Cambridge Tracts in Theoretical Computer Science 62

Initial Algebras and Terminal Coalgebras

Providing an in-depth treatment of an exciting research area, this text's central topics are initial algebras and terminal coalgebras, primary objects of study in all areas of theoretical computer science connected to semantics.

It contains a thorough presentation of iterative constructions, giving both classical and new results on terminal coalgebras obtained by limits of canonical chains and initial algebras obtained by colimits. These constructions are also developed in enriched settings, especially those enriched over complete partial orders and over complete metric spaces, connecting the book to topics such as domain theory. Also included are an extensive treatment of set functors and the first book-length presentation of the rational fixed point of a functor and of lifting results that connect fixed points of set functors with fixed points of endofunctors on other categories.

Representing more than 15 years of work, this will be the leading text on the subject for years to come.

JIŘÍ ADÁMEK is Professor in the Department of Mathematics at Czech Technical University in Prague and Professor Emeritus in the Department of Computer Science at Technical University Braunschweig.

STEFAN MILIUS is Professor in the Department of Computer Science at Friedrich-Alexander-Universität Erlangen-Nürnberg.

LAWRENCE S. MOSS is Professor in the Mathematics Department at Indiana University Bloomington.

CAMBRIDGE TRACTS IN THEORETICAL COMPUTER SCIENCE 62

Titles in the series

A complete list of books in the series can be found at www.cambridge.org/computer-science.
Recent titles include the following:

32. R. Carpenter *The Logic of Typed Feature Structures*
33. E. G. Manes *Predicate Transformer Semantics*
34. F. Nielson & H. R. Nielson *Two-Level Functional Languages*
35. L. M. G. Feijs & H. B. M. Jonkers *Formal Specification and Design*
36. S. Mauw & G. J. Veltink (eds) *Algebraic Specification of Communication Protocols*
37. V. Stavridou *Formal Methods in Circuit Design*
38. N. Shankar *Metamathematics, Machines and Gödel's Proof*
39. J. B. Paris *The Uncertain Reasoner's Companion*
40. J. Desel & J. Esparza *Free Choice Petri Nets*
41. J.-J. Ch. Meyer & W. van der Hoek *Epistemic Logic for AI and Computer Science*
42. J. R. Hindley *Basic Simple Type Theory*
43. A. S. Troelstra & H. Schwichtenberg *Basic Proof Theory*
44. J. Barwise & J. Seligman *Information Flow*
45. A. Asperti & S. Guerrini *The Optimal Implementation of Functional Programming Languages*
46. R. M. Amadio & P.-L. Curien *Domains and Lambda-Calculi*
47. W.-P. de Roever & K. Engelhardt *Data Refinement*
48. H. Kleine Büning & T. Lettmann *Propositional Logic*
49. L. Novak & A. Gibbons *Hybrid Graph Theory and Network Analysis*
50. J. C. M. Baeten, T. Basten & M. A. Reniers *Process Algebra: Equational Theories of Communicating Processes*
51. H. Simmons *Derivation and Computation*
52. D. Sangiorgi & J. Rutten (eds) *Advanced Topics in Bisimulation and Coinduction*
53. P. Blackburn, M. de Rijke & Y. Venema *Modal Logic*
54. W.-P. de Roever et al. *Concurrency Verification*
55. Terese *Term Rewriting Systems*
56. A. Bundy et al. *Rippling: Meta-Level Guidance for Mathematical Reasoning*
57. A. M. Pitts *Nominal Sets*
58. S. Demri, V. Goranko & M. Lange *Temporal Logics in Computer Science*
59. B. Jacobs *Introduction to Coalgebra*
60. S. Mihov & K. U. Schulz *Finite-State Technology*
61. M. Gehrke & S. van Gool *Topological Duality for Distributive Lattices*

Initial Algebras and Terminal Coalgebras

The Theory of Fixed Points of Functors

JIŘÍ ADÁMEK
Czech Technical University in Prague
and
Technical University Braunschweig

STEFAN MILIUS
Friedrich-Alexander-Universität Erlangen-Nürnberg

LAWRENCE S. MOSS
Indiana University Bloomington

CAMBRIDGE
UNIVERSITY PRESS

CAMBRIDGE
UNIVERSITY PRESS

Shaftesbury Road, Cambridge CB2 8EA, United Kingdom

One Liberty Plaza, 20th Floor, New York, NY 10006, USA

477 Williamstown Road, Port Melbourne, VIC 3207, Australia

314–321, 3rd Floor, Plot 3, Splendor Forum, Jasola District Centre,
New Delhi – 110025, India

103 Penang Road, #05–06/07, Visioncrest Commercial, Singapore 238467

Cambridge University Press is part of Cambridge University Press & Assessment,
a department of the University of Cambridge.

We share the University's mission to contribute to society through the pursuit of
education, learning and research at the highest international levels of excellence.

www.cambridge.org
Information on this title: www.cambridge.org/9781108835466

DOI: 10.1017/9781108884112

When citing this work, please include a reference to the DOI 10.1017/9781108884112

First published 2025

A catalogue record for this publication is available from the British Library

Library of Congress Cataloging-in-Publication Data
Names: Adámek, Jiří, 1947– author. | Milius, Stefan, author. |
Moss, Lawrence Stuart, 1959– author.
Title: Initial algebras and terminal coalgebras : the theory of fixed points
of functors / Jiří Adámek (Czech Technical University in Prague),
Stefan Milius (Friedrich-Alexander-Universität Erlangen-Nürnberg, Germany),
Lawrence S. Moss (Indiana University, Bloomington).
Description: Cambridge, United Kingdom ; New York, NY : Cambridge
University Press, 2025. | Series: Cambridge tracts in theoretical
computer science ; 62 | Includes bibliographical references and index.
Identifiers: LCCN 2024011074 | ISBN 9781108835466 (hardback) |
ISBN 9781108884112 (ebook)
Subjects: LCSH: Categories (Mathematics) | Recursion theory. |
Induction (Mathematics) | Fixed point theory. |
Duality theory (Mathematics)
Classification: LCC QA169 .A31994 2025 | DDC 512/.62–dc23/eng/20240628
LC record available at https://lccn.loc.gov/2024011074

ISBN 978-1-108-83546-6 Hardback

To the memory of Věra Trnková.

Contents

	Preface		*page* xi
1	**Introduction**		1
	1.1	Why Are Initial Algebras and Terminal Coalgebras Interesting?	1
	1.2	Expected Background	6
	1.3	Fixed Points and Special Kinds of Algebras and Coalgebras	8
	1.4	Algebraic versus Coalgebraic Concepts	10
2	**Algebras and Coalgebras**		12
	2.1	Algebras	12
	2.2	Initial Algebras	21
	2.3	Recursion and Induction	33
	2.4	Coalgebras	36
	2.5	Terminal Coalgebras	46
	2.6	Corecursion and Bisimulation	54
	2.7	Summary	64
3	**Finitary Iteration**		65
	3.1	Initial-Algebra Chain	65
	3.2	Examples of Initial Algebras	69
	3.3	Terminal-Coalgebra Chain	81
	3.4	Summary	89
4	**Finitary Set Functors**		90
	4.1	Limits and Colimits of Algebras and Coalgebras	91
	4.2	Weakly Terminal Coalgebras	95
	4.3	Presentation of Finitary Set Functors	101
	4.4	Iterating the Terminal-Coalgebra Chain to $\omega + \omega$	113

4.5 Finite Power-Set Functor: The Terminal Coalgebra vs. the ω^{op}-Limit 122
4.6 Summary 132

5 Finitary Iteration in Enriched Settings 133
5.1 Canonical Fixed Points in CPO-Enriched Categories 135
5.2 CMet-Enriched Categories 151
5.3 Solving Domain Equations 161
5.4 Summary 166

6 Transfinite Iteration 167
6.1 The Initial-Algebra Chain 169
6.2 Smooth Monomorphisms 177
6.3 The Initial-Algebra Theorem 185
6.4 Characterization of Varietors on Set 191
6.5 The Terminal-Coalgebra Chain 193
6.6 Subfunctors and Quotient Functors 202
6.7 Canonical Fixed Points in CPO-Enriched Categories 210
6.8 Summary 213

7 Terminal Coalgebras as Algebras, Initial Algebras as Coalgebras 215
7.1 Corecursive Algebras 218
7.2 Terminal Coalgebras from Corecursive Algebras 224
7.3 Completely Iterative Algebras 230
7.4 Recursive Coalgebras 243
7.5 The Initial-Algebra Theorem Revisited 250
7.6 Summary 252

8 Well-Founded Coalgebras 253
8.1 Well-Founded Coalgebras and Well-Founded Graphs 254
8.2 The Next-Time Operator on Coalgebras 263
8.3 The Well-Founded Part of a Coalgebra 270
8.4 Closure Properties of Well-Founded Coalgebras 274
8.5 The General Recursion Theorem 279
8.6 The Converse of the General Recursion Theorem 285
8.7 Summary 292

9 State Minimality and Well-Pointed Coalgebras 293
9.1 Simple Coalgebras 294
9.2 Pointed and Reachable Coalgebras 297
9.3 Well-Pointed Coalgebras 309
9.4 Summary 318

10	**Fixed Points Determined by Finite Behaviour**	319
10.1	Locally Finitely Presentable Categories	320
10.2	The Rational Fixed Point	326
10.3	Iterative Algebras	336
10.4	The Rational Fixed Point of a Set Functor	345
10.5	Full Abstractness and Finitely Generated Objects	347
10.6	Beyond the Rational Fixed Point	360
10.7	Summary	362
11	**Sufficient Conditions for Initial Algebras and Terminal Coalgebras**	363
11.1	Bounded Functors	363
11.2	Accessible Functors	365
11.3	Presentability of Algebras and Coalgebras	376
11.4	Preaccessible Functors	383
11.5	Adjunctions	391
11.6	Summary	396
12	**Liftings and Extensions from** Set	398
12.1	Complete Partial Orders	399
12.2	Complete Metric Spaces and Pseudometric Spaces	401
12.3	Kleisli Categories	408
12.4	Relation Liftings	426
12.5	Eilenberg–Moore Algebras	428
12.6	Connecting Coalgebraic Trace and Language Semantics	439
12.7	Coalgebraic Language Semantics as a Quotient	445
12.8	Summary	458
13	**Interaction between Initial Algebras and Terminal Coalgebras**	461
13.1	Canonical Morphism	462
13.2	Cauchy Completion	464
13.3	Ideal Completion	472
13.4	Summary	476
14	**Derived Functors**	477
14.1	Coproducts and Products with a Constant Functor	477
14.2	Composite Functors	484
14.3	Relatively Terminal Coalgebras	488
14.4	Mutual Recursion	489
14.5	Parametric Fixed Points	491
14.6	Initial Double-Algebras	501
14.7	Coproducts of Monads on Set	505
14.8	Summary	508

15 Special Topics 510
15.1 Variations on Cantor's Theorem 510
15.2 Vietoris Polynomial Functors on Topological Spaces 514
15.3 Hausdorff Polynomial Functors on Metric Spaces 520
15.4 The Interval [0, 1] as a Terminal Coalgebra 524
15.5 Terminal Coalgebras Related to Subsets of the Reals 527
15.6 Summary 535

Appendix A **Functors with Initial Algebras or Terminal Coalgebras** 537

Appendix B **A Primer on Fixed Points in Ordered and Metric Structures** 541
B.1 Fixed Points in Posets 541
B.2 Pataraia's Theorem 546
B.3 Fixed Points in Complete Metric Spaces 550

Appendix C **Set Functors** 552
C.1 Examples of Set Functors 552
C.2 Basic Facts 553
C.3 Common Preservation Properties 555
C.4 Monoid-Valued Functors 568
C.5 Finitary Set Functors 574
C.6 Trnková Hull 579
C.7 Standard Functors 586
C.8 Accessible and Bounded Functors 589
C.9 Accessible Coreflection 593

References 598
List of Categories 617
Index 618

Preface

Initial algebras for endofunctors on a category have been used since the 1970s in algebraic specification and for the semantics of inductive data types definitions. They provide a generic framework to study notions such as recursive function definitions and proofs by structural induction. This has been developed for example in the work of the ADJ group in the 1970s on initial-algebra semantics of abstract data types [143]. Domain theory is another example. Dana Scott's model of the lambda calculus [273] works with initial algebras for endofunctors on the category of domains or complete partial orders [138]. This usage developed into Michael Smyth and Gordon Plotkin's treatment of the solution of recursive domain equations [279]. The 1980s and 90s saw the further development of such topics and their treatment in textbooks such as Ernest Manes and Michael Arbib's book [220] on algebraic semantics of programming, and Samson Abramsky and Achim Jung's survey of domain theory [2].

At the turn of the new millennium, the dual concept, coalgebras for endofunctors, attracted increased attention. While the study of coalgebras had its roots in earlier work parallel to algebras, it was sparked in earnest by Peter Aczel's book [3] on non-well-founded sets, where coalgebras were mentioned at the end. His observation was that the infinite processes that we see in theoretical computer science may be profitably studied as elements of the terminal coalgebra, with specifications of them coming from other coalgebras. He and Nax Mendler also exhibited Robin Milner and David Park's notion of bisimulation from process algebra as a coalgebraic notion. This then led to the breakthrough essay on coalgebras as a theory of systems, Jan Rutten's seminal paper [264]. It demonstrated that many types of state-based systems studied in fields such as automata theory, concurrency, and verification arise as examples of coalgebras for endofunctors. Moreover, the terminal coalgebra yields a fully abstract domain for the behaviour of states of systems. This laid the basis for the new subject of *universal coalgebra*. In the 2000s the subject then rapidly developed and has unified a host of topics that looked similar but were not always understood that way. These were topics from theoretical computer science and logic like automata theory, process calculi, streams, non-well-founded sets, and modal logic, and also areas of mathematics such as power series. Meanwhile universal coalgebra has become a diverse research field offering a generic framework for the semantics of state-based systems, specification and proof principles such as corecursion and coinduction, and the development of coalgebraic logics. In the last few years, generic methods and algorithms for reasoning, model checking, minimization and learning of coalgebras have become a focus of research. Coalgebra continues to be a lively and active area of research, and we hope to offer our readers a source that will help them to enter the field.

Throughout this development it has been the pull of category theory that has provided the language and conceptual apparatus that is needed to unify topics and pose new questions. Therefore, the aim of our book is to give a category-theoretic account of initial algebras, terminal coalgebras, and, as the title of our book suggests, to pursue the topic of fixed points of functors more generally. We also put a focus on the interplay of algebras and coalgebras, a feature that most other texts on those subjects miss, treating one or the other only. Many of the results in this book are stated and proved for categories that go beyond the category of sets, e.g. complete posets and complete metric spaces. Furthermore, we use special features of the category of sets such as presentations of finitary functors and transfinite recursion. We frequently call on facts about set functors, so much so that we have decided to provide an appendix on this topic. A number of these facts are based on work by the Prague mathematical community, especially Věra Trnková and her colleagues and students Vaclǎv Koubek and Jan Reiterman. To make this more 'accessible' for readers not so familiar with it, we provide some background on those topics as needed.

The central topics of our book are:

(1) The iterative constructions of initial algebras and terminal algebras as colimits of chains and limits of op-chains, respectively. We discuss this at great length, providing hosts of concrete examples and developing generalizations to the transfinite setting. We also develop the enriched setting, in particular, enrichment over the categories of complete partial orders and of complete metric spaces, that has been so useful in topics like domain theory.

(2) Initial algebras and terminal coalgebras of finitary functors, and accessible functors more generally. For example, we develop and extend the work of James Worrell [316, 318] on the terminal coalgebra of accessible set functors.

(3) Categorical recursion theory, especially corecursion, completely iterative algebras and well-founded coalgebras. This not only connects our subject to the

past but also highlights topics that we believe will be ever more important in the future.

(4) The relations between various fixed points. In particular, we treat the *rational fixed point* of a functor, which is a fully abstract domain of 'finite-state' behaviour with instances such as the regular languages from automata theory. Here again we go beyond sets for most results.

It has been said that there never is a good time to write a book about anything, and this is especially true of a book coming from an active research field. We are not only trying to summarize algebra, coalgebra and related fields, we are also contributing to them. Some of the results in this book appear here for the first time, and in many other cases we have revised our text in order to develop a subject that only came into existence a few years ago. We hope that an up-to-the-moment book itself will prove useful to the reader.

We have tried hard to be scholarly about the history of results, and we hope that those whose work was not mentioned correctly, or at all, will forgive us.

We have neglected several topics that are near and dear to the hearts of many in the coalgebra community. We did this partly to keep the book of reasonable length, and partly because those topics are already treated in books. In particular, we are thinking of bisimulation and of coalgebraic modal logic. Bisimulation is treated in Bart Jacob's 2016 book [173]; that book also has a lot of material on another topic which we do not treat: predicate liftings. Coalgebraic modal logic is the topic of Dirk Pattinson and Lutz Schröder's forthcoming book. In addition, one also should read Jan Rutten's *The Method of Coalgebra: Exercises in Coinduction* [268]. These are all excellent resources, and they will be especially useful for newcomers to the subject.

But now we hope our readers will have as much fun with this book as we have had in writing it.

Jiří Adámek
Stefan Milius
Lawrence S. Moss

1
Introduction

This book is about topics in category theory motivated by theoretical computer science. The study of initial algebras and terminal coalgebras is a lively area rather than a well-established field. We present a topical picture but do not aim for the last word on the subject. Our goal in this opening chapter is to motivate the book and its overall field for readers who open it without understanding even the title. What are initial algebras, terminal coalgebras, and fixed points of functors? Why do the authors care about them? What do we hope you will learn? What background will you need?

1.1 Why Are Initial Algebras and Terminal Coalgebras Interesting?

Recursion and induction are important tools in mathematics and computer science. Recursion is a definition principle for functions over (inductive) structures of data types such as natural numbers, lists, or trees. Induction is the corresponding proof principle used to prove properties of functions defined by recursion. An important question of theoretical computer science concerns the semantics of such definitions. *Initial Algebra Semantics*, studied since the 1970s, uses the tools of category theory to unify recursion and induction at the appropriate abstract conceptual level. In this approach, the type of data on which one wants to define functions recursively and to prove properties inductively is captured by an *endofunctor* F on the category of sets (or another appropriate base category such as posets, metric spaces, or vector spaces). This functor describes the signature of the data type constructors. An F*-algebra* is an object A together with a morphism $\alpha\colon FA \to A$. An initial algebra for the functor F provides a canonical model of a data type with the desired constructors. Typically the initial algebra is the intended model, and then one tries to use initiality to prove properties of various definitions, such as equivalence of programs or specifications.

Let us illustrate this with the simplest concrete example: Consider the endofunctor on sets given by $FX = X + 1$. This is the construction adding a fresh element to the set X, made into a functor in the obvious way. An algebra for F is just a set A equipped with a unary operation $u\colon X \to X$ and an element $a \in A$. We combine u and a into a function $\alpha\colon A + 1 \to A$. We are going to properly define algebras and their morphisms below along with initial algebras. For now, we simply assert that the initial algebra of our functor F is the algebra of natural numbers $\mathbb{N}$ with the successor function $s(n) = n + 1$ and the constant 0. The initiality of this algebra is equivalent to the usual principle of *recursion on natural numbers*: given an F-algebra (A, α), there exists precisely one function $h\colon \mathbb{N} \to A$ with $h(0) = a$ and $h(s(n)) = u(h(n))$ for all natural numbers n. In other words, functions from the initial algebra to another algebra can be defined by recursion.

As a second example, consider the set functor $FX = X \times X + 1$. An algebra $\alpha\colon A \times A + 1 \to A$ consists of a set equipped with a binary operation and a constant. This time, the initial algebra is the algebra of finite binary trees, and initiality yields a tree-recursion principle which is quite parallel to the induction principle for $\mathbb{N}$.

A collection of objects 'built from below' (such as the natural numbers or the finite trees) is typically described as an initial algebra of some set functor. Addressing our opening question: initial algebras are interesting because they are the natural objects with which to capture recursion and induction.

A *coalgebra* for a functor F is the dual concept of an F-algebra: it consists of an object A and a morphism $\alpha\colon A \to FA$. We think of A as the set of states in a system, and α assigning to a given state a structured collection formed by its successor states modelling the one-step behaviour of the state.

For example, a deterministic automaton with input alphabet Σ can be described by the set A of its states together with a function

$$\alpha\colon A \to \{0, 1\} \times A^{\Sigma}$$

whose first component $\mathsf{acc}\colon A \to \{0, 1\}$ describes the yes/no predicate of being an accepting state. The second component $A \to A^{\Sigma}$ may be canonically recast as $\mathsf{next}\colon A \times \Sigma \to A$. This is the next-state function of the automaton. Thus, a deterministic automaton (with no initial state specified) is a coalgebra for the set functor F given by

$$FX = \{0, 1\} \times X^{\Sigma}.$$

As with initial algebras, we mention an example of a terminal coalgebra before giving the definitions. In the case of deterministic automata, let Σ^* be the set of words on the alphabet Σ, and let $\mathcal{L} = \mathscr{P}(\Sigma^*)$ be the set of (formal) languages over Σ. Then $\mathcal{L}$ itself has a natural automaton structure: the next-state function assigns to a language L and an input symbol σ the language $\partial_\sigma L$ of all words w with $\sigma w \in L$. The accepting states are the languages containing the empty word. This automaton turns out to be the terminal coalgebra. To explain this, we must mention the concept of a *coalgebra homomorphism*. In general, this is a morphism in the base category that respect the coalgebra structure. In the special case of deterministic automata, this would be a function between the state sets that respects the accepting-state predicate and the next-state function. Such functions are called *strong simulations* in automata theory [96]. For a given automaton (A, α), we consider the map $\mathsf{lang}\colon A \to \mathcal{L}$ assigning to a state a the language L_a of words accepted by A when started in a. Then $\mathsf{acc}(a) = 1$ iff $\varepsilon \in L_a$ iff L is an accepting state of $\mathcal{L}$. Furthermore, $\partial_\sigma(L_a) = L_{\mathsf{next}(a,\sigma)}$. This verifies that lang is a coalgebra homomorphism. With further work, one verifies that lang is the unique such function. This is what it means to say that $\mathcal{L}$ is a *terminal coalgebra*.

For another example, take an input alphabet Σ and an output alphabet Γ. A transducer is an automaton which reads in infinite words on Σ and outputs infinite words on Γ according to a next-state function, which may be described as a function of type $A \times \Sigma \to A \times \Gamma$. Closer to our purposes, we transpose it into a function $\alpha\colon A \to (A \times \Gamma)^\Sigma$. Thus, we are considering the set functor $FX = (X \times \Gamma)^\Sigma$. Transducers are precisely coalgebras for F. Let Σ^ω and Δ^ω be the set of infinite sequences of elements of Σ and Δ, respectively. Let C be the set of functions $f\colon \Sigma^\omega \to \Delta^\omega$ which are *causal*: the nth entry in $f(\sigma_1\sigma_2\cdots)$ depends only on $\sigma_1\sigma_2\cdots\sigma_n$. The terminal coalgebra is carried by this set C. We leave it to the reader to guess the terminal coalgebra structure $C \to (C \times \Gamma)^\Sigma$.

The point is that various examples of state-based systems of various sorts may be taken to be coalgebras for set functors. Then the terminal coalgebra appears as the 'intended semantic space'. Terminal coalgebras for set functors have an infinitary aspect that contrasts with what we saw in the most basic initial algebras. For example, languages $L \in \mathcal{L}$ usually are infinite sets of words, and the elements of Σ^ω and Δ^ω are infinite sequences. Dually to recursion and induction, there are principles of function definition by *corecursion* and proof by *coinduction*. Proofs by coinduction are a bit more difficult to illustrate; we present examples in Section 2.6. For now, let us circle back to the title of this section. Terminal coalgebras are interesting because the intended semantic spaces associated with various forms of automata appear as terminal coalgebras. In addition, they are tied up with corecursion, an interesting dual to recursion.

So to the extent that we are interested in recursion (hence in initial algebras), we should also be at least curious about its cousin, corecursion.

We turn next to the formal definitions of algebra and coalgebra, presented for all categories $\mathscr{A}$ and endofunctors $F\colon \mathscr{A} \to \mathscr{A}$.[1] (Incidentally, when $\mathscr{A}$ is the category Set of all sets and functions, we speak of *set functors* in lieu of endofunctors on Set.) An *F-algebra* is a pair (A, α) consisting of an object A of $\mathscr{A}$ and a morphism $\alpha\colon FA \to A$; the *F-algebra homomorphisms* (shortly *homomorphisms*) from (A, α) to (B, β) are those morphisms $h\colon A \to B$ of $\mathscr{A}$ for which the square below commutes:

$$\begin{array}{ccc} FA & \xrightarrow{\alpha} & A \\ {\scriptstyle Fh}\big\downarrow & & \big\downarrow{\scriptstyle h} \\ FB & \xrightarrow{\beta} & B \end{array}$$

The category of algebras is denoted by $\mathsf{Alg}\, F$. By an *initial algebra* μF of F is meant the initial object in $\mathsf{Alg}\, F$ (if it exists): this is an algebra such that for every algebra, there exists a unique homomorphism from μF.

When we mentioned $\mathbb{N}$ as an initial algebra for the set functor $FX = X + 1$, we stated the initiality property in concrete terms. Now we wish to view the initiality of $\mathbb{N}$ as a special case of the general definitions which we have just seen. Suppose we are given as set A and a function $\alpha\colon A + 1 \to A$. From α we can extract a function $u\colon A \to A$ and an element $a\colon 1 \to A$. Then the initiality of $\mathbb{N}$ means that there is a unique function $h\colon \mathbb{N} \to A$ as follows:

$$\begin{array}{ccc} \mathbb{N} + 1 & \xrightarrow{[s,0]} & \mathbb{N} \\ {\scriptstyle h+\mathsf{id}}\big\downarrow & & \big\downarrow{\scriptstyle h} \\ A + 1 & \xrightarrow{\alpha=[u,a]} & A \end{array}$$

How is initiality connected to definition by recursion? Following the element of the singleton set 1 around the square shows that $h(0) = a$, and following a number n in $\mathbb{N}$ shows that $h(s(n)) = u(h(n))$. So h satisfies the recursion equation driven by u and a. Conversely, any function h' which happens to satisfy those recursion equations is an algebra morphism. So by the uniqueness part of initiality, h and h' must be the same. This translates back to the fact that functions which satisfy recursion equations do so uniquely. The overall point is that recursion on the natural numbers is tantamount to initiality of $(\mathbb{N}, [s, 0])$ in $\mathsf{Alg}(F)$, where $FX = X + 1$.

For the coalgebraic concepts, we turn the structure morphisms around. An *F-coalgebra* is a pair (A, α) with $\alpha\colon A \to FA$. An *F-coalgebra homomorphism*

[1] This book properly begins in Chapter 2. What we are doing here is presenting just a few definitions so that we can discuss the content of the book in a very general way.

(shortly: *homomorphism*) from (A, α) to (B, β) is a morphism h of $\mathscr{A}$ such that $Fh \cdot \alpha = \beta \cdot h$:

$$\begin{array}{ccc} A & \xrightarrow{\alpha} & FA \\ {\scriptstyle h}\downarrow & & \downarrow{\scriptstyle Fh} \\ B & \xrightarrow{\beta} & FB \end{array}$$

This defines a category $\mathsf{Coalg}\, F$ of coalgebras. The *terminal coalgebra*νF is (if it exists) the terminal object of this category.

We have already seen a homomorphism when we discussed $\mathsf{lang}\colon A \to \mathcal{L}$ in connection with deterministic automata. The unique homomorphism from an automaton into the terminal coalgebra assigns to every state its language. We see again an instance of a general trend: the unique homomorphism from a coalgebra to νF assigns to every state its 'behaviour'. This underscores the importance of terminal coalgebras. We make a junction between computer science and category theory when we see that concepts like language acceptance satisfy universal properties every bit as compelling, beautiful, and useful as those in other areas of mathematics. This is one of the reasons to be interested in the coalgebraic modelling of dynamical systems of all sorts: it leads to a generic description of behaviour-preserving maps between those systems and an ensuing notion of behavioural equivalence of states. Two states are *behaviourally equivalent* if they are merged by some coalgebra homomorphism. This specializes to well-known notions of behavioural equivalence, for example language equivalence in the case of deterministic automata or Milner's (strong) bisimilarity for labelled transition systems (modelled as coalgebras). This observation was made forcefully by Jan Rutten [264]; he presented a persuasive survey of applications of coalgebra to the theory of discrete dynamical systems. Aczel and Mendler [4] made another decisive contribution by proposing a categorical formulation of bisimulation which revised our conception of everything that went before it in this area and attracted the interest of a new generation of computer scientists. One of the motivations for this book is to present a selection of the large body of work inspired by the papers of Rutten and Aczel–Mendler.

Let us return to homomorphisms, this time of algebras. These are the maps preserving operations, so they are well-known from abstract algebra. The interest of initial algebras for computer science is bolstered because they provide minimal realizations of abstract data type specifications as terms possibly modulo equations.

We shall see in Section 1.3 that initial algebras and terminal coalgebras do not always exists. When do they exist? How can we describe them when they exist? If one exists, does the other? These questions turn out to be difficult, and there are no easily-quoted or complete answers. A large part of this book addresses them. The most basic construction is discussed in Chapter 3. It applies to both initial algebras and terminal coalgebras. This leads to a more in-depth look at certain terminal coalgebras in sets in Chapter 4, and to variations that work in complete partial orders and complete metric spaces in Chapter 5. These variations provide settings where μF and νF are the same. That is, we have an initial algebra (A, α) with the property that (A, α^{-1}) is a terminal coalgebra.

Other chapters which take up the theme of constructing and describing initial algebras and terminal coalgebras include Chapters 6, 11, 12, 13, 15, and Appendix A.

1.2 Expected Background

We intend the book to be our part of the deep connection of category theory to computer science. The root of the matter is that category theory is the mathematics of *structure* par excellence and the semantic side of computer science is full of issues that revolve around mathematical structure.

What background is needed to read this book? We are mainly writing to readers who already know about basic concepts in category theory: initial and terminal objects, duality, products and coproducts (and limits/colimits more generally), monomorphisms, and epimorphisms. Very little else is required. Later chapters in the book require more, of course. The topics that would be most helpful for later work include adjunctions, monads, factorization systems, (co)complete categories, and locally finitely presentable categories.

Even more than knowledge of the specific topics, what we assume is a level of comfort with the presentational style that comes from category theory. Let us illustrate this point by presenting a very simple result having to do with an abstract formulation of induction. The rest of this section contains that result and, in addition, it is a comment on it and the tone and reasoning style of this book.

We mentioned at the outset that initiality is connected with recursion and induction, but we have in fact not said a word about the connection to induction. Now that we have the category $\mathsf{Alg}(F)$, we can connect initiality and induction. Let $\mathscr{A}$ be a category, and let $F\colon \mathscr{A} \to \mathscr{A}$. If (A, α) is an algebra, a *subalgebra* of it is represented by a homomorphism $m\colon (B, \beta) \rightarrowtail (A, \alpha)$ which is a monomorphism in $\mathscr{A}$ (hence the notation $\rightarrowtail$). An algebra (A, α) is *minimal* if every $m\colon (B, \beta) \rightarrowtail (A, \alpha)$ is an isomorphism.

We recall from category theory that a *split epimorphism* is a morphism $e\colon X \to Y$ for which $m\colon Y \to X$ exists with $e \cdot m = \mathsf{id}_Y$. The reader will recall that if a morphism is both a split epimorphism and a monomorphism, then it is an isomorphism.

Proposition 1.2.1. *Every initial algebra $(\mu F, \iota)$ is minimal.*

Proof By initiality, there is an algebra homomorphism $f\colon (\mu F, \iota) \to (B, \beta)$. The composition $m \cdot f$ is a homomorphism from $(\mu F, \iota)$ to itself, so by initiality again, $m \cdot f = \mathrm{id}_{\mu F}$. Hence m is a split epimorphism and a monomorphism, so it is an isomorphism. □

This general result specializes to well-known cases. For the set functor $FX = X + 1$, it corresponds to *weak induction*: every subset of $\mathbb{N}$ which contains 0 and is closed under successors contains all natural numbers.

There are other ways to formulate induction. For a parallel pair $f_1, f_2\colon \mu F \to A$ of morphisms in the base category $\mathscr{A}$ with domain μF, in order to prove $f_1 = f_2$ it is sufficient to present a morphism $\alpha\colon FA \to A$ for which f_1 and f_2 are algebra homomorphisms from $(\mu F, \iota)$ to (A, α). Notice that this is not the same formulation that we saw in Proposition 1.2.1. It is very much in the spirit of this book to investigate different categorial formulations of concepts like weak induction, strong induction, etc. *Coinduction* is the dual principle which for the terminal coalgebra νF allows us to prove equality of morphisms of the form $f_1, f_2\colon A \to \nu F$.

We return to the general point about the tone and reasoning style of the book. In introducing Proposition 1.2.1, we mentioned what a split epimorphism is and also one fact about it. Our brief mention is barely enough to read further. In other words, we want to jog the memories of people who have seen the definition, and we trust that others will acquire some background on it. The proof of Proposition 1.2.1 is shorter than almost any result in the book, but the reasoning style is indicative: it uses point-free category-theoretic arguments and makes use of universal properties.

We generally assume that readers know about automata, terms and signatures, equational logic, and other topics used in theoretical computer science. We do not assume much about any of those topics, but we take for granted that the reader is interested in them.

Several chapters require background on topics related to metric spaces or partially ordered sets. We have more background on that material. For example, we provide an appendix on fixed points in settings that come from complete metric spaces or complete partially ordered sets. We would like readers to come away from the book with appreciation of two things: the general theory, and results about particular concrete categories.

Let us comment on the aspect of the book connected to set theory. Our set theory is the standard one using the Zermelo–Fraenkel axioms, including the Axiom of Choice. At times we do need definitions and results from basic set theory, touching on matters such as ordinal numbers, cardinal numbers, cardinal arithmetic, and regular infinite cardinals. We have provided a more thorough introduction to those topics in Chapter 6. We also collect a number of facts concerning set functors into a single place, Appendix C. Material from this appendix is used in most of the chapters. In a few other places, we mention some other concept in set theory (e.g. filters and ultrafilters), and readers who do not know about such things may skip those passages.

When speaking about a category $\mathscr{C}$ we assume that the collection $\mathscr{C}(X,Y)$ of all morphisms from X to Y is a set (not a proper class) for every pair of objects X, Y. We also assume that these hom-sets are pairwise disjoint.

1.3 Fixed Points and Special Kinds of Algebras and Coalgebras

As stated in its subtitle, our book is about the theory of fixed points of functors.

A *fixed point* of an endofunctor F is an object A together with an isomorphism of FA and A. One general reason to be interested in fixed points comes from the following basic result:

Lambek's Lemma. *Every initial algebra is a fixed point.*

Dually, every terminal coalgebra is a fixed point. Besides these two, there are other interesting fixed points, one of which we now mention.

In the study of state-based systems, one is mostly interested in *finite* systems (e.g. finite-state automata), or ones whose state space has a finite representation (e.g. a finite-dimensional vector space or an orbit-finite nominal set). For example, in classical automata theory the regular languages are those accepted by finite automata, and rational streams over a field are precisely those behaviours obtained from systems whose state space is a finite-dimensional vector space over the field.

We present a general treatment of regularity of system behaviour and the semantics of 'finite' coalgebras. For this we assume that the functor F is *finitary* (see Definition 4.3.1). For example, finitary set functors are those determined by their action on finite sets; for them the behaviour of all finite coalgebras yields a subcoalgebra of the terminal coalgebra νF (consisting of the behaviour of *all* coalgebras). This subcoalgebra turns out to be a fixed point of F which we call the *rational fixed point* and denote by ϱF. For example, for the functor $FX = \{0, 1\} \times X^{\Sigma}$ for deterministic automata, this is the coalgebra of all regular languages over the alphabet Σ. For $FX = X \times X + 1$, the rational fixed point ϱF is the subcoalgebra of νF (of all binary trees) consisting of precisely those binary trees that have only finitely many subtrees (up to isomorphism); these are precisely Courcelle's *regular trees* [107]. In addition to the finite binary trees, the complete (infinite) binary tree is an example of a tree contained in ϱF.

The concept of a rational fixed point makes sense for endofunctors on categories other than Set, too. For example, for the functor F on the category of vector spaces whose coalgebras are linear weighted automata, we obtain the rational formal power series, known from weighted automata theory [113], as ϱF. We study rational fixed points in Chapter 10.

Coming back to initial algebras, it follows from Lambek's Lemma that an initial algebra can be considered as a coalgebra by taking the inverse of its structure morphism. It is interesting to study properties of this coalgebra, especially ones that characterize it in classes of coalgebras with computational relevance. We mention two properties here: *recursive* coalgebras and *well-founded* ones. Without going into the technical definitions, let us just say here that recursive coalgebras are coalgebras which admit function definitions by structural recursion; a good intuition here is to think about 'divide-and-conquer' algorithms. Well-founded coalgebras present the notion of a well-founded relation at a general coalgebraic level. So they are coalgebras admitting an induction principle. An initial algebra turns out to be the terminal recursive or well-founded coalgebra; the latter holds under mild assumptions. We shall study recursivity and well-foundedness in Chapters 7 and 8.

Yet another application of Lambek's Lemma, this time in dual form, lets us consider a terminal coalgebra νF as an algebra for F using the inverse structure. We shall be interested in studying properties which characterize νF in classes of algebras with computational relevance. *Completely iterative* algebras are connected to the matter of solving recursive equations. Again exemplifying via automata, we can solve systems of equations for languages in a unique way, such as $L = aM$, $M = bL + 1$. Completely iterative algebras are even tied up with areas of mathematics that at first blush have little to do with our subject: fractal sets, for example. A terminal coalgebra is, equivalently, an initial completely iterative algebra. Complete iterativity shall be studied in Chapter 7.

Finally, Lambek's Lemma also has a negative consequence for our study. It implies that even for $\mathscr{A}$ = Set, there are important endofunctors that do not have an initial algebra: the power-set functor $\mathscr{P}$, for example. A fundamental result of set theory known as Cantor's Theorem [99] states that no set A is in bijective correspondence with $\mathscr{P}A$. (The short proof may be found in Example 2.2.7(1).) So no set is a fixed point of $\mathscr{P}$.

algebra for a functor	coalgebra for a functor
initial algebra	terminal coalgebra
least fixed point	greatest fixed point
congruence relation	bisimulation equivalence relation
equational logic	modal logic
recursion: map out of an initial algebra	corecursion: map into a terminal coalgebra
iterative conception of set	coiterative conception of set
set with operations	set with transitions and observations
useful in syntax	useful in semantics
bottom-up	top-down

Figure 1.1 Conceptual comparison.

However, this negative point does raise questions about conditions on functors that ensure the existence of initial algebras and terminal coalgebras and results showing how to construct them. For example, what about functors related to $\mathscr{P}$, such as the finite power-set functor, or the countable power-set functor? Do these have fixed points? It is known that in other areas such as metric spaces or posets, fixed points do exist for important classes of functions. We review this general topic in Appendix B.

1.4 Algebraic versus Coalgebraic Concepts

Although *algebra* and *coalgebra* are dual terms, and although this duality persists to the level of *initial algebra* and *terminal coalgebra*, $\mathsf{Alg}\, F$ is not dual to $\mathsf{Coalg}\, F$. There are easy examples of this; here is a very simple one: for the constant functor with value 1, a one-element set, $\mathsf{Alg}\, F$ is the category of pointed sets, while $\mathsf{Coalg}\, F$ is (isomorphic to) the category of sets.

What is true is the following: every functor $F\colon \mathscr{A} \to \mathscr{A}$ defines a functor $F^{\mathsf{op}}\colon \mathscr{A}^{\mathsf{op}} \to \mathscr{A}^{\mathsf{op}}$ by the same rule as F. The category of algebras for F (in $\mathscr{A}$) is dual to the category of coalgebras for F^{op} (in $\mathscr{A}^{\mathsf{op}}$). Shortly,

$$(\mathsf{Alg}\, F)^{\mathsf{op}} = \mathsf{Coalg}(F^{\mathsf{op}}).$$

We present in Figure 1.1 a comparison between concepts and ideas in algebra and coalgebra. The algebraic concepts on the left are connected to *sets with operations* and the coalgebraic ones on the right are about sets with structure corresponding to *transitions* and *observations*. The two columns are duals, we are speaking of 'duals' here in an informal way. This book will not have much to say about most of this chart: to explore it in any depth would require several additional chapters at the least, and perhaps an entire book of its own. However, we hope that readers find this chart enticing and that they will want to explore the book in order to construct their own conceptual charts. Our claim in this book is that the mathematical tools which we study will be useful for anyone interested in the chart.

To sum up, this book is concerned with fundamental notions in theoretical computer science related to recursion and induction. These notions are dualized and generalized, leading to a wealth of questions and results. We bring the viewpoint of category theory to our subject, therefore we adopt the vocabulary and tool set from that theory. In addition to general work, we frequently study particular concrete categories and particular endofunctors.

2
Algebras and Coalgebras

As the title of our book suggests, we are mainly interested in *initial algebras* and *terminal coalgebras*. But to understand what these are, we must discuss the more general concepts of *algebras* and *coalgebras* first. This is the purpose of this chapter.

This chapter has two halves: one devoted to algebras and the other to coalgebras. For algebras, we first discuss them in general, primarily using examples from the categories Set of sets, Set^S of sorted sets, and $\mathsf{CPO}_\perp$ of complete partial orders with a least element (see Example 2.1.7(2)). Then we turn to initial algebras. The main results are detailed constructions of initial algebras, mainly for polynomial endofunctors on these categories. Following this, we examine two concepts closely related to initial algebras: recursion and induction. Then we turn to the parallel topics for coalgebras: examples from the same categories, terminal coalgebras for polynomial functors, and finally corecursion and bisimulation.

The overall treatment in this chapter is in a sense preliminary: at numerous points we mention discussions later in the book which amplify what we do here.

2.1 Algebras

To specify algebras, we must have an *underlying* or *base category*, say $\mathscr{A}$, and an endofunctor F, that is, a functor $F\colon \mathscr{A} \to \mathscr{A}$. With this in mind, we introduce the concepts of *algebra* and *homomorphism*. Often the action of F on morphisms is quite obvious, so one suppresses it and states only its action on objects. For example, $FX = X \times X + A$ is given for objects X, and it is understood that for every morphism $f\colon X \to Y$, we have $Ff = f \times f + \mathsf{id}_A$.

Definition 2.1.1. An *algebra* for an endofunctor F (or an F-*algebra*) consists of an object A and a morphism $\alpha\colon FA \to A$. We sometimes call A the *carrier* and α the *structure*. So technically the algebra is a pair (A, α). But, as usual we shorten this to A whenever we can. Usually, we use upper case letters for algebras and Greek letters for structures.

A *homomorphism* of F-algebras from (A, α) to (B, β) (or a *morphism* of F-algebras) is a morphism $h: A \to B$ of $\mathscr{A}$ such that the square below commutes:

$$\begin{array}{ccc} FA & \xrightarrow{\alpha} & A \\ {\scriptstyle Fh}\downarrow & & \downarrow{\scriptstyle h} \\ FB & \xrightarrow{\beta} & B \end{array} \tag{2.1}$$

It should be clear that if A, B, and C are algebras, and $g: A \to B$ and $h: B \to C$ are homomorphisms, then $h \cdot g: A \to C$ is an algebra homomorphism. In other words, composition of homomorphisms works in the same way as in the base category $\mathscr{A}$. Similarly, the identity morphism from the base category is always an algebra homomorphism. This endows the collection of algebras for a fixed endofunctor with a category structure.

Notation 2.1.2. The category of F-algebras and homomorphisms is denoted by $\mathsf{Alg}\, F$. The base category $\mathscr{A}$ is suppressed in our notation.

Example 2.1.3. Algebras in $\mathscr{A} = \mathsf{Set}$ for various endofunctors.

(1) Algebras for $FX = X + 1$. Recall that 1 denotes a terminal object, here a singleton set. The coproduct operation + is disjoint union. An algebra consists of a set A and a function $\alpha: A + 1 \to A$. That is, on a set A (of data) a unary operation $\alpha_1: A \to A$ is given together with an 'initial datum': an element expressed as $\alpha_0: 1 \to A$. Indeed, specifying $\alpha: A + 1 \to A$ is the same as specifying α_1 and α_0, the two components of α.

A homomorphism from (A, α) to (B, β) is a function $h: A \to B$ between these sets which preserves the unary operations and the initial data: $h \cdot \alpha_1 = \beta_1 \cdot h$ and $h \cdot \alpha_0 = \beta_0$. Indeed, this is equivalent to the commutativity of the following square:

$$\begin{array}{ccc} A + 1 & \xrightarrow{[\alpha_1, \alpha_0]} & A \\ {\scriptstyle h+\mathsf{id}}\downarrow & & \downarrow{\scriptstyle h} \\ B + 1 & \xrightarrow{[\beta_1, \beta_0]} & B \end{array}$$

(2) Analogously, algebras for $FX = X \times X + 1$ are sets (of data) with a binary operation and a constant, and homomorphisms are functions preserving the operation and the constant.

(3) For a given set Σ, the algebras for $FX = \Sigma \times X$ are sets A endowed with unary operations $\alpha_s: A \to A$, where s ranges through Σ. Indeed, this is the

same as specifying a function $\alpha\colon A \times \Sigma \to A$. Homomorphisms from (A, α) and (B, β) are the functions $h\colon A \to B$ preserving all the unary operations: $h \cdot \alpha_s = \beta_s \cdot h$ for every $s \in \Sigma$.

(4) The *power-set functor* is defined by $\mathscr{P}X = \{M : M \subseteq X\}$, and for $f\colon X \to Y$ and $M \subseteq X$, $\mathscr{P}f(M)$ is the image set $f[M] = \{f(x) \colon x \in M\}$. For this functor, an algebra consists of a set A and a function $\alpha\colon \mathscr{P}A \to A$. For example, joins in a complete lattice A form an algebra for $\mathscr{P}$.

A homomorphism from (A, α) to (B, β) is a function $h\colon A \to B$ such that for every set $M \subseteq A$ we have

$$h(\alpha(M)) = \beta(h[M]).$$

If α and β are both joins (of complete lattices), this states that h is join-preserving.

Next, we generalize Example 2.1.3, parts (1)–(3).

Definition 2.1.4. A *(finitary) signature* is a collection $\Sigma = (\Sigma_n)_{n \in \mathbb{N}}$ of sets Σ_n, where Σ_n is called the set of *n-ary operation symbols*. In particular, the elements of Σ_0 are the *nullary* symbols, or *constants*. We associate with the signature Σ the *polynomial functor*

$$H_\Sigma\colon \mathsf{Set} \to \mathsf{Set}$$

given by

$$H_\Sigma X = \coprod_{n \in \mathbb{N}} \Sigma_n \times X^n. \tag{2.2}$$

The operation symbols from the signature are 'inscribed in' the functor. This functor assigns to a function $f\colon X \to Y$ the function

$$H_\Sigma f = \coprod_{n \in \mathbb{N}} \mathsf{id}_{\Sigma_n} \times \underbrace{f \times \cdots \times f}_{n \text{ times}}.$$

Example 2.1.5. Let Σ be a signature. A Σ-*algebra* is a set A together with interpretations of the operation symbols; a symbol σ of arity n is interpreted as a function $\sigma^A\colon A^n \to A$. Moreover, a homomorphism of Σ-algebras is a function $h\colon A \to B$ preserving every operation: for all $\sigma \in \Sigma_n$ and $x_0, \ldots, x_{n-1} \in A$,

$$h(\sigma^A(x_0, \ldots, x_{n-1})) = \sigma^B(h(x_0), \ldots, h(x_{n-1})). \tag{2.3}$$

This notion of a Σ-algebra is from general algebra.

We check that Σ-algebras are the same as algebras for the set functor H_Σ and that homomorphisms, as we have defined them in (2.3), are the same as homomorphisms of H_Σ-algebras.

Given a Σ-algebra A, we define $\alpha\colon H_\Sigma A \to A$ by coproduct components: on the summand $\Sigma_n \times A^n$ of $H_\Sigma A$ we let

$$\alpha(\sigma, (x_0, \ldots, x_{n-1})) = \sigma^A(x_0, \ldots, x_{n-1}). \tag{2.4}$$

In the other direction, given $\alpha\colon H_\Sigma A \to A$, we obtain a Σ-algebra A in the following way. For every $\sigma \in \Sigma_n$, define σ^A using (2.4) from right to left. This recovers algebras, and a map $h\colon A \to B$ is a homomorphism of Σ-algebras iff the square below commutes:

$$\begin{array}{ccc} \coprod_{n\in\mathbb{N}} \Sigma_n \times A^n & \xrightarrow{\alpha} & A \\ {\scriptstyle H_\Sigma h}\big\downarrow & & \big\downarrow{\scriptstyle h} \\ \coprod_{n\in\mathbb{N}} \Sigma_n \times B^n & \xrightarrow{\beta} & B \end{array}$$

Again, given a signature Σ, the category of Σ-algebras is the same as the category of H_Σ-algebras.

Example 2.1.6. Many-sorted algebras. Fix a set S. An *S-sorted set* is just a family of sets $X = (X_s)_{s\in S}$ indexed by S. Given $X = (X_s)_{s\in S}$ and $Y = (Y_s)_{s\in S}$, an *S-sorted function* is a family $f = (f_s)_{s\in S}$ of functions, where $f_s\colon A_s \to B_s$. We thus work in the category

$$\mathsf{Set}^S$$

of S-sorted sets and S-sorted functions. Products and coproducts in Set^S are formed sort-wise. An *S-sorted signature* is a set Σ of operation symbols together with arities of members σ of Σ. An *arity* has the form

$$s_0 \times \cdots \times s_{n-1} \to s$$

for an operation symbol whose n variables have sorts $s_0, \ldots, s_{n-1}$, respectively, and whose result has sort s.

Incidentally, in all of our discussions of many-sorted sets in this book, S can be arbitrary. Most of the time, taking $S = 2 = \{0, 1\}$ illustrates our point: as categories, Set and Set^2 have enough differences to be interesting for us.

A many-sorted algebra is an S-sorted set $(A_s)_{s\in S}$ together with an interpretation of the operation symbols; a symbol σ of arity $s_0 \times \cdots \times s_{n-1} \to s$ is interpreted as a function $\sigma^A\colon A_{s_1} \times \cdots A_{s_n} \to A_s$. Equivalently, a many-sorted algebra is an algebra for a polynomial endofunctor on Set^S. If Σ consists of a single operation symbol of arity $s_0 \times \cdots \times s_{n-1} \to s$, then the corresponding polynomial functor H_Σ has empty sorts $(H_\Sigma X)_t$ for all $t \neq s$, and its sort s is

$$(H_\Sigma X)_s = X_{s_0} \times \cdots \times X_{s_{n-1}}.$$

Its action on morphisms is analogous:

$$(H_\Sigma f)_s = f_{s_0} \times \cdots \times f_{s_{n-1}}.$$

For general S-sorted signatures, the *polynomial functor* $H_\Sigma\colon \mathsf{Set}^S \to \mathsf{Set}^S$ is defined by

$$(H_\Sigma X)_s = \coprod_{\sigma \in \Sigma} X_{s_0} \times \cdots \times X_{s_{n-1}},$$

where the coproduct ranges over all $\sigma \in \Sigma$ with the result sort s and the arity $s_1 \times \cdots \times s_n \to s$. It is easy to verify that many-sorted Σ-algebras and homomorphisms in the above sense form precisely the category $\mathsf{Alg}\, H_\Sigma$.

Example 2.1.7. *Continuous algebras.* These are algebras with a complete partial order on them and where operations are continuous. A *complete partial order,*[1] or *cpo* for short, is a poset $(A, \sqsubseteq)$ in which all ω-chains $a_0 \sqsubseteq a_1 \sqsubseteq a_2 \sqsubseteq \cdots$ have a join. Examples include all complete lattices (these have joins of all sets) and finite posets. Posets like the rational or real numbers are not complete. Incidentally, the empty sequence is not an ω-sequence, and so cpo's need not have a bottom element. Also, the empty poset is a cpo. For every cpo A, the poset A^n ordered pointwise is again a cpo.

A *continuous function* from a cpo A to a cpo B is a monotone function $f\colon A \to B$ preserving joins of ω-chains. An operation $\sigma^A\colon A^n \to A$ is continuous if it is a continuous function on the cpo A^n. Continuous algebras can be represented as F-algebras over one of the following categories:

(1) The category CPO of all cpos and all continuous maps between them. Products of cpos are formed as in Set, that is, they are cartesian products with the coordinatewise order and joins. Similarly, coproducts are, as in Set, the disjoint unions with the given order of the summands and with elements in different summands incomparable. Hence, a number of examples of F-algebras work analogously as for Set:

(a) The algebras for $FX = X + 1$ are cpos A together with a constant and a unary continuous operation.

(b) The algebra for $FX = X \times X + 1$ are cpos A with a constant and a binary continuous operation $A \times A \to A$.

(c) For every cpo X we denote by $X_\perp$ its *lifting*, obtained by adding a (new) least element $\perp$ to X. This defines an endofunctor $FX = X_\perp$ assigning to every continuous function f its extension $f_\perp$ preserving the least element $\perp$. An algebra for this endofunctor is given by a cpo A, a continuous unary

[1] This is often called an ω-complete partial order, in contrast to the directed-complete partial orders (dcpos) that we introduce in Example 5.1.2(3).

operation $\alpha\colon A \to A$ and a constant $c_\perp \in A$ satisfying $c_\perp \sqsubseteq \alpha(x)$ for all $x \in A$.

(2) The picture is more interesting for the (non-full sub-)category $\mathsf{CPO}_\perp$ of all cpos with a least element $\perp$, and all *strict* continuous maps (i.e. those preserving $\perp$). This subcategory is closed under products in CPO. But its coproducts are not Set-based: a coproduct in $\mathsf{CPO}_\perp$ is obtained from that in CPO by identifying the least elements of all summands to one (least) element. For example, here $FX = X + 1$ is simply the identity functor on $\mathsf{CPO}_\perp$.

(a) Strict continuous algebras with one unary operation and one constant are the algebras for the functor $FX = X_\perp + 1_\perp$ forming the coproduct of the lifting of X and the 2-chain $1_\perp = \{\perp, *\}$. On morphisms, F acts as expected: $Ff = f_\perp + \mathsf{id}_{1_\perp}$. Giving an algebra $\alpha\colon A_\perp + 1_\perp \to A$ in $\mathsf{CPO}_\perp$ is equivalent to giving a cpo A with $\perp$, a constant $\alpha(*) \in A$ and a strict continuous unary operation $A \to A$.

(b) Analogously, algebras for the endofunctor on $\mathsf{CPO}_\perp$ defined by $FX = (X \times X)_\perp + 1_\perp$ are the strict continuous algebras with one binary operation and one constant.

Algebraic operations binding variables, for example λ-abstraction in the λ-calculus, or name abstraction in the π-calculus, turn structural induction principles into a more delicate issue. It was the idea of Fiore et al. [124] to deal with these issues by working in the category of sets in context.

Example 2.1.8. *Algebras over sets in context.* Denote by $\mathscr{F}$ the category of finite sets and maps, and form the category $\mathsf{Set}^{\mathscr{F}}$ of (covariant) *presheaves*. This category has as objects functors $\mathscr{F} \to \mathsf{Set}$ and as morphisms the natural transformations between them. Products and coproducts in $\mathsf{Set}^{\mathscr{F}}$ are formed objectwise. One thinks of a finite set as a context Γ of variables, and a presheaf $X\colon \mathscr{F} \to \mathsf{Set}$ assigns to every context Γ a set $X(\Gamma)$ of 'terms in context Γ'. For example, let L be the presheaf of λ-terms, that is, for a context $\Gamma = \{x_1, \ldots, x_n\}$, $L(\Gamma)$ is the set of α-equivalence classes of λ-terms with free variables included in Γ. A variable-binding operation, such as λ-abstraction, may be encoded by an endofunctor on $\mathsf{Set}^{\mathscr{F}}$. For example, consider the endofunctor $F\colon \mathsf{Set}^{\mathscr{F}} \to \mathsf{Set}^{\mathscr{F}}$ that maps a presheaf $X\colon \mathscr{F} \to \mathsf{Set}$ to

$$FX = V + X \times X + \delta X,$$

where $V\colon \mathscr{F} \hookrightarrow \mathsf{Set}$ is the inclusion functor and $\delta X(\Gamma) = X(\Gamma + 1)$. The shape of this functor F reflects the constructors of λ-terms. In fact, the presheaf L of α-equivalence classes of λ-terms is an algebra for F: consider the algebra structure $\alpha\colon FL \to L$ whose three coproduct components are given by

(1) the inclusion of variables $V(\Gamma) = \Gamma \hookrightarrow L(\Gamma)$ as λ-terms;

(2) the operation $L(\Gamma) \times L(\Gamma) \to L(\Gamma)$ given, for two λ-terms t_1 and t_2 in context Γ, by the application $t_1 t_2$ of t_1 to t_2; and

(3) the operation $\delta L(\Gamma) = L(\Gamma + \{x\}) \to L(\Gamma)$ of λ-abstraction, which takes a λ-term t in context $\Gamma + \{x\}$ and assigns to it the λ-term $\lambda x.t$ in context Γ.

Remark 2.1.9. Similarly, one can deal with variable binding operators of higher arity: Fiore et al. [124] introduce the notion of a *binding signature*, that is, a pair (Σ, ar), where Σ is a set of operation symbols and $\mathsf{ar}\colon \Sigma \to \mathbb{N}^*$ assigns to each operation symbol $\sigma \in \Sigma$ its *binding arity*. If $\mathsf{ar}(\sigma) = \langle n_1, \dots, n_k \rangle$, then σ has k arguments, and it binds n_i variables in its ith argument $(1 \le i \le k)$. For each binding signature, we have an associated binding polynomial endofunctor on $\mathsf{Set}^{\mathcal{F}}$ defined by

$$F_\Sigma X = V + \coprod_{\substack{\sigma \in \Sigma \\ \mathsf{ar}(\sigma) = \langle n_1, \dots, n_k \rangle}} \delta^{n_1} X \times \cdots \times \delta^{n_k} X,$$

where δ^n denotes the n-fold composition of δ (cf. Example 2.1.8).

Example 2.1.10. *Algebras over nominal sets.* An alternative way to deal with the subtle issues of induction in the presence of variable binding is to use nominal sets (also known as sets with atoms). Let us recall the definition of the category Nom of nominal sets (see e.g. Pitts' book [250]). We fix a countably infinite set $\mathbb{A}$ of *atomic names*. Let $\mathsf{Perm}(\mathbb{A})$ denote the group of all *finite* permutations on $\mathbb{A}$ (generated by all transpositions $(a\,b)$ with $a, b \in \mathbb{A}$). Consider a set X with an action of this group, denoted by $\pi \cdot x$ for a finite permutation π and $x \in X$. A subset $A \subseteq \mathbb{A}$ is called a *support* of an element $x \in X$ provided that every permutation $\pi \in \mathsf{Perm}(\mathbb{A})$ that fixes all elements of A also fixes x:

$$\pi(a) = a \text{ for all } a \in A \implies \pi \cdot x = x.$$

A *nominal set* is a set with an action of the group $\mathsf{Perm}(\mathbb{A})$ where every element has a finite support. The category Nom is formed by nominal sets and *equivariant maps*, that is, maps f preserving the given group action: $f(\pi \cdot x) = \pi \cdot f(x)$.

Finite products and coproducts of nominal sets are given by cartesian products and disjoint union, respectively, with the evident ensuing nominal structure. It is a standard result that every element x of a nominal set has the least support, denoted by $\mathsf{supp}(x)$. The support plays the role of the set of free variables or names of the elements of the nominal set. For example, λ-terms (modulo α-equivalence) form a nominal set L, where the support of a λ-term is the set of its free variables, and the group action is given by renaming free variables.

An important example of an endofunctor on Nom is that assigning to every nominal set X the *abstraction set* [132], which is defined by

$$[\mathbb{A}]X = (\mathbb{A} \times X)/\sim,$$

where the equivalence relation $\sim$ abstracts the notion of α-equivalence known from the λ-calculus: $(a, x) \sim (b, y)$ holds iff $(c\,a) \cdot x = (c\,b) \cdot y$ for some *fresh* name c, which means that c is distinct from a and b and is not contained in the support of x or y. One writes $\langle a \rangle x$ for the equivalence class of (a, x). The action of $[\mathbb{A}]X$ is given by $\pi \cdot \langle a \rangle x = \langle \pi(a) \rangle (\pi \cdot x)$. Given an equivariant map $f\colon X \to Y$, one has

$$[\mathbb{A}]f\colon [\mathbb{A}]X \to [\mathbb{A}]Y \qquad \text{defined by} \qquad [\mathbb{A}]f(\langle a \rangle x) = \langle a \rangle (f(x)).$$

This is easily seen to yield an endofunctor on Nom.

The operations of the λ-calculus are modelled by the functor $FX = \mathbb{A} + X \times X + [\mathbb{A}]X$ on Nom. In fact, the nominal set L of λ-terms modulo α-equivalence is an algebra for F in an obvious way.

Remark 2.1.11. Similarly, for every binding signature Σ (Remark 2.1.9), one obtains the corresponding *binding polynomial functor* on Nom defined by

$$F_\Sigma X = \mathbb{A} + \coprod_{\substack{\sigma \in \Sigma \\ \mathsf{ar}(\sigma) = \langle n_1, \ldots, n_k \rangle}} [\mathbb{A}]^{n_1} X \times \cdots \times [\mathbb{A}]^{n_k} X.$$

Here $[\mathbb{A}]^n(-)$ denotes the n-fold composition of the abstraction functor as usual: $[\mathbb{A}]^0 X = X$ and $[\mathbb{A}]^{n+1} X = [\mathbb{A}]([\mathbb{A}]^n X)$.

The terms over a given binding signature modulo α-equivalence form an algebra for F_Σ (see e.g. Kurz et al. [209, Sec. 5.2] or cf. Pitts [250, Ch. 8]).

Remark 2.1.12. Historically, nominal sets go back to Fraenkel and Mostowski's permutation models for set theory devised in the 1920s and 1930s. Gabbay and Pitts [132] rediscovered the ideas and have shown how to employ nominal sets to deal abstractly with syntax with variable binding operations.

General properties of the category Alg F The foregoing part of this chapter was devoted to examples of functors and algebras. We close this section with a discussion of a few general properties of Alg F. These facts hold for all endofunctors on all categories.

Proposition 2.1.13. *If the base category $\mathscr{A}$ has finite products, then a product of algebras $\alpha\colon FA \to A$ and $\beta\colon FB \to B$ is formed on the level of $\mathscr{A}$. That is, there exists a unique algebra structure on $A \times B$*

$$\gamma\colon F(A \times B) \to A \times B$$

for which both projections (in $\mathscr{A}$) become homomorphisms. And $(A \times B, \gamma)$ is a product of the given algebras in $\mathsf{Alg}\, F$.

Proof Let π_A and π_B denote the projections in $\mathscr{A}$. Given γ for which π_A and π_B are homomorphisms; that is, the following diagram commutes:

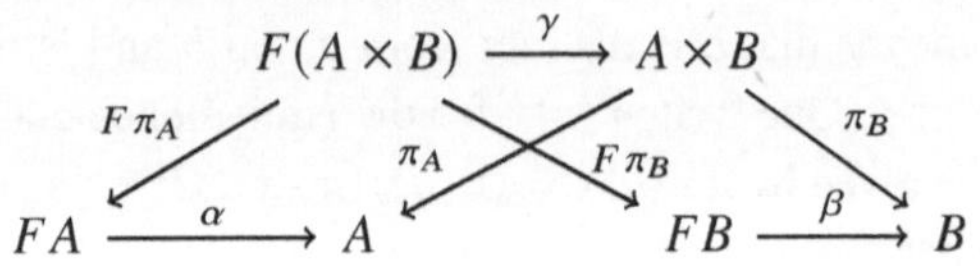

It follows that

$$\gamma = \langle \alpha \cdot F\pi_A, \beta \cdot F_{\pi_B} \rangle.$$

So the uniqueness of γ is clear. Now define γ by the last equality. Then π_A and π_B are both homomorphisms. Consider an algebra $\delta\colon FD \to D$ and homomorphisms $h_A\colon D \to A$ and $h_B\colon D \to B$. Let $h\colon D \to A \times B$ be the $\mathscr{A}$-morphism $\langle h_A, h_B \rangle$; we check that h is a morphism in $\mathsf{Alg}\, F$ as well. That is, the square below commutes:

$$\begin{array}{ccc} FD & \xrightarrow{\ \delta\ } & D \\ {\scriptstyle Fh}\downarrow & & \downarrow{\scriptstyle h} \\ F(A \times B) & \xrightarrow{\ \gamma\ } & A \times B \end{array}$$

Indeed, π_A and π_B are collectively monic; thus, we only need to prove that the square above commutes when post-composed with π_A (and π_B – this follows by symmetry). We have

$$\pi_A \cdot (h \cdot d) = h_A \cdot d = a \cdot Fh_A \quad \text{and} \quad \pi_A \cdot (c \cdot Fh) = a \cdot F\pi_A \cdot Fh = a \cdot Fh_A.$$

To complete the proof, we show that h is the unique algebra morphism. Suppose that in $\mathsf{Alg}\, F$, a morphism $h'\colon D \to A \times B$ satisfies $\pi_A \cdot h' = h_A$ and $\pi_B \cdot h' = h_B$. Then these all hold in the base category, so h' must be $\langle h_A, h_B \rangle = h$. This completes the proof. □

Remark 2.1.14. Similarly terminal objects and pullbacks – in fact, all limits in $\mathsf{Alg}\, F$ – are formed on the level of $\mathscr{A}$. This is discussed in detail in Section 4.1 (see Remark 4.1.4).

Corollary 2.1.15. *If $\mathscr{A}$ has a terminal object* 1, *then the terminal algebra is* $F1 \to 1$.

So there is not much to say about terminal algebras. Initial algebras, on the contrary, are really interesting.

Corollary 2.1.16. *A homomorphism $h\colon (A,\alpha) \to (B,\beta)$ is monic in* $\mathsf{Alg}\, F$ *iff h is monic in $\mathscr{A}$.*

Indeed, recall that h is monic iff the following commutative square is a pullback

$$\begin{array}{ccc} A & \xrightarrow{\ \mathsf{id}\ } & A \\ {\scriptstyle \mathsf{id}}\big\downarrow & & \big\downarrow{\scriptstyle h} \\ A & \xrightarrow{\ h\ } & B \end{array}$$

and use that pullbacks of F-algebras are formed on the level of $\mathscr{A}$.

Remark 2.1.17. For epimorphisms, the situation is not as simple. However, every algebra homomorphism $h\colon (A,\alpha) \to (B,\beta)$, with h being epic in $\mathscr{A}$, is epic in $\mathsf{Alg}\, F$.

Remark 2.1.18. We assume that the reader is familiar with the general notion of a subobject. For an object A, a *subobject* is represented by a monomorphism $m_B\colon B \to A$. If $m_B\colon B \to A$ and $m_C\colon C \to A$ are monomorphisms, we write

$$m_B \le m_C$$

if m_B factorizes through m_C. (That is, for some $k\colon B \to C$, $m_B = m_C \cdot k$.) Whenever $m_B \le m_C \le m_B$, then k above is an isomorphism. In this case, we identify m_B and m_C as subobject representatives. Thus, the poset of subobjects of A is formed by equivalence classes of monomorphisms with codomain A. As usual, we will refer to the monomorphisms m_B as if it were the subobject it represents. A category $\mathscr{A}$ is *well-powered* if, for every object A, only a set of subobjects exist.

Subalgebras of an algebra are understood to be subobjects in the category $\mathsf{Alg}\, F$ of algebras for F. As we will see in Remark 4.1.4, monomorphisms in $\mathsf{Alg}\, F$ are precisely the homomorphisms which are monic in the base category.

2.2 Initial Algebras

We have motivated initial algebras in Section 1.1. This section presents our first examples, mainly using endofunctors on Set and $\mathsf{CPO}_\perp$. In addition, we present some basic properties of initial algebras.

Definition 2.2.1. An algebra for F is *initial* if it admits a unique homomorphism into every F-algebra. Shortly: it is the initial object of $\mathsf{Alg}\, F$.

Example 2.2.2. If F preserves the initial object 0 of $\mathscr{A}$, then $(0, \mathsf{id})$ is the initial algebra.

Notation 2.2.3. We write μF or $\mu X.\ FX$ for (the underlying object of) the initial algebra. The algebra structure is often denoted by

$$\iota\colon F(\mu F) \to \mu F.$$

Examples will follow soon. But even before presenting those, we would like to mention an important result.

Definition 2.2.4. By a *fixed point* of an endofunctor F is meant an object A together with an isomorphism $FA \xrightarrow{\cong} A$.

Lambek's Lemma 2.2.5 [210]. *An initial algebra of an endofunctor is a fixed point.*

Proof Let (A, α) be initial. Since $(FA, F\alpha)$ is an algebra, we have a homomorphism h:

$$\begin{array}{ccc} FA & \xrightarrow{\ \alpha\ } & A \\ {\scriptstyle Fh}\downarrow & & \downarrow{\scriptstyle h} \\ FFA & \xrightarrow{\ F\alpha\ } & FA \\ {\scriptstyle F\alpha}\downarrow & & \downarrow{\scriptstyle \alpha} \\ FA & \xrightarrow{\ \alpha\ } & A \end{array}$$

The square at the bottom obviously commutes. Consequently the outside commutes, showing that $\alpha \cdot h$ is an endomorphism of (A, α) in $\mathsf{Alg}\,F$. Thus $\alpha \cdot h = \mathrm{id}_A$ by initiality. Then $h \cdot \alpha = F\alpha \cdot Fh = \mathrm{id}_{FA}$. Thus $h = \alpha^{-1}$. □

For endofunctors on Set there is a converse:

Theorem 2.2.6 [290, Thm. II.4]. *A set functor has an initial algebra iff it has a fixed point.*

We shall discuss this result in more detail and greater generality in Chapter 6 (see Theorem 6.3.4).

Example 2.2.7. (1) The power-set functor $\mathscr{P}$: Set $\to$ Set does not have an initial algebra: Cantor's Theorem [99] tells us that for all sets A, there is no map of A onto $\mathscr{P}A$. For the proof, suppose towards a contradiction that $f: A \to \mathscr{P}A$ is a surjection. Let $X = \{s \in S : s \notin f(s)\}$. Since f is surjective, let s_0 be such that $f(s_0) = X$. Then we have a contradiction: $s_0 \in X$ iff $s_0 \in f(s_0)$ iff $s_0 \notin X$. Therefore, there exists no fixed point of $\mathscr{P}$.

(2) Let $(P, \leq)$ be a poset, considered as a category. An endofunctor $F: P \to P$ is just a monotone function, and a fixed point is an element $p \in P$ with $F(p) = p$ as usual. An algebra is an element $p \in P$ such that $Fp \leq p$. Such an element p is a *pre-fixed point*. Lambek's Lemma then says that if p is the least pre-fixed

point, then $p = Fp$. In Appendix B we collect results about fixed points of monotone functions.

(3) The initial algebra of the set functor $FX = X + 1$ is the algebra of natural numbers:

$$\mu X.\,(X+1) = \mathbb{N}.$$

The algebra structure $a = [a_1, a_0]$, compare Example 2.1.3(1), consists of the unary operation a_1 which is the successor function $s\colon \mathbb{N} \to \mathbb{N}$, and the constant $a_0 = 0$. Shortly: $\iota = [s, 0]$. Indeed, given an algebra $[\beta_1, \beta_0]\colon B{+}1 \to B$ the unique homomorphism $h\colon \mathbb{N} \to B$ is determined by

$$\begin{aligned} h(0) &= \beta_0, \\ h(sn) &= \beta_1(h(n)) \qquad\qquad \text{for all } n \in \mathbb{N}. \end{aligned}$$

See Example 2.2.9 for a generalization of this example to general base categories.

(4) Let $\mathscr{P}_{\mathsf{f}}$, *the finite power-set functor*, be the variant version of $\mathscr{P}$ from Example 2.1.3(4) given by

$$\mathscr{P}_{\mathsf{f}}A = \{M \subseteq A : M \text{ is finite}\}.$$

The initial algebra $\mu\mathscr{P}_{\mathsf{f}}$ can be described as the set of all *hereditarily finite sets*. These are finite sets all elements of which are hereditarily finite sets again.

In concrete terms,

$$V_\omega = \emptyset \cup \mathscr{P}\emptyset \cup \mathscr{P}\mathscr{P}\emptyset \cup \cdots \cup \mathscr{P}^n\emptyset \cup \cdots.$$

Each hereditarily finite set is a finite subset of V_ω, and conversely. Thus $\mathscr{P}_{\mathsf{f}}V_\omega = V_\omega$. The proof that (V_ω, id) is initial for $\mathscr{P}_{\mathsf{f}}$ can be found in Example 3.2.9.

Remark 2.2.8. Although we use notation μF, initial algebras are only unique up to isomorphism. Indeed, given an isomorphism $i\colon A \to \mu F$ in the base category, then A with the algebra structure

$$\alpha = (FA \xrightarrow{Fi} F(\mu F) \xrightarrow{\iota} \mu F \xrightarrow{i^{-1}} A)$$

is also initial. This follows from $i\colon (A, \alpha) \to (\mu F, \iota)$ being an isomorphism in $\mathsf{Alg}\,F$.

Nevertheless, we shall always speak of *the* initial algebra for an endofunctor (whenever it exists).

Frequently, there are different ways to present an initial algebra; we shall see some examples in this chapter.

Example 2.2.9. We now present an example that applies to all of the base categories which we have seen so far in this chapter. In fact, it applies to every base category that has countable copowers

$$\mathbb{N} \bullet A = A + A + A + \cdots .$$

Then for every object A the endofunctor $FX = X + A$ has the initial algebra

$$\mu X.\,(X + A) = \mathbb{N} \bullet A.$$

If $\mathsf{in}_k \colon A \to \mathbb{N} \bullet A$ are the coproduct injections for $k \in \mathbb{N}$, then the algebra structure is

$$\iota = [\alpha_1, \mathsf{in}_0] \colon (\mathbb{N} \bullet A) + A \to \mathbb{N} \bullet A,$$

where α_1 is determined by the following commutative triangles

$$\begin{array}{ccc} \mathbb{N} \bullet A & \xrightarrow{\quad \alpha_1 \quad} & \mathbb{N} \bullet A \\ {\scriptstyle \mathsf{in}_k} \nwarrow & & \nearrow {\scriptstyle \mathsf{in}_{k+1}} \\ & \mathbb{N} & \end{array} \qquad (k \in \mathbb{N})$$

Indeed, let $[\beta_1, \beta_0] \colon B + A \to B$ be an algebra. Then a homomorphism h:

$$\begin{array}{ccc} \mathbb{N} \bullet A + A & \xrightarrow{[\alpha_1, \alpha_0]} & \mathbb{N} \bullet A \\ {\scriptstyle h+\mathsf{id}} \downarrow & & \downarrow {\scriptstyle h} \\ B + A & \xrightarrow{[\beta_1, \beta_0]} & B \end{array}$$

is uniquely determined by $h \cdot \mathsf{in}_0$ and $h \cdot \mathsf{in}_{k+1} = \beta_1 \cdot (h \cdot \mathsf{in}_k)$: we have

$$h = [\beta_0, \beta_1\beta_0, \beta_1\beta_1\beta_0, \ldots].$$

In the special case where $A = 1$ is a terminal object, the initial algebra $\mathbb{N} \bullet 1$ for $FX = X + 1$ is called a *natural numbers object* (or *NNO*, for short).

Remark 2.2.10. (1) In the next example, and at many later points, we use trees to describe algebras of special interest. Let us recall that a *tree* is a directed graph with a distinguished node called the *root* from which every other node can be reached by a unique directed path. A tree might well be infinite. We always identify isomorphic trees.

(2) We distinguish between *unordered trees*, defined as above, and *ordered trees*. An ordered tree comes with a linear order on the children of every node. In pictures, this linear order is the left-to-right order. In this chapter, all the trees will be ordered. But later in the book we shall also consider unordered trees.

A node is a *leaf* if it has no children. A tree is *binary* if every node which is not a leaf has precisely two children. The *depth* of a node of a tree is its distance from the root.

(3) A *labelled tree* (ordered or unordered) is a tree together with a function assigning to every node an element of a given set M (of labels). We also consider labelled trees up to (label preserving) isomorphism.

(4) Let us also recall that the *complete n-ary tree* (in which every node has precisely n children) can be encoded as A^*, the set of words over $A = \{a_0, \ldots, a_{n-1}\}$, for any n-element set A. The root is ε, the empty word. And the n children of a word w are the words $wa_0, \ldots, wa_{n-1}$.

Example 2.2.11. Initial Σ-algebra. It is well known from general algebra that, for a signature Σ, the initial algebra is formed as follows:

$$\mu H_\Sigma = \text{all ground terms, that is, terms without variables.}$$

This is the smallest set such that

(1) every constant symbol is a term, i.e. $\Sigma_0 \subseteq \mu H_\Sigma$, and

(2) given an n-ary symbol $\sigma \in \Sigma_n$ and n ground terms $t_0, \ldots, t_{n-1}$, then $\sigma(t_0, \ldots, t_{n-1})$ is a ground term.

The operations of the algebra μH_Σ are given by term formation.

Given a Σ-algebra A, every ground term computes in A to a (unique) element, and the resulting mapping from μH_Σ to A is the unique homomorphism from μH_Σ to A.

We now provide an alternative description of μH_Σ using trees.

Definition 2.2.12. Given a signature Σ, by a Σ-*tree* is meant an ordered tree labelled in $\coprod_{n\in\mathbb{N}} \Sigma_n$ in such a way that every node of k children is labelled by a k-ary symbol. Every n-ary symbol $\sigma \in \Sigma_n$ defines an n-ary operation of *tree-tupling* constant on the set of all Σ-trees: it sends an n-tuple $t_0, \ldots, t_{n-1}$ to the following Σ-tree

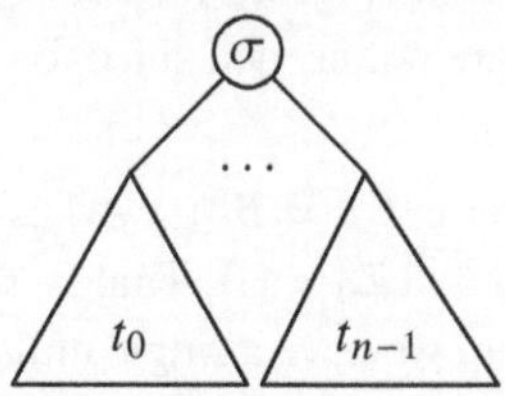

Remark 2.2.13. In most of the classical literature (e.g. Courcelle's paper [107]), Σ-trees are described in the following equivalent way: a Σ-tree is a partial function $t\colon \mathbb{N}^* \to \coprod_{n\in\mathbb{N}} \Sigma_n$ such that

(1) the domain of definition of t is nonempty and *prefix-closed*, i.e. whenever a word wi lies in the domain, so does w (this entails that ε lies in the domain), and

(2) whenever $t(w) \in \Sigma_n$ then $t(wi)$ is defined if and only if $i = 0, \ldots, n-1$.

Note that a Σ-tree may be infinite.

We will use this description of Σ-trees in the proof of Theorem 2.5.9. It is equivalent to the above one: given a partial function t as above, the nodes of the corresponding Σ-tree are the elements of the domain of definition of t, ε is the root, the label of a node w is $t(w)$, and the children of w with an n-ary label are $w0, \ldots, w(n-1)$.

Proposition 2.2.14. *The algebra of finite Σ-trees is an initial algebra μH_Σ.*

Proof It is sufficient to find an isomorphism i between the above algebras of terms and of finite Σ-trees in $\mathsf{Alg}\, H_\Sigma$. It is defined by structural induction on terms as follows: for all $\sigma \in \Sigma_0$ let $i(\sigma)$ be the single-node tree labelled by σ:

and for all $\sigma \in \Sigma_n$, $n > 0$, use tree-tupling: the term $\sigma(t_0, \ldots, t_{n-1})$ is mapped to the following Σ-tree

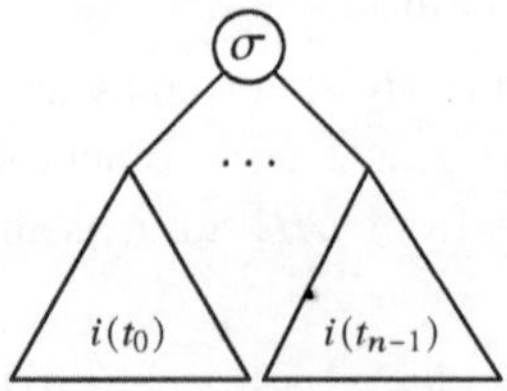

This is a bijection: the inverse function i^{-1} assigns to a tree t with root labelled by $\sigma \in \Sigma_n$ the term $\sigma(t_0, \ldots, t_{n-1})$, where t_k is the term such that $i(t_k)$ is the kth maximum subtree of t.

It is clear that i preserves the operations. □

Example 2.2.15. (1) The initial algebra for $FX = X \times X + 1$ is formed by all finite binary trees. (Labels are not needed since for every arity there is at most one operation symbol.)

(2) For a set B, the set functor $FX = B \times X + 1$ is the polynomial functor of the signature Σ with $\Sigma_1 = B$ and $\Sigma_0 = \{\cdot\}$. That is, the elements of B appear as unary operation symbols, and we have a single nullary symbol, here written as a dot $\cdot$ symbol. Here are three examples of finite Σ-trees:

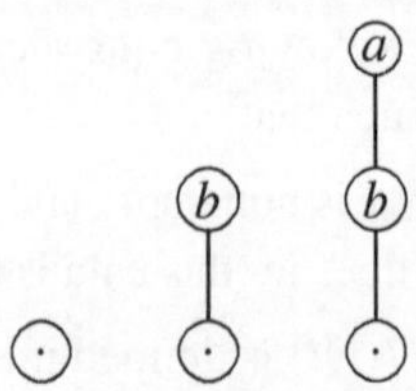

We thus conclude that $\mu F = B^*$. Indeed, the set of all Σ-trees for this signature is isomorphic to the set B^* of all words over B.

Example 2.2.16. For many-sorted algebras, see Example 2.1.6, we also get that μH_Σ consists of all ground terms. The sorted set of all ground terms is defined as the set $T = (T_s)_{s\in S}$ where each T_s (terms of output-sort s) is the smallest set such that

(1) every constant of sort s is a term in T_s, and

(2) given an operation symbol σ of arity $s_0 \times \cdots \times s_{k-1} \to s$ and k terms $t_0 \in T_{s_0}, \dots, t_{k-1} \in T_{s_{k-1}}$, then $\sigma(t_0, \dots, t_{k-1})$ is a term in T_s.

Again, we can instead work with Σ-*trees* which are ordered trees labelled in $S\times\Sigma$, where the label (s, σ) means that the node has sort s and corresponds to the opertation σ. This labelling is such that every node labelled by (s, σ) with a symbol σ of arity $s_1 \times \cdots \times s_n \to s$ has n children, and the ith child has sort s_i for all $i = 0, \dots, n-1$. The *sort* of a Σ-tree is defined to be the sort in its root label. We have the following description of μH_Σ: for every sort s,

$$(\mu H_\Sigma)_s = \text{all finite } \Sigma\text{-trees of output sort } s.$$

The proof is completely analogous to that of Proposition 2.2.14.

Example 2.2.17. Let us consider initial algebras in $\mathsf{CPO}_\perp$ (see Example 2.1.7). Recall that coproducts are disjoint unions with bottom elements smashed to one.

(1) The initial algebra for $FX = X_\perp$ is the algebra

$$\mu X.X_\perp = \mathbb{N}^\top$$

of natural numbers extended by a largest element ∞. The unary operation is the successor function $s\colon (\mathbb{N}^\top)_\perp \to \mathbb{N}^\top$ where we put $s(\perp) = 0$ and $s(\infty) = \infty$.

Indeed, given an algebra $\alpha\colon A_\perp \to A$, the unique homomorphism $h\colon \mathbb{N}^\top \to A_\perp$ fulfils $h(0) = \perp$ (since h is strict) and $h(sn) = \alpha(h(n))$. This determines h on $\mathbb{N}$, and continuity determines the value

$$h(\infty) = h(\textstyle\bigsqcup_{n\in\mathbb{N}} n) = \bigsqcup_{n\in\mathbb{N}} h(n).$$

(2) For the functor $FX = (X \times X)_\perp + 1_\perp$ of one binary continuous operation and one constant we have

$$\mu X.(X \times X)_\perp + 1_\perp = \text{all binary trees}$$

(finite and infinite!). The constant is the root-only tree. The binary operation is tree-tupling. And the ordering is the least one for which the root-only tree is the smallest element, and tree-tupling is continuous. Example:

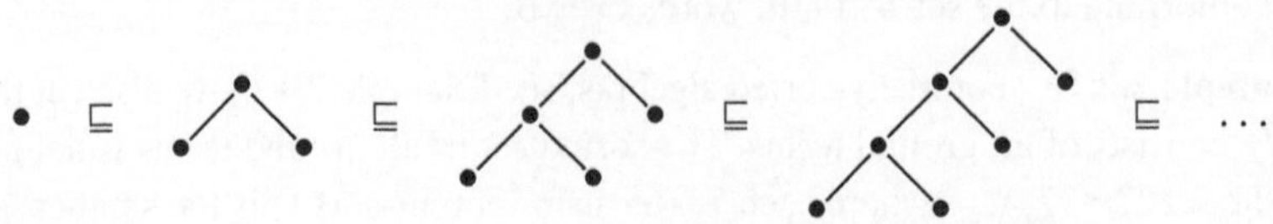

The above is an ω-chain whose join is an (obvious) infinite tree. This explains the presence of infinite trees in μH_Σ. Indeed, every infinite binary tree t has the form $t = \bigsqcup_{n\in\mathbb{N}} t_n$, where t_n is the finite tree obtained by cutting t at the level n. Clearly, $t_0 \sqsubseteq t_1 \sqsubseteq t_2 \cdots$.

Example 2.2.18. (1) Let us come back to the category Nom of nominal sets and the functor $FX = \mathbb{A} + X \times X + [\mathbb{A}]X$ (see Example 2.1.10). We have seen that the nominal set L of all λ-terms modulo α-equivalence is an algebra for F. In fact, Gabbay and Pitts [133, Thm. 6.2] have shown that it is the initial algebra for F:

$$\mu F = L = \text{all } \lambda\text{-terms modulo } \alpha\text{-equivalence.}$$

The initiality of L provides a structural induction principle. For example, it allows to define the usual (capture avoiding) substitution of λ-terms seamlessly. Indeed, let

$$\iota = [\iota_1, \iota_2, \iota_3] : \mathbb{A} + L \times L + [\mathbb{A}]L \to L$$

be the structure of the initial algebra, and suppose we would like to substitute a variable $a \in \mathbb{A}$ by the λ-term t. Let $s : \mathbb{A} \to L$ be the unique equivariant map with $s(a) = t_\alpha$, where t_α denotes the α-equivalence class of t. Then one forms the algebra

$$\sigma = [s, \iota_2, \iota_3] : \mathbb{A} + L \times L \times [\mathbb{A}]L \to L.$$

It is now easy to verify that the unique algebra homomorphism $(L, \iota) \to (L, \sigma)$ maps every $t'_\alpha \in L$ to $t'[t/a]_\alpha$.

(2) Similarly, terms over a binding signature (Remark 2.1.9) modulo α-equivalence form an initial algebra (see Gabbay and Pitts [133, Rem. 6.13] and cf. Pitts [250, Thm. 8.15]). Initiality can be used to define capture-avoiding substitution for such terms.

(3) Note that λ-terms (but *not* modulo α-equivalence) form the initial algebra for a different endofunctor on Nom, $F'X = \mathbb{A} + X \times X + \mathbb{A} \times X$. However, initiality then does not yield the desired capture-avoiding substitution of λ-terms.

(4) We also have seen that λ-terms modulo α-equivalence form an algebra for the functor $FX = V + X \times X + \delta X$ on the category $\mathsf{Set}^{\mathcal{F}}$ of sets in context (Example 2.1.8). In fact, Fiore et al. [124, Thm. 2.1] have shown that this is the

initial F-algebra. Again, initiality can be used to give a convenient definition of capture-avoiding substitution of λ-terms.

More generally, for the functor F_Σ associated to a binding signature Σ (Remark 2.1.9) μF_Σ assigns to a context Γ the α-equivalence classes of terms over Σ with variables from Γ.

Recall that subalgebras are subobjects in $\mathsf{Alg}\, F$: given an algebra (A, α), subalgebras are represented by homomorphisms $h\colon (B, \beta) \to (A, \alpha)$ which are monic in the base category.

Definition 2.2.19. An algebra is *minimal* if it has no proper subalgebra. That is, every subalgebra is carried by an isomorphism.

We have seen in Proposition 1.2.1 that initial algebras are minimal.

Remark 2.2.20. (1) In contrast with initial algebras, minimal algebras are plentiful: every set functor has a minimal algebra. Indeed, either $\alpha\colon F1 \to 1$ is minimal, or it has a proper subalgebra. In the second case, the carrier is $\emptyset$, and this implies that $F\emptyset = \emptyset$. Then $\mu F = \emptyset$ is minimal.

(2) We will see later that under hypotheses, minimal algebras which are fixed points are initial.

Proposition 2.2.21. *Let F be a set functor with an initial algebra. Its minimal algebras are the quotient algebras of μF, which are those with a surjective homomorphism from μF.*

Proof (1) Given a surjective homomorphism $e\colon (\mu F, \tau) \to (A, \alpha)$ and a subalgebra $m\colon (B, \beta) \to (A, \alpha)$, we prove that m is invertible. We have a unique homomorphism $h\colon (\mu F, \tau) \to (B, \beta)$, and the universal property of μF implies that $m \cdot h = e$. Thus, m is surjective and hence invertible.

(2) Given a minimal algebra (A, α), we factorize the unique homomorphism $h\colon (\mu F, \tau) \to (A, \alpha)$ as surjective map e followed by an injective map m (in Set). This gives the left- and right-hand parts of the following diagram:

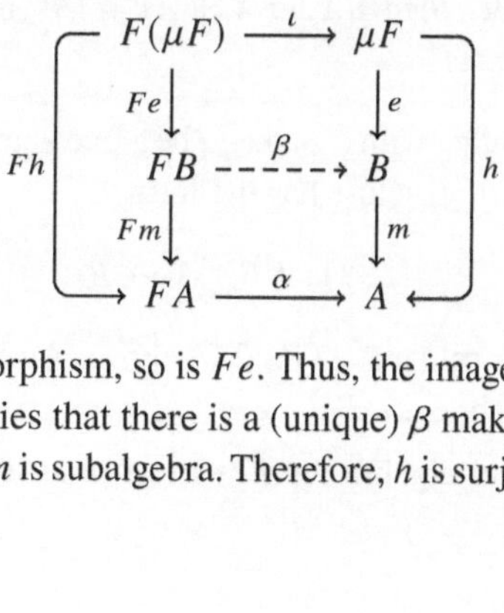

Since e is a split epimorphism, so is Fe. Thus, the image of $\alpha \cdot Fm$ is included in that of m. This implies that there is a (unique) β making the diagram above commute. We see that m is subalgebra. Therefore, h is surjective, as required. $\square$

Remark 2.2.22. We shall give the definition of a *strong monomorphism* in Definition 11.4.2. Readers familiar with this notion will recognize that our argument above reviews the proof that all monomorphisms in Set are strong, and that β is obtained by appealing to the strength of the monomorphism m.

Initial algebras and free algebras

Remark 2.2.23. Initial algebras are closely related to free algebras. By a *free F-algebra* on an object A (of generators) in $\mathscr{A}$ is meant an algebra

$$\varphi_A \colon FA^\sharp \to A^\sharp$$

together with a universal arrow $\eta_A \colon A \to A^\sharp$. Universality means that for every algebra $\beta \colon FB \to B$ and every morphism $f \colon A \to B$ in $\mathscr{A}$, there exists a unique homomorphism $\bar{f} \colon A^\sharp \to B$ extending f, that is, a unique morphism of $\mathscr{A}$ for which the diagram below commutes:

$$\begin{array}{ccccc} FA^\sharp & \xrightarrow{\varphi_A} & A^\sharp & \xleftarrow{\eta_A} & A \\ {\scriptstyle F\bar{f}}\downarrow & & \downarrow{\scriptstyle \bar{f}} & \swarrow{\scriptstyle f} & \\ FB & \xrightarrow{\beta} & B & & \end{array} \tag{2.5}$$

If $\mathscr{A}$ has binary coproducts, then we can reduce (2.5) to a square as follows:

$$\begin{array}{ccc} FA^\sharp + A & \xrightarrow{[\varphi_A, \eta_A]} & A^\sharp \\ {\scriptstyle F\bar{f}+\mathrm{id}_A}\downarrow & & \downarrow{\scriptstyle \bar{f}} \\ FB + A & \xrightarrow{[\beta, f]} & B \end{array} \tag{2.6}$$

Note that for an endofunctor F and fixed object A, we get a new functor $F(-) + A$ defined in the obvious way.

Proposition 2.2.24. *Let $\mathscr{A}$ be a category with finite coproducts. For every endofunctor F, the free F-algebra on A is precisely the initial algebra for $F(-) + A$.*

That is, if $A^\sharp$ is free, then $\mu X.(FX + A) = A^\sharp$ with the algebra structure $[\varphi_A, \eta_A]$, and vice versa.

Proof (1) Let $A^\sharp$ together with φ_A and η_A be a free algebra. Given an algebra B for $F(-) + A$, its algebra structure has the form

$$[\beta, f] \colon FB + A \to B$$

for a unique pair of morphisms as in (2.6) above. And $\bar{f}$ above is the unique homomorphism of algebras for $F(-) + A$.

(2) Let $A^\sharp$ be an initial algebra for $F(-) + A$, and let $\varphi_A \colon FA^\sharp \to A^\sharp$ and

$\eta_A\colon A \to A^\sharp$ denote the components of its algebra structure. This is a free F-algebra on A. Indeed, given an F-algebra $\beta\colon FB \to B$ and a morphism $f\colon B \to A$, we get a unique homomorphism $\bar{f}$ of algebras for $F(-) + A$ from $A^\sharp$ to B. That is, a unique morphism for which the above square commutes. But this is equivalent to the commutativity of the diagram (2.5). Hence, $(A^\sharp, \varphi_A)$ is a free F-algebra on A. □

Example 2.2.25. (1) The identity functor on Set has free algebras

$$A^\sharp = \mathbb{N} \times A$$

with the algebra structure $\varphi_A\colon \mathbb{N} \times A \to \mathbb{N} \times A$ given by $\varphi(n, a) = (n + 1, a)$ and the universal morphism $\eta_A\colon A \to \mathbb{N} \times A$ taking a to $(0, a)$.

More generally, if $\mathscr{A}$ is a category with countable coproducts, then a free algebra for Id on an object A is $A^\sharp = \mathbb{N} \bullet A$ (Example 2.2.9).

(2) Let Σ be a (one-sorted) signature. A free Σ-algebra (in Set) on a set A is precisely an initial Σ_A-algebra, where Σ_A is the signature Σ expanded by nullary operations indexed by A. Indeed, from the formula $H_\Sigma X = \coprod_{n\in\mathbb{N}} \Sigma_n \times X^n$, we see that $H_\Sigma(-) + A = H_{\Sigma_A}$.

We can describe $A^\sharp$ as the algebra of all finite Σ_A-trees. That is, leaves are labelled either by nullary symbols of Σ or by elements of A.

(3) Analogously for S-sorted signatures: the free S-sorted Σ-algebra on a set $A = (A_s)_{s\in S}$ is the algebra of all finite Σ_A-trees. Here Σ_A is the signature Σ with additional nullary operations of sorts s indexed by A_s for every sort s.

Definition 2.2.26. An endofunctor F is called a *varietor* if every object generates a free algebra for F.

Thus, for $\mathscr{A}$ = Set we have seen that H_Σ is a varietor and $\mathscr{P}$ is not. A characterization of varietors on Set will be presented in Corollary 6.4.1.

Corollary 2.2.27. *An endofunctor F on a category with finite coproducts is a varietor iff for every object A an initial algebra for $F(-) + A$ exists.*

Remark 2.2.28. (1) We recall that a *monad* on a category $\mathscr{A}$ consists of an endofunctor M on $\mathscr{A}$ and a pair of natural transformations $\eta\colon \mathsf{Id} \to M$ (the *unit*) and $\mu\colon MM \to M$ (the *multiplication*) such that the following diagrams (called the *unit laws* and *associativity*, respectively) commute:

$$\begin{array}{ccccc} M & \xrightarrow{\eta M} & MM & \xleftarrow{M\eta} & M \\ & \searrow\!\!\!\!= & \downarrow{\scriptstyle \mu} & =\!\!\!\!\swarrow & \\ & & M & & \end{array} \qquad \begin{array}{ccc} MMM & \xrightarrow{\mu M} & MM \\ {\scriptstyle M\mu}\downarrow & & \downarrow{\scriptstyle \mu} \\ MM & \longrightarrow & M \end{array}$$

We often just write M for a monad in lieu of the triple (M, η, μ). The dual notion is that of a *comonad*.

(2) Recall (e.g. from MacLane [217]) that a functor $L\colon \mathscr{A} \to \mathscr{B}$ is a *left adjoint* to a functor $R\colon \mathscr{B} \to \mathscr{A}$ if there are natural transformations $\eta\colon \mathsf{id}_{\mathscr{B}} \to RL$ and $\varepsilon\colon LR \to \mathsf{Id}_{\mathscr{A}}$ such that the following *triangular identities* hold:

$$\varepsilon L \cdot L\eta = \mathsf{id}_L \qquad \text{and} \qquad R\varepsilon \cdot \eta R = \mathsf{id}_R.$$

We say that R is the *right adjoint* to L, and we write $L \dashv R$.

(3) Every adjoint situation $L \dashv R\colon \mathscr{B} \to \mathscr{A}$ with unit $\eta\colon \mathsf{Id} \to RL$ and counit $\varepsilon\colon LR \to \mathsf{Id}$ gives rise to a monad $M = RL$ with unit η and multiplication $\mu = R\varepsilon L$.

(4) A functor F is a varietor iff the forgetful functor from $\mathsf{Alg}\, F$ to $\mathscr{A}$ has a left adjoint $A \mapsto A^\sharp$. This defines a monad (M, η, μ) on $\mathscr{A}$ by

$$MA = A^\sharp.$$

For a morphism $f\colon A \to B$, we have a unique homomorphism of F-algebras $Mf\colon A^\sharp \to B^\sharp$ with $Mf \cdot \eta_A = \eta_B \cdot f$. The unit $\eta\colon \mathsf{Id} \to M$ has as its components the universal arrows $\eta_A\colon A \to A^\sharp$, and the monad multiplication $\mu\colon MM \to M$ has as its components the unique F-algebra homomorphism

$$\mu_A\colon (A^\sharp)^\sharp \to A^\sharp \qquad \text{with} \qquad \mu_A \cdot \eta_{A^\sharp} = \mathsf{id}_{A^\sharp}.$$

This monad is free on F with respect to the universal natural transformation $F \to M$ with components

$$FA \xrightarrow{F\eta_A} FA^\sharp \xrightarrow{\varphi_A} A^\sharp.$$

This was proved by Barr [69]. For a complete and well-powered category, he also proved the converse which was later strengthened by Kelly [187] to categories with powers:

Theorem 2.2.29. *Given a category with powers, an endofunctor F generates a free monad iff it is a varietor.*

Example 2.2.30. (1) If $\mathscr{A}$ has countable coproducts, then the free monad on Id is $MA = \mathbb{N} \bullet A$, see Example 2.2.9.

(2) The free monad on the set functor $FX = X \times X$ is given by

$$MA = \text{ finite binary ordered trees with leaves labelled in } A.$$

Indeed, the functor $F(-) + A$ is the polynomial functor of the signature of one binary operation and constants indexed by A.

2.3 Recursion and Induction

The most basic form of *recursion* concerns functions on the natural numbers: we specify a function f by specifying its value at 0, and for all n, deriving the value $f(n+1)$ from $f(n)$. As we have seen, the natural numbers form the initial algebra of $FX = X + 1$, see Example 2.2.7(2). This is a special case of the following.

Definition 2.3.1. Let F be an endofunctor with an initial algebra. Given an object A, we say that a morphism $f\colon \mu F \to A$ is *recursively specified* if there exists an algebra structure $\alpha\colon FA \to A$ (a specification of f) turning f into a homomorphism.

Example 2.3.2. (1) The height of a finite tree. Here we work with finite binary trees as an initial algebra

$$T \quad = \quad \mu X.\,(X \times X + 1) \ ,$$

see Example 2.2.15. The function $\mathsf{ht}\colon T \to \mathbb{N}$ assigns 0 to the root-only tree, and given a tree t with maximum subtrees t_1, t_2, then

$$\mathsf{ht}(t) = 1 + \max(\mathsf{ht}(t_1), \mathsf{ht}(t_2)).$$

In order to specify height recursively, we need an algebra structure on $\mathbb{N}$ of the form $[\alpha_1, \alpha_0]\colon \mathbb{N} \times \mathbb{N} + 1 \to \mathbb{N}$. The obvious candidate is

$$\alpha_1(n, m) = 1 + \max(n, m) \qquad \text{and} \qquad \alpha_0 = 0.$$

Indeed, the following square (where ι_1 is the tree-tupling and ι_0 represents the root-only tree) clearly commutes:

$$\begin{array}{ccc} T \times T + 1 & \xrightarrow{[\iota_1, \iota_0]} & T \\ {\scriptstyle \mathsf{ht}\times\mathsf{ht}+\mathsf{id}}\downarrow & & \downarrow{\scriptstyle \mathsf{ht}} \\ \mathbb{N} \times \mathbb{N} + 1 & \xrightarrow{[\alpha_1, \alpha_0]} & \mathbb{N} \end{array}$$

It is easy to see by structural induction that ht is indeed the unique homomorphism from T to $\mathbb{N}$.

(2) The same example but now played with potentially infinite trees: put $\mathscr{A} = \mathsf{CPO}_\perp$ and $FX = (X \times X)_\perp + 1_\perp$, the functor with the initial algebra

$$\bar{T} = \text{(finite and infinite) binary trees;}$$

see Example 2.2.17(2). We want to recursively specify the function

$$\mathsf{ht}\colon \bar{T} \to \mathbb{N}^\top$$

assigning ∞ to every infinite tree. This requires an algebra structure for F (in $\mathsf{CPO}_\perp$) on the set $\mathbb{N}^\top$. Analogously to the preceding example we put

$$[\overline{\alpha}_1, \overline{\alpha}_0] \colon \mathbb{N}^\top \times \mathbb{N}^\top + 1 \to \mathbb{N}^\top,$$

where $\overline{\alpha}_1(n, m) = 1 + \max(n, m)$, which is understood as expected ($1 + \infty = \infty$ and $\max(n, \infty) = \infty$) and $\overline{\alpha}_0 = 0$.

The leading idea in Example 2.3.2 and many similar results is to use the initiality principle to establish a principle of definition by recursion. We now present an elegant categorical generalization of this idea, which generalizes recursion with parameters and is often called 'primitive recursion' in the literature:

Theorem 2.3.3 (Primitive Recursion)**.** *Let F be a functor with an initial algebra on a category with finite products. Then for every morphism $\alpha \colon F(A \times \mu F) \to A$ there exists a unique morphism $h \colon \mu F \to A$ such that the square below commutes:*

$$\begin{array}{ccc} F(\mu F) & \xrightarrow{\;\iota\;} & \mu F \\ {\scriptstyle F\langle h, \mathsf{id}_{\mu F}\rangle}\downarrow & & \downarrow{\scriptstyle h} \\ F(A \times \mu F) & \xrightarrow{\;\alpha\;} & A \end{array} \tag{2.7}$$

Proof Let π_1 and π_2 denote the projections of the product $A \times \mu F$. Form the following algebra with carrier $A \times \mu F$:

$$\bar{\alpha} = \big(F(A \times \mu F) \xrightarrow{\langle \mathsf{id}, F\pi_2\rangle} F(A \times \mu F) \times F(\mu F) \xrightarrow{\alpha \times \iota} A \times \mu F\big).$$

Let $\bar{h} \colon \mu F \to A \times \mu F$ be the unique homomorphism. The right-hand component of $\bar{h}$ is $\mathsf{id}_{\mu F}$. Indeed, the diagram below shows that $\pi_2 \cdot \bar{h}$ is a homomorphism from μF to itself:

$$\begin{array}{ccccc} F(\mu F) & & \xrightarrow{\qquad\qquad\iota\qquad\qquad} & & \mu F \\ {\scriptstyle F\bar{h}}\downarrow & & & & \downarrow{\scriptstyle \bar{h}} \\ F(A \times \mu F) & \xrightarrow{\langle \mathsf{id}, F\pi_2\rangle} & F(A \times \mu F) \times F(\mu F) & \xrightarrow{\alpha \times \iota} & A \times \mu F \\ {\scriptstyle F\pi_2}\downarrow & \swarrow{\scriptstyle \pi_2} & & & \downarrow{\scriptstyle \pi_2} \\ F(\mu F) & & \xrightarrow{\qquad\qquad\iota\qquad\qquad} & & \mu F \end{array} \tag{2.8}$$

Thus, putting $h = \pi_1 \cdot \bar{h}$ we have $\bar{h} = \langle h, \mathsf{id}_{\mu F}\rangle$, and extending the upper square above by π_1 we obtain the desired square (2.7).

For the uniqueness, suppose that h makes (2.7) commutative. Then we see that for $\bar{h} = \langle h, \mathsf{id}_{\mu F}\rangle$ the upper square in diagram (2.8) commutes. We know that its left-hand component is commutative, and the right-hand one trivially commmutes. □

Example 2.3.4. Recursion with parameters and the classical primitive recursion for functions on natural numbers are special cases of Theorem 2.3.3: choose $\mathscr{A}$ = Set and $FX = X + 1$.

(1) Recursion with parameters. Given $a \in A$ and $f\colon A \times \mathbb{N} \to A$ we form

$$\alpha = \big(F(A \times \mu F) = A \times \mathbb{N} + 1 \xrightarrow{[a,f]} \mathbb{N} = \mu F\big).$$

Spelling out the commutativity of (2.7), we see that this uniquely determines the function $h\colon \mathbb{N} \to A$ such that

$$\begin{aligned} h(0) &= a, \\ h(n+1) &= f(h(n), n). \end{aligned}$$

For a concrete example, choose $A = \mathbb{N}$, $a = 0 \in \mathbb{N}$ and let $f\colon \mathbb{N} \times \mathbb{N}$ be given by $f(m, n) = m \cdot (n + 1)$. Then $h\colon \mathbb{N} \to \mathbb{N}$ is the factorial function given by its usual recursive definition $h(0) = 0$ and $h(n + 1) = h(n) \cdot (n + 1)$.

(2) Classically, primitive recursion defines a function $h\colon \mathbb{N}^{k+1} \to \mathbb{N}$ by recursion on the first variable, taking the remaining k variables as parameters. More precisely, given $f\colon \mathbb{N}^k \to \mathbb{N}$ and $g\colon \mathbb{N}^{k+2} \to \mathbb{N}$ the function h is uniquely determined by

$$\begin{aligned} h(0, n_1, \dots, n_k) &= f(n_1, \dots, n_k), \\ h(n+1, n_1, \dots, n_k) &= g(f(n, n_1, \dots, n_k), n, n_1, \dots, n_k). \end{aligned}$$

In order to obtain this as a special instance of (2.7) choose $A = [\mathbb{N}^k, \mathbb{N}]$ the set of functions from $\mathbb{N}^k$ to $\mathbb{N}$ and define $\alpha\colon \mathbb{N} \times A + 1 \to A$ componentwise using currying: the left-hand component is curry $f\colon 1 \to [\mathbb{N}^k, \mathbb{N}]$ and the right-hand one is given by currying the function

$$([\mathbb{N}^k, \mathbb{N}] \times \mathbb{N}) \times \mathbb{N}^k \cong [\mathbb{N}^k, \mathbb{N}] \times \mathbb{N}^k \times \mathbb{N} \xrightarrow{\langle \mathrm{ev}, \pi_3, \pi_2 \rangle} \mathbb{N} \times \mathbb{N} \times \mathbb{N}^k \cong \mathbb{N}^{k+2} \xrightarrow{g} \mathbb{N},$$

where π_2 and π_3 are the second and third product projections, respectively. This function clearly maps the triple given by the function $f(n, -, \dots, -)$ in $[\mathbb{N}^k, \mathbb{N}]$, $n_1, \dots, n_k \in \mathbb{N}^k$, and $n \in \mathbb{N}$ to the right-hand side of the second equation above. It is therefore easy to see that the commutativity of (2.7) states precisely that h is defined by primitive recursion from f and g.

Remark 2.3.5. A classical induction proof takes a subset A of $\mathbb{N} = \mu X.X + 1$. If $0 \in A$ and $n \in A \Rightarrow sn \in A$, then it concludes $A = \mathbb{N}$. In other words, if A is a subalgebra of $(\mathbb{N}, \iota)$, where $\iota = [s, 0]$, then $A = \mathbb{N}$.

We now formulate the corresponding principle for F-algebras. Recall the concept of subalgebra from Remark 2.1.18.

Lemma 2.3.6 (Induction Principle [212]). *If F is an endofunctor with an initial algebra, then every subalgebra $m\colon (B,\beta) \to (\mu F, \iota)$ is an isomorphism.*

Proof We have a unique homomorphism $h\colon (\mu F, \iota) \to (B, \beta)$. Then $m \cdot h$ is an endomorphism of the initial object Alg F. But an initial object has only one endomorphism, the identity. Therefore $m \cdot h = \mathsf{id}$. Since m is monic and split epic, it is an isomorphism. □

Example 2.3.7. (1) The case $\mathbb{N} = \mu X.X + 1$ is the most basic form of mathematical induction. To show that $A \subseteq \mathbb{N}$ is all of $\mathbb{N}$, one shows that $0 \in A$ and that A is closed under the successor function.

(2) For the set $T = \mu X.X \times X + 1$ of all finite binary trees (cf. Example 2.2.15), the 'tree-induction principle' states that to prove that a set $A \subseteq T$ contains all trees, one needs only to verify that

(a) A contains the root-only tree, and
(b) with every pair A contains its tree-tupling.

(3) For $\Sigma^* = \mu X.X \times \Sigma + 1$ (cf. Example 2.2.15), a set $A \subseteq X^*$ contains all words provided that

(a) A contains ε, and
(b) with every word w it contains all sw, $s \in \Sigma$.

2.4 Coalgebras

The concept of a coalgebra for an endofunctor F is formally just the dual of an algebra: it consists of an object A and a morphism $\alpha\colon A \to FA$. A number of interesting types of state-based systems can be formalized as coalgebras, thinking of the object A as the collection of states and the structure α as the dynamics of the system.

Definition 2.4.1. A *coalgebra* for an endofunctor F (or just F-coalgebra) consists of an object A and a morphism $\alpha\colon A \to FA$.

Given another coalgebra $\beta\colon B \to FB$, a *homomorphism* (or a morphism of F-coalgebras) is a morphism $h\colon A \to B$ of $\mathscr{A}$ such that the square below commutes:

$$\begin{array}{ccc} A & \xrightarrow{\alpha} & FA \\ {\scriptstyle h}\downarrow & & \downarrow{\scriptstyle Fh} \\ B & \xrightarrow{\beta} & FB \end{array} \tag{2.9}$$

The category of coalgebras and homomorphisms for F is denoted by $\mathsf{Coalg}\, F$.

Example 2.4.2. The following examples of coalgebras over Set stem essentially from Rutten's fundamental paper [264].

(1) *Systems with termination* are given by a set S of states; for every state either a unique next state is specified, or the state is terminal. Such systems are precisely coalgebras for the set functor

$$FX = X + 1.$$

Indeed, define $\alpha\colon S \to S + 1$ by assigning to a nonterminal state x its next state $\alpha(x)$ (in the left-hand summand) and to a terminal state the unique element of the right-hand summand.

Homomorphisms $h\colon (S, \alpha) \to (S', \alpha')$ between systems with termination are functions $h\colon S \to S'$ between their state sets preserving next states: if $\alpha(x) = x'$ in S', then $h \cdot \alpha(x) = \alpha' \cdot h(x)$. Moreover, x is terminal iff $h(x)$ is.

(2) *Binary input and termination.* These are coalgebras for

$$FX = X \times X + 1.$$

A coalgebra $\alpha\colon A \to A \times A + 1$ is given by a set A of states x which are either terminal ($\alpha(x)$ in the right-hand summand) or have, for every input $i = 0, 1$ precisely one next state x_i. Then $\alpha(x) = (x_0, x_1)$.

Coalgebra homomorphisms preserve next states and preserve and reflect termination.

(3) *Deterministic automata.* Let Σ be a set of inputs. A deterministic automaton on a set S of states is given by a state transition function $\delta_s\colon S \to S$ for every $s \in \Sigma$ and a set $T \subseteq S$ of *accepting states*. Usually an initial state is also specified. We ignore this here, but in Example 2.5.5 we will see that initial states are to some extent recovered. We can represent the state transition functions in curried form by

$$\delta\colon S \to S^{\Sigma},$$

and the subset T via the characteristic function

$$\gamma\colon S \to \{0, 1\}, \qquad T = \gamma^{-1}(1).$$

Thus, deterministic automata are precisely the coalgebras for the set functor

$$FX = \{0, 1\} \times X^{\Sigma}.$$

Indeed, to specify a coalgebra $\alpha\colon S \to \{0, 1\} \times S^{\Sigma}$ means precisely to specify a pair $\delta\colon S \to S^{\Sigma}$ and $\gamma\colon S \to \{0, 1\}$ and then to set $\alpha = \langle \gamma, \delta \rangle$.

Homomorphisms of coalgebras

$$\begin{array}{ccc} S & \xrightarrow{\langle \gamma, \delta \rangle} & \{0,1\} \times S^{\Sigma} \\ {\scriptstyle h}\downarrow & & \downarrow{\scriptstyle \mathrm{id} \times h} \\ S' & \xrightarrow{\langle \gamma', \delta' \rangle} & \{0,1\} \times (S')^{\Sigma} \end{array}$$

are precisely the functions preserving transitions:

$$h \cdot \delta_s = \delta'_s \cdot h \qquad \text{for all } s \in \Sigma$$

and preserving and reflecting accepting states:

$$\gamma = \gamma' \cdot h.$$

We have not mentioned initial states here, but we 'recover' them in Example 2.5.5.

(4) *Directed graphs.* A *directed graph* (or just graph) is a set V (of vertices) equipped with a binary relation $E \subseteq V \times V$ (of edges). That is, it is a pair (V, E). Equivalently, it is a coalgebra for the power-set functor $\mathscr{P}$. More precisely, given a coalgebra $\alpha\colon V \to \mathscr{P}V$, take V as the set of vertices of the graph, and

$$E = \{(u, v) \mid u \in V, v \in \alpha(u)\}$$

as its edges. Conversely, every graph (V, E) defines a coalgebra $\alpha\colon V \to \mathscr{P}V$ with $\alpha(u) = \{v \mid (u, v) \in E\}$. We denote by Gra the category of graphs and graph morphisms $f\colon (V, E) \to (V', E')$. These are functions preserving edges: $f \times f[E] \subseteq E'$.

However, coalgebra homomorphisms are more special than graph morphisms, which are precisely the edge-preserving functions. The square

$$\begin{array}{ccc} V & \xrightarrow{\alpha} & \mathscr{P}V \\ {\scriptstyle h}\downarrow & & \downarrow{\scriptstyle \mathscr{P}h} \\ V' & \xrightarrow{\alpha'} & \mathscr{P}V' \end{array}$$

commutes iff

(a) h preserves edges: $(u, v) \in E$ implies $(h(u), h(v)) \in E'$, and
(b) for every edge $(h(u), v') \in E'$ there exists $v \in V$ with $(u, v) \in E$ and $v' = h(v)$.

Thus Coalg $\mathscr{P}$ is a non-full subcategory of Gra (with the same objects).

We sometimes distinguish between graphs (on a set V) and *relations* on V: given a relation $R \subseteq V \times V$ we also consider it as a coalgebra $\alpha\colon V \to \mathscr{P}V$, but this time we define $\alpha(v) = \{u : (u, v) \in R\}$.

(5) *Non-Deterministic automata.* Here transitions are given by relations on the state set S, or, equivalently, by functions

$$\delta_s : S \to \mathscr{P}S \qquad \text{for all } s \in \Sigma,$$

where $\mathscr{P}$ is the power-set functor. These can be represented by a single function

$$\delta : S \to (\mathscr{P}S)^{\Sigma}.$$

We conclude that non-deterministic automata are coalgebras for the set functor

$$FX = \{0,1\} \times (\mathscr{P}X)^{\Sigma}.$$

Homomorphisms of coalgebras

$$\begin{array}{ccc} S & \xrightarrow{\langle \gamma, \delta \rangle} & \{0,1\} \times (\mathscr{P}S)^{\Sigma} \\ {\scriptstyle h}\big\downarrow & & \big\downarrow{\scriptstyle \mathrm{id} \times \mathscr{P}h} \\ S' & \xrightarrow{\langle \gamma', \delta' \rangle} & \{0,1\} \times (\mathscr{P}S')^{\Sigma} \end{array}$$

are precisely the functions that

(a) preserve transitions: for every $s \in \Sigma$, $y \in \delta_s(x)$ implies $h(y) \in \delta_s(h(x))$,
(b) reflect them: for every $y' \in \delta_s(h(x))$ there exists a $y \in \delta_s(x)$ with $h(y) = y'$, and
(c) preserve and reflect accepting states: $\gamma = \gamma' \cdot h$.

(6) *Moore and Mealy automata.* Here a set Σ of inputs and a set Γ of outputs are specified. In a Moore automaton, every state emits an output, thus a function $\gamma : S \to \Gamma$ is given (generalizing the case $\Gamma = \{0,1\}$ in point (3)). We see that Moore automata are coalgebras for the set functor

$$FX = \Gamma \times X^{\Sigma}.$$

In a Mealy automaton the output depends not only on the state but also on the input. Thus, the output function has the form $\gamma : \Sigma \times S \to \Gamma$ that we consider curried as $S \to \Gamma^{\Sigma}$. Thus Mealy automata are coalgebras for the set functor

$$FX = (\Gamma \times X)^{\Sigma}.$$

(7) *Labelled transition systems (LTS).* Given a set Σ of actions, an LTS consists of a set S of states and a binary relation $\xrightarrow{s}$ on S for every action s. We consider, again, relations as functions from S to $\mathscr{P}S$, then the above Σ-tuple of relations forms a function

$$\alpha : S \to (\mathscr{P}S)^{\Sigma}.$$

Thus an LTS is precisely a coalgebra for the functor $FX = (\mathscr{P}X)^{\Sigma}$ which is composed of $\mathscr{P}$ and the polynomial functor $X \mapsto X^\Sigma$. An equivalent way to present a labelled transition system is by one relation $\to \subseteq S \times \Sigma \times S$ which is equivalent to giving a coalgebra $S \to \mathscr{P}(\Sigma \times S)$. Thus, an LTS is, equivalently, a coalgebra for the set functor $FX = \mathscr{P}(\Sigma \times X)$.

Example 2.4.3. (1) *Weighted transition systems.* Given a commutative monoid $(M, +, 0)$, a *finite multiset with values in* M is a pair (X, m) where X is a set and $m\colon X \to M$ is a function with *finite support*; this means that for all but finitely many x we have $m(x) = 0$. (The number $m(x)$ is called the *multiplicity of* x in the multiset.)

We have the set functor $M^{(-)}$ assigning to a set X the set $M^{(X)}$ of all finite multisets over X with values in M. It is defined on maps $f\colon X \to Y$ by assigning to a multiset $m\colon X \to M$ the multiset

$$M^{(f)}(m)\colon Y \to M \quad \text{given by} \quad y \mapsto \sum_{x \in f^{-1}(y)} m(x). \tag{2.10}$$

This set functor was introduced by Gumm and Schröder [152]. A coalgebra for $M^{(-)}$ is, equivalently, a finitely branching transition system with transitions labelled in M, where a transition label 0 means no transition.

By a *weighted bisimulation* on a coalgebra (A, α) is meant an equivalence relation $\sim$ on A such that for every $x \sim y$ one has

$$\sum_{z' \sim z} \alpha(x)(z') = \sum_{z' \sim z} \alpha(y)(z') \qquad \text{for all } z \in A.$$

A coalgebra homomorphism h from $\alpha\colon A \to M^{(A)}$ to $\beta\colon B \to M^{(B)}$ is, equivalently, a map $h\colon A \to B$ whose kernel $\{(x, y) : f(x) = f(y)\} \subseteq A \times A$ is a weighted bisimulation (cf. Buchholz [94] and Sassone and Klin [193, Def. 4]).

(2) Note that the finite power-set functor is (naturally isomorphic to) the functor $M^{(-)}$ for the Boolean monoid $(\{\bot, \top\}, \vee, \bot)$.

(3) *Bag functor.* Another important special case uses the monoid $(\mathbb{N}, +, 0)$. Finite multisets with values in $\mathbb{N}$ are called *bags* and the corresponding functor is called the *bag functor* and denoted by $\mathscr{B}\colon \mathsf{Set} \to \mathsf{Set}$. For further use we record that the *size* of a bag $b\colon X \to \mathbb{N}$ is $\sum_{x \in X} b(x)$. A morphism of bags $f\colon (X, b) \to (Y, c)$ is a function $f\colon X \to Y$ such that the multiplicity of every $y \in Y$ is the sum of multiplicities in $f^{-1}(y)$; in symbols: $c(y) = \sum_{f(x)=y} b(x)$ for all $y \in Y$.

(4) *Markov chains.* The *finite distribution functor* $\mathscr{D}_\mathrm{f}$ $\mathscr{D}_\mathrm{f}$ is a subfunctor of the functor $\mathbb{R}^{(-)}$ for the usual monoid of addition on the real numbers. It maps a set X to the set of all finite probability distributions on the set X: $\mathscr{D}_\mathrm{f}X$ is the set of all maps $d\colon X \to [0, 1]$ of finite support such that $\sum_{x \in X} d(x) = 1$. The action of $\mathscr{D}_\mathrm{f}$ on maps $f\colon X \to Y$ is the same as that of $\mathbb{R}^{(-)}$ (cf. (2.10)).

As shown by Rutten and deVink [269], coalgebras $\alpha\colon A \to (\mathscr{D}_f A + 1)^\Sigma$ are precisely Larsen and Skou's probabilistic transition systems [211] (aka. labelled Markov chains [112]) with the label alphabet Σ. In fact, for each state $x \in A$ and action label s that state either cannot perform an s-action (when $\alpha(x)(s) \in 1$) or the distribution $\alpha(x)(s)$ determines for every state $y \in A$ the probability with which x transitions to y with an s-action.

Coalgebra homomorphisms between probabilistic transition systems are precisely those functions whose kernel is a *probabilistic bisimulation* in the sense of Larsen and Skou (as shown by Rutten and de Vink [269, Cor. 4.7]); the definition is similar to the one of a weighted bisimulation in the previous item.

Example 2.4.4. Given a signature Σ (see Example 2.1.5), Σ-*coalgebras* are coalgebras for the polynomial endofunctor H_Σ on Set. A coalgebra

$$\alpha\colon A \to H_\Sigma A = \coprod_{n\in\mathbb{N}} \Sigma_n \times A^n$$

can be viewed as a system with the state set A and a dynamics consisting of two functions: one, denoted by $\mathsf{head}\colon A \to \coprod_n \Sigma_n$, states which of the operation symbols is assigned to the state $x \in A$. The other function $\mathsf{body}(x,i)$ is defined for $i = 0,\dots,n-1$ where n is the arity of $\mathsf{head}(x)$. If $\alpha(x) = (\sigma,(y_0,\dots,y_{n-1}))$ and $i < n$, then $\mathsf{body}(x,i) = y_i$.

This example subsumes the following systems of Example 2.4.2:

(1) systems with termination: Σ has a unary operation and a constant,

(2) systems with binary input and termination: Σ has a binary operation and a constant,

(3) deterministic automata: consider the signature of two n-ary operations if the input set Σ has n elements (and observe that $\{0,1\} \times X^\Sigma \cong X^n + X^n$),

(4) Moore automata (n-ary operations indexed by Γ) and Mealy automata (n-ary operations indexed by Γ^n).

Example 2.4.5. *Multigraphs.* Let us now move from the base category Set to the category Set × Set of two-sorted sets. This is a special case of many-sorted sets; see Example 2.1.6. We recall that a *multigraph*, where multiple edges between a pair of vertices are allowed, can be represented by two sets V (vertices) and E (edges) and two functions $s, t\colon E \to V$ (source and target of an edge). Define an endofunctor

$$F\colon \mathsf{Set}\times\mathsf{Set} \to \mathsf{Set}\times\mathsf{Set} \qquad \text{by} \qquad F\langle V,E\rangle = \langle 1, V\times V\rangle$$

on objects, and $F\langle f,g\rangle = \langle \mathsf{id}, g\times g\rangle$ on morphisms. Then a multigraph is precisely a coalgebra for F: it consists of a two-sorted set $A = \langle V,E\rangle$ and a two-sorted function

$$\langle \alpha_v, \alpha_e \rangle \colon \langle V, E \rangle \to \langle 1, V \times V \rangle.$$

Indeed, we can ignore α_v. And $\alpha_e \colon E \to V \times V$ is precisely a pair of functions from E to V, as desired.

This time, coalgebra homomorphisms *are* precisely the usual graph homomorphisms, that is, pairs $h = \langle h_v, h_e \rangle$ of functions $h_v \colon V \to V'$ and $h_e \colon E \to E'$ such that for every edge $x \in E$, the edge $h_e(x)$ has the corresponding source and target:

$$h_v \cdot s(x) = s' \cdot h_e(x) \qquad \text{and} \qquad h_v \cdot t(x) = t' \cdot h_e(x). \tag{2.11}$$

Indeed, this condition is equivalent to the commutativity of the following square:

$$\begin{array}{ccc} \langle V, E \rangle & \xrightarrow{\alpha} & \langle 1, V \times V \rangle \\ {\scriptstyle \langle h_v, h_e \rangle}\downarrow & & \downarrow{\scriptstyle \langle \mathsf{id}, h_v \times h_v \rangle} \\ \langle V', E' \rangle & \xrightarrow{\alpha'} & \langle 1, V \times V \rangle \end{array}$$

We can clearly ignore the left-hand component, thus, this commutativity condition is equivalent to the commutativity of the square below:

$$\begin{array}{ccc} E & \xrightarrow{\langle s, t \rangle} & V \times V \\ {\scriptstyle h_e}\downarrow & & \downarrow{\scriptstyle h_v \times h_v} \\ E' & \xrightarrow{\langle s', t' \rangle} & V' \times V'. \end{array}$$

This is precisely (2.11).

Example 2.4.6. Weighted automata are non-deterministic automata with outputs and transition multiplicities in a semiring modelling, for example, costs or rewards (see e.g. [113]). A *semiring* $(\mathbb{S}, +, 0, \cdot, 1)$ consists of two monoids $(\mathbb{S}, +, 0)$ and $(\mathbb{S}, \cdot, 1)$, the first of which is commutative, obeying the usual distributivity laws: $r \cdot 0 = 0 = 0 \cdot r$, $r \cdot (s + t) = r \cdot s + r \cdot t$, and $(r + s) \cdot t = r \cdot t + s \cdot t$. A *weighted automaton* on n states with input alphabet Σ is a tuple $(i, (M_a)_{a \in \Sigma}, o)$, where $i, o \in \mathbb{S}^n$ are the input and output vectors, and each M_a is an $(n \times n)$-matrix over $\mathbb{S}$. The (i, j)-entry of the matrix M_a is the weight of the transition from state i to state j, $1 \le i, j \le n$, by input a.

Weighted automata are coalgebras in the category $\mathbb{S}$-Mod of (left) $\mathbb{S}$-semimodules. An $\mathbb{S}$-*semimodule* is a commutative monoid $(M, +, 0)$ equipped with a (left)-action $\mathbb{S} \times M \to M$ denoted by juxtaposition rm (for $r \in \mathbb{S}$ and $m \in M$), such that the following laws hold for all $r, s \in \mathbb{S}$ and $m, n \in M$:

$$(r+s)m = rm + sm, \qquad r(m+n) = rm + rn,$$
$$0m = 0, \qquad r0 = 0,$$
$$1m = m, \qquad r(sm) = (r \cdot s)m.$$

A morphism of $\mathbb{S}$-semimodules is a monoid morphism $h\colon M_1 \to M_2$ such that $h(rm) = rh(m)$ for each $r \in \mathbb{S}$ and $m \in M_1$. Products of semimodules are defined using the cartesian product with component-wise addition. The free semimodule on n generators is $\mathbb{S}^n$.

Consider the endofunctor $FX = \mathbb{S} \times X^\Sigma$. Then a weighted automaton with n states (ignoring the initial vector) is precisely a coalgebra for F on the (free) $\mathbb{S}$-semimodule $\mathbb{S}^n$. Indeed, the output vector corresponds to a semimodule morphism $o\colon \mathbb{S}^n \to \mathbb{S}$ and each matrix M_a represents a semimodule morphism $\delta_a\colon \mathbb{S}^n \to \mathbb{S}^n$. Pairing the latter for $a \in \Sigma$ yields, equivalently, a semimodule morphism $\delta\colon \mathbb{S}^n \to (\mathbb{S}^n)^\Sigma$. Together with the output morphism we obtain an F-coalgebra $\alpha = \langle o, \delta\rangle\colon \mathbb{S}^n \to \mathbb{S} \times (\mathbb{S}^n)^\Sigma$.

Coalgebra homomorphisms between weighted automata are precisely the *simulations* in the sense of weighted automata theory [84, 119]. Indeed, a coalgebra homomorphism

$$\begin{array}{ccc} \mathbb{S}^n & \xrightarrow{\langle o,\delta\rangle} & \mathbb{S}\times(\mathbb{S}^n)^\Sigma \\ {\scriptstyle h}\downarrow & & \downarrow{\scriptstyle \mathsf{id}\times h^\Sigma} \\ \mathbb{S}^m & \xrightarrow{\langle o',\delta'\rangle} & \mathbb{S}\times(\mathbb{S}^m)^\Sigma \end{array}$$

corresponds to an $(n \times m)$-matrix M_h such that

(1) $oM_h = o'$ (i.e. o considered as a row vector of length n multiplied by the matrix M_h yields o' considered as a row vector of length m); and

(2) $M_a M_h = M_h M'_a$ for every $a \in \Sigma$, where M_a and M'_a are the matrices representing $\delta_a\colon \mathbb{S}^n \to \mathbb{S}^n$ and $\delta'_a\colon \mathbb{S}^m \to \mathbb{S}^m$, respectively.

Example 2.4.7. Here we consider coalgebras for endofunctors of $\mathsf{CPO}_\perp$, see Example 2.2.17. First, we need a definition. By an *ideal* of a cpo P is meant a nonempty subset closed both downwards and under ω-joins. For example, let $\mathbb{2} = \{\perp, \top\}$ the two-element chain with $\perp < \top$. For every morphism $f\colon P \to \mathbb{2}$ in $\mathsf{CPO}_\perp$, the set $f^{-1}(\perp)$ is an ideal of P. Conversely, given an ideal I, we have $I = f^{-1}(\perp)$ for a unique morphism f.

(1) A coalgebra for $FX = X_\perp$ is a system with termination whose state space A has the structure of a cpo such that

(a) all terminal states form an ideal, and

(b) the next-state function is continuous.

Indeed, this yields a strict and continuous function $\alpha\colon A \to A_\perp$ sending all terminal states to the (new) $\perp$ element.

(2) A coalgebra for $FX = X_\perp + 1_\perp$ can be viewed as a system with termination and one deadlock state $\perp$. Every state is either a terminal state or $\perp$, or it has a (unique) next state. In the cpo A of states the terminal states and the states with a next state are always incomparable. Moreover, the conditions (a) and (b) in item (1) hold.

Indeed, this yields a strict and continuous function $\alpha\colon A \to A_\perp + 1_\perp$ mapping deadlock states to the non-bottom element of the right-hand summand.

(3) Continuous deterministic automata. These are automata with a cpo A as a state space and whose transition self-maps are strict and continuous. Moreover, all non-accepting states form an ideal.

Such automata are precisely the coalgebras for the endofunctor of $\mathsf{CPO}_\perp$ defined (as in the case of Set) by

$$FX = \mathbb{2} \times X^\Sigma.$$

(This is a product of the chain $0 \le 1$ and of n copies of X if $|\Sigma| = n$.) Indeed, a coalgebra $\alpha\colon A \to \mathbb{2} \times A^\Sigma$ defines two strict and continuous functions

$$\delta\colon A \to A^\Sigma \quad \text{and} \quad \gamma\colon A \to \mathbb{2}.$$

Here δ represents the transitions and γ the accepting states.

General Properties of the category $\mathsf{Coalg}\, F$

Notation 2.4.8. Given an endofunctor F of $\mathscr{A}$ the corresponding endofunctor of $\mathscr{A}^{\mathrm{op}}$ is denoted by F^{op}: it is defined by $F^{\mathrm{op}}A = FA$ on objects and it sends a morphism $f\colon X \to Y$ of $\mathscr{A}^{\mathrm{op}}$ to $Ff\colon FY \to FX$ in $\mathscr{A}$, i.e. to $Ff\colon FX \to FY$ in $\mathscr{A}^{\mathrm{op}}$.

Lemma 2.4.9. *The category of coalgebras for $F\colon \mathscr{A} \to \mathscr{A}$ is dual to the category of algebras for $F^{\mathrm{op}}\colon \mathscr{A}^{\mathrm{op}} \to \mathscr{A}^{\mathrm{op}}$: $(\mathsf{Coalg}\, F)^{\mathrm{op}} = \mathsf{Alg}\, F^{\mathrm{op}}$.*

Proof Indeed, the objects are the same: an F-coalgebra $\alpha\colon A \to FA$ in $\mathscr{A}$ is precisely an F^{op}-algebra $\alpha\colon FA \to A$ in $\mathscr{A}^{\mathrm{op}}$. Given another coalgebra $\beta\colon B \to FA$, a homomorphism of coalgebras $h\colon (A,\alpha) \to (B,\beta)$ for F is precisely a homomorphism of algebras $h\colon (B,\beta) \to (A,\alpha)$ for F^{op}. □

Remark 2.4.10. This lemma is trivial, but important: it implies that every statement about algebras has a dual statement about coalgebras. For example, Proposition 2.1.13 yields the following fact:

If $\mathscr{A}$ has finite coproducts, then finite coproducts of coalgebras are formed on the level of $\mathscr{A}$. In particular, the initial coalgebra is simply $0 \to F0$. (Terminal coalgebras are much more interesting!)

Lemma 2.4.9 does *not* say that the category of coalgebras for $F\colon \mathscr{A} \to \mathscr{A}$ is dual to the category of algebras for the same functor F. In fact, this will almost always turn out to be false! However, dually to Corollary 2.1.15 and Corollary 2.1.16 we have the following fact.

Corollary 2.4.11. *A homomorphism $h\colon (A,\alpha) \to (B,\beta)$ is epic in* $\mathsf{Coalg}\, F$ *iff h is epic in $\mathscr{A}$.*

Remark 2.4.12. (1) For monomorphisms, the situation is not as simple in $\mathsf{Coalg}\, F$ as in $\mathsf{Alg}\, F$ (cf. Corollary 2.1.16). However, every coalgebra homomorphism $h\colon (A,\alpha) \to (B,\beta)$ with h monic in $\mathscr{A}$ is monic in $\mathsf{Coalg}\, F$. Thus it represents a subobject of (B, b) (cf. Remark 2.1.18). When we speak about *subcoalgebras* of a coalgebra (B,β), we always understand those represented by coalgebra homomorphisms carried by monic morphisms of $\mathscr{A}$.

(2) Suppose that $F\colon \mathscr{A} \to \mathscr{A}$ preserves monomorphisms. Then given a coalgebra $\alpha\colon A \to FA$, every subobject in $\mathscr{A}$ carries at most one coalgebra structure making m a subcoalgebra. Indeed, if the following two squares

$$
\begin{array}{ccc}
B & \xrightarrow{\beta_i} & FB \\
{\scriptstyle m}\downarrow & & \downarrow{\scriptstyle Fm} \\
A & \xrightarrow{\alpha} & FA
\end{array}
\qquad \text{for } i = 1, 2,
$$

commute, then $\beta_1 = \beta_2$ since Fm is monic and $Fm \cdot \beta_1 = \alpha \cdot m = Fm \cdot \beta_2$.

(3) Moreover, the order on subcoalgebras is the same as that of subobjects we saw in Remark 2.1.18. Indeed, let $n\colon C \rightarrowtail A$ be another subobject that carries a (unique) subcoalgebra of (A,α). Then $m \leq n$ holds in $\mathscr{A}$ iff there exists $k\colon B \to C$ such that $m = n \cdot k$. But then k is automatically a coalgebra homomorphism, as we see from the following diagram:

$$
m \quad
\begin{array}{ccc}
B & \xrightarrow{\beta} & FB \\
{\scriptstyle k}\downarrow & & \downarrow{\scriptstyle Fk} \\
C & \xrightarrow{\gamma} & FC \\
{\scriptstyle n}\downarrow & & \downarrow{\scriptstyle Fn} \\
A & \xrightarrow{\alpha} & FA
\end{array}
\quad Fm
$$

Indeed, the left- and right-hand parts commute, and so do the lower square and the outside. Thus, the upper square commutes when post-composed by Fn, which implies that this part commutes as desired.

2.5 Terminal Coalgebras

Due to the duality between coalgebras in $\mathscr{A}$ and algebras in $\mathscr{A}^{\mathrm{op}}$ which we saw in Lemma 2.4.9, everything we said about initial algebras in Section 2.2 dualizes:

Definition 2.5.1. A coalgebra for F is *terminal* if it admits a unique homomorphism from every coalgebra. We denote it by νF or $\nu X.FX$ and the coalgebra structure by

$$\tau\colon \nu F \to F(\nu F).$$

Lemma 2.5.2 [210]. *Every terminal coalgebra of an endofunctor is a fixed point.*

A converse similar to Theorem 2.2.6 for set functors is not known for terminal coalgebras, and it is unlikely that one is possible. In fact, there exist two set functors that coincide on all sets (but not all maps) such that one has a terminal coalgebra and one does not, see [50, Ex. 3.14].

Example 2.5.3. (1) The power-set functor does not have a terminal coalgebra, since it has no fixed points (see Example 2.2.7(1)).

(2) The terminal coalgebra for $FX = X + 1$ (cf. Example 2.4.2(1)) is the set $\mathbb{N}^\top$, with the 'predecessor' map as coalgebra structure:

$$* \longleftarrow\!\shortmid\; 0 \longleftarrow\!\shortmid\; 1 \longleftarrow\!\shortmid\; 2 \longleftarrow\!\shortmid\; \cdots \qquad\qquad \infty \circlearrowright$$

where $*$ denotes the element of 1. Indeed, for every F-coalgebra, that is, a system A with termination, the behaviour defines a homomorphism $b_A\colon A \to \mathbb{N}^\top$: If a state x has behaviour $n > 0$, then its next state x' has behaviour $n - 1 = \tau(n)$. If x has behaviour ∞, then so does the next state. Finally, x is terminal iff $b_A(x) = 0$. This shows that the square defining homomorphisms commutes:

$$\begin{array}{ccc} A & \xrightarrow{\alpha} & A + 1 \\ {\scriptstyle b_A}\downarrow & & \downarrow{\scriptstyle b_A+\mathrm{id}} \\ \mathbb{N}^\top & \xrightarrow{\tau} & \mathbb{N}^\top + 1 \end{array}$$

It is further easy to see that b_A is the unique homomorphism.

(3) Coalgebras for the functor

$$FX = \Sigma \times X$$

can be viewed as dynamical systems with outputs in Σ: Given a coalgebra

$$\alpha = \langle \alpha_0, \alpha_1 \rangle : A \to \Sigma \times A,$$

the function $\alpha_1 : A \to A$ determines next states and $\alpha_0 : A \to \Sigma$ determines outputs.

Every state $q \in A$ emits an infinite *stream* of outputs

$$x_0 = \alpha_0(q), x_1 = \alpha_0\alpha_1(q), x_2 = \alpha_0\alpha_1\alpha_1(q), \ldots .$$

This defines a function $b_A : A \to \Sigma^\omega$ into the set Σ^ω of all infinite streams. This set carries itself the structure of a coalgebra

$$\tau = \langle \mathsf{head}, \mathsf{tail} \rangle : \Sigma^\omega \to \Sigma \times \Sigma^\omega.$$

Here head assigns to a stream $(x_n)_{n \in \mathbb{N}}$ its head x_0, and tail its tail $(x_{n+1})_{n \in \mathbb{N}}$. The following square commutes:

$$\begin{array}{ccc} A & \xrightarrow{\langle \alpha_1, \alpha_0 \rangle} & \Sigma \times A^\omega \\ {\scriptstyle b_A}\downarrow & & \downarrow{\scriptstyle \mathsf{id} \times (b_A)^\omega} \\ \Sigma^\omega & \xrightarrow{\langle \mathsf{head}, \mathsf{tail} \rangle} & \Sigma \times \Sigma^\omega \end{array}$$

So b_A is a homomorphism.

It is easy to verify that no other homomorphism exists from A to Σ^ω. We thus proved that Σ^ω is the terminal coalgebra:

$$\nu X. (\Sigma \times X) = \Sigma^\omega.$$

(4) A slight variation: adding terminal states means that we work with the functor

$$FX = \Sigma \times X + 1.$$

In a coalgebra, every state which is not terminal has a next state and emits an output in Σ. The behaviour of a state is either a finite word (if the execution path encounters a terminal state) or an infinite stream. The set of all these streams is denoted by

$$\Sigma^\infty = \Sigma^* + \Sigma^\omega.$$

Again, we have a coalgebra structure $\tau : \Sigma^\infty \to \Sigma \times \Sigma^\infty + 1$ taking the empty word to the right-hand summand, and all nonempty strings to the pair $\langle \mathsf{head}, \mathsf{tail} \rangle$.

And again behaviour defines a unique homomorphism into (Σ^∞, τ). Thus we have proved that

$$\nu X.\,(\Sigma \times X + 1) = \Sigma^\infty.$$

Remark 2.5.4. The preceding examples demonstrate the important role that the terminal coalgebra νF has: for systems 'of type F' the elements of νF are all possible behaviours of states. Given a system with the state set A, the unique coalgebra homomorphism from A to νF assigns to every state its behaviour. The next example is a classical case:

Example 2.5.5. *Terminal automaton.* The terminal coalgebra for the functor $FX = \{0,1\} \times X^\Sigma$ of deterministic automata, see Example 2.4.2(4) is

$$\nu X.\,(\{0,1\} \times X^\Sigma) = \mathscr{P}\Sigma^*,$$

the set of all *formal languages* over Σ. Recall that $\mathscr{P}\Sigma^*$ can itself be considered as an automaton: a language $L \subseteq \Sigma^*$ is accepting iff it contains the empty word ε. Given a state (i.e. a language) L and an input $s \in \Sigma$ the next state is defined by the *Brzozowski derivative*

$$s^{-1}L = \{w \in \Sigma^* : sw \in L\}.$$

This yields a coalgebra structure $\tau \colon \mathscr{P}\Sigma^* \to F(\mathscr{P}\Sigma^*)$.

For every automaton A and every state $x \in A$ we form the language $L(x)$ accepted by x in A. Observe that

(1) x is accepting iff $L(x)$ contains ε, that is, iff $L(x)$ is accepting in $\mathscr{P}\Sigma^*$, and

(2) for every input $s \in \Sigma$ if y denotes the next state of x, then

$$L(y) = s^{-1}L(x).$$

From (a) and (b) it is easy to see that the above language-function $L \colon A \to \mathscr{P}\Sigma^*$ is a coalgebra homomorphism, that is, the square below commutes:

$$\begin{array}{ccc} A & \xrightarrow{\ \alpha\ } & \{0,1\} \times A^\Sigma \\ {\scriptstyle L}\big\downarrow & & \big\downarrow{\scriptstyle \mathrm{id} \times L^\Sigma} \\ \mathscr{P}\Sigma^* & \xrightarrow{\ \tau\ } & \{0,1\} \times (\mathscr{P}\Sigma^*)^\Sigma \end{array}$$

Here L^Σ is the function $p \mapsto L \cdot p$, for all $p \in A^\Sigma$. It is also easy to verify the uniqueness of L. A different (element-free) proof that formal languages form the terminal coalgebra is presented in Example 11.5.8(1). Thus, although our definition of an automaton as a coalgebra did not introduce an initial state (as is usual), the homomorphism L above assigns to every state x the language of the automaton augmented by x being the initial state.

Remark 2.5.6. The functor $FX = \{0, 1\} \times (\mathscr{P}X)^\Sigma$ used for modelling non-deterministic automata as coalgebras does not have a terminal coalgebra, just as $\mathscr{P}$ does not have one. However, since in applications one is usually interested in *finite* non-deterministic automata, one may of course replace $\mathscr{P}$ above by the finite power-set functor $\mathscr{P}_\mathsf{f}$. Then the resulting functor does have a terminal coalgebra, and we will discuss it when we present the terminal coalgebra for $\mathscr{P}_\mathsf{f}$ in Theorem 4.5.15. For the moment let us just note that the terminal semantics does not provide the usual language semantics of non-deterministic automata. This can be achieved by working over the category Rel of sets and relations, see Example 5.1.27.

Example 2.5.7. (1) Terminal Moore automata. Moore automata are coalgebras for $FX = \Gamma \times X^\Sigma$, see Example 2.4.2(6). The terminal coalgebra is the coalgebra Γ^{Σ^*} of all functions from Σ^* to Γ (formal power series). Again Γ^{Σ^*} is a Moore automaton: given a state $f\colon \Sigma^* \to \Gamma$ and an input $s \in \Sigma$, the corresponding output is $f(\varepsilon)$ and the next state is $\lambda w.f(sw)$.

(2) Analogously, for Mealy automata presented by the functor $FX = (\Gamma \times X)^\Sigma$: here terminal coalgebra can be described as the set of all *causal stream functions* from Σ^ω (the set of all streams in Σ) to Γ^ω. These are functions $f\colon \Sigma^\omega \to \Gamma^\omega$ such that for every stream $x \in \Sigma^\omega$ the nth element of $f(x)$ depends only on the first n elements of x. The coalgebra structure assigns to every causal stream function f and every input $a \in \Sigma$

(a) the causal stream function f_a given by $f_a(x) = \mathsf{tail}(f(ax))$ and

(b) the output $\mathsf{head}(f(ay))$ for any stream y.

(Note that this is well-defined since the head of the causal stream function f only depends on the head of its argument.) This is how Rutten [266] described the terminal coalgebra. Another description is presented in Example 2.5.11(6) below.

Example 2.5.8 (Terminal Σ-coalgebra)**.** Generalizing all of the above examples, consider coalgebras for a polynomial endofunctor H_Σ, see Example 2.4.4. The terminal coalgebra is pleasantly analogous to the initial algebra (consisting of all finite Σ-trees, see Example 2.2.11). Recall that trees are always considered up to isomorphism. We shall show that

$$\nu H_\Sigma = \text{all } \Sigma\text{-trees}.$$

The coalgebra structure $\tau\colon \nu H_\Sigma \to H_\Sigma(\nu H_\Sigma)$ is the inverse of the tree-tupling map from Definition 2.2.12. It assigns to every Σ-tree

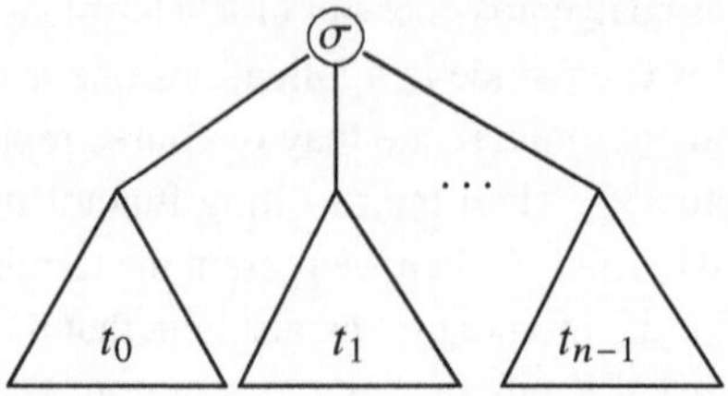

the pair $(\sigma, (t_0, \ldots, t_{n-1}))$.

Theorem 2.5.9. *The terminal coalgebra for a polynomial functor H_Σ is the coalgebra of all Σ-trees with coalgebra structure inverse to the tree-tupling.*

Proof Let us denote by T_Σ the set of all Σ-trees, and let (A, α) be a coalgebra for H_Σ. We are to define a homomorphism h as shown below:

$$\begin{array}{ccc} A & \xrightarrow{\alpha} & H_\Sigma A \\ {\scriptstyle h}\downarrow & & \downarrow{\scriptstyle H_\Sigma h} \\ T_\Sigma & \xrightarrow{\tau} & H_\Sigma T_\Sigma \end{array}$$

Since τ is the inverse of tree-tupling, for each $x \in A$ the Σ-tree $t_x = h(x)$ must fulfil the following equation

$t_x =$ 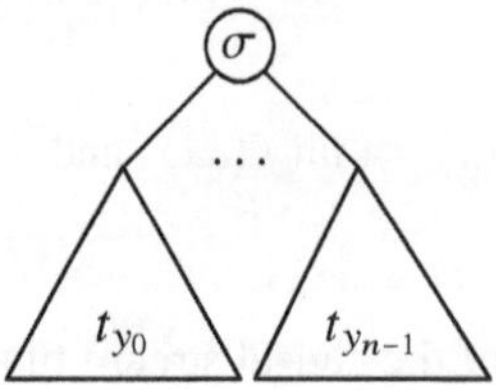 where $\alpha(x) = (\sigma, y_0, \ldots, y_{n-1})$. (2.12)

This determines the trees t_x uniquely. Indeed, let us use their description as partial functions on $\mathbb{N}^*$ as in Remark 2.2.13. The values $t_x(w)$ are defined simultaneously for all x by recursion over $w \in \mathbb{N}^*$ as follows:

$$t_x(\varepsilon) = \sigma, \qquad t_x(iv) = \begin{cases} t_{y_i}(v) & \text{if } i < n, \\ \text{undefined} & \text{else.} \end{cases} \tag{2.13}$$

This recursion determines a unique function $h\colon A \to T_\Sigma$, $h(x) = t_x$ for which the equations in (2.12) hold. Thus, h is the desired unique homomorphism. □

Remark 2.5.10. Note that the unique homomorphism $h\colon (A, \alpha) \to (\nu H_\Sigma, \tau)$ assigns to every element $x \in A$ its tree expansion as a Σ-tree using the coalgebra structure α. This tree has a root labelled by $\mathsf{head}(x)$, and the children of a node y are the tree expansions of $\mathsf{body}(y, i)$ for $i = 0, 1, \ldots, n$, where n is the arity of $\mathsf{head}(y)$ (recall the notation from Example 2.4.4). In fact, this is precisely the contents of the recursive definition (2.13).

Example 2.5.11. (1) $FX = X + 1$ corresponds to one constant and one unary operation. A Σ-tree is either a single path of length $n \in \mathbb{N}$, or a single infinite path. This yields $\nu X.\,(X + 1) = \mathbb{N}^\top$, see Example 2.5.3(2).

(2) $FX = \Sigma \times X$ corresponds to the signature Σ with unary operations only. The corresponding trees have no leaves, thus, they are simply infinite paths labelled in Σ. This yields $\nu X.\,(\Sigma \times X) = \Sigma^\omega$, see Example 2.5.3(3).

(3) Analogously $FX = \Sigma \times X + 1$: the additional constant is a (unique) labelling of a leaf. Thus the corresponding tree is a finite or infinite path labelled in Σ, therefore $\nu X.\,(\Sigma \times X + 1) = \Sigma^\infty$ as in Example 2.5.3(4).

(4) The functor $FX = X \times X + 1$ is the polynomial functor associated to the signature with one binary operation symbol and one constant. Thus νF may be identified with the set of all (finite and infinite) binary trees.

(5) Consider $FX = B \times X^A$ where A and B are fixed sets, and assume that A has size n. This functor F corresponds to H_Σ where the signature Σ consists of n-ary operations indexed by B. An F-coalgebra is a dynamical system with n inputs and with outputs using symbols of B. Recall from Remark 2.2.10(4) that A^* represents the complete n-ary tree. Therefore, the set of Σ-trees for this signature is exactly B^{A^*}. (This is the set of functions from words on A into B.) Then νF is (B^{A^*}, τ), where the structure $\tau\colon B^{A^*} \to B \times (B^{A^*})^A$ is described as follows: τ assigns to $f\colon A^* \to B$ the pair $(f(\varepsilon), g)$, where $g\colon A \to B^{A^*}$ is given by $g(a)(w) = f(aw)$.

(6) As a special case of the last item, consider deterministic automata as coalgebras for $FX = \{0, 1\} \times X^\Sigma$ from Example 2.4.2(3). As we have just seen, the terminal coalgebra is $\{0, 1\}^{\Sigma^*}$. Owing to the general correspondence between subsets of a given sets and maps into $\{0, 1\}$, the terminal coalgebra may be described as the set $\mathscr{P}\Sigma^*$ of formal languages on Σ as in Example 2.5.5.

(7) For Mealy automata regarded as coalgebras for $FX = (\Gamma\times X)^\Sigma$ we have seen a description of the terminal coalgebra in Example 2.5.7(2). Another description is obtained as follows: the functor $FX \cong \Gamma^\Sigma \times X^\Sigma$ is polynomial for the signature of n-ary operations (for Σ of size n) indexed by Γ^Σ. Thus the terminal coalgebra consists of all labellings of the complete n-ary tree Σ^* by labels in Γ^Σ. Every such tree is given by a function from Σ^* to Γ^Σ, or, equivalently, a function from $\Sigma^+ = \Sigma \times \Sigma^*$ to Γ:

$$\nu X.\,(\Gamma^\Sigma \times X^\Sigma) = \Gamma^{\Sigma^+}.$$

The coalgebra structure assigns to every function $g\colon \Sigma^+ \to \Gamma$ and every input $a \in \Sigma$ the function $g(a-)\colon \Sigma^+ \to \Gamma$ and the output $g(a) \in \Gamma$. This coalgebra Γ^{Σ^+} is indeed isomorphic to that of all causal stream functions (see Example 2.5.7(2)) by the following canonical isomorphism. Given a function $g\colon \Sigma^+ \to \Gamma$, define $\bar{g}\colon \Sigma^\omega \to \Gamma^\omega$ by assigning to every stream $a_0a_1a_2\cdots$ in Σ the stream

$$g(a_0)\, g(a_0a_1)\, g(a_0a_1a_2)\, \ldots .$$

Then $\bar{g}$ is obviously a causal stream function, and it is easy to verify that this yields the desired isomorphism.

(8) Let A be a set and consider the endofunctor on Set given by $FX = A \times X \times X$. This is the polynomial functor for the signature with a binary operation symbol for every $a \in A$. So we know from Theorem 2.5.9 that νF is carried by all labellings of the infinite binary tree by elements of $\mathscr{A}$.

A different (and very useful) isomorphic description of the terminal coalgebra was given by Grabmayer et al. [148, Prop. 26] (see also Kupke and Rutten [204] for a related result). Consider the set A^ω of streams over A and equip it with the coalgebra structure

$$\langle \mathsf{head}, \mathsf{odd}, \mathsf{even}\rangle\colon A^\omega \to A \times A^\omega \times A^\omega,$$

where for a stream $s = (s_0, s_1, s_2, s_3, \ldots)$ we have

$$\mathsf{head}(s) = s_0, \quad \mathsf{odd}(s) = (s_1, s_3, s_5, \ldots) \quad \text{and} \quad \mathsf{even}(s) = (s_2, s_4, s_6, \ldots).$$

This is a terminal F-coalgebra. To see this consider the bijection $\mathsf{z}\colon \omega \to \{0,1\}^*$ recursively defined by

$$\mathsf{z}(0) = \varepsilon \qquad \mathsf{z}(2n+1) = 0\mathsf{z}(n) \qquad \mathsf{z}(2n+2) = 1\mathsf{z}(n).$$

More explicitly, to obtain $\mathsf{z}(n)$ for $n > 0$, take the binary representation of $n+1$, drop the leading 1, and reverse the sequence. This map is Eilenberg's *reversed bijective interpretation*; see [114, p. 116].

Now labellings of the infinite binary tree in A may be identified with functions $\{0,1\}^* \to A$, and streams over A are functions $\omega \to A$. Thus, precomposition with z yields an isomorphism between the sets νF above and A^ω. One then verifies that this yields an isomorphism of coalgebras. Analogously, A^ω carries a terminal coalgebra for every functor $FX = A \times X^k$ for $k \in \mathbb{N}$ [148]. Its structure is $\langle \mathsf{head}, \mathsf{p}\rangle\colon A^\omega \to A \times (A^\omega)^k$, where $\mathsf{p}(s)(i) = (s_i, s_{k+i}, s_{2k+i}, \ldots)$ for $1 \le i \le k$.

We shall revisit this example in Chapter 10 when we study the behaviour of finite coalgebras, which in this case turn out to be given by the so-called *automatic sequences* in A^ω [58].

Example 2.5.12. Terminal weighted automaton. Weighted automata are coalgebras for the functor $FX = \mathbb{S} \times X^{\Sigma}$ on the category $\mathbb{S}$-Mod of semimodules for the semiring $\mathbb{S}$ (see Example 2.4.6). The terminal coalgebra for F is carried by the set

$$\nu F = \mathbb{S}^{\Sigma^*}$$

of *formal power series* (or *weighted languages*) with the obvious componentwise $\mathbb{S}$-semimodule structure. The coalgebra structure $\langle o, t\rangle : \mathbb{S}^{\Sigma^*} \to \mathbb{S} \times (\mathbb{S}^{\Sigma^*})^{\Sigma}$ is given by $o(L) = L(\varepsilon)$ and $t(L)(a) = \lambda w.L(aw)$.

Given a coalgebra $\alpha : \mathbb{S}^n \to \mathbb{S} \times (\mathbb{S}^n)^{\Sigma}$ the unique coalgebra homomorphism is the map $h : \mathbb{S}^n \to \mathbb{S}^{\Sigma^*}$ mapping each $i \in \mathbb{S}^n$ to the weighted language accepted by the weighted automaton $(i, (M_a)_{a\in\Sigma}, o)$ corresponding to $(\mathbb{S}^n, \alpha)$ with initial vector i. Indeed, it is easy to prove that $h(i)$ is the weighted language $L : \Sigma^* \to \mathbb{S}$ with $L(w) = iM_w o$ where i is considered as a row vector, o as a column vector, and M_w is the obvious extension of the M_a to words: M_ε is the $n \times n$ identity matrix and $M_{aw} = M_a M_w$.

Example 2.5.13. Let us consider terminal coalgebras in $\mathsf{CPO}_\perp$.

(1) The terminal coalgebra for $FX = X_\perp$ is the cpo

$$\nu X.X_\perp = \mathbb{N}^\top$$

of natural numbers extended by the top element ∞ with coalgebra structure inverse to the algebra structure of Example 2.2.17(1).

(2) The terminal coalgebra for $FX = (X \times X)_\perp + 1_\perp$ consists of all binary trees. Again, the coalgebra structure is inverse to tree-tupling of Example 2.2.17(2).

Let us consider the terminal coalgebras in Nom, the category of nominal sets (see Example 2.1.10) for functors arising from binding signatures (Remark 2.1.11). Later, in Example 10.5.11 items (2) and (3), we will mention terminal coalgebras for the type functors of deterministic nominal automata.

Example 2.5.14. The initial algebra for the functor $FX = \mathbb{A} + X \times X + [\mathbb{A}]X$ on Nom is carried by the nominal set L of λ-terms modulo α-equivalence (see Example 2.2.18(1)). Recall from Proposition 2.2.14 that for the set functor H_Σ the initial algebra is formed by the set of finite Σ-trees, and the terminal coalgebra is formed by all (finite and infinite) Σ-trees (Theorem 2.5.9). Similarly, one moves from λ-terms to (finite and infinite) λ-trees considering variables from $\mathbb{A}$ as constant operation symbols, application as a binary operation symbol and $\lambda x.-$, for every variable x, as a unary operation symbol. Kurz et al. [209] have shown that the terminal coalgebra νF is carried by the nominal set of all α-equivalence classes of λ-trees with finitely many free variables (but, possibly, infinitely many bound ones).

More generally, for every countable binding signature Σ the terminal coalgebra for its associated binding polynomial functor F_Σ on Nom (Remark 2.1.11) is carried by the nominal set of α-equivalence classes of infinite trees for Σ with finitely many free variables (see Kurz et al. [209, Cor. 5.35]).

Example 2.5.15. Consider the category $\mathsf{Set}^{\mathscr{F}}$ of sets in context (see Example 2.1.8) and the endofunctor $FX = V + X \times X + \delta X$. Similarly to Example 2.5.14, its terminal coalgebra is carried by the presheaf of all λ-trees modulo α-equivalence [44, Thm. 2.10]: for every context $\Gamma = \{x_1, \dots, x_n\}$ one has

$$\nu F(\Gamma) = \text{all } \alpha\text{-equivalence classes of finite and infinite } \lambda\text{-trees with free variables } x_1, \dots, x_n \text{ (but, possibly, infinitely many bound variables).}$$

Similarly, for a binding signature Σ (Remark 2.1.9) the terminal coalgebra νF_Σ assigns to Γ the set of α-equivalence classes of finite and infinite trees over Σ; this is a straightforward extension of results on λ-Σ-terms (for an ordinary signature Σ) [44, Thms. 3.11 & 4.18].

Remark 2.5.16. Initial algebras are related to terminal coalgebras in a canonical way: suppose that both exist for an endofunctor F. Then the inverse of the coalgebra structure $\tau\colon \nu F \to F(\nu F)$ makes νF an algebra. Hence, we have a unique algebra homomorphism $k\colon (\mu F, \iota) \to (\nu F, \tau^{-1})$. (We can play the dual game and consider μF as a coalgebra: that would lead to the same morphism k, which is indeed also a coalgebra homomorphism.)

We will see later in Corollary 6.3.9 that for every set functor F if νF exists, so does μF. In this case, the morphism k is monic. Thus, an initial algebra is always a subalgebra of the terminal coalgebra (when the latter is considered as an algebra). Observe that this holds in all the examples of the present section.

2.6 Corecursion and Bisimulation

We have seen the connection of *initiality*, *recursion*, and *induction*: recursion in its most basic form on the natural numbers may be considered as an application of initiality. Proof by induction is thus an application of the minimality of initial algebras.

In this section, we study the dual concepts. Whereas recursion deals with functions out of an initial algebra, corecursion is a definition principle for functions *into* a terminal coalgebra. This is the first topic in this section. After seeing examples, we then turn to *bisimulation principles* allowing us to prove interesting assertions about corecursively defined functions.

Definition 2.6.1. Let F be an endofunctor with a terminal coalgebra. Given an object A, we say that a morphism $f\colon A \to \nu F$ is *corecursively specified* if there exists a (specification) morphism $\alpha\colon A \to FA$ such that f is a coalgebra homomorphism: $\tau \cdot f = Ff \cdot \alpha$.

Example 2.6.2. Corecursive specification of addition [264]. We consider systems with termination as coalgebras for the set functor $FX = X + 1$. The terminal coalgebra is given by natural numbers extended by ∞:

$$\nu F = \mathbb{N}^\top \qquad \text{with coalgebra structure } \tau = \mathsf{pred},$$

the predecessor function, see Example 2.5.3(2). In order to recursively specify addition (extended by $\infty + x = \infty = x + \infty$) we need a coalgebra structure on the set $\mathbb{N}^\top \times \mathbb{N}^\top$. Its terminating states must sum to 0, whence $(0, 0)$ is the only terminating state. The next state function of $\mathbb{N}^\top \times \mathbb{N}^\top$ expresses a move from (x, y) to (x', y') such that $\tau(x+y) = x'+y'$. Thus, the coalgebra structure α has $(0, 0)$ as the only terminating state, and the next state of $(x, y) \neq (0, 0)$ is $(0, \tau y)$ if $x = 0$, else $(\tau x, y)$.

The corresponding coalgebra homomorphism $\mathsf{add}\colon \mathbb{N}^\top \times \mathbb{N}^\top \to \mathbb{N}^\top$ is the desired addition function, since the square below commutes:

$$\begin{array}{ccc} \mathbb{N}^\top \times \mathbb{N}^\top & \xrightarrow{\alpha} & \mathbb{N}^\top \times \mathbb{N}^\top + 1 \\ {\scriptstyle \mathsf{add}}\downarrow & & \downarrow{\scriptstyle \mathsf{add}+\mathsf{id}} \\ \mathbb{N}^\top & \xrightarrow{\tau} & \mathbb{N}^\top + 1 \end{array} \tag{2.14}$$

Thus, α is a corecursive specification of addition.

Example 2.6.3. Corecursive specification of streams. We recall from Example 2.5.3(3) that

$$\nu X.\, \Sigma \times X = \Sigma^\omega.$$

(1) To specify corecursively a concrete stream as a function $f\colon 1 \to \Sigma^\omega$ we need a specification $\alpha\colon 1 \to F(1) \cong \Sigma$, viz. an element $x \in \Sigma$. It is easy to see that the corresponding stream is simply $(x, x, x, \ldots)$ as a homomorphism $1 \to \Sigma^\omega$.

(2) What about non-constant streams, say,

$$(x, y, x, y, x, y, \ldots)?$$

We can corecursively specify this stream by using pairs of streams, $f\colon 2 \to \Sigma^\omega$ where $2 = \{0, 1\}$. The specification

$$\alpha\colon \{0, 1\} \to \Sigma \times \{0, 1\} \quad \text{with} \quad \alpha(0) = (x, 1) \quad \text{and} \quad \alpha(1) = (y, 0)$$

defines the following coalgebra homomorphism

$$\begin{array}{ccc} \{0,1\} & \xrightarrow{\alpha} & \Sigma \times \{0,1\} \\ {\scriptstyle h}\downarrow & & \downarrow{\scriptstyle \mathsf{id}\times h} \\ \Sigma^\omega & \xrightarrow{\langle \mathsf{head}, \mathsf{tail} \rangle} & \Sigma \times \Sigma^\omega \end{array}$$

such that the streams $h(0)$ and $h(1)$ have heads

$$\mathsf{head} \cdot h(0) = x \qquad \text{and} \qquad \mathsf{head} \cdot h(1) = y$$

and tails that satisfy

$$\mathsf{tail} \cdot h(0) = h(1) \qquad \text{and} \qquad \mathsf{tail} \cdot h(1) = h(0).$$

Obviously, this determines h as follows:

$$h(0) = (x, y, x, y, \ldots) \qquad \text{and} \qquad h(1) = (y, x, y, x, \ldots).$$

(3) We next want to specify the zipping function which assigns to a pair of streams $\bar{x} = (x_0, x_1, x_2 \ldots)$ and $\bar{y} = (y_0, y_1, y_2 \ldots)$ the stream

$$\mathsf{zip}(\bar{x}, \bar{y}) = (x_0, y_0, x_1, y_1, \ldots).$$

The corecursive specification describes the head and tail of $\mathsf{zip}(\bar{x}, \bar{y})$. Clearly,

$$\mathsf{head} \cdot \mathsf{zip}(\bar{x}, \bar{y}) = x_0 = \mathsf{head}(\bar{x}). \tag{2.15}$$

Moreover, $\mathsf{tail} \cdot \mathsf{zip}(\bar{x}, \bar{y}) = (y_0, x_1, y_1, x_2, y_2 \ldots)$ fulfils

$$\begin{aligned} \mathsf{head}(\mathsf{tail} \cdot \mathsf{zip}(\bar{x}, \bar{y})) &= \mathsf{head}(\bar{y}) \\ \mathsf{tail}(\mathsf{tail} \cdot \mathsf{zip}(\bar{x}, \bar{y})) &= \mathsf{zip}(\mathsf{tail}\bar{x}, \mathsf{tail}\bar{y}). \end{aligned} \tag{2.16}$$

The desired coalgebra structure $\alpha \colon \Sigma^\omega \times \Sigma^\omega \to \Sigma \times \Sigma^\omega \times \Sigma^\omega$ will have the form $\alpha = \langle \alpha_{\mathsf{head}}, \alpha_{\mathsf{tail}} \rangle$ for the following functions

$$\alpha_{\mathsf{head}} \colon \Sigma^\omega \times \Sigma^\omega \to \Sigma \qquad \text{and} \qquad \alpha_{\mathsf{tail}} \colon \Sigma^\omega \times \Sigma^\omega \to \Sigma^\omega \times \Sigma^\omega,$$

where (2.15) and (2.16) indicate that

$$\alpha_{\mathsf{head}}(\bar{x}, \bar{y}) = \mathsf{head}(\bar{x}) \qquad \text{and} \qquad \alpha_{\mathsf{tail}}(\bar{x}, \bar{y}) = (\bar{y}, \mathsf{tail}(\bar{x})).$$

Indeed, the function zip clearly makes the following squares commutative:

$$\begin{array}{ccc} \Sigma^\omega \times \Sigma^\omega & \xrightarrow{\alpha_{\mathsf{head}}} & \Sigma \\ {\scriptstyle \mathsf{zip}}\downarrow & & \downarrow{\scriptstyle \mathsf{id}} \\ \Sigma^\omega & \xrightarrow{\mathsf{head}} & \Sigma \end{array} \qquad \begin{array}{ccc} \Sigma^\omega \times \Sigma^\omega & \xrightarrow{\alpha_{\mathsf{tail}}} & \Sigma^\omega \times \Sigma^\omega \\ {\scriptstyle \mathsf{zip}}\downarrow & & \downarrow{\scriptstyle \mathsf{zip}} \\ \Sigma^\omega & \xrightarrow{\mathsf{tail}} & \Sigma^\omega \end{array}$$

Thus, zip is uniquely determined by $\alpha = \langle \alpha_{\mathsf{head}}, \alpha_{\mathsf{tail}} \rangle$ since the pairing of the above two squares yields the commutative square below:

$$\begin{array}{ccc} \Sigma^\omega \times \Sigma^\omega & \xrightarrow{\alpha} & \Sigma \times \Sigma^\omega \times \Sigma^\omega \\ {\scriptstyle \mathsf{zip}}\downarrow & & \downarrow{\scriptstyle \mathsf{id} \times \mathsf{zip}} \\ \Sigma^\omega & \xrightarrow{\langle \mathsf{head}, \mathsf{tail} \rangle} & \Sigma \times \Sigma^\omega \end{array}$$

We have seen primitive recursion in its categorical form in Theorem 2.3.3. Dually, we have

Theorem 2.6.4 (Primitive Corecursion). *Let F be an endofunctor with a terminal coalgebra on a category with finite coproducts. Then for every morphism $\alpha \colon X \to F(X + \nu F)$ there exists a unique morphism $h \colon X \to \nu F$ such that the square below commutes:*

$$\begin{array}{ccc} X & \xrightarrow{\alpha} & F(X + \nu F) \\ {\scriptstyle h}\downarrow & & \downarrow{\scriptstyle F[h, \mathsf{id}_{\nu F}]} \\ \nu F & \xrightarrow{\tau} & F(\nu F) \end{array}$$

Example 2.6.5. For the set functor $FX = X \times X + 1$, the terminal coalgebra νF is carried by the set of all binary trees (see Example 2.5.11(4)). Fix a tree $t \in \nu F$. We use primitive corecursion to define the function $h \colon \nu F \to \nu F$ that takes a tree u and substitutes t for all the leaves in u. To achieve this we define the morphism $\alpha \colon \nu F \to F(\nu F + \nu F)$ by

$$\alpha = \big(\nu F \xrightarrow{\tau} \nu F \times \nu F + 1 \xrightarrow{[\mathsf{inl} \times \mathsf{inl}, f]} (\nu F + \nu F) \times (\nu F + \nu F) + 1\big),$$

where f is the function given by $f(*) = \mathsf{inl} \cdot \langle \mathsf{inr}(t_1), \mathsf{inr}(t_2) \rangle$ if t has the maximum subtrees t_1 and t_2, and $f = \mathsf{inr}$ if t is the single-node tree. Then from the commutativity of the square in Theorem 2.6.4 we see that h is the desired substitution function.

The point of all of this is that we have solution principles that go beyond what we get from terminal coalgebras alone.

Coinduction Principle By dualizing the Induction Principle (see Lemma 2.3.6) we could formulate a coinduction principle: homomorphisms into the terminal coalgebra exist uniquely. Instead, we introduce the concept of a *bisimulation* here and formulate that principle accordingly.

Bisimulation is an important concept in the theory of coalgebra, somewhat dual to that of congruence for algebras. In the case of set functors, a bisimulation between coalgebras $\alpha \colon A \to FA$ and $\beta \colon B \to FB$ is a relation $R \subseteq A \times B$

such that both projections $\pi_A\colon R \to A$ and $\pi_B\colon R \to B$ become coalgebra homomorphisms for some coalgebra structure on R.

In general categories, *relations* between two objects A and B are subobjects of $A \times B$. As explained in Remark 2.1.18, they are represented by monomorphisms $r\colon R \to A \times B$. Or, equivalently, by collectively monic spans of morphisms

$$A \xleftarrow{\;r_A\;} R \xrightarrow{\;r_B\;} B$$

Indeed, the span is collectively monic iff $\langle r_A, r_B\rangle\colon R \to A \times B$ is monic.

Definition 2.6.6. Let (A, α) and (B, β) be coalgebras for F. By a *bisimulation* between them is meant a relation $\langle r_A, r_B\rangle\colon R \to A \times B$ for which a coalgebra structure on R exists, turning both r_A and r_B into coalgebra homomorphisms. If the two coalgebras are equal, we speak about a *bisimulation on* (A, α).

The importance of this notion is that pairs of states related by some bisimulation intuitively have the same behaviour. And continuing with this intuition, the largest bisimulation between two coalgebras should be exactly the relation that relates two states iff they have the same behaviour. We mention some examples.

Example 2.6.7. (1) Given two systems with termination (Example 2.4.2(1))

$$\alpha\colon A \to A + 1 \qquad \text{and} \qquad \beta\colon B \to B + 1,$$

a bisimulation is a relation $R \subseteq A \times B$ such that whenever $x\ R\ y$ then

(a) x is terminal iff y is, and
(b) if x and y are nonterminal, then $x'\ R\ y'$ for the next states x' of x and $y'$of y.

Indeed, if (a) and (b) hold, we can define a coalgebra structure $\gamma\colon R \to R + 1$ by sending a pair (x, y) in R to the right-hand summand of $FR = R + 1$ if x is terminal, else $\gamma(x, y) = (x', y')$ for the above next states. The projections of R into A and B are clearly homomorphisms. Conversely, every bisimulation is easily seen to satisfy (a) and (b).

The *behaviour* function $b_A\colon A \to \mathbb{N}^\top$ of Example 2.5.3(2) fulfils for every bisimulation R that

$$x\ R\ y \quad \text{implies} \quad b_A(x) = b_A(y).$$

Conversely, define a relation between coalgebras A and B by 'having the same behaviour'. This is an example of a bisimulation. Consequently, this is the largest bisimulation between (A, α) and (B, β).

(2) Let A and B be deterministic automata as coalgebras for the set functor given by $FX = \{0,1\} \times X^{\Sigma}$, see Example 2.4.2(3). A bisimulation $R \subseteq A \times B$ is precisely a relation such that for $x\ R\ y$ we have

(a) x is accepting iff y is, and
(b) $x'\ R\ y'$ whenever for some input $s \in \Sigma$, x' is the next state of x and y' the next state of y under s.

It is easy to see that this implies that the language accepting by x in A is equal to that accepting by y in B.

Conversely, the relation 'accepting the same language' is a bisimulation between A and B – thus, it is characterized as the largest bisimulation.

(3) The name bisimulation stems from the theory of labelled transition systems (see Example 2.4.2(6)) where we recover precisely Milner's notion of a strong bisimulation [237]: given a set Σ of actions, an LTS is a coalgebra for $(\mathscr{P}(-))^{\Sigma}$. Given LTSs A and B, a relation R between A and B is a bisimulation iff for every $x\ R\ y$ and every action $s \in \Sigma$ we have

(a) given a transition $x \xrightarrow{s} x'$ in A, there exists a transition $y \xrightarrow{s} y'$ in B with $x'\ R\ y'$, and
(b) given a transition $y \xrightarrow{s} y'$ in B, there exists a transition $x \xrightarrow{s} x'$ in A with $x'\ R\ y'$.

It is easy to see that this coincides, for the functor $FX = (\mathscr{P}X)^{\Sigma}$ (equivalently, for $FX = \mathscr{P}(\Sigma \times X)$), with above concept of bisimulation.

(4) Given graphs (V_1, E_1) and (V_2, E_2) as coalgebras for $\mathscr{P}$ (see Example 2.4.5), a bisimulation is a relation $R \subseteq V_1 \times V_2$ with the following properties: whenever $x\ R\ y$,

(a) for every edge $(x, x') \in E_1$ there exists an edge $(y, y') \in E'$ with $y\ R\ y'$, and
(b) for every edge $(y, y') \in E_2$ there exists an edge $(x, x') \in E_1$ with $x\ R\ x'$.

Remark 2.6.8. (1) The *diagonal subobject* of $A \times A$ is the relation Δ_A represented by $\mathsf{id}_A, \mathsf{id}_A \colon A \to A$. For every coalgebra (A, α) this relation is clearly a bisimulation on (A, a). Any bisimulation contained in Δ_A is called *trivial*. Observe that a bisimulation $r_1, r_2 \colon R \to A$ is non-trivial iff $r_1 \neq r_2$.

(2) For a set functor, another bisimulation on (A, α) is given by an arbitrary parallel pair of coalgebra homomorphisms

$$h_1, h_2 \colon (B, \beta) \to (A, \alpha).$$

More precisely, given such a pair in $\mathsf{Coalg}\, F$, the relation $R \subseteq A \times A$ of all the pairs $(h_1(x), h_2(x))$, $x \in B$, is a bisimulation.

Indeed, for the epimorphism $e\colon B \to R$ given by $e(x) = (h_1(x), h_2(x))$ choose a splitting $m\colon R \to B$. Then we have $e \cdot m = \mathsf{id}$. Since the projections $r_1, r_2\colon R \to B$ fulfil $h_i = r_i \cdot e$, thus, $h_i \cdot m = r_i$, the following diagram commutes for $i = 1$ and 2:

$$\begin{array}{ccccccc} R & \xrightarrow{m} & B & \xrightarrow{\beta} & FB & \xrightarrow{Fe} & FR \\ & \searrow^{r_i} & \downarrow{\scriptstyle h_i} & & \downarrow{\scriptstyle Fh_i} & \swarrow_{Fr_i} & \\ & & A & \xrightarrow{\alpha} & FA & & \end{array}$$

Consequently, r_i are homomorphisms for the coalgebra structure $Fe \cdot \beta \cdot m$ on R.

(3) A *functional bisimulation* is a bisimulation $R \subseteq A \times B$ whose graph of a function $f\colon A \to B$. Observe that, for example, coalgebra homomorphisms between automata are precisely the functional bisimulations: see the description of homomorphisms in Example 2.4.2(3). The same holds for non-deterministic automata, systems with termination and graphs (as coalgebras for $\mathscr{P}$). In all these examples coalgebra homomorphisms are precisely the functional bisimulations.

We now formulate the Coinduction Principle by using the concept of trivial bisimulation.

Lemma 2.6.9 (Coinduction Principle). *If F is an endofunctor with a terminal coalgebra, then every bisimulation on νF is trivial.*

Proof Let $r = \langle r_1, r_2\rangle\colon R \to \nu F \times \nu F$ be a bisimulation. Then R has a coalgebra structure such that $r_1, r_2\colon R \to \nu F$ are coalgebra homomorphisms. Since νF is terminal, this proves $r_1 = r_2$. Equivalently, R is trivial. □

Remark 2.6.10. Lemma 2.6.9 shows that bisimulations are a sound proof principle to establish behavioural equivalence of states. The latter notion is defined for set functors F as follows: given two F-coalgebras (A, α) and (B, β) a pair of element $x \in A$ and $y \in B$ is called *behaviourally equivalent* if there exists a pair of coalgebra homomorphisms $h\colon (A, \gamma) \to (C, \gamma)$ and $k\colon (B, \beta) \to (C, \gamma)$ such that $h(x) = k(y)$. Equivalently, x and y are merged by the unique coalgebra homomorphisms into νF (if it exists).

As bisimulations are not a core topic of this book, we do not discuss the completeness of the bisimulation proof method at length. Let us just remark that it fails in general: there exist functors where behaviourally equivalent states need not be related by a bisimulation. However, if the functor F preserves weak pullbacks (see Remark C.3.3) then every pair of behaviourally equivalent states is related by a bisimulation. For details see [4, 264].

Example 2.6.11. Addition on $\mathbb{N}^\top = \nu X.\, X + 1$ specified in Example 2.6.2 has the usual properties. We present coinductive proofs of some of them. Recall the coalgebra structure $\tau\colon \mathbb{N}^\top \to \mathbb{N}^\top + 1$ given by predecessor. We denote by $s\colon \mathbb{N}^\top \to \mathbb{N}^\top$ the successor function, which is the restriction of the inverse of τ. Let us write $x \oplus y$ in lieu of $\mathsf{add}(x, y)$.

(1) We prove by coinduction that $\oplus$ satisfies the equations by which addition on $\mathbb{N}$ is usually defined by recursion:

$$0 \oplus k = k,$$
$$s(n) \oplus k = s(n \oplus k).$$

For the first equation we define the relation $R = \{(0 \oplus k, k) \mid k \in \mathbb{N}^\top\}$. If we prove that R is a bisimulation, we are ready: then $R = \Delta$ and our equality is proved. We verify the two conditions in Example 2.6.7(1). Indeed, we have that $0 \oplus k$ is terminating iff k is so (equivalently, $k = 0$). For $k \neq 0$ we have $\tau(0 \oplus k) = 0 \oplus \tau(k) \; R \; \tau(k)$.

To prove the second equation, let R be the relation on $\mathbb{N}^\top$ consisting of all pairs $(s(n) \oplus k, s(n \oplus k))$ and all (n, n). We show that R is a bisimulation. First observe that the desired conditions hold for each pair (n, n) in R. The remaining pairs do not contain any terminal states, and we have

$$\tau(s(n) \oplus k) = \tau(s(n)) \oplus k = (n \oplus k) \; R \; (n \oplus k) = \tau(s(n \oplus k)).$$

This proves that R is a bisimulation.

(2) Let us verify the following equation

$$n \oplus s(k) = s(n) \oplus k.$$

To prove this by coinduction, let R be the relation on $\mathbb{N}^\top$ containing the pairs $(n \oplus s(k), s(n) \oplus k)$ and all (n, n). Again, the two conditions are trivially true for all pairs (n, n), and the remaining ones are non-terminating. We use that $\tau(s(n)) = n$ for all $n \in \mathbb{N}^\top$ and that $s(\tau(n)) = n$ for all $n \neq 0$ to verify that, for $n = 0$ we have

$$\tau(n \oplus s(k)) = n \oplus \tau(s(k)) = n \oplus k \; R \; n \oplus k = \tau(s(n)) \oplus k = \tau(s(n) \oplus k),$$

and similarly, for $n \neq 0$ we have

$$\tau(n \oplus s(k)) = (\tau(n) \oplus s(k)) \; R \; (s(\tau(n)) \oplus k) = \tau(s(n)) \oplus k = \tau(s(n) \oplus k).$$

This proves that R is a bisimulation.

(3) Addition is commutative. Indeed, let R be the relation of all pairs $(n \oplus k, k \oplus n)$. To see that R is a bisimulation, note first that the only pair in R containing terminating states is $(0 \oplus 0, 0 \oplus 0)$. For the remaining states we have two cases. If $n = 0$ and $k \neq 0$, then

$$\tau(n \oplus k) = (n \oplus \tau(k)) \; R \; (\tau(k) \oplus n) = \tau(k \oplus n).$$

If $n \neq 0 \neq k$, then we have

$$\begin{aligned} \tau(n \oplus k) &= \tau(n) \oplus k \\ &\qquad R\,(k \oplus \tau(n)) \\ &\qquad = s(\tau(k)) \oplus \tau(n) = \tau(k) \oplus s(\tau(n)) = \tau(k) \oplus n = \tau(k \oplus n), \end{aligned}$$

where we use item (2) as a lemma. Thus R is a bisimulation, and the commutativity is proved.

Example 2.6.12 [264]. Recall from Example 2.6.3 the corecursively specified streams $(x, x, x \ldots)$, $(y, y, y, \ldots)$, and $(x, y, x, y, \ldots)$ that we now denote by $\bar{x}$, $\bar{y}$, and $\overline{xy}$, respectively. We will prove

$$\mathsf{zip}(\bar{x}, \bar{y}) = \overline{xy}$$

by coinduction. Let R be the relation on $\Sigma^\omega = \nu X.X \times \Sigma$ consisting of two pairs:

$$\mathsf{zip}(\bar{x}, \bar{y})\ R\ \overline{xy} \qquad \text{and} \qquad \mathsf{zip}(\bar{y}, \bar{x})\ R\ \overline{yx}.$$

We prove that this is a bisimulation. Define a coalgebra structure on R by the following dynamical system $\langle \delta, \gamma \rangle : R \to \Sigma \times R$ with outputs, see Example 2.5.3(3), where the output is denoted by $\mapsto$:

$$\begin{array}{ccc} \langle \mathsf{zip}(\bar{x}, \bar{y}), \overline{xy} \rangle & \underset{\delta}{\overset{\delta}{\rightleftarrows}} & \langle \mathsf{zip}(\bar{y}, \bar{x}), \overline{yx} \rangle \\ \gamma \big\downarrow & & \big\downarrow \gamma \\ x & & y \end{array}$$

Both projections $r_1, r_2 \colon R \to \Sigma^\omega$ are easily seen to be coalgebra homomorphisms. Thus, $R \subseteq \Delta$ which proves the desired equality.

Cofree coalgebras At the close of Section 2.2, we have seen that free algebras generalize initial algebras, and also that the free algebra construction leads to a monad. We now pursue the same topic, this time on the dual side. Here is the motivation for this. We often view a coalgebra $\alpha \colon A \to FA$ as a 'black box': we cannot observe its states directly and therefore do not have a full information about next states. A terminal coalgebra gives us the observable part: for the given homomorphism $h \colon A \to \nu F$ we do not know the state x, but we know its behaviour $h(x)$. An improved view is provided by an additional output functions $g \colon A \to \Gamma$. Here Γ is a set of colours, and whenever a cofree coalgebra $\Gamma_\sharp$ (as defined below) exists, then the corresponding homomorphism $\bar{g} \colon A \to \Gamma_\sharp$ gives us visible information about the colour of current state x.

Definition 2.6.13. By a *cofree F-coalgebra* on an object Γ (of colours) in $\mathscr{A}$ is meant a coalgebra

$$\tau_\Gamma \colon \Gamma_\sharp \to F\Gamma_\sharp$$

together with a couniversal (colouring) morphism $\varepsilon_\Gamma \colon \Gamma_\sharp \to \Gamma$. Couniversality means that for every coalgebra $\alpha \colon A \to FA$ and every morphism $g \colon A \to \Gamma$ in $\mathscr{A}$ there exists a unique homomorphism $\bar{g} \colon A \to \Gamma_\sharp$ with $g = \varepsilon_\Gamma \cdot \bar{g}$

$$\begin{array}{ccccc} \Gamma & \xleftarrow{\;g\;} & A & \xrightarrow{\;\alpha\;} & FA \\ & \nwarrow^{\varepsilon_\Gamma} & \downarrow{\scriptstyle \bar{g}} & & \downarrow{\scriptstyle F\bar{g}} \\ & & \Gamma_\sharp & \xrightarrow{\;\tau_\Gamma\;} & F\Gamma_\sharp \end{array}$$

Example 2.6.14. Consider dynamical systems as coalgebras of $FX = X + 1$. Let A be a system with an additional output function $g \colon A \to \{0, 1\}$. Every state $x \in B$ defines a finite or infinite sequence of states: we start in x and continue moving, stopping only when a terminating state is reached. We do not know the sequence $x = x_0, x_1, x_2 \ldots$ of states, but we do see the sequence of colours $g(x_0), g(x_1), g(x_2) \ldots$.

So if our interpretation of $A_\sharp$ above is correct, we should have

$$\{0, 1\}_\sharp = \{0, 1\}^\infty$$

the set of finite and infinite binary streams. This is indeed true, and it is a consequence of the following.

Proposition 2.6.15. *The cofree F-coalgebra on Γ is precisely the terminal coalgebra for the endofunctor $F(-) \times \Gamma$.*

This is dual to Proposition 2.2.24. Applied to $FX = X + 1$ this tells us that the cofree coalgebra

$$\{0, 1\}_\sharp = \nu X.\, (X \times \{0, 1\} + \{0, 1\})$$

is the terminal Σ-coalgebra, where Σ has two unary operations and two constants. This terminal Σ-coalgebra consists of all Σ-trees. These are unary trees in which all nodes (inner ones as well as leaves) are labelled by 0 or 1. And this is isomorphic to $\{0, 1\}^\infty$.

Definition 2.6.16. An endofunctor F is a *covarietor* if every object generates a cofree coalgebra for F. The corresponding comonad is denoted by $F_\sharp$.

Example 2.6.17. (1) $FX = X + 1$ is a covarietor with

$$\Gamma_\sharp = \Gamma^\infty.$$

This follows from $F(-) \times \Gamma = H_\Sigma$ where $\Sigma_0 = \Sigma_1 = \Gamma$ and $\Sigma_n = \emptyset$ for $n > 1$.

(2) $FX = X \times X + 1$ is a covarietor. The coalgebra $\Gamma_\sharp$ consists of all binary trees with nodes labelled in Γ. This follows from $F(-) \times \Gamma = H_\Sigma$ where $\Sigma_0 = \Gamma = \Sigma_2$ and otherwise $\Sigma_n = \emptyset$.

Proposition 2.6.18. *For every covarietor F, the cofree comonad on F is $F_\sharp$. Conversely, if $\mathscr{A}$ is a cocomplete category, then every endofunctor generating a cofree comonad is a covarietor.*

This is dual to Theorem 2.2.29.

2.7 Summary

As this long chapter draws to a close, here are the points that readers should be most confident of, and let us also raise some general points about the material.

The main theme of this chapter has been the description of initial algebras and terminal coalgebras of endofunctors F on Set and some other categories (Set^S, $\mathsf{CPO}_\perp$, Nom, and $\mathsf{Set}^{\mathscr{F}}$). We have seen that these are fixed points of F: this is Lambek's lemma [210]. We were particularly interested in initial algebras and terminal coalgebras for polynomial set functors. Initial algebras μF are important for a generic form of induction and recursion. Terminal coalgebras νF formalize the concept of behaviour of a state: given a state-based system on a set A of states (whose dynamics is expressed as a coalgebra for F) then the unique homomorphism from A to νF assigns to each state its behaviour. Terminal coalgebras are also important as the codomains for corecursive definitions. For the corresponding proof principle, coinduction, the concept of a bisimulation plays a crucial rôle. These themes will persist through the book.

3

Finitary Iteration

We show how to obtain an initial algebra for an endofunctor by iterating that endofunctor on the initial object. This can be seen as the categorical version of Kleene's famous fixed point theorem for continuous functions on cpos that we recall first below. (For more about fixed points of endomaps of ordered sets, see Appendix B.) The iterations considered here are countable, the more general case (generalizing the Knaster–Tarski Fixed Point Theorem) is treated in Chapter 6. We concentrate more on the dual construction: the terminal coalgebra as a limit of iterations of the endofunctor on the terminal object. Most of this chapter is devoted to examples. In later chapters, we take up other topics: representations of terminal coalgebras, and constructions that either use uncountable iterations or else use countable iteration in connection with extra order-theoretic or metric structure.

3.1 Initial-Algebra Chain

We recall Kleene's Fixed Point Theorem and present its categorical version: the initial algebra for an endofunctor as a colimit of iterations on the initial object. We show with concrete examples how this yields initial algebras for 'well-behaved' endofunctors. For example, this yields a new view on the initial algebra for a polynomial functor associated with a signature Σ as the algebra of all finite Σ-labelled trees (Proposition 2.2.14).

Throughout this section we assume that a category $\mathscr{A}$ and an endofunctor F on $\mathscr{A}$ are given. We assume that $\mathscr{A}$ has an initial object 0 and denote by $!: 0 \to X$ the unique morphism to a given object X.

Recall from Example 2.1.7 that a cpo is a poset with joins of ω-chains.

Theorem 3.1.1 (Kleene). *Let A be a cpo with a least element $\bot$. Then every* *continuous endofunction F has a least fixed point*

$$\mu F = \bigvee_{n<\omega} F^n(\perp).$$

Proof First, an induction on $n < \omega$ shows that $F^n(\perp) \le F^{n+1}(\perp)$. So the sequence $(F^n(\perp))_{n<\omega}$ is an ω-chain. Write μF for its join. By continuity, $F(\mu F) = \bigvee_n F(F^n(\perp))$. It is easy to check that $\bigvee_n F^n(\perp) = \bigvee_n F^{n+1}(\perp)$, and hence $F(\mu F) = \mu F$. Thus, we have a fixed point of F. If $Fx \le x$, then an easy induction on n shows that $F^n(\perp) \le x$; hence $\mu F \le x$ as well. □

Remark 3.1.2. More generally, a *pre-fixed point* of F is an element x with $Fx \le x$; the argument above shows that μF is also the least pre-fixed point of F.

For future use, we record the following induction principle. It follows from the proof of Theorem 3.1.1.

Corollary 3.1.3 (Scott induction)**.** *Let P be a cpo with a least element $\perp$, and let S be a subset which contains $\perp$ and is closed under joins of ω-chains. If a continuous function $f\colon P \to P$ preserves S (in symbols: $f[S] \subseteq S$), then $\mu f \in S$.*

In this book, we are not really interested in Kleene's Theorem but in generalizations of it and in dualizations of those generalizations. Figure 3.1 shows how we generalize the order-theoretic concepts in Kleene's Theorem to the level of categories. In each line, the order-theoretic concept on the left is a special case of the category-theoretic concept to its right. (To see this, recall that a preorder $(P, \le)$ is exactly a category in which every homset is either empty or a singleton set.) Of special interest is the generalization of the completeness condition on the poset and the continuity condition on the function.

An *ω-chain* in a category $\mathscr{A}$ is a functor from ω (the linearly ordered set of natural numbers) to $\mathscr{A}$. An *ω-colimit* is a colimit of an ω-chain.

Definition 3.1.4. By the *initial-algebra ω-chain* of an endofunctor $F\colon \mathscr{A} \to \mathscr{A}$ is meant the ω-chain

$$0 \xrightarrow{!} F0 \xrightarrow{F!} F^2 0 \xrightarrow{F^2!} \cdots \xrightarrow{F^{n-1}!} F^n 0 \xrightarrow{F^n!} F^{n+1}0 \longrightarrow \cdots \qquad (3.1)$$

Incidentally, the reason we use a hyphen in *initial-algebra chain* is to avoid the suggestion that this chain is initial in any category of 'algebra chains.' (In fact, we have no such chains in this book.)

In Chapter 6, we shall need *transfinite* iterations of the initial-algebra chain. But in this section, we only consider the finite iterations as in (3.1).

Remark 3.1.5. A *cocone* of the initial-algebra ω-chain is an object A of $\mathscr{A}$

order-theoretic concept	category-theoretic generalization
preorder $(A, \leq)$	category $\mathscr{A}$
$x \leq y$ and $y \leq x$	A and B are isomorphic objects
least element 0	initial object 0
monotone $F\colon A \to A$	functor $F\colon \mathscr{A} \to \mathscr{A}$
pre-fixed point: $Fx \leq x$	F-algebra: $f\colon FA \to A$
ω-chain	functor from $(\omega, \leq)$ to $\mathscr{A}$
F is continuous	F preserves ω-colimits
least pre-fixed point: $Fx \leq x$	initial F-algebra: $\iota\colon F(\mu F) \to \mu F$

Figure 3.1 Generalizing Kleene's Theorem to categories.

together with a family of morphisms $a_n\colon F^n0 \to A$ such that the triangles below commute:

$$\begin{array}{ccc} F^n0 & \xrightarrow{F^n!} & F^{n+1}0 \\ & {\scriptstyle a_n}\searrow \quad \swarrow{\scriptstyle a_{n+1}} & \\ & A & \end{array} \qquad \text{for all } n < \omega.$$

We often denote cocones with the notation (a_n) or $(a_n)_n$. A *colimit* of the initial-algebra ω-chain is a cocone $c_n\colon F^n0 \to C$ over it with the universal property that if $a_n\colon F^n0 \to A$ is any cocone, then there is a unique *factorizing morphism* $f\colon C \to A$, which means that f is unique such that for all $n < \omega$ we have $a_n = f \cdot c_n$.

Construction 3.1.6. Every F-algebra (A, α) induces the *canonical cocone* $\alpha_n\colon F^n0 \to A$ of the initial-algebra ω-chain as follows: $\alpha_0\colon 0 \to A$ is unique (since 0 is initial) and

$$\alpha_{n+1} = (FF^n0 \xrightarrow{F\alpha_n} FA \xrightarrow{\alpha} A). \tag{3.2}$$

The cocone property, $\alpha_n = \alpha_{n+1} \cdot F^n!$, is easy to verify by induction.

Remark 3.1.7. (1) Homomorphisms of algebras preserve the induced cocones: given a homomorphism h from (A, α) to (B, β), that is, we have $h \cdot \alpha = \beta \cdot Fh$, it follows that

$$h \cdot \alpha_n = \beta_n.$$

This is trivial for $n = 0$, and the induction step is easy:

$$h \cdot \alpha_{n+1} = (h \cdot \alpha) \cdot F\alpha_n = (\beta \cdot Fh) \cdot F\alpha_n = \beta \cdot F\beta_n = \beta_{n+1}.$$

(2) Let $c_n\colon F^n0 \to \mu F$ be the colimit of the initial-algebra ω-chain. Applying F

to each object and morphism in (3.1) yields another ω-chain

$$F0 \xrightarrow{F!} F^2 0 \xrightarrow{F^2!} F^3 0 \xrightarrow{F^3!} \cdots \xrightarrow{F^n!} F^{n+1}0 \xrightarrow{F^{n+1}!} F^{n+2}0 \longrightarrow \cdots, \tag{3.3}$$

which obviously has the same colimit as (3.1).

Theorem 3.1.8 [7]. *Let $\mathscr{A}$ be a category with initial object* 0 *and with colimits of ω-chains. If $F\colon \mathscr{A} \to \mathscr{A}$ preserves ω-colimits, then it has an initial algebra*

$$\mu F = \operatorname*{colim}_{n<\omega} F^n 0.$$

Remark 3.1.9. We have the colimit cocone $Fc_n\colon F^{n+1}0 \to F(\mu F)$ of the chain (3.3). Let $\iota\colon F(\mu F) \to \mu F$ be the unique morphism such that the following triangles commute:

$$\begin{array}{ccc} F(F^n 0) = F^{n+1}0 & & \\ {\scriptstyle Fc_n}\downarrow & \searrow^{c_{n+1}} & \\ F(\mu F) & \xrightarrow{\iota} & \mu F \end{array} \qquad (n < \omega). \tag{3.4}$$

We are going to prove that $(\mu F, \iota)$ is an initial F-algebra. Moreover, for any algebra (A, α), the unique algebra morphism $h\colon \mu F \to A$ is the factorization induced by the canonical cocone:

$$h \cdot c_n = \alpha_n \qquad \text{for all } n < \omega.$$

Proof In this proof we denote by $c_n\colon F^n 0 \to \mu F$ the colimit above. The hypothesis that F preserves ω-colimits implies that $Fc_n\colon F^{n+1}0 \to F(\mu F)$ is a colimit cocone of (3.3). We have another cocone of (3.3), namely $c_{n+1}\colon F^{n+1}0 \to F(\nu F)$. Hence there is a unique morphism $\iota\colon F(\mu F) \to \mu F$ with the property (3.4).

We claim that the F-algebra $(\mu F, \iota)$ is initial. To check this, let (A, α) be an F-algebra. Consider the induced cocone $\alpha_n\colon F^n 0 \to A$ from Construction 3.1.6. It is easy to check that $F\alpha_n\colon F^{n+1}0 \to FA$ is a cocone of (3.3). Let $h\colon \mu F \to A$ be the unique morphism satisfying $h \cdot c_n = \alpha_n$ for all $n < \omega$.

(1) We prove that $h\colon (\mu F, \iota) \to (A, \alpha)$ is a homomorphism. In order to see that $h \cdot \iota = \alpha \cdot Fh$, we check that both are factorizing morphisms for the cocone $\alpha_{n+1}\colon F^{n+1}0 \to A$. That is, we check that for all n,

$$(h \cdot \iota) \cdot Fc_n = \alpha_{n+1} = (\alpha \cdot Fh) \cdot Fc_n.$$

For the first assertion, $h \cdot (\iota \cdot Fc_n) = h \cdot c_{n+1} = \alpha_{n+1}$. For the second one, consider the diagram below:

$$\begin{array}{ccc} F^{n+1}0 & \xrightarrow{\alpha_{n+1}} & A \\ {\scriptstyle Fc_n}\downarrow & \searrow^{F\alpha_n} & \uparrow{\scriptstyle \alpha} \\ F(\mu F) & \xrightarrow{Fh} & FA \end{array}$$

The upper triangle commutes by (3.2). The lower one commutes by definition of h. So the outside commutes, showing that indeed $(\alpha \cdot Fh) \cdot Fc_n = \alpha_{n+1}$. This verifies that h is a homomorphism.

(2) To prove that h is unique, suppose that $k\colon \mu F \to A$ is also a homomorphism: $k \cdot \iota = \alpha \cdot Fk$. We show that k is also the factorization of the induced cocone: $k \cdot c_n = \alpha_n$. Then the uniqueness of h implies that $k = h$. For $n = 0$, we see that $k \cdot c_0$ and α_0 are both morphisms with domain 0, so they are the same. Assuming that $k \cdot c_n = \alpha_n$, we see that

$$\begin{aligned} k \cdot c_{n+1} &= k \cdot \iota \cdot Fc_n && \text{by definition of } \iota \\ &= \alpha \cdot F(k \cdot c_n) && \text{since } k \text{ is a homomorphism} \\ &= \alpha \cdot F\alpha_n && \text{by induction hypothesis} \\ &= \alpha_{n+1} && \text{by (3.2).} \end{aligned}$$

□

Remark 3.1.10. To obtain an initial algebra for an endofunctor $F\colon \mathscr{A} \to \mathscr{A}$, it is not really necessary that $\mathscr{A}$ has colimits of all ω-chains or that F preserves all ω-colimits. It is sufficient to assume that the colimit of the initial-algebra ω-chain (3.1) exists and that F preserve this colimit. That is, these are the only facts about $\mathscr{A}$ and F that are used in the proof of Theorem 3.1.8. In many cases, it is just as easy to verify the stronger requirements that we stated in Theorem 3.1.8 than it is to verify the special cases used in the proof. But this is not always the case, and we shall see examples where the sufficient conditions hold, but the stronger ones do not. For one such example (in dual form) near the end of the book, see Theorem 15.3.10.

3.2 Examples of Initial Algebras

We present examples of initial algebras obtained by finite iteration as in Theorem 3.1.8. We discuss these at some length because the same functors will appear throughout this book.

Example 3.2.1. (1) The functor $FX = X + 1$. Here 1 is a terminal object. In Example 2.2.7(3), we considered this as a functor on Set and found that μF is the set of natural numbers.

In an arbitrary category $\mathscr{A}$ with binary coproducts and a terminal object 1, the initial algebra for $FX = X + 1$ is a natural numbers object (NNO) given by the copower $\mathbb{N} \bullet 1$ (Example 2.2.9), assuming that it exists. The functor $FX = X + 1$ is easily seen to preserve ω-colimits. Thus it has an initial algebra obtained as the colimit of the initial-algebra ω-chain formed by the right-hand coproduct injections

$$0 \longrightarrow 1 \longrightarrow 1 + 1 \longrightarrow 1 + (1 + 1) \longrightarrow \cdots \tag{3.5}$$

It is easy to show that $\mathscr{A}$ has an NNO iff this colimit exists. When $\mathscr{A} = \mathsf{Set}$, we may identify its nth term in (3.5) with the natural number $n = \{0, \dots, n-1\}$ and obtain an ω-chain of inclusion maps. The colimit is the set $\mathbb{N}$ of natural numbers, with the colimit morphisms $c_n : n \to \mathbb{N}$ inclusion maps. The algebra structure obtained from (3.4)

$$\iota : \mathbb{N} + 1 \to \mathbb{N} \tag{3.6}$$

is the isomorphism taking 0 in the right-hand summand, 1, to $0 \in \mathbb{N}$, and the left summand $\mathbb{N}$ to itself via the successor function. This follows easily from $\iota \cdot Fc_n = c_{n+1}$ for all $n < \omega$. This is a new explanation of Example 2.2.7(3).

(2) Consider the category Pos of posets and monotone functions. Here coproducts are disjoints unions of partially ordered sets, where elements of different coproduct components incomparable. The above endofunctor takes a poset and adds a fresh element incomparable to the elements of X. Then $F^n 0$ can be identified with the set $\{0, \dots, n-1\}$ discretely ordered, and the NNO is again $\mathbb{N}$ with the same structure as in (3.6) and discretely ordered. Observe that, in accordance with Example 2.2.9, $\mathbb{N} = \mathbb{N} \bullet 1$ in Pos.

There is also the functor $FX = X_\perp$ adding a new least element to a poset. Here the initial-algebra ω-chain has the following form (where the connecting morphisms are the obvious inclusion maps):

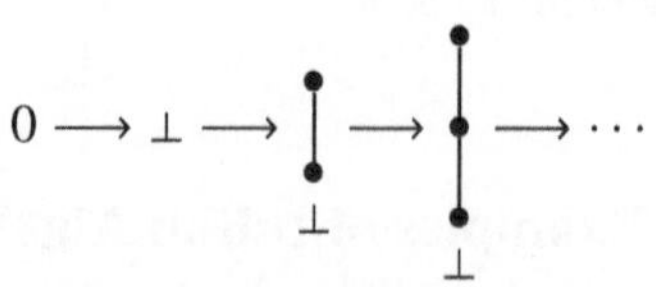

(3.7)

We can describe $F^n 0$ as the linear order $0 < 1 < 2 < \cdots < n-1$, and μF is $\mathbb{N}$ with the usual order and the algebra structure ι with $\iota(\perp) = 0$ and $\iota(n) = n + 1$.

Similarly, we have the endofunctor $FX = X^\top$ adding a new top element. Here μF is the set of natural numbers with the reverse of the usual order: $0 > 1 > 2 > \cdots$.

(3) For the analogous endofunctors on CPO, see Example 2.2.17. In contrast,

the NNO in this category is uninteresting: the functor $FX = X + 1$ is the identity functor, which has the initial algebra $0 = \{\bot\}$.

Notation 3.2.2. (1) Let

$$\mathsf{Met}$$

be the category of 1-bounded metric spaces and non-expanding maps. That is, objects are sets endowed with a *metric* $d\colon X \times X \to [0, 1]$, i.e. a function which satisfies

(a) $d(x, y) = 0$ iff $x = y$
(b) $d(x, y) = d(y, x)$
(c) $d(x, z) \le d(x, y) + d(x, z)$ (the triangle inequality).

Morphisms are *non-expanding functions*: these are functions $f\colon X \to X'$ such that for all $x, y \in X$,

$$d'(f(x), f(y)) \le d(x, y).$$

(2) We also have the full subcategory

$$\mathsf{CMet}$$

of Met of *complete* metric spaces, in which every Cauchy sequence has a limit.

(3) Every metric space X has a *Cauchy completion*, which is a complete metric space $\bar{X}$ containing X as a dense subspace. (This determines $\bar{X}$ uniquely up to isomorphism.) The inclusion map $X \hookrightarrow \bar{X}$ is a reflection of X in CMet: for every non-expanding function $f\colon X \to Y$, with Y being complete, there exists a unique non-expanding extension $\bar{f}\colon \bar{X} \to Y$.

Example 3.2.3. We now consider $FX = X + 1$ as an endofunctor of Met. Coproducts are disjoint unions with distance 1 between points in different summands. The functor $FX = X + 1$ now has as its initial-algebra ω-chain the chain in (3.5). Thus, the NNO is the discrete metric space of natural numbers (with all distances 0 or 1). The same is true for $FX = X + 1$ as an endofunctor of CMet. Indeed, in the initial-algebra ω-chain, we see that $F^n 0$ is the discrete space of n elements, and the NNO, which is the colimit, is the discrete space on $\mathbb{N}$. The algebra structure (3.4) is the same as in (3.6).

Example 3.2.4. The situation changes when we scale the metric by $\frac{1}{2}$ (or by any other constant between 0 and 1). Let $\frac{1}{2}\colon \mathsf{Met} \to \mathsf{Met}$ scale all distances by $\frac{1}{2}$. We now consider $GX = \frac{1}{2}X + 1$.

The initial-algebra ω-chain of G is the following chain of inclusions:

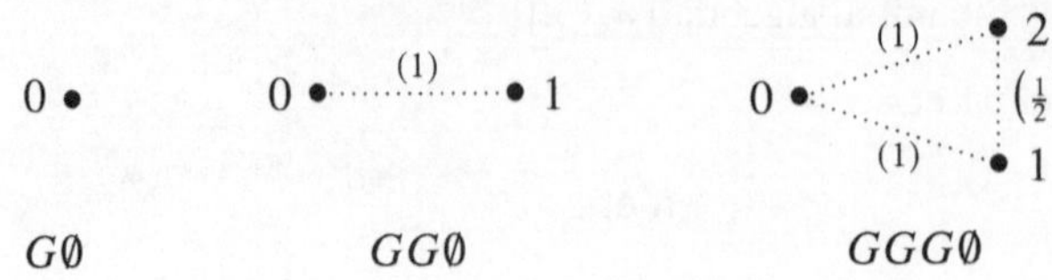

Here 0 represents the unique element of the right-hand summand of

$$G^{n+1}0 = \frac{1}{2}(G^n 0) + 1,$$

and the element i of $G^n 0$ is represented by $i + 1$ in $G^{n+1}0$. We see that $G^n 0$ is the space $\{0, 1, \ldots, n - 1\}$ with the following metric:

$$d(i, i) = 0 \qquad \text{and} \qquad d(i, j) = 2^{-\min(i,j)}, \quad \text{if } i \neq j.$$

The colimit of this chain is $\mu G = \mathbb{N}$ with the above metric.

Example 3.2.5. Let us consider the set functor $FX = X \times X + 1$. When dealing with this functor and related ones, it is often useful to adopt a graphical notation. We shall draw $(x, y) \in X \times X \hookrightarrow FX$ as the ordered tree

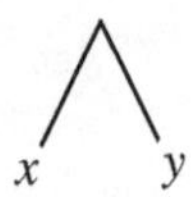

We start with $F^0 0 = \emptyset$. Now we picture the elements of $F^i 0$ for $i = 1, 2$, and 3 using • to denote the element of 1:

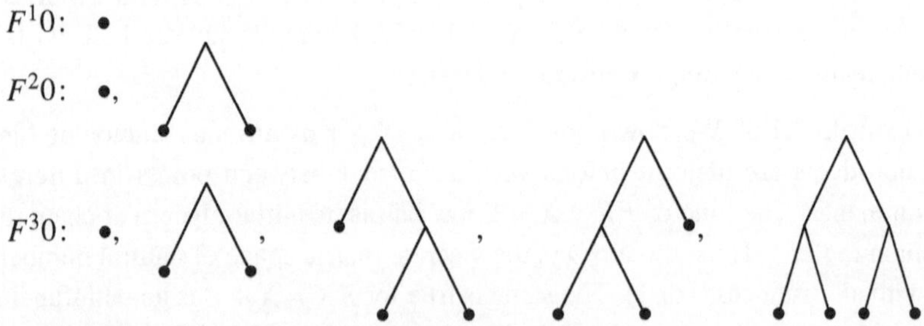

In general, $F^n 0$ is the set of all binary trees of height less than n, and the connecting maps $F^n 0 \to F^{n+1}0$ are the inclusions.

Then the carrier of the initial algebra μF may be taken to be the union $\bigcup_{n<\omega} F^n 0$:

$$\mu F = \text{all finite binary ordered trees.}$$

The structure map

$$\iota\colon (\mu F \times \mu F) + 1 \longrightarrow \mu F$$

takes the unique element of 1 to the single-node tree, and the left-hand component is given by *tree-tupling*:

$$(t_1, t_2) \longmapsto$$

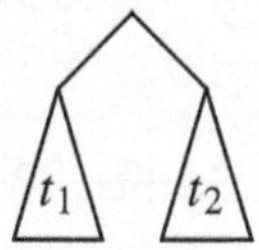

This is a new explanation of Example 2.2.15(1).

Example 3.2.6. (1) We return to the category Met introduced in Notation 3.2.2. In this category, products $X \times X'$ are cartesian products with the *maximum metric*:

$$d((x, x'), (y, y')) = \max\{d(x, y), d'(x', y')\}.$$

Also, recall from Notation 3.2.2 how coproducts work on Met. Thus, the functor $FX = X \times X + 1$ can be considered as an endofunctor on Met. Its initial-algebra ω-chain is

$$0 \longrightarrow 1 \longrightarrow (1 \times 1) + 1 \longrightarrow ((1 \times 1) + 1) \times ((1 \times 1) + 1) + 1 \longrightarrow \cdots$$

the same as that of Example 3.2.5, but with each set $F^n 0$ taken to be the *discrete space*: all distances between distinct points are 1. The colimit μF is then, not surprisingly, the space of all binary trees equipped with the discrete metric.

(2) The situation changes when we scale the distances by half. This leads to the functor

$$GX = \frac{1}{2}(X \times X) + 1.$$

Here $G^{n+1}0$ consists of pairs of ordered trees $(t_1, t_2) \in G^n 0 \times G^n 0$ and of the single-node tree, whose distance from any (t_1, t_2) is 1. And the distance between (t_1, t_2) and (s_1, s_2) is one half of the maximum of the distances $d(t_i, s_i)$, for $i = 1, 2$, in $G^n 0$. From this it follows that $G^n 0$ can be described as the set of all binary trees of height less than n with the metric

$$d(t, u) = \begin{cases} 2^{-k} & \text{if } t \neq u, \\ 0 & \text{if } t = u \end{cases} \tag{3.8}$$

for the least number k such that t and u have different cuttings at level k.

The colimit in Met is the space of all finite binary ordered trees with the above metric (3.8).

(3) When we turn from Met to CMet, the situation changes: for G above, the colimit in Met is not complete. In fact, under the metric (3.8) every infinite binary tree t yields a Cauchy sequence $t_0, t_1, \ldots$, where t_k cuts t at level k. It turns out that the colimit of the initial-algebra ω-chain G^n0 (which is formed by *isometric embeddings*, i.e. distance preserving maps) in CMet is obtained from the colimit in Met by forming the Cauchy completion. Moreover, this Cauchy completion is

$$\mu G = \text{all (finite and infinite) ordered binary trees,}$$

with the metric (3.8) above. We shall see a general reason for this in Section 5.2 below.

Example 3.2.7. Consider the endofunctor $FX = M \times X$, where M is a fixed object in one of our categories.

(1) As an endofunctor on Set, $\mu F = \emptyset$. The same holds in Pos, Met, and CMet.

(2) We next consider the situation in $\mathsf{CPO}_\perp$. For a cpo M the initial algebra ω-chain has the following form:

$$\{\perp\} \longrightarrow M \times \{\perp\} \longrightarrow M \times M \times \{\perp\} \longrightarrow \cdots$$

The tuples in each factor are ordered componentwise, and the connecting maps add to every n-tuple the element $\perp$ in the $(n+1)$-st coordinate. The colimit in $\mathsf{CPO}_\perp$ can be described as the cpo M^ω of all streams of elements of M ordered componentwise. Shortly:

$$\mu X.\,(M \times X) = M^\omega.$$

The algebra structure $M \times M^\omega \to M^\omega$ adjoins a new head to a stream in the evident manner.

Example 3.2.8. (1) For finitary signatures Σ, the polynomial set functor H_Σ of Example 2.1.3(4) preserves colimits of ω-chains. We apply Theorem 3.1.8 to obtain a short proof that the initial chain yields

$$\mu H_\Sigma = \text{the algebra of finite } \Sigma\text{-trees;}$$

cf. Proposition 2.2.14. (For infinitary signatures we describe μH_Σ in Example 6.1.16.) In fact, we can identify $H_\Sigma^1 0 = H_\Sigma \emptyset = \Sigma_0$ with the set of all Σ-trees of height 0, which are one-node trees labelled by an element of Σ_0. We further identify

$$H_\Sigma^2 0 = H_\Sigma \Sigma_0 = \Sigma_0 + \coprod_{k>0} \Sigma_k \times \Sigma_0^k$$

with the set of all Σ-trees of height at most 1. More generally, $H_\Sigma^n 0$ can be identified with the set of Σ-trees of height less than n. We obtain the chain of inclusion maps

$$H^0_\Sigma 0 \hookrightarrow H^1_\Sigma 0 \hookrightarrow H^2_\Sigma 0 \hookrightarrow \cdots$$

whose colimit (union) is the set of all finite Σ-trees.

The algebra structure $\iota\colon H_\Sigma(\mu H_\Sigma) \to \mu H_\Sigma$ of (3.4) is tree-tupling: ι maps an element $(\sigma, (t_i)_{i<k})$ of $\Sigma_k \times (\mu H_\Sigma)^k$ to the Σ-tree with root labelled by σ (k-ary) and with the children $t_0, \ldots, t_{k-1}$ from left to right (cf. Definition 2.2.12).

(2) As a concrete example, consider a finite set $V = \{v_1, \ldots, v_n\}$ (of Boolean variables). Recall the concept of a *binary decision tree*: it is a finite binary ordered tree whose inner nodes are labelled by variables v_i and leaves are labelled by 1 (for 'true') or 0 (for 'false'). We see that the polynomial functor

$$FX = \{0, 1\} + V \times X \times X$$

has the initial algebra

$$\mu F = \text{all binary decision trees.}$$

For a connection with binary decision diagrams see Example 9.3.20(5).

Example 3.2.9. The finite power-set functor $\mathscr{P}_\mathsf{f}\colon \mathsf{Set} \to \mathsf{Set}$ will be of special interest in this book. It preserves ω-colimits, thus we can apply Theorem 3.1.8.

(1) Its initial sequence is given by the inclusion functions of the following sets

$$\emptyset \longrightarrow \{\emptyset\} \longrightarrow \{\emptyset, \{\emptyset\}\} \longrightarrow \cdots$$

where $V_0 = \emptyset$, and $V_{n+1} = \mathscr{P}V_n$. The colimit of this sequence is the union $V_\omega = \bigcup_{n<\omega} V_n$, see Example 2.2.7(4).

(2) An important alternative representation of $\mu\mathscr{P}_\mathsf{f}$ is by finite extensional trees. (Recall from Remark 2.2.10 that trees are considered up to isomorphism.) An unordered tree is *extensional tree* if for every pair of children of a given node, the two corresponding subtrees are different (non-isomorphic). Every unordered tree has an *extensional quotient* obtained by successively identifying children of any given node with the same subtrees. See also Section 4.5 for more on extensional trees.

Observe that if $\mathscr{P}_\mathsf{f}\emptyset$ is represented by the singleton tree, then we have a natural representation of the elements of $\mathscr{P}^n_\mathsf{f}\emptyset$ by all finite extensional trees of height less than n: the tree representing of a set $\{x_1, \ldots, x_n\}$ has n children (representing x_i):

$\mathscr{P}_f^1\emptyset = \{\emptyset\}$ is represented by •

$\mathscr{P}_f^2\emptyset = \{\emptyset, \{\emptyset\}\}$ is represented by 

$\mathscr{P}_f^3\emptyset = \{\emptyset, \{\emptyset\}, \{\{\emptyset\}\}, \{\emptyset, \{\emptyset\}\}\}$ is represented by 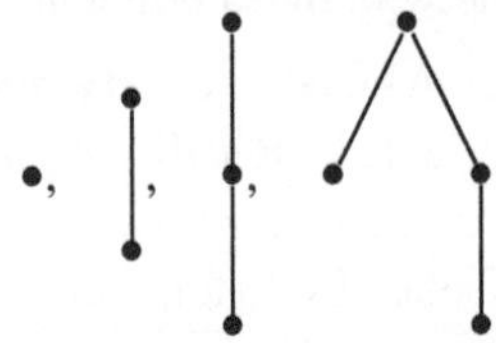

and so on. Therefore, we obtain

$$\mu\mathscr{P}_f = \text{all finite extensional trees}$$

with the algebra structure $\mathscr{P}_f(\mu\mathscr{P}_f) \to \mu\mathscr{P}_f$ given by tree-tupling.

Example 3.2.10. Monoid-valued functors $M^{(-)}$ (see Example 2.4.3(1)) preserve ω-colimits and so Theorem 3.1.8 applies. We obtain a similar description of $\mu M^{(-)}$ as in Example 3.2.9(2). Here we consider unordered M-labelled trees, which means that edges are labelled in M. Similarly as before, an M-labelled tree is *extensional* if distinct children of any vertex determine non-isomorphic M-labelled subtrees.

For $F = M^{(-)}$, the set $F\emptyset$ has a single element which we represent as the single node tree. Given a representation of $F^n\emptyset$ by M-labelled trees, represent $F^{n+1}\emptyset$ as follows: every finite multiset consisting of trees $t_1, \ldots, t_k \in F^n\emptyset$ with multiplicities $m_1, \ldots, m_k \in M$ is represented by the following tree:

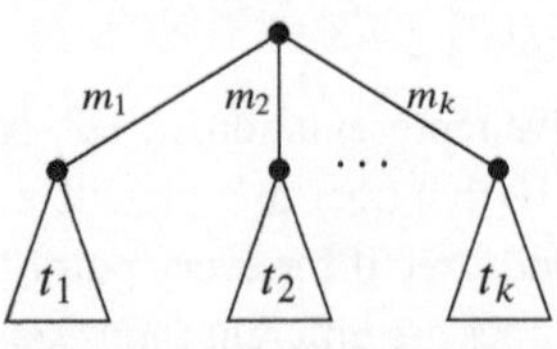

(3.9)

For example, $F^2\emptyset$ is the set of trees shown below:

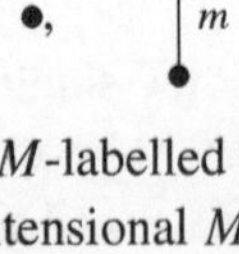

for $m \in M$.

These are all extensional M-labelled trees of height at most 1. It is easy to see that $F^3\emptyset$ consists of all extensional M-labelled trees of height at most 2, etc.

The colimit of the initial-algebra ω-chain is again the union. The initial algebra $\mu M^{(-)}$ can thus be described as follows:

$$\mu M^{(-)} = \text{all finite extensional } M\text{-labelled trees.}$$

The algebra structure $\iota\colon M^{(\mu M^{(-)})} \to \mu M^{(-)}$ is given by tree-tupling of M-labelled trees. More precisely, ι maps a finite multiset of trees t_i with multiplicities m_i $(i = 1, \ldots, k)$ to the tree (3.9).

Example 3.2.11. For the special case of the bag functor from Example 2.4.3(3), we have a related description which does not need labels or extensionality. The set $\mathscr{B}\emptyset$ has a single element which we represent as the single-node tree. Given a representation of $\mathscr{B}^n\emptyset$ by (unordered) trees, represent $\mathscr{B}^{n+1}\emptyset$ as follows: every bag consisting of trees $t_1, \ldots, t_n \in \mathscr{B}^n\emptyset$ with multiplicities $k_1, \ldots, k_n$ is represented by the tree having k_i children given by t_i for $i = 1, \ldots, n$. Thus, the set $\mathscr{B}\mathscr{B}\emptyset$ consists of the following trees

These are all unordered trees of height at most 1. It is easy to see that $\mathscr{B}^3\emptyset$ are all finite unordered trees of height at most 2, etc.

The colimit of the initial-algebra ω-chain is again the union. The initial algebra $\mu\mathscr{B}$ can thus be described as follows:

$$\mu\mathscr{B} = \text{all finite unordered trees.}$$

The algebra structure $\iota\colon \mathscr{B}(\mu\mathscr{B}) \to \mu\mathscr{B}$ is *tree-tupling with multiplicities*, that is, ι maps a bag $(\{t_1, \ldots, t_n\}, b)$ of finite unordered trees to the unordered tree that joins $b(t_i)$ copies of t_i, $i = 1, \ldots, n$, with a new common root.

Example 3.2.12. We return to λ-terms modulo α-equivalence as an initial algebra in Nom for $FX = \mathbb{A} + X \times X + [\mathbb{A}]X$ (Example 2.2.18(1)). This description can be derived from the initial-algebra ω-chain as follows. Let L_n be the set of equivalence classes $[M]_\alpha$ of λ-terms modulo α-equivalence, where M has height less than n. Using recursion we obtain canonical bijections

$$f_n\colon L_n \to F^n 0 \qquad (n < \omega).$$

The base case $n = 0$ is clear since $L_0 = \emptyset = F^0 0$. Given f_n, we recursively define the map $f_{n+1}\colon L_{n+1} \to \mathbb{A} + F^n 0 \times F^n 0 + [\mathbb{A}]F^n 0$ by

$$\begin{aligned}
[a]_\alpha &\mapsto a, \\
[MN]_\alpha &\mapsto (f_n([M]_\alpha), f_n([N]_\alpha)), \\
[\lambda a.\, M]_\alpha &\mapsto \langle a \rangle f_n([M]_\alpha).
\end{aligned}$$

One readily verifies that this is well-defined and yields the desired bijections in Nom. Using these bijections we then see that $\mathsf{colim}_{n<\omega} F^n 0 = \bigcup_{n<\omega} L_n$. We leave the (somewhat tedious) details to the reader.

Example 3.2.13. Generalizing several of the above examples, we recall the concept of an *analytic functor* introduced by Joyal [179, 180]. Let a group G of permutations on $k = \{0, \dots, k-1\}$ be given. We denote by

$$X^k/G$$

the set of orbits under the action of G on X^k by coordinate interchange; this is the set of equivalence classes of X^k modulo the equivalence relating $(x_1, \dots, x_k)$ with $(x_{p(1)}, \dots, x_{p(k)})$ for each $p \in G$. This defines a set functor taking $f\colon X \to Y$ to the obvious function $X^k/G \to Y^k/G$ derived from f^k. The analytic functors are the coproducts of functors of the form X^k/G. More precisely, an analytic functor F is of the form

$$FX = \coprod_{\sigma\in\Sigma} X^k/G_k, \tag{3.10}$$

where Σ is a finitary signature, k the arity of $\sigma \in \Sigma$, and G_k a group of permutations on k. Thus, every polynomial functor H_Σ is analytic: each group G_k is trivial. An important example of an analytic functor is the bag functor, which can be expressed as

$$\mathscr{B}X = \coprod_{k\in\mathbb{N}} X^k/S_k, \tag{3.11}$$

where S_k is the symmetric group of all permutations on k. In contrast, $\mathscr{P}_f$ is not analytic.

For every analytic functor F, the initial-algebra ω-chain yields a quotient of the algebra μH_Σ of Example 3.2.8.

$$\mu F = \mu H_\Sigma/{\sim},$$

where $\sim$ is the equivalence such that for every tree $t \in \mu H_\Sigma$, every node x of t labelled by $\sigma \in \Sigma$ and every permutation g in the group G_σ we have $t \sim t'$, where t' is obtained from t by permuting the children of x according to g.

In fact, the proof is completely analogous to Example 3.2.8. We identify $F^1 0 = \Sigma_0$ with one-point trees labelled in Σ_0. Since $F^{n+1}0 = \coprod_{\sigma\in\Sigma}(F^n 0)^k/G_\sigma$, where σ is a k-ary symbol, we can identify, for every $n \in \mathbb{N}$, the set $F^n 0$ with the quotient of the set of all Σ-trees of height less than n modulo the above equivalence $\sim$. We again obtain a chain of inclusion maps whose colimit (union) is $F_\Sigma/{\sim}$. This is illustrated by the bag functor $\mathscr{B}$ in Example 3.2.11.

Example 3.2.14. Here is an example of an endofunctor F on a category with a rather interesting initial algebra. Let BiP be the category of *bipointed* sets: objects are sets with distinguished elements $\bot$ and $\top$ that are required to be different. A morphism in BiP is a function preserving the distinguished points. There is a binary operation $\oplus$ on BiP taking $(X, \bot_X, \top_X)$ and $(Y, \bot_Y, \top_Y)$ to the disjoint union $X + Y$, identifying $\top_X$ and $\bot_Y$, and then using as distinguished points $\bot_X$ and $\top_Y$. Define an endofunctor on BiP by

$$FX = X \oplus X.$$

Its initial-algebra ω-chain starts with the initial object $\mathbf{0} = \{0, 1\}$ where $0 = \bot_0$ and $1 = \top_0$. Then $F\mathbf{0}$ can be represented by $\{0, \frac{1}{2}, 1\}$ where $\frac{1}{2}$ denotes the equivalence class $\{\bot_1, \top_0\}$. In this manner the initial chain is the following chain of injections

$$\underset{\mathbf{0}}{\boxed{\begin{array}{c}\top = 1\\ \bot = 0\end{array}}} \hookrightarrow \underset{F\mathbf{0}}{\boxed{\begin{array}{c}\top = 1\\ 1/2\\ \bot = 0\end{array}}} \hookrightarrow \underset{FF\mathbf{0}}{\boxed{\begin{array}{c}\top = 1\\ 3/4\\ 1/2\\ 1/4\\ \bot = 0\end{array}}} \hookrightarrow \cdots$$

The colimit of this chain is its union

$$D = \text{all dyadic rational numbers in } [0, 1].$$

This means that D consists of all numbers $\frac{k}{2^n}$, $k = 0, \dots, 2^n$. The algebra structure $\iota\colon D \otimes D \to D$ takes y in the left-hand part to $\frac{y}{2}$ and in the right-hand one to $\frac{y+1}{2}$. Indeed, F preserves the above colimit, thus D is the initial algebra for F. To see this, observe that the connecting map $j\colon D \to FD$ from $\operatorname{colim} F^n\mathbf{0}$ to $F(\operatorname{colim} F^n\mathbf{0})$ is $x \mapsto 2x$ on $[0, \frac{1}{2}]$ and $x \mapsto 2x - 1$ on $[\frac{1}{2}, 1]$. This maps j is bijective, and its inverse $j^{-1}\colon FD \to D$ is the structure of the initial algebra. Related examples are discussed in Section 15.4.

Example 3.2.15. Let us mention also some negative examples. An endofunctor of Set need not have an initial algebra, as we have seen in Example 2.2.7. The endofunctor $F\colon X \mapsto X^{\mathbb{N}} + 1$ does have an initial algebra, as we will see in Chapter 4, but its initial-algebra ω-chain does not converge. In a similar manner to what we have seen in Example 3.2.5, $F^n 0$ can be represented by all countably branching ordered trees of height less than n. But $\bigcup F^n 0$ is not a fixed point of F. We will see in Chapter 6 that an initial-algebra chain of uncountable length is needed for this functor.

Continuing with a discussion of examples of initial algebras obtained by ω-iteration on categories other than Set, we mention a result implying that for some special categories, *every* endofunctor has an initial algebra obtained that way.

Notation 3.2.16. The category Rel has sets as objects, and morphisms from X to Y are relations $R \subseteq X \times Y$. Composition is the usual composition of relations, and the identity morphism on X is the diagonal of $X \times X$.

Definition 3.2.17 [128]**.** A category is *algebraically complete* if every endofunctor on it has an initial algebra.

Theorem 3.2.18 [10]**.** *The categories* Set_c *(countable sets and functions),* Rel_c *(countable sets and relations), and* K-Vec_c *(countably-dimensional vector spaces over a field* K *and linear functions) are algebraically complete. Moreover, every endofunctor* F *has the initial algebra*

$$\mu F = \operatorname*{colim}_{n<\omega} F^n 0.$$

Remark 3.2.19. (1) Every complete lattice is algebraically complete by the well-known Knaster–Tarski Fixed Point Theorem (Theorem B.1.7).

(2) Among categories with products, there are essentially no other examples:

Theorem 3.2.20 [16]**.** *Every algebraically complete category with products is a preorder.*

Proof The following beautiful proof was provided by Peter Freyd [128]: suppose $\mathscr{A}$ has products but is not a preorder. That is, some hom-set $\mathscr{A}(A, B)$ has at least two elements. Consider the endofunctor F obtained by the following composite

$$F = (\mathscr{A} \xrightarrow{\mathscr{A}(A,-)} \mathsf{Set} \xrightarrow{\mathsf{Set}(-,2)} \mathsf{Set}^{\mathsf{op}} \xrightarrow{B^{(-)}} \mathscr{A}).$$

(Here $B^{(-)}$ denotes the functor taking a set M to the power B^M.) This functor F can be expressed as $FD = B^{S(D)}$ where $S(D) = 2^{\mathscr{A}(A,D)}$. It does not have fixed points. In fact, assuming $D \cong FD$, we see that

$$\mathscr{A}(A, D) \cong \mathscr{A}(A, FD) \cong \mathscr{A}(A, B^{S(D)}) \cong \mathscr{A}(A, B)^{S(D)}.$$

But the cardinality of the right-hand side is at least

$$2^{S(D)} \cong 2^{2^{\mathscr{A}(A,D)}},$$

a contradiction: for every set X, the cardinality of 2^X is larger than that of X. Now apply this observation to $X = \mathscr{A}(A, D)$. □

3.3 Terminal-Coalgebra Chain

The construction of the terminal coalgebra in this section dualizes the ω-chain of Section 3.1. This is another categorical version of Kleene's Fixed Point Theorem: a terminal coalgebra for an endofunctor is constructed as a limit of iterations of the functor on the terminal object. An important example is the terminal coalgebra for a polynomial functor which is described as the algebra of all (finite and infinite) Σ-trees. Throughout this section we assume that a category $\mathscr{A}$ with a terminal object 1 is given. We denote by $!\colon X \to 1$ the unique morphism from an object X.

It is straightforward to dualize the general results of Section 3.1. One dualizes initial objects 0 to terminal objects 1, ω-chains to ω^{op}-chains (i.e. functors from ω^{op} to $\mathscr{A}$), ω-colimits to ω^{op}-limits, and the initial-algebra ω-chain of Definition 3.1.4 as follows:

Definition 3.3.1. By the *terminal-coalgebra* ω^{op}*-chain* of an endofunctor $F\colon \mathscr{A} \to \mathscr{A}$ is meant the ω^{op}-chain

$$1 \xleftarrow{\;!\;} F1 \xleftarrow{F!} F^2 1 \xleftarrow{F^2!} \cdots \xleftarrow{F^{n-1}!} F^n 1 \xleftarrow{F^n!} F^{n+1}! \longleftarrow \cdots \tag{3.12}$$

Construction 3.3.2. Dually to Construction 3.1.6, every coalgebra $\alpha\colon A \to FA$ induces a canonical cone over the terminal-coalgebra ω^{op}-chain of F by induction: $\alpha_0\colon A \to 1$ is uniquely determined and $\alpha_{n+1} = F\alpha_n \cdot \alpha\colon A \to F^{n+1}1$.

Remark 3.3.3. Dually to Remark 3.1.7, homomorphisms $h\colon (A, \alpha) \to (B, \beta)$ of coalgebras preserve the induced cones:

$$\alpha_n = \beta_n \cdot h \qquad \text{for all } n < \omega.$$

It is worthwhile putting down the dual statement of Theorem 3.1.8. This was first explicitly formulated by Barr [72]:

Theorem 3.3.4. *Let $\mathscr{A}$ be a category with terminal object* 1 *and with limits of ω^{op}-chains. If $F\colon \mathscr{A} \to \mathscr{A}$ preserves limits of ω^{op}-chains, then it has the terminal coalgebra*

$$\nu F = \lim_{n \in \omega^{\mathrm{op}}} F^n 1.$$

Remark 3.3.5. (1) It is sufficient to assume that F preserves the limit of its terminal-coalgebra chain (cf. Remark 3.1.10).

(2) Let $\ell_n\colon \nu F \to F^n 1 (n < \omega)$ be the limit cone. The coalgebra structure

$$\tau\colon \nu F \to F(\nu F)$$

is the unique morphism with

$$F\ell_n \cdot \tau = \ell_{n+1} \qquad (n < \omega).$$

This is dual to Remark 8.1.12.

We revisit the examples from Section 3.2.

Example 3.3.6. The functor $FX = X + 1$ on Set. The terminal-coalgebra ω^{op}-chain is

$$1 \longleftarrow 1+1 \longleftarrow 1+1+1 \longleftarrow \cdots$$

The nth object may be identified with $n + 1 = \{0, 1, \ldots, n\}$, and the connecting function $F^n! = f_n\colon n + 2 \to n + 1$ is then given by $f_n(i) = i$ for $i \leq n$, and $f_n(n+1) = n$. The limit of this chain is the set of all ω-tuples $(x_0, \ldots, x_n, \ldots)$ with $f_n(x_{n+1}) = x_n$ for every n. One such tuple is $\top = (0, 1, 2, \ldots)$. Every other tuple is of the form $(0, 1, \ldots, k, k, \ldots, k, \ldots)$ which we may identify with k. Thus we describe the terminal coalgebra as

$$\nu X.\,(X + 1) = \mathbb{N}^\top,$$

the set of natural numbers with an element $\top$ added. The coalgebra structure $\tau\colon \mathbb{N}^\top \to \mathbb{N}^\top + 1$ has $\top$ as a fixed point, sends 0 to the point in the right-hand summand of $\mathbb{N}^\top + 1$, and is otherwise the predecessor function on $\mathbb{N}$. This is a new explanation of Example 2.5.3(2).

Example 3.3.7. (1) The functor $FX = X + 1$ on Pos. Using the same argument as above, we see that the terminal coalgebra is $\mathbb{N}^\top$ with the discrete order. The structure is the same map as in Example 3.3.6 above.

The functor $FX = X_\perp$ (cf. Example 2.1.7(1)) has $\nu F = \mathbb{N}^\top$ with the usual order $0 < 1 < 2 < \cdots < \infty$ and the same structure as before, the argument is analogous.

(2) The functor $FX = X + 1$ on $\mathsf{CPO}_\perp$ is essentially the identity functor, whence $\nu F = 1 = \{\perp\}$. The functor $FX = X_\perp$ has as terminal coalgebra the same poset as in point (1) above. This is not surprising: $\mathsf{CPO}_\perp$ is closed in Pos under limits (but not under colimits).

(3) On the category Met of 1-bounded metric spaces, the above argument shows that the terminal coalgebra of the same functor $FX = X + 1$ is again $\mathbb{N}^\top$, this time with a discrete metric. The structure is the same as we have seen. The same is true for CMet.

When we change the functor to $GX = \frac{1}{2}X + 1$ as in Example 3.2.4, the terminal-coalgebra ω^{op}-chain has essentially the same spaces that we saw in that example since $G\emptyset = 1$, and the connecting map $G^{n+1}1 \to G^n 1$ merges n

and $n+1$ and is identity else. We again get $\mathbb{N}^{\top}$ as the terminal coalgebra but with the metric shown in Example 3.2.4.

Example 3.3.8. (1) Let $\mathscr{A}$ have countable products and a terminal object. Consider the functor $FX = M \times X$ where M is a fixed (but arbitrary) object of $\mathscr{A}$. We identify each object with its product with 1. Thus we can write the terminal-coalgebra chain as

$$1 \longleftarrow M \longleftarrow M \times M \longleftarrow M \times M \times M \longleftarrow \cdots$$

The connecting morphisms are all the projections onto the left-most factors. We obtain as a limit the countable power M^{ω} with the limit projections $\ell_n : M^{\omega} \to M^n$ given by the left-hand projections of $M^{\omega} \cong M^n \times \coprod_{i \geq n} M$. For example, if $\mathscr{A} = \mathsf{Set}$, then M^{ω} is the set of all *streams* on M, also known as the *infinite words* on M:

$$\nu X.\,(M \times X) = M^{\omega}.$$

The coalgebra structure $M^{\omega} \to M \times M^{\omega}$ is $\langle \mathsf{head}, \mathsf{tail} \rangle$, where

$$\mathsf{head}(a_1, a_2, a_3 \ldots) = a_1 \qquad \text{and} \qquad \mathsf{tail}(a_1, a_2, a_3 \ldots) = (a_2, a_3 \ldots).$$

This is a special case of Theorem 2.5.9 for Σ consisting of unary operations indexed by M: a Σ-tree is an infinite path labelled in M, equivalently, a stream on M.

(2) As a concrete example, take $\mathscr{A} = \mathsf{Set}$ and $M = \mathbb{R}$, the set of real numbers. A coalgebra for $\mathbb{R} \times X$ is a stream automaton, equivalently a dynamical system with real outputs, see Example 2.5.3(3).

Here are two examples in which the output is represented by a label of the output arrow:

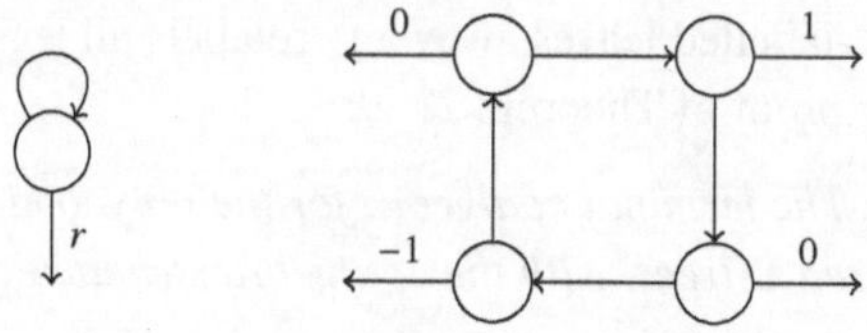

One way to represent real-valued streams is by taking the set A of *real analytic functions*, see [248]. Here one considers $f : \mathbb{R} \to \mathbb{R}$ such that for every x, there is an open interval around 0 on which the nth derivative $f^{(n)}$ is defined, and f agrees with its Taylor series. The coalgebra structure on A is given by

$$f(x) \mapsto (f(0), f'(x))$$

for every analytic function. This coalgebra is isomorphic to a subcoalgebra of the stream coalgebra that we saw above; to every analytic function $f(x)$ one associates the stream of coefficients of the Taylor series of $f(x)$, which implies that A is isomorphic to the subcoalgebra of those streams σ in $\mathbb{R}^\omega$ such that $\sum_{i=0}^{\infty} \frac{\sigma_i}{i!} x_i < \infty$.

The above two stream automata thus present analytic functions by corecursion. In the coalgebra on the left, we have a function whose value at 0 is r and equal to its own derivative: $f(x) = re^x$. On the right, we obtain four functions:

$$\sin x, \cos x, -\sin x, -\cos x.$$

Example 3.3.9. For every signature Σ, the functor H_Σ of Example 2.1.3 preserves limits of ω^{op}-chains. Indeed, these limits commute with coproducts in Set. The functor H_Σ is a coproduct of functors of the form $(-)^n$ which preserve all limits. The terminal-coalgebra ω^{op}-chain for H_Σ can be described as follows. Denote by $\Sigma_\perp = \Sigma + \{\perp\}$ the signature Σ extended by a new constant symbol $\perp$. For every $n < \omega$ we have:

$$H_\Sigma^n 1 = \text{all } \Sigma_\perp\text{-trees of height at most } n \text{ with leaves at height } n \text{ labelled by } \perp.$$

To see this, let us identify $1 = H_\Sigma^0 1$ with the set consisting of the singleton tree labelled by $\perp$. Then $H_\Sigma^1 1 = \coprod \Sigma_k \times 1^k$ can be identified with the set of all trees of height ≤ 1 whose root is labelled in Σ and (for a root label of arity k) whose k leaves are labelled by $\perp$. Analogously $H_\Sigma^2 1 = \coprod \Sigma_k \times (H_\Sigma 1)^k$ are trees t with a root labelled in Σ_k and the k subtrees are trees in $H_\Sigma^1 1$; thus all leaves of t of depth 2 have label $\perp$ etc.

Moreover, the connecting map

$$H_\Sigma^{n+1} 1 \longrightarrow H_\Sigma^n 1$$

simply cuts the $\perp$-labelled leaves away and relabels all leaves of depth n by $\perp$. This yields a new proof of Theorem 2.5.9:

Theorem 3.3.10. *The terminal coalgebra for the polynomial functor H_Σ is the set T_Σ of all ordered Σ-trees, with the coalgebra structure given by the inverse to tree-tupling.*

Proof We have described $H_\Sigma^n 1$ above as the set of all $\Sigma_\perp$-trees of height at most n with leaves of depth n labelled by $\perp$. Since H_Σ preserves limits of ω^{op}-chains, we know that

$$\nu H_\Sigma = \lim_{n<\omega} H_\Sigma^n 1.$$

(1) We prove that the limit of the chain $H_\Sigma^n 1$ (with the connecting

maps $v_{n+1,n}: H_\Sigma^{n+1}1 \to H_\Sigma^n 1$ given by cutting at level n as above) is the set T_Σ with the limit cone $\ell_n: T_\Sigma \to H_\Sigma^n 1$ given, again, by cutting at level n and relabelling all leaves at level n by $\perp$. In other words, we need to show that for every collection of trees $t_n \in H_\Sigma^n 1$ which is compatible:

$$v_{n,m}(t_n) = t_m \qquad \text{for all } n \geq m,$$

there exists a unique tree $t \in T_\Sigma$ with $\ell_n(t) = t_n$ for all $n < \omega$. Existence: let t be the unique Σ-tree which on levels smaller than n agrees with t_n for every n. More precisely, the root of t has the same label as t_n for all $n \geq 1$, the level one labels are the same as in t_n for all $n \geq 2$, etc. Obviously $\ell_n(t) = t_n$ for all n. Uniqueness: if $\ell_n(t) = t_n$, then t and t_n agree on all levels smaller than n.

(2) The coalgebra structure $\tau: T_\Sigma \to H_\Sigma T_\Sigma$ is, following Remark 3.3.5, the unique morphism with $\ell_{m+1} = H_\Sigma \ell_m \cdot \tau$ for all m. Since the tree-tupling $\bar{\tau}: H_\Sigma T_\Sigma \to T_\Sigma$ clearly fulfils $\ell_{m+1} \cdot \bar{\tau} = H_\Sigma \ell_m$ (because to cut a tree-tupling at level $m + 1$ is the same as cutting the maximum subtrees at level m and then performing a tree-tupling), we conclude $\tau = \bar{\tau}^{-1}$. □

Example 3.3.11. Coalgebras of the functor $FX = \{0, 1\} \times X^A$ are deterministic automata with A as the input alphabet (see Example 2.5.11). For this functor,

$$F^n 1 \cong \text{sets of words in } A^* \text{ of length} < n.$$

More precisely, let A have k elements and let Σ be the signature with two k-ary operation symbols. Then $H_\Sigma X = X^A + X^A = FX$. To give a tree in $F^n 1$, that is, a $\Sigma_\perp$-tree of height $\leq n$ with leaves labelled in by $\perp$, means precisely to label the nodes of a complete k-ary (see Remark 2.2.10(4)) tree of height $\leq n - 1$ by 0 or 1. This is equivalent to giving a set of words of length $< n$. The above connecting maps $v_{n,m}$ take a set M of words of length $< n$ to the set formed by the prefixes of length $< m$ of words from M.

A limit of the resulting terminal-coalgebra ω^{op}-chain is the set of all formal languages $\mathscr{P}A^*$. The limit cone $\ell_n: \mathscr{P}A^* \to F^n 1$ takes a formal language $L \subseteq A^*$ to the set of all prefixes of length $n - 1$ of words in L.

Remark 3.3.12. In Chapter 6, we shall consider a more general notion of signature which allows for *infinitary* operations, see Example 6.1.16. Everything which we said in Example 3.3.9 holds for this more general notion of a signature.

Example 3.3.13. For every analytic functor F (see Example 3.2.13), the terminal coalgebra is obtained by the terminal-coalgebra ω^{op}-chain. In fact, it is easy to derive from the formula (3.10) that analytic functors preserve limits of ω^{op}-chains, actually they preserve cofiltered limits [180].

We will see a description of this terminal coalgebra in Example 4.3.31(3). The following example is a special case.

Example 3.3.14. The bag functor $\mathscr{B}$ of Example 2.4.3(3) has the terminal coalgebra

$$\nu\mathscr{B} = \text{all finitely branching unordered trees.}$$

In fact, we can identify $\mathscr{B}^n 1$, analogously as in Example 3.3.9, with all unordered finitely branching trees of height $\leq n$ whose leaves at level n are labelled by $\bot$. The limit $\nu\mathscr{B}$ is then as stated. For more general monoid-valued functors $M^{(-)}$ (Example 2.4.3(1)) descriptions of the terminal coalgebra are more involved (see Example 4.2.12 and Remark 4.5.24).

Example 3.3.15. The finite power-set functor $\mathscr{P}_{\mathsf{f}}$ is an example of a set functor which does not preserve limits of ω^{op}-chains, and, moreover, as we demonstrate in Section 4.4, the limit of its terminal-coalgebra ω^{op}-chain is not the terminal $\mathscr{P}_{\mathsf{f}}$-coalgebra.

We will describe the terminal coalgebra for $\mathscr{P}_{\mathsf{f}}$ below in several different ways (see Section 4.5).

Example 3.3.16. Consider the functor $FX = X \oplus X$ on the category of BiP of bipointed sets. Recall from Example 3.2.14 that the initial F-algebra is carried by the set of all dyadic rationals in $[0, 1]$. Freyd [130] proved that the terminal F-coalgebra is the real unit interval $[0, 1]$ with the structure $t\colon [0,1] \to [0,1] \oplus [0,1]$ given by $x \mapsto 2x$ on $[0, \frac{1}{2}]$ and $x \mapsto 2x - 1$ on $[\frac{1}{2}, 1]$. We discuss this example and related ones in more detail in Section 15.4.

Remark 3.3.17. *Many-sorted sets* are used in many applications. These lead to endofunctors of the category Set^S of S-sorted sets. In Example 2.1.6, we have seen the corresponding polynomial endofunctors on Set^S. Theorem 3.3.10 (and its proof) hold for these more general polynomial functors. Let us illustrate the initial-algebra ω-chain and the terminal-coalgebra ω^{op}-chain for a concrete two-sorted signature.

Example 3.3.18. Consider finite lists of symbols over (arbitrary) alphabets. Here we work with two sorts a (alphabet) and l (list) and consider the signature Σ with the following operation symbols

$$\begin{aligned} &\varepsilon\colon l && \text{(empty list)},\\ &\gamma\colon al \to l && \text{(concatenation)},\\ &0, 1\colon \to a && \text{(constants 0 and 1)}. \end{aligned}$$

The corresponding polynomial functor H_Σ has sorts

$$(H_\Sigma X)_a = \{0,1\} \qquad \text{and} \qquad (H_\Sigma X)_l = X_a \times X_l + \{\varepsilon\}.$$

(1) Here is its initial-algebra chain:

	sort a	sort l
$(\emptyset,\emptyset)$	$\emptyset$	$\emptyset$
$H_\Sigma(\emptyset,\emptyset)$	$0,1$	ε
$H_\Sigma^2(\emptyset,\emptyset)$	$0,1$	$0\varepsilon, 1\varepsilon, \varepsilon$
$H_\Sigma^3(\emptyset,\emptyset)$	$0,1$	$00\varepsilon, 01\varepsilon, 10\varepsilon, 11\varepsilon, \varepsilon$

etc. If we drop the superfluous ε at the ending of words, then we see that for $n \geq 1$,

$$H_\Sigma^n(\emptyset,\emptyset) = \Big(\{0,1\}, \coprod_{k<n} \{0,1\}^k\Big).$$

The connecting maps are inclusions. Thus, the colimit is

$$\mu H_\Sigma = (\{0,1\}, \{0,1\}^*).$$

(2) We next describe the terminal-coalgebra ω^{op}-chain:

	sort a	sort l
$(1,1)$	$\perp_a$	$\perp_l$
$H_\Sigma(1,1)$	$0,1$	$\perp_a\perp_l, \varepsilon$
$H_\Sigma^2(1,1)$	$0,1$	$0\perp_a\perp_l, 1\perp_a\perp_l, 0, 1, \varepsilon$
$H_\Sigma^3(1,1)$	$0,1$	$00\perp_a\perp_l, 01\perp_a\perp_l, 10\perp_a\perp_l, 11\perp_a\perp_l,$ $00, 01, 10, 11, 01, 1, \varepsilon$
$\vdots$	$\vdots$	$\vdots$

Again dropping ε at the end of words, we see that sort l of $H_\Sigma^n(1,1)$ consists of all words $w\perp_a\perp_l$ where w is a binary word of length $n-1$, and all binary words w of lengths $0, 1, \ldots, n-1$. The connecting map $v_{n+1,n}\colon H_\Sigma^{n+1}1 \to H_\Sigma^n 1$ maps each word $w\perp_a\perp_l$ by dropping the last letter of w; analogously for binary words $w \neq \varepsilon$, it drops the last letter. The limit of this ω^{op}-chain has the sort l which consists of (a) all $u\perp_a\perp_l$ where u is an infinite binary word and (b) all finite binary words. By dropping the superfluous $\perp_a\perp_l$ here, we get

$$\nu H_\Sigma = (\{0,1\}, \{0,1\}^* + \{0,1\}^\omega).$$

We thus see that a Σ-tree has one of the following forms: for sort a we just have two trees

$(a, 0)$ and $(a, 1)$

For sort l we have the finite trees

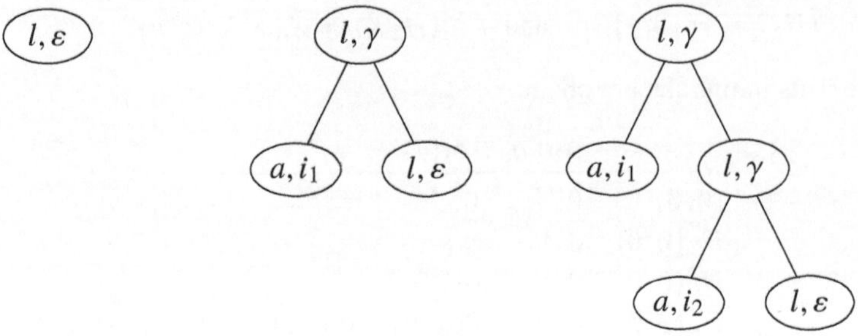

representing the finite binary words $w = i_1 \cdots i_n$, plus the infinite trees

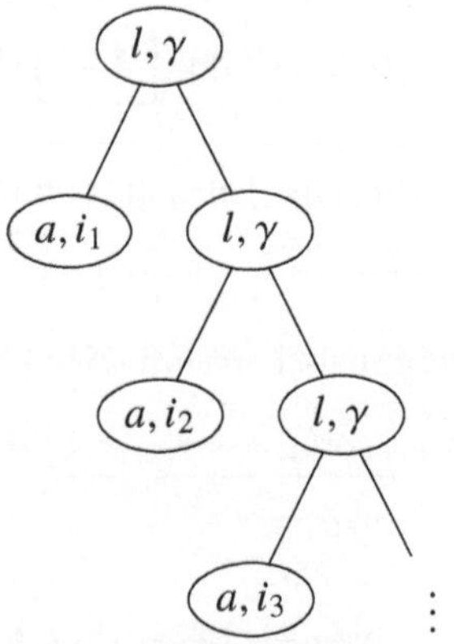

representing the infinite binary words $i_1 i_2 i_3 \cdots$.

Example 3.3.19. Recursive domain equations involving mutual recursion can often be solved by using polynomial endofunctors on many-sorted sets. For example, suppose that we need sets X and Y satisfying

$$X \cong X \times Y + 1 \qquad \text{and} \qquad Y \cong X + X.$$

We form the polynomial endofunctor F on Set $\times$ Set given by

$$F(X, Y) = (X \times Y + 1, X + X).$$

It clearly preserves colimits of ω-chains, thus, it has an initial algebra which is a colimit of $F^n(0, 0)$. By Lambek's Lemma 2.2.5, the components of the initial algebra form an (initial) solution of the above recursive equations.

Theorem 3.3.20. *For every many-sorted signature Σ, the polynomial endofunctor H_Σ has*

(1) *the initial algebra $\mu H_\Sigma = F_\Sigma$, all finite Σ-trees with the algebra structure given by tree-tupling, and*

(2) *the terminal coalgebra $\nu H_\Sigma = T_\Sigma$, all Σ-trees with the coalgebra structure given by the inverse of tree-tupling.*

The proof is completely analogous to that of Proposition 2.2.14 and Theorem 3.3.10.

Remark 3.3.21. Kozen [202] introduces coalgebras for a special class of polynomial functors F on the category of many-sorted sets that arise from graphs representing type declarations of a programming language with sum and product types. He also presents an interesting equivalent description of the category of F-coalgebras and of the terminal F-coalgebra, in particular. However, the type declarations considered do not allow one to capture all polynomial functors on (many-sorted) sets.

3.4 Summary

We started the chapter with a construction of the initial algebra μF of an endofunctor as a colimit of the ω-chain $F^n 0$. In short, $\mu F = \mathsf{colim}\, F^n 0$. This holds whenever F preserves ω-colimits. We saw many examples. To mention one by way of review, for a polynomial endofunctor $F = H_\Sigma$ on Set, $F^n 0$ can be represented by Σ-trees of height less than or equal to n, and this yields μH_Σ = all finite Σ-trees.

Dually, a construction of the terminal coalgebra νF as a limit of the ω^{op}-chain of iterated application of F to the terminal object 1 is obtained whenever F preserves ω^{op}-limits. For polynomial functors, this shows that νH_Σ consists of all Σ-trees, and for the functor F representing deterministic automata with input alphabet Σ, we see that νF consists of all formal languages over Σ. Note, however, that preservation of ω^{op}-limits appears to be much more restrictive in practice than preservation ω-colimits (while logically independent): the latter holds for many everyday functors, the former is more rare; for example, the finite power-set functor satisfies the latter but not the former.

4
Finitary Set Functors

The focus of Chapters 1 to 3 has been initial algebras, terminal algebras, and their construction via finite iteration. A lot of the rest of our book builds on the results in those chapters, and the examples there are critical for our subject. But we pause in this chapter to discuss terminal coalgebras which are *not obtainable* via finitary iteration.

The first example of this phenomenon is the *finite power-set* functor $\mathscr{P}_{\mathsf{f}}\colon \mathsf{Set} \to \mathsf{Set}$ (see Example 2.2.7(4)). We shall see in Example 4.4.6 that the limit of the terminal-coalgebra ω^{op}-chain from Definition 3.3.1 is not preserved by the functor. Therefore Theorem 3.3.4 does not apply. We present several descriptions of the terminal coalgebra for $\mathscr{P}_{\mathsf{f}}$ and of the limit of the terminal-coalgebra ω^{op}-chain in Section 4.5. More generally, we shall study *finitary* set functors in this chapter, since this is the natural setting for a lot of results. This is the central motivation of the chapter. We shall see several lines of investigation, leading to several descriptions of the terminal coalgebras of finitary set functors.[1]

We start in Section 4.1 with general material on limits in a base category and a coalgebra category. We continue with weakly terminal coalgebras in Section 4.2: the word 'weakly' indicates that we weaken the definition of *terminal* by dropping the uniqueness of homomorphisms. We also discuss congruences and precongruences. We pass from weakly terminal coalgebras to terminal coalgebras by taking the quotient modulo the largest congruence. This congruence is related to *equational presentations* of set functors, which we study in Section 4.3.

Section 4.4 goes back to the analysis of the terminal-coalgebra ω^{op}-chain and to an extension of that chain. We present Worrell's construction of the

[1] For finitary functors, the initial algebra construction from Chapter 3 does apply. We thus have very little to say about initial algebras in this chapter; Proposition 4.3.22 is the only relevant result.

terminal coalgebra as a limit of 'twice' the finitary iterations. That is, we iterate to $\omega + \omega$ rather than ω. We also present some results relating the limit of the terminal-coalgebra ω^{op}-chain to νF.

4.1 Limits and Colimits of Algebras and Coalgebras

The main material in this chapter begins with Section 4.3. But we have two preliminary sections for use there and elsewhere in the remainder of this book. In this section, we present some basic facts about algebras and coalgebras for arbitrary endofunctors. We prove that the forgetful functor of the category of F-coalgebras creates (1) all colimits and (2) all limits that are preserved by F.

Recall, e.g. from Mac Lane's book [217], that a *diagram scheme* $\mathscr{D}$ is a small category, and a *a diagram* in a category $\mathscr{B}$ is a functor $D\colon \mathscr{D} \to \mathscr{B}$. A functor $G\colon \mathscr{B} \to \mathscr{A}$ is said to *create colimits* of scheme $\mathscr{D}$ provided that for every diagram $D\colon \mathscr{D} \to \mathscr{B}$ the following holds: given a colimit cocone of GD in $\mathscr{A}$, say

$$c_d\colon GDd \to C \qquad (d \in \mathsf{obj}\,\mathscr{D}),$$

then there exists a unique cocone of D in $\mathscr{B}$ that G maps to (c_d), and, moreover, that cocone is a colimit of D in $\mathscr{B}$.

Given $F\colon \mathscr{A} \to \mathscr{A}$, we apply this to the forgetful functor

$$U\colon \mathsf{Coalg}\,F \to \mathscr{A}, \qquad U(A,\alpha) = A.$$

That U creates colimits implies that colimits of F-coalgebras are formed on the level of $\mathscr{A}$. More precisely, given a diagram $D\colon \mathscr{D} \to \mathsf{Coalg}\,F$ of coalgebras with $Dd = (A_d, \alpha_d)$, the functor U creates the colimit of D if for a colimit cocone $c_d\colon A_d \to C$ of UD in $\mathscr{A}$ we have a unique coalgebra structure $\gamma\colon C \to FC$ such that every c_d is a coalgebra homomorphism $c_d\colon (A_d, \alpha_a) \to (C, \gamma)$. Moreover, the last cocone is a colimit of D in $\mathsf{Coalg}\,F$.

Proposition 4.1.1. *For every endofunctor F on $\mathscr{A}$, the forgetful functor $U\colon \mathsf{Coalg}\,F \to \mathscr{A}$ creates colimits.*

Proof Given a diagram $D\colon \mathscr{D} \to \mathsf{Coalg}\,F$ and a colimit cocone $(c_d\colon A_d \to C)$ of UD in $\mathscr{A}$, we need to find a unique coalgebra structure $\gamma\colon C \to FC$ such that each of the following squares commute:

$$\begin{array}{ccc} A_d & \xrightarrow{\alpha_d} & FA_d \\ {\scriptstyle c_d}\big\downarrow & & \big\downarrow{\scriptstyle Fc_d} \\ C & \overset{\gamma}{\dashrightarrow} & FC \end{array}$$

For that it is sufficient to verify that the family of morphisms $\alpha_d \cdot Fc_d$ forms a compatible cocone of the diagram UD; then γ is uniquely determined by the universal property of the colimit cocone (c_d). Indeed, for every morphism $h\colon d \to d'$ in $\mathscr{D}$ we have $c_d = c_{d'} \cdot UDh$ by the compatibility of the cocone (c_d). This implies the desired compatibility property of (Fc_d, α_d):

$$Fc_{d'} \cdot \alpha_{d'} \cdot UDh = Fc_{d'} \cdot F(UDh) \cdot \alpha_d = Fc_d \cdot \alpha_d,$$

using that $Dh\colon (A_d, \alpha_d) \to (A_{d'}, \alpha_{d'})$ is a coalgebra homomorphism. Thus we obtain the desired unique coalgebra structure γ.

It remains to verify the universal property of the cocone $c_d\colon (A_d, \alpha_d) \to (C, \gamma)$ in $\mathsf{Coalg}\, F$. Let $f_d\colon (A_d, \alpha_d) \to (B, \beta)$ be a cocone of D. Then $f_d\colon A_d \to B$ is a cocone of UD. Thus there exists a unique morphism $f\colon C \to B$ of $\mathscr{A}$ with $f \cdot c_d = f_d$ for all $d \in \mathsf{obj}\,\mathscr{D}$. To see that f is a homomorphism, consider the following diagram:

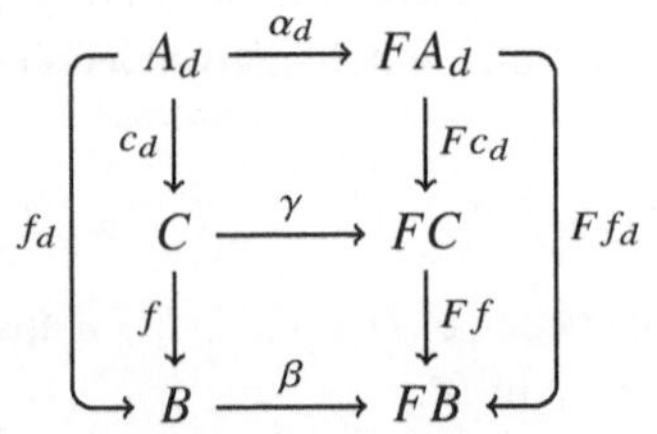

(4.1)

The left-hand and right-hand parts, the upper square, and the outside of the diagram clearly commute. Thus so does the lower square when pre-composed by every colimit injection c_d. Since the colimit injections c_d are collectively epic, this proves that f is a coalgebra homomorphism, as desired. □

Example 4.1.2. (1) As a first (easy) application of Proposition 4.1.1, suppose that $\mathscr{A}$ has coproducts. Given two coalgebras $\alpha\colon A \to FA$ and $\beta\colon B \to FB$, there is a unique coalgebra structure on $A + B$ such that the coproduct injections $\mathsf{inl}\colon A \to A + B$ and $\mathsf{inr}\colon B \to A + B$ are coalgebra homomorphisms:

$$A + B \xrightarrow{\alpha+\beta} FA + FB \xrightarrow{[F\mathsf{inl}, F\mathsf{inr}]} F(A + B).$$

This is dual to Proposition 2.1.13.

(2) Also, let us mention one application that we shall need in Section 4.2. Suppose that $\mathscr{A}$ has coequalizers, and that we have a parallel pair $f, g\colon (A, \alpha) \to (B, \beta)$ of coalgebra homomorphisms. Let $c\colon B \to C$ be the coequalizer in the base category. Then there is a unique coalgebra structure $C \to FC$ so that c becomes a coalgebra homomorphism and (the main point) that we have a coequalizer in $\mathsf{Coalg}\, F$.

(3) Isomorphisms in $\mathsf{Coalg}\,F$ are precisely the homomorphisms that are invertible in the base category. This is easy to prove directly from first principles. It follows from Proposition 4.1.1 since in any category $\mathscr{A}$, a morphism $f\colon A \to B$ is an isomorphism iff the diagram consisting of A and id_A has a colimit cocone $f\colon A \to B$.

(4) A homomorphism in $\mathsf{Coalg}\,F$ is epic iff the underlying morphism is epic (in the base category). This follows from the fact that in any category, e is epic iff the pushout along itself is formed by isomorphisms.

We emphasize that creation of colimits does not make any assumptions on the existence of colimits in $\mathscr{A}$, and it holds for all endofunctors and diagrams. But we do have the following easy consequence of Proposition 4.1.1, which is worth stating on its own. We recall that a category is co-well-powered if for every object only a set of quotient objects exist.

Corollary 4.1.3. *Let $F\colon \mathscr{A} \to \mathscr{A}$ be given.*

(1) *If $\mathscr{A}$ is cocomplete, then so is* $\mathsf{Coalg}\,F$.

(2) *If $\mathscr{A}$ is co-well-powered, then so is* $\mathsf{Coalg}\,F$.

Proof The first part is immediate from Proposition 4.1.1. The second follows from Example 4.1.2(4). □

Remark 4.1.4. Dually, all limits of F-algebras are formed on the level of $\mathscr{A}$. More precisely, creation of limits is the dual concept of creation of colimits, and for every endofunctor $F\colon \mathscr{A} \to \mathscr{A}$ the forgetful functor $\mathsf{Alg}\,F \to \mathscr{A}$ creates limits. In particular, monomorphisms in $\mathsf{Alg}\,F$ are precisely those homomorphisms whose underlying morphisms are monic in $\mathscr{A}$.

For limits of coalgebras the situation is more involved, as they are usually not formed on the level of $\mathscr{A}$. However, limits which F preserves are:

Proposition 4.1.5. *For every endofunctor F on $\mathscr{A}$ preserving limits of the diagram scheme $\mathscr{D}$, the forgetful functor $U\colon \mathsf{Coalg}\,F \to \mathscr{A}$ creates limits of that scheme.*

Proof Suppose that we are given a diagram $D\colon \mathscr{D} \to \mathsf{Coalg}\,F$ with $Dd = (A_d, \alpha_d)$ and a limit of UD with limit projections

$$\pi_d\colon L \to A_d \qquad (d \in \mathsf{obj}\,\mathscr{D}).$$

We need to prove that there exists a unique coalgebra structure $\lambda\colon L \to FL$

making each π_d a homomorphism

$$\begin{array}{ccc} L & \overset{\lambda}{\dashrightarrow} & FL \\ {\scriptstyle \pi_d}\big\downarrow & & \big\downarrow{\scriptstyle F\pi_d} \\ A_d & \xrightarrow{\alpha_d} & FA_d \end{array}$$

and that, moreover, the morphisms $\pi_d\colon (L,\lambda) \to (A_d, \alpha_d)$ form the limit of D. Since F preserves the limit of UD, the diagram FUD has the limit cone given by the morphisms $F\pi_d\colon FL \to FA_d$, $d \in \mathsf{obj}\,\mathscr{D}$. We show that $\alpha_d \cdot \pi_d$ form a compatible cone of that diagram. The argument is similar to what we saw in the proof of Proposition 4.1.1. For every morphism $f\colon d \to d'$ of $\mathscr{D}$ we have $UDh \cdot \pi_d = \pi_{d'}$ by the compatibility of the cone (π_d). This implies the desired compatibility of $(\alpha_d \cdot \pi_d)$:

$$FUDh \cdot \alpha_d \cdot \pi_d = \alpha_{d'} \cdot UDh \cdot \pi_d = \alpha_{d'} \cdot \pi_{d'}$$

using that $Df\colon (A_d, \alpha_d) \to (A_{d'}, \alpha_{d'})$ is a coalgebra homomorphism. Thus we obtain the desired unique coalgebra structure λ.

It remains to verify that the cone $\pi_d\colon (L,\lambda) \to (A_d,\alpha_d)$ is a limit of D in $\mathsf{Coalg}\,F$. Let $f_d\colon (B,\beta) \to (A_d,\alpha_d)$ be a cone of D. Then $f_d\colon B \to A_d$ is a cone of UD. Thus there exists a unique morphism $f\colon B \to L$ such that $\pi_d \cdot f = f_d$ for all $d \in \mathsf{obj}\,\mathscr{D}$. To see that f is a homomorphism, we argue dually as for Diagram (4.1). Consider the following diagram:

$$\begin{array}{ccccc} & L & \xrightarrow{\lambda} & FL & \\ & {\scriptstyle f}\big\downarrow & & \big\downarrow{\scriptstyle Ff} & \\ f_d & B & \xrightarrow{\beta} & FB & Ff_d \\ & {\scriptstyle \pi_d}\big\downarrow & & \big\downarrow{\scriptstyle F\pi_d} & \\ & A_d & \xrightarrow{\alpha_d} & FA_d & \end{array}$$

and use that the family of morphisms $F\pi_d$ is collectively monic. □

Example 4.1.6. Let us underscore the need for F to preserve the limit in Proposition 4.1.5. Consider $FX = X + 1$ on Set. The limit of the empty diagram is the terminal object, 1. Yet the terminal coalgebra is not of the form $1 \to F1$.

Remark 4.1.7. Dually, colimits of algebras are not always formed on the level of $\mathscr{A}$. However, for every endofunctor F preserving colimits of a diagram scheme $\mathscr{D}$, the forgetful functor $\mathsf{Alg}\,F \to \mathscr{A}$ creates colimits of that scheme.

4.2 Weakly Terminal Coalgebras

Let $F: \mathscr{A} \to \mathscr{A}$ be an endofunctor. An F-coalgebra (A, α) is *weakly terminal* if for every coalgebra (B, β) there is *at least one* homomorphism $(B, \beta) \to (A, \alpha)$. Such coalgebras are the topic of this section. It turns out that the terminal coalgebra for a functor F is the quotient of (any) weakly terminal coalgebra. A beautiful example is the coalgebra of all unordered finitely branching trees, which is weakly terminal for $\mathscr{P}_f$.

Example 4.2.1. (1) The terminal coalgebra for the identity functor $\mathsf{Id}: \mathsf{Set} \to \mathsf{Set}$ is the singleton coalgebra. A coalgebra $\alpha: A \to A$ is weakly terminal iff α has a fixed point. Indeed, given $\alpha(a) = a$, for every coalgebra (B, β) the constant function $f: B \to A$ with value a is a coalgebra morphism. Conversely, if (A, α) is weakly terminal, consider homomorphisms from $\mathsf{id}: 1 \to 1$: they correspond to fixed points of α.
(2) If (A, α) is a weakly terminal coalgebra and $f: (A, \alpha) \to (B, \beta)$ a coalgebra homomorphism, then (B, β) is again weakly terminal.
(3) We next consider the finite power-set functor $\mathscr{P}_\mathsf{f}: \mathsf{Set} \to \mathsf{Set}$. Recall from Example 2.4.2(4) that coalgebras for $\mathscr{P}_\mathsf{f}$ are the finitely branching graphs. We denote by D the set of all finitely branching, rooted, and unordered trees. As always we take trees up to isomorphism. This ensures that the trees under consideration indeed form a set rather than a proper class. We endow D with the coalgebra structure $\delta: D \to \mathscr{P}_\mathsf{f}D$ which takes a tree t to the (finite) set of all maximum proper subtrees of t.

To clarify matters, here is an example. Let $b \in D$ be the full infinite binary tree. Then in the coalgebra D, b is its own child; in fact, it is its only child: $\delta(b) = \{b\}$. The same remark applies to the infinite chain u, or to any infinite tree having the same number of children at every node.

We argue that the coalgebra (D, δ) is weakly terminal for $\mathscr{P}_\mathsf{f}$. Every finitely branching graph, considered as a coalgebra (A, α), admits a canonical coalgebra homomorphism to $(A, \alpha) \to (D, \delta)$. It assigns to every node $x \in A$ the *tree unfolding* of x, which we denote by A_x. The nodes of this tree are the (finite) paths $\vec{p} = p_0 \to p_1 \to \cdots \to p_n$ starting at $p_0 = x$ and following the edges in A. We have an edge from $\vec{p}$ to $\vec{q}$ in A_x iff $\vec{q}$ extends the path $\vec{p}$ by an edge of A:

$$\vec{q} = p_0 \to p_1 \to \cdots \to p_n \to p_{n+1} \qquad \text{for an edge } p_n \to p_{n+1} \text{ of } A.$$

The root of the tree A_x is given by the node x, considered as a path of length 0. For example, if $A = \mathbb{1}$ consists of a single loop, then A_x is an infinite path. To see that (D, δ) is weakly terminal, note that the trees A_y for edges $x \to y$ in A

are precisely the maximum proper subtrees of A_x:

$$\{t : t \in \delta(A_x) \text{ in } D\} = \{A_y : y \in \alpha(x) \text{ in } A\}.$$

This states that $x \mapsto A_x$ is a coalgebra homomorphism $(A, \alpha) \to (D, \delta)$.

However, (D, δ) is not terminal: $\mathbb{1}$ has many homomorphisms to (D, δ); for example, we can map it to either of the infinite path or the complete binary tree. In Example 4.2.10 below, we shall see a quotient of (D, δ) which is a terminal $\mathscr{P}_\mathsf{f}$-coalgebra.

(4) Monoid-valued functors $M^{(-)}$ (Example 2.4.3(1)) have a weakly terminal coalgebra carried by the set of all finitely branching M-labelled trees [48, Proof of Thm. 6.6].

(5) The power-set functor $\mathscr{P}$: Set $\to$ Set does not have a weakly terminal coalgebra. Indeed, we have seen that $\mathscr{P}$ has no terminal coalgebra, and from a weakly terminal coalgebra we can always construct a terminal one, as we prove in Theorem 4.2.8.

The main work in this section concerns quotients, particularly quotients of weakly terminal coalgebras. We begin with some general points.

Remark 4.2.2. (1) Dually to Remark 2.1.18, quotient objects of an object A of a category $\mathscr{A}$ are represented (uniquely up to isomorphism of the codomain) by epimorphisms $e\colon A \to \bar{A}$. As usual, we will speak about the quotient e when we mean the quotient object of A represented by e.

(2) For an object A in a category $\mathscr{A}$, the quotients of A are ordered by factorization: given quotients $e\colon A \twoheadrightarrow \bar{A}$ and $f\colon A \twoheadrightarrow A^*$ we write $e \le f$ iff e factorizes through f:

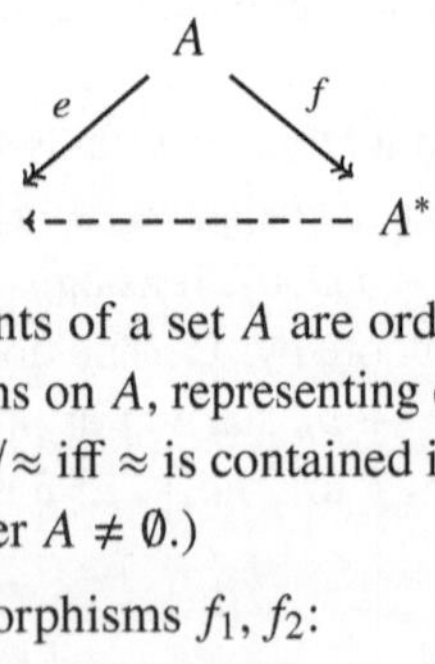

Thus for $\mathscr{A}$ = Set the quotients of a set A are ordered by their cardinality as usual, but equivalence relations on A, representing quotients are ordered dually: $A/{\sim}$ is smaller or equal to $A/{\approx}$ iff $\approx$ is contained in $\sim$. (The smallest quotient thus has codomain 1 whenever $A \ne \emptyset$.)

(3) Given a pushout of epimorphisms f_1, f_2:

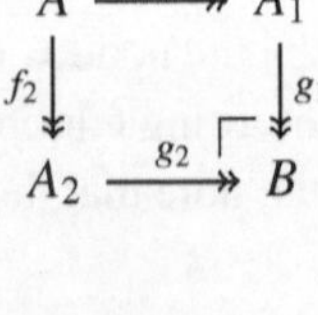

it follows that g_1, g_2 are epimorphisms, too. Thus, the quotient of A represented by $e = g_i \cdot f_i \colon A \twoheadrightarrow B$ is the meet of f_1 and f_2 in the poset of quotients of A.

(4) More generally, a meet of a collection of quotients $f_i \colon A \to A_i$ $(i \in I)$ is represented by their wide pushout.

Remark 4.2.3. We apply Remark 4.2.2(1) to Coalg F: a *quotient coalgebra* is represented by a homomorphism carried by an epimorphism in the base category (cf. Example 4.1.2(4)).

In other words, a quotient $e \colon A \twoheadrightarrow \bar{A}$ in the base category $\mathscr{A}$ represents a quotient coalgebra of (A, α) iff there exists a (necessarily unique) coalgebra structure $\bar{\alpha} \colon \bar{A} \to F\bar{A}$ making e a coalgebra homomorphism:

$$\begin{array}{ccc} A & \xrightarrow{\alpha} & FA \\ {\scriptstyle e}\big\downarrow & & \big\downarrow{\scriptstyle Fe} \\ \bar{A} & \xrightarrow{\bar{\alpha}} & F\bar{A} \end{array}$$

Notation 4.2.4. Let $f \colon X \to Y$ in Set. The *kernel equivalence* of f is the equivalence relation

$$\ker(f) = \{(x, x') \colon f(x) = f(x')\} \subseteq X \times X.$$

Categorically speaking $\ker(f)$ is the (domain of the) pullback of f along itself.

In case $\mathscr{A}$ = Set, we often identify the epimorphism e with its kernel equivalence. The kernel equivalence of a quotient coalgebra structure is called a *congruence*. We write $A/\!\sim$ for the coalgebra $(\bar{A}, \bar{\alpha})$ above.

Example 4.2.5. Let G be a finitely branching graph, considered as a coalgebra for $\mathscr{P}_f$. A congruence on G is then precisely an equivalence relation on G which is also a *graph bisimulation* (cf. Example 2.6.7(4)). This is an equivalence relation $R \subseteq G \times G$ such that whenever xRy, then

for every child x' of x, there is some child y' of y such that $x'Ry'$, and
for every child y' of y, there is some child x' of x such that $x'Ry'$.

Remark 4.2.6. We will not be using the general concept of a bisimulation on an F-coalgebra in this section. But we remind readers familiar with that concept that whenever F preserves weak pullbacks, the largest bisimulation on a coalgebra is always an equivalence relation, and hence a congruence (see Rutten [264, Corollary 5.6]).

Remark 4.2.7. (1) Quotient coalgebras of (A, α) are ordered as we have seen in Remark 4.2.2(2). For quotient coalgebras $e \colon (A, \alpha) \to (\bar{A}, \bar{\alpha})$ and

$f: (A, \alpha) \to (A^*, \alpha^*)$ we write $e \leq f$ if e factorizes through f; that is, there is $g: A^* \to \bar{A}$ with $e = g \cdot f$. It follows that g is a coalgebra homomorphism:

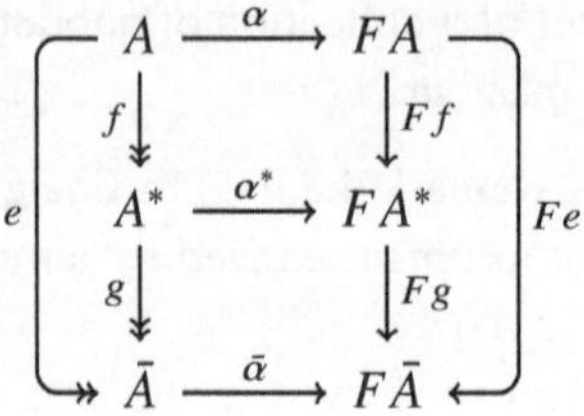

Indeed, the lower square commutes because it does when pre-composed with the epimorphism f.

(2) Given a coalgebra (A, α), a meet of its quotient coalgebras (in the poset of quotients of A in the base category) is a quotient coalgebra (see Remark 4.2.2(4)).

Theorem 4.2.8. *An endofunctor on a cocomplete category has a terminal coalgebra if and only if it has a weakly terminal one. Moreover, the terminal coalgebra is the smallest quotient coalgebra of any weakly terminal one.*

Proof (1) We first prove that a cocomplete category $\mathscr{C}$ has a terminal object if and only if it has a weakly terminal object.

The 'if' direction is clear. For the 'only if' direction let W be a weakly terminal object of $\mathscr{C}$. Form the coequalizer $c: W \to T$ of all endomorphisms $W \to W$. Since W is weakly terminal so is T (cf. Example 4.2.1(2)).

Thus for every object X we have a morphism $t: X \to T$. Now suppose that $t': X \to T$ is another morphism. We prove that $t = t'$. Take the coequalizer $d: T \to T'$ of t and t'. We have a morphism $w: T' \to W$ since W is weakly terminal, and therefore we obtain an endomorphism

$$W \xrightarrow{c} T \xrightarrow{d} T' \xrightarrow{w} W.$$

Since c merges all endomorphisms on W, $c \cdot (w \cdot d \cdot c) = c \cdot \mathsf{id}_W = c$. Therefore $c \cdot w \cdot d = \mathsf{id}_T$, since the coequalizer c is epic. Thus, d is a split monomorphism. It is also epic, being a coequalizer. Therefore it is an isomorphism. From $d \cdot t = d \cdot t'$ it follows that $t = t'$ as desired.

It follows that $c: W \twoheadrightarrow T$ is the smallest quotient of W; given any smaller quotient yielding $d: T \twoheadrightarrow T'$, then d is an isomorphism by the previous argument.

(2) Now for an endofunctor $F: \mathscr{C} \to C$, apply (1) to $\mathscr{C} = \mathsf{Coalg}\, F$, using that this category is cocomplete by Corollary 4.1.3. □

Remark 4.2.9. Note that the statement in (1) is the (dual of) a simplified form of Freyd's Adjoint Functor Theorem [217, Theorem V.6.1]. In fact, this states that a cocomplete category $\mathscr{C}$ has a terminal object if and only if it has a weakly terminal set $\mathcal{S}$ of objects (meaning that every object of $\mathscr{C}$ has a morphism into some object from $\mathcal{S}$). The proof yields a construction of the terminal object: one takes the coproduct C of all objects of $\mathcal{S}$, which is clearly a weakly terminal object. Then one proceeds as in (1) above by forming the coequalizer of all endomorphisms of C, which yields the terminal object.

Instantiated for $\mathscr{C} = \mathsf{Coalg}\,F$, we obtain Theorem 4.2.8. This instance was first noted by Lambek [210, Prop. 2.3] and later by Barr [72].

Example 4.2.10. The terminal coalgebra for $\mathscr{P}_\mathsf{f}$ is the quotient $D/\!\approx$ of the coalgebra D of all finitely branching trees (see Example 4.2.1(3)) modulo the greatest congruence. To describe $\approx$, first recall the initial-algebra ω-chain of $\mathscr{P}_\mathsf{f}$ from Example 3.2.9(2):

$$\mathscr{P}_\mathsf{f}^n\emptyset = \text{extensional trees of height} < n.$$

The terminal-coalgebra ω^{op}-chain is quite analogous. Indeed, since $1 \cong \mathscr{P}\emptyset$, we have $\mathscr{P}_\mathsf{f}^n 1 \cong \mathscr{P}_\mathsf{f}^{n+1}0$. We thus have

$$\mathscr{P}_f^n 1 = \text{all extensional trees of height} \leq n.$$

Given a tree $t \in D$, the canonical cone from D to $\mathscr{P}_\mathsf{f}^n 1$ of Construction 3.3.2 assigns to t the following tree $\partial_n t$.[2] First we take *the cutting of t at level n*. This is the set of nodes whose depth is at most n, considered as an induced subtree of t. Then we take the extensional quotient $\partial_n t$ of this cutting. For an illustration, here is a tree t and four of its cuttings below:

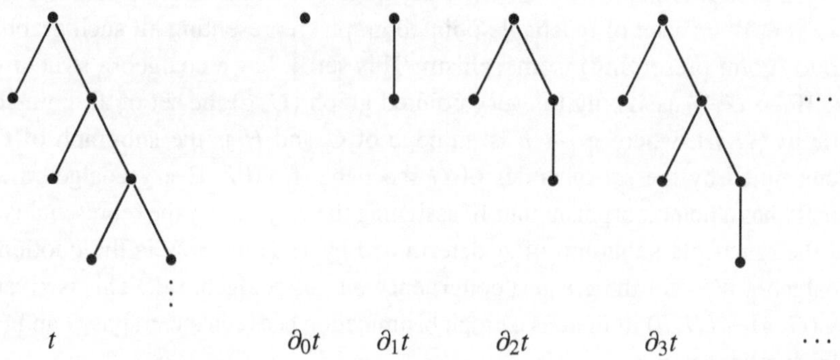

The greatest congruence $\approx$ of D is defined by $t \approx s$ iff $\partial_n t = \partial_n s$ holds for all n (cf. Example 4.3.27). This description, $\nu\mathscr{P}_\mathsf{f} = D/\!\approx$, is due to Barr [72, Ex. 4.6]. To see that it is correct, suppose that $\sim$ is any congruence on D and let $q\colon (D,\delta) \twoheadrightarrow (D/\!\sim, \bar{\delta})$ denote the corresponding quotient coalgebra homomorphism. Our task is to prove that $\sim$ is contained in $\approx$, that is, $q(t) = q(s)$

[2] This means that $\partial_n : D \to \mathscr{P}_\mathsf{f}^n 1$ is the nth map in the canonical cone for D.

implies $t \approx s$. Since (D, δ) is weakly terminal we have a colgebra homomorphism $h\colon (D/{\sim}, \bar{\delta}) \to (D, \delta)$. We prove by induction that

$$\partial_n = \partial_n \cdot h \cdot q \qquad \text{for all } n \in \mathbb{N}.$$

The desired result then follows: from $q(t) = q(s)$ we obtain $\partial_n(t) = \partial_n(s)$ for all n, whence $t \approx s$. The base case $n = 0$ is trivial, and for the induction step we use the following diagram:

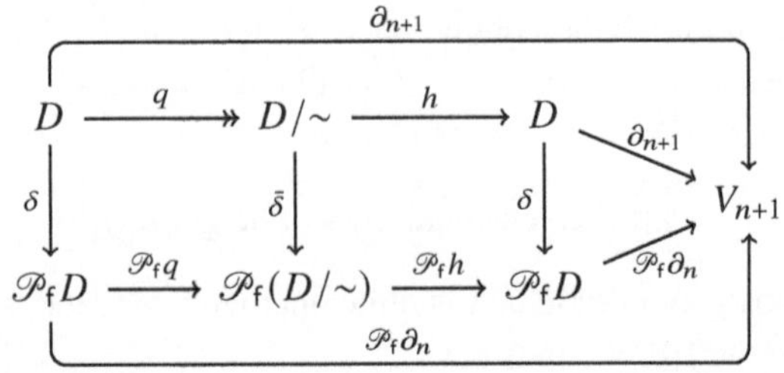

The two inner squares commutes since q and h are coalgebra homomorphisms, and the right-hand triangle and the outside do by the definition of the canonical cone. The lower part commutes by the induction hypothesis. Hence, the upper part commutes, as desired.

Example 4.2.11. Here is another weakly terminal coalgebra for $\mathscr{P}_\mathsf{f}$, due to Aczel [3]. A *pointed graph* is a pair (G, g) consisting of a graph G and a node g in it. It is called *reachable* if every node x admits a directed path from g to x. Let W be a set of reachable pointed graphs representing all such graphs up to (point preserving) isomorphism. This set W has a coalgebra structure $\alpha\colon W \to \mathscr{P}_\mathsf{f} W$ assigning to every pointed graph (G, g) the set of all pointed graphs (H, h), where $g \to h$ is an edge of G and H is the subgraph of G determined by the set of nodes of G reachable from h. Every coalgebra A for $\mathscr{P}_\mathsf{f}$ has a homomorphism into W assigning to every node x the representative of the reachable subgraph of A determined by x. Thus, $\nu\mathscr{P}_\mathsf{f}$ is the quotient coalgebra $W/{\sim}$ for the greatest congruence on this coalgebra W. This is given by $(G, g) \sim (H, h)$ iff there is a graph bisimulation between the graphs G and H which relates g to h.

Example 4.2.12. For monoid-valued functors $M^{(-)}$ (Example 2.4.3(1)) we have a similar description of $\nu M^{(-)}$ as for $\mathscr{P}_\mathsf{f}$ in Example 4.2.10. The terminal coalgebra is the quotient of the coalgebra of all finitely branching M-labelled trees (Example 4.2.1(4)) modulo the greatest congruence. This has a similar description to $\approx$ in Example 4.2.10. Indeed, the formation of extensional quotients is easily generalized to M-labelled trees involving the summation of edge labels in M when identifying isomorphic subtrees of a given node [48, Def. 6.4 and Thm. 6.6].

At this point, we have presented a construction of the terminal coalgebra that differs from the iterative construction which we saw in Chapter 3. Although taking a quotient of a weakly terminal coalgebra seems easier than taking a limit, we frequently want to have information about the terminal coalgebra which is not readily available from its description as a quotient. This is addressed in Section 4.3, where we turn to presentations of finitary set functors and their terminal coalgebras.

4.3 Presentation of Finitary Set Functors

In this section we turn to the main notion of this chapter, *finitary* endofunctors on Set. Intuitively, a functor on sets is finitary if its behaviour is completely determined by its action on *finite* sets and functions. For a general functor, this intuition is captured by the following definition concerning functors between arbitrary categories:

Definition 4.3.1. A *directed diagram* in a category $\mathscr{A}$ is a diagram $D\colon \mathscr{D} \to \mathscr{A}$ where $\mathscr{D}$ is a directed poset: $\mathscr{D}$ is nonempty, and every pair of elements has an upper bound.

A functor is *finitary* if it preserves directed colimits, that is, colimits of directed diagrams.

Remark 4.3.2. (1) Equivalently, filtered colimits can be used in lieu of directed ones, see Theorem C.5.1.

(2) Note that an ω-chain is a special kind of directed diagram. Hence, every finitary functor preserves ω-colimits.

When $\mathscr{A} = \mathsf{Set}$, a more intuitive and equivalent concept is finite boundedness:

Definition 4.3.3 [52]**.** A set functor F is *finitely bounded* if for each set X and each $x \in FX$, there exists a finite subset $M \subseteq X$ such that $x \in Fi[FM]$, where $i\colon M \hookrightarrow X$ is the inclusion map.

Proposition 4.3.4 [52, Cor. 3.3]**.** *A set functor is finitely bounded iff it is finitary.*

A proof is presented in Section C.5. In this chapter, the finite boundedness condition is more central. We also will see in Theorem C.5.9 that a set functor is finitary iff all sufficiently large sets X are *pre-fixed points*: $|FX| \leq |X|$.

Example 4.3.5. (1) Constant functors and the identity functor are finitary.

(2) Finitary set functors are closed under taking coproducts and colimits, composition, and finite products.

(3) It follows that polynomial set functors H_Σ for (finitary) signatures Σ are finitary. (To say that a signature is finitary means that all of the arities in Σ are finite. But Σ might have infinitely many operation symbols, see Definition 2.1.4.)

(4) The finite power-set functor $\mathscr{P}_\mathrm{f}$ is finitary, but the full power-set functor $\mathscr{P}$ is not.

Just as quotient objects of an object A are represented by epimorphisms with domain A, *quotient functors* of a given endofunctor H are represented by natural *epi-transformations* with domain H. That is, natural transformations with all components epic. We use the notation $\varepsilon\colon H \twoheadrightarrow F$ to say that F is a quotient functor of H via ε. An important special case are the set functors that are quotients of a polynomial functor H_Σ associated signature Σ (see Definition 2.1.4): these functors have an equational presentation using the signature Σ.

The main results of this section deal with functors F for which there is an epi-transformation from a polynomial functor, so we have $\varepsilon\colon H_\Sigma \twoheadrightarrow F$. This turns out to be an important situation to study.

The following was proved for set functors by Gumm and Schröder [153, Lemma 2.3(iii)]. Recall that a *split epimorphism* is a morphism $e\colon X \to Y$ for which $m\colon Y \to X$ exists with $e \cdot m = \mathsf{id}_Y$.

Lemma 4.3.6. *Let $H\colon \mathscr{A} \to \mathscr{A}$ be an endofunctor with a weakly terminal coalgebra $\tau\colon T \to HT$. Let $\varepsilon\colon H \twoheadrightarrow F$ be a quotient functor, and assume that every component ε_X is a split epimorphism. Then F has the following weakly terminal coalgebra:*

$$T \xrightarrow{\tau} HT \xrightarrow{\varepsilon_T} FT.$$

Proof Let $\alpha\colon A \to FA$ be an F-coalgebra. We can choose a morphism $m\colon FA \to HA$ such that $\varepsilon_A \cdot m = \mathsf{id}$. So we obtain an H-coalgebra $m \cdot \alpha\colon A \to HA$. We thus have an H-coalgebra homomorphism $h\colon (A, m \cdot \alpha) \to (T, \tau)$, and we show that h is also an F-coalgebra homomorphism from (A, α) to $(T, \varepsilon_T \cdot \tau)$. Indeed, consider the diagram below:

$$\begin{array}{ccccc}
 & & \overset{\alpha}{\curvearrowright} & & \\
A & \xrightarrow{m\cdot\alpha} & HA & \xrightarrow{\varepsilon_A} & FA \\
{\scriptstyle h}\downarrow & & \downarrow{\scriptstyle Hh} & & \downarrow{\scriptstyle Fh} \\
T & \xrightarrow{\tau} & HT & \xrightarrow{\varepsilon_T} & FT
\end{array}$$

The top commutes because $\varepsilon_A \cdot m = \mathsf{id}$, the square on the left does since h is a homomorphism, and the one on the right does by naturality of ε. Thus the outside commutes, showing h to be an F-coalgebra homomorphism. □

Recall the concept of a (finitary) signature Σ from Definition 2.1.4. Recall also that the terminal coalgebra of H_Σ consists of all Σ-trees (Theorem 2.5.9).

Definition 4.3.7. By a *finitary presentation* of a set functor F we mean a (finitary) signature Σ and a natural epi-transformation $\varepsilon\colon H_\Sigma \twoheadrightarrow F$.

Remark 4.3.8. Note that every natural transformation $\varepsilon\colon H \to F$ induces a functor $\mathsf{Coalg}\, H \to \mathsf{Coalg}\, F$ by post-composition:

$$(A \xrightarrow{\alpha} HA) \quad \longmapsto \quad (A \xrightarrow{\alpha} HA \xrightarrow{\varepsilon_A} FA).$$

In particular, for every presentation $\varepsilon\colon H_\Sigma \twoheadrightarrow F$ a coalgebra (A, α) for H_Σ can be considered as the coalgebra $(A, \varepsilon_A \cdot \alpha)$ for F.

Corollary 4.3.9. *For every set functor F with a presentation $\varepsilon\colon H_\Sigma \twoheadrightarrow F$, the terminal coalgebra for H_Σ is weakly terminal for F. Consequently, F has a terminal coalgebra given by the smallest quotient coalgebra of the F-coalgebra of all Σ-trees.*

Indeed, this follows from Lemma 4.3.6 and Theorem 4.2.8.

Remark 4.3.10. Corollary 4.3.9 above stems from Barr [72] and also Gumm and Schröder [153, Lemma 2.3(iv)]. Both consider a more general situation: an epi-transformation $\varepsilon\colon H \twoheadrightarrow F$ where H is any set functor which has a terminal coalgebra. We use this in the next example.

Example 4.3.11. The coalgebra D of all unordered finitely branching trees is weakly terminal for $\mathscr{P}_\mathsf{f}$ (see Example 4.2.1(3)), and here is a different proof. Let $\mathscr{B}$ be the bag functor (see Example 3.2.11). There is an epi-transformation $\varepsilon\colon \mathscr{B} \twoheadrightarrow \mathscr{P}_\mathsf{f}$ mapping a bag to the set of its elements. Naturality is easy to see, as is the surjectivity of each ε_X. As a splitting of any ε_X choose the map that maps a finite subset of $S \subseteq X$ to the bag of S (each element occurring once). Now by Lemma 4.3.6, we obtain that D, being the terminal coalgebra for $\mathscr{B}$, is weakly terminal for $\mathscr{P}_\mathsf{f}$.

In addition, we can also apply Lemma 4.3.6 to the presentation of $\mathscr{P}_\mathsf{f}$ in Example 4.3.14(2) to see that the algebra T_Σ of all *ordered* finitely branching trees is weakly terminal.

Corollary 4.3.12. *Every finitary set functor has an initial algebra and a terminal coalgebra.*

This result highlights our interest in finitary functors in this chapter. The initial algebra part comes from Proposition 4.3.22, and the terminal coalgebra result comes from Theorem 4.3.26.

Proposition 4.3.13. *A set functor has a finitary presentation iff it is finitely bounded.*

Proof (1) Let F be a finitely bounded functor and let Σ be the finitary signature defined by

$$\Sigma_n = Fn \qquad (n < \omega)$$

(where n is, as usual, considered to be the set $\{0, 1, \ldots, n-1\}$ for every $n < \omega$). Define a natural transformation $\varepsilon\colon H_\Sigma \to F$ to have the following components

$$\varepsilon_X\colon \coprod_{n<\omega} \Sigma_n \times X^n \to FX:$$

given a pair $(\sigma, f) \in \Sigma_n \times X^n$, the function $f\colon n \to X$ yields a function $Ff\colon Fn \to FX$ and we put

$$\varepsilon_X(\sigma, f) = Ff(\sigma).$$

Let us verify the naturality square for an arbitrary function $h\colon X \to Y$:

$$\begin{array}{ccc} \coprod_{n<\omega} \Sigma_n \times X^n & \xrightarrow{\varepsilon_X} & FX \\ {\scriptstyle \coprod_{n<\omega} \mathrm{id}\times h^n}\downarrow & & \downarrow{\scriptstyle Fh} \\ \coprod_{n<\omega} \Sigma_n \times Y^n & \xrightarrow{\varepsilon_Y} & FY \end{array}$$

The upper passage takes (σ, f) to $Fh(Ff(\sigma)) = F(h \cdot f)(\sigma)$. The lower one takes it to $\varepsilon_Y(\sigma, h \cdot f) = F(h \cdot f)(\sigma)$.

Moreover, ε_X is an epimorphism since F is finitely bounded. Thus for every element $x \in FX$ there exists a natural number n and an n-element subset $i\colon n \rightarrowtail X$ such that x lies in the image of Fi. Let $\sigma \in Fn$ be such that $x = Fi(\sigma)$. Then we have

$$x = \varepsilon_X(\sigma, i).$$

Thus ε is a finitary presentation of F.

(2) If $\varepsilon\colon H_\Sigma \twoheadrightarrow F$ is a finitary presentation of F, then we prove that F is a finitely bounded functor. Fix a set X. For every element $x \in FX$, choose $y \in H_\Sigma X$ with $\varepsilon_X(y) = x$. Then y lies in the component X^n of $H_\Sigma X$ corresponding to some n-ary operation σ of Σ. Use the naturality square on $y\colon n \to X$:

$$\begin{array}{ccc} H_\Sigma n & \xrightarrow{\varepsilon_n} & Fn \\ {\scriptstyle H_\Sigma y}\downarrow & & \downarrow{\scriptstyle Fy} \\ H_\Sigma X & \xrightarrow{\varepsilon_X} & FX \end{array}$$

The summand of $H_\Sigma n$ corresponding to the n-ary operation σ is n^n. Apply the square above to the element id_n of n^n: this yields $Fy(\varepsilon_n(\mathrm{id}_n)) = \varepsilon_X(y \cdot \mathrm{id}_n)$, consequently

$$Fy(a) = \varepsilon_X(y) = x \qquad \text{for } a = \varepsilon_n(\mathrm{id}_n).$$

Finally, factorize y as a (split) epimorphism $e\colon n \to Y$ followed by an inclusion map $i\colon Y \hookrightarrow X$. Then $x = Fi(b)$ where $b = Fe(a)$. Since Y is finite, this proves that F is finitely bounded. □

We have just seen that every finitary set functor F has a finitary presentation $\varepsilon\colon H_\Sigma \twoheadrightarrow F$ with $\Sigma_n = Fn$. However, in particular special cases we can find smaller presentations.

Example 4.3.14. (1) The functor $\mathscr{P}_3$ takes a set X to the set of subsets of X of size less than 3. This functor has a presentation given by the signature Σ with one constant and one binary operation symbol. Hence $H_\Sigma X = X \times X + 1$, and the natural epi-transformation $\varepsilon\colon H_\Sigma \twoheadrightarrow F$ is given by

$$\varepsilon(\bullet) = \emptyset \qquad \text{and} \qquad \varepsilon_X(x, y) = \{x, y\} \qquad \text{for every } x, y \in X,$$

where $\bullet$ denotes the element of 1.

(2) A presentation of $\mathscr{P}_\mathrm{f}$ is given by the signature Σ having, for each natural number n, just one operation symbol σ_n of arity n. Thus,

$$H_\Sigma X = \coprod_{n<\omega} X^n = X^*.$$

The epi-transformation $\varepsilon\colon H_\Sigma \to \mathscr{P}_\mathrm{f}$ is given by

$$\varepsilon_X(x_1, \ldots, x_n) = \{x_1, \ldots, x_n\}.$$

(3) A presentation for the bag functor $\mathscr{B}$ (see Example 3.2.11) can be easily obtained from its representation as an analytic functor, see (3.11). Take the same Σ as in the previous item and let $\varepsilon_X\colon H_\Sigma X \to \mathscr{B}X$ take $(x_1, \ldots, x_n)$ to the bag (X, b), where $b\colon X \to \mathbb{N}$ assigns to x is the number of times x occurs in the sequence $x_1, \ldots, x_n$.

(4) The *Aczel–Mendler functor* [4] is the set functor $(-)^3_2$ given by

$$(X)^3_2 = \{(a, b, c) \in X \times X \times X : (a = b) \text{ or } (a = c) \text{ or } (b = c)\}.$$

Note that if $(a, b, c) \in (X)^3_2$ and $f\colon X \to Y$ is a any function, then $(fa, fb, fc) \in (Y)^3_2$. We therefore take the action of $(-)^3_2$ on morphisms to be pointwise application, as expected. Our main interest in this functor is that it provides interesting (counter-)examples; see Appendix C. For a presentation of $(-)^3_2$ we take Σ to have three binary symbols, say σ, τ, and ϱ. Then we define $\varepsilon\colon H_\Sigma \to (-)^3_2$ by: $\varepsilon_X(\sigma(a,b)) = (a,a,b)$, $\varepsilon_X(\tau(a,b)) = (a,b,a)$, and $\varepsilon_X(\varrho(a,b)) = (b,a,a)$.

Remark 4.3.15. Given a presentation $\varepsilon\colon H_\Sigma \twoheadrightarrow F$, the functor F is determined (up to natural isomorphism) by the kernel equivalences of $\varepsilon_X\colon H_\Sigma X \to FX$ for all sets X. We can describe the kernel equivalence of ε_X as a set of pairs of elements of $H_\Sigma X = \coprod_{n\in\mathbb{N}} \Sigma_n \times X^n$ (see Definition 2.1.4). Each such element is, for some $\sigma \in \Sigma_n$, an $(n+1)$-tuple $(\sigma, x_0, \ldots, x_{n-1})$. We write this as the term $\sigma(x_0,\ldots,x_{n-1})$, and we write the pairs contained in the kernel equivalence of ε_X as term equations

$$\sigma(x_0,\ldots,x_{n-1}) = \tau(y_0,\ldots,y_{m-1}),$$

where $\sigma \in \Sigma_n$, $\tau \in \Sigma_m$ and all x_i and y_j are elements of X. These equations have the following special form:

Definition 4.3.16. A term over Σ is called *basic* if it contains precisely one symbol from Σ. That is, it has the form $\sigma(x_0,\ldots,x_{n-1})$ for some $\sigma \in \Sigma_n$ and n variables x_i. The variables might contain repeats.

A *basic equation* over X is a pair of basic terms with variables from the set X.

Note that $H_\Sigma X$ is thus the set of all basic terms over X.

Notation 4.3.17. Fix a countably infinite set V whose elements are called *variables*. Let Σ be a signature, and let $\mathcal{E}$ be a set of basic equations over V. We define a quotient functor $H_\Sigma/\mathcal{E}$ of H_Σ as follows: given a set X, form the smallest equivalence $\equiv_{\mathcal{E},X}$ on $H_\Sigma X$ such that

$$\sigma(s(v_0),\ldots,s(v_{n-1})) \equiv_{\mathcal{E},X} \tau(s(w_0),\ldots,s(w_{m-1})),$$

for every $s\colon V \to X$ and $\sigma(x_0,\ldots,x_{n-1}) = \tau(y_0,\ldots,y_{m-1})$ in $\mathcal{E}$. We can think of s as a *substitution* of elements of X for the variables in V, and we identify all substitution instances of all pairs of basic equations. The functor $H_\Sigma/\mathcal{E}$ maps a set X to the set $H_\Sigma X/{\equiv_{\mathcal{E},X}}$ of equivalence classes $[t]$ of basic terms t over X. To a function $f\colon X \to Y$ it assigns the function defined by

$$[\sigma(s(v_0),\ldots,s(v_{n-1}))] \mapsto [\sigma(f(s(v_0)),\ldots,f(s(v_{n-1})))].$$

Remark 4.3.18. In order to show that two elements of A have the same image in the terminal F-coalgebra, it is enough to exhibit a relation R on A relating them and such that $R(a,b)$ implies $[\alpha(a)] \equiv_{\mathcal{E},A/R} [\alpha(b)]$, where $[\alpha(a)]$ denotes $H_\Sigma q_R(\alpha(a))$ and similarly for $[\alpha(b)]$.

Proposition 4.3.19 [40, Thm. 3.9]**.** *Every finitary set functor F can be presented by a finitary signature Σ and a set $\mathcal{E}$ of basic equations in the sense that F and $H_\Sigma/\mathcal{E}$ are naturally isomorphic functors.*

Example 4.3.20. (1) The functor $\mathscr{P}_3$ is presented by the signature Σ with one constant c and one binary operation symbol $*$ modulo the equation $x * y = y * x$ expressing commutativity of $*$.

(2) $\mathscr{P}_\mathrm{f}$ is presented by the signature $\Sigma = \{\sigma_n\}_{n<\omega}$ with σ_n of arity n (see Example 4.3.14(2)), and all basic equations

$$\sigma_n(x_0, \ldots, x_{n-1}) = \sigma_m(y_0, \ldots, y_{m-1})$$

for which the sets $\{x_0, \ldots, x_{n-1}\}$ and $\{y_0, \ldots, y_{m-1}\}$ are equal. For example, one of the equations is $\sigma_2(x, y) = \sigma_3(x, y, x)$.

(3) Let $FX = \coprod_{\sigma\in\Sigma} X^k/G_k$ be an analytic functor (see Example 3.2.13), where Σ is a signature, k is the arity of σ, and G_k is the given group of permutations on $k = \{0, 1, \ldots, k-1\}$. Then F is presented by the signature Σ and the following basic equations:

$$\sigma(x_1, \ldots, x_k) = \sigma(x_{p(1)}, \ldots, x_{p(k)}) \qquad \text{for all } \sigma \in \Sigma \text{ and all } p \in G_\sigma.$$

(4) The Aczel–Mendler $(-)^3_2$ functor (Example 4.3.14(4)) is presented by the two basic equations

$$\sigma(x, x) = \tau(x, x) \qquad \text{and} \qquad \tau(x, x) = \varrho(x, x).$$

We now consider algebras of set functors in connection with functor presentations.

Remark 4.3.21. (1) Given a presentation $F \cong H_\Sigma/\mathcal{E}$, the category of F-algebras is equivalent to the variety of all Σ-algebras satisfying the equations of $\mathcal{E}$ in the usual sense.

(2) Conversely, every variety of Σ-algebras presented by basic equations is equivalent to $\mathsf{Alg}\, F$ for some finitary set functor F. This is proved by Adámek and Trnková [25, Chapter III].

Given a presentation $\varepsilon\colon H_\Sigma \twoheadrightarrow F$ of a set functor F, we can consider the category $\mathsf{Alg}\, F$ as a full subcategory of $\mathsf{Alg}\, H_\Sigma$, following Remark 4.3.21(1). More precisely, we have a full embedding assigning to every algebra $\alpha\colon FA \to A$ for F the corresponding Σ-algebra $\alpha \cdot \varepsilon_A\colon H_\Sigma A \to A$ and defined on homomorphisms by $f \mapsto f$.

Analogously to quotient coalgebras (see Remark 4.2.3), we have the concept of *quotient algebra* of $\alpha\colon FA \to A$: it is an epimorphism $e\colon A \twoheadrightarrow \bar{A}$ in $\mathscr{A}$ for which an algebra structure $\bar{\alpha}$ exists making it a homomorphism:

$$\begin{array}{ccc} FA & \xrightarrow{\alpha} & A \\ {\scriptstyle Fe}\downarrow & & \downarrow{\scriptstyle e} \\ F\bar{A} & \xrightarrow{\bar{\alpha}} & \bar{A} \end{array}$$

Since all set functors preserve epimorphisms, Fe is an epimorphism, and hence the structure morphism $\bar{\alpha}$ is unique.

Proposition 4.3.22 [19]. *Given a presentation $\varepsilon\colon H_\Sigma \twoheadrightarrow F$, the initial algebra μF is the largest quotient of the initial Σ-algebra lying in* Alg F.

More concretely,

$$\mu F = \mu H_\Sigma/\!\sim,$$

where for two finite Σ-trees s, t we have

$$s \sim t \quad \text{iff } s \text{ and } t \text{ can be equated by finitely many applications of the basic equations for } \varepsilon. \tag{4.2}$$

This means that $s = t$ is derivable using the standard rules of equational logic.

Example 4.3.23. The initial algebra for $\mathscr{P}_3$ is the quotient of $\mu X.(X \times X + 1)$, the algebra of all finite binary ordered trees of Example 2.2.15 modulo the least congruence merging trees $x * y$ and $y * x$. Consequently, the linear ordering of children is simply forgotten:

$$\mu\mathscr{P}_3 = \text{all finite binary unordered trees.}$$

Remark 4.3.24. Given a presentation $\varepsilon\colon H_\Sigma \twoheadrightarrow F$ of a finitary set functor, then a similar description holds for the terminal coalgebra [19, Thm. 3.15]; we have

$$\nu F = T_\Sigma/\!\approx,$$

where $\approx$ is the congruence of finite and *infinite* applications of the basic equations. Of course, we have to make clear what is meant by 'infinite application' of basic equations:

Notation 4.3.25. Recall from Theorem 3.3.10 that the limit projection $\ell_n\colon T_\Sigma = \nu H_\Sigma \to H_\Sigma^n 1$ assigns to every Σ-tree t the cutting of t at level n, where all leaves of that level are relabelled by $\perp$. Thus $\ell_n(s)$ is a finite $\Sigma_\perp$-tree where $\Sigma_\perp$ is the signature Σ with an additional nullary symbol $\perp$. Given a presentation $\varepsilon\colon H_\Sigma \twoheadrightarrow F$, we define an equivalence relation $\approx$ on T_Σ as follows: for two Σ-trees s and t, we put

$$s \approx t \qquad \text{iff for all } n < \omega \text{ we have } \ell_n(s) \sim \ell_n(t). \tag{4.3}$$

Here $\sim$ is the congruence of (4.2) for the quotient functor

$$\varepsilon_X + \{\bot\} \colon H_{\Sigma_\bot} X \to FX + \{\bot\}.$$

Recall from Corollary 4.3.9 that the terminal coalgebra $\tau \colon T_\Sigma \twoheadrightarrow H_\Sigma T_\Sigma$ (of all Σ-trees) is a weakly terminal coalgebra for F.

Theorem 4.3.26 [19]. *Given a presentation $\varepsilon \colon H_\Sigma \twoheadrightarrow F$ of a finitary set functor F, the terminal coalgebra of F is a quotient of the F-coalgebra T_Σ of all Σ-trees modulo the equivalence $\approx$ above; in symbols:*

$$\nu F = T_\Sigma / \approx.$$

Example 4.3.27. For the finite power-set functor $\mathscr{P}_\mathsf{f}$ consider the presentation

$$\varepsilon_X \colon X^* \to \mathscr{P}_\mathsf{f} X$$

of Example 4.3.14. Since Σ has one n-ary operation for every n, we know that the terminal coalgebra T_Σ is (isomorphic to) the coalgebra of all finitely branching ordered trees. Recall our concept of extensional quotient (see Example 3.2.9(2)) and our convention that isomorphic trees are identified. For finite trees t and u, the equivalence $\sim$ of (4.2) is easily seen to be

$$t \sim u \qquad \text{iff the extensional quotients of } t \text{ and } u \text{ are equal.}$$

By applying (4.3) we conclude that for arbitrary trees t and u in T_Σ we have in the notation of Example 4.2.10

$$t \approx u \qquad \text{iff } \partial_n t = \partial_n u \text{ for all } n < \omega.$$

We conclude that $\nu \mathscr{P}_\mathsf{f} = T_\Sigma / \approx$.

Remark 4.3.28. The equivalence $\approx$ ignores the linear order on the children of nodes. Therefore, the example above can be formulated using the coalgebra D of all unordered finitely branching trees of Example 4.2.1(3): denoting by $t \approx u$ the above equivalence for unordered trees, we have

$$\nu \mathscr{P}_\mathsf{f} = D / \approx.$$

This description of $\nu \mathscr{P}_\mathsf{f}$ is due to Barr [72, Ex. 4.6].

Proposition 4.3.29. *The terminal coalgebra for $\mathscr{P}_\mathsf{f}$ is the quotient $D/\approx$ of the coalgebra D of all finitely branching trees (see Example 4.2.1(3)) modulo the greatest congruence.*

Remark 4.3.30. There is an interesting connection of Theorem 4.3.26 to the terminal-coalgebra ω^{op}-chain. Let $\varepsilon \colon H_\Sigma \twoheadrightarrow F$ be a presentation. Firstly, ε

induces a natural transformation γ from the terminal-coalgebra ω^{op}-chain of H_Σ to that of F by induction: $\gamma_0 = \mathsf{id}_1$ and

$$\gamma_{n+1} = (H_\Sigma H_\Sigma^n 1 \xrightarrow{\varepsilon_{H_\Sigma^n 1}} F H_\Sigma^n 1 \xrightarrow{F\gamma_n} F F^n 1). \tag{4.4}$$

Hence we obtain the following commutative diagram:

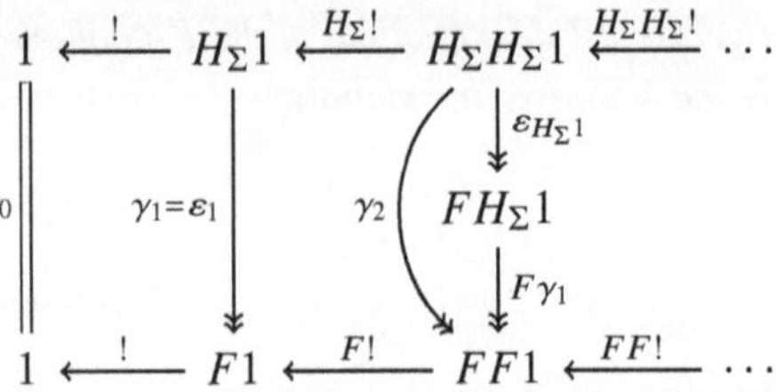

It is not difficult to prove that $\gamma_n \colon H_\Sigma^n 1 \twoheadrightarrow F^n 1$ is an epimorphism whose kernel equivalence is given by the congruence $\sim$ of finite applications of basic equations from (4.2) restricted to trees of height at most n. Thus, for every $s, t \in T_\Sigma$ we obtain that

$$s \approx t \qquad \text{iff } \gamma_n \cdot \ell_n(s) = \gamma_n \cdot \ell_n(t) \text{ for all } n < \omega,$$

where ℓ_n cuts trees at level n and γ_n is the quotient of finite application of basic equations.

Example 4.3.31. (1) We continue Example 4.3.23 where $F = \mathscr{P}_3$. Here is a pair of infinite trees which are equivalent in T_Σ:

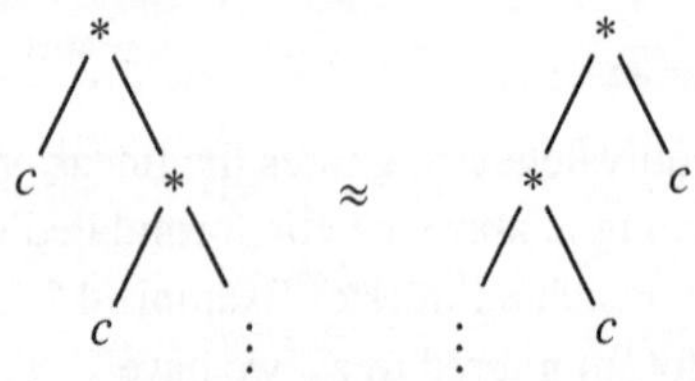

Indeed, we can check that this pair is related by $\approx$, since we have

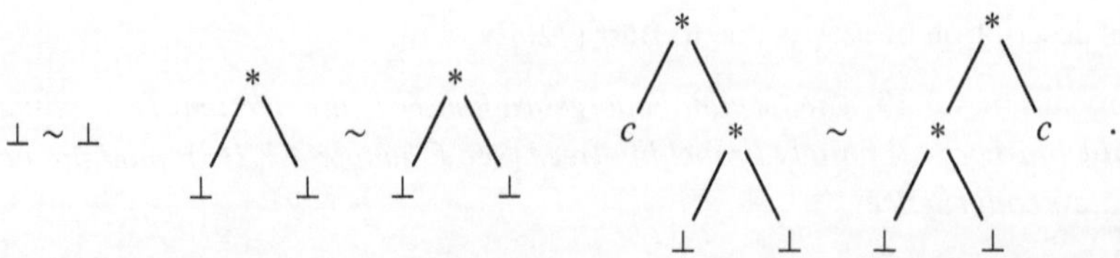

(2) For the finite power-set functor $\mathscr{P}_\mathsf{f}$ recall the presentation $\varepsilon_X \colon X^* \to \mathscr{P}_\mathsf{f} X$

of Example 4.3.14(1). Here we have, for example,

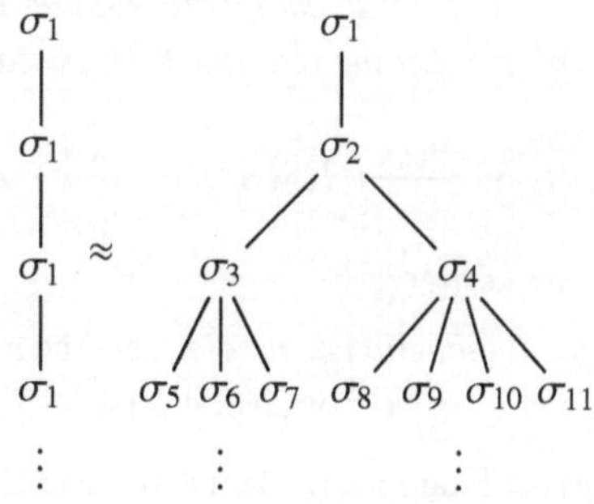

similarly as in (1) above.

(3) We have introduced analytic functors F on Set in Example 3.2.13. By Theorem 4.3.26 we have a direct description of νF: Let $FX = \coprod_{\sigma\in\Sigma} X^k/G_k$ be an analytic functor, where k is the arity of σ and G_k is the given group of permutations on k. Then the terminal coalgebra is the quotient

$$\nu F = T_\Sigma/\approx$$

of the Σ-tree coalgebra modulo the equivalence $\approx$ analogous to $\sim$ of (4.2) but allowing infinitely many permutations of children of nodes; in other words, νF is the coalgebra of all Σ-trees modulo permutations of children of any σ-labelled node (using elements of the permutation group asssociated with σ).

(4) In particular, for the bag functor $\mathscr{B}X = \coprod_{n\in\mathbb{N}} X^n/S_n$ (cf. Example 3.2.11), the terminal coalgebra $\nu\mathscr{B}$ is carried by the set of all unordered finitely branching trees (we saw in Example 4.2.1(3) that the same set carries a weakly terminal $\mathscr{P}_\mathsf{f}$-coalgebra). In fact, the corresponding polynomial functor is $H_\Sigma X = X^*$. We know that νH_Σ is the coalgebra of all finitely branching trees. Hence $\nu\mathscr{B}$ is the quotient coalgebra of νH_Σ given by allowing arbitrary permutations of children. This means that the carrier consists of all unordered trees.

Remark 4.3.32. We have seen congruences on a coalgebra in Remark 4.2.3. To determine whether or not a given relation is a congruence is often tedious, as is the task of coming up with a congruence that relates two points in a given coalgebra. The point of Theorem 4.3.34 below is to make it easier to give

Notation 4.3.33. For every set A and every relation $R \subseteq A \times A$, we write $A \xrightarrow{q_R} A/R$ for the coequalizer of the evident projections $R \rightrightarrows A$. Thus, A/R is the quotient set of A by the the smallest equivalence relation which includes R, and q_R is the canonical map. Notice that with this notation every presentation of a functor F has a component $\varepsilon_{A/R}\colon H_\Sigma(A/R) \twoheadrightarrow F(A/R)$.

Theorem 4.3.34 [283]. *Let $\varepsilon\colon H_\Sigma \twoheadrightarrow F$ be a presentation. Let (A,α) be an H_Σ-coalgebra, so that $(A, \varepsilon_A \cdot \alpha)$ is an F-coalgebra, and let $h\colon (A, \varepsilon_A \cdot \alpha) \to \nu F$ be the unique homomorphism. Let the relation R be contained in the kernel of*

$$A \xrightarrow{\ \alpha\ } H_\Sigma A \xrightarrow{H_\Sigma q_R} H_\Sigma(A/R) \xrightarrow{\varepsilon_{A/R}} F(A/R)\ .$$

Then R is contained in the kernel of h.

Theorem 4.3.34 gives a *coinduction principle*: Let $\mathcal{E}$ be a set of basic equations which present F, and let $\equiv_{\mathcal{E},A/R}$ be from Notation 4.3.17.

Example 4.3.35. We revisit Example 4.3.31(1). We take Σ to consist of a binary symbol $*$ (written in infix notation) and a constant c. The set $\mathcal{E}$ has just one basic equation: $x * y = y * x$. Consider the coalgebra is (A,α), where $A = \{a,b,z\}$, $\alpha(a) = a * z$, $\alpha(b) = z * b$, and $\alpha(z) = c$. The images of a and b in νH_Σ are the two infinite trees in Example 4.3.31(1); call these t_1 and t_2. We also showed what it would take to prove that $t_1 \approx t_2$: it would require an inductive argument that has not been presented. But given that $t_1 \approx t_2$, we see that t_1 and t_2 are identified in $\nu\mathscr{P}_3$. The point of this example is to verify this using Remark 4.3.18 and Theorem 4.3.34.

Consider the relation $R = \{(a,b)\}$ on A. Let us verify that $[\alpha(a)] \equiv_{\mathcal{E},A/R} [\alpha(b)]$. We have

$$[a * z] = [a] * [z] \equiv_{\mathcal{E},A/R} [z] * [a] = [z] * [b] = [z * b].$$

The '$\equiv_{\mathcal{E},A/R}$' is due to the equation $x * y = y * x$. For the second '=', note that A/R has two equivalence classes, and $q_R(a) = q_R(b)$. That is, $[a] = [b]$.

We conclude from this and Theorem 4.3.34 that for the $\mathscr{P}_3$-coalgebra $(A, \varepsilon_A \cdot \alpha)$, a and b have the same image in $\nu\mathscr{P}_3$.

Example 4.3.36. We return to analytic functors, see Examples 3.2.13 and 4.3.20(3). Consider the analytic functor $FX = X^3/G_3$ where $G_3 = A_3$, the three permutations of $\{1,2,3\}$ which are not transpositions. (For example, the permutation p given by $1 \mapsto 2, 2 \mapsto 3, 3 \mapsto 1$ belongs to A_3, but $1 \mapsto 3, 2 \mapsto 2, 3 \mapsto 1$ does not.) So F has a presentation $\varepsilon\colon H_\Sigma \twoheadrightarrow F$, where Σ consists of a single ternary operation symbols σ.

Consider the H_Σ-coalgebra (A,α), where $A = \{a,b,c\}$, $\alpha(a) = \alpha(b) = \sigma(a,b,c)$, and $\alpha(c) = \sigma(c,b,a)$. We claim that b and c have the same image in the terminal F-coalgebra. We use $R = \{(b,c)\}$, and then we verify:

$$\begin{aligned} \sigma(a,b,c) &\equiv \sigma(b,c,a) && \text{(using } p) \\ &\equiv \sigma(c,b,a) && \text{(since } b \equiv c,\ c \equiv b, \text{ and } a \equiv a). \end{aligned}$$

The point again is that this argument is simpler than one which uses the cuttings of the images of b and c in the terminal F-coalgebra.

4.4 Iterating the Terminal-Coalgebra Chain to $\omega + \omega$

When a functor F preserves the limit of its terminal coalgebra ω^{op}-chain, then this limit is a terminal coalgebra (see Theorem 3.3.4). The starting point of this section is that this limit preservation does not always happen, even for finitary set functors such as $\mathscr{P}_{\mathsf{f}}$.

However, Worrell [318] provided a construction of the terminal coalgebra that works for every finitary set functor. The proof presented here is simpler than Worrel's original one. Actually, he proved a more general result about *accessible* set functors that we present later (cf. Theorem 6.5.11).

We begin with a general point about set functors due to Trnková, and then an application to finitary set functors. Recall that an *intersection* of a family of subobjects $s\colon S_i \rightarrowtail X$ $(i \in I)$ is their wide pullback. We use the suggestive notation $\bigcap_{i\in I} s_i$ for intersections. In Set we call an intersection *nonempty* if the domain of the wide pullback is nonempty.

Proposition 4.4.1 [291]. *Every set functor preserves finite nonempty intersections.*

We prove this in Theorem C.2.2 on page 554.

Example 4.4.2. Empty intersections need not be preserved. For example, let $\mathscr{P}'_{\mathsf{f}}$ be the functor obtained from $\mathscr{P}_{\mathsf{f}}$ by changing just the value at $\emptyset$ to $\mathscr{P}'_{\mathsf{f}}\emptyset = \emptyset$. Then $\mathscr{P}'_{\mathsf{f}}$ does not preserve the (empty) intersection of the two maps $t, f\colon 1 \to 2$.

Theorem 4.4.3. *Every finitary set functor preserves nonempty intersections.*

Proof Given a set A, its subobjects are represented by the inclusion maps of subsets $S \subseteq A$. We use lower-case letters and write $s\colon S \hookrightarrow A$ for the inclusion map of S into A.

Let $\mathcal{S}$ be a collection of subsets of A with intersection $T \neq \emptyset$. Then $t\colon T \hookrightarrow A$ is a split monomorphism, hence so is Ft. Thus, in order to prove that F preserves this intersection, we just need to verify that $x \in Ft[FT]$ holds for every element $x \in FA$ such that $x \in Fs[FS]$ for all $S \in \mathcal{S}$.

(1) Suppose that $\mathcal{S}$ contains a finite member S_0, and let us write the finite set $S_0 \setminus T$ as $\{a_1, \ldots, a_k\}$. From $a_i \notin T$ we conclude that there exists an $S_i \in \mathcal{S}$ such that $a_i \notin S_i$ $(i = 1, \ldots, k)$. This implies that $T = S_0 \cap S_1 \cap \cdots \cap S_k$. Since F preserves this intersection by Proposition 4.4.1, we obtain that $x \in Ft[FT]$.

(2) Next suppose that all members of $\mathcal{S}$ are infinite. Fix an element $y \in T$.

Given $S \in \mathcal{S}$, we use that F is finitely bounded (Proposition 4.3.4) to deduce from $s \in Fs[FS]$ that there exists a finite subset $S' \subseteq S$ such that $x \in Fs'[FS']$. Moreover, we can assume that $y \in S'$ (otherwise we replace S' by $S' \cup \{y\}$). The collection of all these subsets $S' \subseteq S$ for $S \in \mathcal{S}$ has an intersection $T' \subseteq T$ containing y. By item (1), F preserves that intersection. Thus, $x \in Ft'[FT'] \subseteq Ft[FT]$ as desired. □

Example 4.4.4. (1) There are 'natural' set functors that fail to preserve nonempty intersections; for example, the filter functor. A *filter* on a set X is a collection $\mathbb{F}$ of subsets of X closed under finite intersection and upwards closed. (In particular, $\mathscr{P}X$ is a filter on X. In set theory filters are often requested to consist of *nonempty* subsets, which excludes $\mathscr{P}X$. With this exception, our definition conforms with the set-theoretical one.) The filter functor $\mathscr{F}$ assigns to a set X the set $\mathscr{F}X$ of all filters on X, and for a map $f: X \to Y$ we have

$$\mathscr{F}f(\mathbb{F}) = \{M \subseteq Y : f^{-1}[M] \in \mathbb{F}\}.$$

This functor does not preserve the intersection $\{0\}$ of the subsets $A_n = \{0\} \cup \{k : k \geq n\}$ of $\mathbb{N}$. Denote by $i: \{0\} \hookrightarrow \mathbb{N}$ and $a_n: A_n \hookrightarrow \mathbb{N}$ the inclusion maps. There exists filters in $\mathscr{F}\mathbb{N}$ that lie in the image of each $\mathscr{F}a_n$, but not in the image of $\mathscr{F}i$. For example, the Fréchet filter consisting of all subsets of $\mathbb{N}$ with a finite complement.

(2) Empty intersections are not preserved by finitary set functors in general. For example, the functor $C_{0,1}$ which takes the empty set to itself and every nonempty set to 1 does not preserve the intersection of the coproduct injections of $1 + 1$.

Notation 4.4.5. (1) Recall from Remark 3.3.5 the limit of the terminal ω^{op}-chain and put

$$V_\omega = \lim_{n \in \omega^{\mathrm{op}}} F^n 1 \qquad \text{with limit projections } \ell_n : V_\omega \to F^n 1 \text{ for all } n \in \omega^{\mathrm{op}}. \tag{4.5}$$

We obtain a unique map $m: FV_\omega \to V_\omega$ having the property that for all $n \in \omega^{\mathrm{op}}$, the triangles below commute:

$$\begin{array}{ccc} FV_\omega & \xrightarrow{\ m\ } & V_\omega \\ & {\scriptstyle F\ell_n} \searrow & \downarrow {\scriptstyle \ell_{n+1}} \\ & & F^{n+1}1 \end{array} \tag{4.6}$$

(2) We put $V_{\omega+1} = FV_\omega$, $V_{\omega+2} = FV_{\omega+1}$, and so forth and form the following ω^{op}-chain:

$$V_\omega \xleftarrow{\ m\ } V_{\omega+1} \xleftarrow{\ Fm\ } V_{\omega+2} \xleftarrow{\ FFm\ } \cdots \tag{4.7}$$

Its limit is denoted by $V_{\omega+\omega} = \lim_{n<\omega} V_{\omega+n}$ with the limit cone $\bar{l}_n : V_{\omega+\omega} \to$

$V_{\omega+n}$, for $n < \omega$. (In Definition 6.5.1 we introduce V_i for all ordinal numbers i. The limit projections $\bar{l}_n$ are denoted by $v_{\omega+\omega,\omega+n}$ there.)

We know that V_ω is the terminal F-coalgebra whenever the limit in (4.5) is preserved by F (see Theorem 3.3.4); equivalently, whenever m is invertible. We shall soon prove that for every finitary set functor F, m is a split monomorphism. Hence $V_{\omega+\omega}$ is the intersection of the chain (4.7) of monomorphisms. Moreover, we obtain that $V_{\omega+\omega} = \nu F$. But first let us record that m is not always invertible:

Example 4.4.6 [17]. For the finite power-set functor, the morphism m is not epic (whence not invertible). To see this, consider an arbitrary element $y \in V_{\omega+1} = \mathscr{P}_{\mathsf{f}} V_\omega$, say of cardinality k. For $x = m(y)$ we conclude from $\ell_{n+1} \cdot m = \mathscr{P}_{\mathsf{f}} \ell_n$ that the set $\ell_{n+1}(x)$ satisfies $|\ell_{n+1}(x)| \le k$ (for all $n \in \mathbb{N}$): Indeed, $\ell_{n+1}(x) = \mathscr{P}_{\mathsf{f}} \ell_n(y) = \ell_n[y]$. Thus, we just need to find $x \in V_\omega$ such that the sets $\ell_{n+1}(x)$, $n \in \mathbb{N}$, have unbounded cardinalities to see that m is not epic.

We define a sequence of elements $x_n \in \mathscr{P}_{\mathsf{f}}^n 1$ as follows: x_0 is unique, and x_{n+1} is the largest element of $\mathscr{P}_{\mathsf{f}}(\mathscr{P}_{\mathsf{f}}^n 1)$: $x_{n+1} = \mathscr{P}_{\mathsf{f}}^n 1$ of cardinality 2^n. The connecting maps $p_n = \mathscr{P}_{\mathsf{f}}^n !$ of the terminal-coalgebra ω^{op}-chain are all epic, and for $n \ge 1$ they are monotone with respect to the subset inclusion on $\mathscr{P}_{\mathsf{f}}(\mathscr{P}_{\mathsf{f}}^n 1)$ (because $p_{n+1}(x) = \mathscr{P}_{\mathsf{f}} p_n(x) = p_n[x]$). Consequently, $p_{n+1}(x_{n+1}) = x_n$ for every $n \in \mathbb{N}$. In $V_\omega = \lim \mathscr{P}_{\mathsf{f}}^n 1$ we thus have a unique element x such that $\ell_n(x) = x_n$. The cardinality of $\ell_{n+1}(x)$ is 2^n for every x, as desired.

Lemma 4.4.7. *Let $\ell_n \colon L \to X_n$ be a limit cone of an ω^{op}-chain in* Set. *For every finite subset $s \colon S \hookrightarrow L$ there exists n such that $\ell_n \cdot s$ is a monomorphism.*

Proof Since the limit cone is collectively monic, for every pair of distinct elements in S we have $k \in \omega$ such that l_k distinguishes that pair. Let n be the maximum of all k's (ranging over $S \times S$). □

Lemma 4.4.8. *Let F be a finitary set functor. Then for every ω^{op}-chain $(X_n)_{n<\omega}$ the canonical map $F(\lim X_n) \to \lim FX_n$ is monic.*

Proof Given an ω^{op}-chain $(X_n)_{n<\omega}$ in Set denote by $\ell_n \colon L \to X_n$ its limit cone and by $p_n \colon \lim_{n<\omega} FX_n \to FX_n$ the limit cone of its image under F. The canonical map $m \colon FL \to \lim FX_n$ is the unique map such that $p_n \cdot m = F\ell_n$ for every $n < \omega$. We prove that the cone $F\ell_n \colon FL \to FX_n$ is collectively monic, which implies that m is monic. Assuming that for all k, $F\ell_k(x) = F\ell_k(y)$, we show that $x = y$. By the assumption that F is finitary, there is a finite subset $s \colon S \hookrightarrow L$ and $x', y' \in FS$ such that $x = Fs(x')$ and $y = Fs(y')$. Without loss of generality we may assume that S is nonempty. We see that

$$(F\ell_k \cdot Fs)(x') = F\ell_k(x) = F\ell_k(y) = (F\ell_k \cdot Fs)(y').$$

By Lemma 4.4.7, there exists some n such that $\ell_n \cdot s$ is monic, thus $S \neq \emptyset$ implies that it is a split monomorphism. Hence $F(\ell_n \cdot s)$ is monic, too. So $x' = y'$, and thus $x = y$. □

Corollary 4.4.9 [318]. *Let F be a finitary set functor. Then $m\colon FV_\omega \to V_\omega$ in (4.6) is a split monomorphism.*

Proof (1) If F is constantly $\emptyset$, the statement is trivial. If not, then $FY \neq \emptyset$ for every $Y \neq \emptyset$ (given X such that $FX \neq \emptyset$, for every set Y we have a map from Y to X, thus, one from FY to FX). Consequently, $V_\omega \neq \emptyset$. Indeed, since $F1 \neq \emptyset$ we have a coalgebra $\alpha\colon 1 \to F1$. The canonical cone (Construction 3.3.2) $\alpha_n\colon 1 \to F^n 1$ induces a morphism $1 \to V_\omega$.

(2) From the definition of m in (4.6), Lemma 4.4.8 yields that it is a monomorphism. It splits since $FV_\omega \neq \emptyset$. □

Recall from Remark 3.3.5 that whenever F preserves limits of ω^{op}-chains, its terminal coalgebra $\nu F = V_\omega$ has the structure τ defined by $F\ell_n \cdot \tau = \ell_{n+1}$. Here is an analogous fact.

Theorem 4.4.10 [318]. *Every finitary set functor F has the terminal coalgebra*

$$\nu F = V_{\omega+\omega} = \lim_{n<\omega} V_{\omega+n}.$$

The coalgebra structure τ is determined by the following commutative triangles

$$\begin{array}{ccc} \nu F & \xrightarrow{\ \tau\ } & F(\nu F) \\ & \searrow^{\bar{l}_{n+1}} & \downarrow{\scriptstyle F\bar{l}_n} \\ & & FV_{\omega+n} \end{array} \qquad (n < \omega) \qquad (4.8)$$

Proof (1) We are going to prove in part (2) that F preserves the limit $V_{\omega+\omega}$. It then follows that there is a unique isomorphism τ for which the above triangles commute.

The proof that $(\nu F, \tau)$ is a terminal coalgebra is then analogous to the proof of Theorem 3.3.4. There we used the canonical cone $\alpha_n\colon A \to F^n 1$ for a given coalgebra $\alpha\colon A \to FA$ defined in Construction 3.3.2. We extend this to a cone $\alpha_{\omega+n}\colon A \to V_{\omega+n}$ as follows. The cone α_n induces a unique morphism $\alpha_\omega\colon A \to V_\omega$ with $\ell_n \cdot \alpha_\omega = \alpha_n$ for all $n < \omega$, and given $\alpha_{\omega+n}$ we define $\alpha_{\omega+n+1} = F\alpha_{\omega+n} \cdot \alpha\colon A \to V_{\omega+n+1}$. To see that this is a cone for (4.7) it suffices to show that $\alpha_\omega = m \cdot \alpha_{\omega+1}$; the rest then follows by an easy induction. We prove the desired equation by post-composing it by every limit projection ℓ_{n+1}, $n < \omega$, and then use that those limit projections are collectively monic:

$$
\begin{aligned}
\ell_{n+1} \cdot \alpha_\omega &= \alpha_{n+1} && \text{(def. of } \alpha_\omega\text{)}\\
&= F\alpha_n \cdot \alpha && \text{(def. of } \alpha_{n+1}\text{)}\\
&= F\ell_n \cdot F\alpha_\omega \cdot \alpha && \text{(def. of } \alpha_\omega\text{)}\\
&= F\ell_n \cdot \alpha_{\omega+1} && \text{(def. of } \alpha_{\omega+1}\text{)}\\
&= \ell_{n+1} \cdot m \cdot \alpha_{\omega+1} && \text{(def. of } m\text{).}
\end{aligned}
$$

We now obtain a unique morphism $\bar{\alpha}\colon A \to V_{\omega+\omega}$ such that $\bar{l}_n \cdot \bar{\alpha} = \alpha_{\omega+n}$ for all $n < \omega$. Then $\bar{\alpha}\colon A \to V_{\omega+\omega}$ is a coalgebra homomorphism: in the following diagram

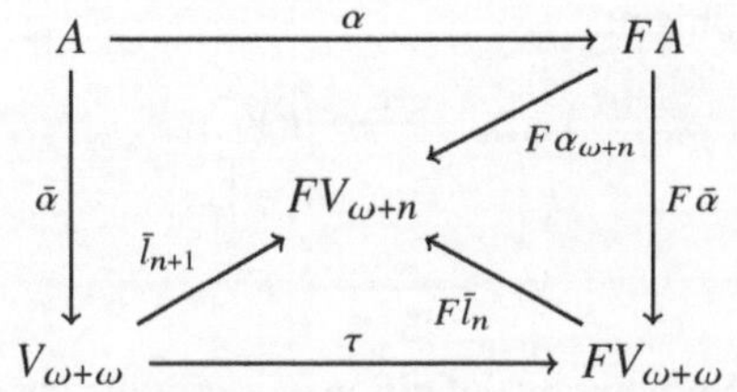

all inner parts commute, thus the outside does: recall that F preserves the above limit, hence the morphisms $F\bar{l}_n$ form a limit cone, which is thus collectively monic. To verify uniqueness, let $\widetilde{\alpha}\colon A \to V_{\omega+\omega}$ be another coalgebra homomorphism. One first proves by induction that for every $n < \omega$ we have

$$\alpha_n = (A \xrightarrow{\tilde{\alpha}} V_{\omega+\omega} \xrightarrow{\bar{l}_0} V_\omega \xrightarrow{\ell_n} F^n 1).$$

This shows that $\bar{l}_0 \cdot \widetilde{\alpha} = \alpha_\omega$ because both sides are merged by all limit projections ℓ_n. Using this as the base case, another induction proof then shows that for all $n < \omega$ we have

$$\alpha_{\omega+n} = (A \xrightarrow{\tilde{\alpha}} V_{\omega+\omega} \xrightarrow{\bar{l}_n} V_{\omega+n}).$$

Thus the family $\bar{l}_n$ merges $\bar{\alpha}$ and $\widetilde{\alpha}$, which proves $\widetilde{\alpha} = \bar{\alpha}$ (since $\bar{l}_n$ forms a limit cone).

(2) The proof that F preserves the limit $V_{\omega+\omega}$ is trivial in case F is constant with value $\emptyset$. Assuming the contrary, we have clearly $F1 \neq \emptyset$, thus there is (at least one) coalgebra $\alpha\colon 1 \to F1$. This implies $V_{\omega+\omega} \neq \emptyset$ due to $\bar{\alpha}\colon 1 \to V_{\omega+\omega}$, whence $V_{\omega+n} \neq \emptyset$ for all n.

From Corollary 4.4.9 we see that m is a split monomorphism, hence so are Fm, FFm, etc. Thus, the chain (4.7) is a decreasing chain of nonempty subobjects $u_n\colon V_{\omega+n} \to V_\omega$ of V_ω where $u_0 = \mathsf{id}_{V_\omega}$ and $u_{n+1} = u_n \cdot F^n m$.

Therefore, the limit $V_{\omega+\omega}$ is simply the intersection of these subobjects, more precisely, for the first projection $\bar{l}_0 \colon V_{\omega+\omega} \to V_\omega$ we have

$$\bar{l}_0 = \bigcap_{n<\omega} u_n.$$

Since $V_{\omega+\omega} \neq \emptyset$ and F preserves nonempty intersections by Theorem 4.4.3, it preserves the limit $V_{\omega+\omega}$. □

Corollary 4.4.11. *For every finitary set functor F, we have a weakly terminal coalgebra $\bar{m} \colon V_\omega \to FV_\omega$ given by any splitting of $m \colon FV_\omega \to V_\omega$, say $\bar{m} \cdot m = \mathsf{id}$.*

Proof We observe that $\bar{l}_0 \colon (V_{\omega+\omega}, \tau) \to (V_\omega, \bar{m})$ of Notation 4.4.5 is a coalgebra homomorphism:

$$\begin{array}{ccc} V_{\omega+\omega} & \xrightarrow{\ \tau\ } & FV_{\omega+\omega} \\ {\scriptstyle \bar{l}_0}\downarrow & \overset{\bar{l}_1}{\searrow} & \downarrow{\scriptstyle F\bar{l}_0} \\ V_\omega & \xrightarrow{\ \bar{m}\ } & FV_\omega \end{array}$$

The upper triangle commutes by definition of τ. For the lower one use $\bar{l}_0 = m \cdot \bar{l}_1$ and post-compose this equation with $\bar{m}$. Since $(V_{\omega+\omega}, \tau)$ is a terminal coalgebra, this implies that $(V_\omega, \bar{m})$ is weakly terminal by Example 4.2.1(2). □

Remark 4.4.12. The limit V_ω is an algebra for F in a canonical sense with the structure $m \colon FV_\omega \to V_\omega$ in (4.6). Theorem 4.4.10 implies that for a finitary set functor the terminal coalgebra, considered as an algebra, is a cartesian subalgebra of V_ω. More precisely, recall that the limit projection $\bar{l}_0 \colon \nu F \to V_\omega$ is monic. The following square is a pullback:

$$\begin{array}{ccc} F(\nu F) & \xrightarrow{\ \tau^{-1}\ } & \nu F \\ {\scriptstyle F\bar{l}_0}\downarrow & \lrcorner & \downarrow{\scriptstyle \bar{l}_0} \\ FV_\omega & \overset{m}{\rightarrowtail} & V_\omega \end{array}$$

Here $\bar{l}_0$ is the first projection of $\nu F = \lim_{n<\omega} V_{\omega+n}$ (see Notation 4.4.5). The square commutes: combine $F\bar{l}_0 \cdot \tau = \bar{l}_1$ (see (4.8)) and $\bar{l}_0 = m \cdot \bar{l}_1$ to obtain

$$m \cdot F\bar{l}_0 = m \cdot \bar{l}_1 \cdot \tau^{-1} = \bar{l}_0 \cdot \tau^{-1}.$$

It is a pullback since m is a monomorphism and τ^{-1} an isomorphism.

Kripke polynomial functors. We turn to an important collection of set functors for which the terminal-coalgebra chain converges in $\omega + \omega$ steps: the Kripke polynomial functors. The name stems from *Kripke structures* used in modal logic: given a set At of atomic propositions a Kripke structure consists of a set X states, a function $X \to \mathscr{P}(\mathsf{At})$ assigning to every state the set of atomic propositions that hold in it, and a function $X \to \mathscr{P}(X)$ assigning to every set the set of successor states. Image-finite Kripke structures, in which every state only has finitely many successor states, can be expressed as coalgebras $\alpha\colon X \to \mathscr{P}(\mathsf{At}) \times \mathscr{P}_\mathsf{f} X$ (cf. Example 2.4.2(4)). Our definition below is a slight generalization of the (finite) Kripke polynomial functors presented by Jacobs [173, Def. 2.2.1]; we admit arbitrary products in lieu of just arbitrary exponents.

Definition 4.4.13. The *Kripke polynomial functors* F are the set functors built from the finite power-set functor, constant functors and the identity functor, by using products, coproducts, and composition. In other words, Kripke polynomial functors are built according to the following grammar:

$$F ::= \mathscr{P}_\mathsf{f} \mid A \mid \mathsf{Id} \mid \textstyle\prod_{i \in I} F_i \mid \coprod_{i \in I} F_i \mid FF,$$

where A ranges over all sets (and is interpreted as a constant functor) and I is an arbitrary index set.

Remark 4.4.14. The constant functors could be omitted from the above grammar since the constant functor on 1 is given by the empty product of functors. Taking a set A as the index set, we then obtain the constant functor on A as a coproduct: $X \mapsto \coprod_{a \in A} 1$.

Example 4.4.15. (1) The type functor $\mathscr{P}(\mathsf{At}) \times \mathscr{P}_\mathsf{f}(-)$ for image-finite Kripke structures is a Kripke polynomial functor.

(2) Every polynomial functor H_Σ for a (possibly infinitary) signature Σ (see Example 6.1.16) is obviously a Kripke polynomial functor: it is a coproduct of functors $F_n X = \Sigma_n \times X^n$, a product of a constant functor and n copies of the identity functor.

(3) Given an input alphabet Σ, the functor $FX = \{0, 1\} \times (\mathscr{P}_\mathsf{f} X)^\Sigma$ is a Kripke polynomial functor. This is the finitary version of the type functor of non-deterministic automata (Example 2.4.2(5)), which has the same finite coalgebras (cf. Remark 2.5.6).

Proposition 4.4.16. *Every Kripke polynomial functor F has the terminal co-algebra given by $\nu F = V_{\omega+\omega}$.*

Proof (1) The functor F preserves monomorphisms and wide intersections. This is clear for constant functors and for Id, and it is easy to see for $\mathscr{P}_\mathsf{f}$. Moreover, these properties are clearly preserved by products, coproducts, and composition.

(2) Let $(X_n)_{n<\omega}$ be an ω^{op}-chain in Set. Then the canonical morphism $F(\lim X_n) \to \lim FX_n$ is monic. This is obvious for constant functors and Id; for $\mathscr{P}_{\mathrm{f}}$ this holds by Lemma 4.4.8. Moreover, this property is preserved by products, coproducts, and composition. In particular, for a composition FG note that the canonical morphism for FG is the composition

$$FG(\lim X_n) \xrightarrow{Fm} F(\lim GX_n) \xrightarrow{m'} \lim FGX_n,$$

where m is the canonical morphism for G with respect to the given ω^{op}-chain and m' the one for F and the ω^{op}-chain $(GX_n)_{n<\omega}$. So this morphism is monic since both m and m' are so and F preserves monomorphisms by item (1).

(3) By item (2) we know $F\ell_n\colon FV_\omega \to F^{n+1}1$ is collectively monic. From the definition of $m\colon FV_\omega \to V_\omega$ in (4.6) it now follows that it is monic. Since F preserves monomorphisms, we know that Fm, FFm etc. are monic. Thus the chain (4.7) is a decreasing chain of subobjects $u_n\colon V_{\omega+n} \rightarrowtail V_\omega$ (like in the proof of Theorem 4.4.10). Therefore, the limit $V_{\omega+\omega} = \lim_{n<\omega} V_{\omega+n}$ is simply the intersection of these subobjects. Thus, from item (1) we know that F preserves this limit. Now proceed as in item (1) of the proof of Theorem 4.4.10. □

Corollary 4.4.17. *Every Kripke polynomial functor is a covarietor.*

Indeed, if F is a Kripke polynomial functor, so is every functor $F(-) \times A$. Now apply Proposition 2.6.15.

We will later see two related classes of endofunctors: the Hausdorff polynomial functors on Met (Definition 15.3.7) and the Vietoris polynomial functors on Top (Definition 15.2.10).

The terminal-coalgebra chains of H_Σ and F. Our final topic in this section returns to presentations of finitary set functors F and to the relation between the terminal-coalgebra chains of H_Σ and F which we have seen in Remark 4.3.30. Recall the natural transformation γ from the first chain to the second:

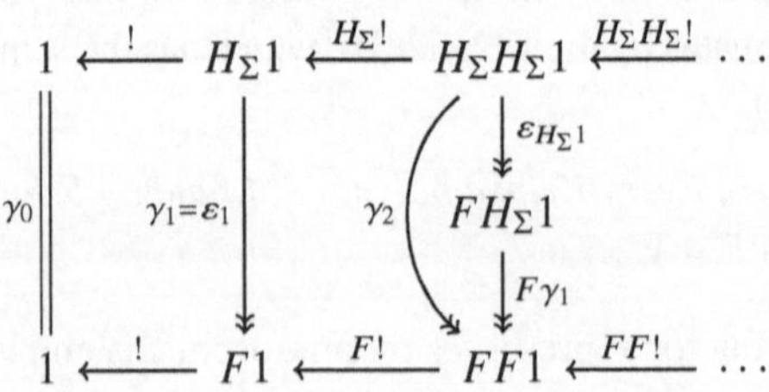

We extend γ to $\gamma^*\colon \nu H_\Sigma \to V_\omega$ so that for all n, $\ell_n \cdot \gamma^* = \gamma_n \cdot p_n$, where $p_n\colon \nu H_\Sigma \to H_\Sigma^n 1$ is the canonical projection. It is easy to check that the following square commutes:

$$\begin{array}{ccc} \nu H_\Sigma & \xleftarrow{\tau^{-1}} & H_\Sigma(\nu H_\Sigma) \\ & & \big\downarrow \varepsilon_{\nu H_\Sigma} \\ \gamma^* \big\downarrow & & F(\nu H_\Sigma) \\ & & \big\downarrow F\gamma^* \\ V_\omega & \xleftarrow[m]{} & FV_\omega \end{array}$$

Indeed, this is yet another argument using the fact that the cone (ℓ_n) is collectively monic. In addition, this fact follows from stronger facts which we shall see below on solution morphisms in corecursive algebras. In general, γ^* is *not epic*. However, there is a natural assumption that implies this condition, as we now present.

Definition 4.4.18. A function $f\colon X \to Y$ is *finite-to-one* if the inverse image of every $y \in Y$ is finite. A set functor F *preserves finite-to-one morphisms* if Ff is finite-to-one whenever f is. A presentation $\varepsilon\colon H_\Sigma \to F$ of a finitary functor is *finite-to-one* if every component ε_X is finite-to-one.

Example 4.4.19. We revisit the presentations in Example 4.3.20.

(1) The presentations of analytic functors are finite-to-one. Indeed, any presentation with the following two properties is finite-to-one: Σ has only finitely many symbols of each arity n, and all basic equations relate terms of the same arties.

(2) The presentation of $\mathscr{P}_\mathsf{f}$ is not finite-to-one: for all sets A and all $a \in A$, ε_A merges $\sigma_1(a)$, $\sigma_2(a, a)$, $\sigma_3(a, a, a)$, and so forth. Moreover, it follows from Corollary 4.4.21 that $\mathscr{P}_\mathsf{f}$ has no finite-to-one presentation.

Lemma 4.4.20. *Let $\varepsilon\colon H_\Sigma \twoheadrightarrow F$ be a finite-to-one presentation of the finitary set functor F. Then the following hold:*

(1) *The functor F preserves finite-to-one functions.*

(2) *Each $\gamma_n\colon H_\Sigma^n 1 \to F^n 1$ is finite-to-one.*

(3) *The map $\gamma^*\colon \nu H_\Sigma \to V_\omega$ is an epimorphism.*

Proof Every polynomial functor clearly preserves finite-to-one functions. A naturality argument shows that F, too, preserves finite-to-one functions. An easy induction on n using (4.4) shows that γ_n is finite-to-one.

The last part is the most important one. Let $x \in V_\omega$ and write x_n for $\ell_n(x)$. Construct a Σ-tree T as follows: for every n we have $\gamma_n^{-1} \subseteq H^n 1$, and we put

$$T = \gamma_0^{-1}(x_0) \cup \gamma_1^{-1}(x_1) \cup \cdots \cup \gamma_n^{-1}(x_n) \cup \cdots .$$

We may assume that the unions above are all pairwise disjoint. So T is the disjoint union of the inverse images of the points x_n under the maps γ_n. Put an arrow from t to u provided that for some n, $t \in \gamma_n^{-1}(x_n)$, $u \in \gamma_{n+1}^{-1}(x_{n+1})$, and $H^n!(u) = t$. We think of T as the tree of attempts to build an element of $y \in \nu H$ such that $\gamma_n(\ell_n(y)) = x_n$ for all n for the limit projections $\ell_n\colon \nu H_\Sigma \to H_\Sigma^n 1$. Notice that T is a finitely branching tree, since each γ_n is finite-to-one. And T has infinitely many nodes, since each ℓ_n is surjective. By Kőnig's Lemma [191], there is some infinite branch through T. This branch corresponds to some $y \in \nu H$ with the property mentioned above: $\gamma_n(\ell_n^H(y)) = x_n$. It follows that $\gamma^*(y) = x$. □

Corollary 4.4.21 [72]. *Let F be a finitary set functor which has a finite-to-one presentation. Then $\nu F = V_\omega$.*

Proof Consider $m\colon FV_\omega \to V_\omega$ from Notation 4.4.5. We have $\gamma^* \cdot \tau^{-1} = m \cdot F\gamma \cdot \varepsilon_{\nu H}$. By Lemma 4.4.20, $\gamma^* \cdot \tau^{-1}$ is an epimorphism. It follows that m, too, is epic. By Corollary 4.4.9, m is an isomorphism, and we are done by Theorem 3.3.4. □

Corollary 4.4.22. *For every analytic set functor F we have $\nu F = V_\omega$.*

Remark 4.4.23. The converse of Corollary 4.4.21 does not hold in general. There are finitary set functors F with $\nu F = V_\omega$ that have no finite-to-one presentation. For example, the nonempty finite power-set functor $\mathscr{P}'_\mathsf{f}$ (see Example 4.4.2) satisfies $\nu\mathscr{P}'_\mathsf{f} = V_\omega = 1$ since $\mathscr{P}'_\mathsf{f}1 = 1$. However, it has no finite-to-one presentation. For suppose that $\varepsilon\colon H_\Sigma \twoheadrightarrow \mathscr{P}'_\mathsf{f}$ is any finite-to-one presentation, let $\bar\Sigma$ extend the signature Σ by the constant symbol c, and extend ε to $\bar\varepsilon\colon H_{\bar\Sigma} \twoheadrightarrow \mathscr{P}_\mathsf{f}$ by $\varepsilon_X(c) = \emptyset$ for every set X. Then $\bar\varepsilon$ is a finite-to-one presentation of $\mathscr{P}_\mathsf{f}$, contradicting Example 4.4.6.

4.5 Finite Power-Set Functor: The Terminal Coalgebra versus the ω^{op}-Limit

The finite power-set functor $\mathscr{P}_\mathsf{f}$ is fascinating when it comes to

(1) its terminal coalgebra $\nu\mathscr{P}_\mathsf{f}$ as well as

(2) the limit V_ω of its terminal-coalgebra ω^{op}-chain, which contains $\nu\mathscr{P}_\mathsf{f}$ as a proper subset (cf. Example 4.4.6).

This can be seen from the number of papers devoted to various descriptions of these two coalgebras, see e.g. [1, 17, 48, 72, 316, 318]. Before turning to our descriptions, let us recall the terminal-coalgebra ω^{op}-chain from Example 4.2.10:

$$\mathscr{P}_f^n 1 = \text{all extensional trees of height} \le n.$$

The connecting maps

$$v_{n+1,n} \colon \mathscr{P}_f^{n+1} 1 \to \mathscr{P}_f^n 1, \qquad t \mapsto \partial_n t$$

are given by cutting the trees at level n and forming the extensional quotient. Indeed, this is trivial for $n = 0$, and the induction step is easy to see.

Remark 4.5.1. Throughout this section trees are understood to be non-ordered. Given a tree t, we continue to write $\partial_n t$ for the extensional quotient of the cutting of t at level n, extending the notation from Example 4.2.10 from finitely branching to all trees.

Similarly, we extend the equivalence $\approx$ from the set D of all finitely branching trees (Remark 4.3.28) to all trees: given trees t and u, we write $t \approx u$ if $\partial_n t = \partial_n u$ for all $n < \omega$.

A first description: V_ω is the set of compatible sequences of finite extensional trees. We can describe V_ω as the set of all sequences $(t_n)_{n<\omega}$ of finite extensional trees, where t_n has height at most n, and for all n we have $t_n = \partial_n(t_{n+1})$. The limit cone is given by the projections $V_\omega \to \mathscr{P}_f^n 1$ taking $(t_k)_{k<\omega}$ to t_n.

The correctness of this description follows from the fact that V_ω is the limit of the ω^{op}-chain $(\mathscr{P}_f^n 1)_{n<\omega}$ above.

Example 4.5.2. (1) The following sequence of finite trees t_n:

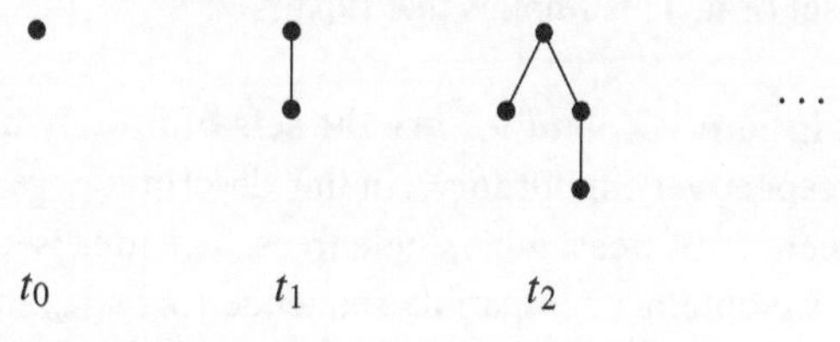

gives an element of V_ω.

(2) Consider the following infinite trees

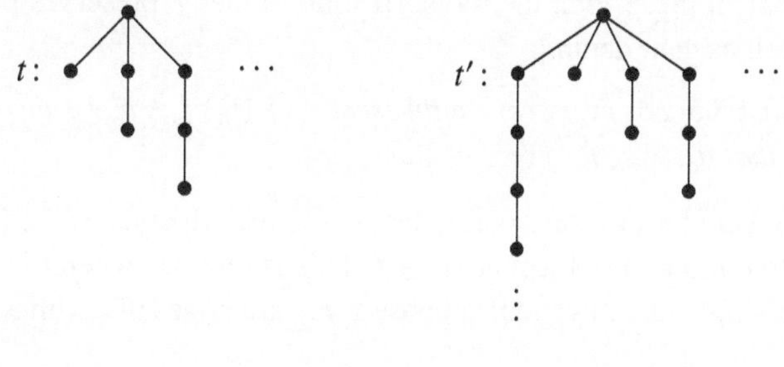

They are related by $\approx$. Indeed, the cuttings $\partial_n t = \partial_n t'$, $n < \omega$, are precisely the trees t_n above.

(3) For every set $M \subseteq \mathbb{N} \setminus \{0\}$ denote by t_M the tree obtained from an infinite path by adding a leaf of depth n for each $n \in M$. For example, for the set M of all odd numbers we obtain the following tree

The elements $(\partial_n t_M)_{n<\omega}$ of V_ω are pairwise distinct: given $M' \neq M$ such that $k \in M \setminus M'$, say, $\partial_{k+1} t_M$ contains a leaf at depth k while $\partial_{k+1} t_{M'}$ does not. Thus $(\partial_n t_M)_{n<\omega} \neq (\partial_n t_{M'})_{n<\omega}$.

Proposition 4.5.3. *The cardinality of* V_ω *is* $2^{\aleph_0}$.

Proof Example 4.5.2(3) demonstrates that $|V_\omega| \geq 2^{\aleph_0}$ since the set $\mathbb{N} \setminus \{0\}$ has $2^{\aleph_0}$ subsets.

To see that the reverse inequality holds, let A_n be the (finite) set of all extensional trees of height at most n. Then $\prod_{n<\omega} A_n$ has cardinality $2^{\aleph_0}$, and V_ω is a subset of it. This implies that $|V_\omega| \leq 2^{\aleph_0}$. □

A second description: $\nu\mathscr{P}_\mathrm{f}$ and V_ω are the sets of finitely branching trees and all trees, respectively, modulo $\approx$. In this description we work, in lieu of compatible sequences of trees, with single trees. The idea is simple: given an arbitrary tree t, we obtain a compatible sequence $(\partial_n t)_{n<\omega}$ in V_ω. We verify that, conversely, every compatible sequence can be obtained in this way.

Definition 4.5.4. Let s and t be trees. A *tree morphism* $f : t \to s$ is a graph homomorphism preserving the roots. (It follows that f preserves parents of nodes as well as their depth.)

Proposition 4.5.5. *For every compatible sequence* $(t_n)_{n<\omega}$ *in* V_ω *there exists a tree* t *such that* $\partial_n t = t_n$ *for every* $n < \omega$.

Proof (1) Let $\bar{t}$ be the extensional quotient of a tree t (Example 3.2.9(2)). Then there is a tree monomorphism $m : \bar{t} \to t$. This is obvious when $\bar{t}$ is obtained from t by a single identification: suppose $x \neq y$ are nodes of t with a common parent and such that $t_x = t_y$. Suppose further that $\bar{t}$ is the quotient of t obtained by identifying t_x and t_y. Then $\bar{t}$ is isomorphic to the subtree of t given by deleting t_y from it. This yields the desired embedding $m\colon \bar{t} \to t$. In the general case the argument is similar.

(2) Let $(t_n)_{n<\omega}$ be a compatible sequence in V_ω. Then for every n the equality $t_n = \partial_n t_{n+1}$ implies that we obtain a tree monomorphism $t_n \to t_{n+1}$: for the cutting s_{n+1} of t_{n+1} at level n compose $m\colon t_n = \overline{s_{n+1}} \to s_n$ with the inclusion $s_{n+1} \hookrightarrow t_{n+1}$.

(3) Now let t be obtained by taking the disjoint union of all t_n and merging all their roots (to obtain a tree). For every $n < \omega$ we verify that $\partial_n t = t_n$: the tree $\partial_n t$ is obtained from the disjoint union of all trees $\partial_n t_k$ $(k < \omega)$, joining them at the root and then forming the extensional quotient. For every $k \geq n$ we know (from the compatibility) that $\partial_n t_k = t_n$. Thus in the extensional quotient $\partial_n t$, all copies of $\partial_n t_k$ for $k \geq n$ are merged. For $k \leq n$ we clearly have $\partial_n t_k = t_k$, which in $\partial_n t$ are all merged with $\partial_n t_n = t_n$ because we have level-preserving monomorphisms $t_k \to t_n$ by item (2). Thus, only t_n remains, as claimed. □

Let $\mathcal{T}$ be the class of all trees. The equivalence $\approx$ has precisely $2^{\aleph_0}$ classes: the trees t_M of Example 4.5.2(3) are pairwise non-equivalent, and we have an injective map

$$b\colon \mathcal{T}/\!\approx\; \to V_\omega$$

(for V_ω from the first description) assigning to (the equivalence class of) a tree t the sequence $(\partial_n t)_{n<\omega}$. Hence, $|\mathcal{T}/\!\approx| = 2^{\aleph_0}$.

Moreover, note that for every tree t we have

$$\partial_n t = \pi_n(b([t])) \qquad \text{for all } n < \omega, \tag{4.9}$$

where $\pi_n\colon V_\omega \to \mathscr{P}_{\mathsf{f}}^n 1$ are the limit projections.

Corollary 4.5.6. *The set $\mathcal{T}/\!\approx$ is the limit of the terminal-coalgebra ω^{op}-chain for $\mathscr{P}_{\mathsf{f}}$ (with the limit projections given by $\pi_n \cdot b$ for $n < \omega$).*

Proof The map $b\colon \mathcal{T}/\!\approx\; \to V_\omega$ is an isomorphism: it is monic since the projections π_n $(n < \omega)$ are collectively monic. By Proposition 4.5.5 and using (4.9), every element of V_ω has the form $b([t])$ for some tree t. Thus b is surjetive, which proves the statement. □

Now let $D \subseteq \mathcal{T}$ be the set of finitely branching trees from Example 4.2.1(3). We know that D is a weakly terminal coalgebra for $\mathscr{P}_{\mathsf{f}}$. We have mentioned in Remark 4.3.28 that the terminal coalgebra for $\mathscr{P}_{\mathsf{f}}$ is $D/\!\approx$.

Corollary 4.5.7. *The cardinality of $\nu\mathscr{P}_{\mathsf{f}}$ is $2^{\aleph_0}$.*

Proof It follows from Example 4.5.2(3) that the cardinality of $\nu\mathscr{P}_\mathrm{f}$ is at least $2^{\aleph_0}$. Since $\nu\mathscr{P}_\mathrm{f}$ is a subset of V_ω we obtain $|\nu\mathscr{P}_\mathrm{f}| \leq |V_\omega| = 2^{\aleph_0}$ from Proposition 4.5.3. □

A third description: $\nu\mathscr{P}_\mathrm{f}$ is the set of strongly extensional trees that are finitely branching. Here we present a description of $\nu\mathscr{P}_\mathrm{f}$ due to Worrell [318], which is based on the notion of a *tree bisimulation*: this is more special than graph bisimulation (Example 4.2.5) since the goal is that by factoring a tree modulo the greatest bisimulation we obtain a tree again. Recall from Remark 2.6.8(1) that a bisimulation on a coalgebra A is *trivial* if it is contained in Δ_A.

We still assume that trees are unordered. We use the notation t_x for the subtree of t rooted in the node x.

Definition 4.5.8 [318]. (1) A *tree bisimulation* between two trees t and u is a graph bisimulation R such that

(a) the roots of t and u are related, and they are not related to other nodes; and
(b) whenever two nodes are related, their parents are also related.

Two trees are *tree bisimilar* if there is a tree bisimulation between them.

(2) A tree t is *strongly extensional* if every tree bisimulation on it is trivial. Said differently: t is strongly extensional iff distinct children x and y of the same node define subtrees t_x and t_y which are *not* tree bisimilar.

Remark 4.5.9. (1) Every strongly extensional tree is clearly extensional (see Example 3.2.9(2)).

(2) Consider the following trees (where the right-hand one has n children for the nth vertex in the breadth-first search):

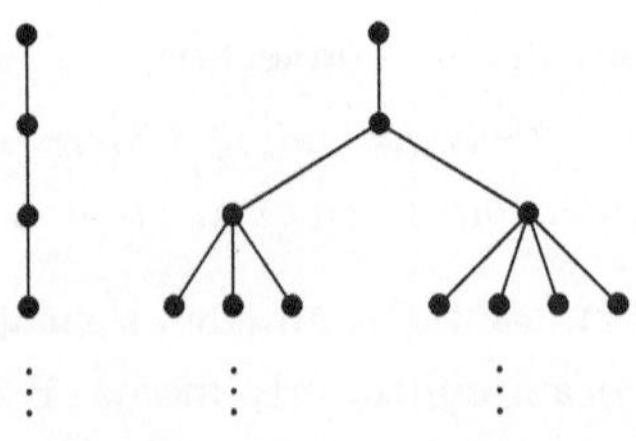

(4.10)

Both are extensional; the left-hand one is strongly extensional, the right-hand one is not, since the relation relating all nodes of the same depth is a tree bisimulation on it.

(3) It is trivial to prove that every composition and every union of tree bisimulations is again a tree bisimulation. In addition, the opposite relation of every tree bisimulation is a tree bisimulation: if R is a tree bisimulation from t to u,

then R^{op} is a tree bisimulation from u to t. Consequently, the largest tree bisimulation on every tree is an equivalence relation.

(4) Observe that the notion of tree bisimulation is different from the usual graph bisimulation. For example, the following tree

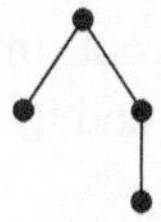

(4.11)

is strongly extensional, but there is a graph bisimulation relating the two leaves.

(5) A subtree s of a strongly extensional tree t is strongly extensional. Indeed, every tree bisimulation R on s can be extended to a tree bisimulation R' on t. Thus, R is trivial since so is R'.

Proposition 4.5.10. *A finite tree is extensional iff it is strongly extensional.*

Proof 'If' is clear. For 'only if' let t be an extensional finite tree, and let R be a tree bisimulation on it. We claim that if $x\,R\,y$, then the corresponding subtrees t_x and t_y are equal. First notice that every node of t_x must be related by R to some node of t_y (to see this, use induction on the depth of nodes, i.e. their distance from the root) and vice versa. Thus, t_x and t_y have the same height, n say. We now prove $t_x = t_y$ by induction on n. For $n = 0$, the result is obvious because the nodes of height 0 are leaves. Assume our result for n, and let x and y be related by R and of height $n + 1$. Then by the induction hypothesis and extensionality of t, for every child x' of x there is a *unique* child y' of y and $t_{x'} = t_{y'}$; and vice-versa. This implies that $t_x = t_y$. It now follows that t is strongly extensional. □

Lemma 4.5.11. *For every tree t the quotient graph t/R modulo the largest tree bisimulation is a tree.*

Proof Let $[x]$ denote the equivalence classes modulo R. If x is the root of t, then $[x]$ is the root of t/R. Moreover, the children of $[y]$ for a node y of t are simply $[z]$ for all children z of y. □

Definition 4.5.12. The *strongly extensional quotient* $\bar{t}$ of a tree t is the quotient tree of t modulo its largest tree bisimulation.

Lemma 4.5.13. *Let t and u be trees.*

(1) *The tree $\bar{t}$ is strongly extensional, and it is tree bisimilar to t.*

(2) *If t and u are strongly extensional and related by a tree bisimulation, then $t = u$.*

Proof In (1), it is easy to see that $\bar{t}$ is strongly extensional. Moreover, the kernel of the canonical quotient map $t \to \bar{t}$ is a tree bisimulation.

For (2), let R be a tree bisimulation between t and u. By Remark 4.5.9, $R^{\mathrm{op}} \cdot R$ is a tree bisimulation on t, whence $R^{\mathrm{op}} \cdot R \subseteq \Delta$ by strong extensionality. But every node of t is related to at least one node of u (use induction on the depth of nodes, i.e. their distance from the root) implying that $R^{\mathrm{op}} \cdot R = \Delta$. Similarly, $R \cdot R^{\mathrm{op}} = \Delta$. Thus, R (is a function and it) is an isomorphism of trees, and we identify such trees. □

Notation 4.5.14. We denote by D_0 the set of all strongly extensional finitely branching trees.

Since a subtree of a strongly extensional tree is strongly extensional (Remark 4.5.9(5)) we see that D_0 is a subcoalgebra of the coalgebra D of all finitely branching trees (Remark 4.3.28); the coalgebra structure takes a tree to the set of its maximum proper subtrees. The proof of the following result is original here.

Theorem 4.5.15 [318]**.** *The coalgebra D_0 is terminal for $\mathscr{P}_\mathsf{f}$.*

Proof We prove that D_0 is isomorphic to the terminal coalgebra $D/{\approx}$ we have seen in Remark 4.3.28. For that we only need to verify that given trees $t, u \in D$, then for their strongly extensional quotients we have

$$t \approx u \qquad \text{iff} \qquad \bar{t} = \bar{u}.$$

Then the map $[t] \mapsto \bar{t}$ is well-defined and is an isomorphism of coalgebras from $D/{\approx}$ to D_0. (Note that surjectivity is obvious since $\bar{t} = t$ for every strongly extensional tree t, and the fact that the map is a coalgebra homomorphism follows from D_0 being a subcoalgebra of D.)

(1) If $t \approx u$ we prove that t and u are tree bisimilar. Then, by Remark 4.5.9, it follows that $\bar{t}$ and $\bar{u}$ are tree bisimilar, which implies that $\bar{t} = \bar{u}$ by Lemma 4.5.13. Define $R \subseteq t \times u$ by relating nodes $x \in t$ and $y \in u$ iff they have the same depth n and for every $m \geq n$ the node of $\partial_m t$ corresponding to x equals the node of $\partial_m u = \partial_m t$ corresponding to y. Using that t and u are finitely branching, it is not difficult to prove that R is a tree bisimulation. We leave the details to the reader.

(2) Conversely, suppose we have $\bar{t} = \bar{u}$. Then we get, by composing with the quotient maps, a tree bisimulation $R \subseteq t \times u$. By restricting R to all pairs of nodes of depth at most n we obtain a tree bisimulation R_n between the cuttings of t and u at level n. Then also the extensional quotients $\partial_n t$ and $\partial_n u$ are tree bisimilar. Since $\partial_n t$ and $\partial_n u$ are strongly extensional by Proposition 4.5.10, we see that $\partial_n t = \partial_n u$. Thus, $t \approx u$. □

A fourth description: V_ω is the set of strongly extensional trees that are compactly branching or saturated. The following two descriptions of the limit V_ω use topology [318] or modal logic [48], respectively. The first description is due to Worrell [318]. He introduced the following pseudometric d on the class of all strongly extensional trees:

$$d(t,u) = \inf\{2^{-n} : n < \omega \text{ with } \partial_n t = \partial_n u\}. \tag{4.12}$$

Consequently t and u have distance 0 iff $t \approx u$, that is $\partial_n t = \partial_n u$ for all $n < \omega$. The resulting pseudometric space is *totally bounded* because for every $\epsilon > 0$ we have a finite number of ϵ-balls covering it. Indeed, choose n with $2^{-n} < \epsilon$ and take the (finite) set A of all extensional trees of height at most n. Then every tree t satisfies $\partial_n t \in A$ and $d(t, \partial_n t) \leq 2^{-n} < \epsilon$. Furthermore, this pseudometric space is complete. We sketch an argument; a detailed proof has been provided by Adámek et al. [6]. Let (t_n) be a Cauchy sequence. For each fixed k, the sequence $(\partial_k(t_n))_{n<\omega}$ is eventually constant. Let $s_k \in V_k$ be such that for all but finitely many n, $\partial_k(t_n) = s_k$. It is easy to check that the sequence $(s_k)_{k<\omega}$ is compatible: $\partial_k(s_{k+1}) = s_k$. Then by Proposition 4.5.5 there exists a single tree t such that $\partial_k(t) = s_k$ for all k. This tree t is a limit of the original sequence $(t_n)_{n<\omega}$.

Consequently, a set M of strongly extensional trees is compact (equivalently, complete and totally bounded) iff it is sequentially compact: every sequence has a convergent subsequence.

Remark 4.5.16. Here is what compactness means in elementary terms: for every sequence (t_n) from M there is a subsequence (s_n) of (t_n) and a tree $t \in M$ such that for all n, and all but finitely many k we have $\partial_n(s_k) = \partial_n(t)$.

Definition 4.5.17 [318]**.** A strongly extensional tree t is *compactly branching* if for every node x the set of all maximum proper subtrees of t_x is compact.

Example 4.5.18. We return to the trees t and t' of Example 4.5.2(2). The tree t is not compactly branching:

t:

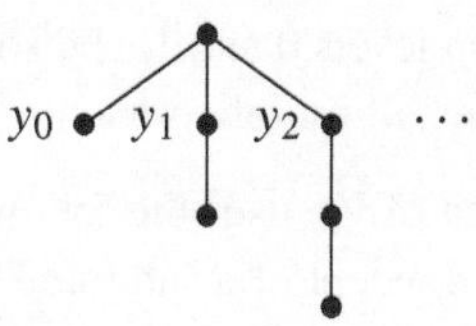

To see this, consider the sequence $t_{y_0}, t_{y_1}, \ldots$. Note that for $n \geq m$, $\partial_n(t_{y_m}) = t_{y_m}$. We claim that this sequence (t_{y_n}) has no convergent subsequence. To simplify the notation, we only verify this for the sequence (t_{y_n}) itself. It does not converge to any fixed element t_{y_m} because for $p > m$, $\partial_p(t_{y_m}) \neq \partial_p(t_{y_p})$.

In contrast, t' is compactly branching.

t':

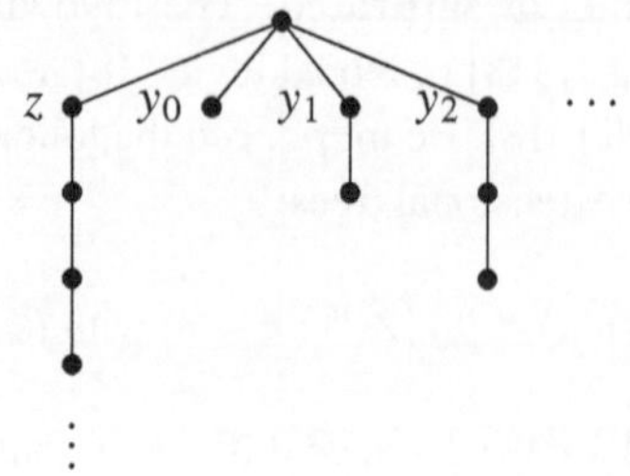

To check the compactness, we only need to consider a sequence of subtrees of the root, say (t_{x_n}). If there is an infinite subsequence which is constant, then of course that sequence converges. If not, then there is a subsequence of (t_{x_n}), say (t_{w_n}), where each w_n is of the form y_k for some $k \geq n$. In this case, (t_{w_n}) converges to t_z. This is because for all but finitely many m, $\partial_n(t_z) = t_{w_n} = \partial_n(t_{w_n})$.

The following result was stated by Worrell [318] without proof; a proof has been presented in later work [6].

Theorem 4.5.19. *The set of all compactly branching strongly extensional trees is a limit of the terminal-coalgebra ω^{op}-chain for $\mathscr{P}_{\mathsf{f}}$ with the limit projections given by ∂_n ($n < \omega$).*

Remark 4.5.20. (1) The pseudometric d in (4.12) is a metric when restricted to the set V_ω of all saturated, strongly extensional trees. Indeed, for such trees t and u we have: $t \approx u$ implies $t = u$. (Recall that trees are considered up to isomorphism.)

(2) When V_ω is described by all saturated, strongly extensional trees and $\nu\mathscr{P}_{\mathsf{f}}$ by the finitely branching ones, what is the canonical monomorphism $\bar{l}_0 : \nu\mathscr{P}_{\mathsf{f}} \to V_\omega$ of Remark 4.4.12? This is just the inclusion map. Indeed, the monomorphism m of Notation 4.4.5 represents the subobject of V_ω of all trees finitely branching at the root. Analogously $\mathscr{P}_{\mathsf{f}} m : \mathscr{P}_f^2 V_\omega \to \mathscr{P}_{\mathsf{f}} V_\omega$ represents the subobject of all trees finitely branching on levels 0 and 1, and so on. The intersection $\bar{l}_0$ of these subobjects thus represents all finitely branching trees in V_ω.

The second description of V_ω uses the following concept of a saturated tree, which is inspired by the concept of a saturated model in modal logic:

Definition 4.5.21 [48]. A tree t is called *saturated* provided that every node x of t has the following property: for every tree s, if the node x has, for every $n < \omega$, a child y_n satisfying $\partial_n(s) = \partial_n(t_{y_n})$, then x also has a child y with $s \approx t_y$.

We observe from the elementary characterization of sequential compactness in Remark 4.5.16 that a strongly extensional tree is saturated iff it is compactly branching. Previously, this was proved directly:

Theorem 4.5.22 [48, Cor. 3.19]. *The limit V_ω can be described as follows:*

$$V_\omega = \text{all saturated, strongly extensional trees.}$$

Other descriptions of V_ω. Let us mention further descriptions due to Abramsky [1]:

(1) Cauchy completion of $\mu\mathscr{P}_{\mathsf{f}}$. In Section 6.2 we will see that the initial algebra carries a canonical metric whose Cauchy completion is a metric space on the set V_ω.

(2) Ideal completion of $\mu\mathscr{P}_{\mathsf{f}}$. In Section 6.2 we will also see a canonical partial order on the initial algebra whose ideal (= free CPO) completion is carried by V_ω.

(3) All maximal consistent theories of the modal logic K. Saturated trees precisely correspond to *modally saturated trees* (when modal logic is taken without atoms) [48, Thm. 5.11]. The latter notion was defined by Fine [123]. Moreover, we have obtained the following descriptions [48, Prop. 5.7, Thm. 5.9]:

$$V_\omega = \text{all maximal consistent theories in } \mathsf{K}\text{, and}$$

$$\nu\mathscr{P}_{\mathsf{f}} = \text{all hereditarily finite theories in } \mathsf{K}.$$

(4) Terminal coalgebra of a modified Hausdorff functor. In Example 5.2.26 we will present an endofunctor on the category CMet of complete metric spaces whose terminal coalgebra is carried by V_ω.

Example 4.5.23. Returning to Theorem 4.5.15, similar descriptions of νF arise for the terminal coalgebras of related set functors F. For example, consider the functor $FX = \{0,1\} \times (\mathscr{P}_{\mathsf{f}}X)^\Sigma$, which is the finitary version of the type functor of non-deterministic automata, see Example 2.4.2(5). In order to describe νF one considers finitely branching trees whose nodes are labelled by 0 or 1 and whose edges are labelled in the input alphabet Σ. The notion of tree bisimulation then needs to be adjusted as follows: it is a bisimulation of labelled transition systems (cf. Example 2.6.7(3)) such that the roots are related, and every pair of related nodes have the same depth and the same label in $\{0,1\}$. The notion of a strongly extensional tree is then defined as before. It is not difficult to adapt the proof of Theorem 4.5.15 to prove that νF is the coalgebra of all strongly extensional finitely branching trees with nodes labelled in $\{0,1\}$ and edges labelled in Σ.

Note that this means that the unique map from an F-coalgebra into the terminal one thus provides the behaviour of states modulo bisimilarity, that is, taking into account the non-deterministic branching of the given automaton. This semantics is therefore different from the usual language semantics of non-deterministic automata. We shall see in Example 5.1.27 that by modelling non-deterministic automata as coalgebras for a functor on the category Rel of sets and relations, one obtains their usual language semantics similarly as for deterministic automata in Example 2.5.5 via the terminal coalgebra.

Remark 4.5.24. Theorem 4.5.15 can be generalized to those monoid-valued functors $M^{(-)}$ (Example 2.4.3(1)) which preserve weak pullbacks. This is the case precisely when the monoid M is

(1) *positive*: $a + b = 0$ implies $a = 0 = b$, and

(2) *refinable*: $a_1 + a_2 = b_1 + b_2$ implies that there exists a 2×2 matrix with row sums a_1, a_2 and columns sums b_1, b_2, respectively.

This fact was proved by Gumm and Schröder [152, Thm. 5.13(iii)] (see Corollary C.4.7 for a full proof).

The concepts of tree bisimulation and strong extensionality are easily generalized to M-labelled trees. One obtains that for a positive and refinable monoid M the terminal coalgebra for the corresponding monoid-valued functor $M^{(-)}$ is carried by the set of all strongly extensional, finitely branching M-labelled trees [48, Thm. 6.11].

4.6 Summary

We have first proved some useful facts about constructions of (co)algebras for general endofunctors F: colimits of coalgebras (and dually limits of algebras) are canonically formed on the level of the base category, and the same is true for those limits of coalgebras that F preserves (and dually for colimits of F-algebras). We then discussed quotient coalgebras and proved that whenever a weakly terminal coalgebra is given, then the smallest quotient coalgebra yields νF.

The main topic of this chapter was a construction of terminal coalgebras for finitary set functors. We presented a new proof of Worrell's result [316] that νF is a limit of the terminal-coalgebra chain after $\omega + \omega$ steps. We also observed that the object V_ω (obtained after the first ω steps of that chain) is a weakly terminal coalgebra. For the finite power-set functor $\mathscr{P}_\mathsf{f}$ that limit was studied by a number of authors, and we presented various descriptions of $\nu\mathscr{P}_\mathsf{f}$ and V_ω in the last section.

5
Finitary Iteration in Enriched Settings

In Chapter 3 we have presented two constructions: the initial algebra as a colimit of (finitary) iterations and the terminal coalgebra as a limit of (finitary) iterations. Then in Chapter 4 we have seen ways to 'get our hands on' some of the terminal coalgebras which are obtained by iteration, namely the terminal coalgebras of finitary functors on Set. In this chapter, we strike out in a different direction by considering categories other than Set. In two important types of base categories, namely those enriched over complete partial orders and over complete metric spaces, it turns out that (under mild conditions on the endofunctor) the two constructions from Chapter 3 coincide. Moreover, they yield a fixed point that is *canonical* in the sense of the following definition (recall from Lambek's Lemma that the structure morphism of an initial algebra has an inverse):

Definition 5.0.1. Let $F\colon \mathscr{A} \to \mathscr{A}$ be an endofunctor. A *canonical fixed point* of F is an initial algebra $\alpha\colon FA \to A$ such that $\alpha^{-1}\colon A \to FA$ is a terminal coalgebra.

Proposition 5.0.2. *If* $F0 = 1 = F1$, *then* $\mathrm{id}\colon F1 \to 1$ *is a canonical fixed point.*

Proof Since F preserves the terminal object 1, $\mathrm{id}_1\colon 1 \to F1$ is a terminal coalgebra. In addition, the initial-algebra chain starts $0 \to 1 \to 1 \to \cdots$. Thus this chain converges to 1, and this convergence is preserved by F. Thus, $F1 \to 1$ is an initial algebra. □

Example 5.0.3. One example of a non-constant set functor with $F0 = 1 = F1$ is as follows: FX is the set of subsets of X which are either empty or of size 2. For $f\colon X \to Y$ and $A \subseteq X$ of size 2: if $f[A]$ has size 1, let $Ff(A) = \emptyset$; otherwise, let $Ff(A) = f[A]$.

The examples of canonical fixed points of set functors coming from Proposition 5.0.2 seem artificial. We have seen much more interesting examples when we considered $\mathsf{CPO}_\perp$ (complete partial orders with least element) and CMet

(complete metric spaces). For example, a canonical fixed point of $FX = X_\perp$ on $\mathsf{CPO}_\perp$ was presented in Examples 2.2.17(1) and 3.3.6. It is $\mathbb{N}^\top$, the natural numbers with an additional largest element $\top$.

What the $\mathsf{CPO}_\perp$ and CMet settings have in common is a nice theory of *approximation*. In the case of $\mathsf{CPO}_\perp$ approximation is given in terms of joins of ω-chains, and in the case of CMet it is given by limits of Cauchy sequences. In this chapter, we are going to work with functors which preserve this additional approximation structure. We also typically work on a *pointed* category: this means that the initial object 0 is the same as the terminal object 1. We call $0 = 1$ a *zero object*. In such a category we may lay the initial ω-chain on top of the terminal ω^{op}-chain, as follows:

$$0 = 1 \underset{\widehat{e}}{\overset{e}{\rightleftarrows}} F1 \underset{F\widehat{e}}{\overset{Fe}{\rightleftarrows}} F^2 1 \underset{F^2\widehat{e}}{\overset{F^2 e}{\rightleftarrows}} \cdots \tag{5.1}$$

Here e is by initiality of 0 and $\widehat{e}$ is by finality of 1. Since clearly $\widehat{e} \cdot e = \mathsf{id}$, the two chains are already related to each other. It turns out that under suitable hypotheses, we have a *limit–colimit coincidence*: the colimit of the initial ω-chain exists and coincides with the limit of the terminal ω^{op}-chain. For $\mathsf{CPO}_\perp$ this is a classical result by Smyth and Plotkin [279] that we shall present in Section 5.1 (see Theorem 5.1.24). The case of CMet was first studied by America and Rutten [59], and we consider it in Section 5.2. In both settings, the 'approximation structure' is taken to apply to the morphisms in the category rather than the objects, and this is why we are working with enriched categories in this chapter. A most appealing part of the theory is the prominence of duality (opposite categories) in various proofs.

This chapter has two themes: the canonical fixed points which we just mentioned, but also fixed points of functors involving *mixed variance*. This topic is related to semantic models for the lambda calculus and other calculi. We are not going to discuss these applications in any detail. But we do want to foreshadow some of the ideas that will be prominent in the chapter. The search for interesting models of the lambda calculus essentially becomes a problem of the following form: can we find a category $\mathscr{A}$ and some object X satisfying $X \cong [X, X]$? In this discussion,

$$[X, Y]$$

is an object in the category corresponding to the hom-set $\mathscr{A}(X, Y)$. We would especially want constructions in concrete categories, and the methods should be general enough that they allow us to solve related equations such as $X \cong [X, X] + A$ for all objects A. It is clear that $\mathscr{A}$ cannot be Set, since the only way to have

$X \cong [X, X]$ in Set is for X to have just one element. So one would want other categories.

Organization of the chapter. The two themes of this chapter are each interesting and will be presented somewhat separately. Section 5.1 discusses canonical fixed points in categories enriched over complete partial orders. We present full detail in this section because our proofs are shorter than those in the literature, and also because this body of material does not seem to have been presented comprehensively in other books. Following this, we turn in Section 5.2 to canonical fixed points in the metric setting rather than the order-theoretic one. We close the chapter with a discussion of domain equations in Section 5.3. Section 5.2 may be read on its own, but some of the ideas in Section 5.1 are used there. Section 5.3 may be read independently of Section 5.2.

5.1 Canonical Fixed Points in CPO-Enriched Categories

Recall the category

$$\mathsf{CPO}$$

of complete partial orders (Example 2.1.7(1)) and continuous maps. The objects in the category need not have a least element $\perp$.

The category CPO is *cartesian closed*; this means that every endofunctor $X \times -$ on that category has a right adjoint. It assigns to every cpo Y the internal hom-object $[X, Y]$ which is the hom-set $\mathsf{CPO}(X, Y)$ ordered pointwise. We denote by $\bigsqcup f_i$ the join of an ω-chain $f_i \colon X \to Y$ $(i < \omega)$ in $[X, Y]$.

Definition 5.1.1. A category $\mathscr{A}$ is CPO-*enriched* if its hom-sets come with a CPO structure, and composition is continuous in both variables. That is, if $f_1 \sqsubseteq f_2 \colon A \to B$, then $h \cdot f_1 \cdot g \sqsubseteq h \cdot f_2 \cdot g$ for all h and g composable with f_1 and f_2. Moreover, given an ω-chain $f_i \colon X \to Y$, $h \cdot \left(\bigsqcup_i f_i\right) \cdot g = \bigsqcup_i h \cdot f_i \cdot g$.

Example 5.1.2. (1) The category CPO is, of course, CPO-enriched using the above pointwise ordering on hom-sets.

(2) The category $\mathsf{CPO}_\perp$ is also CPO-enriched (see Example 2.1.7(2)).

(3) Analogously to a cpo, a poset is called a dcpo (*directed-complete partial order*) if every directed subset has a join. The corresponding morphisms are the functions preserving directed joins. The category $\mathsf{DCPO}_\perp$ of directed-complete partial orders with least element is again CPO-enriched.

(4) The category Pfn of sets and partial functions is CPO-enriched by inclusion: for partial functions $f_1, f_2 \colon X \to Y$ we put $f_1 \sqsubseteq f_2$ iff whenever $f_1(x)$ is defined

so is $f_2(x)$ and $f_1(x) = f_2(x)$, that is, the graph of f_1 is a subset of the graph of f_2. The least element of $\mathsf{Pfn}(X,Y)$ is the nowhere defined function, and $\bigsqcup f_i$ is the set-theoretic union for every ω-chain $(f_i)_{i<\omega}$.

(5) The category Rel of sets and relations (Notation 3.2.16) is also CPO-enriched by inclusion.

(6) CLatis the category of complete lattices and maps preserving joins and meets. This category has two CPO-enriched structures. One is where we use the pointwise order in each hom-set, and the other uses the dual of this order.

(7) If $\mathscr{A}$ is CPO-enriched, then so is $\mathscr{A}^{\mathsf{op}}$. We order $\mathscr{A}^{\mathsf{op}}(X,Y)$ using the same relation as in $\mathscr{A}(Y,X)$.

(8) A product of CPO-enriched categories is CPO-enriched with respect to the componentwise ordering.

Definition 5.1.3 [273]. In a CPO-enriched category, a morphism $e\colon X \to Y$ is an *embedding* if there exists a morphism $\widehat{e}\colon Y \to X$, referred to as the *projection* for e, such that

$$\widehat{e} \cdot e = \mathsf{id}_X \quad \text{and} \quad e \cdot \widehat{e} \sqsubseteq \mathsf{id}_Y. \tag{5.2}$$

The pair $(e, \widehat{e})$ is called an *embedding-projection pair*.

Remark 5.1.4. The morphism $\widehat{e}$ is uniquely determined by (5.2): Suppose that $e^* \cdot e = \mathsf{id}_X$ and $e \cdot e^* \sqsubseteq \mathsf{id}_Y$. Then $\widehat{e} = e^* \cdot e \cdot \widehat{e} \sqsubseteq e^*$. Similarly, $e^* \sqsubseteq \widehat{e}$. Thus $\widehat{e} = e^*$.

Example 5.1.5. (1) In CPO, the embeddings are precisely those monomorphisms $e\colon X \to Y$ such that for every $y \in Y$ there exists a largest $x \in X$ with $e(x) \sqsubseteq y$. In fact, this condition allows us to define $\widehat{e}\colon Y \to X$ by choosing this largest x as $\widehat{e}(y)$. Then $\widehat{e} \cdot e = \mathsf{id}_X$ (since e is one-to-one), and $e \cdot \widehat{e} \sqsubseteq \mathsf{id}_Y$. The verification that the condition is also necessary is trivial.

(2) It follows from (1) that in the category CLat of complete lattices, every monomorphism is an embedding.

(3) Every monomorphism in Pfn is an embedding: monomorphisms are the total injective functions, and for these, $\widehat{e}$ is the (partially defined) inverse function e^{-1}. Identifying functions with their graph relations, $\widehat{e}$ is just the converse of e.

(4) Every split monomorphism in Rel is an embedding: split monomorphisms are precisely those relations $e\colon X \to Y$ which are injective maps, and for these $\widehat{e} = e^{\mathsf{op}}$.

Observation 5.1.6. Suppose that $\mathscr{A}$ is a CPO-enriched category in which every hom-set $\mathscr{A}(X,Y)$ has a least element $\bot_{XY}$.

(1) Given a coproduct $X = \coprod_{i \in I} X_i$, each injection v_i is an embedding. Indeed, the projection $\widehat{v}_i \colon X \to X_i$ has components id_{X_i} and $\perp_{X_i X_j}$ for $j \neq i$. In particular, the morphisms with domain 0 are always embeddings.

(2) The name 'projection' stems from the dual of (1): for every product $X = \prod_{i \in I} X_i$ the projection π_i has the form $\widehat{e}_i$ where $e_i \colon X_i \to X$ has components id_{X_i} and $\perp_{X_i X_j}$.

Notation 5.1.7. The objects of a CPO-enriched category $\mathscr{A}$ together with the embeddings form a category

$$\mathscr{A}^E.$$

The point is that a composite $e \cdot f$ of embeddings is itself an embedding with

$$\widehat{e \cdot f} = \widehat{f} \cdot \widehat{e}. \tag{5.3}$$

Remark 5.1.8. Let $\mathscr{A}$ be a CPO-enriched category in which every hom-set also has meets of decreasing ω-chains. For example, the categories CPO, Pfn, Rel, and CLat in Example 5.1.2 have this property. Then $\mathscr{A}^E$ is a CPO-enriched category, since given an ω-chain e_i of embeddings in $\mathscr{A}(X, Y)$, the join $\bigsqcup_i e_i$ is an embedding with projection given by the meet $\bigsqcap_i \widehat{e}_i$.

Observation 5.1.9. If $e \colon X \to Y$ is an embedding in $\mathscr{A}$, then $\widehat{e} \colon X \to Y$ is an embedding in $\mathscr{A}^{\mathsf{op}}$. So the dual category of $\mathscr{A}^E$ is the category whose objects are those of $\mathscr{A}$ and whose morphisms are the projections.

The following result, due to Smyth and Plotkin [279], could look technical at first glance. It aims to connect a global property of a given cocone of embeddings (being a colimit) and a local property mentioned in (5.5) below. This connection is exploited in all the rest of this section.

Theorem 5.1.10. *Let $\mathscr{A}$ be a CPO-enriched category. Consider an ω-chain of embeddings in $\mathscr{A}$*

$$E_0 \xrightarrow{e_0} E_1 \xrightarrow{e_1} E_2 \xrightarrow{e_2} \cdots \tag{5.4}$$

Let $c_i \colon E_i \to C$ be a cocone. Then the following are equivalent:

(1) (c_i) is a colimit cocone.

(2) Each c_i is an embedding, the composites $c_i \cdot \widehat{c}_i$ form an ω-chain in $\mathscr{A}(C, C)$, and

$$\bigsqcup_i c_i \cdot \widehat{c}_i = \mathsf{id}_C. \tag{5.5}$$

Proof (1) ⇒ (2): For every i we have the following cocone of the shortened chain with codomain E_i:

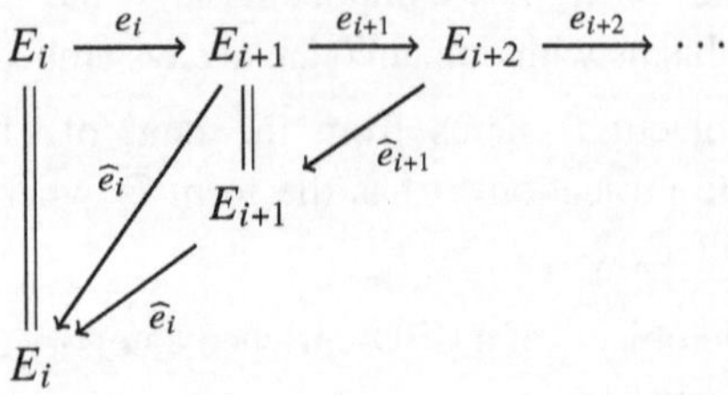

(5.6)

Since the shortened chain has the same colimit as the original one, namely C, there exists a unique morphism $\widehat{c}_i \colon C \to E_i$ such that the left-hand triangle in (5.7) commutes for all $k \geq i \geq 0$:

(5.7)

In particular, we have

$$\widehat{c}_i \cdot c_i = \mathsf{id}_{E_i}. \tag{5.8}$$

We verify the triangles on the right in (5.7) commute by induction on $n = k - i$. Equation (5.8) is the base case $k = i$. Assuming the commutativity when the difference is $n - 1$, we get it when the difference is n:

$$(e_k \cdot e_{k-1} \cdot \dots \cdot e_{i+1}) \cdot e_i = \widehat{c}_k \cdot c_{i+1} \cdot e_i = \widehat{c}_k \cdot c_i.$$

Comparing the left-hand triangles in (5.7) for $\widehat{c}_i$ and $\widehat{c}_{i+1}$, we see that given $k \geq i + 1$ we have

$$\widehat{c}_i \cdot c_k = \widehat{e}_i \cdot \widehat{c}_{i+1} \cdot c_k.$$

Now $(c_k)_{k \geq i+1}$ is a colimit cocone and therefore a jointly epic family. Thus, we conclude

$$\widehat{c}_i = \widehat{e}_i \cdot \widehat{c}_{i+1}. \tag{5.9}$$

Hence, the morphisms $c_i \cdot \widehat{c}_i$ form a chain in $\mathscr{A}(C, C)$:

$$c_i \cdot \widehat{c}_i = (c_{i+1} \cdot e_i) \cdot (\widehat{e}_i \cdot \widehat{c}_{i+1}) \sqsubseteq c_{i+1} \cdot \widehat{c}_{i+1}$$

due to $e_i \cdot \widehat{e}_i \sqsubseteq \mathsf{id}$.

We next prove that the join of the chain $(c_i \cdot \widehat{c}_i)$ is id_C, thus also proving that $c_i \cdot \widehat{c}_i \sqsubseteq \mathsf{id}$; this inequality together with (5.8) establishes that c_i is an embedding. To see that id_C is the desired join, it is sufficient to prove that for every k,

$$\left(\bigsqcup_i c_i \cdot \widehat{c_i}\right) \cdot c_k = c_k.$$

For this, we only need to show that for $i \geq k$, $c_i \cdot \widehat{c_i} \cdot c_k = c_k$. Using the triangle on the right in (5.7) and interchanging k and i, we see that

$$c_i \cdot \widehat{c_i} \cdot c_k = c_i \cdot e_i \cdot e_{i-1} \cdot \cdots \cdot e_k = c_k.$$

We have proved (5.5). This verifies all parts of (2) in our lemma.

(2) $\Rightarrow$ (1): Suppose we are given a cocone $b_i \colon E_i \to B$ of the given chain:

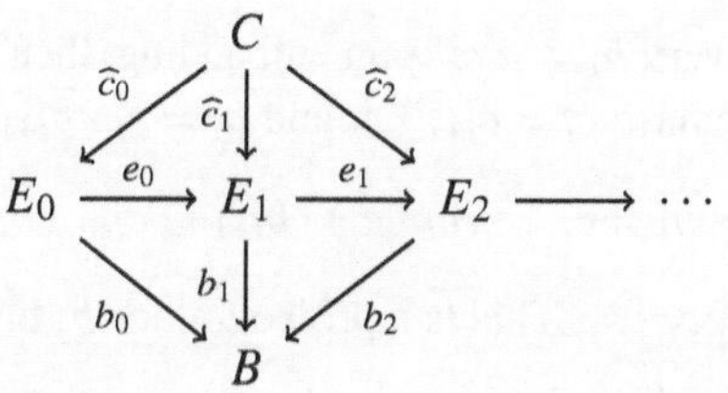

We first observe that the morphisms $b_i \cdot \widehat{c_i}$ form a chain in $\mathscr{A}(C, B)$. Since the c_i's are embeddings and $c_i = c_{i+1} \cdot e_i$, we obtain (5.9) by applying (5.3). Thus

$$b_i \cdot \widehat{c_i} = (b_{i+1} \cdot e_i) \cdot (\widehat{e_i} \cdot \widehat{c_{i+1}}) = b_{i+1} \cdot (e_i \cdot \widehat{e_i}) \cdot \widehat{c_{i+1}} \sqsubseteq b_{i+1} \cdot \widehat{c_{i+1}}.$$

Define

$$b = \bigsqcup_i (b_i \cdot \widehat{c_i}) \colon C \to B. \tag{5.10}$$

We show that this is the desired factorization: we fix k and show that $b \cdot c_k = b_k$. For this, we may restrict the join to $i \geq k$. So we shall show that

$$b \cdot c_k = \bigsqcup_{i \geq k} (b_i \cdot \widehat{c_i} \cdot c_k) = b_k. \tag{5.11}$$

Recall that (c_i) is a cocone. Thus for $i \geq k$, the top triangle below commutes:

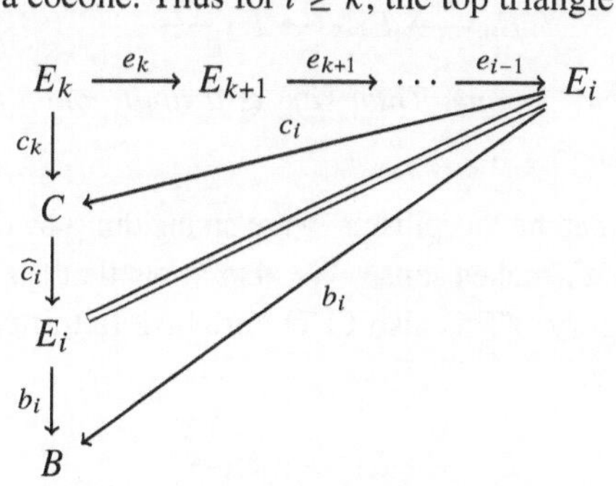

The other triangles clearly commute, and so the outside does too. Since (b_i) is a cocone, we have

$$b_i \cdot \widehat{c_i} \cdot c_k = b_i \cdot (e_{i-1} \cdot \dots \cdot e_k) = b_k.$$

This for all $i \geq k$ proves (5.11). We have shown that b is a morphism through which the cocone (b_i) factorizes.

The factorization is unique: Given $b' \colon C \to B$ with $b' \cdot c_i = b_i$ for all $i < \omega$, we have, due to (5.5)

$$b' = b' \cdot \bigsqcup_i c_i \cdot \widehat{c_i} = \bigsqcup_i b_i \cdot \widehat{c_i} = b,$$

which completes the proof. □

Remark 5.1.11. If every $b_i, i < \omega$ is an embedding, then so is the morphism b of (5.10). Indeed, we have $c_i = c_{i+1} \cdot e_i$ and $\widehat{b_i} = \widehat{e_i} \cdot \widehat{b}_{i+1}$. Thus

$$c_i \cdot \widehat{b_i} = c_{i+1} \cdot (e_i \cdot \widehat{e_i}) \cdot \widehat{b}_{i+1} \sqsubseteq c_{i+1} \cdot \widehat{b}_{i+1}.$$

We define $\widehat{b}$ to be $\bigsqcup_i c_i \cdot \widehat{b_i}$. This is a projection for b: first

$$b \cdot \widehat{b} = \Big(\bigsqcup_i b_i \cdot \widehat{c_i}\Big) \cdot \Big(\bigsqcup_i c_i \cdot \widehat{b_i}\Big) = \bigsqcup_i b_i \cdot \widehat{c_i} \cdot c_i \cdot \widehat{b_i} = \bigsqcup_i b_i \cdot \widehat{b_i} \sqsubseteq \mathsf{id}_B.$$

Second, we have $\widehat{b} \cdot b = \mathsf{id}_C$ due to (5.5):

$$\widehat{b} \cdot b = \bigsqcup_i c_i \cdot \widehat{b_i} \cdot b_i \cdot \widehat{c_i} = \bigsqcup_i c_i \cdot \widehat{c_i} = \mathsf{id}_C.$$

Remark 5.1.12. Although formulated for ω-chains only, a result analogous to Theorem 5.1.10 holds for λ-chains for all limit ordinals λ (in categories enriched so that hom-sets are posets with joins of all chains and composition preserves joins of chains), cf. [54, Thm. 5.2].

We draw several corollaries from Theorem 5.1.10.

Corollary 5.1.13 (Limit–colimit coincidence). *Consider an ω-chain of embeddings in a* CPO*-enriched category*

$$E_0 \xrightarrow{e_0} E_1 \xrightarrow{e_1} E_2 \xrightarrow{e_2} \cdots$$

Let (c_i) be a colimit cocone. Then $(\widehat{c_i})$ is a limit cone of the ω^{op}-chain of projections $\widehat{e_i} \colon E_{i+1} \to E_i$.

Proof First, note that the morphisms c_i are embeddings by the Theorem 5.1.10, and so the notation $\widehat{c_i}$ makes sense. We also know that $\bigsqcup_i c_i \cdot \widehat{c_i} = \mathsf{id}_C$. Now recall that the category $\mathscr{A}^{\mathrm{op}}$ is also CPO-enriched. In this category, we have a chain of embeddings

$$E_0 \xrightarrow{\widehat{e_0}} E_1 \xrightarrow{\widehat{e_1}} E_2 \xrightarrow{\widehat{e_2}} \cdots$$

Each $\widehat{c}_i$ is also an embedding, $\widehat{c}_i \cdot c_i$ is a chain, and $\bigsqcup_i \widehat{c}_i \cdot c_i = \mathrm{id}_C$. Thus by Theorem 5.1.10 again, $(\widehat{c}_i)$ is a colimit cocone in $\mathscr{A}^{\mathrm{op}}$ of $(\widehat{e}_i)$. Moving from $\mathscr{A}^{\mathrm{op}}$ back to $\mathscr{A}$, we see that $(\widehat{c}_i)$ is a limit cone in $\mathscr{A}$ of $(\widehat{e}_i)$, as desired. □

Remark 5.1.14. Thus in a CPO-enriched category we have a limit–colimit coincidence for ω-chains of embeddings (or, equivalently, for ω^{op}-chains of projections in the sense of Definition 5.1.3). A colimit cone of an ω-chain of embeddings yields a limit cone of the corresponding ω^{op}-chain of projections, and vice versa.

Corollary 5.1.15. *If a* CPO*-enriched category has colimits of ω-chains of embeddings, then $\mathscr{A}^{\mathrm{E}}$ has colimits of all ω-chains.*

Proof Let (e_i) be an ω-chain in $\mathscr{A}^E$ and let (c_i) be a colimit cocone in $\mathscr{A}$ for the underlying ω-chain in $\mathscr{A}$. Then by Theorem 5.1.10, each colimit morphism c_i is itself an embedding. Moreover, given a cocone (b_i) for (e_i) in $\mathscr{A}^E$, the factorization morphism b is an embedding by Remark 5.1.11. □

At this point, we need an important definition in this subject. The concept of local continuity is important because it is typically easier to check than the preservation of ω-colimits.

Definition 5.1.16. A functor $F\colon \mathscr{A} \to \mathscr{B}$ between CPO-enriched categories is called *locally continuous* if it is CPO-enriched, that is, it preserves the order of hom-sets, as well as joins of ω-chains: given $f_0 \sqsubseteq f_1 \sqsubseteq f_2 \sqsubseteq \cdots$ in $\mathscr{A}(X,Y)$, then $F(\bigsqcup_n f_n) = \bigsqcup_n F f_n$.

For the following examples, recall from Example 2.1.7(1) that coproducts in CPO are disjoint unions, and products are given by cartesian products with coordinate-wise order. More generally, a limit of a diagram in CPO is the limit of the underlying diagram in Set with the largest order making every limit projection monotone (we leave the easy proof of this fact to the reader).

Example 5.1.17. Here are some examples of locally continuous endofunctors on CPO-enriched categories.

(1) The identity functor Id is locally continuous, and so is every constant functor.

(2) A composite, product or coproduct of locally continuous functors is locally continuous; this is easy to prove, see also Barr [71].

(3) In analogy to the polynomial endofunctors on Set, given a collection $\Sigma = (\Sigma_n)_{n<\omega}$ of cpos, we denote by H_Σ the endofunctor on CPO given by $H_\Sigma X = \coprod_{n<\omega} \Sigma_n \times X^n$. Its local continuity follows from (1) and (2).

(4) The functor $FX = X_\perp$ on CPO from Example 2.1.7(1) is locally continuous.

Example 5.1.18. Ideal completion is an example of a natural endofunctor that is *not* locally continuous. For every poset P the ideal completion is the free dcpo on it. By an *ideal* is meant a nonempty directed down-set in P. The *ideal completion* $\mathsf{Idl}(P)$ is the poset of all ideals ordered by inclusion. In particular, every element $x \in P$ defines the *principal ideal* $\downarrow x = \{y \in P : y \le x\}$. Moreover, $x \le y$ holds in P iff $\downarrow x \subseteq \downarrow y$ in $\mathsf{Idl}(P)$. Therefore, P may be identified with the subposet of $\mathsf{Idl}(P)$ given by the principal ideals. The ideal completion has the following universal property: every monotone map from P to a cpo has a unique continuous extension to a continuous map from $\mathsf{Idl}(P)$ to that cpo, see Abramsky and Jung [2]. As a concrete example consider the cpo $\mathbb{N}^\top$ of Example 3.3.7(1). Its only non-principal ideal is $\mathbb{N}$. Thus, $\mathsf{Idl}(\mathbb{N}^\top)$ is (isomorphic to) the chain $0 < 1 < 2 < \cdots < \top' < \top$, where $\top'$ represents the ideal $\mathbb{N}$.

Denote by Idl the endofunctor on CPO mapping a cpo X to $\mathsf{Idl}(X)$ and a continuous map $f\colon X \to Y$ to the continuous extension taking an ideal M to $\downarrow f[M]$. This endofunctor preserves the order on hom-sets, but it is not locally continuous. To see this, consider the poset $\mathbb{N}^\top$, and let $f_i\colon \mathbb{N}^\top \to \mathbb{N}^\top$ be defined by $n \mapsto \min(n, i)$. Then $\bigsqcup_{i\in\mathbb{N}} f_i = \mathsf{id}$, and Idl does not preserve this join: the map $\mathsf{Idl}(f_i)$ takes $\top$ to i, whence the join $\bigsqcup_{i\in\mathbb{N}} \mathsf{Idl}(f_i)$ takes $\top$ to $\top'$. This example is from [138, Example IV-5.12].

Remark 5.1.19. Most of the functors found in domain theory are locally continuous. Here are some further examples.

(1) Consider

$$F\colon \mathsf{CPO}^{\mathsf{op}} \times \mathsf{CPO} \to \mathsf{CPO}$$

defined as follows: $F(D, E) = [D, E]$ is the set of CPO morphisms as a CPO object. On morphisms, $(f, g)\colon (D, E) \to (X, Y)$, define $F(f, g)$ as the function which takes $u\colon D \to E$ to $f \cdot u \cdot g$. The category $\mathsf{CPO}^{\mathsf{op}} \times \mathsf{CPO}$ is CPO-enriched due to Example 5.1.2(7), (8). Moreover, F is easily seen to be locally continuous.

(2) In contrast, the functor $F\colon \mathsf{Pfn}^{\mathsf{op}} \times \mathsf{Pfn} \to \mathsf{Pfn}$ defined by

$$F(D, E) = \mathsf{Pfn}(D, E) \qquad \text{and} \qquad F(f, g)\colon u \mapsto f \cdot u \cdot g$$

is not locally continuous: $F(f, g)$ is a total function. Thus, if $f \sqsubseteq f'$ but $f' \not\sqsubseteq f$, we have $F(f, g) \not\sqsubseteq F(f', g)$.

(3) On the category of ω-algebraic CPOs, the Plotkin powerdomain operation (i.e. the formation of a free semilattice in that category) extends to a locally continuous functor (see e.g. Knijnenburg [195]).

We have another corollary to Theorem 5.1.10, connecting local continuity to the preservation of ω-colimits and ω^{op}-limits.

Corollary 5.1.20. *Let $\mathscr{A}$ be a* CPO*-enriched category. Then every locally continuous endofunctor on $\mathscr{A}$ preserves colimits of ω-chains of embeddings and limits of ω^{op}-chains of projections.*

Proof We only prove the statement about colimits of ω-chains. The statement about limits of ω^{op}-chains of projections follows by duality (see Example 5.1.2(7) and cf. Observation 5.1.9).

Let $F\colon \mathscr{A} \to \mathscr{A}$ be locally continuous. Consider an ω-chain of embeddings in $\mathscr{A}$

$$E_0 \xrightarrow{e_0} E_1 \xrightarrow{e_1} E_2 \xrightarrow{e_2} \cdots$$

Let $c_i\colon E_i \to C$ be a colimit cocone. By Corollary 5.1.13, all c_i are embeddings (with $\widehat{c_i}$ being the limit projections of the ω^{op}-chain given by the $\widehat{e_i}$). Since F preserves the order on hom-sets, each Fc_i is an embedding and $\widehat{Fc_i} = F\widehat{c_i}$. Moreover, the composites $Fc_i \cdot F\widehat{c_i} = F(c_i \cdot \widehat{c_i})$ form a chain in $\mathscr{A}(FC, FC)$, and their join is id_{FC} using local continuity of F:

$$\bigsqcup_i Fc_i \cdot F\widehat{c_i} = F\Big(\bigsqcup_i c_i \cdot \widehat{c_i}\Big) = F\mathsf{id}_C = \mathsf{id}_{FC}.$$

By Theorem 5.1.10, (Fc_i) is a colimit cocone. □

We now restrict attention to certain CPO-enriched categories pertaining to least elements. We should point out that our next definition has a certain subtlety: this is not the same thing as being *enriched over* $\mathsf{CPO}_\perp$.

Definition 5.1.21. A *strict* CPO*-enriched* category is a nonempty CPO-enriched category $\mathscr{A}$ with a least element $\perp_{XY}$ in every cpo $\mathscr{A}(X,Y)$ such that for all $f\colon Y \to Y'$ we have

$$f \cdot \perp_{XY} = \perp_{XY'} \quad \text{and} \quad \perp_{Y'Z} \cdot f = \perp_{YZ}.$$

Example 5.1.22. The category $\mathsf{CPO}_\perp$ of complete partial orders with least element and continuous functions that are strict, that is, preserve that least element, is a strict CPO-enriched category. Also, $\mathsf{DCPO}_\perp$, Pfn, and Rel are strict CPO-enriched.

Lemma 5.1.23 [71]**.** *Every strict* CPO*-enriched category with ω-colimits or ω^{op}-limits has a zero object* $0 = 1$.

Proof We prove the case where ω-colimits exist; the other case is dual. Choose any object X and form the ω-chain whose objects are all equal to X and whose connecting morphisms are $\perp_{XX}$. The colimit $c_n\colon X \to C$ then fulfils $c_n = c_{n+1} \cdot \perp_{XX} = \perp_{XC}$ for all n. Thus C is initial: for every object Y the unique morphism from C is $\perp_{CY}$. Indeed, given $f\colon C \to Y$, we have

$f \cdot c_n = f \cdot \bot_{CX} = \bot_{XY} = \bot_{CY} \cdot c_n$ for all n, thus, $f = \bot_{CY}$ since the colimit injections are jointly epic. Moreover, C is terminal: the unique morphism from Y to C is $\bot_{YC}$ because clearly $\bot_{CC} = \mathsf{id}_C$, hence, given $f\colon Y \to C$ we have $f = \mathsf{id}_C \cdot f = \bot_{CC} \cdot f = \bot_{YC}$. □

The next result provides a sufficient condition for the existence of canonical fixed points. It is the centrepiece of this section and is inspired by the results of Smyth and Plotkin [279].

Theorem 5.1.24. *Let $\mathscr{A}$ be a strict* CPO*-enriched category with ω-colimits or ω^{op}-limits. Every locally continuous endofunctor $F\colon \mathscr{A} \to \mathscr{A}$ has a canonical fixed point*

$$\mu F = \nu F = \mathsf{colim}\, F^n 0.$$

Proof By Lemma 5.1.23 we know that $\mathscr{A}$ has a zero object. The unique morphism $e\colon 0 \to F0$ is an embedding with projection $\widehat{e} = \bot_{F0,0}$ being the unique morphism to the terminal object. Indeed, $\widehat{e} \cdot e = \mathsf{id}$ is clear, and we have $e \cdot \widehat{e} = e \cdot \bot_{F0,0} = \bot_{F0,F0} \sqsubseteq \mathsf{id}_{F0}$. Going further, we have the initial-algebra chain and the terminal-coalgebra chain of F together in (5.1), and all the pairs of morphisms are embedding-projection pairs. In case $\mathscr{A}$ has ω-colimits, let $c_i\colon F^i 1 \to C$ be a colimit cocone of the initial-algebra chain. By Corollary 5.1.20, F preserves this colimit. Moreover, by Corollaries 5.1.13 and 5.1.20, $(\widehat{c_i})$ is a limit cone of the terminal-coalgebra chain of F, and $(F\widehat{c_i})$ is a limit cone of the same chain, but again without the first term.

If $\mathscr{A}$ has limits of ω^{op}-chains, one takes the limit of the terminal-coalgebra chain and argues dually.

As a result, the initial algebra μF and the terminal coalgebra νF exist. The structure ι of the initial algebra is, by Remark 3.1.9, defined by $\iota \cdot Fc_i = c_{i+1}$. Following equation (5.10), this yields

$$\iota = \bigsqcup\nolimits_{i\geq 1} c_i \cdot F\widehat{c}_{i-1}.$$

Dually, the structure τ of the terminal coalgebra fulfils

$$\tau = \bigsqcup\nolimits_{i\geq 1} Fc_{i-1} \cdot \widehat{c}_i.$$

We conclude that ι is inverse to τ using Theorem 5.1.10 once again:

$$\begin{aligned}
&\iota\cdot\tau && \tau\cdot\iota\\
&= \Big(\bigsqcup c_i\cdot F\widehat{c}_{i-1}\Big)\cdot\Big(\bigsqcup Fc_{i-1}\cdot\widehat{c}_i\Big) && = \Big(\bigsqcup Fc_i\cdot\widehat{c}_{i-1}\Big)\cdot\Big(\bigsqcup c_{i-1}\cdot F\widehat{c}_i\Big)\\
&= \bigsqcup c_i\cdot F(\widehat{c}_{i-1}\cdot c_{i-1})\cdot\widehat{c}_i && = \bigsqcup Fc_i\cdot(\widehat{c}_{i-1}\cdot c_{i-1})\cdot F\widehat{c}_i\\
&= \bigsqcup c_i\cdot\widehat{c}_i && = \bigsqcup Fc_i\cdot F\widehat{c}_i\\
&= \mathsf{id} && = \mathsf{id}
\end{aligned}$$

This completes the proof. □

Remark 5.1.25. (1) It follows from the proof of Theorem 5.1.24 that, in lieu of assuming all ω-colimits in $\mathscr{A}$, it is sufficient to assume a zero object $0 = 1$ and colimits of ω-chains of embeddings. Also, instead of assuming that F be locally continuous, we could assume the weaker condition that it preserve ω-colimits of embeddings.

(2) Similarly, in lieu of assuming all ω^{op}-limits in $\mathscr{A}$, it suffices to assume a zero object, limits of ω^{op}-chains of projections and that F preserves those.

Example 5.1.26. (1) A continuous algebra (Example 2.1.7) with a binary operation and a constant consists of a cpo A, a continuous operation $\alpha_0\colon A\times A\to A$, and an element of A. As we have previously observed, CPO has limits and coproducts. It also has colimits of ω-chains of discrete cpos (which are discrete and formed as in Set). A continuous algebra is given by a single morphism $A\times A+1\to A$, namely the structure of an algebra for the endofunctor $FX = X\times X+1$.

This functor preserves colimits of ω-chains of monomorphism. Consequently, we have $\mu F = \mathsf{colim}_{n<\omega} F^n0$ (see Theorem 3.1.8). As in Example 3.2.5, the underlying set is that of all finite binary trees with the discrete order (since F preserves discrete posets).

The functor F also preserves limits of ω^{op}-chains. Thus, $\nu F = \lim_{n<\omega} F^n1$ (see Theorem 3.3.4). As in Theorem 3.3.10 this is the set of all binary trees, again with the discrete order. The reason why we do not obtain a canonical fixed point in this example is that CPO is not strict CPO-enriched.

(2) Let us now consider continuous algebras with a binary operation and a constant in $\mathsf{CPO}_\perp$, where operations are required to be continuous (but not necessarily strict), and morphisms are the strict continuous homomorphisms. A non-strict binary operation on a poset A can be expressed by a strict continuous function from $(A\times A)_\perp$ to A, where $(-)_\perp$ is the lifting, see Example 2.1.7(2). Recall that coproducts in $\mathsf{CPO}_\perp$ are formed by identifying in the disjoint

union the bottom elements. Thus, the structure of a continuous algebra is expressed by a single morphism $\alpha\colon (A \times A)_\bot + 1_\bot$. This is an algebra for the endofunctor $FX = (X \times X)_\bot + 1_\bot$ from Example 2.1.7(2). It is easy to see that F is locally continuous, so by Theorem 5.1.24 it has a canonical fixed point $\mu F = \nu F = \mathsf{colim}_{n<\omega} F^n 0$.

Let us represent the initial object of $\mathsf{CPO}_\bot$ by $\{\bot\}$ where $\bot$ is considered as a single-node labelled tree. Given a tree representation of $F^n 0$, represent the pair (t_1, t_2) in $F^n 0 \times F^n 0$ as the binary tree with maximum subtrees t_1 and t_2. Then it is easy to see that $F^n 0$ is the cpo of all binary trees of height at most n ordered by

$$t \sqsubseteq t' \quad \text{iff} \quad t \text{ is a cutting of } t' \text{ at some level.}$$

The connecting maps $F^n 0 \hookrightarrow F^{n+1} 0$ are the inclusions. It follows that the canonical fixed point is

$$\mu F = \nu F = \text{all binary trees,}$$

ordered as above.

(3) Generalizing the previous example, for every finitary signature Σ we consider continuous algebras in $\mathsf{CPO}_\bot$. Operations are continuous and homomorphisms are the strict continuous Σ-homomorphisms. These algebras are precisely the F-algebras for the endofunctor given by

$$FX = \coprod_{\sigma \in \Sigma} (X^n)_\bot,$$

where n denotes the arity of $\sigma \in \Sigma$. This functor is locally continuous. Its canonical fixed point is T_Σ, the algebra of all Σ-trees, with the following order [29]: Given $t \in T_\Sigma$ as a partial function on $\mathbb{N}^*$ as explained in Remark 2.2.13, let $\hat{t}\colon \mathbb{N}^* \to (\coprod_{n\in\mathbb{N}} \Sigma_n)_\bot$, where the coproduct on the right is discretely ordered, be the total function with $\hat{t}(w) = \bot$ if $t(w)$ is undefined and $\hat{t}(w) = t(w)$ else. Then for two trees $s, t \in T_\Sigma$ put

$$s \sqsubseteq t \quad \text{iff} \quad \hat{s}(w) \sqsubseteq \hat{t}(w) \text{ for all } w \in \mathbb{N}^*.$$

In our next example we work in the strict CPO-enriched category Rel of sets and relations. Coproducts in this category are disjoint unions (as in Set). Since Rel is self-dual, products are also given by disjoint unions where the projections are the opposite relations to the coproduct injections. Moreover, an embedding in Rel is precisely an injective function $e\colon A \to B$ with the projection $\hat{e}$ the opposite relation. Colimits of ω-chains of embeddings are the unions.

Example 5.1.27 [157]. We have discussed deterministic automata in Examples 2.4.2(3) and 2.5.5, and we have seen non-deterministic automata as coalgebras over Set in Example 2.4.2(5). Here we consider the latter as coalgebras in the category Rel (Notation 3.2.16). Recall that non-deterministic automata (ndas) are defined in the same way as deterministic ones, except that the transition functions $\delta_s : S \to S$ for every $s \in \Sigma$ are replaced by transition relations $\delta_s \subseteq S \times S$. Thus, in the category Rel, an nda is given by an object S of states, a morphism $S \to 1$ relating the accepting states to the single element of 1 and a morphism $\delta : S \to \Sigma \times S$ (where $\times$ denotes cartesian product) relating a state $x \in S$ to all pairs (s, y) with $(x, y) \in \delta_s$. We can combine this to obtain a single morphism

$$\alpha : S \to 1 + \Sigma \times S,$$

whence ndas are precisely the coalgebras for the functor $F : \mathsf{Rel} \to \mathsf{Rel}$ given by $FX = 1 + \Sigma \times X$. The initial-algebra ω-chain

$$\emptyset \xrightarrow{!} 1 \xrightarrow{F!} 1 + \Sigma \xrightarrow{FF!} 1 + \Sigma + \Sigma \times \Sigma \xrightarrow{F^3!} \cdots$$

coincides with that of the corresponding endofunctor on Set, since the connecting relations are simply the inclusion maps. Thus, the colimit is Σ^*. It follows from Theorem 5.1.24 that the canonical fixed point of F is $\mu F = \nu F = \Sigma^*$.

It is easy to see that a coalgebra homomorphism from an nda $\alpha : S \to 1 + \Sigma \times S$ to an nda $\alpha' : S' \to 1 + \Sigma \times S'$ is a relation $R : S \to S'$ such that

(1) a state $x \in S$ is accepting iff $x \, R \, x'$ for some accepting state $x' \in S'$, and
(2) for every triple $(x, s, x') \in S \times \Sigma \times S'$ the following statements are equivalent

- there is a transition $x \xrightarrow{s} y$ in S with $y \, R \, x'$, and
- there is a transition $z \xrightarrow{s} x'$ in S' with $x \, R \, z$.

One readily shows that for every nda (S, α) the unique coalgebra homomorphism $R : S \to \Sigma^*$ to the terminal coalgebra relates every state x to the language $L(x)$ it accepts, that is, we have $x \, R \, w$ iff $w \in L(x)$.

In contrast, we saw in Example 4.5.23 that the semantics of non-deterministic automata considered as coalgebras over Set yields the behaviour of states modulo bisimilarity rather than the usual language semantics.

We have seen the finite power-set functor $\mathscr{P}_{\mathsf{f}} : \mathsf{Set} \to \mathsf{Set}$ at several points, and we gave many descriptions of its terminal coalgebra. So it should be interesting to consider analogs of $\mathscr{P}_{\mathsf{f}}$ on Rel.

Example 5.1.28. (1) The functor $\mathscr{P}_{\mathsf{f}}$ has a lifting $P_{\mathsf{f}} : \mathsf{Rel} \to \mathsf{Rel}$ assigning to every relation $r \subseteq X \times Y$ the relation $P_{\mathsf{f}} r \subseteq \mathscr{P}_{\mathsf{f}} X \times \mathscr{P}_{\mathsf{f}} Y$ that consists of all pairs (A, B) of finite sets such that

$$\text{for every } a \in A \text{ there exist } b \in B \text{ with } (a, b) \in r. \tag{5.12}$$

It is easy to see that this is a well-defined locally continuous functor on Rel. Indeed, suppose that $r_n\colon X \to Y$, $n < \omega$, is an ω-chain of relations and $r = \bigcup_{n<\omega} r_n$. Then a pair $(A, B) \in \mathscr{P}_f X \times \mathscr{P}_f Y$ lies in $P_f r$ iff there exists an $n < \omega$ such that (A, B) lies in $P_f r_n$, using the fact that A and B are finite.

Therefore P_f has a canonical fixed point by Theorem 5.1.24 which is a colimit of its initial-algebra chain. Since this is a chain of embeddings and colimits of such chains in Rel are the same as in Set, we conclude that

$$\mu P_f = \nu P_f = V_\omega,$$

i.e. the hereditarily finite sets (see Example 2.2.7(1)).

(2) There are other locally continuous liftings of $\mathscr{P}_f$ to Rel. For example, the lifting P'_f obtained by changing (5.12) to

$$\text{for every } b \in B \text{ there exists } a \in A \text{ with } (a, b) \in r,$$

or P'_f defined such that $P'_f r$ is the intersection of $P_f r$ and $P'_f r$. The canonical fixed points of P'_f and P'_f are given by V_ω, again.

Remark 5.1.29. In some applications one uses categories such as CPO_* whose objects are cpos with $\bot$ and morphisms are continuous maps (not necessarily preserving $\bot$). Observe that this means that composition of morphisms is *left-strict*: $\bot \cdot f = \bot$, but not necessarily right-strict:

Definition 5.1.30. A CPO-enriched category is called *left-strict* if every hom-set has a least element $\bot$ and every morphism f fulfils $\bot \cdot f = \bot$.

Theorem 5.1.24 does not extend to left-strict CPO-enriched categories as shown by the following.

Example 5.1.31. The identity functor on CPO_* does not have an initial algebra although it is locally continuous. (The 'natural candidate' $\mathsf{id}_{\{\bot\}}$ is not an initial algebra: consider any algebra with two fixed points.) Indeed, CPO_* does not have an initial object.

This shows that Lemma 5.1.23 requires strict CPO-enrichedness. However, Freyd [129] proved a weaker result; it uses preservation of fixed points by strict continuous maps:

Lemma 5.1.32 (Uniformity). *Let P and Q be cpos and let $f\colon P \to P$ and $g\colon Q \to Q$ be continuous. For every strict continuous map $h\colon P \to Q$ such that $g \cdot h = h \cdot f$ we have $h(\mu f) = \mu g$.*

Proof First, $h(\mu f)$ is a fixed point of g: we have $g(h(\mu f)) = h(f(\mu f)) = h(\mu f)$. Therefore $\mu g \leq h(\mu f)$. For the reverse, we use Scott induction (Corollary 3.1.3). Let $S = \{x \in P : h(x) \leq \mu g\}$. Since h is strict, we see that $\bot \in S$. Moreover, S is closed under f, for if $x \in S$ we obtain $h(f(x)) = g(h(x)) \leq g(\mu g) = \mu g$ using monotonicity of g in the second step. Finally, S is closed under joins of ω-chains: for any ω-chain (x_n) in S we obtain $h(\bigvee_{n<\omega} x_n) = \bigvee_{n<\omega} h(x_n) \leq \bigvee_{n<\omega} \mu g = \mu g$, whence $\bigvee_{n<\omega} x_n$ lies in S. Thus, $\mu f \in S$, which means that $h(\mu f) \leq \mu g$. □

Proposition 5.1.33. *Let $\mathscr{A}$ be a left-strict* CPO*-enriched category. If a locally continuous functor F has an initial algebra, then μF is a canonical fixed point.*

Proof Let $\iota\colon FI \to I$ be an initial algebra. For every coalgebra $\alpha\colon A \to FA$, we prove that a unique homomorphism into (I, ι^{-1}) exists.

(1) Existence. The endomap g of $\mathscr{A}(A, I)$ given by $h \mapsto \iota \cdot Fh \cdot \alpha$ is obviously continuous, hence, it has a fixed point $h\colon A \to I$ by Kleene's Theorem 3.1.1. This is a coalgebra homomorphism since $\iota^{-1} \cdot h = Fh \cdot \alpha$.

(2) Uniqueness. First notice that for $\mathscr{A}(I, I)$ we have an analogous endomap f given by $k \mapsto \iota \cdot Fk \cdot \iota^{-1}$. Since I is initial, the only fixed point is $k = \mathsf{id}_I$. Thus $\mu f = \mathsf{id}_I$. Now suppose that $h'\colon (A, \alpha) \to (I, \iota^{-1})$ is any coalgebra homomorphism. We know that $\mathscr{A}(h', I)\colon \mathscr{A}(I, I) \to \mathscr{A}(A, I)$ is a strict continuous map; strictness follows from left-strictness of composition: $\bot_{I,I} \cdot h' = \bot_{A,I}$. We now show that $g \cdot \mathscr{A}(h', I) = \mathscr{A}(h', I) \cdot f$. Indeed, unfolding the definitions, we have for every $k\colon I \to I$:

$$
\begin{aligned}
g \cdot \mathscr{A}(h', I)(k) &= g(k \cdot h') = \iota \cdot F(k \cdot h') \cdot \alpha = \iota \cdot Fk \cdot Fh' \cdot \alpha \\
&= \iota \cdot Fk \cdot \iota^{-1} \cdot h' = f(k) \cdot h' = \mathscr{A}(h', I)(f(k)).
\end{aligned}
$$

Therefore, by Lemma 5.1.32, we obtain $\mathscr{A}(h', I)(\mu f) = \mu g$, which means that $h' = \mathsf{id}_I \cdot h' = h$. □

Remark 5.1.34. There is an analogous result for locally monotone functors on categories $\mathscr{A}$ left-strictly enriched in directed-complete partial orders (Example 5.1.2(3)). This means that each hom-set of $\mathscr{A}$ is a dcpo with bottom and composition preserves directed joins and is left-strict. If a locally monotone functor F on $\mathscr{A}$ has an initial algebra $(\mu F, \iota)$, then $(\mu F, \iota^{-1})$ is a terminal coalgebra [54, Prop. 5.6].

One of the themes of this section has been canonical fixed points. We saw in Theorem 5.1.24 conditions on a category and a functor that ensure the existence of a canonical fixed point. Looking ahead to Section 5.2, we shall see conditions there that insure *unique* fixed points. So one might ask whether the fixed points given by Theorem 5.1.24 are unique. As it happens, they are not (see Example 5.1.28 below). However if we assume an extra condition on the fixed point, we do have a uniqueness result.

Proposition 5.1.35. *The canonical fixed point of a locally continuous endofunctor F on $\mathsf{CPO}_\perp$ is the only fixed point $FX \cong X$ with the property that id_X is the least algebra endomorphism of X.*

Moreover, X is the only fixed point with the property that id_X is the only strict algebra endomorphism. For the proof, see Abramsky and Jung [2, Thm. 5.3.2] or Gierz et al [138, Sec. IV-7].

We also might note another fact about the canonical fixed points.

Theorem 5.1.36. *Under the conditions of Theorem 5.1.24, the algebra morphism from the canonical fixed point μF to any fixed point of F is an embedding.*

Proof Let $\alpha\colon FA \to A$ be an isomorphism. Recall the cocone $\alpha_n\colon F^n0 \to A$ from (3.2): $\alpha_0\colon 0 \to A$ is given by initiality, and $\alpha_{n+1} = \alpha \cdot F\alpha_n$. The algebra morphism $h\colon \mu F \to A$ is the factorizing morphism for this cocone. So our result follows from Corollary 5.1.15 as soon as we show that each α_n is an embedding. We also have a coalgebra $\alpha^{-1}\colon A \to FA$ and hence a canonical cone $\beta_n\colon A \to F^n1 = F^n0$. This cone fulfils $\beta_0\colon A \to 1$ by finality, and $\beta_{n+1} = F\beta_n \cdot \alpha^{-1}$. An easy induction shows that $\widehat{\alpha_n} = \beta_n$ for all n. So indeed each α_n is an embedding. □

Remark 5.1.37. A higher-order variation of Theorem 5.1.24 was established by Adámek [11] where cpos are generalized as follows: A category $\mathscr{A}$ is called *Scott complete* if it has

(1) filtered colimits,

(2) an initial object,

(3) limits of diagrams with a cone, and

(4) a set of finitely presentable objects (i.e. objects X whose hom-functor $\mathscr{A}(X, -)$ is finitary) whose closure under filtered colimits is $\mathscr{A}$.

We obtain a 2-category of Scott-complete categories, using as 1-cells the functors preserving filtered colimits and the initial object, and as 2-cells the natural transformations.

Theorem 5.1.38 [11]. *Every locally continuous 2-endofunctor of the category of Scott-complete categories has a canonical fixed point.*

Remark 5.1.39. We have seen how to obtain initial algebras and terminal coalgebras for locally continuous endofunctors on strict CPO-enriched categories. Surprisingly, for all set functors F with $F\emptyset \neq \emptyset$ and preserving ω^{op}-limits, it turns out that the terminal coalgebra also carries a structure of a cpo (see Section 13.3).

In Section 6.7 we prove a (non-finitary) generalization of the main result:

every stable functor on $\mathsf{DCPO}_\perp$ with a fixed point has a canonical one (see Corollary 6.7.8).

5.2 CMet-Enriched Categories

Complete partial orders, used in Section 5.1, express the idea of how much information is contained in an approximation of a solution of a recursive specification. But we can also measure the distance of various approximations, and apply the idea of a metric space in lieu of a poset.

For example, recall that the terminal coalgebra for the polynomial functor H_Σ is the set of all Σ-labelled trees, as explained by Theorem 3.3.10. Arnold and Nivat [61] studied this set using the following natural metric: the distance of distinct trees t and s is

$$d(s,t) = 2^{-n} \text{ for the least level } n \text{ at which } t \text{ and } s \text{ differ.} \tag{5.13}$$

This metric space is *complete*: every Cauchy sequence converges.

In this section we consider categories enriched over the category CMet of complete metric spaces, and prove a result analogous to that of Section 5.1: every locally contracting endofunctor has a canonical fixed point. In fact, a stronger and surprising result is true: such a functor has, up to isomorphism, a *unique* fixed point. Thus every fixed point is canonical.

Complete metric spaces were first applied by de Bakker and Zucker [111], and their method was further developed by America and Rutten in their seminal paper [59] on this topic. In both of these papers, the morphisms are a little different from the non-expanding maps: they are the *retraction-embedding pairs* $(e, \widehat{e})$ consisting of an isometric embedding $e\colon X \to Y$ (a function that preserves distances) and a non-expanding map $\widehat{e}\colon Y \to X$, such that $\widehat{e} \cdot e = \mathsf{id}_X$.

Notation 5.2.1. We denote by

$$\mathsf{CMet}$$

the category of complete metric spaces with distances bounded by 1 and *non-expanding* functions as morphisms; these are the functions f with the property that $d(fx, fy) \le d(x, y)$ for all x, y.

Definition 5.2.2. A category $\mathscr{A}$ is CMet-*enriched* if its hom-sets come equipped with a complete metric, and composition is non-expanding in both variables. This means that given morphisms $f_1, f_2\colon A \to B$, then for all h and g composable with f_i we have

$$d(h \cdot f_1 \cdot g, h \cdot f_2 \cdot g) \le d(f_1, f_2). \tag{5.14}$$

The category CMet has finite products $A_1 \times A_2$: they are given by cartesian products equipped with the maximum metric:

$$d((x_1, x_2), (y_1, y_2)) = \max\{d_1(x_1, y_1), d_2(x_2, y_2)\}.$$

Unlike CPO, it is not cartesian closed. However, there is a natural complete metric on $\mathsf{CMet}(X, Y)$, the supremum metric which is given by

$$d_{X,Y}(f, g) = \sup_{x \in X} d_Y(f(x), g(x)).$$

Example 5.2.3. (1) The category CMet is CMet-enriched with respect to the supremum metric.

(2) Denote by CMet_* the category of pointed complete metric spaces, that is, objects are pairs (X, x) where X is a complete metric space and $x \in X$, and morphisms are non-expanding maps preserving the distinguished point. This category is also CMet-enriched using the supremum metric.

(3) If $\mathscr{C}$ is CMet-enriched, then so is $\mathscr{C}^{\mathrm{op}}$, where the metric of $\mathscr{C}^{\mathrm{op}}(A, B)$ is that of $\mathscr{C}(B, A)$.

(4) A product $\mathscr{C} \times \mathscr{C}'$ of CMet-enriched categories (with metrics $d_{A,B}$ and $d'_{A',B'}$ on hom-sets, respectively) is CMet-enriched using the maximum metric: the distance of (f, f') and (g, g') is $\max\{d(f, g), d'(f', g')\}$.

At the root of our study is the following classical result in which a function $f : X \to Y$ is *contracting* if for some $\varepsilon < 1$ we have $d(f(x), f(y)) < \varepsilon \cdot d(x, y)$ for all $x, y \in X$.

Theorem 5.2.4 (Banach Fixed Point Theorem [66]). *Every contracting endofunction on a nonempty complete metric space has a unique fixed point.*

A proof can be found in Section B.3.

Definition 5.2.5 [59]. An endofunctor F of a CMet-enriched category $\mathscr{A}$ is *locally contracting*if there exists $\varepsilon < 1$ such that for all parallel pairs $f, g : X \to Y$ we have

$$d(Ff, Fg) \leq \varepsilon \cdot d(f, g).$$

Example 5.2.6. (1) For every $\varepsilon < 1$, the functor which takes a space (X, d) and scales the metric by ε, giving $(X, \varepsilon d)$ is contracting.

(2) Every composite, coproduct, and finite product of locally contracting endofunctors is locally contracting.

(3) Every polynomial functor H_Σ on Set has a *contracting lifting* H'_Σ to CMet.

This means that the following square

$$\begin{array}{ccc} \mathsf{CMet} & \xrightarrow{H'_\Sigma} & \mathsf{CMet} \\ {\scriptstyle U}\downarrow & & \downarrow{\scriptstyle U} \\ \mathsf{Set} & \xrightarrow{H_\Sigma} & \mathsf{Set} \end{array}$$

commutes, where U is the forgetful functor taking a metric space to its set of points. This follows from (1) and (2) since polynomial functors are formed from Id by finite products and coproducts. Using the constant $\varepsilon = \frac{1}{2}$ in (1) above, we can describe the terminal coalgebra for H'_Σ on CMet as the set of all Σ-trees equipped with the metric of (5.13) as we will see in Example 5.2.25(3) below. Moreover, if Σ contains at least one constant symbol, then this is a canonical fixed point of H'_Σ.

(4) The Hausdorff Functor. This is the endofunctor $\mathcal{H}$ of Met assigning to a metric space (X, d) the space $\mathcal{H}(X, d)$ of all nonempty compact subsets. The definition of its metric goes back to Pompeiu [253] and was popularized by Hausdorff [159, p. 293]; it is nowadays often called the *Hausdorff metric*. It is defined by

$$\bar{d}(S, T) = \max\left(\sup_{x\in S} d(x, T), \sup_{y\in T} d(y, S)\right).$$

for every pair S, T of nonempty compact sets. Here $d(x, S)$ denotes $\inf_{y\in T} d(x, y)$. For morphisms $f\colon (X, d) \to (X', d')$ we define $\mathcal{H}f\colon A \mapsto f[A]$. The resulting functor $\mathcal{H}$ is locally non-expanding, but not locally contracting.

It is well-known that for a complete metric space X, the space $\mathcal{H}X$ is complete, too (see e.g. Barnsley [68, Thm. 7.1]).

Thus, $\mathcal{H}$ restricts to an endofunctor on CMet, which we denote by the same symbol. This restricted functor is finitary [47], while $\mathcal{H}$ on Met is not.

By modifying the Hausdorff functor to halve all distances as in (3), we obtain a contracting endofunctor.

We are going to present the proof that every locally contracting endofunctor has a canonical fixed point with all details. As in the CPO-enriched case, this material has not been presented so comprehensively before. Once again, for CMet the main ideas stem from America and Rutten [59]. Their results were later extended [22], and we sharpen them a bit. First a surprising result about canonical fixed points:

Theorem 5.2.7. *Let $\mathscr{A}$ be a* CMet-*enriched category with nonempty hom-sets. Every fixed point of a locally contracting endofunctor on $\mathscr{A}$ is canonical.*

Proof (1) Let F be a locally contracting endofunctor with contraction factor ε. Given a fixed point $\alpha\colon FA \to A$, we prove that this is the initial algebra of F.

For any algebra $\beta\colon FB \to B$ define an endofunction k of $\mathscr{A}(A,B)$ as follows:

$$k\colon (A \xrightarrow{f} B) \longmapsto (A \xrightarrow{\alpha^{-1}} FA \xrightarrow{Ff} FB \xrightarrow{\beta} B).$$

Then k is ε-contracting: for every parallel pair $f, g\colon A \to B$ we have

$$\begin{aligned} d(k(f), k(g)) &= d(\beta \cdot Ff \cdot \alpha^{-1}, \beta \cdot Fg \cdot \alpha^{-1}) \\ &\leq d(Ff, Fg) \\ &\leq \varepsilon \cdot d(f, g) \end{aligned}$$

since composition is non-expanding and F is ε-contracting. Thus k has a fixed point:

$$h\colon A \to B \text{ with } h = \beta \cdot Fh \cdot \alpha^{-1}.$$

This implies that h is an algebra homomorphism. Conversely, every algebra homomorphism ($\alpha \cdot h = \beta \cdot Fh$) is a fixed point of k. By Banach's Theorem 5.2.4, we see that (A, α) is an initial algebra.

(2) The proof that $\alpha^{-1}\colon A \to FA$ is a terminal coalgebra is by duality: if $\mathscr{C}$ is a CMet-enriched category, then so is $\mathscr{C}^{\mathsf{op}}$ (using the same metric on hom-sets). Moreover, F^{op} is a contracting endofunctor. By applying (1) to it we see that (A, α^{-1}) is an initial algebra for F^{op}, that is, a terminal coalgebra for F. □

Notation 5.2.8. Let $\mathscr{A}$ be a CMet-enriched category. We denote by $\mathscr{A}^E$ the category of all objects of $\mathscr{A}$, where the morphisms from A to B are all pairs $(e, \widehat{e})$ of morphisms $e\colon A \to B$ and $\widehat{e}\colon B \to A$ in $\mathscr{A}$ such that $\widehat{e} \cdot e = \mathsf{id}_A$:

$$\mathsf{id} \circlearrowright A \underset{\widehat{e}}{\overset{e}{\rightleftarrows}} B$$

Composition and identity morphisms in $\mathscr{A}^E$ are defined componentwise. If $\mathscr{A} = \mathsf{CMet}$, then $\widehat{e} \cdot e = \mathsf{id}_A$ implies that e is an isometric embedding. Thus CMet^E is precisely the category $\mathscr{C}$ of America and Rutten [59].

One difference between the work in this section and the previous one is that in the case of CPO-enriched categories, each embedding e had a unique projection $\widehat{e}$. The uniqueness is lost in the CMet setting. The morphisms in categories $\mathscr{A}^E$ are pairs in the CMet setting, while in the CPO setting we were able to make due with only the embeddings.

Definition 5.2.9. An ω-chain in the category $\mathscr{A}^E$

$$E_0 \underset{\widehat{e}_0}{\overset{e_0}{\rightleftarrows}} E_1 \underset{\widehat{e}_1}{\overset{e_1}{\rightleftarrows}} E_2 \underset{\widehat{e}_2}{\overset{e_2}{\rightleftarrows}} \cdots \tag{5.15}$$

is *contracting* if there exists $\varepsilon < 1$ such that for all i we have

$$d(e_{i+1} \cdot \widehat{e}_{i+1}, \mathsf{id}_{E_{i+1}}) \leq \varepsilon \cdot d(e_i \cdot \widehat{e}_i, \mathsf{id}_{E_i}). \tag{5.16}$$

Example 5.2.10. Let F be a locally contracting endofunctor on a CMet-enriched category with a terminal object 1 and such that $\mathscr{A}(1, F1)$ is nonempty. Given $e\colon 1 \to F1$, the chain

$$1 \underset{!}{\overset{e}{\rightleftarrows}} F1 \underset{F!}{\overset{Fe}{\rightleftarrows}} F^2 1 \underset{F^2!}{\overset{F^2 e}{\rightleftarrows}} \cdots$$

is contracting. Indeed, if ε is the contraction constant for F, then

$$d(F^{i+1}e \cdot F^{i+1}!, \mathsf{id}) = d(F(F^i e \cdot F^i !), F\mathsf{id}) \leq \varepsilon \cdot d(F^i e \cdot F^i !, \mathsf{id}).$$

Theorem 5.2.11. *Let $\mathscr{A}$ be a* CMet*-enriched category. Consider a contracting chain*

$$E_0 \underset{\widehat{e}_0}{\overset{e_0}{\rightleftarrows}} E_1 \underset{\widehat{e}_1}{\overset{e_1}{\rightleftarrows}} E_2 \underset{\widehat{e}_2}{\overset{e_2}{\rightleftarrows}} \cdots \tag{5.17}$$

in $\mathscr{A}^E$. Let $c_i\colon E_i \to C$ be a cocone of $(e_i)_{i\in\mathbb{N}}$ in $\mathscr{A}$. Then the following are equivalent:

(1) *(c_i) is a colimit cocone in $\mathscr{A}$.*

(2) *Each c_i is a split monomorphism and there exists a family of splittings $(\widehat{c}_i)$ forming a cone of $(\widehat{e}_i)$ in $\mathscr{A}$ with* $\lim_i (c_i \cdot \widehat{c}_i) = \mathsf{id}_C$.

Proof First observe that condition (5.16) implies that for all $i < k$ we have

$$d(e_k \cdot \widehat{e}_k, \mathsf{id}) \leq \varepsilon^{k-i} \cdot d(e_i \cdot \widehat{e}_i, \mathsf{id}) \leq \varepsilon^k.$$

The proof is any easy induction on i where in the first step we use that we have $d(e_0 \cdot \widehat{e}_0, \mathsf{id}) \leq 1$. (Recall that 1 is our bound on all distances.)

To prove (1) $\Rightarrow$ (2) use, for every i, the shortened ω-chain

$$E_i \xrightarrow{e_i} E_{+1} \xrightarrow{e_{i+1}} E_{i+2} \xrightarrow{e_{i+2}} \cdots$$

and define, analogously to Theorem 5.1.10, morphisms $\widehat{c}_i\colon C \to E_i$ by

$$\widehat{c}_i \cdot \widehat{c}_i = \mathsf{id}_{E_i} \qquad \text{and} \qquad \widehat{c}_i \cdot c_k = \widehat{e}_i \cdot \dots \cdot \widehat{e}_{k-1} \quad \text{for } k > i.$$

Again, we obtain the formula (5.9) for all i

$$\widehat{c}_i = \widehat{e}_i \cdot \widehat{c}_{i+1}.$$

The sequence $c_i \cdot \widehat{c}_i : C \to C$ is Cauchy: to prove this, we verify below that

$$d(c_i \cdot \widehat{c}_i, c_{i+1} \cdot \widehat{c}_{i+1}) \leq \varepsilon^i \quad \text{for all } i,$$

from which we conclude that the distance of $c_i \cdot \widehat{c}_i$ and $c_k \cdot \widehat{c}_k$, where $k > i$, is at most

$$\varepsilon^i + \varepsilon^{i+1} + \cdots + \varepsilon^{k-1} \leq \sum_{j=i}^{\infty} \varepsilon^j = \frac{\varepsilon^i}{1-\varepsilon}.$$

This converges to 0 as $i \to \infty$. The desired inequality is clear from $\widehat{e}_i \cdot \widehat{c}_{i+1} = \widehat{c}_i$ and $c_{i+1} \cdot e_i = c_i$:

$$\begin{aligned} d(c_i \cdot \widehat{c}_i, c_{i+1} \cdot \widehat{c}_{i+1}) &= d(c_{i+1} \cdot (e_i \cdot \widehat{e}_i) \cdot \widehat{c}_{i+1}, c_{i+1} \cdot \widehat{c}_{i+1}) \\ &\leq d(e_i \cdot \widehat{e}_i, \mathsf{id}) \qquad \text{by (5.14)} \\ &\leq \varepsilon^i. \end{aligned}$$

The rest is analogous to Theorem 5.1.10 again: we know that $\lim_i (c_i \cdot \widehat{c}_i)$ exists, and we derive that it is id_C.

The proof of (2) $\Rightarrow$ (1) is also analogous to Theorem 5.1.10. Given a cocone $b_i : E_i \to B$ of $(\widehat{e}_i)$, the sequence $b_i \cdot \widehat{c}_i : C \to B$ is Cauchy. This follows from

$$\begin{aligned} d(b_i \cdot \widehat{c}_i, b_{i+1} \cdot \widehat{c}_{i+1}) &= d(b_{i+1} \cdot (e_i \cdot \widehat{e}_i) \cdot \widehat{c}_{i+1}, b_{i+1} \cdot \widehat{c}_{i+1}) \\ &\leq d(e_i \cdot \widehat{e}_i, \mathsf{id}) \qquad \text{by (5.14)} \\ &\leq \varepsilon^i. \end{aligned}$$

Then $\lim_i (b_i \cdot \widehat{c}_i)$ is the desired unique factorization morphism. □

The following corollary yields a limit–colimit coincidence in CMet-enriched categories analogous to the one we have seen in Remark 5.1.14.

Corollary 5.2.12. *Given a chain (5.17) in* $\mathscr{A}^E$, *a cone* $(\widehat{c}_i)$ *of the* ω^{op}*-chain* $(\widehat{e}_i)$ *is a limit cone in* $\mathscr{A}$ *iff each* $\widehat{e}_i$ *is a split epimorphism and there exist splittings* c_i *forming a cocone of* (e_i) *with* $\lim c_i \cdot \widehat{c}_i = \mathsf{id}_C$.

By Example 5.2.3(3), we can apply Theorem 5.2.11 to $\mathscr{A}^{\mathsf{op}}$.

Corollary 5.2.13. *If a* CMet*-enriched category has* ω*-colimits, then so does* $\mathscr{A}^E$.

Remark 5.2.14. (1) Theorem 5.2.11 also holds for the *converging towers* of America and Rutten [59]: These are sequences (5.15) such that for every $\varepsilon > 0$ there exists N where the distance between id_{E_m} and $e_n \cdot \cdots \cdot e_{m+1} \cdot \widehat{e}_{m+1} \cdot \cdots \cdot \widehat{e}_n$ is at most ε for all $m > n \geq N$. The proof is the same.

(2) In analogy to Theorem 5.1.24 we might now expect that contracting endofunctors on CMet-enriched categories have canonical fixed points. Unfortunately, this is not quite true:

(a) A locally contracting endofunctor need not have a terminal coalgebra: take the poset $\mathbb{N}^{op}$ as a category. It is trivially CMet-enriched. Every endofunctor is (trivially) contracting. But not every endofunctor has a terminal coalgebra, for example $s(x) = x + 1$ does not have one.

(b) A locally contracting endofunctor on CMet with a terminal coalgebra need not have a canonical fixed point: take $F(X, d) = (X, \frac{1}{2}d)$. Here $\mu F = \emptyset$ and $\nu F = 1$. (We might note in passing that on Met, this functor has other fixed points, such as the open interval $(0, 1)$.)

Assuming strictness, analogous as required in the Theorem 5.1.24, canonical fixed points exist:

Definition 5.2.15. A *strict* CMet-*enriched category* $\mathscr{A}$ is a nonempty CMet-enriched category in which every hom-object $\mathscr{A}(X,Y)$ has a chosen element $\bot_{XY}$ such that for all morphisms $f\colon A \to B$ we have $f \cdot \bot_{XA} = \bot_{XB}$ and $\bot_{BY} \cdot f = \bot_{AY}$.

Remark 5.2.16. For example, the category CMet_* from Example 5.2.3(2) is strict CMet-enriched. Moreover, the notion of a strict CMet-enriched category is precisely that of a category enriched over CMet_* in the sense of enriched category theory [188].

Theorem 5.2.17. *Every locally contracting endofunctor on a strict* CMet-*enriched category with* ω*-colimits has a canonical fixed point.*

Proof Every strict CMet-enriched category $\mathscr{A}$ has a zero object, the proof is completely analogous to Lemma 5.1.23. The rest of the proof is analogous to the proof of Theorem 5.1.24, using Theorem 5.2.11 and arguing with $\lim_i$ in lieu of $\bigsqcup_i$. □

Corollary 5.2.18. *Every locally contracting endofunctor* F *on* CMet *such that* $F\emptyset \neq \emptyset$ *has a canonical fixed point.*

Indeed, we modify F to an endofunctor $\bar{F}$ on CMet_* as follows: choose $a_0 \in F\emptyset$ and for every space X put $a_X = Fh_X(a_0)$, where $h_X\colon \emptyset \to X$ is the unique morphism. Define $\bar{F}$ on objects by

$$\bar{F}(X,x) = (FX, a_X)$$

and on morphisms $f\colon (X,x) \to (Y,y)$ by $\bar{F}f = Ff$. This is well-defined due to $h_Y = f \cdot h_X$. Then $\bar{F}$ is locally contracting, hence, has a canonical fixed point. This is also the canonical fixed point of F.

Remark 5.2.19. (1) Instead of assuming all ω-colimits in Theorem 5.2.17, it is sufficient to assume a zero object and colimits of ω-chains of split monomorphisms.

(2) For CMet-enriched categories, the theorem above does not hold in general, but the following weaker result has a completely analogous proof:

Theorem 5.2.20. *Let $\mathscr{A}$ be a* CMet*-enriched category with ω-colimits. Let $F\colon \mathscr{A} \to \mathscr{A}$ be locally contracting with $\mathscr{A}(1, F1)$ nonempty. Then F has a terminal coalgebra.*

Remark 5.2.21. In the above proof of the existence of a canonical fixed point, we did not fully use the power of F being locally contracting: we only needed this for pairs consisting of id_A and an endomorphism on A. This led America and Rutten [59] to the concept of a *hom-contracting* endofunctor F, that is, there exists $\varepsilon < 1$ with

$$d(Ff, \mathrm{id}_A) \leq \varepsilon \cdot d(f, \mathrm{id}_A) \qquad \text{for all } f\colon A \to A.$$

However, this is equivalent to being locally contracting whenever the following type of coproducts exists:

Definition 5.2.22. In a CMet-enriched category, given $\varepsilon < 1$, an *ε-contracting copower* of an object A is an object $A +_\varepsilon A$ with morphisms $i_1, i_2\colon A \to A +_\varepsilon A$ universal in the following sense:

(1) $d(i_1, i_2) \leq \varepsilon$;

(2) given morphisms $f_1, f_2\colon A \to B$ with $d(f_1, f_2) \leq \varepsilon$, there exists a unique morphism $[f_1, f_2]\colon A +_\varepsilon A \to B$ with $f_k = [f_1, f_2] \cdot i_k$ for $k = 1, 2$;

(3) the distance of morphisms $u, v\colon A +_\varepsilon A \to C$ is the maximum of $d(u \cdot i_1, v \cdot i_1)$ and $d(u \cdot i_2, v \cdot i_2)$. (This implies that $\mathrm{id}_{A+_\varepsilon A}$ and $[i_2, i_1]$ have distance at most ε: $d(id, [i_2, i_1]) \leq \varepsilon$; to see this choose $C = A +_\varepsilon A$ and $u = \mathrm{id} = [i_1, i_2]$, $v = [i_2, i_1]$.)

Example 5.2.23. The category CMet has contracting copowers. Indeed, put $A + A = A \times \{1, 2\}$ where the metric on $A \times \{i\}$ is that of A and the distance of $(x, 1)$ and $(x, 2)$ is ε, while $(x, 1)$ and $(y, 2)$ have distance 1 for $x \neq y$.

Lemma 5.2.24. *Every hom-contracting endofunctor on a* CMet*-enriched category with contracting copowers for all $\varepsilon < 1$ is locally contracting.*

Proof Let F be hom-contracting with contraction factor ε. Given a parallel pair $f_1, f_2\colon A \to B$, we prove $d(Ff_1, Ff_2) \leq \varepsilon \cdot d(f_1, f_2)$. To this end, consider the function $g = [f_1, f_2]\colon A +_r A \to B$ and put $\sigma = [i_2, i_1]$. Then we have

$$\begin{aligned}
d(Ff_1, Ff_2) &= d(Fg \cdot Fi_1, Fg \cdot Fi_2) \\
&\le d(Fi_1, Fi_2) \\
&= d(Fi_1, F\sigma \cdot Fi_1) && \text{since } \sigma \cdot i_1 = i_2 \\
&\le d(\mathsf{id}, F\sigma) \\
&\le \varepsilon \cdot d(\mathsf{id}, \sigma) && \text{since } F \text{ is contracting} \\
&= \varepsilon \cdot d(f_1, f_2).
\end{aligned}$$

□

Example 5.2.25. (1) Binary streams. Let us consider $F(X, d) = 2 \times (X, \frac{1}{2}d)$ on CMet. Here, 2 is the two-point space $\{0, 1\}$ with the discrete metric. Since $F\emptyset = \emptyset$, the initial algebra of F is also $\emptyset$. Its terminal coalgebra is carried by the set 2^ω of binary streams with the metric given for distinct streams (s_i) and (t_i) by

$$d(s_0 s_1 s_2 \cdots, t_0 t_1 t_2 \cdots) = 2^i, \quad \text{where } i \text{ is least with } s_i \ne t_i.$$

To see this, first note that the set 2^ω is the terminal coalgebra for $FX = 2 \times X$ on Set. Moreover, every space $F^n 1$ is the set 2^n of binary sequences of length n with the metric given similarly. Notice also that the above metric is a special case of the metric (5.13) as F is (isomorphic to) the lifting H'_Σ of the polynomial functor $H_\Sigma X = X + X$.

(2) We have discussed deterministic automata and formal languages in Example 2.5.5, and now we revisit the topic from the metric point of view. This involves the functor $FA = \{0, 1\} \times X^A$ on Set. We lift this to CMet as in our previous examples: $\{0, 1\}$ is the discrete space and X^A carries the half-of-the-maximum metric. We obtain as a terminal coalgebra the set $\mathscr{P}(A^*)$ of formal languages over A. The structure

$$\langle o, t\rangle\colon \mathscr{P}(A^*) \to \{0, 1\} \times \mathscr{P}(A^*)^A$$

is given by $t(L)(a) = \{w : aw \in L\}$, and $o(L) = 1$ iff the empty word ε belongs to L. The metric assigns to two different languages L and M the number 2^{-n}, where n is least so that there is a word of length n in $(L \setminus M) \cup (M \setminus L)$.

(3) Σ-trees. Recall the contracting lifting H'_Σ of the polynomial endofunctor H_Σ from Example 5.2.6(3) with $\varepsilon = \frac{1}{2}$. That is, given elements of $H'_\Sigma X$ of the same summand $\sigma \in \Sigma$, say $x = \sigma(t_i)$ and $x' = \sigma(t'_i)$, then $d(x, x') = \frac{1}{2} \max d(t_i, t'_i)$. It follows that the metric space T_Σ of all Σ-trees with the Arnold–Nivat metric (5.13) is a fixed point of H'_Σ: the usual tree-tupling map $\alpha\colon H'_\Sigma T_\Sigma \to T_\Sigma$ is an isomorphism. Indeed, consider two elements of $H' T_\Sigma = \coprod_{\sigma \in \Sigma} T^n_\Sigma : x = (t_i)_{i<n}$ in the summand σ and $x' = (t'_j)_{j<m}$ in the summand σ'. If $\sigma \ne \sigma'$ the

distances of x, x' and $\alpha(x), \alpha(x')$ are both 1. If $\sigma = \sigma'$, then $d(x, x') = \frac{1}{2}2^{-n}$ where $2^{-n} = \max d(t_i, t_i')$ means that for some i_0 the trees t_{i_0}, t_{i_0}' differ at level n. Then $\alpha(x)$ and $\alpha(x')$ differ first at level $n+1$, thus $d(\sigma(x), \sigma(x')) \leq 2^{-(n+1)}$.

In fact, T_Σ is the terminal coalgebra for H_Σ'. In addition, if Σ contains a constant symbol, T_Σ is the canonical fixed point of H_Σ'.

(4) If Σ in (3) consists of two $|A|$-ary operations, we get (2) above. Here T_Σ is the complete $|A|$-ary tree without leaves, labelled by $\{0, 1\}$. Since that tree can be identified with A^*, we again get $T_\Sigma = \mathscr{P}(\Sigma^*)$ with the metric above.

(5) As another concrete example consider Σ of one binary operation and one constant. The functor $H_\Sigma X = X \times X + 1$ with the maximum metric on $X \times X$ is not contracting: its initial algebra is as in Set given by finite binary trees equipped with the discrete metric, and similarly for the terminal coalgebra. But the scaled functor $H_\Sigma' X$ has T_Σ, the binary trees (with the metric (5.13)) as the unique fixed point.

Example 5.2.26 [1]. Recall the limit $\mathbb{F} = \lim_{n<\omega} \mathscr{P}_\mathsf{f}^n 1$ from Section 4.5, where $\mathscr{P}_\mathsf{f}$ is the finite power-set functor. We present a 'natural' endofunctor on CMet whose terminal coalgebra is $\mathbb{F}$.

Denote by 1 the terminal endofunctor of CMet with value, say, $\{\emptyset\}$, and recall the Hausdorff functor $\mathcal{H}\colon \mathsf{CMet} \to \mathsf{CMet}$ from Example 5.2.6(3). Then the coproduct $\mathcal{H}(-) + 1$ assigns to every finite metric space (X, d) a metric space on the set $\mathscr{P}_\mathsf{f} X$: every nonempty subset is of course compact, and the summand 1 takes care of the empty set. We shall see in Example 15.3.2(2) that $\mathcal{H}(-) + 1$ has a terminal coalgebra given by $\nu\mathscr{P}_\mathsf{f}$ equipped with the discrete metric. Let

$$\widetilde{\mathcal{H}}\colon \mathsf{CMet} \to \mathsf{CMet}$$

be the functor scaling the metric on $\mathcal{H}(-) + 1$ by $\frac{1}{2}$. Then $\widetilde{\mathcal{H}}$ is a locally contracting endofunctor and, again, $\widetilde{\mathcal{H}}(X, d)$ has the underlying set $\mathscr{P}_\mathsf{f} X$ for all finite metric spaces (X, d). Consequently, the terminal ω^{op}-chain of $\widetilde{\mathcal{H}}$ is a lifting of that of $\mathscr{P}_\mathsf{f}$. Since ω^{op}-limits in CMet are preserved by the forgetful functor to Set, and since $\nu\widetilde{\mathcal{H}} = \lim \widetilde{\mathcal{H}}^n 1$, we conclude that $\nu\widetilde{\mathcal{H}}$ is carried by $\mathbb{F}$ using Theorem 5.2.20. More precisely, considering $\nu\widetilde{\mathcal{H}}$ as an algebra for $\widetilde{\mathcal{H}}$ (by inverting the coalgebra structure), the underlying $\mathscr{P}_\mathsf{f}$-algebra is $\mathbb{F}$.

Later, in Theorem 11.2.18, we shall obtain a result that allows us to obtain a terminal coalgebra for the class of functors formed by coproducts, products and compositions of polynomial functors, and the Hausdorff functor without any scaling (see Example 11.2.20(2)).

Example 5.2.27. Every limit of an ω^{op}-chain in Set (with projections $\ell_n: L \to X_n$) carries a canonical complete metric as follows: given elements $x \neq y$ of L, put

$$d(x, y) = 2^{-n} \quad \text{for the least } n \text{ with } \ell_n(x) \neq \ell_n(y). \tag{5.18}$$

For every Cauchy sequence (x_k) in the limit it is easy to see that there exists a subsequence (y_m) satisfying $d(y_m, y_{m+1}) \leq 2^{-(m+1)}$ for every m. This means that the connecting morphism $f_n: X_{n+1} \to X_n$ fulfils

$$f_n \cdot \ell_{n+1}(y_{m+1}) = \ell_n(y_{n+1}) = \ell_n(y_n).$$

Thus, the elements $\ell_n(y_n) \in X_n$ form a compatible sequence of elements of the given chain. Therefore, there is a unique $y \in L$ with $\ell_n(y) = \ell_n(y_n)$ for all n. Hence $d(y, y_n) \leq 2^{-n}$, which means that $\lim y_m = y$; whence $\lim x_k = y$.

The following result was first proved by Barr [72] under the additional assumption that F also preserves ω-colimits:

Theorem 5.2.28 [13]. *Let F: $\mathsf{Set} \to \mathsf{Set}$ fulfil $F\emptyset \neq \emptyset$ and have the terminal coalgebra $\nu F = \lim_{n<\omega} F^n 1$. Then νF with the metric in* (5.18) *is a Cauchy completion of the initial algebra.*

Remark 5.2.29. (1) Birkedal, Støvring, and Thamsborg [82] considered a setting that is quite similar to our work in this section. They work with *ultrametric spaces*, where one strengthens the triangle inequality to the condition that $d(x, z) \leq \max\{d(x, y), d(y, z)\}$. For example, the above metric (5.18) is an ultrametric.

(2) Alessi, Baldan, and Belle [57] study endofunctors of KMet, the full subcategory of CMet on compact metric spaces. This category is obviously CMet-enriched, but fails to have ω^{op}-limits, which makes the situation more complicated.

Theorem 5.2.20 has an analog. Let us call an endofunctor F on KMet *locally contracting* if there exists a constant $\varepsilon < 1$ with $d(Ff, Fg) \leq \varepsilon \cdot d(f, g)$ for every parallel pair f, g of morphisms.

Theorem 5.2.30 [57]. *Let F: $\mathsf{KMet} \to \mathsf{KMet}$ be locally contracting, and assume that $F0 \neq 0$. Then F has a canonical fixed point.*

5.3 Solving Domain Equations

Theorem 5.1.24 allows us to solve the 'equation' $X \cong FX$ whenever F is a locally continuous endofunctor on a strict CPO-enriched category with ω-colimits.

The solutions are not in general unique, but the canonical fixed point gives a workable notion of solution.

We are also interested in solving 'equations with mixed variance' such as $D \cong [D, D] + N_\perp$. Indeed, solving this kind of equation was Scott's original motivation for much of the work on embedding-projection pairs, due to its connection with models of the untyped lambda calculus. So work on this topic is, in effect, the source of embedding-projection pairs in domain theory and also the categorical formulation of the limit–colimit coincidence (see Scott [273] and Smyth and Plotkin [279]). The obstacle is that $[D, D]$ is not a functor of the form $\mathsf{CPO}_\perp \times \mathsf{CPO}_\perp \to \mathsf{CPO}_\perp$. Indeed, it is contravariant in its first argument, so its type is $\mathsf{CPO}_\perp^{\mathsf{op}} \times \mathsf{CPO}_\perp \to \mathsf{CPO}_\perp$. But the theory which we have developed up to this point does not allow us to obtain a fixed point of such a functor, due precisely to the contravariance. As it happens, there are two ways to proceed. They are related, and we are going to present both and briefly touch on their relationship afterwards. Recall from Notation 5.1.7 the category $\mathscr{A}^E$ of objects and embeddings.

Theorem 5.3.1. *Let $\mathscr{A}$ be a strict* CPO*-enriched category with ω-colimits of embeddings. Every locally continuous functor $F\colon \mathscr{A}^{\mathsf{op}} \times \mathscr{A} \to \mathscr{A}$ yields an endofunctor F^E on $\mathscr{A}^E$ defined by*

$$F^E X = F(X, X) \qquad \text{and} \qquad F^E e = F(\widehat{e}, e).$$

Moreover, F^E has an initial algebra $X = \mu F^E$ which satisfies $F(X, X) \cong X$ in $\mathscr{A}$.

Proof It is easy to check the functoriality of F^E. To verify that $F(\widehat{e}, e)$ is an embedding, one checks that its projection is $F(e, \widehat{e})$.

We show that F^E preserves ω-colimits of embeddings (cf. Corollary 5.1.15):

$$E_0 \xrightarrow{e_0} E_1 \xrightarrow{e_1} E_2 \xrightarrow{e_2} \cdots$$

Let $c_i\colon E_i \to C$ be a colimit cocone. By Theorem 5.1.10, $\bigsqcup_i c_i \cdot \widehat{c_i} = \mathsf{id}$. We consider the following ω-chain in $\mathscr{A}$:

$$F(E_0, E_0) \xrightarrow{F(\widehat{e_0}, e_0)} F(E_1, E_1) \xrightarrow{F(\widehat{e_1}, e_1)} F(E_2, E_2) \xrightarrow{F(\widehat{e_2}, e_2)} \cdots,$$

and we show that $(F(\widehat{c_i}, c_i))$ is a colimit cocone. By Theorem 5.1.10 again, we need to show that

$$\bigsqcup_i F(\widehat{c_i}, c_i) \cdot F(c_i, \widehat{c_i}) = \mathsf{id}.$$

For this, we use the local continuity:

$$\begin{aligned}\bigsqcup_i F(\widehat{c_i}, c_i) \cdot F(c_i, \widehat{c_i}) &= \bigsqcup_i F(\widehat{c_i} \cdot c_i, c_i \cdot \widehat{c_i}) \\ &= F\left(\bigsqcup_i \widehat{c_i} \cdot c_i, \bigsqcup_i c_i \cdot \widehat{c_i}\right) \\ &= F(\mathsf{id}, \mathsf{id}) \\ &= \mathsf{id}.\end{aligned}$$

From Theorem 3.1.8, we conclude that $X = \mathsf{colim}(F^E)^n 0$ is an initial F^E-algebra. By Lambek's Lemma, X is a fixed point of F^E. Clearly, $X \cong F(X, X)$ in $\mathscr{A}$. This completes the proof. □

We now turn to the second solution method. The proof of the following theorem is based on an argument by Glimming and Ghani [142].

Theorem 5.3.2 [129]. *Let $\mathscr{A}$ be a strict* CPO*-enriched category with ω-colimits, and let $F\colon \mathscr{A}^{\mathsf{op}} \times \mathscr{A} \to \mathscr{A}$ be locally continuous. We define an endofunctor G on $\mathscr{A}^{\mathsf{op}} \times \mathscr{A}$ by*

$$\begin{aligned} G(X,Y) &= (F(Y,X), F(X,Y)) \\ G(f,g) &= (F(g,f), F(f,g)) \qquad \textit{for } f\colon X' \to X \textit{ and } g\colon Y \to Y'. \end{aligned}$$

Then G has a fixed point of the form (X, X), and for such X we have $F(X, X) \cong X$.

Proof We define an isomorphism of categories

$$I\colon \mathsf{Alg}\, G \to (\mathsf{Coalg}\, G)^{\mathsf{op}}.$$

Given a G-algebra $A = ((X,Y), (\alpha, \beta))$, so that $\alpha\colon X \to F(Y,X)$ and $\beta\colon F(X,Y) \to Y$ are morphisms of $\mathscr{A}$, we define IA to be the G-coalgebra

$$(Y, X) \xrightarrow{(\beta,\alpha)} (F(X,Y), F(Y,X)).$$

Given another G-algebra $A' = ((X', Y'), (\alpha', \beta'))$ and a homomorphism $(f, g)\colon A \to A'$, we put $I(f,g) = (g, f)\colon IA' \to IA$, which is clearly a G-coalgebra homomorphism.

The category $\mathscr{A}^{\mathsf{op}} \times \mathscr{A}$ is strict CPO-enriched and has ω-colimits, and the functor G is locally continuous. By Theorem 5.1.24, we have a canonical fixed point

$$(\alpha, \beta)\colon G(X,Y) \to (X,Y).$$

In particular,

$$(\alpha, \beta)^{-1} = (\alpha^{-1}, \beta^{-1})\colon (X,Y) \to G(X,Y)$$

is a terminal coalgebra. Since I preserves initial objects, it takes the initial algebra to the terminal coalgebra, whence $(\beta, \alpha)\colon (Y, X) \to G(Y, X)$ is also a terminal coalgebra. Thus, there is an isomorphism $(i, j)\colon (X, Y) \to (Y, X)$ of G-coalgebras. In particular, $i\colon Y \to X$ and $j\colon Y \to X$ are isomorphisms in $\mathscr{A}$. All of the morphisms shown below are isomorphisms:

$$\begin{array}{ccccc}
(X,X) & \underset{(\mathsf{id},j)}{\overset{(\mathsf{id},j^{-1})}{\rightleftarrows}} & (X,Y) & \underset{(\alpha,\beta)}{\overset{(\alpha^{-1},\beta^{-1})}{\rightleftarrows}} & (F(Y,X),(F(X,Y)) \\
 & & & & {\scriptstyle (F(i,\mathsf{id}),F(\mathsf{id},j^{-1}))}\ \uparrow\downarrow\ {\scriptstyle (F(i^{-1},\mathsf{id}),F(\mathsf{id},j))} \\
 & & & & (F(X,X),F(X,X)).
\end{array}$$

The composites show that $(X, X) \cong G(X, X)$ and $X \cong F(X, X)$, as desired. □

Example 5.3.3. We illustrate two ways to solve a system of domain equations in $\mathsf{CPO}_\perp$. This category has coproducts given by disjoint union with least elements identified. Consider the following system:

$$X \cong [Y, X] \qquad Y \cong Z + D \qquad Z \cong [Z, X], \tag{5.19}$$

where D is a fixed object of $\mathsf{CPO}_\perp$.

First, we solve (5.19) following the method suggested by Theorem 5.3.1. Let

$$\mathscr{A} = \mathsf{CPO}_\perp \times \mathsf{CPO}_\perp \times \mathsf{CPO}_\perp.$$

This category is strict CPO-enriched. We also have $\mathscr{A}^E = (\mathsf{CPO}_\perp^E)^3$. We express our system in terms of the following endofunctor on $\mathscr{A}^E$:

$$(X, Y, Z) \mapsto ([Y, X], Z + D, [Z, X]).$$

For embeddings $f\colon X \to X'$, $g\colon Y \to Y'$, $h\colon Z \to Z'$, the first component of $F(f, g, h)$ takes $p\colon Y \to X$ to $f \cdot p \cdot \widehat{f}$. In more detail, we take the contravariant functor $[Y, X]$ from Remark 5.1.19(1) and then apply the construction in Theorem 5.3.1. The third component is defined similarly, again using the projection $\widehat{g}$. The second component is $h + \mathsf{id}_D$. Then one forms the colimit of the initial-algebra ω-chain of this functor to obtain the canonical fixed point (X^*, Y^*, Z^*). Then we get the desired solution $X^* \cong [Y^*, X^*]$, $Y^* \cong Z^* + D$, and $Z^* \cong [Z^*, X^*]$.

Second, let us solve (5.19) in a manner inspired by Theorem 5.3.2. The variables X, Y, and Z become doubled: each gets a + version and a − version. Our equations now become

$$\begin{array}{ll}
X^+ \cong [Y^-, X^+], & X^- \cong [Y^+, X^-], \\
Y^+ \cong Z^+ + D, & Y^- \cong Z^- + D, \\
Z^+ \cong [Z^-, X^+], & Z^- \cong [Z^+, X^+].
\end{array}$$

We think of this as a locally continuous endofunctor $G\colon \mathscr{B} \to \mathscr{B}$, where

$$\mathscr{B} = (\mathsf{CPO}_\perp)^3 \times (\mathsf{CPO}_\perp^{\mathrm{op}})^3,$$

and $G(X^+, Y^+, Z^+, X^-, Y^-, Z^-)$ is given by the equations above. Then Theorem 5.3.2 tells us that the canonical fixed point will be of the form $(X^*, Y^*, Z^*, X^*, Y^*, Z^*)$: the solutions for the 'positive' and 'negative' variables will agree. Moreover, the triple (X^*, Y^*, Z^*) solves our original system (5.19).

Example 5.3.4. Scott's model of the untyped λ-calculus. The formulas t of the λ-calculus have the form

$$t ::= k \mid x \mid tt \mid \lambda x.t$$

where k ranges through a nonempty set K of constants, and x through a countable set of variables. The meaning of $t_1 t_2$ is 'application': we evaluate t_1 (a function) in t_2. The meaning of $\lambda x.t$ is 'λ-abstraction': this function takes a value a and responds with $t[a/x]$, the term t in which x is substituted by a. Thus if D is the set of all closed terms, we obtain an isomorphism

$$D \cong K + D \times D + [D, D].$$

No such set D exists because of the third summand which has larger cardinality than D whenever D is not a singleton set.

Scott [273] decided to use the cartesian closed category of continuous lattices and interpret the above equation in that category. But Smyth and Plotkin [279] made it clear that working in $\mathsf{CPO}_\perp$ is sufficient (and simpler). In fact, consider the flat cpo

$$K_\perp = K + \{\perp\}$$

with all pairs in K incomparable, in place of the set K. We obtain a locally continuous functor

$$F\colon \mathsf{CPO}_\perp^{\mathrm{op}} \times \mathsf{CPO}_\perp \to \mathsf{CPO}_\perp \quad \text{with} \quad F(X, Y) = K_\perp + Y \times Y + [X, Y].$$

If D is the initial algebra for F^E (Theorem 5.3.1), then

$$D \cong K_\perp + D \times D + [D, D]$$

is a model of λ-calculus. (Observe that the 'artificial' $\perp$ of K disappears in the formation of coproduct.)

We conclude this section by mentioning a result which is parallel to Theorem 5.3.1 but in the metric setting. It can also be used for getting fixed points of functors of mixed variance.

Theorem 5.3.5 [59]. *Let $\mathscr{A}$ be a strict* CMet*-enriched category with colimits of ω-chains. Every locally contracting endofunctor $F\colon \mathscr{A} \times \mathscr{A}^{\mathsf{op}} \to \mathscr{A}$ yields an endofunctor F^E on $\mathscr{A}^E$ defined by*

$$F^E X = F(X, X) \quad \textit{and} \quad F^E(e, \widehat{e}) = F(\widehat{e}, e).$$

Moreover, it has an initial algebra $X = \mu F^E$ which satisfies $F(X, X) \cong X$ in $\mathscr{A}$.

There is also an obvious parallel to Theorem 5.3.2, and with the same proof, mutatis mutandis.

5.4 Summary

We have presented two closely related types of categories in which the terminal coalgebra and the initial algebra coincide for every 'well-behaved' endofunctor: categories enriched over

- complete partial orders, where 'well-behaved' are the locally continuous endofunctors,
- complete metric spaces, where 'well-behaved' are the locally contracting endofunctors.

In both cases the main technical result was a coincidence of colimits of ω-chains of embeddings (or of contracting ω-chains of split subobjects, respectively) and limits of ω^{op}-chains of the corresponding projections (or splittings, respectively).

Let us point out some of the sources that inspired the material of this chapter. The famous construction of a model of the untyped λ-calculus presented by Scott in [273] applied ω^{op}-limits and it is the starting point of our study. Scott used the category of continuous lattices and embedding-projection maps. Later, Smyth and Plotkin [279] introduced the concept of a locally continuous endofunctor and noticed that in the category of domains, the finitary constructions of the initial algebra and the terminal coalgebra coincide for these endofunctors, yielding a canonical fixed point. For more on this, see Abramsky and Jung [2]. Taylor proved the same result in the category of domains and adjoint pairs [285], and this was generalized further in Velebil's Ph.D. thesis [308]. Adámek [11] proved a related result in which cpos are generalized to finitely accessible categories and locally continuous functions to finitary functors.

6

Transfinite Iteration

This chapter deals with iterations of the initial-algebra chain which extend into the transfinite. We generalize the objects $F^n 0$ for $n \in \mathbb{N}$ to objects W_i for all ordinals i and prove that whenever the connecting map from W_i to W_{i+1} is invertible, W_i is the initial algebra. A dual result holds for terminal coalgebras: we generalize $F^n 1$ to V_i for all ordinals i and obtain the terminal coalgebra from an invertible connecting map.

In order to help the reader understand this material, we provide a very short background discussion on ordinal numbers.

Background on ordinal and cardinal numbers and transfinite induction. Cardinal numbers are representatives of sets up to isomorphism: to every set X one assigns a cardinal number $|X|$, called the *cardinality* of X, so that a set Y is isomorphic (in Set) to X iff $|X| = |Y|$. The set-theoretic choice of such representatives uses ordinal numbers (or, ordinals, for short). Recall that a *well-ordered set* is a poset with a total order (i.e. for every pair x, y of elements in it, either $x \leq y$ or $y \leq x$) such that every nonempty subset contains a least element. For example, $\mathbb{N}$ is a well-ordered set but $\mathbb{Z}$ is not. For every well-ordered set P, one obtains another well-ordered set by adding a new top element. Usually this new poset is written as $P \cup \{P\}$ where $x \leq P$ for all $x \in P$. Thus starting from the smallest (empty) well-ordered set we get the following well-ordered subsets of $\mathbb{N}$:

$$0 = \emptyset, \quad 1 = \{\emptyset\}, \quad 2 = \{\emptyset, \{\emptyset\}\}, \quad 3 = \{\emptyset, \{\emptyset\}, \{\emptyset, \{\emptyset\}\}\}, \quad \ldots \tag{6.1}$$

Recall that, in formal set theory, elements of every set are sets themselves. The proper definition of an ordinal is as follows:

Definition 6.0.1. An *ordinal* is a set which is well-ordered by the relation defined by

$$x \leq y \qquad \text{iff} \qquad x \in y \text{ or } x = y.$$

The first ordinals are the natural numbers in (6.1) above. For every ordinal i we obtain an ordinal $i \cup \{i\}$ denoted by $i + 1$. An ordinal is a *successor ordinal* if it is of the form $i + 1$ for some i. An ordinal which is neither 0 nor a successor ordinal is called a *limit ordinal*. The smallest limit ordinal is denoted by ω; it is the set of natural numbers as defined above

$$\omega = \{0, 1, 2, \ldots\}.$$

The *sum* $i + j$ of two ordinals is the ordinal $i \cup \{i + k : k < j\}$. (Observe that $i + 1$ above is a special case.) Incidentally, the second limit ordinal $\omega + \omega$ plays an interesting role in work on the finitary set functors (see Example 6.5.3). This ordinal is a countable set. The first uncountable ordinal is denoted by ω_1.

Moreover, the ordinals themselves form a proper class (a collection which is not a set), which is itself well-ordered by the inclusion relation: $i \leq j$ iff $i \subseteq j$. Every element of an ordinal is an ordinal, and the union of every set of ordinals is an ordinal. This gives a well-ordered proper class

$$\mathsf{Ord}.$$

Remark 6.0.2. We often use proofs by *transfinite induction*: to prove that a class C of ordinals contains all ordinals, it is sufficient to verify that

(1) 0 lies in C (the *base case*),

(2) if $i \in C$, then $i + 1 \in C$ (the *isolated step*), and

(3) a limit ordinal i lies in C if each ordinal $j < i$ belongs to C (the *limit step*).

Analogously for a given ordinal k (which is the set of all smaller ordinals): to prove that a set $C \subseteq k$ is all of k, one verifies: $0 \in C$; if $i \in C$ and $i + 1 < k$, then $i + 1 \in C$; and the limit step as above.

We have already mentioned that cardinals are canonical representatives of isomorphism classes of sets: every set is in bijective correspondence with a unique cardinal. The set-theoretic definition is as follows:

Definition 6.0.3. A *cardinal* is an ordinal i such that every ordinal j which is isomorphic to i in Set fulfils $i \leq j$.

The smallest cardinals are the natural numbers (the finite cardinals), and then ω. The infinite cardinals are listed in a sequence $(\aleph_i)$, where i ranges through ordinals. So $\aleph_0 = \omega$, then $\aleph_1 = \omega_1$ is the first uncountable cardinal, etc.

For every cardinal λ, the *cardinal successor* (the smallest cardinal larger than λ) is denoted by λ^+. Thus, $n^+ = n + 1$ for $n \in \mathbb{N}$ and if $\lambda = \aleph_i$, then $\lambda^+ = \aleph_{i+1}$.

Definition 6.0.4. (1) A function between ordinals $f: i \to j$ is *cofinal* if for every ordinal $k < j$ there is some $i' < i$ such that $k < f(i')$.

(2) A cardinal λ is *regular* if it is infinite and there is no cofinal map from a smaller ordinal to λ. A cardinal is *singular* if it is infinite and not regular.

The smallest singular cardinal is $\aleph_\omega$, the join of $\aleph_0, \aleph_1, \aleph_2, \ldots$. For every ordinal i, the cardinal $\aleph_{i+1}$ is regular.

Remark 6.0.5. (1) Recall that an ordinal i is the (linearly ordered) set of all ordinals smaller than i. As such it is also a category.

(2) By an *i-chain* in a category $\mathscr{C}$ is meant a functor $C: i \to \mathscr{C}$. It consists of objects C_j for all ordinals $j < i$ and (connecting) morphisms $c_{j,j'}: C_j \to C_{j'}$ for all pairs $j \le j' < i$.

Analogously, an Ord-*chain* in $\mathscr{C}$ is a functor from Ord to $\mathscr{C}$.

In both cases we will drop the domain and simply speak of a (transfinite) *chain* whenever confusion is unlikely.

(3) A category $\mathscr{C}$ *has colimits of chains* if for every ordinal i, colimits of all i-chains exist in $\mathscr{C}$. (This does *not* include Ord-chains.) In particular, $\mathscr{C}$ has an initial object since the ordinal 0 is the empty set.

(4) Dually, an i^{op}-*chain* in $\mathscr{C}$ is a functor $C: i^{\mathrm{op}} \to \mathscr{C}$. It is given by objects C_j ($j < i$) and morphisms $c_{j',j}: C_{j'} \to C_j$ for all $j \le j' < i$. Moreover, an Ord^{op}-*chain* is a functor C: Ord$^{\mathrm{op}} \to \mathscr{C}$.

Again, we will simply speak of a (transfinite) chain whenever confusion about the length or direction is unlikely.

(5) A category $\mathscr{C}$ *has limits of chains* if for every ordinal i, limits of all i^{op}-chains exist in $\mathscr{C}$. This includes the terminal object (the case $i = 0$).

6.1 The Initial-Algebra Chain

In this section, we pursue the *transfinite* iteration of the initial-algebra chain. The terminal-coalgebra chain is treated in Section 6.5. We begin with a famous result on fixed points of monotone maps on a chain-complete poset. See Appendix B for more about fixed points of monotone maps.

Let P be a *chain-complete* poset, that is, every chain in it has a join. In particular, P has a least element $\bot$, the join of the empty chain. Furthermore, let $f: P \to P$ a monotone map. Then we can define an Ord-indexed sequence $f^j(\bot)$

as follows:

$$f^0(\bot) = \bot, \quad f^{j+1}(\bot) = f(f^j(\bot)), \text{ and}$$
$$f^j(\bot) = \bigvee_{i<j} f^i(\bot) \quad \text{for limit ordinals } j. \tag{6.2}$$

It is easy to verify that this is a chain in P: $f^i(\bot) \le f^j(\bot)$ if $i \le j$.

Theorem 6.1.1 (Zermelo). *Let P be a chain-complete poset. Every monotone function $f\colon P \to P$ has a least fixed point μf. Moreover, for some ordinal j,*

$$\mu f = f^j(\bot).$$

Proof Take i to be an ordinal larger than $|P|$. For this i, there must be some $j < i$ such that $f^j(\bot) = f^{j+1}(\bot)$. Indeed, this follows from Hartogs' Lemma, stating that for every set P there exists an ordinal i such that there is no injection $i \rightarrowtail P$. Thus, there are $j < k < i$ such that $f^j(\bot) = f^k(\bot)$. Since $f^j(\bot) \le f^{j+1}(\bot) \le f^k(\bot)$ we have $f^j(\bot) = f^{j+1}(\bot)$. Hence $f^j(\bot)$ is a fixed point of f. Let $f(x) = x$. An easy transfinite induction shows that $x \ge f^i(\bot)$ for all i. This shows that $f^j(\bot)$ is the least fixed point of f. □

We attribute this theorem to Zermelo since the mathematical content of the result appears in his 1904 paper [319] proving the Well-ordering Theorem. Note that the well-known Knaster–Tarski Theorem (Theorem B.1.7) has a stronger hypothesis (complete lattices) and a stronger conclusion (the fixed points form a complete lattice) [194]. In view of Kleene's Theorem 3.1.1 there is a tradeoff: with chain-completeness we have a stronger condition on the existence of joins in P while with monotonicity, we have a weaker condition on f. There are also variations on Theorem 6.1.1, such as the result often called the Bourbaki–Witt Theorem: if $f\colon P \to P$ has the property that $f(x) \ge x$, then f has a fixed point. See Section B.1 for more on this and other fixed point theorems.

Remark 6.1.2. (1) In all of these results, it is critical that P be a set. For consider the successor function on the ordered class Ord of ordinals. It has no fixed point.

(2) Alternatively, one may assume in Theorem 6.1.1 that P be a dcpo (with bottom). This is not a stronger assumption because chain-complete posets are equivalently dcpos, as shown by Markowsky [222] (Theorem B.2.2).

Remark 6.1.3. A *pre-fixed point* of a monotone function f is an element x with $f(x) \le x$. Note that μf in Theorem 6.1.1 is also the least pre-fixed point of f: whenever $x \in P$ fulfils $f(x) \le x$, then $\mu f \le x$. This follows from the proof. Moreover, by Lambek's Lemma 2.2.5, a least pre-fixed point of any monotone function (functor) on any poset is a fixed point.

A category-theoretic generalization of the construction of the least fixed point in the proof of Theorem 6.1.1 was formulated by Adámek [7].

It was applied there to the functor $F(-) + A$; in other words, the free F-algebra on an object A was considered instead of the initial F-algebra (see Proposition 2.2.24).

Definition 6.1.4 [7]. Let $\mathscr{A}$ be a category with colimits of chains (see Remark 6.0.5(3)). For an endofunctor F we define the *initial-algebra chain* $W\colon \mathsf{Ord} \to \mathscr{A}$. Its objects are denoted by W_i and its connecting morphisms by $w_{ij}\colon W_i \to W_j$, $i \le j \in \mathsf{Ord}$. These are defined by transfinite recursion as follows:

$$\begin{aligned} W_0 &= 0, & & \\ W_{j+1} &= FW_j & &\text{for all ordinals } j, \\ W_j &= \mathrm{colim}_{i<j} W_i & &\text{for all limit ordinals } j, \text{ and} \end{aligned}$$

$$\begin{aligned} &w_{0,1}\colon 0 \to W_1 \text{ is unique,} \\ &w_{j+1,k+1} = Fw_{j,k}\colon FW_j \to FW_k, \\ &w_{i,j}\ (i < j) \text{ is the colimit cocone for limit ordinals } j. \end{aligned}$$

Remark 6.1.5. (1) Note that for finite ordinals n we have $W_n = F^n 0$ and the initial-algebra ω-chain from Definition 3.1.4 is the beginning of the transfinite chain (W_i).

(2) We did not mention the connecting morphism

$$w_{\omega,\omega+1}\colon W_\omega \to FW_\omega = W_{\omega+1}$$

because it is uniquely determined by the universal property of the colimit $W_\omega = \mathrm{colim}_{n<\omega} W_n$. We have already seen this (in dual form) in Notation 4.4.5 (see Diagram (4.6)). Indeed, leaving out the first member 0 of the initial-algebra ω-chain we still have the same colimit $W_\omega = \mathrm{colim}_{n+1<\omega} W_{n+1}$ with colimit injections $w_{n+1,\omega}\colon W_{n+1} \to W_\omega$ $(n+1 < \omega)$. We also see that the following morphisms

$$w_{n+1,\omega+1} = Fw_{n,\omega}\colon W_{n+1} = FW_n \to FW_\omega$$

form a cocone. Hence, we obtain $w_{\omega,\omega+1}$ uniquely such that the following triangles commute:

$$\begin{array}{ccc} W_{n+1} = FW_n & & \\ {\scriptstyle w_{n+1,\omega}}\downarrow & \searrow^{Fw_{n,\omega}} & \\ W_\omega & \overset{w_{\omega,\omega+1}}{\dashrightarrow} & FW_\omega = W_{\omega+1} \end{array}$$

(3) Proceeding as in item (2) we obtain $w_{i,j}$ for all pairs of ordinals $i \le j$. In

fact, the initial-algebra chain is determined by Definition 6.1.4 uniquely up to natural isomorphism, as we prove in Remark 6.6.2.

We therefore speak of *the* initial-algebra chain of F. Notice that Definition 6.1.4 generalizes what we have seen in (6.2) above for sequences associated with a given monotone endofunction $f\colon P \to P$: If $\mathscr{A}$ is a chain-complete poset and f a monotone function on $\mathscr{A}$, then W_i is precisely $f^i(\bot)$ from the proof of Theorem 6.1.1.

Remark 6.1.6. Our goal is to provide a category-theoretic generalization of Theorem 6.1.1. Since a monotone function on a poset is an endofunctor, one may expect a result stating that on a category satisfying certain conditions, every endofunctor has an initial algebra, which is to say that the category is algebraically complete (see Definition 3.2.17). However, Theorem 3.2.20 tells us that every algebraically complete category with products is a preorder. Therefore, we should not expect a sweeping categorification of Theorem 6.1.1, the way the main result on iterative constructions of initial algebras (Theorem 3.1.8) generalizes Kleene's Theorem 3.1.1.

Construction 6.1.7. Analogously to Construction 3.1.6 every algebra (A, α) induces a *canonical cocone*

$$\alpha_i\colon W_i \to A \qquad (i \in \mathsf{Ord})$$

on the initial-algebra chain: this is the unique cocone with

$$\alpha_{i+1} = (W_{i+1} = FW_i \xrightarrow{F\alpha_i} FA \xrightarrow{\alpha} A)$$

for all ordinals i.

This is a definition by transfinite recursion. Indeed, α_0 is unque since 0 is the initial object. The isolated step is as above. For a limit ordinal j such that the morphisms α_i $(i < j)$ form a cocone, we use that $W_j = \mathsf{colim}_{i<j} W_i$ to obtain α_i as the unique factorizing morphism with $\alpha_i \cdot w_{i,j} = \alpha_i$ for $i < j$.

Remark 6.1.8. Homomorphisms of F-algebras $h\colon (A, \alpha) \to (B, \beta)$ preserve the canonical cocones: for every ordinal i we have

$$\beta_i = h \cdot \alpha_i\colon W_i \to B.$$

This is easy to verify by transfinite induction.

Definition 6.1.9. We say that the initial-algebra chain of a functor F *converges in λ steps* if $w_{\lambda,\lambda+1}$ is an isomorphism. We say that it converges in *exactly λ steps* if λ is the least such ordinal. In this case, we also call λ the *convergence ordinal* of the initial-algebra chain of F.

If $w_{\lambda,\lambda+1}$ is an isomorphism, then so is $w_{i,j}$, for all $j \geq i > \lambda$. This is easy to prove by transfinite induction. The following proof of the converse is from Barr [71]:

Proposition 6.1.10. *If $i < j$ and $w_{i,j}$ is an isomorphism, then so is $w_{i,i+1}$.*

Proof Since $w_{i,j}$ is an isomorphism, so is $Fw_{i,j} = w_{i+1,j+1}$. Moreover, because $w_{i+1,j+1} = w_{j,j+1} \cdot w_{i+1,j}$, we see that $w_{i+1,j}$ is a split monomorphism. In addition, $w_{i+1,j}$ is a split epimorphism, since $w_{i,j} = w_{i+1,j} \cdot w_{i,i+1}$. Hence $w_{i+1,j}$ is an isomorphism. Thus so is $w_{i,i+1} = w_{i+1,j}^{-1} \cdot w_{i,j}$. □

Theorem 6.1.11 [7]. *Let $\mathscr{A}$ be a category with colimits of chains. If the initial-algebra chain of an endofunctor F converges in j steps, then W_j is the initial algebra with the algebra structure*

$$w_{j,j+1}^{-1} : FW_j \to W_j.$$

The following proof is analogous to that of Theorem 3.1.8. Note that, as in Remark 3.1.9, we do not need to assume all chain colimits exist in Theorem 6.1.11. It is sufficient that those needed in the construction of W_i for $i \leq j$ do.

Proof Let (A, α) be any F-algebra. Denote by $h : W_j \to A$ the component α_j of the canonical cocone $\alpha_i : W_i \to A$ (see Construction 6.1.7).

(1) In order to see that h is a homomorphism, note that the outside of the following diagram commutes:

$$\begin{array}{ccc} FW_j & \xrightarrow{w_{j,j+1}^{-1}} & W_j \\ {\scriptstyle Fh=F\alpha_j}\downarrow & \searrow^{\alpha_{j+1}} & \downarrow{\scriptstyle h=\alpha_j} \\ FA & \xrightarrow{\alpha} & A \end{array} \tag{6.3}$$

Indeed, the left-hand triangle commutes by the definition of α_{j+1}, and the right-hand one does since (α_i) is a cocone.

(2) To prove that h is unique, let $k : W_j \to A$ be a homomorphism. Then we see that $h \cdot w_{i,j} = k \cdot w_{i,j}$ holds for all $i \leq j$: the proof is by transfinite induction as in Theorem 3.1.8. The case $i = j$ yields $k = h$. □

Theorem 6.1.11 is the raison d'être of this section. We turn to the following topics:

(1) sufficient conditions for the convergence of the initial-algebra chain;

(2) examples of specific set functors and of ordinals by which their initial-algebra chains converge;

(3) an examination of when convergence of the initial-algebra chain is necessary for the existence of the initial algebra.

Corollary 6.1.12. *Let $\mathscr{A}$ be a category with colimits of chains. Let $F: \mathscr{A} \to \mathscr{A}$ preserve colimits of λ-chains for some infinite ordinal λ. Then the initial-algebra chain of F converges in λ steps. Therefore,*

$$\mu F = W_\lambda.$$

The proof is analogous to that of Theorem 3.1.8. Since $W_\lambda = \mathrm{colim}_{i<\lambda} W_i$ implies $FW_\lambda = \mathrm{colim}_{i<\lambda} FW_i$ with the colimit cocone $Fw_{i\lambda} = w_{i+1,\lambda+1}$, we have a unique morphism $\iota: FW_\lambda \to W_\lambda$ with $\iota \cdot Fw_{i,\lambda} = w_{i+1,\lambda}$ for all i. This is the inverse of $w_{\lambda,\lambda+i}$.

Recall from Remark 2.2.23 that a free F-algebra on A is precisely the initial algebra for $F(-) + A$. Whenever F preserves colimits of λ-chains, so does $F(-) + A$ for all objects A. Hence, F is a varietor (Definition 2.2.26):

Corollary 6.1.13. *Let $\mathscr{A}$ be a cocomplete category. Every endofunctor preserving colimits of λ-chains for some infinite ordinal λ is a varietor.*

Remark 6.1.14. (1) Subsequently we use the concept of *height* of a tree (as defined in Remark 2.2.10). A tree is *well-founded* if it has no infinite path (cf. Chapter 7, where well-founded coalgebras are studied more generally). The height of a well-founded tree t is defined by

$$h(t) = \sup\{h(t') + 1 : t' \text{ a maximum proper subtree of } t\}.$$

The height of a well-founded tree is an ordinal. If t has an infinite path, then we set $h(t) = \infty$, a fixed value 'larger than all ordinals'.

Every finite tree t is automatically well-founded. For such trees, $h(t)$ is the the length of the longest path from the root to a leaf. For example, the following infinite trees

t: 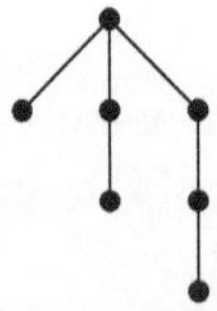 . . . and t': 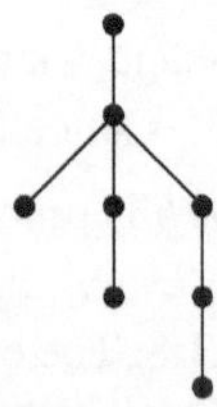 . . .

have height $h(t) = \omega$ and $h(t') = \omega + 1$, respectively.

(2) For a given ordinal i, a tree has height at most i iff all its proper subtrees have heights less than i.

Example 6.1.15. (1) Let us consider the set functor $FX = X^{\mathbb{N}} + 1$ whose algebras have a constant and an $\mathbb{N}$-ary operation. We will show that the initial algebra consists of all well-founded countably branching trees. To this end, we will verify that we can describe the initial-algebra chain as follows:

$$W_i = \text{all countably branching trees of height less than } i.$$

The connecting maps are given by the inclusions $W_i \hookrightarrow W_j$ for all $i \le j$. Recall that trees are always considered up to isomorphism. We will see that this chain converges in ω_1 steps (where ω_1 is the first uncountable ordinal).

In the first steps, $W_0 = \emptyset$ since no tree has height less than 0, and $W_1 = 1$ which we present as the singleton $\{t_0\}$, where t_0 is the root-only tree. This is the unique tree of height 0. Next $W_2 = 1 + W_1^{\mathbb{N}}$ has two elements: the tree t_0 (in the left-hand summand) and the sequence $(t_0, t_0, t_0, \ldots)$, which is represented by the unique countably branching tree of height 1:

t_1 :

In general, the equation $W_{i+1} = 1 + W_i^{\mathbb{N}}$ states that W_{i+1} consists of t_0 and all the trees below:

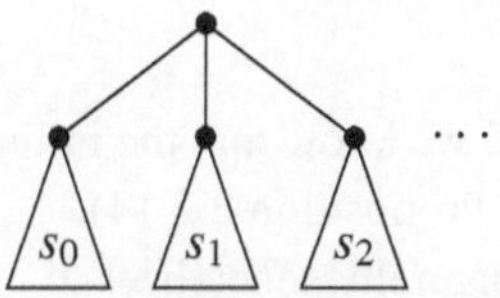

where $s_n \in W_i$ for every $n \in \mathbb{N}$. Assuming that W_i consists of all countably branching trees of height less than i, it follows from Remark 6.1.14(2) that W_{i+1} consists of all such trees of height less than $i + 1$.

The connecting map $w_{01} : \emptyset \to W_1$ is, of course, the (empty) inclusion map. The functor $FX = X^{\mathbb{N}} + 1$ clearly takes every inclusion map to an inclusion map. Thus, $w_{12} = Fw_{01}$, $w_{23} = Fw_{12}$ etc. are all inclusion maps. Consequently the colimit $W_\omega = \mathrm{colim}_{i<\omega} W_i$ is simply the union:

$$W_\omega = \bigcup_{i<\omega} W_i = \text{all countably branching trees of finite height.}$$

The initial-algebra chain continues: $W_{\omega+1}$ is a proper superset of W_ω: for example, the following tree

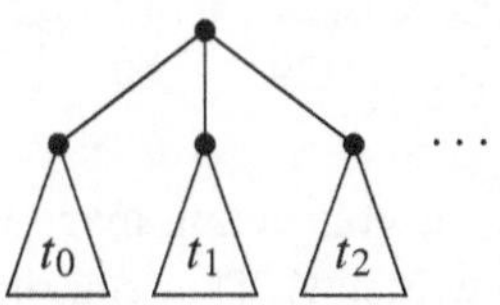

(6.4)

where t_n is the complete countably branching tree of height n, has height ω; thus, it lies in $W_{\omega+1} \setminus W_\omega$.

It is easy to see that for every well-founded countably branching tree t, the height $h(t)$ is a countable ordinal: $h(t) < \omega_1$ for the first uncoutable ordinal ω_1. Consequently, the initial-algebra chain converges at this ordinal: $\mu F = W_{\omega_1}$. This set consists of all well-founded countably branching trees. The algebra structure $W_{\omega_1}^{\omega} + 1 \to W_{\omega_1}$ is given by tree-tupling: to a sequence $(t_0, t_1, \ldots)$ of trees in the left-hand summand it assigns the tree in (6.4).

(2) In contrast $FX = X^{\mathbb{N}} + 1$ as an endofunctor on CPO is locally continuous. Thus its initial algebra is, by Theorem 5.1.24, W_ω. It consists of *all* countably branching trees.

Example 6.1.16. Let us generalize the polynomial functors (Definition 2.2.12) to *infinitary signatures*. Such a signature is a set Σ of operation symbols σ, each with an arity $\mathsf{ar}(\sigma)$, which is a (finite or infinite) cardinal number. For every set X and every cardinal κ, the set X^κ consists of all functions from κ to X. The signature Σ determines a polynomial functor H_Σ. It is defined on sets X by

$$H_\Sigma X = \coprod_{\sigma \in \Sigma} X^{\mathsf{ar}(\sigma)}$$

and similarly on maps.

For a finitary signature Σ we know that the initial algebra μH_Σ is carried by the set of all finite Σ-trees (Proposition 2.2.14).

The concept of a Σ-tree naturally generalizes to infinitary signatures, too: it is a tree labelled in Σ such that every node with k children (where k is a cardinal) is labelled by a k-ary operation symbol.

Analogously to (1), the polynomial functor H_Σ has the initial-algebra chain

$$W_i = \text{all } \Sigma\text{-trees of height less than } i$$

and the initial algebra

$$\mu H_\Sigma = \text{all well-founded } \Sigma\text{-trees.}$$

The algebra structure is given by tree-tupling.

Example 6.1.17. (1) The power-set functor has no fixed point, thus, no initial algebra. Nonetheless, it has an interesting initial-algebra chain given by the sets

$$W_0 = \emptyset, \qquad W_{i+1} = \mathscr{P} W_i, \qquad \text{and} \qquad W_i = \bigcup_{j<i} W_j \quad \text{for limit ordinals } i$$

and the inclusion functions as connecting maps $W_i \hookrightarrow W_j$ for $i \le j$. In set theory these sets are usually written V_i. They constitute a proper hierarchy: every ordinal i is a set which belongs to $W_{i+1} \setminus W_i$. Due to the Foundation Axiom, every set x lies in W_{i+1} for some i. The least such i is called the *rank* of x.

Of special interest is the set

$$W_\omega = HF$$

of all *hereditarily finite sets*. We have seen in Example 2.2.7(4) that HF is the initial algebra of $\mathscr{P}_\mathsf{f}$.

(2) Analogously, denote by

$$\mathscr{P}_\mathsf{c}$$

the subfunctor of $\mathscr{P}$ of all countable subsets. The initial-algebra chain of $\mathscr{P}_\mathsf{c}$ yields the initial algebra in ω_1 steps, and we have the initial algebra $W_{\omega_1} = HC$ of *hereditarily countable* sets. This is the set of sets which are countable and all elements of which are hereditarily countable.

(3) Generalizing $\mathscr{P}_\mathsf{f}$ and $\mathscr{P}_\mathsf{c}$ above, let us define, for every cardinal λ, the subfunctor

$$\mathscr{P}_\lambda$$

of $\mathscr{P}$ by taking all subsets of cardinality less than λ. When λ is regular and infinite, $\mathscr{P}_\lambda$ preserves colimits of λ-chains. Hence,

$$\mu\mathscr{P}_\lambda = W_\lambda,$$

the set of all sets hereditarily smaller than λ. Indeed, the initial-algebra chain converges exactly at λ. To see this, assume it converges exactly at $i < \lambda$. Then for all $j < i$ we have $W_j \subsetneqq W_i$ and we can choose $x_j \in W_i \setminus W_j$. The set $X = \{x_j : j < i\}$ has cardinality less than λ and consists of sets hereditarily smaller than λ. Thus, $X \in \mathscr{P}_\lambda W_i = W_{i+1}$. It is easy to see that $X \notin W_i$. So $W_i \neq W_{i+1}$, a contradiction.

We can also study the functor $\mathscr{P}_\lambda$ when λ is a singular cardinal. This functor does not preserve colimits of λ-chains. But λ^+ is regular, and so the initial-algebra chain converges in λ^+ steps. Indeed, it cannot be less than λ, see the above argument, and it cannot be an ordinal i with $\lambda \leq i < \lambda^+$. Such an ordinal would have a cofinal subset of size $\leq \lambda$, and it is easy to extend the argument above.

6.2 Smooth Monomorphisms

At this point we know that iteration is a powerful way of obtaining initial algebras. Thus we can ask whether it is *necessary*. In other words, are all initial algebras obtainable by iteration? It turns out that the answer is 'No, but in settings that are sufficiently Set-like, Yes'. Spelling this out is our next goal.

Remark 6.2.1. We recall the general definition of a *subobject* from Remark 2.1.18. Generalizing a bit, let $\mathcal{M}$ be a class of monomorphisms. For an object A, an $\mathcal{M}$-*subobject* is a subobject represented by a morphism $m\colon S \to A$ in $\mathcal{M}$. If the object A has only a set of $\mathcal{M}$-subobjects, then we write

$$(\mathsf{Sub}_{\mathcal{M}}(A), \leq)$$

for the poset of $\mathcal{M}$-subobjects of A.

Definition 6.2.2. Let $\mathcal{M}$ be a class of monomorphisms closed under isomorphisms and composition. We say that

(1) an object A has *smooth* $\mathcal{M}$-*subobjects* provided that $\mathsf{Sub}_{\mathcal{M}}(A)$ is a chain-complete poset (equivalently, a dcpo with bottom, see Remark 6.1.2(2)) with joins of chains given by colimits of the corresponding chains of subobjects;

(2) the class $\mathcal{M}$ is *smooth* if every object of $\mathscr{A}$ has smooth $\mathcal{M}$-subobjects.

Moreover, we say that a category has *smooth monomorphisms* if the class of all monomorphisms is smooth. (Similarly, for other classes $\mathcal{M}$ of monomorphisms.)

Remark 6.2.3. (1) In more detail, $\mathscr{A}$ has smooth $\mathcal{M}$-subobjects iff for every ordinal λ the following holds. Let λ be an ordinal, and let $m_i\colon A_i \rightarrowtail A$ $(i < \lambda)$ be a λ-chain of $\mathcal{M}$-subobjects of A. Then the chain has a join in $\mathsf{Sub}_{\mathcal{M}}(A)$, say $m\colon C \rightarrowtail A$. Moreover, consider the λ-chain of objects $(A_i)_{i<\lambda}$ with connecting morphisms $a_{i,j}\colon A_i \rightarrowtail A_j$ for $i \leq j < \lambda$ given by the unique factorizations witnessing $m_i \leq m_j$:

$$\begin{array}{ccc} A_i & \overset{a_{i,j}}{\rightarrowtail} & A_j \\ & {\scriptstyle m_i}\searrow \quad \swarrow{\scriptstyle m_j} & \\ & A & \end{array}$$

(Note that $a_{i,j}$ need not lie in $\mathcal{M}$.) Then for every $i < \lambda$ there exists a monomorphism $c_i\colon A_i \rightarrowtail C$ with $m \cdot c_i = m_i$, since $m_i \leq m$. The smoothness requirement is that these monomorphisms form a colimit cocone. Equivalently, the corresponding details hold for joins of *directed* subsets of $\mathsf{Sub}_{\mathcal{M}}(A)$ (see the proof of Theorem B.2.2).

(2) In particular, a category $\mathscr{A}$ with smooth $\mathcal{M}$-subobjects has an initial object 0. Indeed, 0 is the colimit of the empty chain of monomorphisms in $\mathcal{M}$. Given an object A in $\mathscr{A}$ the unique morphism $0 \to A$ lies in $\mathcal{M}$.

(3) An initial object is *simple* if all morphisms $0 \to A$ are monic. Hence, if a category has a smooth class of monomorphisms, then its initial object is simple.

(4) Finally, note that a category with a smooth class $\mathcal{M}$ of monomorphisms is $\mathcal{M}$-*well-powered*: $\mathsf{Sub}_{\mathcal{M}}(A)$ is a set for every object A.

We are especially interested in two generally defined classes $\mathcal{M}$ of monomorphisms: the class of all monomorphisms, and the class of regular monomorphisms.

Remark 6.2.4. A monomorphism is *regular* if it is the equalizer of a parallel pair of morphisms. Below we show that in a number of categories the classes of monomorphisms and regular monomorphisms are smooth. We also present some counterexamples. Regular monomorphisms usually represent 'embeddings', for example, subposets in Pos or subspaces in Met. (However, we reserve the term 'embedding' for CPO-enriched categories, see Definition 5.1.3.)

Example 6.2.5. (1) Every split monomorphism $m\colon X \to Y$ (with $e \cdot m = \mathsf{id}_X$ for some $e\colon Y \to X$) is regular: it is the equalizer of $m \cdot e$ and id_Y. In particular, embeddings in CPO-enriched categories are regular.

(2) In a number of categories, regular monomorphisms $m\colon X \to Y$ represent precisely the subobjects of Y which inherit their structure from Y. For example, in Pos and Gra, they represent induced subposets and subgraphs, respectively, of Y.

(3) An *isometric embedding* in Met is a map $f\colon X \to Y$ preserving the distances. Isometric embeddings represent the subspaces of Y with the induced metric. One readily proves that regular monomorphisms in Met are precisely the isometric embeddings $m\colon X \to Y$ whose image $m[X]$ is closed in Y; in other words, they represent precisely the closed subspaces.

(4) Regular monomorphisms in CMet are precisely the isometric embeddings. This follows from the fact that a complete subspace of a complete metric spaces is necessarily closed.

We now turn to examples of classes of smooth monomorphisms in various categories. Our work is summarized Table 6.1. Example 6.2.6 presents examples without proof. In contrast, proofs are given afterwards for the categories Met, CMet, and CPO. After that, we turn in Proposition 6.2.12 to an equivalent condition for an object A to have smooth $\mathcal{M}$-subobjects.

Example 6.2.6. (1) The category Set has smooth monomorphisms. Since every monomorphism is regular in Set, the regular monomorphisms are smooth in Set as well.

(2) In the category Pfn of sets and partial functions, the classes of all monomorphisms and of all regular monomorphisms are both smooth. (The latter are precisely the monomorphisms of Set.)

(3) In the category Pos both monomorphisms and embeddings of subposets (equivalently, regular monomorphisms) form smooth classes. The same is true

category	monomorphisms	regular monomorphisms
Set	+	+
Pfn	+	+
Pos	+	+
Gra	+	+
K-Vec	+	+
algebras for $FX = X + 1$	−	−
Ring	−	−
CPO	−	−
Met	+	−
CMet	−	+

Table 6.1 Smooth classes of morphisms

for the category Gra of graphs and graph homomorphisms (functions preserving the edges).

(4) The categories of monoids, groups, Boolean algebras, and vector spaces over a field K have smooth monomorphisms. In fact, all of these categories are varieties, that is, classes of algebras specified by a finitary signature and a set of equations. Monomorphisms in a variety are precisely the homomorphisms whose underlying functions are injective. The smoothness condition is easy to verify, except for the case $\lambda = 0$ in Remark 6.2.3(2). Thus, a variety $\mathcal{V}$ has smooth monomorphisms iff its initial algebra I is simple, that is, all homomorphisms $I \to A$ in $\mathcal{V}$ are injective. (Regular monomorphisms are not easy to characterize in general varieties.)

(5) In the category K-Vec of vector spaces over a fixed field, every monomorphism splits and thus is regular. This gives another reason why K-Vec has smooth monomorphisms. Indeed, given a subspace $m\colon A \rightarrowtail B$ of a vector space B, there exists a complementary subspace $m'\colon A' \rightarrowtail B$ with $B = A + A'$. Then $[\mathsf{id}_A, 0]$, where $0\colon A' \to A$ is the trivial linear map, splits m.

(6) Let F be the set functor $FX = X + 1$. Algebras for F form the variety of algebras with one unary operation and one constant (no equations). In the category $\mathsf{Alg}\, F$, the collection of all monomorphisms is not smooth because the initial object in the category is $\mathbb{N}$ (see Example 2.2.7(3)), and the unique homomorphism into the terminal F-algebra 1 is not injective. Regular monomorphisms and monomorphisms coincide in $\mathsf{Alg}\, F$.

(7) Analogously, monomorphisms and regular monomorphisms are not smooth in the category of rings. Its initial object is the ring $\mathbb{Z}$ of integers, and the unique homomorphism $\mathbb{Z} \to 1$ is not injective.

(8) In CPO-enriched categories with colimits of chains of embeddings, the class of all embeddings is smooth (see Remark 5.1.12 and cf. Corollary 5.1.15). However, the class of all monomorphisms is usually non-smooth. For example, in the category CPO itself consider the cpo $\mathbb{N}^\top$ of natural numbers with a top element $\top$. The inclusion maps of the subposets $C_n = \{0, \dots, n\} \cup \{\top\}$, $n \in \mathbb{N}$, form an ω-chain in CPO. Its colimit is $\mathbb{N}^\top \cup \{\infty\}$ where $n < \infty < \top$ for all $n \in \mathbb{N}$. The cocone formed by the inclusion maps $C_n \hookrightarrow \mathbb{N}^\top$ is formed by monomorphisms. However, the factorizing morphism from colim C_n to $\mathbb{N}^\top$ is not monic, as it merges ∞ and $\top$. Analogously, regular monomorphisms are not smooth.

Remark 6.2.7. Generalizing Example 6.2.6(1)–(7), a locally finitely presentable category (Definition 10.1.5) has smooth monomorphisms iff it has a simple initial object [23, Prop. 1.62] (with the case of the empty chain due to simplicity of the initial object).

We now turn to the categories Met and CMet of metric spaces and complete ones, respectively (Notation 3.2.2).

Proposition 6.2.8. *In the category* Met*:*

(1) *both monomorphisms and isometric embeddings are smooth, and*

(2) *regular monomorphisms are not smooth.*

Proof (1) For isometric embeddings, this is easy to see: Given a λ-chain $(A_i, d_i) \hookrightarrow (A, d)$, $i < \lambda$, of embeddings, their join is the subspace $\bigcup_{i<\lambda} A_i$. This is also the colimit in Met.

Let us prove that monomorphisms are smooth. Fix a space (A, d), and consider a λ-chain of subobjects $m_i \colon (A_i, d_i) \to (A, d)$, $i < \lambda$. Let $B = \bigcup_{i<\lambda} m_i[A_i]$, and let $d' \colon B \to [0, 1]$ be defined as follows:

$$d'(x, y) = \inf\{d_i(x', y') : i \in \lambda, x', y' \in A_i, m_i(x') = x, \text{ and } m_i(y') = y\}. \tag{6.5}$$

We show that d' is a metric. It is clearly symmetric. We verify that distinct points x, y in B have non-zero distance. For each i, x', and y' as in (6.5), we have $d_i(x', y') \geq d(x, y)$, since m_i is non-expanding. Thus $d'(x, y) \geq d(x, y) > 0$.

Finally, we verify that d' satisfies the triangle inequality. To this end it suffices to show that for all $x, y, z \in B$ and every $\varepsilon > 0$, $d'(x, z) \leq d'(x, y) + d'(y, z) + \varepsilon$. Let i, x', y' be as in (6.5). Assume that $d_i(x', y') < d'(x, y) + \varepsilon/2$. Let j, y', z' be such that $d_{A_j}(y', z') < d'(x, y) + \varepsilon/2$. Without loss of generality, $i \leq j$. Since the connecting map $a_{i,j}$ is non-expanding, $d_j(a_{i,j}(x'), a_{i,j}(y')) < d'(x, y) + \varepsilon/2$. Since m_j is injective, $a_{i,j}(y') = y'$. Let $x' = a_{i,j}(x')$, and note that $m_j(x') = x$.

By the triangle inequality in A_j,

$$d_j(x',z') \le d_j(x',y') + d_j(y',z') < d'(x,y) + d'(y,z) + \varepsilon.$$

It follows that $d'(x,z) \le d'(x,y) + d'(y,z) + \varepsilon$, as desired.

As we have just seen, the inclusion $m\colon (B,d') \rightarrowtail (A,d)$ is non-expanding. For $i < \lambda$, we have the codomain restriction $m'_i\colon A_i \rightarrowtail B$. This, too, is non-expanding. It is also easy to check that (B,d') is the join in $\mathsf{Sub}(A,d)$ of the original chain.

We verify that the family $m'_i\colon (A_i,d_i) \rightarrowtail (B,d')$ is a colimit cocone. It clearly is a cocone. Consider a cocone $f_i\colon (A_i,d_i) \to (A^*,d^*)$, $i < \lambda$. We know that the union B is the colimit in Set (see item (1)). Therefore, we have a unique map $f\colon B \to A^*$ such that $f_i = f \cdot m'_i$ for all $i < \lambda$. This is given by $f(x) = f_i(x)$ whenever $x \in m_i[A_i]$. We check that f is non-expanding, and this will conclude our verification. Let $x, y \in B$, and choose i, x', y' as in (6.5). Since f_i is non-expanding,

$$d^*(f(x), f(y)) = d^*(f_i(x'), f_i(y')) \le d_i(x',y') \le d'(x,y).$$

(2) To see that regular monomorphisms are not smooth in Met, consider the ω-chain of closed subspaces $X_n = \{1/k : 1 \le k \le n\}$ of the unit interval $[0,1]$. Their colimit is the union $X = \{1/k : 1 \le k\}$. However, the join of the ω-chain formed by the inclusion maps $X_n \hookrightarrow [0,1]$ in $\mathsf{Sub}_{\mathcal{M}}([0,1])$ is the inclusion of the closure $X \cup \{0\} \hookrightarrow [0,1]$. □

Example 6.2.9 [56, Ex. 3.5]. Analogously, in the category Top of topological spaces and continuous functions, monomorphisms are smooth but regular monomorphisms (subspace embeddings) are not.

Proposition 6.2.10. *In the category* CMet*:*

(1) *monomorphisms are not smooth,*

(2) *regular monomorphisms (= isometric embeddings) are smooth.*

Proof (1) To see that monomorphisms are not smooth in CMet, we will show that the closed interval $I = [-1/2, 1/2]$ does not have smooth subobjects. Let A_k be the subspace (with the induced metric) on the set $\{2^{-n} : n = 1, \ldots, k\}$ and B_k that on $\{-2^{-n} : n = 1, \ldots, k\}$. The injection maps $a_k\colon A_k \hookrightarrow I$ and $b_k\colon B_k \hookrightarrow I$ define an ω-chain of subobjects $[a_k, b_k]\colon A_k + B_k \to I$. The supremum of this chain in $\mathsf{Sub}(I)$ clearly has the property that both sequences (2^{-n}) and (-2^{-n}) are Cauchy, and they have the same limit, 0. However, in the colimit of that ω-chain, the limits of the two sequences have distance 1, since colimits commute with coproducts.

(2) We prove that isometric embeddings are smooth in CMet (cf. Example 6.2.5(4)). Let $A_i \hookrightarrow A$ form a λ-chain of embeddings of complete subspaces. This means that every A_i is a closed subspace of A. Their join $B \hookrightarrow A$ is the closure of their union

$$B = \overline{\bigcup_{i<\lambda} A_i}.$$

We know from Proposition 6.2.8(1) that the union is the colimit of the given λ-chain in Met. Moreover, the colimit of a diagram D in CMet is given by forming the Cauchy completion of the colimit of D in Met. (This follows from the fact that CMet is a reflective subcategory of Met with Cauchy completions as reflections.) Since B is complete and $\bigcup_{i<\lambda} A_i$ is dense in it, B is the Cauchy completion of that union, whence of the λ-chain (A_i) in CMet. □

We have seen that in a number of categories all objects have smooth subobjects, and in the remaining examples no 'non-trivial' object has them. Here is an exception:

Example 6.2.11. In CPO an object A has smooth subobjects iff it contains no strictly increasing ω-chain. To see this, let $a_0 < a_1 < a_2 < \cdots$ be such a chain in A. Then A does not have smooth subobjects. Indeed, for $b = \bigsqcup_{n<\omega} a_n$ we form the following ω-chain in $\mathsf{Sub}(A)$ given by

$$B_n = \{a_0, \ldots, a_n\} \cup \{b\} \qquad \text{for } n < \omega$$

with the order inherited from A. The colimit C of this ω-chain contains $\{a_n : n < \omega\}$, and the join $\bigsqcup_{n<\omega} a_n$ in C is strictly smaller than b. This is different from the join in $\mathsf{Sub}(A)$, which is simply $\{a_n \mid n < \omega\} \cup \{b\}$.

Conversely, suppose that no strictly increasing ω-chain exists in A. Then all subposets are sub-cpos, and the fact that A has smooth subobjects in Pos implies that this also holds in CPO.

The following equivalent formulation is often used in proofs.

Proposition 6.2.12. *An object A has smooth $\mathcal{M}$-subobjects iff for every ordinal λ, every λ-chain $(A_i)_{i<\lambda}$ of monomorphisms in $\mathscr{A}$, and every cocone $m_i\colon A_i \rightarrowtail A$ of $\mathcal{M}$-monomorphisms, the following holds:*

(1) *the λ-chain $(A_i)_{i<\lambda}$ has a colimit, and*

(2) *the factorization morphism of the cocone $(m_i)_{i<\lambda}$ is again an $\mathcal{M}$-monomorphism.*

Proof The 'only if' direction is obvious. For the 'if' direction, suppose we are given a λ-chain $m_i\colon A_i \rightarrowtail A$ of $\mathcal{M}$-subobjects ($i < \lambda$) as in Remark 6.2.3(1). By (1), this chain has a colimit C, and we will prove that this yields the join $\bigvee_{i<\lambda} m_i$. We denote the colimit injections by $c_i\colon A_i \to C, i < \lambda$. By (2), we have a unique $\mathcal{M}$-monomorphism $m\colon C \rightarrowtail A$ such that $m \cdot c_i = m_i$ for all $i < \lambda$.

Now let $s\colon S \rightarrowtail A$ be an $\mathcal{M}$-subobject such that $m_i \leq s$ for all $i < \lambda$. That is, we have monomorphisms $s_i\colon A_i \rightarrowtail S$ with $s \cdot s_i = m_i$ for all $i < \lambda$. Then the family s_i is a cocone because for every $a_{i,j}\colon A_i \rightarrowtail A_j$ we have

$$s \cdot s_j \cdot a_{ij} = m_j \cdot a_{i,j} = m_i = s \cdot s_i$$

whence $s_j \cdot a_{i,j} = s_i$ since s is monic. We therefore obtain a unique $t\colon C \rightarrowtail S$ with $t \cdot c_i = s_i$ for all $i < \lambda$. Consequently, we have

$$s \cdot t \cdot c_i = s \cdot s_i = m_i = m \cdot c_i \qquad \text{for all } i < \lambda.$$

Since the colimit injections c_i form a jointly epic family, we conclude that $s \cdot t = m$, which means that $m \leq s$ in $\mathsf{Sub}_{\mathcal{M}}(A)$, as desired. □

Remark 6.2.13. (1) Note that the conditions for $\mathcal{M}$ to be a smooth class of monomorphisms are essentially a part of the conditions of Taylor's notion of a *locally complete class of supports* [286, Def. 6.1. & 6.3] (see also [288, Assumption 4.18]). However, Taylor worked with directed joins and colimits in lieu of joins/colimits of λ-chains.

(2) Furthermore, our conditions are closely related to the list of conditions for a class of monomorphisms given by Trnková et al. [290]. Note that a class $\mathcal{M}$ of monomorphisms containing the identities and closed under composition can be regarded as the subcategory of $\mathscr{A}$ given by all morphisms in $\mathcal{M}$. The list of conditions in op. cit. is equivalent to stating that the inclusion functor $\mathcal{M} \hookrightarrow \mathscr{A}$ creates colimits of chains. This implies that the class $\mathcal{M}$ is smooth. For the converse, we would need to add that for every chain of $\mathcal{M}$-monomorphisms the colimit cocone consists of $\mathcal{M}$-monomorphisms, too.

Remark 6.2.14. Surprisingly, in a category with smooth monomorphisms,

(1) a chain of monomorphisms need not have a colimit; and

(2) even if a colimit does exist, the colimit injections need not be monic.

To illustrate (1), consider the natural numbers as a category $(\mathbb{N}, \leq)$. Its (mono)morphisms are trivially smooth, but $\mathbb{N}$ as an ω-chain has no colimit.

For (2), consider Met. We have seen in Proposition 6.2.8(1) that monomorphisms in Met are smooth. For $n < \omega$, let A_n be the space $\{a, b\}$ two elements of distance $1/n$. We have an ω-chain with connecting morphisms given by $\mathsf{id}_{\{a,b\}}$. Its colimit is the one-point space 1. The colimit maps are the obvious ones $A_n \to 1$, and they are not monic.

Nevertheless, there is a positive result for categories with smooth monomorphisms: If a chain has a cocone formed by monomorphisms, then it has a colimit. This follows from Remark 6.2.3(1): let A be the codomain of a cocone of monomorphisms on some chain. Then that cocone represents a chain of subobjects of A. The join of that chain exists in $\mathsf{Sub}(A)$ and yields the desired colimit.

The upshot is that the condition in Remark 6.2.13(2) that the inclusion functor $\mathcal{M} \hookrightarrow \mathscr{A}$ creates colimits of chains is stronger than the requirement that $\mathscr{A}$ has smooth monomorphisms.

Remark 6.2.15. (1) Epimorphisms are automatically 'smooth': a colimit of a chain of epimorphisms is formed by epimorphisms. Moreover, given a cocone of that chain formed by epimorphisms, the unique factorizing morphism is epic. This holds in all categories and is easy to prove.

(2) Thus, monomorphisms are automatically 'co-smooth' in the following sense: a limit of a chain of monomorphisms is formed by monomorphisms. Moreover, given a cone of monomorphisms, the unique factorizing morphism is monic.

(3) We have seen in Example 6.2.6 that many everyday categories have smooth monomorphisms. In contrast, the dual property often fails: limits of ω^{op}-chains of epimorphisms need not have epimorphic factorizations for cones formed by epimorphisms. For example, in Set this fails: here is an example of an ω^{op}-chain of epimorphisms $a_n : A_{n+1} \twoheadrightarrow A_n$, $n < \omega$, and a cone of epimorphisms $b_n : B \twoheadrightarrow A_n$, $n < \omega$, such that the factorizing morphism $b : B \to \lim_{n<\omega} A_n$ is not epic.

Let $A_n = n + 1$ be the ω^{op}-chain of Example 3.3.6, and let $B = \mathbb{N}$. We have the limit $\lim A_n = \mathbb{N}^\top$. The following cone

$$b_n : \mathbb{N} \twoheadrightarrow n + 1, \qquad b_n(i) = \min(i, n + 1) \qquad \text{for all } i \in \mathbb{N}$$

consists of epimorphisms. But the factorizing morphism is not epic: this is the inclusion map $b : \mathbb{N} \hookrightarrow \mathbb{N}^\top$.

(4) It is nevertheless true that in Set limits of ω^{op}-chains of epimorphisms are formed by epimorphisms.

6.3 The Initial-Algebra Theorem

In this section we shall see a very useful sufficient condition for the convergence of the initial-algebra chain, based on this concept of smooth classes.

We say that a functor F *preserves* $\mathcal{M}$*-monomorphisms* provided that $m \in \mathcal{M}$ implies $Fm \in \mathcal{M}$. For short, we say F *preserves* $\mathcal{M}$.

Definition 6.3.1. By an $\mathcal{M}$-*pre-fixed point* of F we mean an object A with an $\mathcal{M}$-monomorphism $\alpha\colon FA \rightarrowtail A$.

Theorem 6.3.2. *Let A be an object having smooth $\mathcal{M}$-subobjects. If $\alpha\colon FA \rightarrowtail A$ is an $\mathcal{M}$-pre-fixed point of a functor F preserving $\mathcal{M}$, then the initial-algebra chain for F converges, and the initial algebra is a subalgebra of (A, α).*

Proof Define the function

$$f\colon \mathsf{Sub}_{\mathcal{M}}(A) \to \mathsf{Sub}_{\mathcal{M}}(A) \quad \text{by} \quad f(B \overset{u}{\rightarrowtail} A) = (FA \overset{Fu}{\rightarrowtail} FA \overset{\alpha}{\rightarrowtail} A).$$

It is well-defined and monotone by our assumption that F preserves $\mathcal{M}$. Theorem 6.1.1 applies since $\mathsf{Sub}_{\mathcal{M}}(A)$ is a chain-complete poset by assumption. Therefore f has the least fixed point $\mu f = f^i(\bot)$ for some ordinal i. The canonical cocone $\alpha_j\colon W_j \to A$ (see Construction 6.1.7) satisfies $\alpha_j = f^j(\bot)$ for all $j \in \mathsf{Ord}$. This is easily verified by transfinite induction starting with Remark 6.2.3(2) and using smoothness at the limit steps. Thus, from $f(f^i(\bot)) = f^i(\bot)$ we conclude that α_i and α_{i+1} represent the same subobject of A. Since $\alpha_i = \alpha_{i+1} \cdot w_{i,i+1}$, it follows that $w_{i,i+1}$ is invertible. So the initial-algebra chain converges by Theorem 6.1.11. The initial algebra $(W_i, w_{i,i+1}^{-1})$ is a subalgebra of (A, α) via the homomorphism $\alpha_i = f^i(\bot)$: see (6.3) with i replacing j. □

We present a different proof of Theorem 6.3.2 in Theorem 7.5.2. That proof uses facts about recursive coalgebras developed in Chapter 7.

Example 6.3.3. We present several examples which show, *inter alia*, that none of the hypotheses in Theorem 6.3.2 can be left out. In each case $\mathcal{M}$ is the class of all monomorphisms.

(1) Well-poweredness [71]. The category $\mathsf{Ord}^{\top}$, the ordinals with a new element $\top$ added 'on top' satisfies all assumptions of Theorem 6.3.2 except that it is not well-powered. Let $F\colon \mathsf{Ord}^{\top} \to \mathsf{Ord}^{\top}$ be given by $F(\lambda) = \lambda + 1$, and $F(\top) = \top$. Clearly, $\top$ is the only fixed point of F, and this is not W_i for any ordinal i.

(2) Smooth subobjects. Recall from Example 6.2.6(6) the variety $\mathscr{A}$ of algebras (A, u, c) with one unary operation u and one constant c. Its initial object is $(\mathbb{N}, s, 0)$ with $s(n) = n + 1$, which is not simple. We present an endofunctor having no initial algebra (whence the initial-algebra chain does not converge) even though it has a fixed point and preserves monomorphisms. Let $\mathscr{P}_0$ be the nonempty power-set functor: $\mathscr{P}_0 X = \mathscr{P} X \setminus \{\emptyset\}$. We obtain an analogous endofunctor (in fact, a lifting, cf. Chapter 12) $\bar{\mathscr{P}}_0$ on $\mathscr{A}$ defined by $\bar{\mathscr{P}}_0(A, u, c) = (\mathscr{P}_0 A, \mathscr{P}_0 u, \{c\})$. It clearly preserves monomorphisms. The

terminal object 1 is a fixed point of $\bar{\mathscr{P}}_0$ (since $\mathscr{P}_0 1 \cong 1$). This is, up to isomorphism, the only fixed point. But this is not $\mu\bar{\mathscr{P}}_0$ because given an algebra on (A, u, c) with $u(x) \neq x$ for all $x \in A$, no $\bar{\mathscr{P}}_0$-algebra homomorphism exists from 1 to A.

(3) Preservation of monomorphisms. Here we use the category Set × Set which satisfies all assumptions of Theorem 6.3.2. We define an endofunctor F by

$$F(X,Y) = \begin{cases} (\emptyset, 1) & \text{if } X \neq \emptyset, \\ (\emptyset, \mathscr{P}Y) & \text{if } X = \emptyset. \end{cases}$$

It is defined on morphisms as expected, using $\mathscr{P}$ in case $X = \emptyset$. This functor has many pre-fixed points, for example, $F(1,1) = (\emptyset, 1) \rightarrowtail (1,1)$. But it has no fixed points (thus no initial algebra): first, $(\emptyset, Y)$ and $(\emptyset, \mathscr{P}Y)$ are never isomorphic, by Cantor's Theorem (cf. Example 2.2.7(1)). Second, if $X \neq \emptyset$, then there exists no morphism from (X, Y) to $F(X,Y) = (\emptyset, 1)$.

Theorem 6.3.4 (Initial-Algebra Theorem [290]). *Let $\mathscr{A}$ have a smooth class $\mathcal{M}$ of monomorphisms. Then the following are equivalent for every endofunctor F preserving $\mathcal{M}$:*

(1) *The initial-algebra chain converges.*

(2) *F has an initial algebra.*

(3) *F has a fixed point.*

(4) *F has an $\mathcal{M}$-pre-fixed point.*

Moreover, if (4) holds, then μF is an $\mathcal{M}$-subobject of every $\mathcal{M}$-pre-fixed point of F.

Indeed, we know that (1) implies (2), and Lambek's Lemma tells us that (2) implies (3). Clearly, (3) implies (4). Finally, we have shown that (4) implies (1) in Theorem 6.3.2. It follows from its proof that μF is a subobject of every $\mathcal{M}$-pre-fixed point of F.

As a shorter, more memorable form of the Initial-Algebra Theorem we note the following:

Corollary 6.3.5. *An endofunctor preserving monomorphisms of a smooth class has an initial algebra iff it has a fixed point.*

Corollary 6.3.6. *Let F be an endofunctor preserving monomorphisms of a smooth class and which has a terminal coalgebra. Then the canonical morphism $k\colon \mu F \to \nu F$ of Remark 2.5.16 is monic.*

Proof Apply Theorem 6.3.2 to the (pre-)fixed point $\alpha = \tau^{-1}\colon F(\nu F) \to \nu F$. As we have seen, the canonical cone (α_i) consists of monomorphisms, and we have an ordinal i such that $w_{i,i+1}$ is invertible. Its inverse is the initial algebra $\iota\colon F(\mu F) \to \mu F$. Using that $\alpha_{i+1} = \tau^{-1} \cdot F\alpha_i$ and $\alpha_i = \alpha_{i+1} \cdot w_{i,i+1}$, we see that the following square commutes:

$$\begin{array}{ccc} F(\mu F) & \xrightarrow{\ \iota\ } & \mu F \\ {\scriptstyle F\alpha_i}\big\downarrow & \searrow^{\alpha_{i+1}} & \big\downarrow{\scriptstyle \alpha_i} \\ F(\nu F) & \xrightarrow{\ \tau^{-1}\ } & \nu F \end{array}$$

Thus, $k = \alpha_i$ is monic as desired. □

We now briefly return to a point mentioned in the introduction: minimal algebras. An algebra (A, α) is *minimal* if it has no proper subalgebras: every monomorphism $m\colon (B, \beta) \rightarrowtail (A, \alpha)$ is an isomorphism. We have seen in Proposition 1.2.1 that every initial algebra is minimal. Under the hypothesis of the Initial-Algebra Theorem, initial algebras are precisely the minimal fixed points:

Corollary 6.3.7 [18]. *Under the assumptions of Theorem 6.3.4, every minimal algebra which is a fixed point is initial.*

Proof Suppose that (A, α) is a minimal algebra which is a fixed point. Let i be the convergence ordinal of the initial-algebra chain. We have $\alpha_i = \alpha_{i+1} \cdot w_{i,i+1}$. Since $\alpha_{i+1} = \alpha \cdot F\alpha_i$ this yields

$$\alpha_i \cdot w_{i,i+1}^{-1} = \alpha \cdot F\alpha_i.$$

Thus $\alpha_i\colon (W_i, w_{i,i+1}^{-1}) \to (A, \alpha)$ is a homomorphism. Since it is monic, it is invertible (by minimality). Hence (A, α) is isomorphic to $(\mu F, \iota)$ and is thus initial. □

For every endofunctor on Set, the four conditions of Theorem 6.3.4 are equivalent. This is clear if F preserves monomorphisms. For general set functors we present here a substantially shorter proof than that in the literature [28].

Theorem 6.3.8. *For every set functor with a pre-fixed point the initial-algebra chain converges.*

Proof This is clear if $F\emptyset = \emptyset$. So suppose we have $F\emptyset \neq \emptyset$.

(1) We prove first that for every pre-fixed point $\alpha\colon FA \rightarrowtail A$ with the canonical cocone $\alpha_j\colon W_j \to A$ we have that α_ω is a monomorphism. Since $F\emptyset \neq \emptyset$, we see that FA and therefore A are nonempty. Hence, we can choose $\alpha'\colon A \to FA$ with $\alpha' \cdot \alpha = \mathsf{id}_{FA}$. We next define morphisms $u_n\colon A \to F^{n+1}0$ $(n < \omega)$ by recursion: choose an arbitrary morphism $u_0\colon A \to F\emptyset$ and let $u_{n+1} = Fu_n \cdot \alpha'$.

Then we obtain

$$w_{n,n+1} = u_n \cdot \alpha_n \colon F^n 0 \to F^{n+1} 0 \qquad \text{for all } n < \omega$$

by induction: we clearly have $w_{0,1} = u_0 \cdot \alpha_0 \colon \emptyset \to F\emptyset$ and for the induction step we compute:

$$w_{n+1,n+2} = Fw_{n,n+1} = Fu_n \cdot F\alpha_n = (Fu_n \cdot \alpha') \cdot (\alpha \cdot F\alpha_n) = u_{n+1} \cdot \alpha_{n+1}.$$

Given elements $x, y \in W_\omega$ merged by α_ω, we prove $x = y$. Since $W_\omega = \mathsf{colim}_{n<\omega} F^n\emptyset$, we can choose $n < \omega$ and elements x', y' of $F^n 0$ mapped by $w_{n,\omega}$ to the given pair. Then $\alpha_n = \alpha_\omega \cdot w_{n,\omega}$ merges x' and y', thus, these elements are also merged by

$$(w_{n+1,\omega} \cdot u_n) \cdot \alpha_n = w_{n+1,\omega} \cdot w_{n,n+1} = w_{n,\omega},$$

which proves $x = y$.

(2) Now we prove that the initial-algebra chain converges. Since α_ω is a monomorphism with nonempty domain, it splits. Thus, $F\alpha_\omega$ is also a monomorphism. This proves that $\alpha_{\omega+1} = \alpha \cdot F\alpha_\omega$ is a monomorphism. In this manner we see that all α_j with $j \geq \omega$ are monomorphisms. Thus $\alpha_j[W_j]$ is a chain of subsets of A for $j \in \mathsf{Ord}$. This cannot be strictly increasing. Given j with $\alpha_j[W_j] = \alpha_{j+1}[W_{j+1}]$, we conclude from $\alpha_j = \alpha_{j+1} \cdot w_{j,j+1}$ that $\mu F = W_j$. □

Corollary 6.3.9. *An endofunctor on one of the categories*

$$\mathsf{Set}, \quad \mathsf{Pfn}, \quad \text{or} \quad K\text{-}\mathsf{Vec}$$

has an initial algebra iff it has a pre-fixed point.

Proof For Set, use Theorem 6.3.8. For Pfn and K-Vec, apply Theorem 6.3.4 with $\mathcal{M}$ the class of all split monomorphisms (which every endofunctor preserves). □

Corollary 6.3.10. *For every set functor with a terminal coalgebra, the canonical morphism $k \colon \mu F \to \nu F$ of Remark 2.5.16 is monic.*

The argument is the same as for Corollary 6.3.6 but using Theorem 6.3.8 in lieu of Theorem 6.3.2.

Example 6.3.11. Every Kripke polynomial set functor F (Definition 4.4.13) has an initial algebra. Indeed, by Proposition 4.4.16, we have that F has a terminal coalgebra, and this provides a fixed point $\tau^{-1} \colon F(\nu F) \to \nu F$.

It follows that every Kripke polynomial functor is a varietor (Definition 2.2.26). Indeed, if F is a Kripke polynomial set functor, then so is every functor $F(-) + A$.

Example 6.3.12 [26]. Fixed points do not in general imply initial algebras even if $\mathscr{A}$ satisfies all assumptions in Theorem 6.3.2. Consider the functor

$$F\colon \mathsf{Set}\times\mathsf{Set} \to \mathsf{Set}\times\mathsf{Set} \qquad \text{given by} \qquad F(X,Y) = \begin{cases} (1,1) & \text{if } X \neq \emptyset, \\ (\emptyset, \mathscr{P}Y) & \text{if } X = \emptyset. \end{cases}$$

It is defined on morphisms as expected, using $\mathscr{P}$ in case $X = \emptyset$. Although $(1,1)$ is a fixed point, F has no initial algebra: the F-algebras $(\emptyset, \mathscr{P}Y) \to (\emptyset, Y)$ cannot be initial by Lambek's Lemma, and algebras of the form $(1,1) \to (X,Y)$ admit no homomorphisms to those of the form $(\emptyset, \mathscr{P}Y) \to (\emptyset, Y)$. Again, what is going wrong here is that this functor does not preserve monomorphisms.

Remark 6.3.13. The same general argument shows that in the category Rel (Notation 3.2.16), existence of a pre-fixed point implies the convergence of the initial-algebra chain. More precisely, let $\mathcal{M}$ be the class of all injective functions (considered as morphisms in Rel) and suppose that $F\colon \mathsf{Rel} \to \mathsf{Rel}$ is an *extension* of a set functor $G\colon \mathsf{Set} \to \mathsf{Set}$. This means that for the inclusion functor $J\colon \mathsf{Set} \to \mathsf{Rel}$ we have a commutative square

$$\begin{array}{ccc} \mathsf{Rel} & \xrightarrow{F} & \mathsf{Rel} \\ {\scriptstyle J}\uparrow & & \uparrow{\scriptstyle J} \\ \mathsf{Set} & \xrightarrow{G} & \mathsf{Set} \end{array}$$

Note that J is a left adjoint and so preserves all colimits. (Its right adjoint $R\colon \mathsf{Rel} \to \mathsf{Set}$ is given on objects by $RX = \mathscr{P}X$. It assigns to a relation $R \subseteq X \times Y$ the map $f\colon \mathscr{P}X \to \mathscr{P}Y$ defined by

$$f(A) = \{y \in Y : a\,R\,y \text{ for some } a \in A\}.$$

Thus, $\mathcal{M}$ is a smooth class in Rel. Moreover, the initial-algebra chain for F extends from Set to Rel in the obvious way. Thus, for every endofunctor on Rel which is an extension of a set functor, the conditions of the Initial-Algebra Theorem 6.3.4 are equivalent for the class $\mathcal{M}$ of all injective functions.

The proofs of the following two theorems are rather technical. We therefore state them without proof.

Theorem 6.3.14 [26]. *If the initial-algebra chain for a set functor converges in exactly λ steps, then either $\lambda \leq 3$ or λ is an infinite regular cardinal.*

Example 6.3.15. Conversely, all such convergence ordinals are possible. Convergence in exactly λ steps is illustrated by $\mathscr{P}_\lambda$ for regular cardinals λ. Convergence in exactly $i \leq 3$ steps is illustrated by the following functors F_i: $F_0 = \mathsf{Id}$, $F_1 = C_1$, the constant functor of value 1, and F_2 differing from C_1 only by $F_2\emptyset = 2 = \{0,1\}$.

Define $F_3\emptyset = 3$ and for $X \neq \emptyset$ put $F_3X = \{A \subseteq X : |A| = 3 \text{ or } 0\}$. For a map $f: X \to Y$, the map Ff is constantly $\emptyset$ if $X = \emptyset$; if $X \neq \emptyset$ put $Ff(A) = f[A]$ whenever $f[A]$ has size 3, else $Ff(A) = \emptyset$.

Theorem 6.3.16 [26]. *For the category* Set^S *of many-sorted sets, whenever an endofunctor has an initial algebra, then the initial-algebra chain converges.*

6.4 Characterization of Varietors on Set

In this section we return to varietors and treat results related to what we have seen in the previous sections. Recall from Remark 2.2.28 that F is called a varietor if free F-algebras exist.

The following result is a corollary of Corollary 6.3.9.

Corollary 6.4.1. *Let F be an endofunctor of* Set, Pfn, *or* K-Vec. *Then F is a varietor iff F has arbitrarily large pre-fixed points: for every cardinal there exists a pre-fixed point of a larger cardinality.*

Proof Let F be a varietor on Set. Given a cardinal κ, fix a set A with $|A| = \kappa$. By Proposition 2.2.24, $F(-) + A$ has an initial algebra. So there is some set X with $FX + A \cong X$, by Lambek's Lemma 2.2.5. Hence $|FX + A| \leq |X|$. In particular, $|FX| \leq |X|$ and $|A| \leq |X|$. Analogously for Pfn or K-Vec.

In the other direction, assume that F has arbitrarily large pre-fixed points. In the case of Set or Pfn, given a set A, let X be an infinite pre-fixed point with $|A| \leq |X|$. Then $|FX + A| \leq |X|$. (In the case of K-Vec, we have $FX + A = FX \times A$.) So there is a monomorphism $m: FX + A \rightarrowtail X$. By Corollary 6.3.9, the functor $F(-)+A$ has an initial algebra. By Proposition 2.2.24, F is a varietor. □

Here is an unfortunate fact [25]: the collection of all set functors possessing initial algebras is not well-behaved. There are set functors F, G having initial algebras such that none of $F + G$, $F \times G$, or $G \cdot F$ has a fixed point. We defer the details to Example 6.5.17.

In contrast, post-fixed points of set functors F are much more frequent: unless the restriction of F to nonempty sets is essentially constant (see Definition 6.4.2 below), all sufficiently large sets X are post-fixed points, that is, they fulfil $|X| \leq |FX|$.

Definition 6.4.2. A functor is called *essentially constant* if it is naturally isomorphic to a constant functor.

The following result was first proved by Koubek [198, Thm. 1.3]. We present a new and simpler proof here.

Theorem 6.4.3. *Let F be a set functor. Then one of the following two alternatives holds:*

(1) *All sufficiently large sets are post-fixed points of F: there is a cardinal κ such that $|FY| \geq |Y|$ for all sets Y such that $|Y| \geq \kappa$.*

(2) *The restriction of F to nonempty sets is almost constant.*

Proof For every set X we write $t_X : X \to 1$ for the unique function.

Case 1: For some $X \neq \emptyset$, Ft_X is not injective. We prove that the cardinal $\kappa = \max(|X|, \aleph_0)$ has the desired property. Choose an element $x^* : 1 \to X$. Since Ft_X is not injective, and $Ft_X \cdot Fx^* = \mathrm{id}_{F1}$, Fx^* is not surjective. Let Y be a set with $|Y| \geq \kappa$. Since Y is infinite and $|Y| \geq |X|$, we have $|Y| = |X \times Y|$. Therefore there is a partition of Y into $|Y|$ subsets of cardinality $|X|$:

$$Y = \bigcup_{i \in I} Y_i \qquad \text{with } |I| = |Y| \text{ and } |Y_i| = |X| \text{ for every } i \in I.$$

As Y is infinite, we can fix $y^* \in Y$ and add it to each Y_i without changing cardinality. Choose an injection $m_i : X \rightarrowtail Y$ such that $m_i[X] = Y_i \cup \{y^*\}$ and $m_i(x^*) = y^*$. Then for $i \neq j$, we have $Y_i \cap Y_j = \{y^*\}$. Thus, we have pullbacks as on the left below:

$$\begin{array}{ccc} 1 & \xrightarrow{x^*} & X \\ {\scriptstyle x^*}\downarrow & \lrcorner & \downarrow{\scriptstyle m_j} \\ X & \xrightarrow{m_i} & Y \end{array} \qquad\qquad \begin{array}{ccc} F1 & \xrightarrow{Fx^*} & FX \\ {\scriptstyle Fx^*}\downarrow & \lrcorner & \downarrow{\scriptstyle Fm_j} \\ FX & \xrightarrow{Fm_i} & FY \end{array}$$

Since F preserves finite nonempty intersections (see Proposition 4.4.1), we therefore obtain the pullback squares on the right. As shown above, Fx^* is not surjective. Hence, we can choose $\overline{x} \in FX$ that is not in the image of Fx^*. The squares on the right above being pullbacks, we see that for $i \neq j$, we have $Fm_i(\overline{x}) \neq Fm_j(\overline{x})$. It follows that $|FY| \geq |I| = |Y|$, as desired.

Case 2: For all nonempty X, Ft_X is injective. Then Ft_X is an isomorphism, since t_X is surjective, and F clearly preserves surjections in Set. We conclude that item (2) holds: the assignment $\tau_X = Ft_X$ yields a natural isomorphism $\tau : F' \to G$, where F' is the restriction of F to nonempty sets, and G is the constant functor with value $F1$. Naturality is due to the fact that for all $f : X \to Y$, $t_Y \cdot f = t_X$. □

Corollary 6.4.4. *A set functor is a varietor iff it has arbitrarily large pre-fixed points.*

Proof Let F be a set functor. The statement is trivial in case F is essentially constant when restricted to infinite sets. So let us assume the contrary. Fix a cardinal κ as in Theorem 6.4.3(1).

Assume first that F is a varietor. Let $\lambda \geq \kappa$ be a cardinal. By Corollary 6.4.1, let A be a set such that $|A| \geq \lambda$ and $|FA| \leq |A|$. But by Theorem 6.4.3(1), we have $|FA| \geq A$. So $|FA| = |A|$. Thus A is a fixed point. Since λ is arbitrary, F has arbitrarily large fixed points.

In the other direction, assume that F has arbitrarily large fixed points. Fix a set X. Let Y be an infinite set with $|Y| \leq |X|$ and $|Y| \leq |FY|$. Then $|Y| \leq |FY+X|$. So Y is a pre-fixed point of $F(-) + X$. Thus $F(-) + X$ has an initial algebra, by Corollary 6.3.9. □

6.5 The Terminal-Coalgebra Chain

Recall from Remark 6.0.5(4) that functors from $\mathsf{Ord}^{\mathsf{op}}$, the dually ordered class of all ordinals are called $\mathsf{Ord}^{\mathsf{op}}$*-chains*.

Recall that a category $\mathscr{A}$ is said to *have limits of chains* if for every ordinal i all diagrams $D\colon i^{\mathsf{op}} \to \mathscr{A}$ have limits. This includes the existence of a terminal object 1 (the case $i = 0$).

The following definition is nothing else than the dual of the initial-algebra chain of Definition 6.1.4. It was formulated explicitly by Barr [72].

Definition 6.5.1. Let $\mathscr{A}$ be a category with limits of chains. For an endofunctor F the *terminal-coalgebra chain* is the $\mathsf{Ord}^{\mathsf{op}}$-chain in $\mathscr{A}$, having objects V_j and connecting morphism v_{ji}, $j \geq i$, defined by transfinite recursion as follows:

$$
\begin{aligned}
V_0 &= 1, \\
V_{j+1} &= FV_j && \text{for all ordinals } j, \\
V_j &= \lim_{i<j} V_i && \text{for all limit ordinals } j, \text{ and}
\end{aligned}
$$

$$
\begin{aligned}
&v_{1,0}\colon V_1 \to 1 \text{ is unique}, \\
&v_{k+1,j+1} = Fv_{k,j}\colon FV_k \to FV_j, \\
&v_{j,i}\ (j > i) \text{ is the limit cone for every limit ordinal } j.
\end{aligned}
$$

Just as in Definition 6.1.9, we say that the terminal-coalgebra chain *converges in λ steps* if $v_{\lambda+1,\lambda}$ is an isomorphism. We similarly say that it *converges in exactly λ steps*, if λ is the least such ordinal.

Remark 6.5.2. (1) Dually to the initial-algebra chain, Definition 6.5.1 determines the chain (V_j) uniquely up to natural isomorphism.

(2) For finite ordinals n we have $V_n = F^n 1$, and we obtain the terminal-coalgebra ω^{op}-chain of Definition 3.3.1.

(3) Note that $v_{\omega+1,\omega}\colon FV_\omega \to V_\omega$ is precisely the morphism m in Notation 4.4.5; there the limit cone $v_{\omega+\omega,\omega+n}\colon V_{\omega+\omega} \to V_{\omega+n}$ was denoted by $\bar{\ell}_n$.

(4) Dually to Construction 6.1.7 we have for every coalgebra $\alpha\colon A \to FA$ the *canonical cone*

$$\alpha_i\colon A \to V_i \qquad (i \in \mathsf{Ord}^{\mathsf{op}})$$

of the terminal-coalgebra chain satisfying $\alpha_{i+1} = F\alpha_i \cdot \alpha$ for all ordinals i.

Example 6.5.3. The terminal-coalgebra chain of every finitary endofunctor on Set converges in $\omega + \omega$ steps; see the proof of Theorem 4.4.10.

Theorem 6.5.4 (Dual of Theorem 6.1.11)**.** *Let $\mathscr{A}$ be a category with limits of chains. If the terminal-coalgebra chain of a functor F converges in λ steps, then V_λ is the terminal algebra with coalgebra structure $v^{-1}_{\lambda,\lambda+1}\colon V_\lambda \to FV_\lambda$.*

Corollary 6.5.5 (Dual of Corollary 6.1.12)**.** *Let $\mathscr{A}$ be a category with limits of chains. If an endofunctor F preserves limits of λ^{op}-chains for some infinite ordinal λ, then the terminal-coalgebra chain converges in λ steps, hence,*

$$\nu F = V_\lambda.$$

Example 6.5.6. For polynomial functors H_Σ (even the non-finitary ones of Example 6.1.16) we conclude that the terminal coalgebra is constructed by the terminal-coalgebra ω^{op}-chain:

$$\nu H_\Sigma = V_\omega.$$

Indeed, for all cardinals κ, the functors $X \mapsto X^\kappa$ preserve limits of ω^{op}-chains. Hence, so do all polynomial functors. This follows from the fact that a coproduct of set functors preserving limits of ω^{op}-chains preserves them, too. Recall that νH_Σ is the coalgebra of all Σ-trees with the inverse of tree-tupling as its structure; for finitary signatures we showed this in Theorem 2.5.9, and for non-finitary ones the proof is analogous.

At this point, we have seen one condition, preservation of limits of λ^{op}-chains, which implies an upper bound on the convergence ordinal of the terminal-coalgebra chain. However, this condition is not satisfied by many everyday functors. We therefore now turn to another, rather mild, condition. Recall that finitary set functors have the convergence ordinal $\omega + \omega$ (see Theorem 4.4.10). We are going to generalize this to *accessible* set functors below. Intuitively, a set functor is λ-accessible, where λ is a fixed infinite regular cardinal, if its behaviour is completely determined by its action on sets of cardinality less than λ. This is captured by the following definition.

Definition 6.5.7. Let λ be an infinite regular cardinal. A set functor F is *λ-bounded* if for each set X and each $x \in FX$ there exists a subset $M \subseteq X$

of cardinality $|M| < \lambda$ such that $x \in Fi[FM]$, where $i\colon M \hookrightarrow X$ is the inclusion map.

Remark 6.5.8. A set functor F is λ-bounded iff it is λ-*accessible*. This means that F preserves λ-filtered colimits (see Definition 11.2.2). A proof of this fact is presented in Section C.8. Following most of the recent literature, we will speak of λ-accessible (in lieu of λ-bounded) set functors, but we work with the property in Definition 6.5.7 in the proofs of this chapter.

Example 6.5.9. (1) Finitary functors are precisely the $\aleph_0$-accessible ones (preserving directed colimits).

(2) The set functor $X \to X^n$ is finitary for every natural number n. For a cardinal number n, F is λ-accessible for every $\lambda > n$.

(3) A coproduct of λ-accessible functors is λ-accessible. Thus, a polynomial functor H_Σ (see Example 6.1.16) is λ-accessible iff all arities of operations symbols in Σ are smaller than λ.

(4) The bounded power-set functor $\mathscr{P}_\lambda X = \{M \subseteq X : |M| < \lambda\}$, a subfunctor of $\mathscr{P}$, is λ-accessible. In contrast, $\mathscr{P}$ is not λ-accessible for any λ.

We have seen that for a finitary set functor the terminal coalgebra $\nu F = V_{\omega+\omega}$ is the limit of the ω-chain $V_{\omega+i}$ $(i < \omega)$. The proof of the generalization to λ-accessible functors below is very similar. However, there is a crucial difference: the previous proof made use of the fact that finitary set functors preserve nonempty intersections (Theorem 4.4.3). In general, this fails for λ-accessible functors, and so the proof of the more general theorem below needs an adjustment.

Lemma 6.5.10. *Let F be a set functor which is not constant of value $\emptyset$. Then all V_i in the terminal-coalgebra chain are nonempty.*

Proof We have $F1 \neq \emptyset$ (otherwise $FX = \emptyset$ for all X using the map $X \to 1$). Thus there is a coalgebra $\alpha\colon 1 \to F1$. The canonical cone $\alpha_i\colon 1 \to V_i$ yields $V_i \neq \emptyset$. □

Theorem 6.5.11 [318, Thm. 11]. *The terminal-coalgebra chain of a λ-accessible set functor converges in $\lambda + \lambda$ steps:*

$$\nu F = V_{\lambda+\lambda}.$$

Proof If $F1 = \emptyset$, then F is constant with value $\emptyset$, and so $\nu F = V_0$. We thus assume that $F1 \neq \emptyset$. By Lemma 6.5.10, all objects V_i in terminal-coalgebra chain are nonempty.

(1) We prove that all connecting morphisms $v_{i,\lambda}$, for $i \geq \lambda$, are monic. It is

sufficient to prove that $v_{\lambda+1,\lambda}$ is monic. Indeed then it is a split monomorphism, hence $v_{\lambda+2,\lambda+1} = Fv_{\lambda+1,\lambda}$ is monic, etc. We obtain by transfinite induction that all $v_{i,\lambda}$ with $i \geq \lambda$ are monic.

We verify that two distinct elements x and y of $V_{\lambda+1} = FV_\lambda$ remain distinct under $v_{\lambda+1,\lambda}$. Since F is λ-bounded, there exists a nonempty subset $u\colon U \hookrightarrow V_\lambda$ such that $|U| < \lambda$ and x and y lie in the image of Fu. In addition, $(v_{\lambda,i})_{i<\lambda}$ is a limit cone, thus collectively monic. Hence, every pair of distinct elements of U remains distinct under $v_{\lambda,i}$ for some $i < \lambda$. From the fact that $|U \times U| < \lambda$ we conclude that one $i_0 < \lambda$ can be chosen for *all* distinct pairs in U. (This is precisely where we use the regularity of the cardinal λ.) In other words, $v_{\lambda,i_0} \cdot u$ is a monomorphism. It splits since $U \neq \emptyset$, thus $Fv_{\lambda,i_0} \cdot Fu$ is monic, which implies that Fv_{λ,i_0} keeps x and y distinct. Moreover $v_{\lambda+1,\lambda}$ also keeps them distinct, because

$$Fv_{\lambda,i_0} = v_{\lambda+1,i_0+1} = v_{\lambda,i_0+1} \cdot v_{\lambda+1,\lambda}.$$

(2) In order to prove that $v_{\lambda+\lambda+1,\lambda+\lambda}$ is invertible, it suffices to prove that F preserves the limit $V_{\lambda+\lambda}$ of $(V_i)_{i<\lambda+\lambda}$. We can disregard the first λ members of that chain and obtain the same limit:

$$V_{\lambda+\lambda} = \lim_{i<\lambda} V_{\lambda+i}.$$

The connecting morphisms

$$v_i = v_{\lambda+i,\lambda}\colon V_{\lambda+i} \rightarrowtail V_\lambda$$

of that last λ-chain are monic by (1). So the limit is just the intersection $V_{\lambda+\lambda} = \bigcap_{i<\lambda} V_{\lambda+i}$, and we write $v_\lambda = v_{\lambda+\lambda,\lambda}\colon V_{\lambda+\lambda} \rightarrowtail V_\lambda$ for the monic connecting map. We will now prove that F preserves this intersection.

Given an element $x \in FV_\lambda$ lying in the image of Fv_i, for all $i < \lambda$, we have the task to prove that x lies in the image of $Fv_{\lambda+\lambda,\lambda}$. Using that F is λ-bounded, we choose a subset $u\colon U \hookrightarrow V_\lambda$, $|U| < \lambda$, such that x lies in the image of Fu. Without loss of generality we may assume that $u \cap v_\lambda$ is nonempty; for otherwise we may pick any element from $V_{\lambda+\lambda}$ and add it to U. Since we have the inclusion maps $v_{\lambda+\lambda,\lambda+i}\colon V_{\lambda+\lambda} \hookrightarrow V_{\lambda+i}$ we obtain that $u \cap v_i$ is nonempty for every ordinal number i. Then for $u_i = u \cap v_i$ we know that $x \in \mathrm{im}(Fu_i)$ for all $i < \lambda$ because F preserves nonempty finite intersections by Proposition 4.4.1. However, (u_i) is a decreasing λ-chain of subsets of U. Since $|U| < \lambda$, this chain converges at some $i_0 < \lambda$. It follows that $u_{i_0} = u_\lambda \subseteq v_{\lambda+\lambda,\lambda}$, whence x indeed lies in the image of $Fv_{\lambda+\lambda,\lambda}$. □

Remark 6.5.12. (1) Worrell also showed [318, Cor. 14] that the collection of set functors whose terminal-coalgebra chain converges in $\lambda + \lambda$ steps contains, besides all λ-accessible functors, all set functors preserving limits of λ^{op}-chains. It is closed under composition, coproduct, and limits of functors.

(2) It follows from the proof in [318] that if F preserves countable intersections, then we have $\nu F = V_{\lambda+\omega}$.

We present examples related to the convergence ordinals of various λ-accessible set functors. This contrasts surprisingly with Example 6.3.15.

Example 6.5.13. (1) The condition in Theorem 6.5.11 is not necessary. Worrell [318] shows that the countably supported subprobability distribution functor $\mathscr{D}$ has $\mathscr{D}^{\omega+\omega}1$ as a terminal coalgebra, and yet the functor is not finitary. It does preserve infinite intersections.

(2) For every regular cardinal λ, the functor $\mathscr{P}_\lambda$ of Example 6.5.9(4) has a terminal coalgebra consisting of all strongly extensional λ-branching trees, that is, every node has less than λ successors (see Definition 4.5.8). The coalgebra structure $\nu\mathscr{P}_\lambda \to \mathscr{P}_\lambda(\nu\mathscr{P}_\lambda)$ is the inverse map of tree-tupling. This was proved for $\lambda = \aleph_0$ by Worrell [318] and for general λ by Schwencke [272].

The terminal-coalgebra chain for $\mathscr{P}_\lambda$ may be presented in terms of a generalization of the saturated trees which we saw in Definition 4.5.21. This leads to the following results, proved in [48, Thm. 3.17]: the terminal-coalgebra chain of $\mathscr{P}_\lambda$ converges in exactly $\lambda + \omega$ steps for λ regular, and in exactly λ^+ steps for λ singular.

Example 6.5.14 [50]. (1) We present a set functor whose terminal-coalgebra chain converges in exactly $\lambda + \lambda$ steps. Recall the filter functor $\mathscr{F}$ from Example 4.4.4(1). It has no fixed point, hence no terminal coalgebra. The λ-accessible coreflection $\mathscr{F}_\lambda$ is the subfunctor of all *λ-small filters*; that is, $\mathscr{F}_\lambda X$ consists of those filters containing some member of size smaller than λ. The terminal-coalgebra chain of $\mathscr{F}_\lambda$ converges in $\lambda + \lambda$ steps.

(2) We can also present a set functor whose terminal-coalgebra chain converges in exactly $\lambda + \kappa$ steps for an arbitrary regular cardinal $\kappa < \lambda$. A *base* of a filter $\mathbb{F}$ is a family $\mathbb{F}_0$ of sets such that $\mathbb{F}$ is the smallest filter with $\mathbb{F}_0 \subseteq \mathbb{F}$. Let us call a filter *$\alpha$-based* if it has a base whose cardinality is less than α. The subfunctor $\mathscr{F}_\lambda^\kappa$ of $\mathscr{F}_\lambda$ consisting of all λ-small, κ-based filters has the terminal-coalgebra chain converging in exactly $\lambda + \kappa$ steps.

(3) We present a modification $\overline{\mathscr{P}}_\lambda$ of $\mathscr{P}_\lambda$ whose terminal-coalgebra chain converges in exactly λ steps. No modification is needed on objects, but for a morphism $f\colon X \to Y$, given $M \subseteq X$ with $|M| < \lambda$, we put

$$\overline{\mathscr{P}}_\lambda(M) = \begin{cases} f[M] & \text{if } f|_M \text{ is monic,} \\ \emptyset & \text{otherwise.} \end{cases}$$

We have seen a necessary and sufficient condition for a set functor F to have an initial algebra: the existence of a fixed point (see Corollary 6.3.9). We shall see in Example 6.5.16 that a set functor with a fixed point need not have a terminal coalgebra.

However, whenever a terminal coalgebra exists, it can be constructed by the terminal-coalgebra chain. This holds more for many-sorted sets:

Theorem 6.5.15 [26]. *Whenever an endofunctor of* Set^S *has a terminal coalgebra, then its terminal-coalgebra chain converges.*

This generalizes the previous result for endofunctors on Set in Adámek and Koubek [17]. Both proofs heavily depend on the theory of algebraized chains developed by Jan Reiterman in his PhD thesis and summarized by Koubek and Reiterman [200].

The expected generalization of Theorem 6.5.15 to, say, all presheaf categories does not hold. Adámek and Trnková [26] constructed an endofunctor on the category of graphs that has a terminal coalgebra although the terminal-coalgebra chain does not converge.

Example 6.5.16 [17]. In contrast to what we saw in Theorem 6.3.4, a set functor with a fixed point need not have a terminal coalgebra. Consider the set functor $\mathscr{P}_{-\aleph_0}$ defined by

$$\mathscr{P}_{-\aleph_0}X = \text{the set of all finite or uncountable subsets of } X.$$

For a function $f\colon X \to Y$ and $A \in \mathscr{P}_{-\aleph_0}X$, put

$$\mathscr{P}_{-\aleph_0}f(A) = \begin{cases} f[A] & \text{if either } A \text{ is finite or } f|_A \text{ is injective,} \\ \emptyset & \text{otherwise.} \end{cases}$$

This functor has an initial algebra but not a terminal coalgebra. Obviously, $\mathbb{N}$ is a fixed point of $\mathscr{P}_{-\aleph_0}$: the set $\mathscr{P}_{-\aleph_0}\mathbb{N}$ of all finite subsets is countable. Thus, $\mu\mathscr{P}_{-\aleph_0}$ exists by Corollary 6.3.9.

To verify that $\nu\mathscr{P}_{-\aleph_0}$ does not exist, note that $\mathscr{P}_{\mathsf{f}}$ is a subfunctor of $\mathscr{P}_{-\aleph_0}$. If $\nu\mathscr{P}_{-\aleph_0}$ would exist, then by Proposition 6.6.11 below, it would be at least as large as $\nu\mathscr{P}_{\mathsf{f}}$. Recall from Corollary 4.5.7 that $\nu\mathscr{P}_{\mathsf{f}}$ is uncountable. To conclude the proof, we thus only need to check that $\mathscr{P}_{-\aleph_0}$ has no uncountable fixed points. Let A be an uncountable set. Let C be the set of countably infinite subsets of A, write D for $\mathscr{P}_{-\aleph_0}A$, and note that $|C \cup D| = |\mathscr{P}(A)| > |A|$. The map $X \mapsto A \setminus X$ is an injection $\mathscr{P}A \rightarrowtail \mathscr{P}A$. Since A is uncountable, it restricts to an injection $C \rightarrowtail D$. Therefore $|C| \leq |D|$. Now $|D| = \max(|C|, |D|) = |C| + |D| = |\mathscr{P}(A)|$. In other words, $|A| < |\mathscr{P}_{-\aleph_0}A|$. In particular, A is not a fixed point.

Example 6.5.17 [25]. We now provide set functors F and G having initial algebras such that none of $F \cdot G$, $F + G$, or $F \times G$ has a fixed point. Recall $FX = \mathscr{P}_{-\aleph_0} X$ from Example 6.5.16, and let $GX = X + \mathbb{R}$ for the set $\mathbb{R}$ of real numbers. Then F and G have initial algebras. We have seen in Example 6.5.16 that for every uncountable set A, $|A| < |F(A)|$. It follows that the same holds for $F + G$, $F \times G$, and $G \cdot F$. Indeed, for A countable or finite, $(F + G)A$ is uncountable (easily), and the same for $F \times G$ and $G \cdot F$. As for $F \cdot G$, note that $(F \cdot G)X = \mathscr{P}_{-\aleph_0}(X + \mathbb{R})$. For all X, $X + \mathbb{R}$ is uncountable, and so $|X| \leq |X + \mathbb{R}| < |(F \cdot G)X|$. Thus, these functors all fail to have fixed points.

We next present a sufficient condition for the existence of a terminal coalgebra for a set functor F: the existence of a *fixed-point pair*: this means an infinite regular cardinal λ exists such that F has fixed points of cardinalities λ and λ^+, the successor cardinal of λ. In our result on this, we assume the *Generalized Continuum Hypothesis (GCH)*: for every infinite cardinal λ, $\lambda^+ = 2^\lambda$. Under this hypothesis, for all cardinals κ and all regular cardinals $\lambda > \kappa$, we have $\lambda^\kappa = \lambda$ [176, Thm. 5.15(iii)]. Even without GCH, we have $\lambda^\lambda = 2^\lambda$. Let us also recall that the Continuum Hypothesis (CH) is the statement that $2^{\aleph_0} = \aleph_1$. We prove in Theorem 6.5.18 that (assuming GCH) this existence of a fixed-point pair implies that F has a terminal coalgebra.

Theorem 6.5.18 [17, Thm. 5]. *Assuming GCH, every set functor with a fixed-point pair has a terminal coalgebra.*

Proof Let λ and λ^+ be a fixed-point pair of a set functor F. We prove that all of the connecting morphisms $v_{j,i}$ for $j \geq i \geq \lambda$ are monic. Thus, the sets V_i for $i \geq \lambda$ form a non-increasing chain of subsets of V_λ. It follows that there is some i with $v_{i+1,i} \colon V_{i+1} \to V_i$ invertible. For this i we have $\nu F = V_i$.

We can assume that $F1 \neq \emptyset$, thus, $V_i \neq \emptyset$ for all ordinals i (Lemma 6.5.10).

(1) For every $j < \lambda$ we prove that $|V_j| \leq \lambda$ by transfinite induction. The case $\lambda = 0$ is clear. If this holds for V_j, then there exists an injective map $m \colon V_j \rightarrowtail \lambda$. Since m is a split monomorphism, so is Fm. Thus, $|FV_j| \leq |F\lambda| = \lambda$. This is the isolated step. Finally, let $j < \lambda$ be a limit ordinal. Then $V_j = \lim_{i<j} V_i$ has cardinality at most that of $\prod_{i<j} V_j$ which by the induction hypothesis is at most λ^κ, where κ is the cardinality of j. By GCH, $\lambda^\kappa = \lambda$.

(2) Consequently, $|V_\lambda| \leq \lambda^+$, because

$$|V_\lambda| \leq |\prod_{i<\lambda} V_i| \leq \lambda^\lambda = 2^\lambda.$$

Since $2^\lambda = \lambda^+$ is a fixed point of F, it follows that $|V_{\lambda+1}| = |FV_\lambda| \leq \lambda^+$.

(3) We now prove that every element $x \in V_{\lambda+1}$ lies in $Fm[FM]$ for some subset $m \colon M \hookrightarrow V_\lambda$ with $|M| < \lambda$. For this, we apply Lemma C.5.8 twice.

First use the fixed point λ^+. Given $x \in V_{\lambda+1} = FV_\lambda$ there exists $m_0 : M_0 \hookrightarrow V_\lambda$ with $x \in Fm_0[FM_0]$ and $|M_0| < \lambda^+$. So $|M_0| \le \lambda$. Choose $x_0 \in FM_0$ such that $x = Fm_0(x_0)$. Now we use the fixed point λ: there exists $m_1 : M_1 \hookrightarrow M_0$ with $x_0 \in Fm_1[FM_1]$ and $|M_1| < \lambda$. This concludes the proof as the subset $m = m_0 \cdot m_1 : M_1 \hookrightarrow X$ has cardinality less than λ, and $x \in Fm[FM_1]$.

(4) It follows from item (3) that $v_{\lambda+1,\lambda}$ is monic: see the proof of Theorem 6.5.11. Since $V_i \neq \emptyset$ for all i, it follows by transfinite induction (as in the cited proof) that all $v_{j,i}$ with $j \ge i \ge 1$ are monic. □

Example 6.5.19. A set functor can have a terminal coalgebra without having a fixed-point pair. Define F on objects by

$$FX = X + \{M \subseteq X : X \text{ is uncountable or empty}\}.$$

On morphisms, F is defined as expected, using the formula of Example 6.5.14(3) in the second summand. The terminal coalgebra chain is the same for F and the functor $GX = X + 1$. Thus $\nu F = \nu G = \mathbb{N}^\top$ (Example 3.3.6). But an infinite set is a fixed point of F iff it is countable, so F has no fixed point pairs.

Proposition 6.5.20. *The continuum hypothesis $\aleph_1 = 2^{\aleph_0}$ is equivalent to the statement that every set functor with the fixed-point pair $\aleph_0, \aleph_1$ has a terminal coalgebra.*

Proof If the continuum hypothesis holds, the argument is precisely as in the proof of Theorem 6.5.18 for $\lambda = \aleph_0$.

Conversely, if $\aleph_1 < 2^{\aleph_0}$, we present a set functor with fixed points $\aleph_0$ and $\aleph_1$ which has no terminal coalgebra. This is completely analogous to Example 6.5.16: define $\mathscr{P}_{-\aleph_0,\aleph_1}$ to assign to every set the set of all subsets of X of power different from $\aleph_0$ or $\aleph_1$. For a set X of cardinality $\aleph_0$ or $\aleph_1$ we have $\mathscr{P}_{-\aleph_0,\aleph_1} = \mathscr{P}_{\mathrm{f}}X$, so X is a fixed point. The non-existence of a terminal coalgebra is proved as for $\mathscr{P}_{-\aleph_0}$. □

Open Problem 6.5.21. (1) Give a necessary and sufficient condition for the existence of a terminal coalgebra for set functors.

(2) Characterize the converence ordinals of all terminal-coalgebra chains of set functors.

Remark 6.5.22. The following were proved by Barr [71].

(1) The initial-algebra and terminal-coalgebra chain of a functor F are connected by a unique collection of morphisms $k_i : W_i \to V_i$ ($i \in \mathsf{Ord}$) satisfying $k_{i+1} = Fk_i$ for every ordinal i. (See Section 13.1 for more on these morphisms.)

(2) Assume that some k_i is an isomorphism, and suppose that at least one

algebra-to-coalgebra morphism exists. Then F has a canonical fixed point (Definition 5.0.1). In more detail, the hypothesis is that there is an algebra $\alpha\colon FA \to A$, a coalgebra $\beta\colon B \to FB$, and a morphism $f\colon A \to B$ such that the square below commutes:

$$\begin{array}{ccc} FA & \xrightarrow{\alpha} & A \\ {\scriptstyle Ff}\big\downarrow & & \big\downarrow{\scriptstyle f} \\ B & \xleftarrow{\beta} & FB \end{array}$$

Remark 6.5.23. We have seen canonical fixed points $\mu F = \nu F$ in Chapter 5. For set functors they are very rare. However, every set functor F with $|F\emptyset| = |F1| = 1$ has the canonical fixed point 1. It is clear that $\nu F = 1$. Since the initial-algebra chain is $\emptyset \to 1 \to 1 \to \cdots$, we conclude that $\mu F = 1$. For example, the quotient of the power-set functor $\mathscr{P}$ which merges $\emptyset$ and every singleton set $\{x\}$ has $|F\emptyset| = |F1| = 1$.

Returning to varietors (Definition 2.2.26), it follows from Corollary 6.1.12 that every λ-accessible set functor is a varietor. (This is because $F(-) + A$ is also λ-accessible, for each set A.) We now present a set functor which is a varietor, which implies that it has arbitrarily large pre-fixed points, but is not accessible. This example essentially stems from Koubek and Reiterman [199]. There the description was a bit simpler, but the example needed the Generalized Continuum Hypothesis. The presentation below is in the spirit of Kelly [187, Ex. 21.3], and the example was suggested to him by Monro [187, Rem. 21.2].

Example 6.5.24. We present a non-accessible varietor F on Set. A cardinal $\kappa > 0$ is a *strong limit cardinal* if whenever $\lambda < \mu < \kappa$, then also $\lambda^\mu < \kappa$. It is a basic fact of set theory that there are arbitrarily large strong limit cardinals. To see this, note that for every cardinal λ, the ω-sequence $f(0) = \lambda$, $f(n+1) = 2^{f(n)}$ yields a strong limit cardinal $\bigvee_n f(n)$.

We define F on sets X by

$$FX = \{\emptyset\} \cup \{M \subseteq X \mid |M| \text{ is a strong limit cardinal}\}.$$

For a map $f\colon X \to Y$ we define

$$Ff(M) = \begin{cases} f[M] & \text{if } f|_M \text{ is monic,} \\ \emptyset & \text{else.} \end{cases}$$

It is easy to see that F is a well-defined set functor. We show that F is not λ-bounded (see Definition 6.5.7) for any cardinal λ. Hence F is not accessible; see Proposition C.8.2. For this, fix λ. Let $\kappa > \lambda$ be a strong limit cardinal. Then $\kappa \in F(\kappa)$, and there is no subset $M \subseteq \kappa$ of cardinality λ such that $\kappa \in Fi[M]$, with $i\colon M \to \kappa$ the inclusion.

Furthermore, F has arbitrarily large fixed points: for every infinite strong limit cardinal κ we prove $FA \cong A$ for every set A such that $|A| = 2^\kappa$.

(1) Case $|FA| \geq |A|$. Indeed, FA contains all $M \subseteq A$ such that $|M| = \kappa$, thus we have $|FA| \geq (2^\kappa)^\kappa = 2^\kappa = |A|$.

(2) Case $|FA| \leq |A|$. First observe that 2^κ is clearly not a strong limit cardinal. Consequently, if $F_\lambda A \subseteq FA$ denotes the set $\{M \subseteq A : |M| = \lambda\}$, we have

$$FA = \{\emptyset\} \cup \bigcup_{\lambda < 2^\kappa} F_\lambda A,$$

where the union ranges over strong limit cardinals $\lambda < 2^\kappa$. They fulfil $\lambda \leq \kappa$ (else from $\kappa < \lambda < 2^\kappa$ we deduce $2^\kappa < 2^\kappa$). Thus, we have

$$|F_\lambda A| = (2^\kappa)^\lambda \leq (2^\kappa)^\kappa = 2^\kappa.$$

Thus, $|FA| \leq \kappa \cdot 2^\kappa = 2^\kappa = |A|$.

6.6 Subfunctors and Quotient Functors

This section presents results which transfer initial algebras and terminal coalgebras for a functor to subfunctors and quotient functors. So we are concerned with four situations. In some but not all of these, the results use the chain constructions which we have been studying in earlier parts of this chapter.

We first prove a technical result concerning the initial-algebra chains of pairs of functors related by a natural transformation. In the following we consider a pair of endofunctors F and F' with the initial-algebra chains (W_i) for F and (W'_i) for F'. Note that a natural transformation between the corresponding chains is a collection of morphisms $\psi_i : W_i \to W'_i$ ($i \in \mathsf{Ord}$) making the following square commutative for every pair $i \leq j$ of ordinals:

$$\begin{array}{ccc} W_i & \xrightarrow{w_{i,j}} & W_j \\ {\scriptstyle \psi_i}\downarrow & & \downarrow{\scriptstyle \psi_j} \\ W'_i & \xrightarrow{w'_{i,j}} & W'_j \end{array} \tag{6.6}$$

Proposition 6.6.1. *Every natural transformation $\varphi : F \to F'$ between functors with initial-algebra chains (W_i) and (W'_i), respectively, induces a unique natural transformation $\psi_i : W_i \to W'_i$ ($i \in \mathsf{Ord}$) such that for every ordinal i we have*

$$\psi_{i+1} = (FW_i \xrightarrow{F\psi_i} FW'_i \xrightarrow{\varphi W'_i} F'W'_i). \tag{6.7}$$

Proof (1) Uniqueness. Given a natural transformation $\psi_i\colon W_i \to W'_i$ satisfying (6.7) we verify that each ψ_i is uniquely determined by transfinite induction. The base case is clear since $W_0 = W'_0 = 0$. For the isolated step note that (6.7) determines ψ_{i+1} uniquely from ψ_i. Now let j be a limit ordinal. Then we have $W_j = \mathsf{colim}_{i<j} W_i$ and $W'_j = \mathsf{colim}_{i<j} W'_i$. By hypothesis we know that $(\psi_i)_{i<j}$ form a natural transformation of the corresponding j-chains. Since the morphisms $w'_{i,j} \cdot \psi_i\colon W_i \to W'_j$ clearly form a cocone of (W_i) with vertex $W_{j'}$, we see that $\psi_j\colon W_j \to W'_j$ is determined by the universal property of the colimit W_j uniquely such that the naturality squares (6.6) commute for every $i < j$.

(2) Existence. We proceed by transfinite recursion on j to define ψ_j and simultaneously prove the commutativity of (6.6) for every $i < j$.

(2a) In the base case we put $\psi_0 = \mathsf{id}_0$, and there is nothing to prove.

(2b) For the isolated step we define $\psi_{j+1} = \varphi_{W'_j} \cdot F\psi_j$ as in (6.7). We prove the commutativity of (6.6) for every $i < j+1$ by transfinite induction on i. The base case $i = 0$ is clear since $W_0 = 0$ is the intial object. For the isolated step we know by the induction hypothesis on j that (6.6) commutes. Then the corresponding square for the successors of i and j commutes, too:

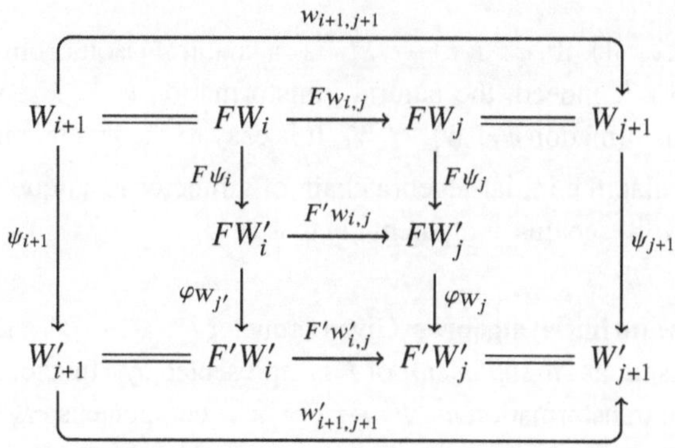

Indeed, the upper and lower parts commute by the definitions of $w_{i+1,j+1}$ and $w'_{i+1,j+1}$, respectively. The left-hand and right-hand squares commute by (6.7), the middle upper square commutes by the induction hypothesis, and the middle lower one by the naturality of φ.

Finally, suppose that i is a limit ordinal. We use that $W_i = \mathsf{colim}_{k<i} W_k$ and show that the desired square commutes when pre-composed by every colimit

injection $w_{k,i}$, where $k < i$:

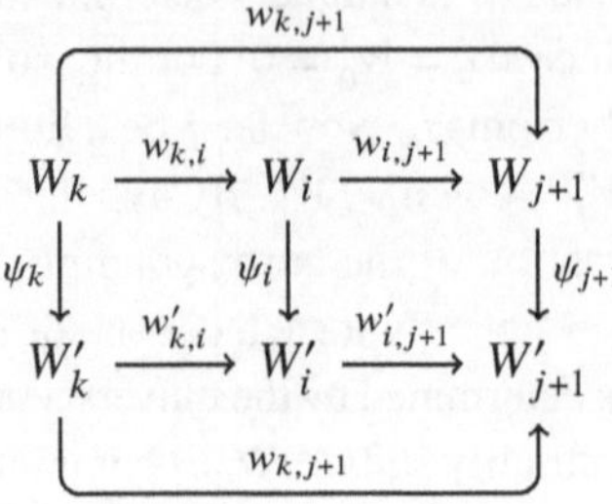

The outside commutes by the induction hypothesis (for our induction on i), and the left-hand inner square does by the induction hypothesis (for the induction on j because $i < j$). Since the upper and lower parts trivially commute, we obtain that the right-hand inner square commutes when pre-composed by $w_{k,i}$ for all $k < i$. This proves that this square commutes since the colimit injections are jointly epic.

(2c) Finally, suppose that j is a limit ordinal. By induction we know that for all $i \leq i' < j$ the squares (6.6) (with i' in lieu of j) commute. Thus, we obtain $\psi_j : W_j \to W'_j$ uniquely such that (6.6) commmutes for every $i < j$, similary as in item (1), by the universal property of the colimit $W_j = \mathsf{colim}_{i<j} W_i$. This completes the proof. □

Remark 6.6.2. (1) If $\varphi : F \to F'$ is a natural isomorphism then so is $\psi_i : W_i \to W'_i$. Indeed, the natural transformation $\varphi^{-1} : F \to F'$ induces a natural transformation $\bar{\psi}_i : W'_i \to W_i$. It is easy to see that it is inverse to ψ_i.

(2) In particular, the initial-algebra chain of a functor is unique up to natural isomorphism. To see this apply item (1) to $\varphi = \mathsf{id}_F$.

Subfunctors and initial algebras Given a functor $F : \mathscr{A} \to \mathscr{B}$, and a class $\mathcal{M}$ of monomorphisms, an $\mathcal{M}$-*subfunctor* of F is represented by a functor $F' : \mathscr{A} \to \mathscr{B}$ with a natural transformation $m : F' \rightarrowtail F$ whose components m_X belong to $\mathcal{M}$.

If an initial algebra of a functor F exists, do all $\mathcal{M}$-subfunctors have an initial algebra? Not in general:

Example 6.6.3. Let Gra be the category of graphs and graph homomorphisms. Let F be the endofunctor mapping every graph $(X, \emptyset)$ without edges to the graph $(\mathscr{P}X, \{(\emptyset, \emptyset)\})$ on the power-set that has just a loop at $\emptyset$. All graphs with edges are mapped to the terminal graph 1. Homomorphisms $f : (X, \emptyset) \to (Y, \emptyset)$ are mapped to $\mathscr{P}f$. Then F has an initial algebra carried by the terminal graph:

$$\mu F = 1.$$

But the subfunctor of F given by $(X, \emptyset) \mapsto (\mathscr{P}X, \emptyset)$ and else constantly 1 does not have an initial algebra, since 1 is its only fixed point, and there exist no algebra homomorphism from it to the algebras $\alpha\colon (\mathscr{P}X, \emptyset) \to (X, \emptyset)$ where $X \neq \emptyset$.

Corollary 6.6.4. *Let $\mathscr{A}$ be a well-powered category, and let $\mathcal{M}$ be a smooth class of monomorphisms. If an endofunctor F on $\mathscr{A}$ has an initial algebra, then so does every $\mathcal{M}$-subfunctor F' of F preserving $\mathcal{M}$-monomorphisms. Moreover, $\mu F'$ is a subobject of μF.*

Proof Let $m\colon F' \to F$ be a natural transformation whose components are in $\mathcal{M}$. The functor F' has the $\mathcal{M}$-pre-fixed point

$$F'(\mu F) \xrightarrow{m_{\mu F}} F(\mu F) \xrightarrow{\cong} \mu F,$$

where the second morphism is the structure of the initial F-algebra. Now apply Theorem 6.3.4 to F' to see that it has an initial algebra which is a subobject of $\mu F'$. □

Remark 6.6.5. Observe that the subfunctor F' preserves $\mathcal{M}$ whenever F does and $\mathcal{M}$ is *left cancellative*: $u \cdot v \in \mathcal{M}$ implies $v \in \mathcal{M}$. (For example, the class of all monomorphisms is left cancellative.) Indeed, let $m\colon F' \rightarrowtail F$ be a natural transformation with components in $\mathcal{M}$. If $u\colon X \rightarrowtail Y$ is an $\mathcal{M}$-monomorphism, then $F'u$ is one, too, since $m_Y \cdot F'u = Fu \cdot m_X$ holds by the naturality of m.

Corollary 6.6.6. *Let $\mathscr{A}$ be well-powered with a left-cancellative and smooth class $\mathcal{M}$ of monomorphisms. If an endofunctor preserves $\mathcal{M}$-monomorphisms and has an initial algebra, then every $\mathcal{M}$-subfunctor has an initial algebra, too.*

Remark 6.6.7. Even if F has no initial algebra, the initial algebra of any $\mathcal{M}$-subfunctor of F has a certain universal property with respect to all algebras for F. Indeed, given $m\colon F' \to F$ as above such that the initial algebra $(\mu F', \iota)$ exists. Then for every F-algebra (A, α) there exists a unique morphism $h\colon \mu F' \to A$ such that the diagram below commutes:

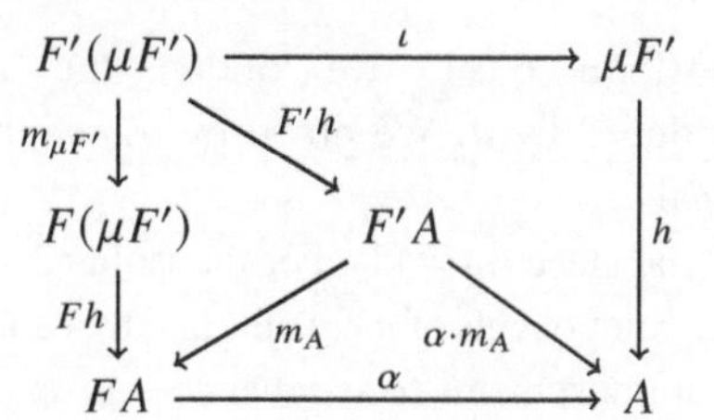

Indeed, note that the left-hand triangle commutes due to the naturality of m. By the initiality of $\mu F'$ there exists a unique h such that the right-hand part commutes. Equivalently, the outside commutes.

Quotient functors and initial algebras Recall that quotients of a functor F are represented by natural transformations $q\colon F \to F'$ with epic components. Quotient functors often inherit the existence of initial algebras, as we prove below. But this does not always hold:

Example 6.6.8. We present an endofunctor F on Gra with an initial algebra and a quotient F' of it which has no fixed points. The functor F assigns to a graph (X, R) the discrete graph on $\mathscr{P}R$ (the set of subsets of the edges):

$$F(X, R) = (\mathscr{P}R, \emptyset).$$

Every homomorphism $f\colon (X, R) \to (Y, S)$ induces a function

$$f^{(2)}\colon \mathscr{P}R \to \mathscr{P}S,$$

and we put

$$Ff = \mathscr{P}f^{(2)}.$$

The initial-algebra chain of F converges in two steps, yielding $\mu F = (\mathscr{P}\emptyset, \emptyset)$.

For the functor F', we modify F by assigning to (X, R) the complete graph on $\mathscr{P}R$:

$$F'(X) = (\mathscr{P}R, \mathscr{P}R \times \mathscr{P}R) \qquad \text{and} \qquad F'f = \mathscr{P}f^{(2)}.$$

Then F' has no fixed points. But it is a quotient of F via the natural transformation $g\colon F \twoheadrightarrow F'$ given by $g_{(X,R)} = \mathsf{id}_{\mathscr{P}R}$.

Proposition 6.6.9. *Let $\mathscr{A}$ be a well-powered and co-well-powered category with a smooth class $\mathcal{M}$ of monomorphisms. Let $F\colon \mathscr{A} \to \mathscr{A}$ preserve $\mathcal{M}$-monomorphisms and epimorphisms. If it has an initial algebra, then so does every quotient functor $q\colon F \twoheadrightarrow F'$, and $\mu F'$ is a quotient object of μF.*

Proof By Theorem 6.3.4, the initial-algebra chain (W_i) of F converges. Thus, there exists an ordinal λ with $\mu F = W_\lambda$. We prove that the initial-algebra chain (W'_i) of F' converges, too.

(1) Applying Proposition 6.6.1, we obtain the natural transformation $\psi_i\colon W_i \to W'_i$ induced by q. We prove by transfinite induction that ψ_i is epic for every ordinal i.

The base case is trivial since $\psi_0 = \mathsf{id}_0$. For the isolated step use the induction hypothesis and that the components of q are epic and that epimorphisms compose and that F preserves epimorphisms to see that ψ_{i+1} is epic:

$$\psi_{i+1} = (W_{i+1} = FW_i \xrightarrow{F\psi_i} FW'_i \xrightarrow{q_{W'_i}} F'W'_i = W'_{i+1}).$$

For a limit ordinal j, observe that $\psi_j\colon W_j \to W'_j$ was (necessarily) defined as the colimit of the chain of epimorphisms $(\psi_i)_{i<j}$ (in the category of $\mathscr{A}^{\to}$ of morphisms of $\mathscr{A}$): $\psi_j = \mathsf{colim}_{i<j}\,\psi_i$. It is easy to see that a colimit in $\mathscr{A}^{\to}$ of epimorphisms of $\mathscr{A}$ is an epimorphism (cf. Remark 6.2.15).

(2) Since $\mu F = W_\lambda$ we know that $w_{\lambda,i}$ is an isomorphism for every $i \geq \lambda$. Therefore the following composites are quotients of W_λ for every $i \geq \lambda$:

$$e_i = \left(W_\lambda \xrightarrow{w_{\lambda,i}} W_i \xrightarrow{\psi_i} W_i'\right).$$

Due to the naturality of ψ_i they form a chain of quotients of W_λ with the connecting morphisms $w'_{i,j}$ for $j \geq i \geq \lambda$:

$$\begin{array}{ccc} & W_\lambda & \\ {}^{w_{\lambda,i}}\swarrow & & \searrow^{w_{\lambda,j}} \\ W_i & \xrightarrow{\quad w_{i,j} \quad} & W_j \\ {\scriptstyle \psi_i}\downarrow & & \downarrow{\scriptstyle \psi_j} \\ W_i' & \xrightarrow{\quad w'_{i,j} \quad} & W_j' \end{array}$$

Since W_λ has only a set of quotients, this chain of quotients converges. Hence there exists an ordinal $i \geq \lambda$ such that e_i and e_{i+1} represent the same quotient of W_λ. In other words, $w'_{i,i+1} : W_i' \to W_{i+1}'$ is an isomorphism, as desired.

(3) The assertion that $\mu F'$ is a quotient of μF follows: $\psi_i : \mu F \twoheadrightarrow \mu F'$ is epic. □

Subfunctors and terminal coalgebras We first show that subfunctors do not inherit the existence of terminal coalgebras in general.

Example 6.6.10. Consider the functor F on Gra from Example 6.6.3 and form the coproduct $GX = FX + A$, where A is a discrete one-vertex graph. The terminal-coalgebra chain for G converges in two steps, yielding $\nu G = 1 + A$.

The subfunctor G' of G defined on objects by

$$G'(X,R) = \begin{cases} (\mathscr{P}A, \emptyset) + A & \text{if } R = \emptyset, \\ A & \text{else} \end{cases}$$

does not have fixed points. Indeed, for a graph (X,R) where $R = \emptyset$ we see that $G'(X,R)$ has greater cardinality than (X,R). If $R \neq \emptyset$, then $G'(X,R)$ is discrete but (X,R) is not.

Proposition 6.6.11. *Let $\mathscr{A}$ be a well-powered category with limits of chains. Let $F' : \mathscr{A} \to \mathscr{A}$ preserve monomorphisms, and assume that the terminal-coalgebra chain of F' converges. Then so does the terminal-coalgebra chain of every subfunctor $m : F \rightarrowtail F'$, and νF is a subobject of $\nu F'$.*

Proof (1) Denote by (V_i) and (V'_i) the respective terminal-coalgebra chains of F and F'. By the dual of Proposition 6.6.1 there exists a unique natural transformation $\psi_i\colon V_i \to V'_i$ ($i \in \mathsf{Ord}$) such that for every ordinal i we have

$$\psi_{i+1} = (V_{i+1} = FV_i \xrightarrow{F\psi_i} FV'_i \xrightarrow{m_{V'_i}} F'V'_i).$$

By transifinite induction on i one proves that every ψ_i is monic. Indeed, the proof is almost identical to what we have seen for epimorphisms in item (1) of the proof of Proposition 6.6.9 except that one uses in the limit step that limits of monomorphisms (in $\mathscr{A}^{\to}$) are monic.

(2) We know that $\nu F' = V'_\lambda$ for some ordinal λ. For every $i > \lambda$ we have the following subobject of νF:

$$s_i = \big(V_i \overset{\psi_i}{\rightarrowtail} V'_i \overset{v'_{i,\lambda}}{\rightarrowtail} \nu F'\big).$$

We are using the fact that $v'_{i,\lambda}$ is an isomorphism. These subobjects form a decreasing chain of subobjects of V'_λ with the connecting morphisms $v_{j,i}$ for $j \geq i \geq \lambda$:

$$\begin{array}{ccc}
V_i & \xleftarrow{\quad v_{j,i}\quad} & V_j \\
{\scriptstyle\psi_i}\downarrow & & \downarrow{\scriptstyle\psi_j} \\
V'_i & \xleftarrow{\quad v'_{j,i}\quad} & V'_j \\
& {\scriptstyle v'_{i,\lambda}}\searrow \quad \swarrow{\scriptstyle v'_{j,\lambda}} & \\
& V'_\lambda &
\end{array}$$

Since $\nu F'$ has only a set of subjobjects, there is an ordinal $i \geq \lambda$ such that the subobjects represented by s_i and s_{i+1} are the same. In other words, $v_{i+1,i}$ is an isomorphism. So as desired, $\nu F = V_i$.

(3) The assertion that $\nu F'$ is a subobject of νF follows from the fact that ψ_i is monic. □

Remark 6.6.12. We will later see that a terminal coalgebra for F' implies that all subfunctors of F' have terminal coalgebras, without assuming that the terminal-coalgebra chain converges (see Corollary 7.2.10).

Quotient functors and terminal coalgebras We again first demonstrate that terminal coalgebras are in general not inherited by quotient functors. On the affirmative side, we present a somewhat weaker result than the previous ones: we will restrict ourselves to quotient functors $q\colon F \to F'$ such that the components q_X are split epimorphisms.

Example 6.6.13. The functor F of Example 6.6.8 has the terminal coalgebra $(\mathscr{P}\emptyset, \emptyset)$. It is obtained in two steps of the terminal-coalgebra chain. The quotient F' has no fixed points.

We now present the following result due to Gumm and Schröder [153]:

Observation 6.6.14. (1) Let $\mathscr{A}$ be a cocomplete category. Suppose that F and G are endofunctors, and let $\varepsilon\colon F \twoheadrightarrow G$ be a natural transformation whose components are split epimorphisms. If F has a terminal coalgebra, then so does G. Indeed, by Lemma 4.3.6, G has a weakly terminal coalgebra. Now use Theorem 4.2.8 to see that νG exists.

(2) Furthermore, we see that νG is a canonical (split) quotient of νF via the unique homomorphism shown below:

$$\begin{array}{ccccc}
\nu F & \xrightarrow{\tau^F} & F(\nu F) & \xrightarrow{\varepsilon_{\nu F}} & G(\nu F) \\
{\scriptstyle e}\big\downarrow\!\!\downarrow & & & & \big\downarrow{\scriptstyle Ge} \\
\nu G & & \xrightarrow{\qquad\tau^G\qquad} & & G(\nu G)
\end{array}$$

To see that e is a split epimorphism, we use that νF is a weakly terminal G-coalgebra to obtain a homomorphism $s\colon (\nu G, \tau^G) \to (\nu F, \varepsilon_{\nu F} \cdot \tau^F)$ (cf. Lemma 4.3.6). Thus, $e \cdot s = \mathsf{id}_{\nu G}$ because this is a coalgebra homomorphism from νG to itself.

Corollary 6.6.15. *If a set functor has an initial algebra (or a terminal coalgebra), then so do all of its subfunctors and all of its quotient functors.*

Proof Let F be a set functor, let G be a subfunctor and H a quotient. Assume that F has an initial algebra, say $FI \cong I$. Then $|GI| \leq |FI| \leq |I|$. Likewise $|HI| \leq |I|$. By Corollary 6.3.9, G and H have initial algebras.

Assume that F has a terminal coalgebra. Then by Theorem 6.5.15 the terminal-coalgebra chain of F converges. We wish to apply Proposition 6.6.11 to see the same for G even if F does not preserve all monomorphisms. But note that in that result, we only need each morphism ψ_i to be monic. If $G1 = \emptyset$, then G is constant with value $\emptyset$. Therefore $V_i = \emptyset$ for every $i > 0$, and we are done. Otherwise all V_i are nonempty (Lemma 6.5.10). Using that nonempty injections are split monomorphisms in Set and are thus preserved by G, we see by transfinite induction that every ψ_i is injective.

As for H, recall that all epimorphisms in Set split. By Observation 6.6.14, H has a terminal coalgebra. □

6.7 Canonical Fixed Points in CPO-Enriched Categories

We now return to CPO-enriched categories and consider sufficient conditions for an endofunctor to have a canonical fixed point. We know from Theorem 5.1.24 that locally continuous endofunctors have canonical fixed points. We introduce a weaker condition called stability and prove that whenever a stable functor has a fixed point, it has a canonical one. (By Observation 6.7.5 below, stability is more or less equivalent to preservation of embeddings.) Recall from Definition 5.1.21 that a category is *strict* CPO*-enriched* if its hom-sets carry the structure of a cpo with $\perp$, and composition is continuous and preserves $\perp$ (in both variables). In the following definition an endomorphism $f\colon X \to X$ is *idempotent* if $f \cdot f = f$.

Definition 6.7.1. A functor F between strict CPO-enriched categories is called *stable* if for every idempotent endomorphism $f\colon X \to X$, we have:

$$f \sqsubseteq \mathsf{id}_X \quad \text{implies} \quad Ff \sqsubseteq \mathsf{id}_{FX}.$$

Example 6.7.2. (1) Every locally monotone functor (see Definition 5.1.16) is stable.

(2) A composite, product or coproduct of stable endofunctors on the category $\mathsf{CPO}_\perp$ (see Example 5.1.22) is stable. (Recall from Example 2.1.7(2) that coproducts in $\mathsf{CPO}_\perp$ are given by disjoint unions with least elements merged.)

(3) For every stable functor $F\colon \mathsf{CPO}_\perp \to \mathsf{CPO}_\perp$ the *lifting* defined by $F_\perp X = FX \cup \{\perp\}$ ($\perp$ a new bottom element) is stable.

(4) It follows that for every signature Σ, we have two stable endofunctors

$$H_\Sigma X = \coprod_{\sigma \in \Sigma} X^n \qquad \text{and} \qquad H'_\Sigma X = \coprod_{\sigma \in \Sigma} X^n_\perp \qquad n = \mathrm{ar}(\sigma).$$

Remark 6.7.3. Let Σ be a signature. The H_Σ-algebras in $\mathsf{CPO}_\perp$ are the *strict continuous* Σ*-algebras*: they are given by a cpo, A, which is a Σ-algebra such that every operation $\sigma_A\colon A^n \to A$ is continuous and strict: $\sigma_A(\perp, \dots, \perp) = \perp$.

The H'_Σ-algebras are Σ-algebras whose operations are continuous but not necessarily strict.

Remark 6.7.4. (1) A category is said to have *split idempotents* if every idempotent endomorphism has a factorization

$$f = s \cdot u \qquad \text{with } u \cdot s = \mathsf{id}_X.$$

(2) Every category with equalizers has split idempotents: given f one takes the equalizer u of f and id_X. Dually, a category with coequalizers has split idempotents.

(3) The above factorization is unique up to isomorphism, for suppose $f = s' \cdot u'$ where $u' \cdot s' = \mathsf{id}_X$. Then $i = u' \cdot s$ is an isomorphism with $s' = s \cdot i^{-1}$ and $u' = i \cdot u$.

Observation 6.7.5. (1) Stable functors F preserve embeddings. To see this, let e be an embedding with projection $\widehat{e}$. Observe that $e \cdot \widehat{e}$ is idempotent. Since $e \cdot \widehat{e} \leq \mathsf{id}$, we also have $Fe \cdot F\widehat{e} \leq \mathsf{id}$. Therefore $F\widehat{e}$ is the projection for Fe.

(2) Let $\mathscr{A}$ be a CPO-enriched category with split idempotents. For every endofunctor $F\colon \mathscr{A} \to \mathscr{A}$ we have:

$$F \text{ stable} \iff F \text{ preserves embedding-projection pairs.}$$

Indeed, let F preserve these pairs and let $f \cdot f = f \sqsubseteq \mathsf{id}_X$. For the factorization in Remark 6.7.4 we have $s \cdot u = f \sqsubseteq \mathsf{id}_X$, thus, s is an embedding with $u = \widehat{s}$. Therefore $Fu = \widehat{Fs}$ and we conclude that

$$Ff = Fs \cdot \widehat{Fs} \sqsubseteq \mathsf{id}_{FX}.$$

Lemma 6.7.6. *Let $\mathscr{A}$ be a nonempty, strict* CPO*-enriched category with colimits of ω-chains. The initial-algebra chain of every stable endofunctor consists of embeddings.*

Proof We prove that if F is a stable endofunctor, then $w_{i,j}\colon W_i \to W_j$ is an embedding by transfinite induction analogous to that in Proposition 6.6.1. First, $w_{0,1}$ is an embedding since, by Lemma 5.1.23, we have a zero object $0 = 1$. The unique morphism $F0 \to 0$ is $\widehat{w}_{0,1}$ because we clearly have $\widehat{w}_{0,1} \cdot w_{0,1} = \mathsf{id}_0$. Since $w_{0,1} = \bot\colon 0 \to F0$, we conclude that $w_{0,1} \cdot \widehat{w}_{0,1} = \bot \leq \mathsf{id}_{F0}$.

If $w_{i,j}$ is an embedding, so is $w_{i+1,j+1} = Fw_{ij}$ by Observation 6.7.5.

For the limit steps, use the fact that a colimit of a chain of embeddings is formed by embeddings (see Theorem 5.1.10 and Remark 5.1.12). □

Remark 6.7.7. In the proof of Lemma 6.7.6, the projections $\widehat{w}_{i,j}\colon W_j \to W_i$ form the terminal-coalgebra chain of F (cf. Section 6.5). This is easily seen by transfinite induction.

Corollary 6.7.8. *Let $\mathscr{A}$ be a strict* CPO*-enriched category with colimits of chains. For every stable endofunctor F the following are equivalent:*

(1) *an initial algebra exists,*

(2) *a terminal coalgebra exists,*

(3) *a fixed point exists.*

Moreover, an initial algebra is always a canonical fixed point.

Item (3) can be strenghthened to state existence of a pre-fixed point carried by an embedding.

Proof We know from Example 6.2.6(8) that the class of all embeddings of $\mathscr{A}$ is smooth, and F preserves embeddings since it is stable. The dual category $\mathscr{A}^{\mathrm{op}}$ is strict CPO-enriched with respect to the same order on hom-sets. But the embeddings in $\mathscr{A}^{\mathrm{op}}$ are precisely the projections in $\mathscr{A}$. Every stable endofunctor F on $\mathscr{A}$ preserving embeddings clearly also preserves projections. Thus, the dual functor F^{op} on $\mathscr{A}^{\mathrm{op}}$ preserves embeddings. Now (1) $\Leftrightarrow$ (3) follows from an application of Theorem 6.3.4 to $\mathscr{A}$ and F, and (2) $\Leftrightarrow$ (3) is an application to $\mathscr{A}^{\mathrm{op}}$ and F^{op}. In each case the class $\mathcal{M}$ consists of all embeddings in $\mathscr{A}$ and $\mathscr{A}^{\mathrm{op}}$, respectively.

Finally, we prove that an initial algebra is a canonical fixed point. Let $\iota\colon FI \to I$ be an initial algebra. From what we have shown above we see that a terminal coalgebra $\tau\colon T \to FT$ exists. Moreover, from the last statement in Theorem 6.3.4 applied to F and its fixed point (T, τ^{-1}) we see that the unique F-algebra homomorphism $e\colon (I, \iota) \to (T, \tau^{-1})$ is an embedding. Another application of Theorem 6.3.4 to F^{op} and its fixed point (I, ι^{-1}) yields that the unique F^{op}-algebra homomorphism $f\colon \mu F^{\mathrm{op}} = (T, \tau) \to (I, \iota^{-1})$ is an embedding in $\mathscr{A}^{\mathrm{op}}$. This means that f is an F-coalgebra homomorphism $f\colon (I, \iota^{-1}) \to (T, \tau)$ which is a projection in $\mathscr{A}$. By the universal properties of (I, ι) and (T, τ), we have $e = f$, and this morphism is both an embedding and a projection, whence an isomorphism. □

Remark 6.7.9. Returning to $\mathsf{CPO}_\perp$-enriched categories, recall that locally continuous endofunctors have canonical fixed points (see Theorem 5.1.24). The concept of a locally continuous endofunctor F can be generalized to that of a *locally λ-continuous* one, where λ is a given infinite cardinal: This means that for every λ-chain $(f_i)_{i<\lambda}$ in $\mathscr{A}(X, Y)$ we have

$$F\big(\textstyle\bigsqcup_{i<\lambda} f_i\big) = \textstyle\bigsqcup_{i<\lambda} Ff_i \qquad \text{in } \mathscr{A}(FX, FY).$$

For these endofunctors we have, whenever $\mathscr{A}$ is strict CPO-enriched and has colimits of chains of cofinality at most λ, the following:

$$F \text{ has a canonical fixed point } \mu F = \nu F = \lim_{i<\lambda} V_i.$$

This is completely analogous to Theorem 5.1.24.

Example 6.7.10 [71]**.** The category of sets and partial one-to-one functions with hom-sets ordered by inclusion is $\mathsf{CPO}_\perp$-enriched. An endomorphism f is smaller than the identity iff it is idempotent. Thus, every endofunctor is stable, and consequently, all endofunctors with a fixed point have a canonical one.

Remark 6.7.11. Corollary 6.7.8 stems from Barr [71], where the slightly stronger assumption that F be order-preserving on homsets was made, and the proof was somewhat more technical. Barr also proved a result establishing canonical fixpoint for endofunctors on a (non-strict) CPO-enriched category (see [71, Theorem 5.1]).

Example 6.7.12. The functor $\mathsf{Idl}\colon \mathsf{DCPO}_\perp \to \mathsf{DCPO}_\perp$ taking an object of $\mathsf{DCPO}_\perp$ to the dcpo of its ideals (see Example 5.1.18) is stable, but it does not have any fixed point (see Corollary 15.1.9).

6.8 Summary

Up to and including this chapter, the main method of constructing initial algebras and terminal coalgebras has been iteration. This was the only method in Chapter 2, where the iterations went on for precisely ω steps. This method worked on Set and a host of other categories. For terminal coalgebras of finitary set functors, we extended the iteration to $\omega + \omega$ in Chapter 4. In the present chapter we take things to the limit, so to speak, allowing the iterations to go as long as needed. Indeed, the results in the chapter provide bounds for the convergence of both the initial-algebra chain and the terminal-coalgebra chain. We also have seen results which address the question of whether iteration is necessary for initial algebras and terminal coalgebras.

We want to emphasize the contrast between the results in Chapter 2 and those in present chapter. We began with Theorem 6.1.1, and this chapter applied that result in key places. But the chapter was not mainly concerned with categorical generalizations. As noted in Remark 6.1.6, such generalizations do not exist. In their place, we have results like Theorem 6.3.4, the Initial-Algebra Theorem.

Some highlights of this chapter: We presented a transfinite construction of initial algebras μF and terminal coalgebras νF for endofunctors on categories with (co)limits of (op-)chains. For a polynomial set functor H_Σ, the initial-algebra chain converges in λ steps to μH_Σ, the algebra of all well-founded Σ-trees, provided that λ is an infinite regular cardinal larger than all arities of the operation symbols in Σ. In contrast, the terminal-coalgebra chain converges already in ω steps to νH_Σ, the coalgebra of all Σ-trees. For every set functor F, an initial algebra exists iff F has a (pre-)fixed point, and it is always constructed by the initial-algebra chain. In contrast, existence of a fixed point X does not guarantee that a terminal coalgebra exists. A fixed-point pair does guarantee the existence of a terminal coalgebra assuming GCH, but that existence is not necessary. For a number of categories, given an endofunctor F preserving monomorphisms, we proved that the existence of a fixed point implies that μF exists.

All terminal coalgebras of endofunctor on Set^S, the category of S-sorted sets, that exist are obtained by the terminal-coalgebra construction. This does not hold for endofunctors on the (presheaf) category of graphs. We also discussed the precise length of the construction until convergence of the initial-algebra chain or the terminal coalgebra chain, respectively, happens.

We use transfinite recursion and induction at many places in the rest of this book. But for the most part, the specific results in this chapter will not reappear. The exception to this concerns the concept of smoothness (Section 6.2) and the application of this concept to initial algebras (Section 6.3). Classes of smooth monomorphisms and Theorem 6.3.4 will be used in many of the remaining chapters.

7

Terminal Coalgebras as Algebras, Initial Algebras as Coalgebras

A fixed point of a functor can be considered both as an algebra or as a coalgebra. For example, the terminal coalgebra can be viewed as an algebra (by inverting its structure map). The theme of this chapter is that the terminal coalgebra is the initial object in a category of algebras that allow one to interpret *structured corecursive* specifications. Dually, we are interested in characterizations of the initial algebra as a terminal object in a category of coalgebras that allow one to interpret specifications by *structured recursion*. The chapter is split into sections that present the various notions of (co)algebras. Each of those notions is interesting in its own right. In fact, they were originally introduced and studied for reasons not connected to our theme.

We begin in Section 7.1 with *corecursive algebras*, and we show that the terminal coalgebra is characterized as the initial corecursive algebra. Section 7.2 shows how terminal coalgebras can be obtained as subalgebras of corecursive algebras. In Section 7.3 we present the stronger notion of a *completely iterative algebra* and again characterize the terminal coalgebra as an initial completely iterative algebra. Dually, we then turn to *recursive* and *parametrically recursive coalgebras*, respectively, in Section 7.4, where the initial algebra is characterized as a terminal object.

Before we turn to Sections 7.1 to 7.4, let us informally motivate corecursive and completely iterative algebras with an example from general algebra. Consider the set functor $FX = X \times X + 1$ expressing the type of algebras with a binary operation $*$ and constant c. In this example, corecursive specifications take the form of systems of formal recursive equations, and we require that those are uniquely solvable. For example, the recursive equation $x \approx x * x$ has a unique solution in an algebra A iff A has a unique idempotent element $i \in A$. The following system of recursive equations

$$x \approx x * x \qquad y \approx x * (y * c) \tag{7.1}$$

requires, in addition, that a unique element $a \in A$ with $a = i *^A (a *^A c^A)$.

Moreover, we would like to have unique solutions of *as many recursive equations as possible*. It will not be possible to allow the right-hand sides to be bare variables because we do not expect equations like $x \approx x$ to have a unique solution. Similarly, 'ungrounded' systems like

$$x_1 \approx x_2 \qquad x_2 \approx x_3 \qquad x_3 \approx x_4 \qquad \cdots$$

are not expected to have unique solutions. So our formal definition of an equation is designed to forbid the right-hand sides of recursive equations to be bare variables. In addition, our definition also includes an important simplification. Even if we wish to solve as many recursive equations as possible, it is sufficient to restrict attention to systems whose right-hand sides are *flat terms*; that is, terms containing precisely one operation symbol. Fortunately, every system of recursive equations in which each right-hand side is a non-variable term (or even an infinite Σ-tree) can be *flattened* by introducing fresh auxiliary variables to represent subterms of non-flat terms. This implies that, as soon as we have solutions for all systems with flat right-hand sides, we also have solutions to systems where the right-hand sides are more complex. For example, for the above system (7.1) we obtain the following flattened system (using fresh variables z and z'):

$$x \approx x * x \quad y \approx x * z \quad z \approx y * z' \quad z' \approx c.$$

Let X denote the set of recursion variables (i.e. the ones on the left-hand sides of the above formal equations). Then a *flat system* of recursive equations assigns to every element of X its right-hand side, which is an element of $FX = X \times X + 1$. Such assignments are simply coalgebras $e\colon X \to FX$. Given an algebra (A, α), a *solution* of e assigns to every recursion variable in X an element of A, that is, we obtain a morphism $e^\dagger\colon X \to A$. The fact that $e^\dagger$ solves e means precisely that the following square commutes:

$$\begin{array}{ccc} X & \xrightarrow{e^\dagger} & A \\ {\scriptstyle e}\downarrow & & \uparrow{\scriptstyle \alpha} \\ FX & \xrightarrow{Fe^\dagger} & FA \end{array}$$

In other words, a solution $e^\dagger$ is just a coalgebra-to-algebra morphism. The F-algebra (A, α) is called *corecursive* if every coalgebra has a unique solution.

Although this notion is interesting, it is not exactly what we are after in this chapter. We are more interested in a stronger notion that allows a wider class of recursive specifications. For example, suppose that we again fix (A, α) and allow elements of A to appear as *parameters* on the right-hand sides of formal recursive equations such as (7.1). This means that *flat equation morphisms* have the form $e\colon X \to FX + A$. We are interested in algebras (A, α) that have a unique solution $e^\dagger$ in the sense that the following square commutes:

$$\begin{array}{ccc} X & \xrightarrow{\quad e^\dagger \quad} & A \\ {\scriptstyle e}\big\downarrow & & \big\uparrow{\scriptstyle [\alpha, \mathrm{id}_A]} \\ FX + A & \xrightarrow{Fe^\dagger + \mathrm{id}_A} & FA + A \end{array}$$

Then we call the algebra (A, α) *completely iterative*, or a *cia*, for short. Again, the commutativity of the above square just expresses the fact that formal equations are turned into identities in A. The main result in Section 7.3 is that a terminal coalgebra for F is precisely the same as an initial cia for F.

Dual to corecursive algebras and cia are notions of coalgebras providing semantics for structured recursion. A coalgebra (A, α) is called *recursive* if for every morphism $e\colon FX \to X$ there is a unique coalgebra-to-algebra morphism $e^\dagger\colon A \to X$. This means that the square below commutes:

$$\begin{array}{ccc} A & \xrightarrow{\ e^\dagger\ } & X \\ {\scriptstyle \alpha}\big\downarrow & & \big\uparrow{\scriptstyle e} \\ FA & \xrightarrow{\ Fe^\dagger\ } & FX \end{array}$$

The dual of complete iterativity is *parametric recursiveness*. A coalgebra (A, α) is *parametrically recursive* if for every morphism $e\colon FX \times A \to X$ there is a unique morphism $e^\dagger\colon A \to X$ so that the square below commutes:

$$\begin{array}{ccc} A & \xrightarrow{\quad e^\dagger \quad} & X \\ {\scriptstyle \langle \alpha, \mathrm{id}_A \rangle}\big\downarrow & & \big\uparrow{\scriptstyle e} \\ FA \times A & \xrightarrow{Fe^\dagger \times \mathrm{id}_A} & FX \times A \end{array} \tag{7.2}$$

Finally, we list the four definitions given above in Table 7.1.

type of equation morphism	a ______ algebra has unique solutions to these	coalgebra in the dual setting
$X \to FX$	corecursive	recursive
$X \to FX + A$	completely iterative	parametrically recursive

Table 7.1 Definitions studied in Sections 7.1–7.4.

7.1 Corecursive Algebras

Corecursive algebras were first studied by Capretta et al. [101] who also compared that notion to a number of related concepts. Adámek et al. [46] investigated corecursive algebras further, and, in particular, they provided a description and construction of *free* corecursive algebras as well as a study of the ensuing monad of free corecursive algebras and its properties. Formally, a corecursive algebra is one that admits a unique coalgebra-to-algebra morphism for every given coalgebra. We use the terminology from the Introduction of this chapter:

Definition 7.1.1. Let F be an endofunctor on a category $\mathscr{A}$. By a *solution* of a coalgebra $e\colon X \to FX$ in an F-algebra (A, α) we mean a morphism $e^\dagger$ such that the following square commutes:

$$\begin{array}{ccc} X & \xrightarrow{e^\dagger} & A \\ {\scriptstyle e}\downarrow & & \uparrow{\scriptstyle \alpha} \\ FX & \xrightarrow{Fe^\dagger} & FA \end{array}$$

The algebra (A, α) is called *corecursive* if every coalgebra has a unique solution in it.

Proposition 7.1.2. (1) *Every terminal coalgebra, considered as an algebra, is corecursive.*

(2) *Every corecursive fixed point is a terminal coalgebra.*

Proof (1) By Lambek's Lemma 2.2.5, the structure $\tau\colon \nu F \to F(\nu F)$ of the terminal coalgebra is invertible, and hence $(\nu F, \tau^{-1})$ is an F-algebra. This algebra is corecursive by the universal property of νF.

(2) This is immediate from Definition 7.1.1 since for an algebra (A, α) with α invertible, the solutions of a coalgebra (X, e) in (A, α) are the same as the coalgebra homomorphisms from (X, e) to (A, α^{-1}).

□

Remark 7.1.3. (1) Fields like general algebra and abstract data types study Σ-algebras, which are precisely the algebras for the functor H_Σ (see Example 2.1.5). Let X be a set of 'recursion variables'. A coalgebra $e\colon X \to H_\Sigma X$ can be understood as a system of formal mutually recursive equations

$$x \approx \sigma(x_1, \dots, x_n), \tag{7.3}$$

one for every recursion variable $x \in X$; here σ is an n-ary operation symbol in Σ and $x_1, \dots, x_n \in X$. Indeed, one just writes $e(x) = \sigma(x_1, \dots, x_n)$ in lieu of the above formal equation. For a Σ-algebra A, a *solution* $e^\dagger : X \to A$ assigns to every recursion variable x an element $e^\dagger(x) \in A$ such that the formal equations in (7.3) become actual identities in A when recursion variables are substituted by their solutions; in symbols: $e^\dagger(x) = \sigma^A(e^\dagger(x_1), \ldots, e^\dagger(x_n))$.

By Proposition 7.1.2(1), a concrete example of a corecursive algebra is the algebra $T_\Sigma = \nu H_\Sigma$ of all Σ-trees.

(2) In Example 7.1.4, we will present further examples of corecursive Σ-algebras. However, let us observe that classical algebras (such as groups, lattices, or Boolean algebras) are almost never corecursive. This is due to the uniqueness requirement for solutions, more precisely, it follows from the fact that a corecursive algebra with a binary operation has a unique idempotent element. Indeed, the formal recursive equation $x \approx x \vee x$ has a unique solution in a lattice or Boolean algebra A iff A is trivial. Similarly, the system $x \approx x \cdot y, y \approx 1$ has a unique solution in a group G iff G is trivial.

Example 7.1.4. (1) Capretta et al. [101] provide the following example of a corecursive algebra for the set functor $FX = E \times X \times X$. Consider the set E^ω of all streams over E with the F-algebra structure $\alpha : E \times E^\omega \times E^\omega \to E^\omega$ where $\alpha(e, s, t)$ is the stream with head e and continuing by the merge of s and t. This is a corecursive F-algebra.

(2) The set functor $FX = X \times X + 1$ is the polynomial functor associated to the signature Σ with a binary operation symbol $*$ and a constant symbol c. By Proposition 7.1.27.1.2, the terminal coalgebra νF consisting of all finite and infinite binary trees (see Example 3.3.9) is corecursive. Let us consider inner nodes of binary trees as being labelled by $*$ and leaves as labelled by c. Let $X = \{x_1, x_2\}$ and consider the following system of mutually recursive equations

$$x_1 \approx c * x_2 \qquad x_2 \approx x_1 * c. \tag{7.4}$$

Then its solution in νF consists of two infinite trees, ${x_1}^\dagger$ and ${x_2}^\dagger$, here written as infinite terms:

$${x_1}^\dagger = c * ((c * (\cdots * c)) * c) \quad \text{and} \quad {x_2}^\dagger = (c * ((c * \cdots) * c)) * c.$$

We invite the reader to draw these as pictures; cf. Example 7.3.7(1).

(3) Here is a different example of a corecursive algebra due to Capretta et al. [101], which we shall revisit when we discuss completely iterative algebras. For the set functor $FX = X \times X$ on Set, let $A = \{0, 1, 2\}$, and let $\alpha : A \times A \to A$ be the algebra defined by $\alpha(1, 2) = 1$, and otherwise $\alpha(i, j) = 0$. Then A is corecursive: given any $e : X \to X \times X$, the unique solution $e^\dagger : X \to A$ is the constant function on 0. Indeed, this is clearly a solution since $\alpha(0, 0) = 0$.

Moreover, since 2 cannot be written as $\alpha(i, j)$, we see that for all x, $e^\dagger(x) \neq 2$. We claim that the same is true for 1. For suppose that $e^\dagger(x) = 1$. Then $e(x)$ must be (y, z) for some y and z such that $e^\dagger(y) = 1$ and $e^\dagger(z) = 2$. However, this contradicts what we have seen about 2. This proves that $e^\dagger$ is the constant function with value 0.

(4) In Section 15.5, we shall see a number of corecursive algebras for set functors whose definitions involve elementary mathematics. Here is one example. Consider $FX = \mathbb{N} \times X$. Then the unit interval $\mathbb{I} = [0, 1]$ is a corecursive algebra for F, via the structure $\tau\colon \mathbb{I} \to \mathbb{N} \times \mathbb{I}$ defined by $\tau(n, r) = (n + r)/(1 + n + r)$ (Theorem 15.5.10).

(5) All algebras $FV_i \to V_i$ in the terminal-coalgebra chain of F (Definition 6.5.1) are corecursive. This is a consequence of Corollary 7.3.12 on completely iterative algebras.

Corecursive algebras for F may be considered as a full subcategory of $\mathsf{Alg}\,F$ because of the following:

Proposition 7.1.5 [46, Lemma 2.9]. *Let (A, α) and (B, β) be corecursive algebras. Every homomorphism $h\colon (A, \alpha) \to (B, \beta)$ preserves solutions. This means that for every $e\colon X \to FX$ we have that*

$$X \xrightarrow{e^\dagger} A \xrightarrow{h} B$$

is the unique solution of e in B.

Proof The following diagram

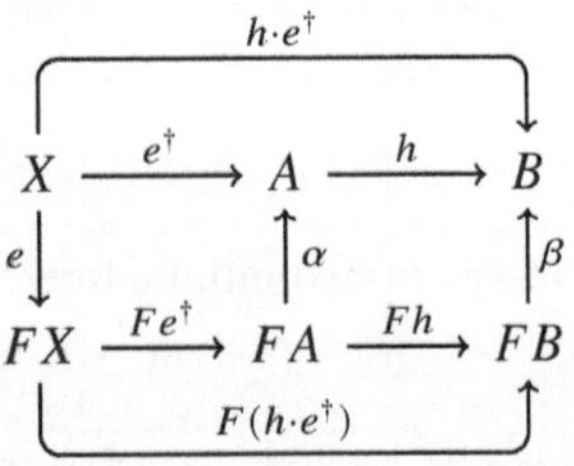

shows that $h \cdot e^\dagger$ is a solution of e in B and therefore is the unique one. □

In order to see that an initial corecursive F-algebra is a fixed point of F, a crucial step is the following result by Capretta, Uustalu, and Vene.

Proposition 7.1.6 [100, dual of Prop. 6]. *If (A, α) is a corecursive algebra, then so is $(FA, F\alpha)$.*

Proof Given a corecursive algebra $\alpha\colon FA \to A$ and $e\colon X \to FX$, denote by $e^\dagger : X \to A$ its unique solution. We will show that

$$e^\ddagger = (X \xrightarrow{e} FX \xrightarrow{Fe^\dagger} FA)$$

is the unique solution of e in FA. First, the following diagram shows that it is a solution:

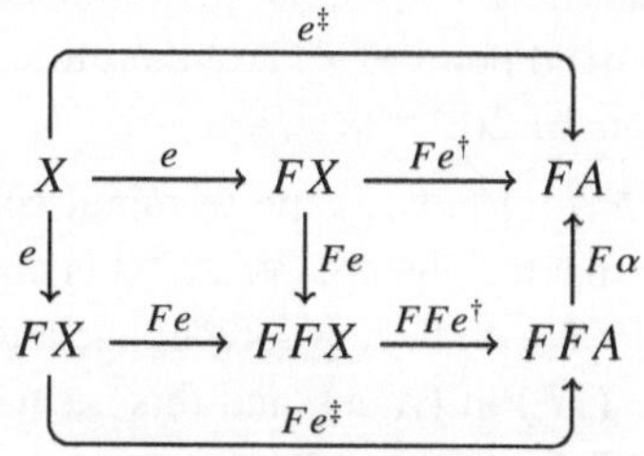

Indeed, the upper and lower parts commute by the definition of $e^\ddagger$, the right-hand square does since $e^\dagger$ solves e, and the left-hand square is trivial.

To see that $e^\ddagger$ is the unique solution in FA, suppose that $s \colon X \to FA$ solves e. Then it follows that $\alpha \cdot s \colon X \to A$ is a solution of e in A, since the following diagram commutes:

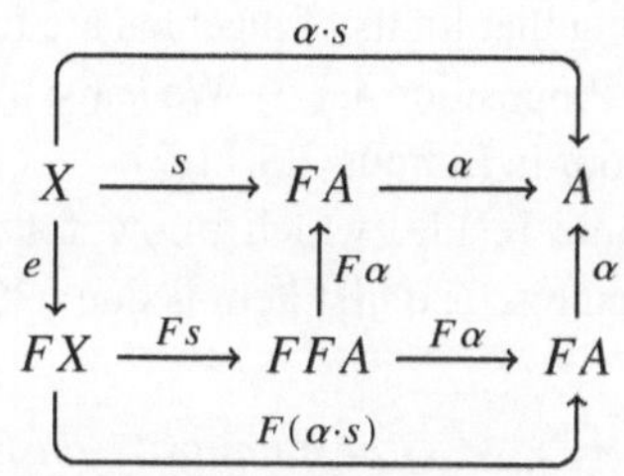

Indeed, the left-hand square commutes since s solves e in FA, and the remaining parts are trivial. Thus, we have $e^\dagger = \alpha \cdot s$, and we conclude that

$$e^\ddagger = Fe^\dagger \cdot e = F\alpha \cdot Fs \cdot e = s,$$

where the last equation holds since, once again, s solves e in FA. □

From this result one obtains Lambek's Lemma for corecursive algebras:

Corollary 7.1.7. *If an initial corecursive F-algebra exists, then it is a fixed point of F.*

Indeed, the proof is the same as the one for Lambek's Lemma 2.2.5, using Proposition 7.1.6 to see that for an initial corecursive algebra (I, ι), $(FI, F\iota)$ is corecursive, too.

Theorem 7.1.8 [100, dual of Prop. 7]**.** *The initial corecursive algebra is precisely the same as the terminal coalgebra.*

In more detail, let $F\colon \mathscr{A} \to \mathscr{A}$ be an endofunctor. Then we have:

(1) If (I, ι) is an initial corecursive algebra, then (I, ι^{-1}) is a terminal coalgebra.

(2) If $(\nu F, \tau)$ is a terminal coalgebra, then $(\nu F, \tau^{-1})$ is an initial corecursive algebra.

Proof (1) Suppose that $\iota\colon FI \to I$ is an initial corecursive algebra. By Corollary 7.1.7, it is a fixed point of F, and therefore (I, ι^{-1}) is the terminal coalgebra by Proposition 7.1.2(2).

(2) Suppose that $\tau\colon \nu F \to F(\nu F)$ is the terminal coalgebra. We have seen in Proposition 7.1.2(1) that the algebra $(\nu F, \tau^{-1})$ is corecursive. It remains to verify its initiality. So let (A, α) be any corecursive algebra. There is a unique solution of $\tau\colon \nu F \to F(\nu F)$ in (A, α), and this means that there is a unique algebra homomorphism h from $(\nu F, \tau^{-1})$ to (A, α). □

Theorem 7.1.8 connects to the theme of this chapter in that it expresses the terminal coalgebra of a functor as the initial object in an interesting category of algebras.

Proposition 7.1.9. *Every limit of corecursive algebras is corecursive.*

This is easy to prove, using that limits of algebras are formed on the level of the underlying category (cf. Proposition 4.1.5). We leave the details to the reader as we shall see a related proof in Lemma 7.3.11(2).

We next mention some results which allow for the propagation of the corecursive algebra structures. The first item is due to Capretta et al. [100, dual of Prop. 10(b)].

Proposition 7.1.10. *Let $n\colon F \to G$ be a natural transformation, and let (A, α) be an algebra for G.*

(1) *If (A, α) is corecursive for G, then $\alpha \cdot n_A\colon FA \to A$ is corecursive for F.*

(2) *If $(\nu G, \tau)$ is the terminal coalgebra, then $\tau^{-1} \cdot n_{\nu G}\colon F(\nu G) \to \nu G$ is corecursive for F.*

Proof (1) Given $e\colon X \to FX$, a morphism $e^\dagger\colon X \to A$ is a solution in the F-algebra $(A, \alpha \cdot n_A)$ iff it is a solution of $n_X \cdot e\colon X \to GX$ in the G-algebra (A, α). To see this consider the diagram below:

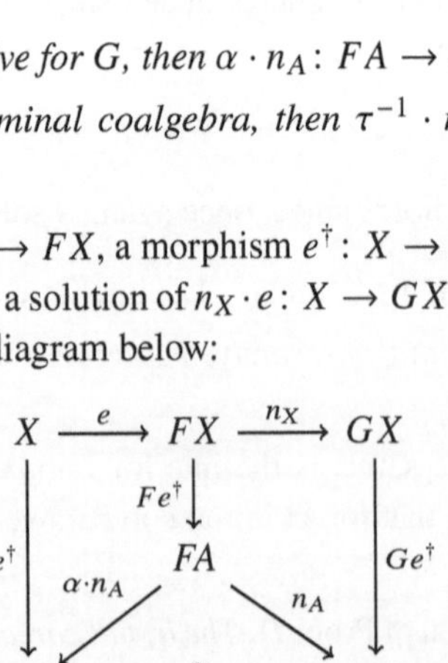

Since the right-hand square and the triangle clearly commute, we see that the outside commutes iff so does the left-hand square.

By assumption, we have a unique $e^\dagger$ such that the outside commutes. Thus, $e^\dagger$ is unique such that the left-hand square commutes, and we are done.

(2) By Theorem 7.1.8 we know that $(\nu G, \tau^{-1})$ is a corecursive G-algebra. Now apply item (1). □

At this point, we would like to briefly recall how *free* corecursive algebras are constructed [46]. Freeness for corecursive algebras is a concept analogous to freeness of ordinary F-algebras (cf. Remark 2.2.23): a free corecursive algebra on an object Y is a corecursive algebra $\gamma_Y : FCY \to CY$ together with a universal morphism $\eta_Y : Y \to CY$. This means that for every corecursive algebra (A, α) and every morphism $f : Y \to A$ there exists a unique algebra homomorphism $f^\sharp : (CY, \gamma_Y) \to (A, \alpha)$ satisfying $f^\sharp \cdot \eta_Y = f$:

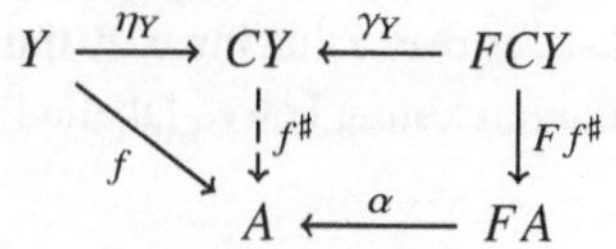

Theorem 7.1.11 [46]. *Assuming that a terminal coalgebra νF and a free F-algebra ΦX on X exist, the free corecursive algebra is their coproduct $\nu F \oplus \Phi X$ in the category of algebras for F.*

Furthermore, op. cit. provides the following iterative construction of free corecursive algebras, somewhat similar to the initial-algebra chain.

Construction 7.1.12 (Free-Corecursive-Algebra Chain). Let $\mathscr{A}$ be a cocomplete category and let $F : \mathscr{A} \to \mathscr{A}$ have a terminal coalgebra. We define for every object Y of $\mathscr{A}$ an essentially unique chain $U : \mathsf{Ord} \to \mathscr{A}$ by the following transfinite recursion:

$$
\begin{aligned}
U_0 &= \nu F \\
U_{i+1} &= FU_i + Y \\
U_j &= \operatorname*{colim}_{i<j} U_i \qquad\qquad \text{for limit ordinals } j.
\end{aligned}
$$

The connecting morphisms $u_{i,j} : U_i \to U_j$ are defined as follows:

$$
\begin{aligned}
u_{0,1} &= (\nu F \xrightarrow{\tau} F(\nu F) \xrightarrow{\mathsf{inl}} F(\nu F) + Y), \\
u_{i+1,j+1} &= F u_{i,j} + \mathsf{id}_Y,
\end{aligned}
$$

and for limit ordinals j, $(u_{i,j})_{i<j}$ is the colimit cocone.

If this chain converges, which means that some $u_{i,i+1}$ is invertible, then the coproduct components of $u_{i,i+1}^{-1} : FU_i + Y \to U_i$ yield an algebra structure on U_i and a morphism $Y \to U_i$. For accessible endofuntors on Set and, more generally, on locally presentable categories (see Definitions 11.2.8 and 11.2.13) the free-corecursive-algebra chain converges:

Theorem 7.1.13 [46, Thm. 4.6]. *Let $\mathscr{A}$ be a locally presentable category. Assume that $F : \mathscr{A} \to \mathscr{A}$ is accessible and preserves monomorphisms. Then the free-corecursive-algebra chain converges to the free corecursive algebra on Y.*

Example 7.1.14 [46]. For a polynomial endofunctor H_Σ on Set, the free corecursive algebra on a set Y can be described as follows. Consider the Σ-algebra $T_\Sigma Y$ of all *Σ-trees over Y*, which are trees for the signature obtained by adding the elements of Y as constant symbols to Σ_0. This means these trees are defined like Σ-trees except that the leaves are labelled by constant symbols in Σ_0 or generators in Y. The free corecursive Σ-algebra on Y is the subalgebra $C_\Sigma Y$ of $T_\Sigma Y$ consisting of all Σ-trees over Y having only finitely many leaves labelled by generators in Y (and the remaining leaves labelled in Σ_0).

7.2 Terminal Coalgebras from Corecursive Algebras

In this section we prove that if a corecursive algebra has a monic structure $\alpha : FA \rightarrowtail A$, then it contains νF as its largest subalgebra which is a fixed point.

We first shortly digress and consider factorizations of coalgebra homomorphisms inherited from the factorizations in the base category, which will also be useful in later chapters.

Remark 7.2.1. (1) We recall that an epimorphism $e : A \to B$ is *strong* if it satisfies the following *diagonal fill-in property*: given a monomorphism $m : C \rightarrowtail D$ and morphisms $f : A \to C$ and $g : B \to D$ such that $m \cdot f = g \cdot e$ (i.e. the outside of the square below commutes) then there exists a unique $d : B \to C$ such that the diagram below commutes:

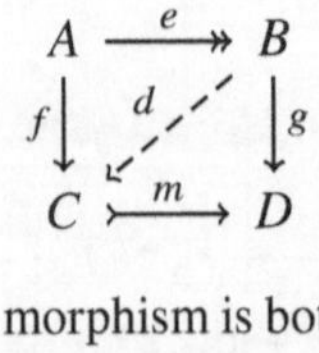

(7.5)

(2) A very basic fact is that if a morphism is both a monomorphism and a strong epimorphism, then it is an isomorphism. For this, take f and g to be identities and $m = e$, and observe that d is the inverse of e.

(3) Every complete and well-powered category has factorizations of morphisms f as $f = m \cdot e$, where e is a strong epimorphism and m is a monomorphism [90, Prop. 4.4.3]. The subobject m is the *image* of f. Also every cocomplete and co-well-powered category has such factorizations; this follows from Adámek et al. [38, dual of Theorem 14.19 and Exercise 14C(e)].

(4) We indicate monomorphisms by $\rightarrowtail$ and strong epimorphisms by $\twoheadrightarrow$.

It follows from a result in Kurz' thesis [207, Prop. 1.3.6] that factorizations of morphisms lift to coalgebras:

Proposition 7.2.2 (Coalg F inherits factorizations from $\mathscr{A}$). *If $\mathscr{A}$ has (strong epi, mono)-factorizations and F preserves monomorphisms, the category* Coalg F *has factorizations of homomorphisms f as $f = m \cdot e$, where e is carried by a strong epimorphism and m by a monomorphism in $\mathscr{A}$. Moreover, the diagonal fill-in property also holds in* Coalg F.

Proof Let $f\colon (A, \alpha) \to (B, \beta)$ be a coalgebra homomorphism and take its (strong epi, mono)-factorization $f = m \cdot e$ in $\mathscr{A}$. Now consider the diagram below:

$$\begin{array}{ccc} A & \xrightarrow{\ \alpha\ } & FA \\ {\scriptstyle e}\big\downarrow & & \big\downarrow{\scriptstyle Fe} \\ C & \overset{\gamma}{\dashrightarrow} & FC \\ {\scriptstyle m}\big\downarrow & & \big\downarrow{\scriptstyle Fm} \\ B & \xrightarrow{\ \beta\ } & FB \end{array} \tag{7.6}$$

Since Fm is monic we can use the diagonal fill-in property to obtain a unique coalgebra structure γ such that e and m are coalgebra homomorphisms.

As for the diagonal fill-in property, note that if m and g in (7.5) are coalgebra homomorphisms, then so is d. To see this we use that F preserves monomorphisms and consider the diagram below:

$$\begin{array}{ccccc} & B & \xrightarrow{\ \beta\ } & FB & \\ & {\scriptstyle d}\big\downarrow & & \big\downarrow{\scriptstyle Fd} & \\ g & C & \xrightarrow{\ \gamma\ } & FC & Fg \\ & {\scriptstyle m}\big\downarrow & & \big\downarrow{\scriptstyle Fm} & \\ & D & \xrightarrow{\ \delta\ } & FD & \end{array}$$

Its outside, left-hand, and right-hand parts commute. Thus so does the upper square when post-composed by the monomorphism Fm, which implies that it commutes. □

Remark 7.2.3. Following Remark 7.2.1(3) we see that C in (7.6) above is the image of f. We therefore say that (C, γ) the *image* of (A, α) under the homomorphism f.

Remark 7.2.4. (1) Proposition 7.2.2 shows that we obtain a *factorization system* on $\mathsf{Coalg}\, F$ (see e.g. [90, Def. 5.5.1] for the definition), where the two morphism classes are homomorphisms carried by monomorphisms and strong epimorphisms in $\mathscr{A}$, respectively. Recall from Corollary 2.4.11 and Remark 2.4.12(1) that coalgebra homomorphisms carried by monomorphisms (or epimorphisms) in $\mathscr{A}$ have the same property in $\mathsf{Coalg}\, F$. Nevertheless, we do not necessarily obtain (strong epi, mono)-factorizations in $\mathsf{Coalg}\, F$ because monomorphisms in $\mathsf{Coalg}\, F$ need not be carried by monos in $\mathscr{A}$.

(2) Note that Proposition 7.2.2 generalizes to an arbitrary factorization system $(\mathcal{E}, \mathcal{M})$ on the base category and endofunctors F such that $m \in \mathcal{M}$ implies $Fm \in \mathcal{M}$. In this case, the proof of the diagonal fill-in property in $\mathsf{Coalg}\, F$ needs more work (see e.g. Wißmann's thesis [313, Lemma 2.4.8]). However, we are not going to work in this generality.

Remark 7.2.5. (1) Note that for $\mathscr{A} = \mathsf{Set}$, Proposition 7.2.2 holds even without the assumption that F preserve monomorphisms. In fact, every set functor preserves epimorphisms and nonempty monomorphisms, and if m in (7.6) is empty, we have $C = \emptyset$ and therefore $A = \emptyset$. Hence, we can take γ to be the empty map in this case.

(2) Milius et al. [235, Lemma 2.5] show that this argument can easily be generalized. Recall [103] that an initial object 0 is called *strict* if every morphism $I \to 0$ is an isomorphism. A *nonempty* monomorphism in a category $\mathscr{A}$ is a monomorphism whose domain is not a strict initial object. Proposition 7.2.2 (and in fact, all ensuing results using it) then hold for endofunctors F on $\mathscr{A}$ preserving nonempty monomorphisms. Note that in categories not having (strict) initial objects, for example, the categories of groups or vector spaces over a field, all monomorphisms are nonempty. In such categories, preservation of nonempty monomorphisms is the same as mono-preservation. In addition, in every category of algebras containing the empty algebra, all functors lifted from Set preserve nonempty monomorphisms.

Notation 7.2.6. (1) We write $\mathsf{Sub}(A)$ for the collection of subobjects of A ordered by factorization (cf. Remark 6.2.1). This is a poset in every well-powered category. The top of this poset is represented by id_A.

(2) The meet of two subobjects $m_1, m_2 \colon M_i \rightarrowtail A$ in $\mathsf{Sub}(A)$ is given by the

pullback of m_1 and m_2 (if it exists):

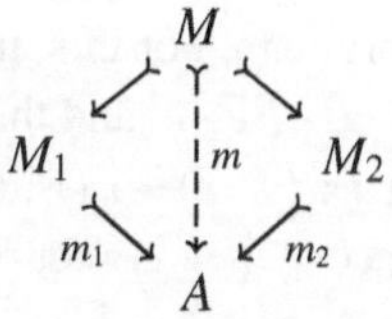

Shortly, $m = m_1 \wedge w_2$, whenever m_1 and m_2 have a pullback.

(3) Analogously, meets of arbitrary collections in $\mathsf{Sub}(A)$ are given by wide pullbacks if those exist. Therefore, $\mathsf{Sub}(A)$ is a complete lattice for every object A of a complete and well-powered category.

(4) The smallest subobject is denoted by $\bot_A$. If the category is complete and well-powered, then this is obtained as the meet of all subobjects. In categories with a *simple* initial object 0, which means that the unique morphisms $0 \to A$ are monic, these represent the smallest subobjects.

We have seen in Chapter 6 that subfunctors 'inherit' terminal coalgebras (Proposition 6.6.11). We now prove a related result which does not require the convergence of the terminal-coalgebra chain.

We call on a preliminary result. It will be used in this section and in Section 15.5 to obtain terminal coalgebras as certain subalgebras of corecursive algebras. Recall from Definition 7.1.1 that an algebra (A, α) is *corecursive* if for every coalgebra $e\colon X \to FX$, there is a unique solution $e^\dagger\colon X \to A$ (this means that $e^\dagger = \alpha \cdot Fe^\dagger \cdot e$).

Lemma 7.2.7. *Let (A, α) be a corecursive algebra for an endofunctor F, and let (B, β) be a fixed point of F which is a subalgebra via $m\colon (B, \beta) \rightarrowtail (A, \alpha)$. Assume that for every coalgebra $e\colon X \to FX$, the unique solution $e^\dagger\colon X \to A$ factorizes through m. Then (B, β^{-1}) is the terminal coalgebra for F.*

Proof Fix a coalgebra $e\colon X \to FX$, and factorize the solution $e^\dagger$ through m:

$$e^\dagger = (X \xrightarrow{\widehat{e}} B \overset{m}{\rightarrowtail} A).$$

By the solution property of $e^\dagger$, the outside of the diagram below commutes:

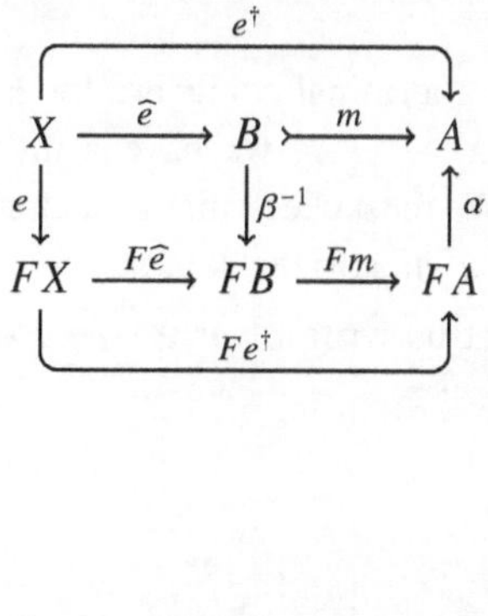

The right-hand square commutes since m is a homomorphism of algebras. The upper part is our factorization of $e^\dagger$, as is the lower part with F applied. We claim that $\widehat{e}$ is a coalgebra morphism. For this, it is enough to deduce from the above diagram that $m \cdot \widehat{e} = m \cdot \beta \cdot F\widehat{e} \cdot e$, and then use the fact that m is monic.

For the uniqueness of $\widehat{e}$, let $f\colon (X, e) \to (B, \beta^{-1})$ be a coalgebra morphism. Then $m \cdot f$ is a solution. Since A is a corecursive algebra, $m \cdot f = e^\dagger$. But then $m \cdot f = m \cdot \widehat{e}$. Since m is monic, $f = \widehat{e}$. □

Proposition 7.2.8. *Let $\mathscr{A}$ be a complete and well-powered category, and let F be an endofunctor on $\mathscr{A}$ preserving monomorphisms. If F has a corecursive pre-fixed point $\alpha\colon FA \rightarrowtail A$, then it has a terminal coalgebra which is the largest fixed point that is a subalgebra of (A, α).*

Proof (1) Recall from Notation 7.2.6(3) that $\mathsf{Sub}(A)$ is a complete lattice. Consider the function $\Phi\colon \mathsf{Sub}(A) \to \mathsf{Sub}(A)$ which maps a given subobject $u\colon X \rightarrowtail A$ to

$$\Phi(u) = \big(FX \overset{Fu}{\rightarrowtail} FA \overset{\alpha}{\rightarrowtail} A\big).$$

We are using that F preserves monomorphisms and that α is monic. It is easy to see that Φ is monotone. By the Knaster–Tarski Theorem B.1.7, Φ has a greatest fixed point $t\colon T \rightarrowtail A$ given by

$$t = \bigvee \{m \in \mathsf{Sub}(A) \mid m \leq \Phi(m)\}. \tag{7.7}$$

Being a fixed point of Φ, we see that there is an isomorphism $\tau\colon T \to FT$ such that (T, τ^{-1}) is a subalgebra of (A, α) via t:

$$\begin{array}{ccc} T & \overset{\cong}{\longleftarrow} & FT \\ {\scriptstyle t}\downarrow & & \downarrow{\scriptstyle Ft} \\ A & \overset{\alpha}{\longleftarrow} & FA \end{array}$$

(2) To see that (T, τ) is a terminal coalgebra for F, we apply Lemma 7.2.7. For each coalgebra $e\colon X \to FX$, we have a unique coalgebra-to-algebra morphism $e^\dagger\colon X \to A$. We must check that $e^\dagger$ factorizes through t.

By Remark 7.2.1(3), $e^\dagger$ factorizes as a strong epimorphism followed by a monomorphism. So let us write $e^\dagger$ as $X \overset{h}{\twoheadrightarrow} I \overset{k}{\rightarrowtail} A$. Then we have the following diagram:

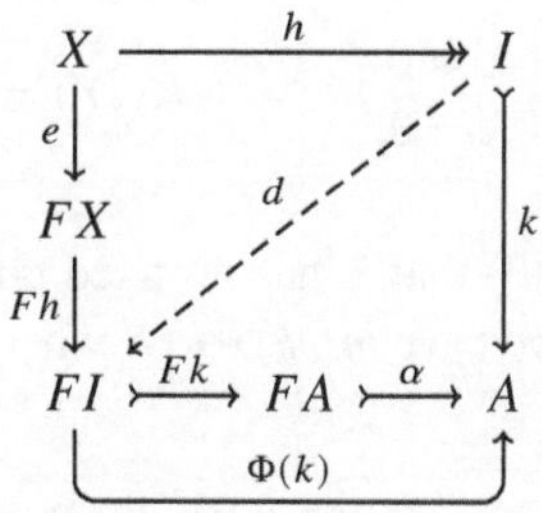

The outside commutes since $k \cdot h = e^\dagger$ is a coalgebra-to-algebra morphism. We thus get d by the diagonal fill-in property (Remark 7.2.1(1)). Since k is monic in the lower triangle, d also is monic. Thus, $k \leq \Phi(k)$ in $\mathsf{Sub}(A)$. By the characterization of $\nu\Phi = t$ in (7.7) above, $k \leq t$. That is, k factorizes through t. Hence so does $k \cdot h = e^\dagger$, as desired.

(3) Finally, suppose that (B, β) is any F-coalgebra with a monic coalgebra-to-algebra morphism m from (B, β) to (A, α) (e.g. a fixed point of F being a subalgebra of (A, α)). Then β witnesses $m \leq \Phi(m)$, whence $m \leq t$ using (7.7). □

Remark 7.2.9. In the last part of the above proof we have actually seen that $t\colon \nu F \rightarrowtail A$ is the largest subobject represented by a monomorphism which is a coalgebra-to-algebra morphism from some F-coalgebra to (A, α).

Corollary 7.2.10. *Let $\mathscr{A}$ be a complete and well-powered category, and let G be an endofunctor on $\mathscr{A}$ preserving monomorphisms. If G has a terminal coalgebra, then every subfunctor of G also has a terminal coalgebra, which is a subobject of νG.*

Proof We are given a subfunctor $m\colon F \rightarrowtail G$. By Remark 6.6.5, F preserves monomorphisms since so does G. Let $\tau\colon \nu G \to G(\nu G)$ be the terminal coalgebra structure, and consider $\alpha = \tau^{-1} \cdot m_{\nu G}$. By Proposition 7.1.10(2), $\alpha\colon F(\nu G) \to \nu G$ is a corecursive algebra. Moreover, α is clearly monic. Thus, the desired result follows from Proposition 7.2.8. □

Example 7.2.11. Here are some examples to show that the hypotheses in Corollary 7.2.10 are needed.

(1) To see that completeness is needed, take $\mathscr{A}$ to be the poset of natural numbers ordered by $\geq$ considered as a category. Let g be the function $g(x) = 2x$, and let f be the identity. Then $f \leq g$. Now $\nu g = 0 \geq 2 \cdot 0$ and νf does not exist.

(2) To see that we need G to preserve monomorphisms, consider Set × Set and

the endofunctors

$$F(X,Y) = \begin{cases} (\emptyset, 1) & \text{if } X \neq \emptyset, \\ (\emptyset, \mathscr{P}Y) & \text{if } X = \emptyset, \end{cases} \qquad G(X,Y) = \begin{cases} (1, 1) & \text{if } X \neq \emptyset, \\ (\emptyset, \mathscr{P}Y) & \text{if } X = \emptyset. \end{cases}$$

We saw in Example 6.3.3(3) that F has no fixed point. But $G(1,1) = (1,1)$, so this is νG. Moreover, we have $m\colon F \rightarrowtail G$, where $m_{(X,Y)}$ for $X \neq \emptyset$ is the inclusion $(\emptyset, 1) \hookrightarrow (1,1)$.

Remark 7.2.12. Under the hypothesis of Proposition 7.2.8, consider the terminal-coalgebra chain of F (Definition 6.5.1). If for some ordinal i, the connecting morphism $v_{i+1,i}\colon FV_i \to V_i$ is monic, then this algebra contains νF. In fact, $(V_i, v_{i+1,i})$ even enjoys a stronger property than being corecursive, as we will see in Corollary 7.3.12.

7.3 Completely Iterative Algebras

We continue our study by looking at a natural strengthening of the defining property of corecursive algebras and verifying that again, the terminal coalgebra is, equivalently, the initial object in the relevant (smaller) subcategory of algebras.

The idea of algebras with unique solutions of recursive equations stems from work in general algebra by Evelyn Nelson [242] and Jerzy Tiuryn [289]. Nelson introduced iterative algebras for a signature as algebras with unique solutions of finite systems of mutually recursive equations such as (7.1). In these systems, parameters from A are allowed to occur on right-hand sides of equations. Dropping the finiteness assumption one arrives at the notion of a *completely iterative algebra*, introduced by Milius [226]. He also investigates the connection to terminal coalgebras.

Although the addition of parameters to formal recursive equations may seem like a small extension, the fact that each flat equation morphism has a unique solution is a strong property of an algebra and has many interesting consequences, as we shall explain in Remark 7.3.23.

Assumption 7.3.1. Throughout this section, we assume that $\mathscr{A}$ is a category with binary coproducts.

Definition 7.3.2 [226]. Let F be an endofunctor on $\mathscr{A}$. A *flat equation morphism* in an object A is a morphism $e\colon X \to FX + A$. A *solution* of e in an algebra (A, α)

is a morphism $e^\dagger\colon X \to A$ such that the square below commutes:

$$\begin{array}{ccc} X & \xrightarrow{\quad e^\dagger \quad} & A \\ {\scriptstyle e}\downarrow & & \uparrow{\scriptstyle [\alpha,\mathsf{id}_A]} \\ FX + A & \xrightarrow{Fe^\dagger + \mathsf{id}_A} & FA + A \end{array} \tag{7.8}$$

The algebra (A, α) is *completely iterative* (or, a *completely iterative algebra (cia)* for short) provided that every flat equation morphism in it has a unique solution.

Remark 7.3.3. (1) Every cia (A, α) is a corecursive algebra. Indeed, for every coalgebra $e\colon X \to FX$ one forms

$$\bar{e} = (X \xrightarrow{e} FX \xrightarrow{\mathsf{inl}} FX + A),$$

and one readily verifies that the unique solution of $\bar{e}$ in the present sense is a unique solution of e in (A, α) in the sense of Definition 7.1.1.

(2) Thus, classical algebras are seldom cias as they are not even corecursive, as explained in Remark 7.1.3(2).

Example 7.3.4. Recall the corecursive algebra A from Example 7.1.4(3). This algebra A is not a cia: the system $x \approx (x, y)$, $y \approx 2$ has two solutions, corresponding to the facts that both $\alpha(1, 2) = 1$ and $\alpha(0, 2) = 0$. This point again comes from Capretta et al. [101].

We take cias to be a full subcategory of the category of all F-algebras. This choice is justified by the following result, analogous to Proposition 7.1.5:

Proposition 7.3.5 [226, Prop. 2.3]. *Let (A, α) and (B, β) be cias for F. Then a morphism $h : A \to B$ is an algebra homomorphism if and only if it preserves solutions: for every flat equation morphism $e : X \to FX + A$ we have that $h \cdot e^\dagger : X \to B$ is the unique solution of*

$$X \xrightarrow{e} FX + A \xrightarrow{FX+h} FX + B.$$

Proof sketch The ‘only if’ direction is very similar to what we have seen in Proposition 7.1.5: one proves with a straightforward diagram chase that $h \cdot e^\dagger$ is a solution for $(FX + h) \cdot e$ in B and then obtains the desired result by the uniqueness of solutions in B.

For the ‘if’ direction one observes that the morphism $[\alpha, \mathsf{id}_A]\colon FA + A \to A$ appears as the unique solution of the flat equation morphism

$$e = F\mathsf{inr} + \mathsf{inr}\colon FA + A \to F(FA + A) + A.$$

Similarly, one proves that $[\beta \cdot Fh, h]\colon FA + A \to B$ is the unique solution of $(F(FA + A) + h) \cdot e$. Since h preserves solutions, one then obtains the following equality

$$h \cdot [\alpha, \mathsf{id}_A] = h \cdot e^\dagger = (F(FA + A) + h) \cdot e)^\dagger = [\beta \cdot Fh, h] \colon FA + A \to B,$$

whose left-hand coproduct component shows that h is a homomorphism. □

Proposition 7.3.6 [226, Ex. 2.5]. *For every endofunctor, the terminal coalgebra is a cia.*

Proof Let $F \colon \mathscr{A} \to \mathscr{A}$ be an endofunctor for which $(\nu F, \tau)$ exists. Given a flat equation morphism $e : X \to FX + \nu F$, form the following F-coalgebra

$$\overline{e} = \big(X + \nu F \xrightarrow{[e,\mathsf{inr}]} FX + \nu F \xrightarrow{FX+\tau} FX + F(\nu F) \xrightarrow{\mathsf{can}} F(X + \nu F)\big),$$

where $\mathsf{can} = [F\mathsf{inl}, F\mathsf{inr}]$. Let $h \colon X + \nu F \to \nu F$ be the corresponding unique coalgebra homomorphism and define

$$e^\dagger = \big(X \xrightarrow{\mathsf{inl}} X + \nu F \xrightarrow{h} \nu F\big).$$

One now shows that $\mathsf{inr} \colon \nu F \to X + \nu F$ is a coalgebra homomorphism from $(\nu F, \tau)$ to $(X + \nu F, \overline{e})$; indeed, the following diagram clearly commutes:

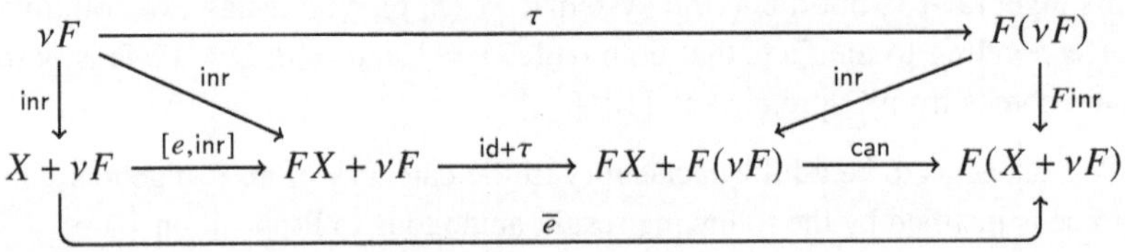

Consequently $h \cdot \mathsf{inr}$ is a coalgebra homomorphism from $(\nu F, \tau)$ to itself, whence $h \cdot \mathsf{inr} = \mathsf{id}$, which implies that $h = [e^\dagger, \mathsf{id}]$.

We now consider the following diagram:

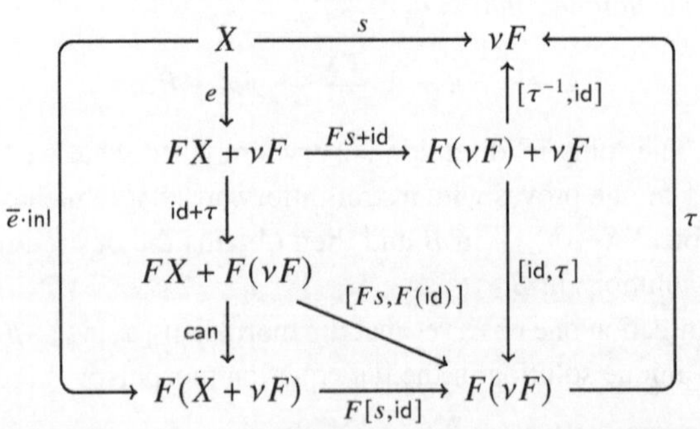

Note that the left- and right-hand parts, as well as the middle one and the lower triangle clearly commute. For $s = e^\dagger$, the left-hand component of h, the outside of the diagram commutes. Thus, so does the upper part, which shows that $e^\dagger$ is a solution of e.

To show its uniqueness, suppose that s is any solution of e. Then the upper square commutes, and therefore so does the outside. This implies that $[s, \mathsf{id}]$ is a coalgebra homomorphism from $(X + \nu F, \overline{e})$ to $(\nu F, \tau)$, and so we have $[s, \mathsf{id}] = h$ by the universal property of the terminal coalgebra. We conclude that

$$e^\dagger = h \cdot \mathsf{inl} = [s, \mathsf{id}] \cdot \mathsf{inl} = s,$$

which completes the proof. □

Example 7.3.7. (1) It follows from Proposition 7.3.6 that the collection of all infinite binary trees is a cia for $FX = X \times X + 1$ (cf. Example 7.1.4(2)). For example, fix an element $t \in \nu F$, and consider the following system of equations:

$$x_1 \approx x_2 * t \qquad x_2 \approx x_1 * c. \tag{7.9}$$

We emphasize that t here is a fixed element of the terminal coalgebra (i.e. a finite or infinite binary tree). The solution of this equation in νF consists of two infinite trees, ${x_1}^\dagger$ and ${x_2}^\dagger$. Here are two different depictions of these trees. First, we may write them as infinite Σ-terms:

$${x_1}^\dagger = (\cdots(\cdots * c) * t) * c) * t) \qquad \text{and} \qquad {x_2}^\dagger = (\cdots(\cdots * t) * c) * t) * c).$$

Second, we may picture them as infinite Σ-trees. For example, here is ${x_1}^\dagger$ shown as such a tree. The small triangles represent the infinite tree t.

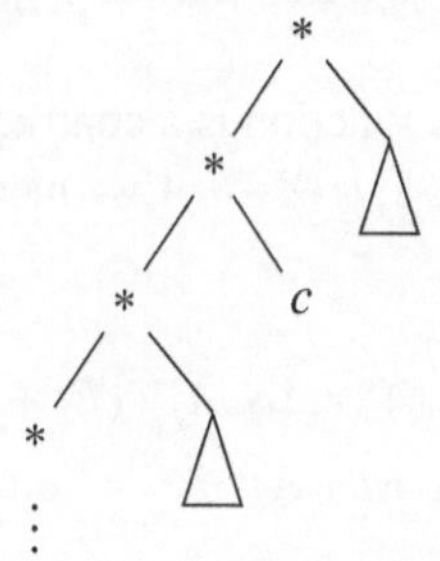

(2) The collection of all finitely branching strongly extensional trees (Definition 4.5.8) is a cia for $\mathscr{P}_\mathsf{f}$ (Theorem 4.5.15). The collection of all unordered finitely branching trees is a terminal coalgebra for the bag functor $\mathscr{B}$ (see Example 3.2.11) and therefore a cia.

(3) The algebra of addition of the extended natural numbers $\widetilde{\mathbb{N}} = \mathbb{N} \cup \{\infty\}$ is a cia for the functor $FX = X \times X$, see Adámek et al. [30].

(4) As shown by Milius [226] a unary algebra $\alpha : A \to A$ (here we consider $F = \mathsf{Id}$ on Set) is a cia if and only if

(a) there exists a unique fixed point $x_0 \in A$ of α, and
(b) if an infinite sequence $y_0, y_1, y_2, \ldots$ fulfils $\alpha(y_{n+1}) = y_n$ $(n < \omega)$, then $y_n = x_0$ for all n.

(5) Free corecursive algebras often fail to be cias. For example, for every signature Σ containing a binary operation symbol $*$ a free corecursive algebra on $Y = \{y\}$ (see Example 7.1.14) does not have a unique solution of recursive equation $x \approx x * y$.

We continue with examples of cias obtained from complete metric spaces. Recall from Notation 3.2.2 that CMet is the category of complete metric spaces with distances at most 1 and non-expanding maps. Recall *locally contracting* endofunctors from Definition 5.2.5.

Proposition 7.3.8 [32, Lemma 2.9]. *Let F be a locally contracting endofunctor on* CMet. *Then every nonempty F-algebra is a cia.*

Proof Suppose that F has a contraction factor $\varepsilon < 1$. Let $\alpha\colon FA \to A$ be a nonempty F-algebra and let $e\colon X \to FX+A$ be a flat equation morphism. Recall that the hom-set $\mathsf{CMet}(X, A)$ is a complete metric space with the supremum metric $d_{X,A}$. Moreover, $\mathsf{CMet}(X, A)$ is nonempty, since A is nonempty. It is clear from Definition 7.3.2 that a solution of e is equivalently a fixed point of the endomap Φ on $\mathsf{CMet}(X, A)$ assigning to every $s\colon X \to A$ the map

$$\Phi(s) = (X \xrightarrow{e} FX + A \xrightarrow{Fs+A} FA + A \xrightarrow{[\alpha,A]} A).$$

We shall now prove that this function is a contraction on $\mathsf{CMet}(X, A)$. Indeed, for two non-expanding maps $s, t\colon X \to A$ we have

$$\begin{aligned}
&d_{X,A}(\Phi s, \Phi t)\\
&\quad = d_{X,A}([\alpha, A] \cdot (Fs + A) \cdot e, [\alpha, A] \cdot (Ft + A) \cdot e) && \text{(definition of } \Phi)\\
&\quad \leq d_{FX+A,FA+A}(Fs + A, Ft + A) && \text{(composition is non-expanding)}\\
&\quad = d_{FX,FA}(Fs, Ft)\\
&\quad \leq \varepsilon d_{X,A}(s,t) && (F \text{ is } \varepsilon\text{-contracting}).
\end{aligned}$$

By Banach's Fixed Point Theorem 5.2.4, there exists a unique fixed point of Φ. □

Remark 7.3.9. Note that Banach's Fixed Point Theorem yields the unique solution $e^\dagger$ as the limit of a Cauchy sequence as follows. Choose some element $a \in A$ and define a sequence $(e^\dagger_n)_{n\in\mathbb{N}}$ in $\mathsf{CMet}(X, A)$ inductively as follows: let $e^\dagger_0 = \mathsf{const}_a$, and given $e^\dagger_n$ define $e^\dagger_{n+1}$ by the commutativity of the following square

$$\begin{array}{ccc} X & \xrightarrow{e^\dagger_{n+1}} & A \\ {\scriptstyle e}\downarrow & & \uparrow{\scriptstyle [\alpha, \mathsf{id}_A]} \\ FX + A & \xrightarrow{Fe^\dagger_n + \mathsf{id}_A} & FA + A \end{array}$$

Then $(e^\dagger_n)_{n\in\mathbb{N}}$ is a Cauchy sequence in $\mathsf{CMet}(X, A)$ whose limit is the unique solution of e:

$$e^\dagger = \lim_{n\to\infty} e^\dagger_n.$$

Example 7.3.10. (1) Many set functors F have a lifting to a locally contracting endofunctor F' on CMet. We have seen this for polynomial endofunctors in Example 5.2.6. We call an F-algebra $\alpha\colon FA \to A$ *completely metrizable* if there exists a complete metric d on A such that the algebra structure is a non-expanding map $\alpha\colon F'(A, d) \to (A, d)$.

Every nonempty completely metrizable F-algebra is a cia. Indeed, to every equation morphism $e\colon X \to FX + A$ its unique solution is the unique solution of $e\colon (X, d_0) \to F'(X, d_0) + (A, d)$ in the F'-algebra (A, α), where d_0 is the discrete metric.

(2) Let $F'\colon \mathsf{CMet} \to \mathsf{CMet}$ be given by

$$F'(X, d) = \left(X, \tfrac{1}{2}d\right) + \left(X, \tfrac{1}{2}d\right),$$

that is, F' takes a space into the disjoint union of two copies of itself, each equipped with a metric shrinking the distance by half, and with the distance between the copies set to 1. Let $A = [0, 1]$, and let $\alpha\colon F'A \to A$ be defined by $x \mapsto \frac{x}{2}$ on the left-hand copy, and $x \mapsto \frac{x}{2} + \frac{1}{2}$ on the right-hand one. Then α is clearly a non-expanding map, and hence (A, α) is a cia. Note that the functor F' is a lifting of $FX = X + X$ on Set. By item (1), $[0, 1]$ with α considered as an F-algebra structure is completely metrizable and therefore a cia for F.

As an illustration, we show how to obtain every real number $r \in [0, 1]$ as $e^\dagger(x)$ for some flat equation morphism $e\colon X \to FX + A$ and $x \in X$. We take an infinite set $X = \{x_0, x_1, \ldots, x_i, \ldots\}$, and we write r in binary notation as $0.b_1b_2b_3 \cdots b_i \cdots$. Our flat equation morphism e is given by the system

$$x_i \approx \frac{b_i}{2} + \frac{1}{2}x_{i+1} \qquad (i \in \mathbb{N}).$$

It is not hard to see that $e^\dagger(x_i) = 0.b_ib_{i+1}b_{i+2}\cdots$ is the solution. So $e^\dagger(x_0)$ is the given real number r with which we started.

(3) Proposition 7.3.8 yields further interesting examples of cias, where solutions

of recursive equations are fractals. The most basic of these examples is the following one due to Milius and Moss [229].

Let A be the set of nonempty closed subsets of the interval $[0,1]$. Then A is a complete metric space equipped with the *Hausdorff metric* (Example 5.2.6(4)): for closed subsets $S, T \subseteq [0,1]$,

$$d(S,T) = \max(\sup_{x \in S} d(x,T), \sup_{y \in T} d(y,S)),$$

where $d(x,T) = \inf_{y \in T} d(x,y)$. We also consider the functor

$$F(X,d) = \left(X \times X, \tfrac{1}{3} d_{\mathsf{max}}\right)$$

sending the complete metric space (X,d) to $X \times X$ equipped with the maximum metric scaled by $\frac{1}{3}$; this functor is clearly locally contracting. Then A is an algebra for F with structure α given by

$$\alpha(S,T) = \tfrac{1}{3} S \cup \left(\tfrac{1}{3} T + \tfrac{2}{3}\right)$$

with the obvious interpretation of addition and multiplication on the closed subsets $S, T \subseteq [0,1]$, for example, $\frac{1}{3}S = \{\frac{1}{3}s : s \in S\}$. Now let $X = \{*\}$ be the one-point space and $e \colon X \to FX + A$ be given by $e(*) = (*,*)$. Then $e^\dagger(*)$ is the famous Cantor set [176, p. 31].

We now turn to constructions of cias as a preparation for the proof that initial cias and terminal coalgebras are the same.

Lemma 7.3.11. (1) *If (A,α) is a cia, then so is $(FA, F\alpha)$.*

(2) *A limit of cias in* $\mathsf{Alg}\, F$ *is a cia. In particular, if $\mathscr{A}$ has a terminal object* 1, *the trivial (terminal) algebra $F1 \to 1$ is a cia.*

Proof sketch (1) Let $e \colon X \to FX + FA$ be a flat equation morphism in FA. Form the equation morphism

$$\overline{e} = (X \xrightarrow{e} FX + FA \xrightarrow{FX+\alpha} FX + A),$$

and obtain $\overline{e}^\dagger \colon X \to A$. Using this, define

$$e^\dagger = (X \xrightarrow{e} FX + FA \xrightarrow{[F\overline{e}^\dagger, FA]} FA).$$

It is not difficult to check that $e^\dagger$ is the unique solution of e in FA. For the details, see Milius [226, Proposition 2.6].

(2) Suppose we have a diagram D of cias (A_i, α_i), $i \in I$, and let (A,α) be a limit in the category of algebras for F with limit projections $p_i \colon (A,\alpha) \to (A_i, \alpha_i)$. Recall from Remark 4.1.4 that this limit is formed on the level of $\mathscr{A}$. We need

to show that (A, α) is a cia. Let $e \colon X \to FX + A$ be a flat equation morphism. For every $i \in I$ one forms the flat equation morphism

$$e_i = (X \xrightarrow{e} FX + A \xrightarrow{FX+p_i} FX + A_i)$$

and takes the unique solution $e_i^\dagger \colon X \to A_i$. Then $e_i^\dagger$ $(i \in I)$ is a cone of D. Indeed, for every connecting morphism $h : (A_i, \alpha_i) \to (A_j, \alpha_j)$ we know from Proposition 7.3.5 that h preserves solutions, thus $h \cdot e_i^\dagger = e_j^\dagger$. We define $e^\dagger \colon X \to A$ to be the unique morphism with $p_i \cdot e^\dagger = e_i^\dagger$ for every $i \in I$. Then $e^\dagger$ is a solution of e: indeed, in order to see that (7.8) commutes, extend this by the limit projections p_i to obtain the following diagram:

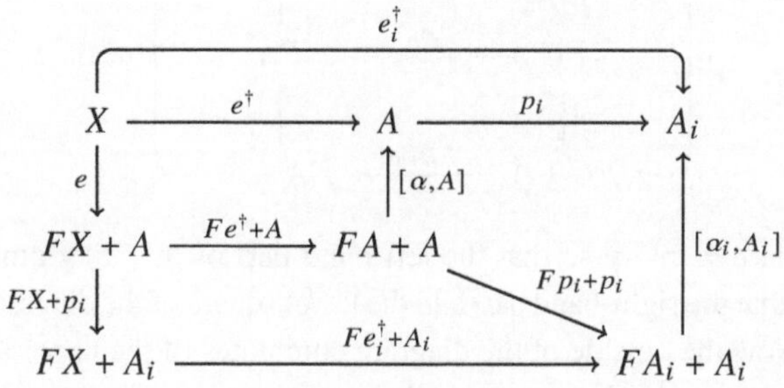

Its outside commutes since $e_i^\dagger$ solves $(FX + p_i) \cdot e$, its right-hand part does because p_i is a homomorphism, and the lower and upper parts commute by the definition of $e^\dagger$. Thus, the upper left-hand square commutes when extended by p_i. Since the limit cone (p_i) is collectively monic, this proves that the desired square commutes.

For the uniqueness of $e^\dagger$ suppose that $s \colon X \to A$ is any solution of e in A. Then we see that for every $i \in I$, $p_i \cdot s$ is a solution of $(FX + p_i) \cdot e$ (to see this, repeat the reasoning in the 'only if' direction of the proof of Proposition 7.3.5). Thus $p_i \cdot s = e_i^\dagger$, whence $s = e^\dagger$, by the universal property of the limit A. □

Corollary 7.3.12. *Let F be an endofunctor on a complete category. Then in the terminal-coalgebra chain of F, all algebras $FV_i \to V_i$ (Definition 6.5.1) are cias.*

Note that this result holds even if the terminal coalgebra itself does not exist. It easily follows from Lemma 7.3.11 by transfinite induction: use item (2) for $i = 0$ and limit steps and item (1) for isolated steps.

We also obtain a version of Lambek's Lemma for cias:

Corollary 7.3.13. *If an initial cia for F exists, then it is a fixed point of F.*

Indeed, the proof is the same as the one for Lambek's Lemma 2.2.5 but using Lemma 7.3.11(1) to see that for an initial cia (I, ι), $(FI, F\iota)$ is a cia, too.

Theorem 7.3.14 [226, Thm. 2.8]. *The initial cia is precisely the same as the terminal coalgebra.*

More precisely, let $F\colon \mathscr{A} \to \mathscr{A}$ be an endofunctor. Then (I, ι) is an initial cia for F iff (I, ι^{-1}) is a terminal coalgebra. The proof is somewhat similar to what we saw in Theorem 7.1.8; we include it for the convenience of the reader.

Proof (1) For a coalgebra $\gamma\colon C \to FC$ and an algebra $\alpha\colon FA \to A$ consider the following diagram:

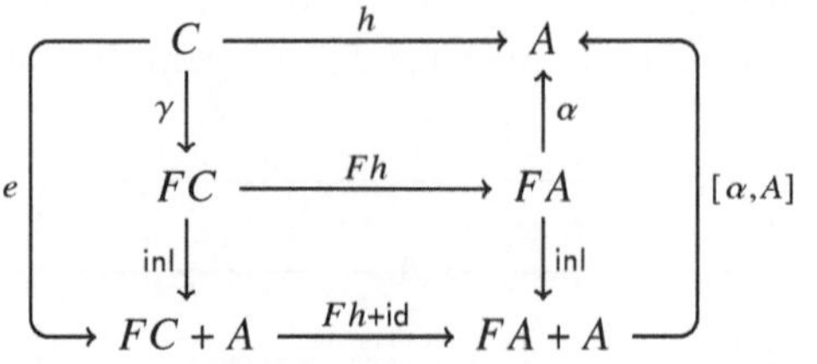

(7.10)

Let us define $e = \mathsf{inl} \cdot \gamma$ so that the left-hand part of this diagram commutes. Notice also that the right-hand part and the lower square of the diagram obviously commute. Now the outside of the diagram commutes iff the upper square does; in other words, h is a solution of γ in the algebra (A, α) iff it is a solution of the flat equation morphism e in that algebra.

(2) Suppose that $\iota : FI \to I$ is an initial cia. Then ι is an isomorphism by Corollary 7.3.13. So we have the coalgebra $\iota^{-1}\colon I \to FI$, and we need to verify that it is terminal. Indeed, for every coalgebra (C, γ) replace (A, α) in Diagram (7.10) by the algebra (I, ι). Then since this algebra is a cia, we have a unique coalgebra-to-algebra homomorphism, equivalently, a unique coalgebra homomorphism from (C, γ) to (I, ι^{-1}).

(3) Conversely, suppose that $\tau : \nu F \to F(\nu F)$ is a terminal coalgebra. Then the corresponding algebra is a cia by Proposition 7.3.6, and it remains to verify its initiality. So let (A, α) be any cia and let (C, γ) in Diagram (7.10) be $(\nu F, \tau)$. Since A is a cia, we have a unique solution of $e = \mathsf{inl} \cdot \tau$ in A, equivalently a unique F-algebra homomorphism from the cia $(\nu F, \tau^{-1})$ to the cia (A, α). □

Remark 7.3.15. The only fixed point among the cias for a functor is the initial cia. This follows from Theorem 7.3.14 and Proposition 7.1.2(2).

More generally, we can characterize *free* cias (cf. Remark 2.2.23): the forgetful functor from the category of all cias for F to the base category $\mathscr{A}$ has a left adjoint whenever all endofunctors $F(-) + Y$ for object Y of $\mathscr{A}$ have a terminal coalgebra:

Theorem 7.3.16 [226, Cor. 2.11]. *A free cia for F on Y is precisely the same as a terminal coalgebra for $F(-) + Y$.*

More precisely, given a terminal coalgebra $\tau_Y : TY \to FTY + Y$, then precomposing τ_Y^{-1} by the two coproduct injections yields an algebra structure $\alpha_Y : FTY \to TY$ and a morphism $\eta_Y : Y \to TY$ which are the structure and universal morphism, respectively, of a free cia on Y. Conversely, given a free cia (TY, α_Y) with universal morphism η_Y then $[\alpha_Y, \eta_Y] : FTY + Y \to TY$ is an isomorphism whose inverse is the structure of a terminal coalgebra for $F(-) + Y$.

Analogously to the notion of a varietor (Definition 2.2.26) we have

Definition 7.3.17 [5]. An endofunctor F is called *iteratable* if every object generates a free cia.

Example 7.3.18. Polynomial set functors H_Σ are iteratable. Indeed, the terminal coalgebra for $H_\Sigma(-) + Y$ is carried by the set $T_\Sigma Y$ of all Σ-trees over Y (see Example 7.1.14) with the coalgebra structure given by the inverse of tree-tupling; this follows from Theorem 2.5.9 applied to the signature Σ_Y (see Example 2.2.25(2)). Thus, $T_\Sigma Y$ is the free cia for H_Σ on Y.

Remark 7.3.19. In general, corecursive algebras need not be cias, see Examples 7.3.4 and 7.3.7(5). But are there interesting functors for which the two notions coincide? The answer is affirmative whenever the base category $\mathscr{A}$ has 'very well-behaved' coproducts (this includes examples such as sets, posets, graphs, and presheaf categories) and the endofunctor has the form $FX = A \times X + B$ for fixed objects A and B [20]. Moreover, among finitary set functors, there are essentially no other functors. In the Appendix (see page 587) we show that for every set functor F there exists a *standard* functor which agrees with F on all nonempty sets and functions. Standard means that the functor preserves finite intersections and inclusions.

Theorem 7.3.20 [20]. *Let F be a standard finitary set functor. If every corecursive F-algebra is a cia, then F is naturally isomorphic to a functor defined by $X \mapsto A \times X + B$.*

Remark 7.3.21. As observed by Paul Levy[1], a cia can, equivalently, be defined as an algebra (A, α) such that for every coalgebra $e : X \to F(X + A)$ there exists a unique morphism $e^\dagger : X \to A$ such that the square below commutes:

$$\begin{array}{ccc} X & \xrightarrow{\quad e^\dagger \quad} & A \\ {\scriptstyle e}\downarrow & & \uparrow{\scriptstyle \alpha} \\ F(X + A) & \xrightarrow{F[e^\dagger, \mathrm{id}_A]} & FA \end{array} \tag{7.11}$$

[1] Private communication.

We present a proof below. Note that (7.11) is very closely related to primitive corecursion (see Theorem 2.6.4); in fact, primitive corecursion is precisely the unique existence of $e^\dagger$ in (7.11) for the algebra $(\nu F, \tau^{-1})$.

However, one advantage of Definition 7.3.2 is that (A, α) is a cia iff $(A, [\alpha, \mathsf{id}_A])$ is corecursive for $F(-) + A$. Moreover, observe that for a polynomial set functor, flat equation morphisms in Definition 7.3.2, regarded as systems of recursive equations, have a slightly simpler syntactic format than systems arising from maps $X \to H_\Sigma(X + A)$ (cf. Remark 7.1.3(1)). Indeed, a flat equation morphism $X\colon X \to H_\Sigma X + A$ can be understood as a recursive system of equations each of whose right-hand sides is either a flat term $\sigma(x_1, \dots, x_n)$ for an n-ary operation symbol σ or a variable from X, or an element of A.

The theory of cias could be developed based on this alternative definition. However, we prefer to use the definition based on flat equation morphism from the literature.

Proposition 7.3.22. *An algebra (A, α) is a cia iff for every $e\colon X \to F(X + A)$ there exists a unique morphism $e^\dagger\colon X \to A$ such that* (7.11) *commutes.*

Proof sketch (1) Given a cia (A, α) and a morphism $e\colon X \to F(X + A)$, form the flat equation morphism

$$X + A \xrightarrow{e + \mathsf{id}_A} F(X + A) + A.$$

Its unique solution is easily seen to satisfy $(e + \mathsf{id}_A)^\dagger \cdot \mathsf{inr} = \mathsf{id}_A$, and one then readily proves that $(e + \mathsf{id}_A)^\dagger \cdot \mathsf{inl}\colon X \to A$ is the desired unique morphism; we omit the details.

(2) Conversely, suppose that (A, α) admits for every $e\colon X \to F(X + A)$ a unique morphism $e^\dagger$ such that (7.11) commutes. Given a flat equation morphism $f\colon X \to FX + A$, form the morphism

$$FX \xrightarrow{Ff} F(FX + A),$$

and let $(Ff)^\dagger\colon FX \to A$ be the unique morphism such that (7.11) commutes (for $e = Ff$). Now let

$$f^\dagger = (X \xrightarrow{f} FX + A \xrightarrow{[(Ff)^\dagger, \mathsf{id}_A]} A).$$

One readily proves that this is a unique solution of f. Again, we omit the details. □

Remark 7.3.23. To conclude this section, let us mention some further work on cias and related structures in the literature.

(1) As we mention in the introduction of this section, one of the reasons why cias are interesting is that one can uniquely solve much more general recursive equations than the flat ones in Definition 7.3.2 in a cia. For example, [5, 226] contain a solution theorem which, when instantiated for a polynomial set functor, states that mutually recursive systems

$$x_i \approx t_i \qquad (i \in I),$$

where I is any set and the right-hand sides t_i are arbitrary (possibly infinite) non-variable Σ-trees have unique solutions.

(2) Assuming that F is iteratable, that is, there exists a free cia TX on every object X of a base category $\mathscr{A}$ (see Theorem 7.3.16), it turns out that T is the object assignment of a monad on $\mathscr{A}$. This monad is characterized by a universal property: it is the free completely iterative monad on F. These results appear in [5, 226] generalizing work on (free) completely iterative theories by Elgot, Bloom, and Tindell [116]. The monad structure on T arises in the usual way from the adjunction between the base category $\mathscr{A}$ and the category of all cias for F whose left adjoint assigns to X the free cia TX.

(3) For further use we note that the above monad structure can also be obtained explicitly as follows. Recall, for example, from Moggi [238], that monads can, equivalently, be presented as *Kleisli triples* $(T, \eta, \widehat{(\_)})$, where T is an object assignment, $\eta_X\colon X \to TX$ is a family of maps indexed by the objects of $\mathscr{A}$ and the *Kleisli extension* $\widehat{(\_)}$ assigns to every morphism $s\colon X \to TY$ a morphism $\hat{s}\colon TX \to TY$ such that the following three axioms hold for every s and $t\colon Y \to TZ$:

$$\hat{s} \cdot \eta_X = s, \qquad \widehat{\eta_X} = \mathsf{id}_{TX}, \qquad \widehat{\hat{s} \cdot t} = \hat{s} \cdot \hat{t}.$$

This notion goes back to Manes [219, Exercise 12], who called it *algebraic theory in extension form*.

Now let T assign to an object X the free cia TX on X. We obtain the structure of a Kleisli triple as follows [5, 226]: the morphism $\eta_X\colon X \to TX$ is the universal morphism; it is obtained by pre-composing the inverse of the terminal coalgebra structure $\tau_X\colon TX \to F(TX) + X$ by the right-hand coproduct injection (cf. Theorem 7.3.16). For a given morphism $s\colon X \to TY$, its Kleisli extension $\hat{s}\colon TX \to TY$ is the unique F-algebra homomorphism such that $\hat{s} \cdot \eta_X = s$. Alternatively, $\hat{s}$ can be defined by primitive corecursion with respect to $F(-) + Y$ (Theorem 2.6.4) as the unique morphism such that the square below commutes (here $\mathsf{can} = [F\mathsf{inl}, F\mathsf{inr}]\colon FTX + FTY \to F(TX + TY)$ is the canonical morphism):

$$
\begin{array}{ccccccc}
TX & \xrightarrow{\tau_X} & FTX + X & \xrightarrow{FTX + \tau_Y \cdot s} & FTX + FTY + Y & \xrightarrow{\mathsf{can}+Y} & F(TX + TY) + Y \\
\Big\downarrow{\scriptstyle \hat{s}} & & & & & & \Big\downarrow{\scriptstyle F[\hat{s}, TY]+Y} \\
TY & & & \xrightarrow{\qquad\qquad\tau_Y\qquad\qquad} & & & FTY + Y
\end{array}
$$

It is not difficult to prove that this satisfies the three axioms of a Kleisli triple, whence T yields a monad.

Note that for a polynomial set functor H_Σ, the map $\hat{s}\colon T_\Sigma X \to T_\Sigma Y$ substitutes in a given Σ-tree over X the leaves labelled by $x \in X$ by the Σ-tree $s(x) \in T_\Sigma Y$ (cf. Example 7.3.18).

(4) In a number of settings, equation morphisms do have a solution albeit not a unique one. Then one is more interested in taking canonical solutions (or fixed points) than requiring their unique existence, for example, least fixed points in complete partial orders. The idea of Bloom and Ésik's *iteration theories* [84] is to study the equational properties that characterize least fixed points in complete partial orders and in related settings. It turns out that similar ideas are important in connection with cias, and they lead to the notion of a *complete Elgot algebra* [32] for an endofunctor F: this is a triple $(A, \alpha, (\_)^\dagger)$ where $\alpha\colon FA \to A$ is an algebra and $(\_)^\dagger$ is an operation taking a flat equation morphism $e : X \to FX + A$ to a solution $e^\dagger : X \to A$ such that two simple and well-motivated properties are satisfied.

Every cia is a complete Elgot algebra. Further examples are continuous algebras (i.e. algebras for locally continuous functors on CPO) or complete lattices; taking binary joins as algebra structure, they are complete Elgot algebras for $FX = X \times X$ on Set.

Theorem 7.3.14 can be augmented to state that the terminal coalgebra is the same as the initial Elgot algebra (in addition, the terminal coalgebra for $F(-) + X$ is equivalently, the free complete Elgot algebra for F on X). The main result [32, Thm. 5.8] is that complete Elgot algebras form precisely the category of Eilenberg–Moore algebras for the monad T of item (2) above.

(5) Completely iterative algebras and complete Elgot algebras as in item (4) play a significant rôle in the category-theoretic semantics of recursive program schemes as presented by Milius and Moss [229]. In their approach to the semantics of recursive function definitions, terminal coalgebras for functors are used in lieu of concepts from general algebra like signatures and infinite trees. In the category-theoretic setting, cias and Elgot algebras serve as those classes of algebras in which one interprets recursive program schemes – in

those algebras one can canonically (or even uniquely, in the case of cias) solve recursive program schemes.

(6) Milius, Moss, and Schwencke [232] demonstrate how Turi and Plotkin's abstract operational rules [252] give rise to new cia structures on the terminal coalgebra. This yields theorems that instantiate to several known unique solution theorems in the literature. For example, Milner's solution theorem for CCS [236], and a unique solution theorem for stream circuits [265], as well as new ones for non-well-founded sets (extending previous results of Barwise and Moss [74]) or formal languages. Besides, one obtains a modular framework for the specification of operations by abstract operational rules [232]; modularity here means that unique solution theorems are preserved when adding operations specified by abstract operational rules to a given cia structure on the terminal coalgebra.

7.4 Recursive Coalgebras

After having looked at algebras with special properties, we now turn to coalgebras with special properties, and we will see the initial algebra appearing as a terminal object in certain categories of coalgebras. The first notion we mention is that of a *recursive coalgebra*, which is formally dual to the notion of a corecursive algebra, thus formulating a unique definition principle for recursive functions. Recursive coalgebras are closely connected to well-founded coalgebras, which we consider in Chapter 8. They provide a categorical formulation of well-founded induction. In his work on categorical set theory, Osius [245] first studied the notions of well-founded and recursive coalgebras (for the power-set functor on sets and, more generally, the power-object functor on an elementary topos). He defined recursive coalgebras as those coalgebras $\alpha\colon A \to \mathscr{P}A$ which have a unique coalgebra-to-algebra homomorphism into every algebra (see Definition 7.4.1).

Taylor [286, 287, 288] considered recursive coalgebras for a general endofunctor under the name 'coalgebras obeying the recursion scheme', and proved the General Recursion Theorem that all well-founded coalgebras are recursive for more general endofunctors (see Chapter 8). Recursive coalgebras were also investigated by Eppendahl [117], who called them algebra-initial coalgebras.

Capretta, Uustalu, and Vene [100] have studied recursive coalgebras, and they have shown how to construct new ones from given ones by using comonads. They have also explained nicely how recursive coalgebras allow for the semantic treatment of (functional) recursive divide-and-conquer programs. In addition, recursive and parametrically recursive coalgebras feature prominently in the unifying investigation of structured recursion schemes by Hinze et al. [163]. Jeannin et al. [175] have proved the general recursion theorem for polynomial functors on the category of many-sorted sets; they have also provided many interesting examples of recursive coalgebras arising in programming.

In this section we will just recall a few basic results, which are dual to what we have seen in Section 7.1 and Section 7.3 on corecursive algebras, and present some examples.

For the dual concept of a solution in a corecursive algebras, we stick to the standard terminology of a colgebra-to-algebra morphism:

Definition 7.4.1. A coalgebra $\gamma\colon C \to FC$ is *recursive* if for every algebra $\alpha\colon FA \to A$ there exists a unique coalgebra-to-algebra morphism $h\colon C \to A$, that is a unique morphism such that the square below commutes:

$$\begin{array}{ccc} C & \xrightarrow{\gamma} & FC \\ {\scriptstyle h}\downarrow & & \downarrow{\scriptstyle Fh} \\ A & \xleftarrow{\alpha} & FA \end{array}$$

Proposition 7.4.2. (1) *An initial algebra, considered as a coalgebra, is recursive.*

(2) *Every recursive fixed point is an initial algebra.*

This is dual to Proposition 7.1.2.

Example 7.4.3. (1) The first examples of recursive coalgebras are well-founded relations. Recall that a binary relation R on a set X is well-founded if there is no infinite descending sequence

$$\cdots R\, x_3\, R\, x_2\, R\, x_1\, R\, x_0.$$

Now a binary relation $R \subseteq X \times X$ is essentially a graph on X, equivalently a coalgebra $\alpha\colon X \to \mathscr{P}X$ defined by $\alpha(x) = \{y \mid y\, R\, x\}$ (cf. Example 2.4.2(4)). Osius [245] showed that for every well-founded relation the associated $\mathscr{P}$-coalgebra is recursive. Shortly: a graph regarded as a coalgebra for $\mathscr{P}$ is recursive iff it has no infinite path.

(2) The (trivial) initial coalgebra $0 \to F0$ is recursive.

(3) If (C, γ) is recursive so is $(FC, F\gamma)$. This is dual to Proposition 7.1.6.

(4) Every colimit of recursive coalgebras in Coalg F is recursive. This is easy to prove, using that colimits of coalgebras are formed on the level of the underlying category. The argument is the 'parameter-less' dual of the proof of Lemma 7.3.11(2).

(5) It follows from items (2)–(4) that in the initial-algebra chain (see Definition 6.1.4) all coalgebras $W_i \to FW_i$ are recursive.

Dually to Corollary 7.1.7, we see that a terminal recursive F-coalgebra is a fixed point of F. Moreover, dually to Theorem 7.1.8 we have

Corollary 7.4.4 [100, Prop. 7]**.** *The initial algebra is precisely the same as the terminal recursive coalgebra.*

The dual notion of a completely iterative algebra (see Definition 7.3.2) is called a *parametrically recursive coalgebra* by Capretta et al. [100] (cf. (7.2)). So the dual statement of Theorem 7.3.14 states that the initial algebra is, equivalently, the terminal parametrically recursive coalgebra.

Remark 7.4.5. Furthermore, dually to Theorem 7.3.16, a cofree parametrically recursive coalgebras on an object Y is precisely the same an initial algebra for $F(-) \times Y$. So assuming that such an initial algebra CY exists for every object Y, we conclude that C is the object assignment of a comonad on the base category $\mathscr{A}$ (dually to what we have seen in Remark 7.3.23(2)). This was shown by Ghani et al. [136] and by Uustalu and Vene [300]. Moreover, they call C the cofree corecursive monad on F [301], owing to its characterization by the universal property dual to the one characterizing the monad T given by free cias. The comonad C also appears more recently in the treatment of course-of-value recursion by Hinze et al. [163, Sec. 5.4].

Every parametrically recursive coalgebra is recursive; this is dual to Remark 7.3.3(1). However, in general the converse fails:

Example 7.4.6 [34]**.** Let $R\colon \mathsf{Set} \to \mathsf{Set}$ be the functor defined by

$$RX = \{(x, y) \in X \times X : x \neq y\} + \{d\}$$

for sets X, and for a function $f\colon X \to Y$ put

$$Rf(d) = d \quad \text{and} \quad Rf(x, y) = \begin{cases} d & \text{if } f(x) = f(y), \\ (f(x), f(y)) & \text{else.} \end{cases}$$

The following coalgebra (C, γ) is recursive: let $C = \{0, 1\}$ and define $\gamma\colon C \to RC$ by $\gamma(0) = \gamma(1) = (0, 1)$. Indeed, for every algebra $\alpha\colon RA \to A$ the constant map $h\colon C \to A$ with value $\alpha(d)$ is the unique coalgebra-to-algebra morphism.

However, (C, γ) is not parametrically recursive. To see this, consider any morphism $e\colon RX \times \{0, 1\} \to X$ such that RX contains more than one pair (x_0, x_1), $x_0 \neq x_1$ with $e((x_0, x_1), i) = x_i$ for $i = 0, 1$. Then each such pair yields $h\colon C \to X$ with $h(i) = x_i$ making (7.2) commutative. Thus, (C, γ) is not parametrically recursive.

The situation in Example 7.4.6 is relatively rare and artificial because for functors preserving inverse images, recursive and parametrically recursive coalgebras coincide (see Corollary 8.6.2 and Corollary 8.6.10). We recall that finite intersections are pullbacks of cospans of monomorphisms, and inverse images are pullbacks of a morphism along a monomorphism.

Proposition 7.4.7. *If a coalgebra (C, γ) has the property that all of its subcoalgebras are recursive, then $(FC, F\gamma)$ has that property, too, provided that*

(1) *F preserves inverse images, or*

(2) *F preserves finite intersections and γ is monic.*

Proof Let $s\colon (S, \sigma) \rightarrowtail (FC, F\gamma)$ be a subcoalgebra and $\alpha\colon FA \to A$ be any algebra.

Existence of a coalgebra-to-algebra morphism from (S, σ) to (A, α) does not need any assumption on F. This follows from the recursivity of $(FC, F\gamma)$ (see Example 7.4.3(3)): we have a coalgebra-to-algebra morphism $h\colon FC \to A$, and therefore $h \cdot s\colon S \to A$ is a coalgebra-to-algebra morphism.

In order to show the uniqueness we distinguish two cases.

(1) Let F preserve inverse images. Suppose that $k\colon S \to A$ is any coalgebra-to-algebra morphism. Now form the pullback of s and γ:

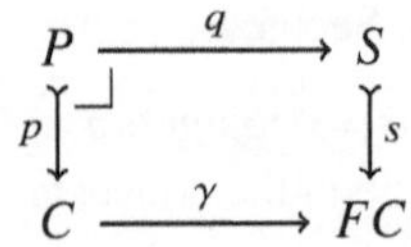

(7.12)

Note that $\gamma\colon (C, \gamma) \to (FC, F\gamma)$ clearly is a coalgebra homomorphism. Thus, since F preserves inverse images, there exists a unique coalgebra structure $\pi\colon P \to FP$ such that $p\colon (P, \pi) \rightarrowtail (C, \gamma)$ and $q\colon (P, \pi) \to (S, \sigma)$ are homomorphisms (cf. Proposition 4.1.5):

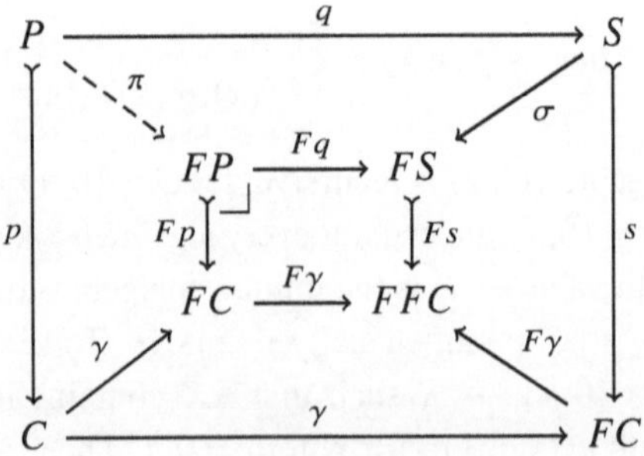

Thus, $k \cdot q$ is a coalgebra-to-algebra morphism from (P, π) to (A, α), and so is $h \cdot s \cdot q$. By hypothesis, (P, π) is recursive since it is a subcoalgebra of (C, γ). Hence we have

$$k \cdot q = h \cdot s \cdot q. \tag{7.13}$$

Furthermore, since the right-hand square in the diagram below is a pullback by assumption (1) or (2) and its outside clearly commutes, we obtain the morphism π_0 such that the upper part and left-hand triangle of the diagram below commute:

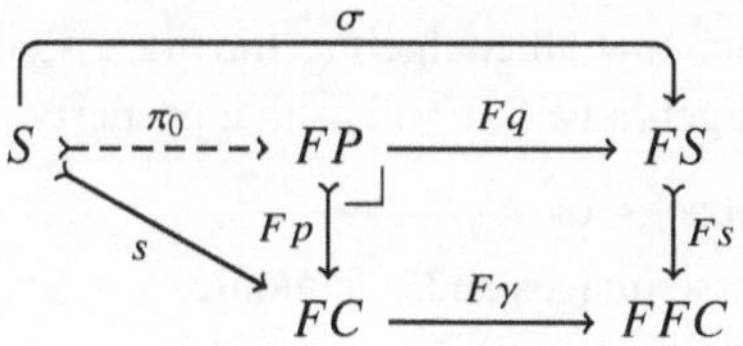

(7.14)

We conclude that

$$
\begin{aligned}
k &= \alpha \cdot Fk \cdot \sigma && \text{(since } k \text{ is a coalgebra-to-algebra morphism)}\\
&= \alpha \cdot Fk \cdot Fq \cdot \pi_0 && \text{(by the upper triangle in (7.14))}\\
&= \alpha \cdot F(k \cdot q) \cdot \pi_0 && \\
&= \alpha \cdot F(h \cdot s \cdot q) \cdot \pi_0 && \text{(by (7.13))}\\
&= \alpha \cdot F(h \cdot s) \cdot Fq \cdot \pi_0 && \\
&= \alpha \cdot F(h \cdot s) \cdot \sigma && \text{(by the upper triangle in (7.14))}\\
&= h \cdot s && \text{(since } h \cdot s \text{ is a coalgebra-to-algebra morphism).}
\end{aligned}
$$

(2) Let F preserve finite intersections and γ be monic. Then F preserves the pullback (7.12), and the rest of the proof is the same. □

In the following corollary we use the notion of a simple initial object 0, which means that the unique morphisms $0 \to X$ are monomorphisms (cf. Notation 7.2.6(4)). This result is essentially due to Urbat and Schröder [298, Prop. A.8].

Corollary 7.4.8. *Every subcoalgebra of a coalgebra $W_n \to FW_n$ in the initial-algebra ω-chain is recursive, provided that*

(1) *F preserves inverse images, or*

(2) *$\mathscr{A}$ has a simple initial object and F preserves finite intersections.*

Proof Let $w_{n,n+1}\colon W_n \to FW_n$, $n \in \mathbb{N}$, be the coalgebras in the initial-algebra ω-chain (cf. Definition 6.1.4).

The desired result follows easily by induction on n. For $n = 0$, the statement clearly holds since every subcoalgebra of $0 \to F0$ is carried by an initial object. Indeed, for every subobject $s\colon S \rightarrowtail 0$ we have $s \cdot u_S = \mathrm{id}_0$ for the unique morphism $u_S\colon 0 \to S$, which shows that s is a split epimorphism, whence an isomorphism.

For the induction step, use Proposition 7.4.7. In the case of assumption (2) we know that since F preserves finite intersections, it preserves monomorphisms.

Hence, using that 0 is simple, an easy induction shows that all coalgebra structures $w_{n,n+1}$ are monomorphisms. □

Remark 7.4.9. (1) The same proof as that for Proposition 7.4.7 shows that for a coalgebra (C, γ) such that all coalgebras having a homomorphism to (C, γ) are recursive, the coalgebra $(FC, F\gamma)$ has that property, too, provided that

(a) F preserves pullbacks, or

(b) F preserves inverse images and γ is monic.

(2) Similarly as in Corollary 7.4.8 one then obtains that for a category $\mathscr{A}$ having a simple initial object and a functor F preserving inverse images, every coalgebra with a homomorphism into a coalgebra $F^n 0 \to F(F^n 0)$ in the initial-algebra ω-chain is recursive.

(3) It is easy to extend the result in Corollary 7.4.8 to all subcoalgebras of coalgebras $(W_i \to FW_i)_{i \in \mathsf{Ord}}$ in the initial-algebra chain provided that $\mathscr{A}$ has universally smooth monomorphisms (see Definition 6.2.2 and Definition 8.4.10).

We conclude this section with a few examples explaining how recursive coalgebras capture familiar recursive function definitions as well as functional divide-and-conquer programs.

Example 7.4.10. (1) The set functor $FX = X + 1$ has unary algebras with a constant as algebras, and coalgebras for F may be identified with partial unary algebras. As we know from Example 2.2.7(3), the initial algebra is the set of natural numbers $\mathbb{N}$ with the structure given by the successor function and the constant 0. The inverse is the coalgebra given by the partial unary operation $n \mapsto n - 1$ (defined iff $n > 0$). This coalgebra is parametrically recursive. Hence every function

$$e = [u, v] \colon FX \times \mathbb{N} \cong X \times \mathbb{N} + \mathbb{N} \to X$$

defines a unique sequence $e^\dagger \colon \mathbb{N} \to X$ such that (7.2) commutes. Putting $x_n = e^\dagger(n)$, this means that $x_0 = v(0)$ and $x_{n+1} = u(x_n, n + 1)$. For example, the factorial function is then given by the choice $X = \mathbb{N}$; $u(n, m) = n \cdot m$ and $v(0) = 1$.

(2) For the set functor F given by $FX = X \times X + 1$, coalgebras $\gamma \colon C \times C + 1$ are deterministic systems with a state set C and a binary input function, and with halting states (expressed by $\gamma^{-1}(1)$).

The coalgebra $\mathbb{N}$ of natural numbers with halting states 0 and 1 and input structure $\gamma \colon n \mapsto (n - 1, n - 2)$ for $n \geq 2$ is parametrically recursive (see Example 8.5.6).

For example, to define the Fibonacci sequence starting with $a_0, a_1 \in \mathbb{N}$, consider the morphism $e\colon F\mathbb{N} \times \mathbb{N} \cong \mathbb{N}^3 + \mathbb{N} \to \mathbb{N}$ defined by

$$e(i,j,k) = i + j \quad \text{and} \quad e(n) = \begin{cases} a_0 & \text{if } n = 0, \\ a_1 & \text{if } n = 1, \\ 0 & \text{if } n \geq 2. \end{cases}$$

We know that there is a unique sequence $e^\dagger\colon \mathbb{N} \to \mathbb{N}$ such that the Diagram (7.2) commutes, which means that $e^\dagger(0) = a_0$, $e^\dagger(1) = a_1$ and $e^\dagger(n+2) = e^\dagger(n+1) + e^\dagger(n)$.

(3) Capretta et al. [100] have shown how to obtain Quicksort using parametric recursivity. Let A be any linearly ordered set (of data elements). Then Quicksort is usually defined as the recursive function $q\colon A^* \to A^*$ given by

$$q(\varepsilon) = \varepsilon \qquad \text{and} \qquad q(aw) = q(w_{\leq a}) \star (aq(w_{>a})),$$

where A^* is the set of all lists on A, ε is the empty list, $\star$ is the concatenation of lists, and $w_{\leq a}$ denotes the list of those elements of w which are less than or equal to a; analogously for $w_{>a}$.

Now consider the functor $FX = A \times X \times X + 1$ on Set, where $1 = \{\bullet\}$, and form the coalgebra $s\colon A^* \to A \times A^* \times A^* + 1$ given by

$$s(\varepsilon) = \bullet \qquad \text{and} \qquad s(aw) = (a, w_{\leq a}, w_{>a}) \qquad \text{for } a \in A \text{ and } w \in A^*. \tag{7.15}$$

Again, we shall see that this coalgebra is recursive in Example 8.5.6. Thus, for the F-algebra $m\colon A \times A^* \times A^* + 1 \to A^*$ given by

$$m(\bullet) = \varepsilon \qquad \text{and} \qquad m(a, w, v) = w \star (av),$$

there exists a unique function q on A^* such that $q = m \cdot Fq \cdot s$.

Notice that the last equation reflects the idea that Quicksort is a divide-and-conquer algorithm. The coalgebra structure s divides a list into two parts $w_{\leq a}$ and $w_{>a}$. Then Fq sorts these two smaller lists, and finally in the combine (or conquer) step, the algebra structure m merges the two sorted parts to obtain the desired whole sorted list.

Similarly, functions defined by parametric recursivity (cf. Diagram (7.2)) can be understood as divide-and-conquer algorithms, where the combine step is allowed to access the original parameter additionally. For instance, in the current example the divide step $\langle s, id_{A^*}\rangle$ produces the pair consisting of $(a, w_{\leq a}, w_{>a})$ and the original parameter aw, and the combine step, which is given by an algebra $FX \times A^* \to X$ will, by the commutativity of (7.2), get aw as its right-hand input.

Jeannin et al. [175, Sec. 4] provide a number of recursive functions arising in programming that are determined by recursivity of a coalgebra, for example, the Euclidean algorithm for the gcd of integers, Ackermann's function, and the Towers of Hanoi.

7.5 The Initial-Algebra Theorem Revisited

We conclude this chapter with a new proof of Theorem 6.3.4 using the topic of this section, recursive coalgebras. While the proof works for a smooth class $\mathcal{M}$, we present it for smooth monomorphisms only since this makes for easier reading.

We need several preliminary results which are also used in later chapters, especially Chapter 8. We begin with an extension of Scott induction (Corollary 3.1.3) to monotone functions on chain-complete posets. It easily follows from the proof of Theorem 6.1.1.

Corollary 7.5.1. *Let P be a chain-complete poset, and let S be a subset which is closed under joins of chains. If a monotone function $f: P \to P$ preserves S (in symbols: $f[S] \subseteq S$), then $\mu f \in S$.*

Indeed, S contains $\bot$, the join of the empty chain. By transfinite induction we have $f^i(\bot) \in S$ for every ordinal i.

In Theorem 6.3.4 we proved that, under hypotheses, the existence of a pre-fixed point implies the existence of an initial algebra. We now present an alternative proof that does not use transfinite recursion. Instead, it uses fixed points and recursive coalgebras. Moreover, the assumption that monomorphisms are smooth in that theorem can be slightly weakened as follows. The notion of smooth subobjects of an object is the case of Definition 6.2.2(1) where $\mathcal{M}$ is the class of all monomorphisms.

Theorem 7.5.2. *Let $m: FA \rightarrowtail A$ be a pre-fixed point of a functor F preserving monomorphisms. If A has smooth subobjects, then F has an initial algebra. Moreover, μF is a subalgebra of (A, m).*

Proof We have the following endomap

$$f: \mathsf{Sub}(A) \to \mathsf{Sub}(A) \text{ defined by } f(B \overset{u}{\rightarrowtail} A) = (FB \overset{Fu}{\rightarrowtail} FA \overset{m}{\rightarrowtail} A).$$

This map is clearly monotone. We are going to apply Corollary 7.5.1 to it. We take the subset $S \subseteq \mathsf{Sub}(A)$ of all $u: B \rightarrowtail A$ such that $u \le fu$ via some recursive coalgebra. More precisely,

$$S = \{u: B \rightarrowtail A : u = m \cdot Fu \cdot \beta \text{ where } \beta: B \rightarrowtail FB \text{ is recursive}\}.$$

Note that if β exists for u, then it is unique. Moreover, $u \in S$ is a coalgebra-to-algebra morphism from (B, β) to (A, m).

The set S is closed under f since $(FB, F\beta)$ is a recursive coalgebra by Example 7.4.3(3) and since for $u \in S$ we have

$$f(u) = m \cdot Fu = m \cdot F(m \cdot Fu \cdot \beta) = m \cdot F(f(u)) \cdot F\beta.$$

We now prove that S is closed under joins of chains. Given an ordinal λ and a chain $u_i\colon B_i \rightarrowtail A$ $(i < \lambda)$ in S with recursive coalgebras $\beta_i\colon B_i \rightarrowtail FB_i$ witnessing $u_i \le f(u_i)$, we show that these recursive coalgebras form a chain. Indeed, suppose that $v_{i,j}\colon B_i \rightarrowtail B_j$ witnesses $u_i \le u_j$; that is, we have $u_j \cdot v_{i,j} = u_i$. Then we have the following diagram:

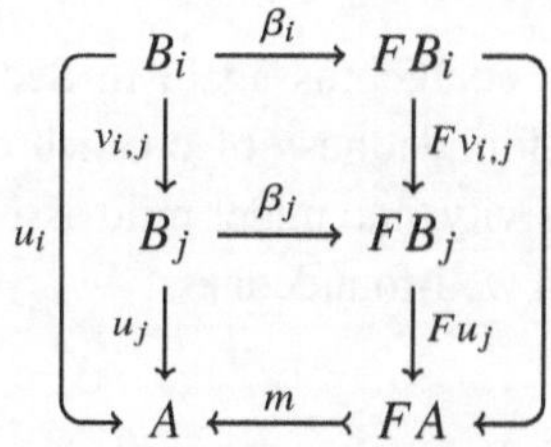

Since the outside, the lower square, and the left-hand and right-hand parts commute, we see that the upper square commutes when extendend by the monomorphism $m \cdot Fu_j$. Thus it commutes, proving that $v_{i,j}$ is a coalgebra homomorphism.

Denote by $u\colon B \rightarrowtail A$ the join of the chain (u_i) in $\mathsf{Sub}(A)$. Since A has smooth subobjects, it is given by the colimit of the chain of the objects B_i and morphisms $v_{i,j}$ in $\mathscr{A}$ (cf. Definition 6.2.2(1)). By Example 7.4.3(4), we obtain a canonical structure $\beta\colon B \to FB$ of a recursive coalgebra. Moreover, since $u_i \in S$ we know that u_i is the (unique) coalgebra-to-algebra morphism from (B_i, β_i) to (A, m), and u is the unique morphism induced by the cocone formed by the u_i. Thus, u is the unique coalgebra-to-algebra morphism from (B, β) to (A, m) (cf. the dual of the proof of Lemma 7.3.11(2)). In symbols, we have $u = m \cdot Fu \cdot \beta$, which states that u lies in S, as desired.

By Theorem 6.1.1, f has a least fixed point, and by Corollary 7.5.1, $\mu f \in S$. We write this subobject as $u\colon I \rightarrowtail A$. We know that u and $f(u) = m \cdot Fu$ represent the same subobject of A. Since $u \in S$, this is witnessed by a recursive coalgebra structure $\iota\colon I \to FI$, which is an isomorphism. Thus (I, ι^{-1}) is an initial algebra, by Proposition 7.4.2(2). □

Corollary 7.5.3. *Let $m\colon FA \rightarrowtail A$ be a pre-fixed point of a functor F preserving monomorphisms on a complete and well-powered category with smooth monomorphisms. Then the following hold:*

(1) *the functor F has an initial algebra which is a subalgebra of (A, α),*

(2) *if (A, α) is corecursive, then F has a terminal coalgebra which is a subalgebra of (A, α).*

Indeed, item (1) follows from Theorem 7.5.2 while in item (2) we restate Proposition 7.2.8 for contrast.

Remark 7.5.4. In the proof of Theorem 6.3.4 we have worked, more generally, with a class $\mathcal{M}$ of monomorphisms. Theorem 7.5.2 also holds in that setting: in lieu of $\mathsf{Sub}(A)$ we work with the collection $\mathsf{Sub}_{\mathcal{M}}(A)$ of all subobjects of A represented by monomorphisms in $\mathcal{M}$ (cf. Definition 6.2.2). We conclude that if F preserves $\mathcal{M}$ and A is an $\mathcal{M}$-pre-fixed point having smooth $\mathcal{M}$-subobjects, then F has an initial algebra.

We shall meet recursive coalgebras again in Section 8.5, where we see that the condition of well-foundedness of coalgebras studied in Chapter 8 implies parametric corecursivity, and under mild assumptions, the parametric corecursivity coincides with well-foundedness.

7.6 Summary

Our theme in this chapter was characterizations of the initial algebra as a terminal object in a category of coalgebras and of the terminal coalgebra as an initial object in a category of algebras.

We have presented corecursive and completely iterative algebras (cias), which allow for the unique solution of recursive equation systems, in other words, they allow for the interpretation of definitions by structured corecursion. The terminal coalgebra turns out to be the initial corecursive algebra as well as the initial cia. Dually, the initial algebra is the initial (parametrically) recursive coalgebra. The latter notions allow for the unique interpretation of definitions by structured recursion.

Finally, we have applied recursive coalgebras in order to obtain a new proof of the Initial-Algebra Theorem.

8
Well-Founded Coalgebras

In the last chapter we have seen recursive coalgebras, which have the property that functions out of them may be specified by structured recursion. This is a desirable property, but it may not be straightforward to establish in general. In this chapter we consider *well-foundedness* of coalgebras, which captures well-founded induction and is usually much easier to verify for a given coalgebra. As mentioned at the beginning of Section 7.4, recursive and well-founded coalgebras were first studied by Osius [245] in connection with categorical set theory. He studied them for the power-object functor on an elementary topos.

Taylor [286, 287] took Osius' ideas much further. He introduced well-founded coalgebras for a general endofunctor, capturing the notion of a well-founded relation categorically, and considered recursive coalgebras under the name 'coalgebras obeying the recursion scheme'. He then proved the General Recursion Theorem that all well-founded coalgebras are (parametrically) recursive for every endofunctor on sets (and on more general categories) preserving inverse images. In subsequent work the latter condition was lifted [53] (see also the later preprint [288]).

We will present a proof of Taylor's General Recursion Theorem for endofunctors preserving monomorphisms on a complete and well-powered category having smooth monomorphisms (cf. Definition 6.2.2). For the category of sets, this implies that 'well-founded $\Rightarrow$ recursive' holds for all endofunctors. We then discuss the converse: is every recursive coalgebra well-founded? Here the assumption that the endofunctor preserves inverse images cannot be lifted, and one needs additional assumptions. In fact, we present two proofs: one assumes the category has universally smooth monomorphisms and the functor has a pre-fixed point (see Theorem 8.6.1). Under these assumptions we also give a new equivalent characterization of recursiveness and well-foundedness of a coalgebra, namely that the coalgebra admits a coalgebra-to-algebra morphism into the initial algebra (which exists under our assumptions), see Corollary 8.6.2.

This characterization was previously established for finitary functors on sets [34]. The other proof of the converse mentioned above is due to Taylor [286], and it uses the concept of a subobject classifier (Theorem 8.6.8). It implies that 'recursive' and 'well-founded' are equivalent concepts for all set functors preserving inverse images. We also prove that a similar result holds for the category of vector spaces over a fixed field (Corollary 8.6.14).

Finally, we shall see that for every set functor, the initial algebra is characterized as the terminal well-founded coalgebra. This is connected to the theme of Chapter 7, where we have seen in Corollary 7.4.4 that the initial algebra is the terminal (parametric) recursive coalgebra.

In Chapter 9 we will see that well-founded coalgebras lead to a surprising description of the initial algebra for a set functor preserving intersections.

On a more technical note, we shall once again make use of *transfinite iteration* that we already studied in Chapter 6. Here we make use of the induction principle in Corollary 7.5.1, when we prove the General Recursion Theorem 8.5.3.

8.1 Well-Founded Coalgebras and Well-Founded Graphs

The concept of well-foundedness is well-known for directed graphs: it means that the graph has no infinite directed paths. For relations, the concept is similar: a relation is well-founded if its *converse* has no infinite directed paths (Example 7.4.3(1)). For example, the membership relation $\in$ of set theory is well-founded; this is precisely the Foundation Axiom.

Taylor [287, Def. 6.2.3] gave a more general category-theoretic formulation of well-foundedness. His definition can be presented based on the concept of subcoalgebra; recall that this means a coalgebra homomorphism into the given coalgebra carried by a monomorphism in the base category (see Remark 2.4.12(1)). For example, consider a graph as a coalgebra $\alpha\colon A \to \mathcal{P}A$ for the power-set functor (see Example 2.4.2(4)). A subcoalgebra is a subset $m\colon B \hookrightarrow A$ such that with every vertex it contains all the neighbours. The coalgebra structure $\beta\colon B \to \mathcal{P}B$ is then the domain–codomain restriction of α. A subcoalgebra $m\colon (B,\beta) \hookrightarrow (A,\alpha)$ *cartesian* if the square below expressing that m is a homomorphism is a pullback:

$$\begin{array}{ccc} B & \xrightarrow{\ \beta\ } & \mathcal{P}B \\ {\scriptstyle m}\big\downarrow & \lrcorner & \big\downarrow{\scriptstyle \mathcal{P}m} \\ A & \xrightarrow{\ \alpha\ } & \mathcal{P}A \end{array}$$

(Being a pullback is indicated by the 'corner' symbol.) This means that

$$\text{whenever a vertex of } A \text{ has all neighbours in } B, \text{ it also lies in } B. \tag{8.1}$$

This implies that a graph is well-founded iff no proper subcoalgebra is cartesian. That is, whenever the above square is a pullback, m is an isomorphism.

Definition 8.1.1. Let $\alpha\colon A \to FA$ be a coalgebra.

(1) A subcoalgebra $m\colon (B,\beta) \rightarrowtail (A,\alpha)$ is *cartesian* provided that the square below is a pullback.

$$\begin{array}{ccc} B & \xrightarrow{\ \beta\ } & FB \\ {\scriptstyle m}\big\downarrow & \lrcorner & \big\downarrow{\scriptstyle Fm} \\ A & \xrightarrow{\ \alpha\ } & FA \end{array} \tag{8.2}$$

(2) (A,α) is *well-founded* if whenever $m\colon (B,\beta) \rightarrowtail (A,\alpha)$ is a cartesian subcoalgebra, then m is an isomorphism.

Lemma 8.1.2 [287, Example 6.3.3]. *The well-founded coalgebras for the power-set $\mathscr{P}$ functor are precisely the graphs without infinite paths.*

Proof Let $\alpha\colon A \to \mathscr{P}A$ be a graph without any infinite paths. Given a cartesian subcoalgebra $m\colon (B,\beta) \hookrightarrow (A,\alpha)$, we prove that $B = A$ by contradiction. Every node $a_0 \in A \setminus B$ has, by (8.1), a neighbour $a_1 \in A \setminus B$. By the same argument, a_1 has neighbour $a_2 \in A \setminus B$, etc. This yields an infinite path $a_0, a_1, a_2, \ldots$, which is a contradiction.

Conversely, if (A,α) is a well-founded coalgebra, then it has no infinite path. Indeed, the subset $B \subseteq A$ of all vertices that do not lie on an infinite path satisfies (8.1). Thus, B is a cartesian subcoalgebra of A, whence $B = A$. □

Remark 8.1.3. Recall from Example 7.4.3(1) that a binary relation $R \subseteq X \times X$ can be represented as a coalgebra for $\mathscr{P}$ by defining $\alpha\colon X \to \mathscr{P}X$ by $\alpha(x) = \{y \in X \mid y\,R\,x\}$. This is the coalgebra representing the graph (X, R^{op}). It is well-founded iff it is well-founded as a relation, which means that there is no sequence $x_0, x_1, x_2, \ldots$ of elements of A such that $x_{n+1}\,R\,x_n$ for all $n \in \mathbb{N}$.

Example 8.1.4. (1) If an initial algebra μF exists, then (considered as a coalgebra) it is well-founded. Indeed, in every pullback (8.2), since α is invertible, so is β. The unique algebra homomorphism from μF to the algebra $\beta^{-1} : FB \to B$ is clearly inverse to m.

(2) If a set functor F fulfils $F\emptyset = \emptyset$, then the only well-founded coalgebra is the empty one; this follows from the fact that the empty coalgebra is a cartesian subcoalgebra of every coalgebra for F.

For example, a deterministic automaton, as a coalgebra for $FX = \{0,1\} \times X^{\Sigma}$ (see Example 2.4.2(3)), is well-founded iff it is empty.

(3) Non-deterministic automata were discussed in Example 5.1.27 as coalgebras for an endofunctor on Rel. But they can also be considered as coalgebras for the set functor $FX = \{0,1\} \times (\mathscr{P}X)^{\Sigma}$ (see Example 2.4.2(5)). Recall that the *state transition graph* of an automaton has the states as its nodes, and an edge leads from a state to each of its successor states. An F-coalgebra is well-founded iff the state transition graph of the corresponding non-deterministic automaton is well-founded (i.e. has no infinite path). This follows immediately from Corollary 8.2.10 below.

(4) We have discussed weighted automata in Example 2.4.6 as coalgebras for endofunctors $FX = \mathbb{S} \times X^{\Sigma}$, where $\mathbb{S}$ is a semiring and the underlying category is the category of $\mathbb{S}$-semimodules. For the case that $\mathbb{S}$ is a field K, this category K-Vec is the category of vector spaces over K and linear functions, and we speak of *linear weighted automata*. We characterize the well-founded coalgebras for $FX = K \times X^{\Sigma}$ in Proposition 8.3.8: a linear weighted automaton is a well-founded coalgebra iff every path in its state transition graph eventually leads to 0.

In the next result we assume that our functor preserves inverse images, that is, which are pullbacks of a morphism and a monomorphism. (See also Remark C.3.7.)

Proposition 8.1.5. *Let $\mathscr{A}$ have and $F\colon \mathscr{A} \to \mathscr{A}$ preserve inverse images, and let $\alpha\colon A \to FA$ be a coalgebra.*

(1) *If (A, α) is a well-founded, then so is $(FA, F\alpha)$.*

(2) *If every subcoalgebra of (A, α) is well-founded, then $(FA, F\alpha)$ has the same property.*

Proof (1) Let $s\colon (S, \sigma) \to (FA, F\alpha)$ be a cartesian subcoalgebra; we prove that s is invertible. Form its inverse image under α:

$$\begin{array}{ccccc} T & \xrightarrow{\ q\ } & S & \xrightarrow{\ \sigma\ } & FS \\ {\scriptstyle t}\downarrow & \lrcorner & {\scriptstyle s}\downarrow & \lrcorner & \downarrow{\scriptstyle Fs} \\ A & \xrightarrow{\ \alpha\ } & FA & \xrightarrow{\ F\alpha\ } & FFA \end{array}$$

Since F preserves inverse images, we see that Ft is the inverse image of Fs under $F\alpha$. In other words, the pullback square on the right above can be chosen as follows:

$$\begin{CD} FT @>{Fq}>> FS \\ @V{Ft}VV @VV{Fs}V \\ FA @>{F\alpha}>> FFA \end{CD}$$

This means that $s = Ft$ as subobjects of FA. Hence T carries a cartesian subcoalgebra of (A, α). Due to the well-foundedness of (A, α) we conclude that t is an isomorphism, thus so is $Ft = s$, as required.

(2) Suppose that every subcoalgebra of (A, α) is well-founded, and consider a subcoalgebra $m\colon (B, \beta) \rightarrowtail (FA, F\alpha)$; we prove that (B, β) is well-founded, too. Form the inverse image (P, p, q) of m under α:

$$\begin{CD} P @>{q}>> B \\ @V{p}VV @VV{m}V \\ A @>{\alpha}>> FA \end{CD}$$

Using that F preserves inverse images, we then obtain the following commutative diagram, where we obtain the dashed arrow β' using the universal property of the right-hand pullback:

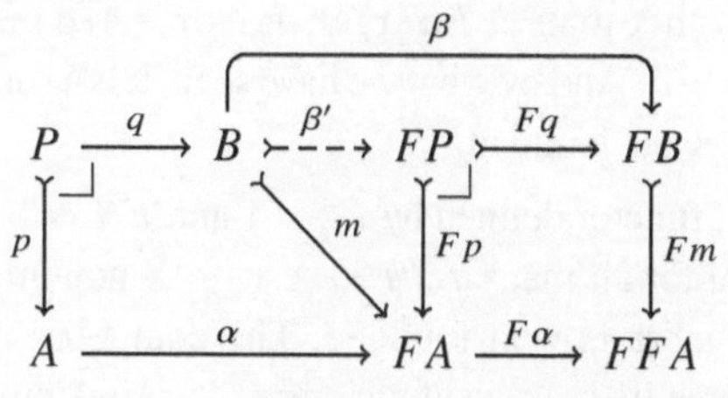

We see that $(P, \beta' \cdot q)$ is a subcoalgebra of (A, α) and is therefore well-founded by hypothesis.

Now consider a cartesian subcoalgebra $s\colon (S, \sigma) \rightarrowtail (B, \beta)$. We prove that s is an isomorphism. Take its inverse image under $q\colon P \to B$:

$$\begin{CD} T @>{r}>> S \\ @V{t}VV @VV{s}V \\ P @>{q}>> B \end{CD}$$

Note that since $\beta = Fq \cdot \beta'$, we can obtain s as the pullback of Fs first along β' and further along Fq. Using that F preserves inverse images, we thus obtain the following commutative diagram (with a uniquely determined morphism σ'):

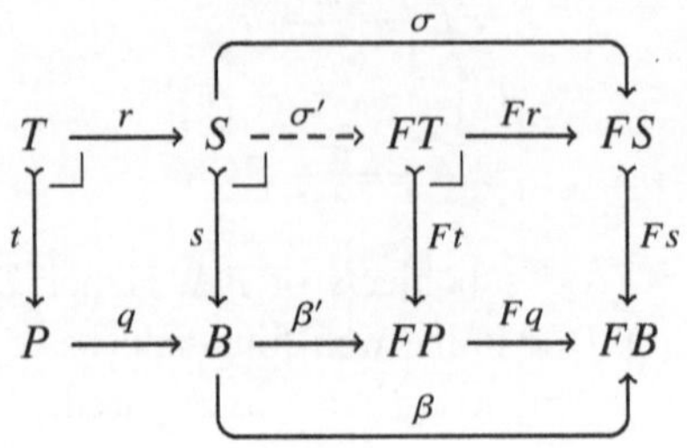

This shows that $(T, \sigma' \cdot r)$ is a cartesian subcoalgebra of $(P, \beta' \cdot q)$. Thus, t is an isomorphism by well-foundedness of $(P, \beta' \cdot q)$. Therefore so are Ft and its inverse image s under β'. This completes the proof. □

Example 8.1.6. Without the preservation of inverse images, the statements of Proposition 8.1.5 may fail.

(1) Let R be the set functor from Example 7.4.6, and let $FX = R(X + 1)$. Take $A = 1 = \{*\}$, and let $\alpha \colon A \to FA$ be $\alpha(*) = d$. Note that FA has three points, say a, b, and d. Aside from (A, α) itself, the only subcoalgebra is $! \colon \emptyset \to F(\emptyset)$. This last is not cartesian, since $\alpha(d) = F!(d)$. So (A, α) is well-founded. Consider $(FA, F\alpha)$. We have $F\alpha(d) = d$, by definition of R, but for all $x \in FA$ other than d, $F\alpha(x) \neq d$. Now (A, α) is a subcoalgebra of $(FA, F\alpha)$ via $m(*) = d$. And by what we have seen, it is a cartesian subcoalgebra. So $(FA, F\alpha)$ is not well-founded.

(2) Let F be the set functor defined by $F\emptyset = 1$ and $FX = 1 + 1$ for all nonempty sets X and $Ff = \mathsf{inl}$ for all maps $f \colon \emptyset \to X$ with X nonempty.

The empty coalgebra is well-founded. The coalgebra $\mathsf{inr} \colon 1 \to F1$ is not well-founded because its empty subcoalgebra is cartesian. On the other hand, consider $\mathsf{inl} \colon 1 \to F1$. It and all its subcoalgebras are well-founded: these are itself and the empty coalgebra (not cartesian). Applying F to the last coalgebra gives $\mathsf{id} \colon 1 + 1 \to 1 + 1$. This coalgebra is the initial algebra and is thus well-founded, but it has a subcoalgebra which is not, namely $\mathsf{inr} \colon 1 \to F1$ via the embedding inr.

These examples show that a well-founded coalgebra for a set functor can have non-well-founded subcoalgebras, and also that even if all subcoalgebras of a given coalgebra are well-founded, the same need not hold after applying the functor. But both of these are very rare; see Proposition 8.1.5 just below.

Remark 8.1.7. (1) Observe that when $\alpha \colon A \rightarrowtail FA$ in Proposition 8.1.5 is a monomorphism, we only need to assume that $\mathscr{A}$ has and F preserves finite intersections since all the pullback squares are then intersections, that is, pullbacks of two monomorphisms.

(2) We obtain a version of Corollary 7.4.8 for well-founded coalgebras as follows. Suppose that, in addition to the above assumptions, $\mathscr{A}$ has a simple initial object, that is, the unique morphisms $0 \to X$ are monomorphisms. Then it follows from Proposition 8.1.5 that every subcoalgebra of a coalgebra $F^n 0 \to F(F^n 0)$ in the initial-algebra ω-chain for F is well-founded. This is easy to see by induction. Indeed, for the unique coalgebra $0 \to F0$ this clearly holds since every subcoalgebra yields a subobject $m\colon S \rightarrowtail 0$, and we have $m \cdot u = \mathrm{id}_0$ for the unique morphism $u\colon 0 \to S$. Thus, m is a split epimorphism, whence an isomorphism. For the induction step use Proposition 8.1.5 and the observation in point (1).

We next show that to every coalgebra for a set functor F one may associate a graph, in a canonical way. Moreover, if F preserves intersections, then a coalgebra is well-founded if and only if so is its canonical graph.

Notation 8.1.8. Given a set functor F, we define for every set X the map $\tau_X\colon FX \to \mathscr{P}X$ assigning to every element $x \in FX$ the intersection of all subsets $m\colon M \hookrightarrow X$ such that x lies in the image of Fm:

$$\tau_X(x) = \bigcap \{m \mid m\colon M \hookrightarrow X \text{ satisfies } x \in Fm[FM]\}. \tag{8.3}$$

Definition 8.1.9. Let F be a set functor. For every coalgebra $\alpha\colon A \to FA$ its *canonical graph* is the following coalgebra for $\mathscr{P}$:

$$A \xrightarrow{\alpha} FA \xrightarrow{\tau_A} \mathscr{P}A.$$

Example 8.1.10. (1) Given a graph as a coalgebra $\alpha\colon A \to \mathscr{P}A$, the condition $\alpha(x) \in \mathscr{P}m[\mathscr{P}M]$ states precisely that all successors of x lie in the set M. The least such set is $\alpha(x)$. Therefore, the canonical graph of (A, α) is itself (see [287, Example 6.3.3]).

(2) For the type functor of $FX = \{0, 1\} \times X^\Sigma$ of deterministic automata, we have

$$\tau_X(i, t) = \{t(s) : s \in \Sigma\} \qquad \text{for } i = 0, 1 \text{ and } t\colon \Sigma \to X.$$

Thus, the canonical graph of a deterministic automaton A is precisely its state transition graph (forgetting the labels of transitions and the finality of states): we have an edge (a, a') iff $a' = \delta(a, s)$ for some $s \in \Sigma$, where δ is the next state function of A.

Similarly, for the type functor $FX = \{0, 1\} \times (\mathscr{P}X)^\Sigma$ of non-deterministic automata we have

$$\tau_X(i, g) = \bigcup_{s \in \Sigma} t(s) \qquad \text{for } i = 0, 1 \text{ and } t\colon \Sigma \to \mathscr{P}X,$$

and we again obtain the state transition graph, forgetting the labels.

(3) For the functor $FX = \mathscr{P}(\Sigma \times X)$ whose coalgebras are labelled transition systems we have

$$\tau_X = (\mathscr{P}(\Sigma \times X) \xrightarrow{\mathscr{P}\pi_X} \mathscr{P}X),$$

where $\pi_X\colon \Sigma \times X \to X$ is the projection. Again, the canonical graph of a labelled transition system is its state transition graph: (a, a') is an edge iff some action leads from state a to a'.

Recall that a functor *preserves intersections* if it preserves (wide) pullbacks of families of monomorphisms. For these functors the well-founded coalgebras will be proved to be precisely those with a well-founded canonical graph (Corollary 8.2.10).

Gumm [150, Theorem 7.3] observed that for a set functor preserving intersections, the maps $\tau_X\colon FX \to \mathscr{P}X$ in (8.3) form a 'subnatural' transformation from F to the power-set functor $\mathscr{P}$. Subnaturality means that (although these maps do not form a natural transformation in general) for every monomorphism $i\colon X \to Y$ we have a commutative square:

$$\begin{array}{ccc} FX & \xrightarrow{\tau_X} & \mathscr{P}X \\ {\scriptstyle Fi}\downarrow & \lrcorner & \downarrow{\scriptstyle \mathscr{P}i} \\ FY & \xrightarrow{\tau_Y} & \mathscr{P}Y \end{array} \qquad (8.4)$$

For such set functors this is even a pullback square:

Theorem 8.1.11 [150, Thm. 7.4] and [286, Prop. 7.5]**.** *A set functor F preserves intersections iff the above squares* (8.4) *are pullbacks.*

Remark 8.1.12. The condition that F preserves intersections is extremely mild for set functors. We discuss this further in Section C.2. Furthermore, the collection of set functors which preserve intersections is closed under products, coproducts, and compositions. A subfunctor $m\colon G \rightarrowtail F$ of an intersection preserving functor F preserves intersections whenever m is a *cartesian* natural transformation, that is, all naturality squares are pullbacks:

$$\begin{array}{ccc} GX & \overset{m_X}{\rightarrowtail} & FX \\ {\scriptstyle Gf}\downarrow & \lrcorner & \downarrow{\scriptstyle Ff} \\ GY & \underset{m_Y}{\rightarrowtail} & FY \end{array}$$

Observe that all the functors in Example 8.1.10 preserve intersections.

As we show next, 'almost all' finitary set functors preserve intersections. This calls on an important construction due to Trnková [292]. We have seen in Proposition 4.4.1 that every set functor preserves finite *nonempty* intersections. By changing a set functor only at the empty set, one can ensure preservation of finite intersections:

Proposition 8.1.13 [292]. *For every set functor F there exists an essentially unique set functor $\bar{F}$ which coincides with F on nonempty sets and functions and preserves finite intersections (whence monomorphisms).*

See Theorem C.6.6 for the proof.

Definition 8.1.14. We say that the functor $\bar{F}$ in Proposition 8.1.13 is the *Trnková hull* of F.

Note that it is the reflection of F into the full subcategory of set functors preserving finite intersections (see Section C.6 for details).

Example 8.1.15. (1) Let F be the functor in Example 8.1.6(2). Its Trnková hull is constant functor with value $1 + 1$.

(2) Let $\mathscr{P}'$ be the subfunctor of the power-set functor $\mathscr{P}$ consisting of all nonempty subsets. Its Trnková hull is $\mathscr{P}$.

Remark 8.1.16. Note that the Trnková hull $\bar{F}$ has the same coalgebras as F. More precisely, the categories $\mathrm{Coalg}\, F$ and $\mathrm{Coalg}\, \bar{F}$ are isomorphic: the initial coalgebra $0 \to F0$ of the former category is related to $0 \to \bar{F}0$ in the latter category, and the nonempty coalgebras are the same for F and $\bar{F}$. Similary, the Trnková hull has the same nonempty algebras as F.

Remark 8.1.17. If F is a finitary set functor, then so is its Trnková hull $\bar{F}$. This follows immediately from the characterization of finitary set functors in terms of finite boundedness in Proposition 4.3.4.

Proposition 8.1.18. *The Trnková hull of a finitary set functor preserves all intersections.*

The proof is a slightly simpler version of the proof of Theorem 4.4.3. In fact, in lieu of Proposition 4.4.1 one uses that $\bar{F}$ preserves finite intersections so that one need not worry about intersections which might be nonempty.

Theorem 8.1.19 [150, Thm. 8.1]. *Let F be a set functor which preserves inverse images and intersections. Then $\tau\colon F \to \mathscr{P}$ is a natural transformation.*

Example 8.1.20. To see that τ is not a natural transformation in general, we use the set functor R from Example 7.4.6. Let $X = \{0, 1\}$, $Y = \{0\}$, and $f\colon X \to Y$

the evident function. Then $(0, 1) \in FX$, and $\tau_X(0, 1) = X$. Further, $\mathscr{P}f(X) = Y$. But $Rf(0, 1) = d$, and $\tau_Y(d) = \emptyset$.

Example 8.1.21. (1) A coalgebra for the identity functor $FX = X$ on Set is a set A equipped with a function $\alpha\colon A \to A$. The canonical graph of (A, α) is the graph of the function α, which is the graph with edges $(a, \alpha(a))$ for all $a \in A$. Hence, (A, α) is well-founded iff it is empty (see Example 8.1.4(2)).

(2) For $FX = X + 1$ coalgebras are sets A equipped with a *partial* function $\alpha\colon A \to A$, and the canonical graph is the graph of α. This functor has many nonempty well-founded coalgebras: by Corollary 8.2.10 below a coalgebra $\alpha\colon A \to A + 1$ is well-founded iff its canonical graph given by $\{(a, \alpha(x)) : x \in A, \alpha(x) \in A\}$ is well-founded. For example, the initial F-algebra, considered as the coalgebra on $\mathbb{N}$ with the structure given by the partial function $n \mapsto n - 1$, for $n > 0$ (cf. Example 7.4.10(1)), is well-founded.

(3) Consider the functor $FX = X \times X + 1$ and a coalgebra $\alpha\colon A \to A \times A + 1$. The edges in its canonical graph are all of the pairs (a, a_1) and (a, a_2) such that $a \in A$ and $\alpha(a) = (a_1, a_2)$. For example, the coalgebra $(\mathbb{N}, \gamma)$ from Example 7.4.10(2) has the canonical graph with edge set $\{(n, n - 1), (n, n - 2) : n \geq 2\}$, which is clearly well-founded, and therefore so is the coalgebra.

Similarly, for the functor $FX = A \times X \times X + 1$, the coalgebra (A^*, s) in Example 7.4.10(3) is easily seen to be well-founded via its canonical graph. Indeed, this graph has for every list w one outgoing edge to the list $w_{\leq a}$ and one to $w_{>a}$, for every $a \in A$. Hence, this is a well-founded graph, and therefore (A^*, s) is a well-founded coalgebra.

(4) More generally, for a polynomial functor $H_\Sigma\colon \mathsf{Set} \to \mathsf{Set}$ associated to a finitary signature Σ, a coalgebra $\alpha\colon A \to \coprod_{n \in \mathbb{N}} \Sigma_n \times A^n$ has the canonical graph in which vertex $a \in A$ has an outgoing edge (a, a') iff $a' \in A$ occurs in some tuple $\alpha(a) \in \Sigma_n \times A^n$ for some $n < \omega$.

Thus, the coalgebra (A, α) is well-founded iff for every $a \in A$ its tree unfolding, that is, its image under the unique homomorphism $h : A \to \nu F$, is a finite Σ-tree.

In particular, if the signature Σ does not contain any constant symbols, then the only well-founded H_Σ-coalgebra is $A = \emptyset$.

For further use we now compare well-founded and recursive coalgebras for a given set functor F with those of its Trnková hull $\bar{F}$ (see Proposition 8.1.13). Since empty coalgebras are (trivially) well-founded and recursive, we can restrict ourselves to the nonempty ones. We have observed in Remark 8.1.16 that $\mathsf{Coalg}\, F$ and $\mathsf{Coalg}\, \bar{F}$ are isomorphic.

Lemma 8.1.22. *Let (A, α) be a nonempty coalgebra for a set functor F. If it is well-founded or (parametrically) recursive, then it also has those properties as a coalgebra for the Trnková hull $\bar{F}$.*

Proof (1) Let (A, α) be well-founded for F. Nonempty subcoalgebras of (A, α) for F and $\bar{F}$ coincide. Thus, we only need to show that the left-hand square below, where $r_X : \emptyset \to X$ denotes the empty map, is not a pullback:

$$\begin{array}{ccc} \emptyset & \xrightarrow{r_{\bar{F}\emptyset}} & \bar{F}\emptyset \\ {\scriptstyle r_A}\downarrow & & \downarrow{\scriptstyle \bar{F}r_A} \\ A & \xrightarrow{\alpha} & \bar{F}A = FA \end{array} \qquad \begin{array}{ccc} \emptyset & \xrightarrow{r_{F\emptyset}} & F\emptyset \\ {\scriptstyle r_A}\downarrow & & \downarrow{\scriptstyle Fr_A} \\ A & \xrightarrow{\alpha} & FA \end{array}$$

Since (A, α) is well-founded for F, we know that the right-hand square is not a pullback. Thus, there exist $a \in A$ and $x \in F\emptyset$ such that $\alpha(a) = Fr_A(x)$. In particular, we have $F\emptyset \neq \emptyset$. But then the two squares above are the same, so the one on the left is not a pullback, either.

(2) Let (A, α) be a nonempty recursive coalgebra for F. Given an algebra $e : \bar{F}X \to X$ we know that $X \neq \emptyset$, for otherwise the existence of a unique coalgebra-to-algebra morphism $A \to X$ would force A to be empty. But then the unique coalgebra-to-algebra morphism from (A, α) to (X, e) for F is also one for $\bar{F}$. □

8.2 The Next-Time Operator on Coalgebras

A compact characterization of well-foundedness can be given by using an operator that generalizes the semantics of the 'next-time' operator of temporal logics for non-deterministic (or even probabilistic) automata and transitions systems (see e.g. Manna and Pnüeli [221]). This operator is also strongly related to the algebraic semantics of modal logic, where one passes from a graph G to a function on $\mathscr{P}G$. Jacobs [170] defined and studied the 'next-time' operator on coalgebras for Kripke polynomial set functors. We generalize his definition to arbitrary functors, obtaining an operator $\bigcirc$ on the subobjects of the carrier of a coalgebra. In Section 8.4, we will see that a coalgebra is well-founded iff none of its proper subcoalgebras is a fixed point of $\bigcirc$ (Proposition 8.3.9).

Assumption 8.2.1. Throughout the remainder of this chapter we work with a category $\mathscr{A}$ which is complete and well-powered. We also assume that $F : \mathscr{A} \to \mathscr{A}$ preserves monomorphisms. (See Remark 7.2.5 for a weakening of the last assumption.)

Remark 8.2.2. Recall from Notation 7.2.6 that for every object A we have a poset $\mathsf{Sub}(A)$ of subobjects of A. Furthermore, since $\mathscr{A}$ is complete and well-powered we know that $\mathsf{Sub}(A)$ is a complete lattice: it is small since $\mathscr{A}$ is well-powered, and a meet of subobjects $m_i \colon A_i \rightarrowtail A$, $i \in I$, is their intersection, obtained by forming their wide pullback. For example, $\bot_A$ is the intersection of all subobjects of A.

It follows that $\mathsf{Sub}(A)$ has all joins as well.

We shall need the fact that forming inverse images, that is, pulling back along a morphism, is a right adjoint:

Notation 8.2.3. For every morphism $f \colon B \to A$ we have two operators:

(1) The inverse-image operator

$$\overleftarrow{f} \colon \mathsf{Sub}(A) \to \mathsf{Sub}(B),$$

assigning to every subobject $s \colon S \rightarrowtail A$ its inverse image under f obtained by the following pullback

$$\begin{array}{ccc} P & \longrightarrow & S \\ {\scriptstyle \overleftarrow{f}(s)}\downarrow & & \downarrow{\scriptstyle s} \\ B & \xrightarrow{\ f\ } & A \end{array}$$

(2) The (direct) image operator

$$\vec{f} \colon \mathsf{Sub}(B) \to \mathsf{Sub}(A),$$

assigning to every subobject $t \colon T \rightarrowtail B$ the image of $f \cdot t$ (see Remark 7.2.1(3)):

$$\begin{array}{ccc} T & \twoheadrightarrow & S \\ {\scriptstyle t}\downarrow & & \downarrow{\scriptstyle \vec{f}(t)} \\ B & \xrightarrow{\ f\ } & A \end{array}$$

Remark 8.2.4. (1) Let X and Y be posets, considered as categories. A monotone map $r \colon X \to Y$ is a right adjoint iff there exists a monotone map $\ell \colon Y \to X$ such that

$$\ell(y) \le x \quad \text{iff} \quad y \le r(x) \qquad \text{for every } x \in X \text{ and } y \in Y.$$

(2) If in addition X is a complete lattice, then r is a right adjoint iff it preserves meets. This in particular holds for $r \colon \mathsf{Sub}(B) \to \mathsf{Sub}(A)$. Indeed, the necessity follows since right adjoints preserve limits. For the sufficiency, suppose that r preserves intersections, and define $\ell \colon Y \to X$ by

$$\ell(y) = \bigwedge_{y \le r(x)} x \qquad \text{for every } y \in Y.$$

Then ℓ is clearly monotone, and for every $x \in X$, $y \in Y$ we have

$$\ell(y) \le x \qquad \text{iff} \qquad y \le r(x).$$

Thus, ℓ is the desired left adjoint of r.

Lemma 8.2.5. *For every morphism $f: B \to A$ we have an adjoint situation:*

$$\mathsf{Sub}(A) \underset{\overleftarrow{f}}{\overset{\vec{f}}{\underset{\longrightarrow}{\overset{\longleftarrow}{\bot}}}} \mathsf{Sub}(B).$$

In other words: $\vec{f}(t) \le s$ iff $t \le \overleftarrow{f}(s)$ for all subobjects $s: S \rightarrowtail A$ and $t: T \rightarrowtail B$.

Proof In order to see this, we consider the following diagram:

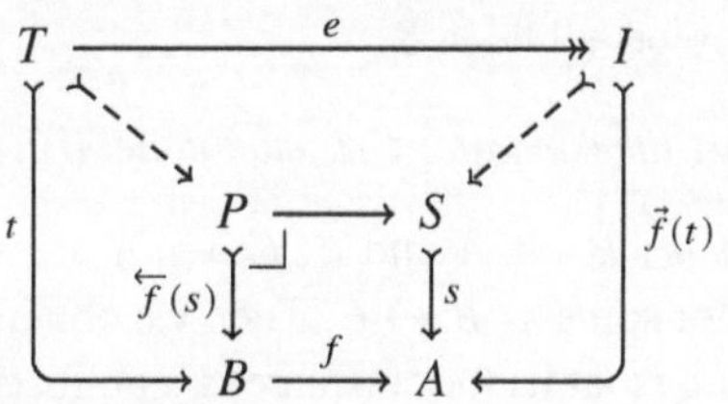

By the universal property of the lower middle pullback square and the diagonal fill-in property, we have the dashed morphism on the left iff we have the one on the right. Thus, $t \le \overleftarrow{f}(s)$ iff $\vec{f}(t) \le s$, as desired. □

We now come to the main definition in this section. It was introduced by Jacobs [170] for Kripke polynomial functors on sets. He presented a number of interesting properties and examples of the next-time operator.

Definition 8.2.6. Every coalgebra $\alpha: A \to FA$ induces the following endofunction $\bigcirc$ on $\mathsf{Sub}(A)$, called the *next-time operator*:

$$\bigcirc: \mathsf{Sub}(A) \to \mathsf{Sub}(A), \qquad \bigcirc(s) = \overleftarrow{\alpha}(Fs) \quad \text{for } s \in \mathsf{Sub}(A).$$

In more detail: for a subobject $s: S \rightarrowtail A$, we define $\bigcirc s$ and $\alpha(s)$ by the following pullback:

$$\begin{array}{ccc} \bigcirc S & \xrightarrow{\alpha(s)} & FS \\ {\scriptstyle \bigcirc s}\downarrow & \lrcorner & \downarrow{\scriptstyle Fs} \\ A & \xrightarrow{\alpha} & FA \end{array} \tag{8.5}$$

Since Fs is a monomorphism (Assumption 8.2.1), $\bigcirc s$ is a monomorphism and $\alpha(s)$ is (for every representation $\bigcirc s$ of that subobject of A) uniquely determined.

Example 8.2.7. (1) Let A be a graph, considered as a coalgebra for the functor $\mathscr{P}$: Set $\to$ Set. If $S \subseteq A$ is a set of vertices, then $\bigcirc S$ is the set of those vertices all of whose successors belong to S.

(2) Consider the set functor $FX = \mathscr{P}(\Sigma \times X)$ expressing labelled transition systems. The operator $\bigcirc$ for a coalgebra $\alpha\colon A \to FA$ is the semantic counterpart of the next time operator of classical linear temporal logic, see e.g. Manna and Pnüeli [221]. In fact, for a subset $S \subseteq A$ of a labelled transition system $\alpha\colon A \to \mathscr{P}(\Sigma \times A)$ and we have that $\bigcirc S$ consists of those states whose next states lie in S, in symbols:

$$\bigcirc S = \{x \in A \mid (s, y) \in \alpha(x) \text{ implies } y \in S, \text{ for all } s \in \Sigma\}.$$

(3) For the endofunctor $FX = K \times X^\Sigma$ on K-Vec whose coalgebras are linear weighted automata, given a subspace S of a linear automaton, $\bigcirc S$ consists of all states whose successors all lie in S.

Lemma 8.2.8. *The next-time operator* $\bigcirc$ *is monotone: if* $m \le n$, *then* $\bigcirc m \le \bigcirc n$.

Proof Suppose that $m\colon B \rightarrowtail A$ and $n\colon C \rightarrowtail A$ are subobjects such that $m \le n$, i.e. $n \cdot s = m$ for some $s\colon B \rightarrowtail C$. Then we obtain the dashed arrow in the diagram below using that its lower square is a pullback:

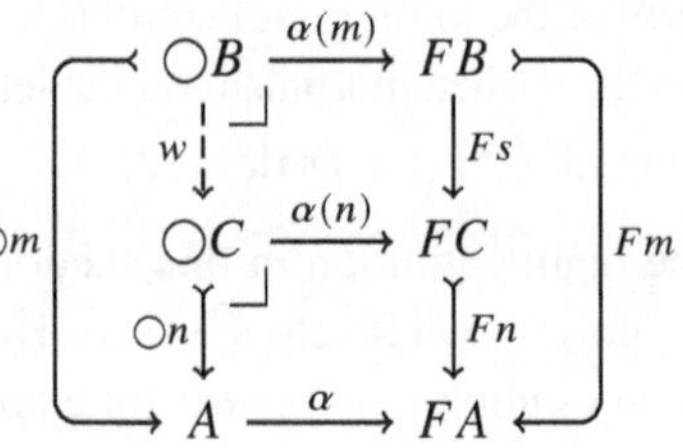

(8.6)

This shows that $\bigcirc m \le \bigcirc n$. □

Lemma 8.2.9. *For every set functor preserving intersections, the next-time operator of a coalgebra coincides with that of its canonical graph (see Definition 8.1.9).*

Proof In the diagram below the outside is a pullback if and only if so is the left-hand square:

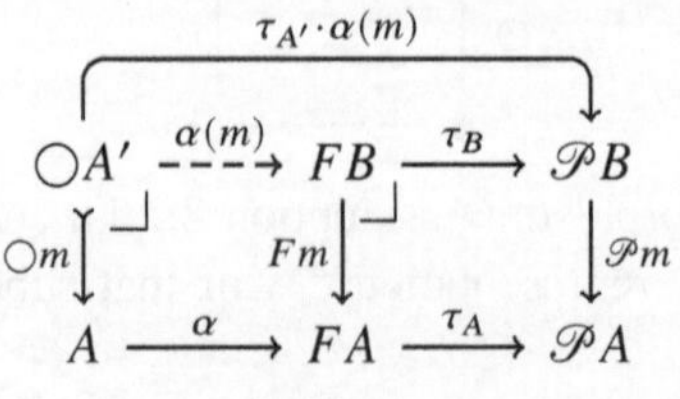

□

Corollary 8.2.10 [287, Rem. 6.3.4]. *A coalgebra for a set functor preserving intersections is well-founded iff its canonical graph is a well-founded graph.*

The following lemma will be useful when we establish the universal property of the well-founded part of a coalgebra in Section 8.3.

Lemma 8.2.11. *For every coalgebra homomorphism $f\colon (B,\beta) \to (A,\alpha)$ we have*

$$\bigcirc_\beta \cdot \overleftarrow{f} \le \overleftarrow{f} \cdot \bigcirc_\alpha,$$

where $\bigcirc_\alpha$ and $\bigcirc_\beta$ denote the next-time operators of the coalgebras (A,α) and (B,β), respectively.

Proof Let $s\colon S \rightarrowtail A$ be a subobject. We see that $\overleftarrow{f}(\bigcirc_\alpha s)$ is obtained by pasting two pullback squares as shown below:

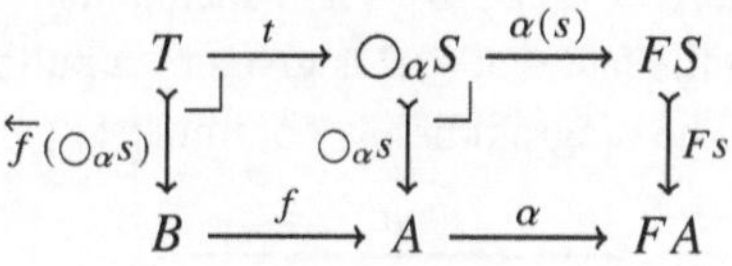

(8.7)

In order to show that $\bigcirc_\beta(\overleftarrow{f}(s)) \le \overleftarrow{f}(\bigcirc_\alpha s)$, we consider the following diagram:

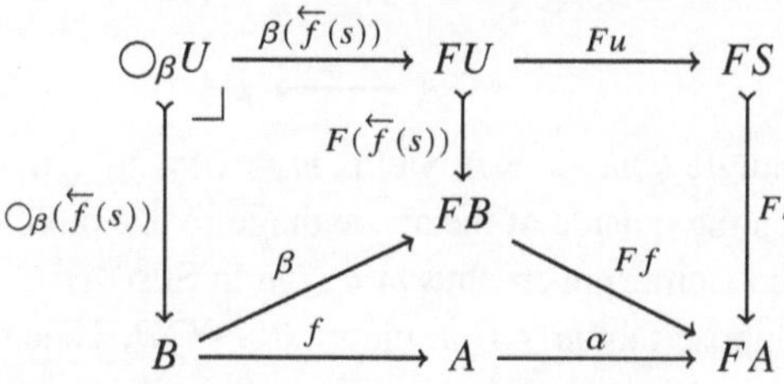

(8.8)

The left-hand part is the pullback square defining $\bigcirc_\beta(\overleftarrow{f}(s))$, and the right-hand one is that defining $\overleftarrow{f}(s)$, with F applied. On the bottom, we use that f is a coalgebra homomorphism. Thus, the outside of the diagram commutes. Since the outside of the diagram in (8.7) is a pullback, we have some $g\colon\ \bigcirc_\beta U \to T$ such that $\bigcirc_\beta(\overleftarrow{f}(s)) = \overleftarrow{f}(\bigcirc_\alpha(s)) \cdot g$, which proves the desired inequality. □

Corollary 8.2.12. *For every coalgebra homomorphism $f\colon (B,\beta) \to (A,\alpha)$ we have $\bigcirc_\beta \cdot \overleftarrow{f} = \overleftarrow{f} \cdot \bigcirc_\alpha$ provided that either*

(1) *f is a monomorphism in $\mathscr{A}$ and F preserves finite intersections, or*

(2) *F preserves inverse images.*

Proof In case (1), the right-hand part in Diagram (8.8) is clearly a pullback. In case (2), this holds because $F(\overleftarrow{f}(s)) = \overleftarrow{Ff}(Fs)$. Either way, pasting this part with the pullback in the upper left of (8.8) and using that the lower part commutes, we see that $\bigcirc_\beta(\overleftarrow{f}(s))$ is obtained by pulling back Fs along $f \cdot \alpha$. This implies the desired equality since this is how $\overleftarrow{f}(\bigcirc_\alpha s)$ is obtained (see (8.7)). □

Lemma 8.2.13. *Let $\alpha\colon A \to FA$ be a coalgebra and $m\colon B \rightarrowtail A$ be a monomorphism.*

(1) *There is a coalgebra structure $\beta\colon B \to FB$ for which m gives a subcoalgebra of (A, α) iff $m \le \bigcirc m$.*

(2) *There is a coalgebra structure $\beta\colon B \to FB$ for which m gives a cartesian subcoalgebra of (A, α) iff $m = \bigcirc m$.*

Proof We prove the left-to-right directions of both assertions first, and then the right-to-left ones.

Suppose first that there exists $\beta\colon B \to FB$ such that $m\colon (B, \beta) \rightarrowtail (A, \alpha)$ is a homomorphism. Then the fact that $\bigcirc B$ is given by a pullback yields the unique morphism x such that the diagram below commutes:

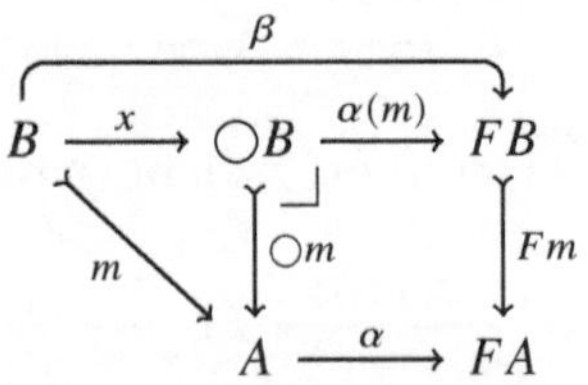

(8.9)

The left-hand triangle $\bigcirc m \cdot x = m$ yields $m \le \bigcirc m$. If (B, β) is a cartesian subcoalgebra, then the outside of the above diagram is a pullback square. This implies that x is an isomorphism, thus $m = \bigcirc m$ in $\mathsf{Sub}(A)$.

Conversely, suppose that $m \le \bigcirc m$ via $x\colon B \rightarrowtail \bigcirc B$. Then (8.9) shows that $\beta = \alpha(m) \cdot x\colon B \to FB$ is a coalgebra, and $m\colon B \rightarrowtail A$ is a homomorphism. If in addition $m = \bigcirc m$, that is, x is an isomorphism, we see that m is a cartesian subcoalgebra. □

Remark 8.2.14. The next-time operator of a coalgebra restricts to its subcoalgebras. Given a subcoalgebra $m\colon (B, \beta) \rightarrowtail (A, \alpha)$, we use the morphism x from (8.9) to obtain a (unique) coalgebra structure on $\bigcirc B$ turning it into a subcoalgebra of (A, α):

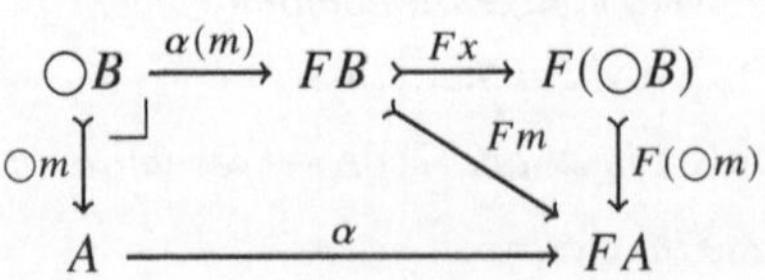

(8.10)

Recall from Definition 4.3.1 that a functor is called finitary if it preserves directed colimits.

Proposition 8.2.15. *Suppose that F is a finitary functor preserving monomorphisms and that A is an object having smooth subobjects. Let (A, α) be a coalgebra such that $\overleftarrow{\alpha}\colon \mathsf{Sub}(FA) \to \mathsf{Sub}(A)$ is continuous: it preserves joins of ω-chains. Then the next-time operator of (A, α) is continuous, too.*

Proof Since F preserves monomorphisms we have a function

$$\mathsf{Sub}(A) \to \mathsf{Sub}(FA) \qquad (s\colon S \rightarrowtail A) \mapsto (Fs\colon FS \rightarrowtail FA). \tag{8.11}$$

This function is continuous because for a given ω-chain in $\mathsf{Sub}(A)$

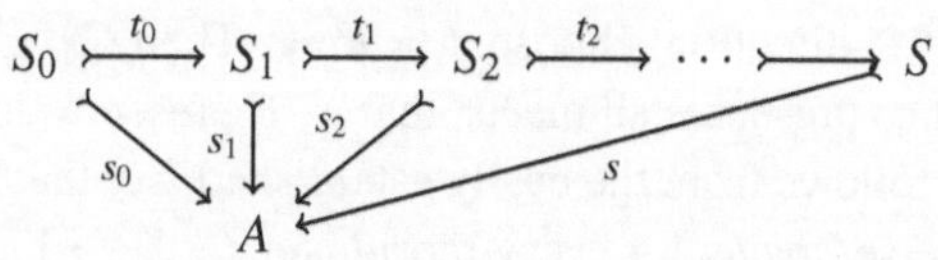

its join $s\colon S \rightarrowtail A$ is, equivalently, a colimit of the above ω-chain (t_i) (cf. Proposition 6.2.12). Since F preserves this colimit, we see that the subobject Fs is the join of all $Fs_i\colon FS_i \rightarrowtail FS$, $i \in \mathbb{N}$.

Now note that the next-time operator of a coalgebra $\alpha\colon A \to FA$ is the composite of the function in (8.11) and $\overleftarrow{\alpha}\colon \mathsf{Sub}(FA) \to \mathsf{Sub}(A)$. The latter is continuous by assumption. □

Remark 8.2.16. (1) The result in Proposition 8.2.15 holds for every coalgebra for a finitary set functor. Indeed, we know from Example 6.2.6(1) that monomorphisms are smooth, and taking inverse images clearly commutes with unions of ω-chains (of subobjects). The assumption that F preserves monomorphisms may be dropped by working with the Trnková hull (cf. Remark 8.1.16).

(2) Similarly, this holds for every coalgebra for a finitary functor on K-Vec, e.g. the functor $FX = K \times X^{\Sigma}$ whose coalgebras are linear weighted automata.

(3) More generally, this also holds for every coalgebra for a finitary endofunctor preserving monomorphisms on a locally finitely presentable category $\mathscr{A}$ with a simple initial object (Remark 6.2.7). Examples of such categories are, besides Set, K-Vec, every finitary variety of algebras with a simple initial object, and the categories of posets and graphs.

Indeed, finitary functors clearly preserve ω-colimits, $\mathscr{A}$ has smooth monomorphisms according to Remark 6.2.7, and that $\overleftarrow{\alpha}$ is continuous follows since joins are given by colimits (cf. Proposition 6.2.12) and the fact the colimit of ω-chains are universal (cf. Example 8.4.13(5)).

We close this section with a characterization result: F preserves intersections if and only if the following 'generalized next-time' operators are right adjoints. Given a morphism $f\colon A \to FB$, we have the operator $\bigcirc_f\colon \mathsf{Sub}(B) \to \mathsf{Sub}(A)$ that maps $m\colon B' \rightarrowtail B$ to the pullback of Fm along f:

$$\begin{array}{ccc} \bigcirc_f A' & \xrightarrow{f(m)} & FB' \\ {\scriptstyle \bigcirc_f m}\big\downarrow & \lrcorner & \big\downarrow{\scriptstyle Fm} \\ A & \xrightarrow{\ f\ } & FB \end{array}$$

Proposition 8.2.17 [314]**.** *An endofunctor F preserves intersections if and only if every generalized next-time operator $\bigcirc_f$ is a right adjoint.*

Proof For the 'if'-direction, choose $f = \mathsf{id}_{FY}$. Then $\bigcirc_{\mathsf{id}_{FY}}\colon m \mapsto Fm$ is a right adjoint and so preserves all meets, that is, F preserves intersections.

The converse follows from the easily established fact that intersections are stable under inverse image: for every morphism $f\colon X \to Y$ and every family $m_i\colon S_i \rightarrowtail Y$ of subobjects, the intersection $m\colon P \rightarrowtail X$ of the inverse images of the m_i under f yields a pullback

$$\begin{array}{ccc} P & \longrightarrow & \bigcap S_i \\ {\scriptstyle m}\big\downarrow & \lrcorner & \big\downarrow{\scriptstyle \bigcap m_i} \\ X & \xrightarrow{\ f\ } & Y \end{array}$$

Hence, if F preserves intersections, then so does every operator $\bigcirc_f$. Equivalently, $\bigcirc_f$ is a right adjoint. □

We will mention another equivalent characterizaton of intersection preservation in Remark 9.2.30.

8.3 The Well-Founded Part of a Coalgebra

We have introduced well-founded coalgebras in Section 8.1. We now discuss the *well-founded part* of a coalgebra, that is, its largest well-founded subcoalgebra. We shall prove that the well-founded part of a coalgebra always exists (Proposition 8.3.6) and is the least fixed point of the next-time operator. We will also see that a coalgebra is well-founded iff it is its own well-founded part. Then we prove that the well-founded part is the coreflection of a coalgebra in the category of well-founded coalgebras (Proposition 8.3.11).

Definition 8.3.1. The *well-founded part* of a coalgebra is its largest well-founded subcoalgebra.

Example 8.3.2. The well-founded part of a graph (as a coalgebra for $\mathscr{P}$) is the subgraph formed by all points which do not lie on any infinite path.

Construction 8.3.3. Let $\alpha\colon A \to FA$ be a coalgebra. By Assumption 8.2.1, we know that $\mathsf{Sub}(A)$ is a complete lattice and that the next-time operator $\bigcirc$ is monotone (see Lemma 8.2.8). Hence, by Theorem 6.1.1, $\bigcirc$ has a least fixed point, which we denote by

$$a^*\colon A^* \rightarrowtail A.$$

Moreover, by Lemma 8.2.13(2), we know that there is a coalgebra structure $\alpha^*\colon A^* \to FA^*$ such that $a^*\colon (A^*, \alpha^*) \rightarrowtail (A, \alpha)$ is the smallest cartesian subcoalgebra of (A, α).

Remark 8.3.4. (1) From the proof of Theorem 6.1.1 we know how to construct a^*, the least fixed point of $\bigcirc$, as the join of the following chain of subobjects $a_i\colon A_i \rightarrowtail A$, $i \in \mathsf{Ord}$. First, put $a_0 = \bot_A$, the least subobject of A. Given $a_i\colon A_i \rightarrowtail A$, put $a_{i+1} = \bigcirc a_i\colon A_{i+1} = \bigcirc A_i \rightarrowtail A$. For every limit ordinal j, put $a_j = \bigvee_{i<j} a_i$. There exists an ordinal i such that $a_i = a^*\colon A^* \rightarrowtail A$.
(2) Under the conditions in Proposition 8.2.15, the least ordinal i with $a^* = a_i$ is at most ω. Indeed, since $\bigcirc$ is continuous we have $a^* = \bigvee_{i\in\mathbb{N}} a_i$ by Kleene's Fixed Point Theorem.

In particular, by Remark 8.2.16, this holds for every coalgebra for a finitary functor on Set or K-Vec, and more generally, for every finitary functor preserving monomorphisms on a locally finitely presentable category with a simple initial object.

Example 8.3.5. For a graph $\alpha\colon A \to \mathscr{P}A$ we have $a_0\colon \emptyset \hookrightarrow A$, $a_1\colon A_1 \hookrightarrow A$ is the set of nodes with no successors, $a_2\colon A_2 \hookrightarrow A$ contains the nodes from A_1 plus those nodes all of whose successors have no successors, etc. Hence, $a_i\colon A_i \hookrightarrow A$ is the sets of nodes x such that every path from x has length at most i. The union $a^*\colon A^* \hookrightarrow A$ is the well-founded part.

Proposition 8.3.6. *For every coalgebra (A, α), the coalgebra (A^*, α^*) is well-founded.*

Proof Let $m\colon (B, \beta) \rightarrowtail (A^*, \alpha^*)$ be a cartesian subcoalgebra. By Lemma 8.2.13, $a^* \cdot m\colon B \to A$ is a fixed point of $\bigcirc$. Since a^* is the least fixed point, we have $a^* \leq a^* \cdot m$, that is, $a^* = a^* \cdot m \cdot x$ for some $x\colon A^* \rightarrowtail B$. Since a^* is monic, we thus obtain $m \cdot x = \mathsf{id}_{A^*}$. So m is a monomorphism and a split epimorphism, whence an isomorphism. □

Example 8.3.7. Consider the coalgebra G for $\mathscr{P}$ depicted as the graph:

$$a \longrightarrow b \qquad c \rightleftarrows d$$

We list all subcoalgebras below (the structures are the obvious ones given by the picture of G). Those are $\emptyset$, $\{b\}$, $\{a,b\}$, $\{c,d\}$, $\{b,c,d\}$, and $\{a,b,c,d\}$. Of these, the cartesian subcoalgebras of G are $\{a,b\}$, and $\{a,b,c,d\}$. The well-founded part of G is the least cartesian subcoalgebra, namely $\{a,b\}$.

Let us come back to a loose end from Example 8.1.4(4):

Proposition 8.3.8. *A linear weighted automaton is well-founded iff every path in its state transition graph eventually leads to* 0.

Proof Given a linear weighted automaton, that is, a coalgebra (A,α) for $FX = K \times X^{\Sigma}$ on K-Vec, denote by $a^*\colon A^* \rightarrowtail A$ the subset of all states with the property in the statement of the proposition. Clearly, A^* is a subspace of A. Furthermore, we obtain from Remark 8.3.4(2) that the least fixed point of $\bigcirc$ is $\bigvee_{n\in\mathbb{N}} \bigcirc^n(\bot_A)$. We also know that $\bot_A$ is the trivial subspace $\{0\}$, and for every subspace $s\colon S \rightarrowtail A$, $\bigcirc s$ is the space of all nodes whose successors lie in S (see Example 8.2.7(3)). Therefore $\bigcirc^n(\bot_A)$ consists of precisely those states from which every path reaches 0 in at most n steps. Since $A^* = \bigvee_{n\in\mathbb{N}} \bigcirc^n(\bot_A)$ and since (A,α) is well-founded iff $A = A^*$, our result follows. □

For the next result, cf. Taylor [287, Exercise VI.17] or Adámek et al. [47, Corollary 2.19]; item (3) is Taylor's definition of a well-founded coalgebra [287, Def. 6.3.2]:

Proposition 8.3.9. *For every coalgebra (A,α), the following are equivalent:*

(1) *(A,α) is well-founded,*

(2) *id_A is the only fixed point of $\bigcirc$,*

(3) *id_A is the only pre-fixed point of $\bigcirc$.*

Proof Since $\bigcirc$ is monotone, (2) and (3) state the same fact, see Remark 6.1.3.

(1) ⇒ (2). Let (A,α) be well-founded. Since A is a cartesian subcoalgebra of itself via id_A, we have $\mathsf{id}_A = \bigcirc\mathsf{id}_A$ by Lemma 8.2.13. Consider an arbitrary subobject $m\colon A' \rightarrowtail A$ such that $m = \bigcirc m$. By Lemma 8.2.13, there is a coalgebra structure $\alpha\colon A' \to FA'$ giving a cartesian subcoalgebra. By well-foundedness, m is an isomorphism. Thus it represents the same subobject as id_A.

(2) ⇒ (1). We know that the least fixed point a^* in Construction 8.3.3 is id_A. Hence, (A,α) is isomorphic to (A^*,α^*) and therefore well-founded by Proposition 8.3.6. □

Remark 8.3.10. Given a coalgebra (A, α) for the power-set functor $\mathscr{P}$, consider the corresponding relation R on A given by $y \mathrel{R} x$ iff $y \in \alpha(x)$ (Remark 8.1.3). In this instance, item (3) states precisely that the relation R satisfies the principle of *well-founded induction* (aka. Noetherian induction). This principle states that a property $P(x)$ of elements $x \in A$ holds for all elements of A provided that the following implication is true for every $x \in A$: if $P(y)$ holds for all $y \in A$ such that $y \mathrel{R} x$, then $P(x)$ holds. In symbols: $\forall x \in A. \forall y \in A. (y \mathrel{R} x \implies P(y)) \implies P(x)$. Indeed, for every subset $S \subseteq A$ we have $\bigcirc S = \{x \in A : \forall y. \, y \mathrel{R} x \implies y \in S\}$ (cf. Example 8.2.7(1)). Hence, given $P(x)$, the corresponding subset $S = \{x \in A : P(x)\}$ is a pre-fixed point of $\bigcirc$ iff the above formula holds.

We know from Proposition 8.3.6 that for every coalgebra (A, α) its subcoalgebra represented by $a^* : A^* \rightarrowtail A$ is well-founded. We now prove that, categorically, this subcoalgebra is characterized uniquely up to isomorphism by the following universal property: every homomorphism from a well-founded coalgebra into (A, α) factorizes uniquely through a^*. In particular, this implies that $a^* : A^* \to A$ is the largest well-founded subcoalgebra of A, namely the well-founded part of A.

Proposition 8.3.11. *The full subcategory of* $\mathsf{Coalg}\, F$ *given by well-founded coalgebras is coreflective. In fact, the well-founded coreflection of a coalgebra* (A, α) *is its well-founded part* (A^*, α^*).

Proof We are to prove that for every coalgebra homomorphism $f : (B, \beta) \to (A, \alpha)$, where (B, β) is well-founded, there exists a coalgebra homomorphism $f^\sharp : (B, \beta) \to (A^*, \alpha^*)$ such that $a^* \cdot f^\sharp = f$. It is unique since $a^* : A^* \rightarrowtail A$ is a monomorphism. It then follows that $a^* : (A^*, \alpha^*) \rightarrowtail (A, \alpha)$ is the largest well-founded subcoalgebra.

For the existence of $f^\sharp$, we first observe that $\overleftarrow{f}(a^*)$ is a pre-fixed point of $\bigcirc_\beta$: indeed, using Lemma 8.2.11 we get

$$\bigcirc_\beta(\overleftarrow{f}(a^*)) \leq \overleftarrow{f}(\bigcirc_\alpha(a^*)) = \overleftarrow{f}(a^*).$$

By Proposition 8.3.9, we therefore have $id_B = b^* \leq \overleftarrow{f}(a^*)$ in $\mathsf{Sub}(B)$. Using the adjunction in Lemma 8.2.5, we obtain $\vec{f}(\mathsf{id}_B) \leq a^*$ in $\mathsf{Sub}(A)$. Now let

$$f = (B \overset{e}{\twoheadrightarrow} C \overset{m}{\rightarrowtail} A)$$

be the factorization of f as in Remark 7.2.1(3). This implies that $\vec{f}(\mathsf{id}_B) = m$. Hence, we obtain

$$m = \vec{f}(\mathsf{id}_B) \leq a^*,$$

that is, there exists a morphism $h\colon C \rightarrowtail A^*$ such that $a^* \cdot h = m$. Thus, $f^\sharp = h \cdot e\colon B \to A^*$ is a morphism satisfying

$$a^* \cdot f^\sharp = a^* \cdot h \cdot e = m \cdot e = f.$$

It follows that $f^\sharp$ is a coalgebra homomorphism from (B, β) to (A^*, α^*) since f and a^* are and F preserves monomorphisms (cf. Remark 7.2.4(1)). □

8.4 Closure Properties of Well-Founded Coalgebras

In this section we will see that strong quotients and subcoalgebras of well-founded coalgebras are well-founded. For subcoalgebras we need to assume more about $\mathscr{A}$ and F than what we mentioned in Assumption 8.2.1. We present two variants in Proposition 8.4.7 and Corollary 8.4.15.

Remark 8.4.1. Recall from Remark 4.2.3 that quotient coalgebras of (A, α) are represented by homomorphisms with domain (A, α) carried by epimorphisms in $\mathscr{A}$. In this chapter we will consider a stronger notion of quotient coalgebra.

Definition 8.4.2. By a *strong quotient* of a coalgebra (A, α) is meant a quotient coalgebra of (A, α) represented by a strong epimorphism in $\mathscr{A}$ (see Remark 7.2.1(1)).

Later, in Section 9.3, we will make use of the following corollary to Proposition 8.3.11. For set functors preserving inverse images, this was stated by Taylor [287, Exercise VI.16]:

Corollary 8.4.3. *The subcategory of* Coalg F *formed by all well-founded coalgebras is closed under strong quotients and coproducts in* Coalg F.

This follows from a general result on coreflective subcategories: the category Coalg F has the factorization system of Proposition 7.2.2 (cf. Remark 7.2.4(1)), and its full subcategory of well-founded coalgebras is coreflective with monic coreflections (see Proposition 8.3.11). Thus, it is closed under strong quotients and colimits [38, dual of Thm. 16.8].

Remark 8.4.4. Let $\mathscr{A}$ have a simple initial object (Remark 6.2.3(3)) and colimits of chains. If F preserves finite intersections, then every coalgebra $W_i \to FW_i$, $i \in \mathsf{Ord}$, in the initial-algebra chain (see Definition 6.1.4) is well-founded. This is shown by transfinite induction. The base case and isolated step are proved as in Remark 8.1.7(2) using Proposition 8.1.5(1), and for the limit step one uses that well-founded coalgebras are closed under colimits.

Remark 8.4.5. We prove next that, for an endofunctor preserving finite intersections, well-founded coalgebras are closed under subcoalgebras, provided that $\mathsf{Sub}(A)$ forms a *frame*. Recall that this means that for every subobject $m: B \rightarrowtail A$ and every family m_i $(i \in I)$ of subobjects of A we have

$$m \wedge \bigvee_{i \in I} m_i = \bigvee_{i \in I} (m \wedge m_i).$$

Equivalently, the inverse image operator $\overleftarrow{m}\colon \mathsf{Sub}(A) \to \mathsf{Sub}(B)$ from Notation 8.2.3 has a right adjoint $m_*\colon \mathsf{Sub}(B) \to \mathsf{Sub}(A)$ (use the dual of Remark 8.2.4).

Example 8.4.6. (1) Set has the property that all $\mathsf{Sub}(A)$ are frames. In fact, given subsets S and S_i $(i \in I)$ of A the equality $S \cap (\bigcup_{i \in I} S_i) = \bigcup_{i \in I}(S \cap S_i)$ clearly holds.

(2) This property is shared by categories such as posets and monotone maps, graphs and homomorphisms, unary algebras and homomorphisms, and presheaf categories $\mathsf{Set}^{\mathscr{C}^{\mathsf{op}}}$, with $\mathscr{C}$ small. This follows from the fact that joins and meets of subobjects of an object A are formed on the level of subsets of the underlying set of A.

(3) The category CPO does not have the above property: for the cpo $A = \mathbb{N}^\top$ of natural numbers with a top element $\top$ (linearly ordered) the lattice $\mathsf{Sub}(A)$ is not a frame. Consider the subobjects given by inclusion maps $m_i\colon \{0, \dots, i\} \hookrightarrow \mathbb{N}^\top$ for $i \in \mathbb{N}$, with domains linearly ordered. It is easy to see that $\bigvee_{i \in \mathbb{N}} m_i = \mathsf{id}_A$. For the inclusion map $m\colon \{\top\} \hookrightarrow \mathbb{N}^\top$ we have $m \wedge m_i = 0$ $(i \in I)$, the empty subobject. Thus, $\bigvee_{i \in \mathbb{N}}(m \wedge m_i) = 0 \neq m = m \wedge \bigvee_{i \in I} m_i$.

(4) For every Grothendieck topos, the posets $\mathsf{Sub}(A)$ are frames. In fact, it is sufficient for a topos to have all coproducts or intersections to satisfy this requirement.

Proposition 8.4.7. *Suppose that F preserves finite intersections, and let (A, α) be a well-founded coalgebra such that $\mathsf{Sub}(A)$ is a frame. Then every subcoalgebra of (A, α) is well-founded.*

Proof Let $m\colon (B, \beta) \rightarrowtail (A, \alpha)$ be a subcoalgebra. We will show that the only pre-fixed point of $\bigcirc_\beta$ is id_B (cf. Proposition 8.3.9). Suppose $s\colon S \rightarrowtail B$ fulfils $\bigcirc_\beta(s) \leq s$. Since F preserves finite intersections, we have

$$\overleftarrow{m} \cdot \bigcirc_\alpha = \bigcirc_\beta \cdot \overleftarrow{m}$$

by Corollary 8.2.12(1). The counit of the adjunction $\overleftarrow{m} \dashv m_*$ yields the relation $\overleftarrow{m}(m_*(s)) \leq s$, so that we obtain

$$\overleftarrow{m}(\bigcirc_\alpha(m_*(s))) = \bigcirc_\beta(\overleftarrow{m}(m_*(s))) \leq \bigcirc_\beta(s) \leq s.$$

Equivalently, we have that $\bigcirc_\alpha(m_*(s)) \leq m_*(s)$, using again the adjunction

$\overleftarrow{m} \dashv m_*$. Thus, $m_*(s)$ is a pre-fixed point of $\bigcirc_\alpha$. Since (A, α) is well-founded, Corollary 8.2.12(1) implies that $m_*(s) = \mathsf{id}_A$. Since $\overleftarrow{m}$ is also a right adjoint (Lemma 8.2.5) and therefore preserves the top element of $\mathsf{Sub}(B)$, we thus obtain

$$\mathsf{id}_B = \overleftarrow{m}(\mathsf{id}_A) = \overleftarrow{m}(m_*(s)) \leq s. \qquad \square$$

Remark 8.4.8. For a set functor F preserving inverse images, a much better result was proved by Taylor [287, Cor. 6.3.6]: for every coalgebra homomorphism $f\colon (B, \beta) \to (A, \alpha)$ with (A, α) well-founded so is (B, β). In fact, our proof above is essentially Taylor's who (implicitly) uses Corollary 8.2.12(2) instead.

Corollary 8.4.9. *If a set functor preserves finite intersections, then subcoalgebras of well-founded coalgebras are well-founded.*

We know from Trnková's result (Proposition 4.4.1) that every set functor preserves all *nonempty* finite intersections. However, this does not suffice for Corollary 8.4.9, as we saw in Example 8.1.6(2).

The property that subcoalgebras of a given well-founded coalgebra are well-founded does not necessarily need the assumption that $\mathsf{Sub}(A)$ is a frame. Using the construction of the least fixed point a^* of $\bigcirc$ provided by the (proof of the) Knaster–Tarski Fixed Point Theorem, it is essentially sufficient that $\overleftarrow{m}$ in the proof of Proposition 8.4.7 preserves joins of unions of chains in $\mathsf{Sub}(A)$. We now discuss this in more detail.

Recall the notion of smooth monomorphisms from Definition 6.2.2.

Definition 8.4.10. Let $\mathscr{A}$ be a category with pullbacks. We say that $\mathscr{A}$ has *universally smooth monomorphisms* if it has smooth monomorphisms, and colimits of chains of monomorphisms are *universal*: for every morphism $f\colon X \to Y$, the functor $\mathscr{A}/Y \to \mathscr{A}/X$ forming pullbacks along f preserves those colimits.

Remark 8.4.11. That $\mathscr{A}$ has universally smooth monomorphisms means precisely that the following property holds for every λ-chain of monomorphisms $a_{i,j}\colon A_i \rightarrowtail A_j$ $(i \leq j < \lambda)$ in $\mathscr{A}$ with colimit cocone $a_i\colon A_i \to A$: Given a morphism $f\colon X \to A$, form the pullback of a_i along f:

$$\begin{array}{ccc} X_i & \xrightarrow{f_i} & A_i \\ {\scriptstyle x_i}\downarrow & \lrcorner & \downarrow{\scriptstyle a_i} \\ X & \xrightarrow{f} & A \end{array} \qquad \text{for every } i < \lambda.$$

Then the objects X_i $(i < \lambda)$ form a λ-chain, using the universal property of pullbacks. This has the colimit cocone $x_i\colon X_i \to X$.

Remark 8.4.12. Let $\mathscr{A}$ have universally smooth monomorphisms.

(1) The initial object 0 is strict (see Remark 7.2.5(2)). Indeed, consider the empty chain in Definition 8.4.10.

(2) For every morphism $f\colon A \to B$, the operator $\overleftarrow{f}\colon \mathsf{Sub}(B) \to \mathsf{Sub}(A)$ preserves unions of chains. Indeed, this follows from universal smoothness using that monomorphisms are stable under pullback and that unions of chains are given by the corresponding colimits (cf. Proposition 6.2.12).

Example 8.4.13. (1) Set has universally smooth monomorphisms.

(2) K-Vec has smooth monomorphisms, but not universally so because the initial object is not strict: we have a morphism from every vector space to 0.

(3) CPO does not have universally smooth monomorphisms. In fact, they are not even smooth (see Example 6.2.6(8)).

(4) Categories in which colimits of chains and pullbacks are formed 'set-like' have universally smooth monomorphisms. These include the categories of posets, graphs, topological spaces, presheaf categories, and many varieties, such as monoids, groups, and unary algebras.

(5) Every locally finitely presentable category $\mathscr{A}$ (Definition 10.1.5) with a simple and strict initial object (Remark 8.4.12(1)) has universally smooth monomorphisms. To see this use that $\mathscr{A}$ has smooth monomorphisms (Remark 6.2.7) and the fact that filtered colimits (whence colimits of nonempty chains) are universal as we now prove. Indeed, suppose that $c_i\colon C_i \to C, i \in I$ is a filtered colimit, and let $f\colon B \to C$ be a morphism. Form the pullback of every c_i along f:

$$\begin{array}{ccc} B_i & \xrightarrow{f_i} & C_i \\ {\scriptstyle b_i}\big\downarrow & \lrcorner & \big\downarrow{\scriptstyle c_i} \\ B & \xrightarrow{f} & C \end{array}$$

Then $b_i\colon B_i \to B$ is a colimit cocone. Indeed, in the category of commmutative squares in $\mathscr{A}$, the filtered diagram formed by the above pullbacks squares has as a colimit a pullback square

$$\begin{array}{ccc} \mathsf{colim}\, B_i & \xrightarrow{\mathsf{colim}\, f_i} & \mathsf{colim}\, C_i = C \\ {\scriptstyle \cong}\big\downarrow & \lrcorner & \big\| \\ B & \xrightarrow{f} & C \end{array}$$

Unfortunately, the example of rings demonstrates that the assumption of strictness of 0 cannot be lifted. In fact, monomorphisms are not smooth in this category (see Example 6.2.6(7)).

Our proof in Theorem 8.6.1 of the converse of the General Recursion Theorem hinges on Proposition 8.4.14, which shows that well-foundedness is reflected by coalgebra homomorphisms. For categories in which $\mathsf{Sub}(A)$ is a frame, this has been proved by Taylor [287, Cor. 6.3.6] and [286, Prop. 7.3].

Our proof makes use of the induction principle which we have seen in Corollary 7.5.1.

As stated in Assumption 8.2.1, our standing assumption at this point is that the underlying category $\mathscr{A}$ be complete and well-powered.

Proposition 8.4.14. *Let $\mathscr{A}$ have universally smooth monomorphisms, and suppose that $F\colon \mathscr{A} \to \mathscr{A}$ preserves inverse images. Then for every coalgebra homomorphism $f\colon (B,\beta) \to (A,\alpha)$ with (A,α) well-founded, (B,β) is well-founded, too.*

Proof Let $s\colon S \rightarrowtail B$ be a pre-fixed point of $\bigcirc_\beta$, that is, $\bigcirc_\beta(s) \le s$. It is our task to prove that $\mathsf{id}_B \le s$ (see Proposition 8.3.9(3)).

We apply the induction principle in Corollary 7.5.1 to the subset of $\mathsf{Sub}(A)$ given by

$$\mathcal{S} = \{t\colon T \rightarrowtail A \mid \overleftarrow{f}(t) \le s\}.$$

We observe that $\bigcirc_\alpha$ preserves $\mathcal{S}$: given $t\colon T \rightarrowtail A$ in $\mathcal{S}$ we have

$$\overleftarrow{f}(\bigcirc_\alpha(t)) = \bigcirc_\beta(\overleftarrow{f}(t)) \le \bigcirc(s) \le s,$$

where the left-hand equation holds by Corollary 8.2.12(2). In addition, $\mathcal{S}$ is closed under joins of chains. For suppose that $(t_i\colon T_i \rightarrowtail A)$, $i \in I$, is a chain in $\mathcal{S}$. Then we have

$$\overleftarrow{f}(\bigvee_{i\in I} t_i) = \bigvee_{i\in I} \overleftarrow{f}(t_i) \le \bigvee_{i\in I} s = s,$$

where the first equation holds due to Remark 8.4.12(2). Thus, by Corollary 7.5.1, the least fixed point of $\bigcirc_\alpha$ lies in $\mathcal{S}$, in symbols $\mathsf{id}_A \in \mathcal{S}$, since (A,α) is well-founded. Consequently, we obtain

$$\mathsf{id}_B = \overleftarrow{f}(\mathsf{id}_A) \le s,$$

which completes the proof. □

Corollary 8.4.15. *Suppose that $\mathscr{A}$ satisfies the assumptions in Proposition 8.4.14 and that F preserves finite intersections. Then every subcoalgebra of a well-founded coalgebra is well-founded.*

Indeed, for a monomorphism $f: (B, \beta) \rightarrowtail (A, \alpha)$ we can use Corollary 8.2.12(1) in the above proof of Proposition 8.4.14.

Example 8.4.16. Suppose that $\mathscr{A}$ has colimits of chains and a simple initial object. Let $F: \mathscr{A} \to \mathscr{A}$ preserve finite intersections, and assume that $\mathsf{Sub}(W_i)$ is a frame for all objects W_i in the initial-algebra chain. Then by Remark 8.4.4 and Proposition 8.4.7 we have that every subcoalgebra of $W_i \to FW_i$, $i \in \mathsf{Ord}$, is well-founded. In lieu of all $\mathsf{Sub}(W_i)$ being frames and 0 being simple, one may also assume that $\mathscr{A}$ is complete and well-powered with universally smooth monomorphisms (cf. Corollary 8.4.15).

8.5 The General Recursion Theorem

The main consequence of well-foundedness is parametric recursivity. This is Taylor's General Recursion Theorem [287, Theorem 6.3.13]. Taylor originally assumed that F preserves inverse images. We present a proof based on subsequent work [53] for which it is sufficient that F preserves monomorphisms, assuming those are smooth. In Section 8.6, we discuss the converse of the General Recursion Theorem in Theorems 8.6.1 and 8.6.8.

Assumption 8.5.1. For the rest of this section, we assume that in our complete and well-powered category $\mathscr{A}$ the class of all monomorphisms is smooth (see Definition 6.2.2).

Remark 8.5.2. (1) In this setting, the bottom element of each poset $\mathsf{Sub}(A)$ is the unique morphism $\perp_A: 0 \to A$, where 0 is the initial object of $\mathscr{A}$ (cf. Remark 6.2.3(2)).

(2) Furthermore, a join of a chain in $\mathsf{Sub}(A)$ is obtained by forming the colimit of the corresponding chain (Remark 6.2.3(1)).

Theorem 8.5.3. *For $F: \mathscr{A} \to \mathscr{A}$ preserving monomorphisms, every well-founded coalgebra is parametrically recursive.*

Proof (1) Let (A, α) be well-founded. We first prove that it is recursive. Recall from Remark 2.4.12(3) that the subcoalgebras of (A, α) form a subposet of $\mathsf{Sub}(A)$. Let S be the subset of the poset $\mathsf{Sub}(A)$ given by the recursive subcoalgebras of (A, α). It follows from Example 7.4.3(4) and Remark 8.5.2(2) that S is closed under joins of chains in $\mathsf{Sub}(A)$. We prove below that the next-time operator $\bigcirc$ preserves S. Using Corollary 7.5.1, we conclude that the least fixed point of $\bigcirc$ is contained in S, and since (A, α) is well-founded, this fixed point is id_A (see Proposition 8.3.9). In other words, (A, α) is recursive.

To complete the proof, we now show that for every recursive subcoalgebra (B, β) the subcoalgebra on $\bigcirc B$ in Remark 8.2.14 is recursive, too.

First, we know from Example 7.4.3(3) that $(FB, F\beta)$ is recursive. Given an algebra $e\colon FX \to X$ we have the coalgebra-to-algebra morphism h so that the lower part of the following diagram commutes:

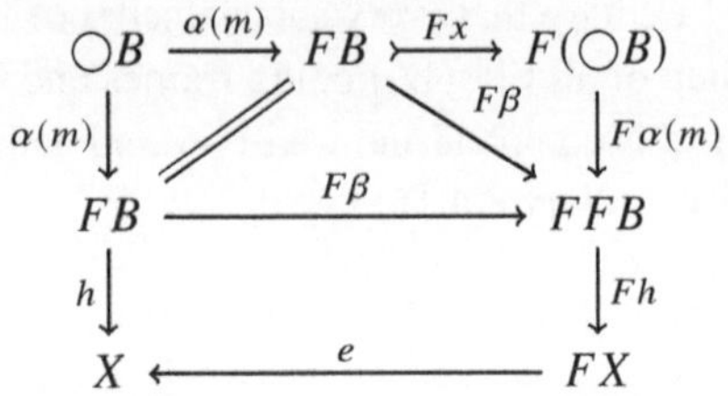

Moreover, in the upper part the left-hand and middle triangles commute trivially, and the right-hand one does by the definition of x (see Remark 8.2.14). Thus, $h \cdot \alpha(m)$ is a coalgebra-to-algebra morphism.

For the uniqueness suppose that $g\colon \bigcirc B \to X$ is a coalgebra-to-algebra morphism:

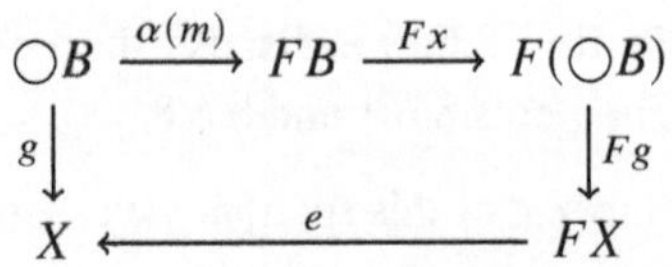

(8.12)

Then $e \cdot Fg \cdot Fx$ is a coalgebra-to-algebra morphism from $(FB, F\beta)$ to (X, e):

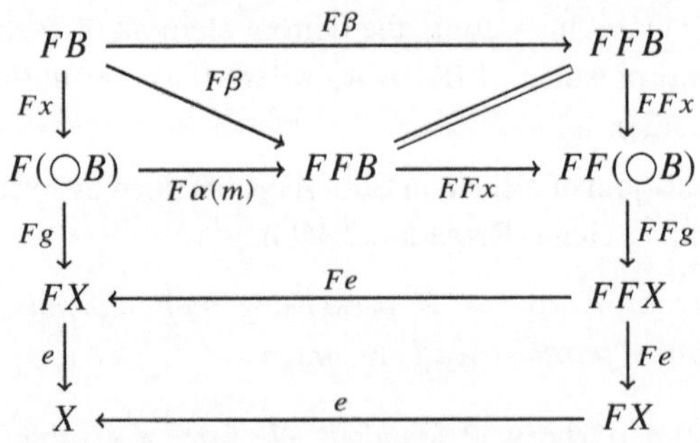

The left-hand triangle commutes by the definition of x, the middle and right-hand triangles are trivial, the middle square is the F-image of (8.12), and the lower part is trivial again.

Thus, we conclude that $h = e \cdot Fg \cdot Fx$ and therefore

$$g = e \cdot Fg \cdot Fx \cdot \alpha(m) = h \cdot \alpha(m),$$

as desired.

(2) Finally, we prove that the coalgebra (A, α) is parametrically recursive.

Consider the coalgebra $\langle \alpha, \mathrm{id}_A \rangle \colon A \to FA \times A$ for $F(-) \times A$. This functor preserves monomorphisms since F does and monomorphisms are closed under products. The next-time operator $\bigcirc$ on $\mathsf{Sub}(A)$ is the same for both coalgebras, since the square (8.5) is a pullback if and only if the square below is one:

$$\begin{array}{ccc} \bigcirc A' & \xrightarrow{\langle \alpha(m), \bigcirc(m) \rangle} & FA' \times A \\ {\scriptstyle \bigcirc m} \downarrow & \lrcorner & \downarrow {\scriptstyle Fm \times A} \\ A & \xrightarrow{\langle \alpha, A \rangle} & FA \times A \end{array}$$

Since id_A is the unique fixed point of $\bigcirc$ with respect to F (see Proposition 8.3.9), it is also the unique fixed point of $\bigcirc$ with respect to $F(-) \times A$. Thus $(A, \langle \alpha, A \rangle)$ is a well-founded coalgebra for $F(-) \times A$. By point (1), it is thus recursive for $F(-) \times A$. This states equivalently that (A, α) is a parametrically recursive coalgebra for F. □

Remark 8.5.4. In the category K-Vec of vector spaces over the field K and linear maps every monomorphism splits: Given a subspace $s \colon V \rightarrowtail W$, choose a base $(v_i)_{i \in I}$ of V and extend it to a basis of W using the vectors $(w_j)_{j \in J}$. The linear map $e \colon W \to V$ determined by $e(v_i) = v_i$ and $e(w_j) = 0$ fulfils $e \cdot s = \mathrm{id}_V$.

Analogously, all epimorphisms split in K-Vec: Given a surjective linear map $e \colon W \twoheadrightarrow V$, let $(v_i)_{i \in I}$ be a base of V and choose, for every $i \in I$, an element $w_i \in W$ with $e(w_i) = v_i$. The linear map $s \colon V \to W$ determined by $s(v_i) = w_i$ for $i \in I$ fulfils $e \cdot s = \mathrm{id}_V$.

Corollary 8.5.5. *For every endofunctor on* Set *or* K-Vec, *every well-founded coalgebra is parametrically recursive.*

Proof For Set, we apply Theorem 8.5.3 to the Trnková hull $\bar{F}$ (see Proposition 8.1.13), noting that F and $\bar{F}$ have the same (nonempty) coalgebras. By Lemma 8.1.22 the desired result follows. For K-Vec, use that monomorphisms split and are therefore preserved by every endofunctor F. □

Example 8.5.6. For the set functor $FX = X \times X + 1$ the coalgebra $(\mathbb{N}, \gamma)$ from Example 8.1.21(3) is well-founded. Hence it is parametrically recursive.

Similarly, we saw that for $FX = A \times X \times X + 1$ the coalgebra (A, s) from Example 7.4.10(3) is well-founded, and therefore it is (parametrically) recursive.

Example 8.5.7. Well-founded coalgebras need not be recursive when F does not preserve monomorphisms. We take $\mathscr{A}$ to be the category Pred of *sets with a predicate*. Objects are pairs (X, A), where $A \subseteq X$. Morphisms $f \colon (X, A) \to (Y, B)$ are functions that satisfy $f[A] \subseteq B$. Pred clearly fulfils

Assumption 8.5.1. Denote by $\mathbb{1}$ the terminal object $(1,1)$. We define an endofunctor F by $F(X,\emptyset) = (X+1,\emptyset)$, and for $A \neq \emptyset$, $F(X,A) = \mathbb{1}$. For a morphism $f\colon (X,A) \to (Y,B)$, put $F = f + \mathsf{id}$ if $A = \emptyset$; if $A \neq \emptyset$, then also $B \neq \emptyset$ and Ff is $\mathsf{id}\colon \mathbb{1} \to \mathbb{1}$.

The terminal coalgebra is $\mathsf{id}\colon \mathbb{1} \to \mathbb{1}$, and it is easy to see that it is well-founded. But it is not recursive: there are no coalgebra-to-algebra morphisms into an algebra of the form $F(X,\emptyset) \to (X,\emptyset)$.

We know that the terminal recursive coalgebra is precisely the initial algebra (Theorem 7.1.8). Under our current assumptions the same holds for the terminal well-founded coalgebra:

Corollary 8.5.8. *For $F\colon \mathscr{A} \to \mathscr{A}$ preserving intersections, the terminal well-founded coalgebra is precisely the initial algebra.*

Proof First note that F preserves monomorphisms since it preserves intersections and since a morphism m is monic if and only the pullback of m along itself is formed by identity morphisms. Furthermore, recall from Remark 7.2.1(3) that $\mathscr{A}$ has factorizations of morphisms into strong epimorphisms followed by monomorphisms, and that $\mathsf{Coalg}\,F$ inherits these factorizations by Proposition 7.2.2.

Now given a terminal well-founded coalgebra (A,α) take the factorization of the homomorphism $\alpha\colon (A,\alpha) \to (FA, F\alpha)$ into a homomorphism carried by a strong epimorphism e followed by one carried by a monomorphism m:

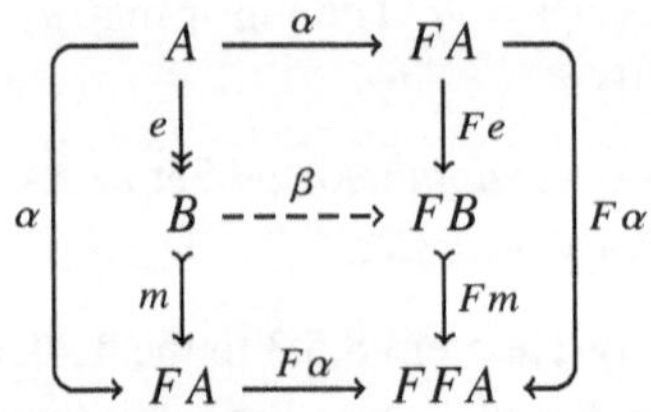

We see that (B,β) is a strong quotient of the coalgebra (A,α). By Corollary 8.4.3, (B,β) is well-founded. Since (A,α) is terminal, we obtain a homomorphism $f\colon (B,\beta) \to (A,\alpha)$. Then $f \cdot e$ is an endomorphism of the terminal well-founded coalgebra, whence $f \cdot e = \mathsf{id}_A$. This proves that e is an isomorphism; in other words, α is a monomorphism. Using Proposition 8.1.5(1) and Remark 8.1.7(1), we see that $(FA, F\alpha)$ is well-founded. Hence an argument precisely as in Lambek's Lemma 2.2.5 proves that α is an isomorphism. Since (A,α) is recursive by Theorem 8.5.3, we obtain that (A,α^{-1}) is an initial algebra by Proposition 7.4.2(2).

Conversely, if (A,α^{-1}) is an initial algebra, then (A,α) is well-founded (see Example 8.1.4(1)). By Theorem 7.1.8, (A,α) is a terminal recursive coalgebra.

Hence, Theorem 8.5.3 implies that it is a terminal well-founded coalgebra, for if (C, γ) is any well-founded coalgebra it is recursive, whence there exists a unique coalgebra homomorphism from (C, γ) to (A, α). □

For $\mathscr{A}$ = Set the previous result even holds for *all* endofunctors (even those not preserving monomorphisms or intersections):

Theorem 8.5.9 [47, Thm. 2.46]**.** *For every set functor, a terminal well-founded coalgebra is precisely an initial algebra.*

The fact that no assumptions on F are needed seems very special to Set. Without any assumptions on F, Theorem 8.5.9 does not even generalize from Set to the category of multigraphs, as the following example shows.

Example 8.5.10. A multigraph is a generalization of the notion of a graph allowing for multiple edges between nodes. Categorically, we take multigraphs to be the category of presheaves over the category $\{\bullet \rightrightarrows \bullet\}$given by two parallel morphisms. Here is a simple endofunctor F whose initial algebra is infinite and whose terminal well-founded coalgebra is a singleton graph: On objects A put $FA = 1$ (the terminal multigraph) if A has edges. For a multigraph A without edges, let FA be A with an additional vertex (and no edges). The definition of F on morphisms $h\colon A \to B$ is as expected: Fh maps the additional vertex of A to that of B in the case where B has no edges. Then μF is the natural numbers without edges. However, the terminal well-founded coalgebra is $F1 \xrightarrow{\cong} 1$.

We close this section with a general fact on well-founded parts of fixed points of functors:

Theorem 8.5.11. *Let $\mathscr{A}$ be a complete and well-powered category with smooth monomorphisms. For F preserving monomorphisms, the well-founded part of every fixed point is the initial algebra. In particular, the only well-founded fixed point is the initial algebra.*

Proof Let $\alpha\colon A \to FA$ be a fixed point of F. By Theorem 6.3.4 we know that the initial algebra $(\mu F, \iota)$ exists. Now let $a^*\colon (A^*, \alpha^*) \rightarrowtail (A, \alpha)$ be the well-founded part of A given in Proposition 8.3.6. This is a cartesian subcoalgebra: we have a pullback square

$$\begin{array}{ccc} A^* & \xrightarrow{\alpha^*} & FA^* \\ {\scriptstyle a^*}\downarrow & \lrcorner & \downarrow{\scriptstyle Fa^*} \\ A & \xrightarrow{\alpha} & FA \end{array}$$

Since α is an isomorphism, so is α^*. By Theorem 8.5.3, (A^*, α^*) is recursive. Finally, by Proposition 7.4.2(2), (A^*, α^*) is the initial algebra. □

Corollary 8.5.12. *If F in Theorem 8.5.11 has a terminal coalgebra νF, it also has an initial algebra which is the well-founded part of νF.*

Remark 8.5.13. For set functors, the assumption that F preserves monomorphisms may be dropped in Theorem 8.5.11 and Corollary 8.5.12. For Theorem 8.5.11, we modify the proof above, replacing Theorem 6.3.4 by Corollary 6.3.9, and Theorem 8.5.3 by Corollary 8.5.5.

Example 8.5.14. We illustrate that for a set functor F, the well-founded part of the terminal coalgebra is the initial algebra.

(1) Consider $FX = \Sigma \times X + 1$. The terminal coalgebra is carried by the set $\Sigma^* + \Sigma^\infty$ of finite and infinite sequences in Σ. The initial algebra is Σ^*. It is easy to check that Σ^* is the well-founded part of $\Sigma^* + \Sigma^\infty$.

(2) For a polynomial set functor H_Σ the terminal coalgebra νH_Σ is formed by all Σ-trees (see Example 6.5.6). Its well-founded part is given by all well-founded Σ-trees (see Example 6.1.16).

Example 8.5.15. When the functor F does not preserve monomorphisms, the well-founded part of the terminal coalgebra need not be the initial algebra. Indeed, for the functor F in Example 8.5.7, the initial algebra is carried by $(\mathbb{N}, \emptyset)$, which is not even a subcoalgebra of $\mathbb{1}$.

We have seen in Corollary 6.3.7 that under the assumptions of the Initial-Algebra Theorem (Theorem 6.3.4), the initial algebra is the only minimal fixed point. There is another way of verifying this:

Proposition 8.5.16. *Every minimal algebra (A, α) which is a fixed point is, considered as a coalgebra, well-founded.*

Proof We show that (A, α^{-1}) is well-founded. Let (B, β) be a cartesian subcoalgebra of (A, α^{-1}). So we have a monic m and a pullback square as shown in the following diagram:

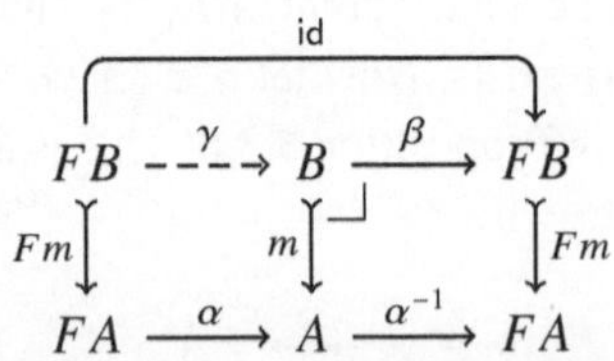

The outside clearly commutes, giving γ as shown. So (B, γ) is a subalgebra of (A, α). By minimality, m is an isomorphism, as desired. □

Corollary 8.5.17. *Under the assumption of Theorem 8.5.11, every minimal algebra which is a fixed point is initial.*

8.6 The Converse of the General Recursion Theorem

Our last topic in this chapter is a converse to Theorem 8.5.3, under various hypotheses. This generalizes results by Adámek et al. [34] and Jeannin et al. [175]. In order to prove the implication

$$\text{recursive} \implies \text{well-founded},$$

one needs to assume more than preservation of finite intersections. In fact, we will assume that F preserves inverse images. But even this is not enough. We additionally assume that

(1) either the underlying category $\mathscr{A}$ has universally smooth monomorphisms and the endofunctor F has a *pre-fixed point*, that is, an object A with a monomorphism $\alpha\colon FA \rightarrowtail A$.

(2) or the underlying category $\mathscr{A}$ has a subobject classifier.

The first case leads to Theorem 8.6.1 which appears in work by Adámek et al. [53], the second is a theorem derived from Taylor's preprints [286, 288]. Finally, at the end of this section we prove the above converse implication for every functor on vector spaces preserving inverse images (see Theorem 8.6.13). This last result is not covered by the previous two, since K-Vec neither has universally smooth monomorphisms nor a subobject classifier. We have not found a unifying proof that would cover the cases we consider.

Theorem 8.6.1 [53, Thm. 5.5]. *Let $\mathscr{A}$ be a complete and well-powered category with universally smooth monomorphisms, and suppose that $F\colon \mathscr{A} \to \mathscr{A}$ preserves inverse images and has a pre-fixed point. Then every recursive F-coalgebra is well-founded.*

Proof By Theorem 6.3.4, we know that the initial algebra μF exists. Let (A, α) be a recursive coalgebra. We have a unique coalgebra-to-algebra morphism f from (A, α) to $(\mu F, \iota)$, that is, a unique coalgebra homomorphism $f\colon (A, \alpha) \to (\mu F, \iota^{-1})$. By Example 8.1.4(1), the coalgebra $(\mu F, \iota^{-1})$ is well-founded. Hence, so is (A, α) by Proposition 8.4.14. □

Corollary 8.6.2. *Let $\mathscr{A}$ and F satisfy the assumptions of Theorem 8.6.1. Then the following properties of a coalgebra are equivalent:*

(1) *well-foundedness,*

(2) *parametric recursiveness,*

(3) *recursiveness,*

(4) *existence of a homomorphism into $(\mu F, \iota^{-1})$,*

(5) *existence of a homomorphism into a well-founded coalgebra.*

Proof The assumption that F preserves inverse images implies that F preserves monomorphisms. Thus Theorem 8.5.3 applies and shows that (1) $\Rightarrow$ (2) $\Rightarrow$ (3) by Theorem 8.5.3. We have also seen (3) $\Rightarrow$ (4) $\Rightarrow$ (1) in the proof of Theorem 8.6.1. The implication (4) $\Rightarrow$ (5) holds since $(\mu F, \iota^{-1})$ is well-founded (see Example 8.1.4(1)). Finally, by Corollary 8.5.8, $(\mu F, \iota^{-1})$ is a terminal well-founded coalgebra. Thus, (5) $\Rightarrow$ (4), which completes the proof. □

Remark 8.6.3. Observe that one can replace universally smooth monomorphisms in Corollary 8.6.2 by a universally smooth class $\mathcal{M}$ of monomorphisms. Taking as $\mathcal{M}$ the class of all strong monomorphisms, this yields examples such as posets, graphs, and topological spaces (where the strong monomorphisms represent subposets, subgraphs, and subspaces, respectively).

Example 8.6.4. (1) The category of many-sorted sets satisfies the assumptions of Theorem 8.6.1, and polynomial endofunctors on that category preserve inverse images. Thus, we obtain as a special instance of the result due to Jeannin et al. [175, Thm. 3.3] that conditions (1)–(4) in Corollary 8.6.2 are equivalent.

(2) Vector spaces fail to have universally smooth monomorphisms (see Example 8.4.13(2)). We will see in Theorem 8.6.13 that nevertheless, 'recursive implies well-founded' holds whenever F preserves inverse images. On the other hand, the implication (4) $\Rightarrow$ (3) in Corollary 8.6.2 does not even hold for the identity functor on K-Vec. To see this, note that we have $\mu\mathsf{Id} = (0, \mathrm{id})$. Hence, every coalgebra has a homomorphism into $\mu\mathsf{Id}$. However, not every coalgebra is recursive, for example the coalgebra (K, id) admits many coalgebra-to-algebra morphisms to the algebra (K, id). Similarly, the implication (4) $\Rightarrow$ (1) does not hold. In fact, a coalgebra $\alpha\colon A \to A$ is well-founded iff for every $x \in A$ there exists a natural number n with $\alpha^n(x) = 0$ (cf. Example 8.1.4(4)). Clearly, not every coalgebra satisfies this property.

We mention in passing a consequence of the general recursion theorem which uses universally smooth monomorphisms but not preservation of inverse images.

Corollary 8.6.5. *Let $\mathscr{A}$ satisfy the assumptions of Theorem 8.6.1, let $F\colon \mathscr{A} \to \mathscr{A}$ preserve intersections and have an initial algebra. Then every subcoalgebra of μF is recursive.*

Indeed, every subcoalgebra of μF is well-founded by Proposition 8.4.7 and Corollary 8.4.15 and therefore recursive by Theorem 8.5.3.

In Theorem 8.6.1, we assumed that the endofunctor has a pre-fixed point. For set functors, this assumption may be lifted. Indeed, whenever a category has a subobject classifier, then every recursive coalgebra is well-founded, as shown by

Taylor [286, Rem. 3.8] (see also [288, Prop. 2.10]). We present this in all details, beginning with a recollection of the following definition from topos theory.

Definition 8.6.6. Given a category with a terminal object 1, a *subobject classifier* is an object Ω with a subobject $\mathrm{t}\colon 1 \rightarrowtail \Omega$ such that for every subobject $b\colon B \rightarrowtail A$ there is a unique morphism $\hat{b}\colon A \to \Omega$ such that the square below is a pullback:

$$\begin{array}{ccc} B & \xrightarrow{\;!\;} & 1 \\ {\scriptstyle b}\big\downarrow & \lrcorner & \big\downarrow{\scriptstyle t} \\ A & \xrightarrow{\;\hat{b}\;} & \Omega \end{array} \tag{8.13}$$

Example 8.6.7. (1) Set has a subobject classifier given by $\Omega = \{\bot, \top\}$ with the evident morphism $\mathrm{t}\colon 1 = \{\top\} \hookrightarrow \Omega$. Indeed, subsets $b\colon B \hookrightarrow A$ are in one-to-one correspondence with characteristic maps $\hat{b}\colon B \to \{\bot, \top\}$.

(2) The category of multigraphs (Example 8.5.10) has a subobject classifier Ω given by the complete graph on $\{\bot, \top\}$ with the morphism $\mathrm{t}\colon 1 \to \Omega$ where 1 is the terminal multigraph given by one node with a loop.

(3) More generally, every elementary topos has a subjobject classifier; for example, every category $\mathsf{Set}^{\mathscr{C}}$ with $\mathscr{C}$ small.

Our standing assumption that $\mathscr{A}$ is a complete and well-powered category is not needed for the next result: finite limits are sufficient. The following proof is based on Taylor's work [286, Rem. 3.8].

Theorem 8.6.8 [286]. *Let F be an endofunctor on a finitely complete category with a subobject classifier. If F preserves inverse images, then every recursive coalgebra is well-founded.*

Proof Let (A, α) be a recursive coalgebra. Clearly, id_A is a fixed point of $\bigcirc$, and we prove below that it is the unique one. Thus, (A, α) is well-founded.

Let $b\colon B \to A$ be any fixed point of $\bigcirc$. Consider the following diagram:

$$\begin{array}{ccccccc} & & & \overset{!}{\frown} & & & \\ B & \xrightarrow{\;\alpha(b)\;} & FB & \xrightarrow{\;F!\;} & F1 & \xrightarrow{\;!\;} & 1 \\ {\scriptstyle b}\big\downarrow & \lrcorner & {\scriptstyle Fb}\big\downarrow & \lrcorner & {\scriptstyle F\mathrm{t}}\big\downarrow & \lrcorner & \big\downarrow{\scriptstyle \mathrm{t}} \\ A & \xrightarrow{\;\alpha\;} & FA & \xrightarrow{\;F\widehat{b}\;} & F\Omega & \xrightarrow{\;\widehat{F\mathrm{t}}\;} & \Omega \\ & & & \underset{\widehat{b}}{\smile} & & & \end{array}$$

The square on the left is a pullback because $b = \bigcirc b$. The middle square is F applied to the pullback square (8.13). The square on the right is the pullback square (8.13) for b chosen as $F\mathrm{t}\colon F1 \to F\Omega$. The upper composite morphism is $!\colon B \to 1$, and so the lower one is $\widehat{b}$. Thus the outside rectangle is again a pullback. In particular,

$$\widehat{b} = \widehat{F\mathsf{t}} \cdot F\widehat{b} \cdot \alpha.$$

So $\widehat{b}$ is a coalgebra-to-algebra morphism

$$\begin{array}{ccc} A & \xrightarrow{\ \alpha\ } & FA \\ {\scriptstyle \widehat{b}}\Big\downarrow & & \Big\downarrow{\scriptstyle F\widehat{b}} \\ \Omega & \xleftarrow{\ \widehat{Ft}\ } & F\Omega \end{array}$$

Since (A, α) is recursive, this means that $\widehat{b}$ is uniquely determined from α, independent of which fixed point b of $\bigcirc$ was used in our argument. Thus, $\widehat{b} = \widehat{\mathsf{id}_A}$, which implies $b = \mathsf{id}_A$ in $\mathsf{Sub}(A)$, as desired. □

Corollary 8.6.9. *For every set functor preserving inverse images, recursive coalgebras are well-founded.*

Corollary 8.6.10. *Let $\mathscr{A}$ and F satisfy the assumptions of Theorem 8.6.8. Then the following properties of a coalgebra are equivalent:*

$$\textit{well-foundedness} \iff \textit{recursiveness} \iff \textit{parametric recursiveness}.$$

Example 8.6.11. The hypothesis in Theorems 8.6.1 and 8.6.8 that F preserves inverse images cannot be lifted. To demonstrate this, we consider the functor $R\colon \mathsf{Set} \to \mathsf{Set}$ of Example 7.4.6. It preserves monomorphisms but not inverse images (Example C.3.8(6)). The coalgebra $A = \{0, 1\}$ with the structure α constant to $(0, 1)$ is recursive: given an algebra $\beta\colon RB \to B$, the unique coalgebra-to-algebra homomorphism $h\colon \{0, 1\} \to B$ is given by

$$h(0) = h(1) = \beta(d).$$

But A is not well-founded: $\emptyset$ is a cartesian subcoalgebra.

As the last result of this section we prove the implication 'recursive $\Rightarrow$ well-founded' for endofunctors on the category K-Vec preserving inverse images. This follows neither from Theorem 8.6.1 (since monomorphisms are not universally smooth in K-Vec) nor from Theorem 8.6.8 (since K-Vec does not have a subobject classifier).

Remark 8.6.12. (1) Given a vector space X, we denote by $z\colon 0 \to X$ the zero map. We also use this notation for composites $z\colon X \to 0 \to Y$.

(2) Recall that the *kernel* of a linear map $f\colon X \to Y$ is the equalizer of f and the zero map $z\colon X \to 0 \to Y$ represented by the subspace $\ker f = \{x \in X : f(x) = 0\}$.

(3) A functor $F\colon K\text{-Vec} \to K\text{-Vec}$ *preserves kernels* if for every linear map $f\colon X \to Y$ its kernel $s\colon \ker f \rightarrowtail X$ is mapped to the kernel of Ff, shortly $Fs = \ker Ff$.

(4) Observe that for every linear map $f\colon X \to Y$ its kernel $s\colon \ker f \rightarrowtail X$ is the inverse image of the zero map $z\colon 0 \rightarrowtail Y$.

(5) If $F\colon K\text{-Vec} \to K\text{-Vec}$ preserves inverse images and $F0 = 0$, then it preserves kernels. Indeed, Fs is then the inverse image of Fz under Ff, and $Fz\colon 0 = F0 \rightarrowtail FY$ is the zero map. Thus Fs is the kernel of Ff as desired.

(6) Conversely, if F preserves kernels, then $F0 = 0$ (the terminal object) and F preserves inverse images. In fact, F preserves finite limits: by [131, Thm. 3.12], a functor preserving kernels is additive, and for an additive functor preservation of kernels is equivalent to preservation of finite limits [91, Prop. 1.11.2].

(7) Every subspace $s\colon S \rightarrowtail X$ induces a quotient space X/S with vectors $x + S$ for $x \in X$. We denote the corresponding canonical quotient map $x \mapsto x + S$ by $\operatorname{coker} s\colon S \to X/S$. Its kernel is S.

(8) Every linear map $f\colon X \to Y$ induces an isomorphism $X \cong \ker f + f[X]$, where $f[X]$ denotes the image of f in Y.

(9) For a given linear map $f\colon X \to Y$ and a subspace $s\colon S \rightarrowtail Y$ let $t\colon T = f^{-1}[S] \rightarrowtail Y$. By the universal property of $\operatorname{coker} t$ there exists a unique monomorphism $u\colon X/T \rightarrowtail Y/S$ such that the following diagram commutes:

$$
\begin{array}{ccc}
T & \longrightarrow & S \\
{\scriptstyle t}\big\downarrow & \lrcorner & \big\downarrow{\scriptstyle s} \\
X & \xrightarrow{\;f\;} & Y \\
{\scriptstyle \operatorname{coker} t}\big\downarrow & & \big\downarrow{\scriptstyle \operatorname{coker} s} \\
X/T & \overset{u}{\dashrightarrow} & Y/S
\end{array}
$$

Moreover, we see that u is injective: if $x + T$ satisfies $u(x + T) = 0$, that is, $f(x) + S = 0$, then we have $f(x) \in S$, thus $x \in T$.

(10) A product $X \times Y$ in K-Vec is also a coproduct: the coproduct injections are $\langle \mathrm{id}_X, z\rangle\colon X \to X \times Y$ and $\langle z, \mathrm{id}_Y\rangle\colon Y \to X \times Y$.

Theorem 8.6.13 [53, Thm. 5.12]. *Let F be an endofunctor on K-Vec preserving inverse images. Then every recursive coalgebra is well-founded.*

Proof Let $\alpha\colon A \to FA$ be a recursive coalgebra and let $a^*\colon (A^*, \alpha^*) \rightarrowtail (A, \alpha)$ be its well-founded part.

(1) Assume first that $F0 = 0$. Then F preserves zero maps and kernels by Remark 8.6.12(5). Therefore Fa^* is the kernel of $F(\operatorname{coker} a^*)$ as shown in the

following diagram:

$$\begin{array}{ccc} A^* & \xrightarrow{\alpha^*} & FA^* \\ {\scriptstyle a^*}\big\downarrow & \lrcorner & \big\downarrow{\scriptstyle Fa^*} \\ A & \xrightarrow{\alpha} & FA \\ {\scriptstyle \operatorname{coker} a^*}\big\downarrow & & \big\downarrow{\scriptstyle F(\operatorname{coker} a^*)} \\ A/A^* & \underset{e}{\overset{u}{\rightleftarrows}} & F(A/A^*) \end{array}$$

Since $\operatorname{coker} a^*$ is epic, so is $F(\operatorname{coker} a^*)$ because epimorphisms split in K-Vec (Remark 8.5.4). Thus, we have $F(\operatorname{coker} a^*) = \operatorname{coker}(Fa^*)$, and by Remark 8.6.12(9) we obtain the unique monomorphism $u\colon A/A^* \rightarrowtail F(A/A^*)$ such that the diagram above commutes. Choose a splitting $e\colon F(A/A^*) \twoheadrightarrow A/A^*$, so that $e \cdot u = \mathrm{id}$. It follows that $q = \operatorname{coker} a^*$ is a coalgebra-to-algebra morphism from (A, α) to $(A/A^*, e)$. Indeed, we obtain

$$e \cdot Fq \cdot \alpha = e \cdot u \cdot q = q.$$

Since F preserves zero morphisms, the zero morphism $z\colon A \to A/A^*$ is also a coalgebra-to-algebra morphism. Consequently, $q = z$, which is equivalent to a^* being an isomorphism $A \cong A^*$, as desired.

(2) Let F be arbitrary, and put $R = F0$. We verify that there is an endofunctor G on K-Vec with $G0 = 0$ and preserving inverse images such that $FX = R \times GX$. Indeed, for every vector space X, let $t_X\colon X \to 0$ denote the zero map, and let $k_X\colon GX \rightarrowtail FX$ be the kernel of Ft_X. For every linear map $f\colon X \to Y$ the equality $t_X = t_Y \cdot f$ implies that Ff yields a linear map Gf making the following square commutative:

$$\begin{array}{ccc} GX & \overset{k_X}{\rightarrowtail} & FX \\ {\scriptstyle Gf}\big\downarrow & & \big\downarrow{\scriptstyle Ff} \\ GY & \underset{}{\overset{k_Y}{\rightarrowtail}} & FY \end{array}$$

It is easy to verify that this defines an endofunctor G and $k\colon G \rightarrowtail F$ is a natural transformation. Observe that t_X is a split epimorphism (whose splitting is the zero morphism $z_X\colon 0 \to X$), whence Ft_X is a split epimorphism with splitting $Fz_X\colon R \to FX$. Using Remark 8.6.12(9), this implies that $FX \cong R + GX$ with coproduct injections Fs_X and k_X. Since $+$ is also product (Remark 8.6.12(10)), we obtain $FX \cong R \times GX$ as desired.

(3) We prove that G preserves kernels. By Remark 8.6.12(6), G then preserves finite limits, whence inverse images. Suppose that $s = \ker f$ so that we have the

pullback on the left below

$$
\begin{array}{ccc}
S & \longrightarrow & 0 \\
{\scriptstyle s}\big\downarrow & \lrcorner & \big\downarrow \\
X & \xrightarrow{\ f\ } & Y
\end{array}
\qquad\qquad
\begin{array}{ccc}
GS & \longrightarrow & 0 \\
{\scriptstyle Gs}\big\downarrow & \lrcorner & \big\downarrow \\
GX & \xrightarrow{\ Gf\ } & GY
\end{array}
$$

It is our task to prove that the square on the right above is a pullback. Since F preserves inverse images, applying it to left-hand square yields the pullback square below: note here that $G0 = 0$ implies that $F(S \to 0)$ is the product projection $\pi\colon R \times GS \to R$ and that $F(0 \to Y)$ is the coproduct injection $i\colon R \to R \times GY$:

$$
\begin{array}{ccc}
R \times GS & \xrightarrow{\quad \pi \quad} & R \\
{\scriptstyle R\times Gs}\big\downarrow & \lrcorner & \big\downarrow{\scriptstyle i} \\
R \times GX & \xrightarrow{\ R\times Gf\ } & R \times GY
\end{array}
$$

Now suppose we have $g\colon Z \to GX$ with $Gf \cdot g = z$, where $z\colon Z \to 0 \to GY$ is the zero morphism. We prove that g factorizes through Gs. For the zero morphism $z'\colon Z \to 0 \to R$ we clearly have

$$(R \times Gf) \cdot \langle z', g \rangle = \langle z', z \rangle = i \cdot z',$$

since the latter two are both the zero morphism $Z \to R \times GY$. Therefore, there is a unique morphism $h\colon Z \to R \times GS$ with $(R \times Gs) \cdot h = \langle z', g \rangle$ and $\pi \cdot h = z'$. This implies that $h = \langle z', h' \rangle$ for a unique morphism $h'\colon Z \to GS$ such that $Gs \cdot h' = g$, which proves the claim.

(4) Our recursive coalgebra has the form $\alpha = \langle \alpha_1, \alpha_2 \rangle\colon A \to R \times GA$. It yields a coalgebra $\alpha_2\colon A \to GA$, and we prove that it is recursive, too. Indeed, given any algebra $\beta\colon GB \to B$, we use the zero morphism $z\colon R \to GB$ to get the following algebra for F:

$$R \times GB \cong R + GB \xrightarrow{[z,\beta]} GB.$$

Now observe that a morphism $h\colon A \to B$ is a coalgebra-to-algebra morphism for F

$$
\begin{array}{ccc}
A & \xrightarrow{\langle \alpha_1, \alpha_2 \rangle} & R \times GA \\
{\scriptstyle h}\big\downarrow & & \big\downarrow{\scriptstyle R\times Gh} \\
B & \xleftarrow{\ [0,\beta]\ } & R \times GB
\end{array}
$$

iff it is a coalgebra-to-algebra morphism from (A, α_2) to (B, β) for G. Since the former exists uniquely, so does the latter. This proves that (A, α_2) is recursive for G.

Since $G0 = 0$, we use item (1) to see that the coalgebra (A, α_2) is well-founded for G. Its next-time operator $\bigcirc$ is the same as that of the F-coalgebra (A, α) because in the diagram below the outside is a pullback iff the left-hand square is:

$$\begin{array}{ccccc} \bigcirc S & \xrightarrow{\alpha(s)} & R \times GS & \xrightarrow{\pi_r} & GS \\ {\scriptstyle \bigcirc s}\downarrow \lrcorner & & \downarrow{\scriptstyle R\times Gs} & & \downarrow{\scriptstyle Gs} \\ At & \xrightarrow{\alpha} & R \times GA & \xrightarrow{\pi_r} & GA \end{array} \qquad \alpha_2 : At \to GA$$

Since id_A is the unique fixed point of $\bigcirc$ with respect to G, it is also the unique fixed point with respect to F. Thus (A, α) is well-founded for F, as desired. □

Corollary 8.6.14. *For every functor on K-*Vec *preserving inverse images, the following properties of a coalgebra are equivalent:*

$$\textit{well-foundedness} \iff \textit{parametric recursiveness} \iff \textit{recursiveness}.$$

8.7 Summary

We have presented well-founded coalgebras, thereby capturing the concept of well-founded induction on an abstract level. We have also provided a new proof of Taylor's General Recursion Theorem stating that every well-founded coalgebra is parametrically recursive. This holds for functors preserving monomorphisms on a complete and well-founded category with smooth monomorphisms. This even holds for *every* set functor, and the converse holds for set functors preserving inverse images. Moreover, for every set functor the initial algebra is, equivalently, the terminal well-founded coalgebra.

We have also provided an iterative construction of the well-founded part of a given coalgebra. It is carried by the least fixed point of Jacobs' next-time operator. In addition, the well-founded part yields the coreflection of a coalgebra in the category of well-founded coalgebras.

In Chapter 9 we will meet well-founded coalgebras again in a different role. There we will describe the initial algebra of set a functor as the algebra consisting of all well-founded, well-pointed coalgebras up to isomorphism.

Finally, let us remark that a dual notion to well-foundedness has been studied by Capretta, Uustalu, and Vene [101].

9
State Minimality and Well-Pointed Coalgebras

The notions and constructions we present in this chapter are inspired by the well-known concept of *minimality* for deterministic automata. Recall that a *minimal* deterministic automaton is one that is *reachable* and *observable*: every state is reachable from the initial state, and distinct states accept distinct languages. Both notions turn out to have equivalent element-free characterizations: a deterministic automaton is observable iff it has no proper quotient (via a coalgebra homomorphism, see Example 2.4.2(3)), and it is reachable iff it has no proper subcoalgebra containing the initial state. Both of these element-free characterizations can be formulated for coalgebras in general. These formulations are the main topic of this chapter.

We first turn to the study of *simple* coalgebras, that is, those having no proper quotient coalgebra, in Section 9.1. Under rather mild assumptions, we show how to construct for every coalgebra its simple quotient.

In Section 9.2 we study reachable coalgebras. The coalgebraic formulation of that property requires us to consider *pointed* coalgebras. These are coalgebras equipped with a *point*, a morphism from the terminal object, modelling an initial state. Again under mild assumptions, every pointed coalgebra has what we call a *reachable part*: the largest reachable subcoalgebra. Moreover, we construct this reachable part in a way parallel to what we have seen for the well-founded part in Chapter 8.

Finally, in Section 9.3, we use the foregoing work to provide new descriptions of initial algebras and terminal coalgebras for set functors preserving intersections. We define *well-pointed* coalgebras as those which are both simple and reachable and thus model state minimality of coalgebras. We present a construction of the well-pointed modification of a given coalgebra much in the spirit of the well-known minimization of deterministic automata. We shall see that the set of all well-pointed F-coalgebras, considered up to isomorphism, carries the terminal coalgebra νF. Moreover, the subset given by the well-founded and well-pointed F-coalgebras carries the initial algebra μF.

9.1 Simple Coalgebras

In this section we study coalgebras having no proper quotient. This property generalizes the notion of an *observable* deterministic automaton, one where distinct states accept distinct languages.

Assumption 9.1.1. Throughout this section, $\mathscr{A}$ denotes a cocomplete, well-powered, and co-well-powered category.

Remark 9.1.2. (1) For every endofunctor F on $\mathscr{A}$, Coalg F is cocomplete with colimits formed on the level of $\mathscr{A}$. This follows from Proposition 4.1.1, since the forgetful functor of Coalg F creates all colimits.

(2) We work with factorizations of morphisms f as $f = m \cdot e$, where e is a strong epimorphism and m is a monomorphism, which exist by Remark 7.2.1(3). Furthermore, recall from Proposition 7.2.2 that these factorizations lift to homomorphisms in Coalg F for every endofunctor F preserving monomorphisms (a condition that can be dropped for set functors, see Remark 7.2.5(1)).

(3) In this chapter we work with *strong quotient coalgebras* (see Definition 8.4.2), that is, quotients represented by strong epimorphisms in the base category.

(4) We recall some basic facts about strong epimorphisms:

(a) A morphism which is both a strong epimorphism and a monomorphism is an isomorphism (Remark 7.2.1(2)).

(b) Strong epimorphisms are stable under pushout [90, Prop. 4.3.8]. That is, in the diagram below e' is a strong epimorphism if e is one:

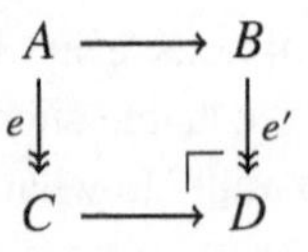

Analogously, a wide pushout of a collection $e_i \colon A \twoheadrightarrow B_i$ $(i \in I)$ of strong epimorphisms is formed by strong epimorphisms.

(c) Strong epimorphisms are *extremal*: if a strong epimorphism $e \colon A \twoheadrightarrow B$ factorizes through a monomorphism $m \colon S \rightarrowtail B$, then m is an isomorphism [90, Prop. 4.3.6].

(d) Every coequalizer is a strong epimorphism [90, Prop. 4.3.6].

(5) A wide pushout of a span of strong epimorphisms is a *cointersection*. Thus,

our assumptions imply that all cointersections exist in $\mathscr{A}$. Moreover, for every endofunctor F all cointersections exist in Coalg F, and they are formed on the level of $\mathscr{A}$.

Definition 9.1.3. A coalgebra (A, α) is called *simple* if it has no proper strong quotient coalgebra: every strong quotient coalgebra of (A, α) is an isomorphism.

Example 9.1.4. (1) If F has a terminal coalgebra, then it is simple. Indeed, given any strong quotient coalgebra $e\colon \nu F \twoheadrightarrow A$, the unique homomorphism $f\colon A \to \nu F$ satisfies $f \cdot e = \mathsf{id}_{\nu F}$. Thus, e is a split monomorphism, and since it is also an epimorphism, it is an isomorphism.

(2) A deterministic automaton, considered as a coalgebra for $FX = \{0, 1\} \times X^{\Sigma}$ is simple iff it is observable. To see this, let (A, α) be simple and consider the unique coalgebra homomorphism $f\colon A \to \nu F$ assigning to every state the formal language it accepts (see Example 2.5.5). Take a (strong epi, mono)-factorization

$$f = \big(A \xrightarrow{e} C \xrightarrow{m} \nu F\big)$$

to obtain its image C in νF. Then C is a quotient of A via e, and therefore it is an isomorphism by simplicity. It follows that two states x and y in A accept the same language (i.e. $f(x) = f(y)$) iff $e(x) = e(y)$. Equivalently, we have $x = y$.

(3) A graph considered as a coalgebra for $\mathscr{P}$ is simple iff no distinct nodes are bisimilar. Indeed, for every graph (A, α) the largest bisimulation $\sim$ on it is a congruence (see Remark 4.2.6): the quotient $A/{\sim}$ carries the structure of a $\mathscr{P}$-coalgebra such that the canonical quotient map $e\colon A \twoheadrightarrow A/{\sim}$ is a coalgebra homomorphism. Thus, (A, α) is simple iff e is an isomorphism, which is equivalent to saying that for every bisimilar pair x, y in A we have $x = y$.

Proposition 9.1.5 [168]. *Every coalgebra has a unique simple quotient represented by the cointersection*

$$e_{(A,\alpha)}\colon (A, \alpha) \twoheadrightarrow (\bar{A}, \bar{\alpha})$$

of all strong quotient coalgebras of (A, α). This is the reflection of (A, α) in the full subcategory of Coalg F *given by all simple coalgebras.*

Proof (1) The above cointersection $e_{(A,\alpha)}$ is a strong quotient coalgebra by Remark 9.1.2(1) and (4). To see that $(\bar{A}, \bar{\alpha})$ is simple, consider any strong quotient coalgebra $q\colon (\bar{A}, \bar{\alpha}) \twoheadrightarrow (B, \beta)$. Then $q \cdot e_{(A,\alpha)}\colon (A, \alpha) \twoheadrightarrow (B, \beta)$ is a strong quotient coalgebra. Hence, by the construction of $e_{(A,\alpha)}$ we have a homomorphism $f\colon (B, \beta) \to (\bar{A}, \bar{\alpha})$ such that $f \cdot q \cdot e_{(A,\alpha)} = e_{(A,\alpha)}$. Since $e_{(A,\alpha)}$ is epic, we obtain $f \cdot q = \mathsf{id}_B$. Thus, q is a strong epimorphism which is also monic, hence an isomorphism.

(2) In order to prove that $e_{(A,\alpha)}$ is a reflection, let $h\colon (A,\alpha) \to (B,\beta)$ be any coalgebra homomorphism with (B,β) simple. Form a pushout in $\mathsf{Coalg}\, F$ as shown below:

$$\begin{array}{ccc} A & \xrightarrow{h} & B \\ {\scriptstyle e_{(A,\alpha)}}\big\downarrow & & \big\downarrow{\scriptstyle \bar{e}} \\ \bar{A} & \xrightarrow{\bar{h}} & \bar{B} \end{array}$$

By Remark 9.1.2(4)(b) $\bar{e}$ is strongly epic in the base category. Since (B,β) is simple, we thus know that $\bar{e}$ is an isomorphism and therefore $k = \bar{e}^{-1} \cdot \bar{h}$ is the desired coalgebra homomorphism satisfying $k \cdot e_{(A,\alpha)} = h$. Uniqueness of k is clear, since $e_{(A,\alpha)}$ is epic. □

Observation 9.1.6. (1) If F preserves monomorphisms, then a coalgebra (A,α) is simple iff every homomorphism with domain (A,α) is monic in $\mathscr{A}$. Indeed, given a coalgebra homomorphism $f\colon (A,\alpha) \to (B,\beta)$ with (A,α) simple, we obtain its image by taking the (strong epi, mono) factorization $f = m \cdot e$ in $\mathscr{A}$ (see Remark 7.2.3). Then (A,α) must be isomorphic to its image via e, which implies that f is monic.

Conversely, suppose that every coalgebra homomorphism with domain (A,α) is monic. Then, in particular the simple quotient $e_{(A,\alpha)}\colon (A,\alpha) \twoheadrightarrow (\bar{A},\bar{\alpha})$ from Proposition 9.1.5 is monic. Hence it is an isomorphism.

(2) For every pair of coalgebra homomorphisms $f_1, f_2\colon (A,\alpha) \to (B,\beta)$ where (B,β) is simple we have $f_1 = f_2$. Indeed, take the coequalizer e of f_1, f_2 in $\mathsf{Coalg}\, F$. Then e is also a coequalizer in the base category, thus it is an isomorphism by simplicity (Remark 9.1.2(4)(d)). It follows that $f_1 = f_2$.

Remark 9.1.7. If the terminal coalgebra νF exists, then the simple reflection of a coalgebra A is given by the image of the unique coalgebra homomorphism $A \to \nu F$. Consequently, an F-coalgebra A is simple iff the unique homomorphism $h\colon A \to \nu F$ is a monomorphism. We leave the easy proof to the reader.

Theorem 9.1.8. *If F preserves monomorphisms, then the following are equivalent:*

(1) *F has a terminal coalgebra, and*

(2) *F has only a set of simple coalgebras up to isomorphism (of* $\mathsf{Coalg}\, F$*).*

Proof Suppose that a terminal coalgebra νF exists. Since $\mathscr{A}$ is well-powered, νF has only a set of subobjects in $\mathscr{A}$. Every simple coalgebra A gives such a subobject, since the coalgebra homomorphism $m_A\colon A \to \nu F$ is, by Observation 9.1.6(1), monic. Moreover, if A and B determine the same subobject, then m_A and m_B factor through each other and thus represent the same subobject.

Conversely, if (A_i, α_i) $(i \in I)$ is a representative set of all simple coalgebras for F, then we prove that their coproduct

$$(B, \beta) = \coprod_{i \in I} (A_i, \alpha_i)$$

is weakly terminal. The existence of νF then follows from Theorem 4.2.8. Let (A, α) be a coalgebra. Since $(\bar{A}, \bar{\alpha})$ in Proposition 9.1.5 is simple, we have an isomorphism $u\colon (\bar{A}, \bar{\alpha}) \to (A_i, \alpha_i)$ for some $i \in I$. Denote by $v_i\colon (A_i, \alpha_i) \to (B, \beta)$ the coproduct injection. Then $v_i \cdot u \cdot e_{(A,\alpha)}\colon (A, \alpha) \to (B, \beta)$ is the desired homomorphism. □

Remark 9.1.9. Note that for $\mathscr{A}$ = Set we may drop the assumption in Observation 9.1.6 and Theorem 9.1.8 that F preserves monomorphisms. This follows from an argument analogous to Remark 7.2.5(1).

Proposition 9.1.10. *Every subcoalgebra of a simple coalgebra is simple.*

Proof Suppose that (A, α) is simple, and let $m\colon (B, \beta) \rightarrowtail (A, \alpha)$ be a subcoalgebra. Given a strong quotient $e\colon (B, \beta) \twoheadrightarrow (C, \gamma)$, we prove that e is an isomorphism. Form the following pushout in Coalg F:

$$\begin{array}{ccc} (B, \beta) & \overset{m}{\rightarrowtail} & (A, \alpha) \\ {\scriptstyle e}\downarrow & & \downarrow{\scriptstyle \bar{e}} \\ (C, \gamma) & \xrightarrow{\bar{m}} & (\bar{C}, \bar{\gamma}) \end{array}$$

Since e is a strong epimorphism in $\mathscr{A}$, so is $\bar{e}$ by Remark 9.1.2(4), that is, $\bar{e}$ represents a strong quotient coalgebra in Coalg F. Hence, $\bar{e}$ is an isomorphism by simplicity. Thus, $\bar{m} \cdot e = \bar{e} \cdot m$ is monic, whence e is monic. Since e is a strong epimorphism which is also monic, it is an isomorphism, and we are done. □

9.2 Pointed and Reachable Coalgebras

In this section we study reachability on the level of coalgebras. For automata, this is the second ingredient of minimality, besides observability. Throughout this section we work with pointed coalgebras, which are coalgebras equipped with a point (a morphism from the terminal object 1) thought of as an initial state.

Reachability has an easy formulation: a pointed coalgebra is reachable iff it does not have any proper pointed subcoalgebras. We will also see an iterative construction of the reachable part of a given pointed coalgebra, which is reminiscent of the usual breadth-first search algorithm for computing the reachable part of a pointed graph. This construction appears in work by Barlocco et al. [67] and Wißmann et al. [314].

Assumption 9.2.1. Throughout this section $\mathscr{A}$ denotes a complete and well-powered category, and F an endofunctor preserving intersections.

We denote by 1 the terminal object of $\mathscr{A}$. Morphisms $1 \to A$ are called *points* (a.k.a. *global elements*) of the object A.

Definition 9.2.2 (Adámek et al. [47]). (1) By a *pointed coalgebra* is meant a triple (A, α, x) consisting of a coalgebra $\alpha\colon A \to FA$ and a point $x\colon 1 \to A$. The category

$$\mathsf{Coalg_p}\, F$$

of pointed coalgebras has as morphisms from (A, α, x) to (B, β, y) those coalgebra homomorphisms $f\colon (A, \alpha) \to (B, \beta)$ which preserve the point: $f \cdot x = y$.

(2) A *pointed subcoalgebra* of a pointed coalgebra is (represented by) a morphism in $\mathsf{Coalg_p}\, F$ carried by a monomorphism in $\mathscr{A}$.

(3) A pointed coalgebra (A, α, x) is called *reachable* if it has no proper pointed subcoalgebra. That is, for every pointed subcoalgebra $m\colon (A', \alpha', x') \rightarrowtail (A, \alpha, x)$ the morphism m is an isomorphism.

Remark 9.2.3. Since F preserves intersections, $\mathsf{Coalg_p}\, F$ has intersections formed on the level of $\mathscr{A}$. Indeed, the forgetful functor $U_p\colon \mathsf{Coalg_p}\, F \to \mathscr{A}$ creates intersections (and, more generally, all limits).

The proof is the same as that of Proposition 4.1.5. One just needs to observe that the given points $x_d\colon 1 \to A_d$ form a cone of $U_p D$. Thus, one obtains a unique point $x\colon 1 \to L$ making all limit projections π_d morphisms of pointed coalgebras.

Example 9.2.4. (1) A deterministic automaton with a given initial state is a pointed coalgebra for the set functor $FX = \{0, 1\} \times X^\Sigma$. Reachability means that every state can be reached (in finitely many steps) from the initial state.

(2) For the power-set functor $\mathscr{P}$ the pointed coalgebras are the pointed directed graphs. Reachability here means that every vertex can be reached by a directed path from the distinguished vertex. This property of a pointed graph is sometimes called *accessibility* [3].

(3) A labelled transition system with a given initial state is a pointed coalgebra

for $FX = \mathscr{P}(\Sigma \times X)$ on Set. It is reachable iff every state can be reached from the initial one by a finite sequence of transitions.

Definition 9.2.5. By the *reachable part* of a pointed coalgebra we mean its smallest pointed subcoalgebra.

Example 9.2.6. (1) For a deterministic automaton as a pointed coalgebra for $FX = \{0,1\} \times X^{\Sigma}$ the reachable part is the subautomaton given by all states reachable (by input words from Σ) from the initial state.
(2) Similarly, a labelled transition system as a pointed coalgebra for $\mathscr{P}(\Sigma \times A)$ has as reachable part the states reachable (by some sequence of actions from Σ) from the initial state.

Proposition 9.2.7. *Every pointed coalgebra has a reachable part. It is its unique reachable subcoalgebra.*

Proof (1) Fix a pointed F-coalgebra (A, α, x). Since $\mathscr{A}$ is well-powered we have a set I indexing all pointed subcoalgebras

$$m_i : (A_i, \alpha_i, x_i) \rightarrowtail (A, \alpha, x).$$

Let m be the intersection of all m_i in $\mathscr{A}$, for $i \in I$. By Remark 9.2.3 this yields the intersection in $\mathsf{Coalg_p}\, F$, which is a pointed subcoalgebra

$$m : (B, \beta, y) \rightarrowtail (A, \alpha, x).$$

We check that (B, β, y) is reachable. Let $i \in I$ and m' be such that $m' : (A_i, \alpha_i, x_i) \rightarrowtail (B, \beta, y)$ is a pointed subcoalgebra. By definition of m we have $m \leq m \cdot m'$, thus there is a monomorphism $n : B \rightarrowtail A'$ such that $m = m \cdot m' \cdot n$. Since m is monic, we obtain $\mathrm{id} = m' \cdot n$. Thus, m' is a split epimorphism and a monomorphism, whence it is an isomorphism, as desired.
(2) For the uniqueness, use Remark 9.2.3: given a reachable pointed coalgebra of (A, α, x) form its intersection with (B, β, y) to obtain a pointed subcoalgebra of (B, β, y). But the construction of B makes it clear that no proper pointed subcoalgebra of (B, β, y) exists. □

Remark 9.2.8. Note that for $\mathscr{A} = \mathsf{Set}$ it suffices for Proposition 9.2.7 that F preserves nonempty intersections. The following result is the converse:

Proposition 9.2.9. *Let G be a set functor such that every pointed coalgebra has a reachable part. Then G preserves nonempty intersections.*

Proof Let $m_i : B_i \hookrightarrow A$ $(i \in I)$ have intersection $m : B \hookrightarrow A$ with $B \neq \emptyset$. Since Gm and Gm_i are (split) monomorphisms, it is sufficient to take a collection $b_i \in GB_i$ so that all $Gm_i(b_i)$ are equal, say to $a \in GA$, and to find $b \in GB$

such that $Gm(b) = a$. We assume without loss of generality that $B_i \subseteq B \subseteq A$ for all $i \in I$ and that m and m_i are the inclusion maps. Choose $x \in B$ and form the pointed coalgebra (A, α, x), where α is the constant function of value a. We have a reachable subcoalgebra $n\colon (C, \gamma, x) \hookrightarrow (A, \alpha, x)$.

(1) The first case is when $m_i \cap n$ is nonempty for all i. Then $C \neq \emptyset$ as well, and it follows that Gn is a monomorphism. The composite $Gn \cdot \gamma = \alpha \cdot n$ is constant with value $Gn \cdot \gamma(x) = a$. Since Gn is monic, γ is constant with value $c = \gamma(x)$. For every i, the intersection of m_i and n is formed by the inclusion maps of $B_i^* = B_i \cap C$ as follows:

$$\begin{array}{ccc} B_i^* & \xrightarrow{m_i^*} & C \\ {\scriptstyle n_i^*}\downarrow & \lrcorner & \downarrow{\scriptstyle n} \\ B_i & \xrightarrow{m_i} & A \end{array}$$

Clearly, B_i^* contains the point x. By Proposition 4.4.1, G preserves this intersection. Consequently, from $Gm_i(b_i) = a = Gn(\gamma(x)) = Gn(c)$ we conclude that there exists $b_i^* \in GB_i^*$ with $Gm_i^*(b_i^*) = c$. Let $\beta_i\colon B_i \to GB_i$ be constant with value b_i^*. Since γ is constant with value $c = Gm_i^*(b_i^*)$ we obtain a pointed subcoalgebra

$$m_i^*\colon (B_i^*, \beta_i, x) \hookrightarrow (C, \gamma, x).$$

From the reachability of (C, γ, x) we conclude that m_i^* is an isomorphism, whence $n \subseteq m_i$. Since this holds for all i, we conclude that $n \subseteq m$, which means that we have $f\colon C \hookrightarrow B$ with $n = m \cdot f$. The desired element is therefore $b = Gf(c)$: we clearly have $Gm(b) = Gm(Gf(c)) = Gn(c) = a$.

(2) The second case is when $m_i \cap n$ is empty for some i. For every map $f\colon A \to A$, put $a_f = Gf(a)$. We claim that a_f is independent of the choice of f. To see this, let $g\colon A \to A$ be given. Since $m_i \cap n = \emptyset$, we can find $h\colon A \to A$ be which agrees with f on B_i and with g on C:

$$h \cdot m_i = f \cdot m_i \qquad \text{and} \qquad h \cdot n = g \cdot n. \tag{9.1}$$

This implies $Gf(a) = Gg(a)$ because using $Gn(c) = a = Gm_i(b_i)$ and (9.1) we have

$$\begin{aligned} Gf(a) &= Gf(Gm_i(b_i)) = Gh(Gm_i(b_i)) = Gh(a) = Gh(Gn(c)) \\ &= Gg(Gn(c)) = Gg(a). \end{aligned}$$

In particular, choose $f = \mathrm{id}_A$ and $g = m \cdot e$, where $e\colon A \to B$ splits m: $e \cdot m = \mathrm{id}_B$. Then $a = Gf(a) = Gm(Ge(a))$. Thus $b = Ge(a)$ is the desired point such that $Gm(b) = a$. □

Remark 9.2.10. Note that Proposition 9.2.9 does not hold with intersections in lieu of nonempty ones. In fact, the functor C_{01} with $C_{01}\emptyset = \emptyset$ and mapping nonempty sets to 1 is a counterexample. Every pointed coalgebra (A, α, x) contains a reachable subcoalgebra carried by $\{x\}$. However, C_{01} does not preserve the intersection of a pair of disjoint nonempty subsets.

Proposition 9.2.11. *Let F be an accessible set functor. Then F has only a set T of reachable coalgebras up to isomorphism.*

Proof Let λ be an infinite cardinal such that F is λ-accessible. Then F is also λ^+-accessible, where λ^+ is the successor cardinal of λ. By Proposition C.8.2 and Lemma C.8.3, given a coalgebra $\alpha\colon A \to FA$, every singleton subset $\{a\}$ is included in the carrier set of a subcoalgebra of size $\leq \lambda$. It follows that if (A, α, x) is reachable, then A is its own reachable part and thus itself has cardinality $\leq \lambda$. Let X be any set of size λ. For the set T as required in our result, we may take the set of coalgebras (Y, β), where $Y \subseteq X$. □

Remark 9.2.12. (1) Recall from Remark 8.2.2 that for every object A of $\mathscr{A}$ the poset $\mathsf{Sub}(A)$ of subobjects of A is a complete lattice with meets obtained by forming intersections.

(2) Since F preserves intersections, we know that for every F-coalgebra (A, α) the next-time operator $\bigcirc$ has a left adjoint by Proposition 8.2.17.

We are ready to proceed to an iterative construction of the reachable part of a coalgebra.

Definition 9.2.13. For every F-coalgebra (A, α) the *previous-time* operator is the left adjoint of the next-time operator. We denote it by

$$\ominus\colon \mathsf{Sub}(A) \to \mathsf{Sub}(A).$$

Remark 9.2.14. (1) More explicitly, $\ominus$ assigns to every subobject $s\colon S \rightarrowtail A$ in $\mathscr{A}$ the intersection of all subobjects m with $s \leq \bigcirc m$. This follows from the formula for the left adjoint given in Remark 8.2.4:

$$\ominus s = \bigwedge_{s \leq \bigcirc m} m.$$

(2) Thus, for the least subobject $\bot$ of A, we have $\ominus\bot = \bot$.

Proposition 9.2.15. *Let (A, α) be a coalgebra. The previous-time operator assigns to every subobject s of A the least subobject m of A for which $\alpha \cdot s$*

factorizes through Fm:

$$\begin{array}{ccc} S & \dashrightarrow & F(\ominus S) \\ {\scriptstyle s}\downarrow & & \downarrow{\scriptstyle F(\ominus s)} \\ A & \xrightarrow{\ \alpha\ } & FA \end{array}$$

Proof (1) We prove that $\alpha \cdot s$ factorizes through $F(\ominus s)$. Indeed, since F preserves intersections, $F(\ominus s) = \bigwedge_{s \leq \bigcirc m} Fm$. Thus, we just need to observe that for every $m \colon M \rightarrowtail A$ with $s \leq \bigcirc m$, $\alpha \cdot s$ factorizes through Fm:

$$\begin{array}{ccccc} S & \rightarrowtail\dashrightarrow & \bigcirc M & \xrightarrow{\ \alpha(m)\ } & FM \\ & {\scriptstyle s}\searrow & \downarrow{\scriptstyle \bigcirc m} \lrcorner & & \downarrow{\scriptstyle Fm} \\ & & A & \xrightarrow{\ \alpha\ } & FA \end{array}$$

(2) Conversely, given a subobject $m \colon M \rightarrowtail A$ with such a factorization, that is, we have $g \colon S \to FM$ with $\alpha \cdot s = Fm \cdot g$, it is our task to prove that $\ominus s \leq m$. One uses the universal property of the pullback to obtain the dashed arrow in the diagram below:

$$\begin{array}{ccccc} S & \rightarrowtail\dashrightarrow & \bigcirc M & \xrightarrow{\ \alpha(m)\ } & FM \\ & {\scriptstyle s}\searrow & \downarrow{\scriptstyle \bigcirc m} \lrcorner & & \downarrow{\scriptstyle Fm} \\ & & A & \xrightarrow{\ \alpha\ } & FA \end{array} \qquad (g \colon S \to FM \text{ along the top})$$

This proves that $s \leq \bigcirc m$, equivalently, $\ominus s \leq m$. □

The name of the operator $\ominus$ stems from the fact that it is a generalized semantic counterpart of the previous-time operator of classical linear temporal logic (see e.g. Manna and Pnüeli [221]), as we will now illustrate.

Example 9.2.16. (1) For the functor $FX = \mathscr{P}(\Sigma \times X)$ on Set, consider a coalgebra $\alpha \colon A \to \mathscr{P}(\Sigma \times A)$, that is, a labelled transition system. Then for every subset $S \subseteq A$, the set $\ominus S$ consists of those states which are reachable from S by a single transition:

$$\ominus S = \{ y \in A \mid y \in \alpha(s, x) \text{ for some } s \in \Sigma \text{ and } x \in S \}.$$

Cf. Example 8.2.7(2) on the next-time operator $\bigcirc$.

(2) Analogously for graphs as coalgebras for $\mathscr{P}$: given a set S of vertices, $\ominus S$ consists of all successor vertices of S.

(3) Finally, deterministic automata A considered as coalgebras for $FX = \{0, 1\} \times X^\Sigma$ have an analogous description of the previous-time operator: for a set $S \subseteq A$ of states, $\ominus S$ are the states reachable from a state in S by a single transition.

In the case where $\mathscr{A}$ = Set, the canonical graph (see Definition 8.1.9) of a given pointed F-coalgebra is a pointed graph. Moreover, the operator $\ominus$ can be computed on the canonical graph:

Corollary 9.2.17. *The previous-time operators of a coalgebra and its canonical pointed graph are the same.*

Indeed, this follows from Lemma 8.2.9 and the fact that left adjoints are unique.

Corollary 9.2.18. *A pointed coalgebra for an intersection preserving set functor is reachable iff so is its canonical pointed graph.*

This result can be used to prove a coalgebraic version of the well-known consequence of Kőnig's Lemma [191] that every finitely branching, well-founded and reachable graph (in the sense of Definition 9.2.2(3)) is finite. (Finitely branching means that every node has only finitely many neighbours.) This statement holds more generally for every finitary endofunctor on sets.

Proposition 9.2.19. *Every well-founded, reachable pointed coalgebra for a finitary set functor is finite.*

Note that in this proposition we do not assume that the given set functor preserves intersections.

Proof Let (A, α, x) be a well-founded, reachable coalgebra for the finitary set funtor F. Suppose first that F preserves intersections. Then we construct the canonical graph $G = (A, \tau_A \cdot \alpha)$. We see that G is finitely branching using that F is finitely bounded (see Proposition 4.3.4). Indeed, for every $x \in A$ we have a finite subset $s\colon S \hookrightarrow A$ and $a \in FS$ such that $Fs(a) = \alpha(x)$. Using that $\tau\colon F \to \mathscr{P}$ is 'subnatural' (see (8.4)), we obtain that $\tau_A(\alpha(x))$ lies in $\mathscr{P}S \subseteq \mathscr{P}A$, that is, the set of successor nodes of x is finite.

Moreover, G is well-founded by Corollary 8.2.10. When pointed by $x \in A$ it is also reachable by Corollary 9.2.18. It follows from Kőnig's Lemma that G is finite.

In general, a finitary functor F need not preserve intersections. In this case, consider the Trnková hull $\bar{F}$ (see Proposition 8.1.13), whose category of coalgebras is clearly isomorphic to that of F-coalgebras. From Proposition 8.1.18 we know that $\bar{F}$ preserves intersections.

Thus, the only fact we need to prove is that the given well-founded and reachable coalgebra (A, α, x) is well-founded and reachable as an $\bar{F}$-coalgebra. We have $FA = \bar{F}A$ since A is nonempty due to $x \in A$.

For reachability the above is clear because pointed subcoalgebras of (A, α, x) are nonempty, and therefore they are the same for F and $\bar{F}$. For well-foundedness, see Lemma 8.1.22. □

Barlocco et al. [67] proved the following fact. Here we obtain it as a corollary of Lemma 8.2.13(1).

Corollary 9.2.20. *Let (A, α) be a coalgebra. A subobject $m\colon A' \rightarrowtail A$ carries a subcoalgebra of (A, α) if and only if m is a pre-fixed point of $\ominus$, that is, $\ominus m \leq m$.*

Indeed, the above inequality is equivalent to $m \leq \bigcirc m$.

Remark 9.2.21. For every pointed coalgebra (A, α, x) we have, besides the previous-time operator $\ominus$ the constant endomap on $\mathsf{Sub}(A)$ with value $x\colon 1 \to A$. Indeed, since 1 is a terminal object, x is a monomorphism. We can then form the (pointwise) join

$$x \vee \ominus(-)\colon \mathsf{Sub}(A) \to \mathsf{Sub}(A),$$

which assigns to every subobject $m\colon S \rightarrowtail A$ the join of x and $\ominus m$ in the complete lattice $\mathsf{Sub}(A)$.

Theorem 9.2.22 [67]**.** *A pointed coalgebra (A, α, x) is reachable if and only if the operator $x \vee \ominus(-)$ on $\mathsf{Sub}(A)$ has the unique fixed point $\top = \mathsf{id}_A$.*

Proof The operator $x \vee \ominus(-)$ is monotone since $\ominus$ is. Thus it has a least fixed point by Theorem 6.1.1. Suppose that (A, α, x) is reachable, and let $r\colon R \rightarrowtail A$ be the least fixed point of $x \vee \ominus(-)$. Then we have $\ominus r \leq x \vee \ominus r = r$. Hence, by Corollary 9.2.20, R carries a subcoalgebra of (A, α) via r. Moreover, we have $x \leq x \vee \ominus r = r$. So there exists a morphism $x_0\colon 1 \to R$ such that $r \cdot x_0 = x$. Thus R is a pointed subcoalgebra of A via r, which implies that r is an isomorphism.

Conversely, suppose that id_A is the only fixed point of $x \vee \ominus(-)$. Let $m\colon (A', \alpha', x') \rightarrowtail (A, \alpha, x)$ be a pointed subcoalgebra. By Corollary 9.2.20, we know that $\ominus m \leq m$. Since $m \cdot x' = x$, we also know that $x \leq m$. Thus, $x \vee \ominus m \leq m$; that is, m is a pre-fixed point of $x \vee \ominus(-)$. By Remark 6.1.3, the least fixed point is also the least pre-fixed point. Thus, we obtain $\mathsf{id}_A \leq m$, which implies that m is an isomorphism. □

Corollary 9.2.23. *For every pointed coalgebra (A, α, x) its reachable part is the least fixed point of $x \vee \ominus(-)$.*

A related construction of the reachable part was given independently by Wißmann et al. [314]. It can be obtained as a consequence of Corollary 9.2.23 using the following general fact from lattice theory.

Remark 9.2.24. For every join-preserving map $\varphi\colon L \to L$ on a complete lattice L, and every $\ell \in L$, the least fixed point of $\ell \vee \varphi(-)$ is given by the following join

$$\bigvee_{i<\omega} \varphi^i(\ell). \tag{9.2}$$

To see this, note that $\ell \vee \varphi(-)$ preserves joins. Hence, by the proof of Kleene's Fixed Point Theorem 3.1.1, the least fixed point of $\ell \vee \varphi(-)$ is the join of the ω-chain given by $x_0 = \bot$, the least element of L, and $x_{n+1} = \ell \vee \varphi(x_n)$. Since φ preserves joins, we know that $\varphi(\bot) = \bot$ and furthermore that this ω-chain is the following one:

$$\bot, \quad \ell \vee \varphi(\bot) = \ell, \quad \ell \vee \varphi(\ell), \quad x \vee \varphi(\ell) \vee \varphi^2(\ell), \quad \ldots,$$

whose join clearly is the one in (9.2).

Since the left adjoint $\ominus$ is a join-preserving map on the complete lattice $\mathsf{Sub}(A)$, we obtain:

Corollary 9.2.25 [314]. *For every pointed coalgebra (A, α, x) its reachable part is carried by the following join of subobjects of A:*

$$\bigvee_{i<\omega} \ominus^i x = x \vee \ominus x \vee \ominus(\ominus x) \vee \cdots. \tag{9.3}$$

More explicitly, one defines subobjects $m_i\colon A_i \rightarrowtail A$ as follows: $m_0 = x$, and given $m_i\colon A_i \rightarrowtail A$, $m_{i+1} = \ominus m_i\colon A_{i+1} = \ominus A_i \rightarrowtail A$ is the least subobject such that $\alpha \cdot m_i$ factorizes through Fm_{i+1}:

$$\begin{array}{ccc} A_i & \longrightarrow & A_{i+1} = \ominus A_i \\ {\scriptstyle m_i}\downarrow & & \downarrow{\scriptstyle m_{i+1} = \ominus m_i} \\ A & \xrightarrow{\ \alpha\ } & FA \end{array}$$

The reachable part is then the union of all $m_i\colon A_i \rightarrowtail A$ for $i < \omega$.

Remark 9.2.26. Note that the above subobjects m_i yield a precise connection to standard algorithms for computing reachability. Indeed, we may compute the subsets $m_i\colon A_i \hookrightarrow A$ as subgraphs of the canonical graph of (A, α, x). This follows by an easy induction from Corollary 9.2.17. Furthermore, observe that in this case the subset $m_i\colon A_i \hookrightarrow A$ consists of precisely those states of A that are reachable by a directed path of length precisely i from the initial state x.

Consequently, one can compute the reachable part of a given pointed coalgebra by a standard graph algorithm such as breadth-first search.

Theorem 9.2.27. *Suppose that F preserves inverse images. Then the full subcategory of* $\mathsf{Coalg}_\mathsf{p}\, F$ *given by all reachable coalgebras is coreflective in* $\mathsf{Coalg}\, F$.

Proof Let (A, α, x) be a pointed coalgebra and $m\colon (A_0, \alpha_0, x_0) \rightarrowtail (A, \alpha, x)$ its reachable part. We will show that this is a coreflection.

Given a homomorphism $h\colon (B, \beta, y) \to (A, \alpha, x)$ where (B, β, y) is reachable, we need to prove that h factorizes uniquely through m in $\mathsf{Coalg}_\mathsf{p}\, F$. Uniqueness is clear since m is monic. For the existence, we form the inverse image of m under h:

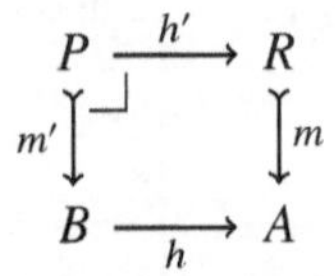

(9.4)

Since $m \cdot x_0 = x = h \cdot y$, we obtain a point $z\colon 1 \to P$. Since F preserves inverse images, we also obtain a coalgebra structure $\pi\colon P \to FP$ making m' and h' homomorphisms in $\mathsf{Coalg}\, F$ (Proposition 4.1.5):

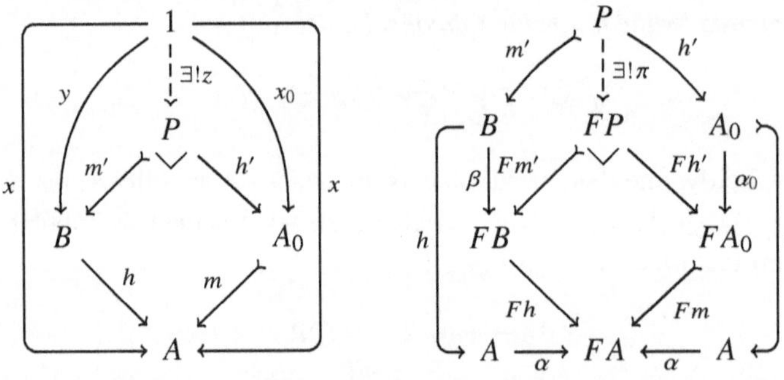

This defines a pointed coalgebra (P, π, z) making m' and h' pointed coalgebra homomorphisms. Since (B, β, y) is reachable and m' is monic, the latter must be an isomorphism. Thus, $h' \cdot (m')^{-1}$ is the desired factorization of h through m, cf. (9.4). □

Corollary 9.2.28. *If F preserves inverse images, then all strong quotients of a reachable coalgebra are reachable.*

Proof Suppose we have a reachable coalgebra (A, α, x) and a quotient $e\colon (A, \alpha, x) \twoheadrightarrow (B, \beta, y)$ in $\mathsf{Coalg}_\mathsf{p}\, F$. Denote by $m\colon (B_0, \beta_0, y_0) \rightarrowtail (B, \beta, y)$

the reachable part of (B, β, y). By Theorem 9.2.27, e factorizes through m:

$$\begin{array}{ccc} (B_0, \beta_0, y_0) & \overset{m}{\rightarrowtail} & (B, \beta, y) \\ {\scriptstyle h}\uparrow & \nearrow_{e} & \\ (A, \alpha, x) & & \end{array}$$

Thus, m is an isomorphism by Remark 9.1.2(4). □

Example 9.2.29. For functors not preserving inverse images, reachable coalgebras need not be closed under quotients. For example, recall the functor $R\colon \mathsf{Set} \to \mathsf{Set}$ from Example 7.4.6, which preserves intersections but not inverse images. Consider the coalgebras $\gamma\colon C \to RC$ with $C = \{x, y, z\}$ and $\gamma(x) = (y, z)$ and $\gamma(y) = \gamma(z) = d$ and $\delta\colon D \to RD$ with $D = \{x', y'\}$ and $\delta(x') = \delta(y') = d$. Then (D, δ) is a quotient of (C, γ) via the coalgebra homomorphism q with $q(x) = x'$ and $q(y) = q(z) = y'$. However, (C, γ, x) is reachable, whereas (D, δ, x') is not.

Note that, in light of the proof of Corollary 9.2.28, this example also shows that reachable coalgebras need not form a coreflective subcategory if F does not preserve inverse images.

We have seen that preservation of intersections by F plays an important role in the results above on the reachable part of a pointed coalgebra. Let us remark that this property has an equivalent characterization in terms of the least subobjects that the previous-time operator $\ominus$ delivers (cf. Proposition 9.2.15):

Remark 9.2.30. (1) Let $F\colon \mathscr{A} \to \mathscr{A}$ preserve monomorphisms. Then F preserves intersections if and only if

for every morphism $f\colon X \to FY$ there is a *least* subobject $m\colon Z \rightarrowtail Y$ such that f factorizes through Fm. (9.5)

This means that there exists a morphism $g\colon X \to FZ$ with

$$\begin{array}{ccc} X & \xrightarrow{f} & FY \\ & {\scriptstyle g}\searrow & \uparrow{\scriptstyle Fm} \\ & & FZ \end{array}$$

and satisfying the following universal property: For every subobject $m'\colon Z' \rightarrowtail Y$ and every morphism $g'\colon X \to FZ'$ with $Fm' \cdot g' = f$ there exists a (necessarily unique) $h\colon Z \rightarrowtail Z'$ with $m' \cdot h = m$.

For the base category $\mathscr{A} = \mathsf{Set}$, the above equivalence was established by Gumm [150, Cor. 4.8]. For a proof for the current setting of a complete and well-powered category, see Wißmann et al. [314, Prop. 5.9].

(2) For a morphism $f\colon X \to FY$, the above factor g provides the part of Y used by f. As an example, consider a polynomial functor $F = H_\Sigma$ on Set. Then for every $f\colon X \to H_\Sigma Y$ the set Z above consists of all those elements occuring in the flat terms $f(x)$ for $x \in X$.

We finish this section with a few remarks on extensions of the results on reachability presented here.

Remark 9.2.31. (1) The results presented in this section hold more generally in a category $\mathscr{A}$ equipped with a class $\mathcal{M}$ of monomorphisms. One then works with $\mathsf{Sub}_{\mathcal{M}}(A)$, the class of subobjects of A represented by $\mathcal{M}$-morphisms $m\colon S \rightarrowtail A$ (cf. Remark 6.2.1). One requires that $\mathsf{Sub}_{\mathcal{M}}(A)$ is a complete lattice for every object A, that inverse images (i.e. pullbacks along monomorphisms in $\mathcal{M}$) exist, and that for every morphism $f\colon A \to B$ the map $f^*\colon\ \mathsf{Sub}(B) \to \mathsf{Sub}(A)$ given by forming inverse images is a right adjoint.

(2) Of course, these requirements hold in our present setting (see Assumption 9.2.1) where $\mathcal{M}$ is the class of all monomorphisms. Another class of examples are well-powered categories $\mathscr{A}$ having coproducts and an $(\mathcal{E}, \mathcal{M})$-factorization system where $\mathcal{M}$ is a class of monomorphisms. It is then easy to see that inverse images exist (see Wißmann et al. [314, Remark 5.2(3)]). Moreover, for every $f\colon X \to Y$, the left adjoint to f^* is $f_*\colon\ \mathsf{Sub}(X) \to \mathsf{Sub}(Y)$, mapping an $\mathcal{M}$-subobject $s\colon S \rightarrowtail X$ to the *image* m of $f \cdot s$, that is, m is given by taking an $(\mathcal{E}, \mathcal{M})$ factorization $m \cdot e$ of $f \cdot s$.

As explained in *loc. cit.*, this setting allows one to construct the reachable part of coalgebras for base categories which are not complete, such as Rel, the category of sets and relations (viz. the Kleisli category for the power-set monad $\mathscr{P}$), and other Kleisli categories, for example, that for the distribution monad.

(3) In some applications, and notably in the setting outlined in point (2), it is desirable to replace the terminal object in a point $1 \to A$ by a different object I. This leads to the notion of an *I-pointed F-coalgebra*: an F-coalgebra (A, α) equipped with a morphism $i\colon I \to A$. In fact, Wißmann et al. [314] formulated and proved the results of this section for I-pointed F-coalgebras over a category satisfying the assumptions in (2) above.

(4) Finally, let us come back to Kleisli categories for a monad. They are used as base categories when one wants to obtain the finite trace or language semantics of state-based systems modelled as coalgebras (cf. Example 5.1.27, and for further related examples see Section 12.3).

Wißmann et al. [314, Section 6] show how to obtain the reachable part of a coalgebra over a Kleisli category for a monad, generalizing point (2), even if this category is not equipped with a factorization system. One works under an additional assumptions on the base category $\mathscr{A}$. In addition one assumes that the monad T on $\mathscr{A}$ as well as the coalgebraic type functor $F: \mathscr{A} \to \mathscr{A}$ preserve intersections, and that F has an extension $\bar{F}$ to the Kleisli category $\mathscr{A}_T$. One then obtains that the reachable part of a given pointed $\bar{F}$-coalgebra is the reachable part of a related coalgebra for F in $\mathscr{A}$.

9.3 Well-Pointed Coalgebras

We now present a notion that captures well the concept of minimality from automata theory: *well-pointed coalgebras*. We also present a new description of the initial algebra and the terminal coalgebra for set functors preserving intersections: we describe νF as the set of all well-pointed F-coalgebras up to isomorphism, and μF as the set of all well-pointed, well-founded F-coalgebras.

Although we restrict our presentation of these descriptions to endofunctors on Set, the same description holds for endofunctors on concrete categories $\mathscr{A}$ satisfying a number of natural assumptions (see Adámek et al. [47]). In particular, the results hold for endofunctors on varieties, as we mention at the end of this section.

Assumption 9.3.1. Throughout this section F denotes a set functor preserving intersections.

In particular, F preserves monomorphisms. We also know that Set is complete, cocomplete, well-powered, and co-well-powered. Thus, all of the results in Sections 9.1 and 9.2 apply.

Definition 9.3.2 [47]**.** A pointed coalgebra is called *well-pointed* if it is reachable and simple (i.e. it has no proper strong quotient coalgebra and no proper pointed subcoalgebra).

Example 9.3.3. We combine Example 9.1.4 and Example 9.2.4:

(1) A deterministic automaton with a given initial state is a well-pointed coalgebra for $FX = \{0, 1\} \times X^{\Sigma}$ iff it is reachable and observable (= simple), that is, iff it is a minimal automaton.

(2) A well-pointed $\mathscr{P}$-coalgebra is a pointed graph which is reachable in the usual sense (see Example 9.2.4(2)) and no distinct nodes are bisimilar (see Example 9.1.4(3)).

(3) A labelled transition systems with a given initial state is a well-pointed coalgebra for $FX = (\mathscr{P}X)^{\Sigma}$ iff it is reachable in the usual sense (see Example 9.2.4(3)) and no distinct states are bisimilar.

Notation 9.3.4. There is a canonical construction for turning an arbitrary pointed coalgebra (A, α, x) into a well-pointed one: first form the simple quotient $(\bar{A}, \bar{\alpha})$ (see Proposition 9.1.5) pointed by $e_{(A,\alpha)}(x) \in \bar{A}$, and then form the reachable part of $(\bar{A}, \bar{\alpha}, e_{(A,\alpha)}(x))$ (see Proposition 9.2.7).

$$
\begin{array}{ccccc}
 & & & & \bar{A}_0 \\
 & & & \overset{x_0}{\nearrow} & \big\downarrow m \\
1 & \xrightarrow{x} & A & \xrightarrow{e_{(A,\alpha)}} & \bar{A}
\end{array}
\qquad
\begin{array}{ccc}
\bar{A}_0 & \xrightarrow{\bar{\alpha}_0} & F\bar{A}_0 \\
{\scriptstyle m}\big\downarrow & & \big\downarrow{\scriptstyle Fm} \\
\bar{A} & \xrightarrow{\bar{\alpha}} & F\bar{A}
\end{array}
$$

We have $m\colon (\bar{A}_0, \bar{\alpha}_0, x_0) \rightarrowtail (\bar{A}, \bar{\alpha}, e_{(A,\alpha)}(x))$. Then $(\bar{A}_0, \bar{\alpha}_0, x_0)$ is well-pointed, and we denote it by

$$\mathsf{wp}(A, \alpha, x).$$

We call it the *well-pointed modification* of (A, α, x).

For functors preserving inverse images, the well-pointed modification may be obtained by first forming the reachable part and then its simple quotient, that is the usual order of the two operations in a minimization procedure.

Proposition 9.3.5. *Suppose that F preserves inverse images. Then* $\mathsf{wp}(A, \alpha, x)$ *is the simple quotient of the reachable part of* (A, α, x).

Proof Given the pointed coalgebra (A, α, x), form first its reachable part $m_0\colon (A_0, \alpha_0, x_0) \rightarrowtail (A, \alpha, x)$ and then the simple quotient $e_{(A,\alpha)}\colon (A, \alpha) \twoheadrightarrow (\bar{A}, \bar{\alpha})$. Now take the factorization of $e_{(A,\alpha)} \cdot m_0$ into a surjective homomorphism followed by an injective one (cf. Proposition 7.2.2):

$$
\begin{array}{ccc}
(A_0, \alpha_0) & \xrightarrow{m_0} & (A, \alpha) \\
{\scriptstyle \bar{e}}\big\downarrow & & \big\downarrow{\scriptstyle e_{(A,\alpha)}} \\
(\bar{A}_0, \bar{\alpha}_0) & \xrightarrow{\bar{m}_0} & (\bar{A}, \bar{\alpha})
\end{array}
$$

By Corollary 9.2.28 the pointed coalgebra $(\bar{A}_0, \bar{\alpha}_0, \bar{e}(x_0))$ is reachable, and by Proposition 9.1.10, $(\bar{A}_0, \bar{\alpha}_0)$ is simple. Hence, it is the well-pointed modification of (A, α, x) but it is also given as the simple quotient of its reachable part (A_0, α_0, x_0), as desired. □

Remark 9.3.6. (1) Note that Proposition 9.3.5 can be proven for more general categories than Set and beyond the setting of coalgebras (see Wißmann [313, Section 7.2]).

(2) Even if F does not preserve inverse images, the well-pointed modification might still be given by the simple quotient of the reachable part. For example, for the functor R from Example 7.4.6, which does not preserve inverse images,

this holds because the simple quotient of every pointed coalgebra is the terminal coalgebra $id\colon 1 \to R1 = 1$.

(3) In general, the well-pointed modification is not given by the simple quotient of the reachable part. In fact, Wißmann [313, Ex. 5.3.27] observed that for a functor not preserving inverse images, reachable coalgebras are not closed under quotients. To see this consider the monoid-valued set functor $FX = \mathbb{Z}^{(X)}$ for monoid of addition on the set of all integers (Example 2.4.3). Coalgebras for F are finitely branching transition systems with transitions labelled by non-zero integers. The following pointed F-coalgebra

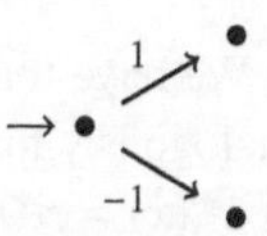

is reachable (by Corollary 9.2.18; the canonical graph being the above graph without the labels). However, its simple quotient consist of two nodes with no transitions and is therefore not reachable.

(4) As in the case of deterministic automata, well-pointed coalgebras are rather special in general. In fact, for many functors F there exists, up to isomorphism of $\mathsf{Coalg}_\mathsf{p}\, F$, only a set of well-pointed coalgebras. As we have seen in Proposition 9.2.11, this is the case for all bounded set functors.

Notation 9.3.7. Let F be a set functor for which there is a representative set T of well-pointed coalgebras. In other words, every well-pointed coalgebra for F is isomorphic in $\mathsf{Coalg}_\mathsf{p}\, F$ to precisely one coalgebra in T. Then we can assume, without loss of generality, that all well-pointed modifications are chosen to be elements of T. For every coalgebra $\alpha\colon A \to FA$ we have a function

$$\alpha^+\colon A \to T \qquad \text{defined by} \qquad \alpha^+(x) = \mathsf{wp}(A, \alpha, x).$$

Lemma 9.3.8 [47]. *For every coalgebra homomorphism $h\colon (A,\alpha) \to (B,\beta)$ the following triangle commutes:*

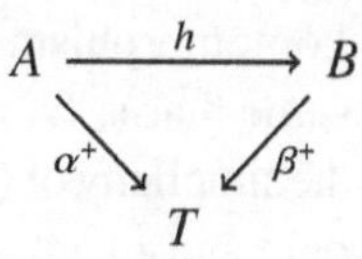

Proof (1) Assume that (A,α) and (B,β) are simple. In particular, h is a monomorphism in $\mathscr{A}$, by simplicity of (A,α); see Observation 9.1.6(1). Fix $x \in A$. By simplicity again, we know that $\alpha^+(x)$ is the reachable part $m\colon (A_0,\alpha_0,x_0) \rightarrowtail (A,\alpha,x)$. Analogously, $\beta^+(h(x))$ is the reachable part $n\colon (B_0,\beta_0,y_0) \rightarrowtail (B,\beta,h(x))$. We have a monomorphism

$h \cdot m\colon (A_0, \alpha_0, x_0) \rightarrowtail (B, \beta, h(x))$. By Proposition 9.2.7, (A_0, α_0, x_0) is isomorphic to (B_0, β_0, y_0). Now T contains just one representative of every well-pointed coalgebra up to isomorphism, consequently, $\beta^+(h(x)) = \alpha^+(x)$.

(2) In general, form simple reflections $(\bar{A}, \bar{\alpha})$ and $(\bar{B}, \bar{\beta})$. By Proposition 9.1.5, there is a unique homomorphism $\bar{h}$ such that the diagram below commutes:

$$\begin{array}{ccc} (A, \alpha) & \xrightarrow{h} & (B, \beta) \\ {\scriptstyle e_{(A,\alpha)}} \big\downarrow & & \big\downarrow {\scriptstyle e_{(B,\beta)}} \\ (\bar{A}, \bar{\alpha}) & \xrightarrow{\bar{h}} & (\bar{B}, \bar{\beta}) \end{array}$$

Let $x \in A$ and put $\bar{x} = e_{(A,\alpha)}(x)$. We have that $\alpha^+(x)$ is the reachable part of $(\bar{A}, \bar{\alpha}, \bar{x})$. Thus $\alpha^+(x) = \bar{\alpha}^+(\bar{x})$; analogously for $\beta^+(h(x))$. By applying (1) to $\bar{h}$ in lieu of h, we conclude $\alpha^+(x) = \bar{\alpha}^+(\bar{x}) = \bar{\beta}^+(\bar{h}(\bar{x})) = \beta^+(h(x))$. □

Notation 9.3.9. Let F be a set functor preserving intersections and having a set T representing all its well-pointed coalgebras up to isomorphism. We denote by $\tau\colon T \to FT$ the function which assigns to every element (A, α, x) of T the following element of FT:

$$\tau(A, \alpha, x) = F\alpha^+ \cdot \alpha(x). \tag{9.6}$$

Theorem 9.3.10. *A set functor F preserving intersections has a terminal coalgebra iff it has only a set T of well-pointed coalgebras up to isomorphism. Moreover, T is the terminal coalgebra with the structure $\tau\colon T \to FT$ in* (9.6). For every coalgebra (A, α), the unique homomorphism to T is $\alpha^+\colon A \to T$.

Proof Necessity follows from Theorem 9.1.8: since νF exists, there is only a set of simple coalgebras up to isomorphism. For each simple coalgebra (A, α) there is only a set of (subcoalgebras of) pointed coalgebras (A, α, x). Thus, the set T exists.

Let us prove sufficiency.

(1) We first show that the coalgebra (T, τ) is weakly terminal. Let (A, α) be a coalgebra. Since there is a homomorphism of A into a simple coalgebra (Proposition 9.1.5), we can assume that is (A, α) is simple. Fix $x \in A$. Write $\mathsf{wp}(A, \alpha, x)$ as (A_0, α_0, x_0). By the simplicity of (A, α) we know that (A_0, α_0, x_0) is the reachable pointed subcoalgebra of (A, α, x). Thus, we have a monic coalgebra homomorphism $m\colon (A_0, \alpha_0) \rightarrowtail (A, \alpha)$ with $m(x_0) = x$.

We first prove that the following equation holds:

$$F\alpha^+ \cdot \alpha(x) = F\alpha_0^+ \cdot \alpha_0(x_0). \tag{9.7}$$

Indeed, $\alpha^+ \cdot m = \alpha_0^+$ by Lemma 9.3.8. Since m is a coalgebra homomorphism, we have $Fm \cdot \alpha_0 = \alpha \cdot m$. Thus

$$F\alpha_0^+ \cdot \alpha_0 = F\alpha^+ \cdot Fm \cdot \alpha_0 = F\alpha^+ \cdot \alpha \cdot m.$$

Apply both sides to x_0 to obtain (9.7). Using this, we prove that α^+ is a homomorphism: we have $\tau \cdot \alpha^+ = F\alpha^+ \cdot \alpha$. Indeed, for every $x \in A$ we have

$$\tau \cdot \alpha^+(x) = \tau(\mathsf{wp}(A, \alpha, x)) = \tau(A_0, \alpha_0, x_0) = F\alpha_0^+ \cdot \alpha_0(x_0) = F\alpha^+ \cdot \alpha(x).$$

This verifies that $\alpha^+ \colon (A, \alpha) \to (T, \tau)$ is a homomorphism.

(2) We conclude the proof by showing that for every homomorphism $h \colon (A, \alpha) \to (T, \tau)$ we have $\alpha^+ = h$. By Lemma 9.3.8, $\tau^+ \cdot h = \alpha^+$. We only need to show that $\tau^+ = \mathrm{id}_T$. Every element (A, α, x) of T is well-pointed, and so we have $\alpha^+(x) = (A, \alpha, x)$. By Lemma 9.3.8 again, $\tau^+ \cdot \alpha^+ = \alpha^+$. This proves that

$$\tau^+(A, \alpha, x) = \tau^+ \cdot \alpha^+(x) = \alpha^+(x) = (A, \alpha, x). \qquad \square$$

Corollary 9.3.11. *Every finitary set functor has a terminal coalgebra carried by a set T representing all well-pointed coalgebras.*

Proof If F preserves intersections, then this follows from Theorem 9.3.10. If it does not, then we use the Trnková hull $\bar{F}$, which preserves intersections by Proposition 8.1.18. Since F and $\bar{F}$ coincide on all nonempty sets, they either have the same terminal coalgebra or $F = C_0$ (the constant functor with value $\emptyset$). In the latter case $\nu F = \emptyset$ and no pointed coalgebras exists, so the desired result holds, too. $\square$

Example 9.3.12. The power-set functor preserves intersections and does not have a terminal coalgebra. Thus it has a proper class of well-pointed coalgebras that are pairwide non-isomorphic (cf. Proposition 9.2.11).

Remark 9.3.13. We have seen that if T is a *set* which contains representatives of all well-pointed coalgebras, then T is a terminal coalgebra. Even if T is a proper class (of representatives), it is a terminal coalgebra for a functor related to F. More precisely, consider Set as the full subcategory of the category Classof all classes and functions, then F has an canonical extension to an endofunctor $\widehat{F}$ on the latter (see Adámek et al. [31]). Moreover, T is a terminal coalgebra for $\widehat{F}$. (The proof is completely analogous to that of Theorem 9.3.10.)

Remark 9.3.14. An analogous description of νF is possible for all endofunctors on a variety $\mathscr{A}$ preserving intersections. Here $\mathscr{A}$ is a category of algebras specified by a (finitary) signature Σ and a set of equations. The concept of a pointed and a well-pointed coalgebra is defined analogously to the case $\mathscr{A}$ = Set: a pointed coalgebra is (A, α, x) where x is an element of A. Thus, we work here with the generalization mentioned in Remark 9.2.31(3): we choose I to be the free algebra on one generator in the variety $\mathscr{A}$. A pointed coalgebra (A, α, x) is well-pointed if (A, α) is simple and has no proper subcoalgebra containing x. Since F preserves intersections, it preserves monomorphisms. Thus, by the proof of Theorem 9.1.8, νF exists iff F has only a set of simple coalgebras (up to isomorphism). Now let T be a representative set of all well-pointed F-coalgebras. Then T carries a canonical structure of an object of $\mathscr{A}$ such that $\tau\colon T \to FT$ in (9.6) is a morphism of $\mathscr{A}$. Moreover, (T, τ) is the terminal coalgebra.

Initial algebras and well-founded, well-pointed coalgebras We have seen in Theorem 9.3.10 that the terminal coalgebra νF for a set functor preserving intersections is carried by a representative set of well-pointed coalgebras. We now show that the initial algebra μF can be described similarly as the algebra of all well-founded, well-pointed coalgebras (up to isomorphism). For this result we recall from Theorem 8.5.9 that $(\mu F, \iota^{-1})$ is the terminal well-founded coalgebra.

Remark 9.3.15. The collection of well-founded coalgebras is closed under subcoalgebras and strong quotients by Corollary 8.4.9 and Corollary 8.4.3. Hence, if a coalgebra is well-founded, then so is its well-pointed modification.

Notation 9.3.16. Suppose that F has only a set of well-founded, well-pointed coalgebras up to isomorphism, and choose a set W of representatives. Analogously to Notation 9.3.7, we have a function

$$\alpha^+\colon A \to W \qquad \text{defined by} \qquad \alpha^+(x) = \mathsf{wp}(A, \alpha, x).$$

We obtain a coalgebra structure $\sigma\colon W \to FW$ analogously to τ in Notation 9.3.9: it assigns to (A, α, x) in W the element $F\alpha^+ \cdot \alpha(x) \in FW$.

Remark 9.3.17. For $\alpha = \sigma\colon W \to FW$ we have $\alpha^+ = \mathsf{id}_W$. This is analogous to $\tau^+ = \mathsf{id}_T$ at the end of the proof of Theorem 9.3.10.

Theorem 9.3.18 [47, Thm. 3.48]**.** *A set functor preserving intersections has an initial algebra iff it has only a set W of well-founded, well-pointed coalgebras up to isomorphism. Moreover, W carries the initial algebra with the structure $\sigma^{-1}\colon FW \to W$.*

Proof For the necessity, suppose that μF exists. By Theorem 8.5.9, $(\mu F, \iota^{-1})$ is the terminal well-founded coalgebra. For every well-founded, well-pointed coalgebra (A, α, x) the unique homomorphism $f\colon (A, \alpha) \to (\mu F, \iota^{-1})$ is injective (see Observation 9.1.6(1)). Since μF has only a set of subsets, this implies that there is only a set of well-founded coalgebras up to isomorphism. Thus, the same holds for well-founded, well-pointed coalgebras.

For the sufficiency, suppose that the set W is given. By Theorem 8.5.9, we only need to prove that (W, σ) is a terminal well-founded coalgebra. This is done in a number of steps, and the order diverges from what we saw in Theorem 9.3.10.

(1) We first prove a version of Lemma 9.3.8 saying that if $h\colon (A, \alpha) \to (B, \beta)$ is a homomorphism of *well-founded* coalgebras, then $\beta^+ \cdot h = \alpha^+$. The proof is analogous: recall that a strong quotient of a well-founded coalgebra is again well-founded (Corollary 8.4.3).

(2) For every well-founded coalgebra (A, α), the function α^+ is a coalgebra homomorphism $\alpha^+\colon (A, \alpha) \to (W, \sigma)$. This is entirely as in the part of the proof of Theorem 9.3.10 where we showed (T, τ) to be weakly terminal.

(3) We prove that (W, σ) is a well-founded coalgebra. To this end notice that for every well-pointed, well-founded coalgebra (A, α, x) in I we have that

$$\alpha^+(x) = (A, \alpha, x).$$

By Proposition 4.1.1 we can take the coproduct (in $\mathsf{Coalg}\, F$) of all (A, α) for which there is an $x \in A$ such that (A, α, x) lies in W. This coproduct is a well-founded coalgebra by Corollary 8.4.3. Moreover, as we have just seen, the unique homomorphism from the coproduct to (W, σ) is surjective, whence (W, σ) is a strong quotient coalgebra of the coproduct. Thus, another application of Corollary 8.4.3 shows that (W, σ) is a well-founded coalgebra as desired.

(4) At this point, we know that (W, σ) is well-founded, and indeed it is a weakly terminal well-founded coalgebra. We conclude as in the proof of Theorem 9.3.10. Let (A, α) be any well-founded coalgebra. Consider a homomorphism $h\colon (A, \alpha) \to (W, \sigma)$. By item (1), $\alpha^+ = \sigma^+ \cdot h$ and $\alpha^+ = \sigma^+ \cdot \alpha^+$. By Remark 9.3.17, $\sigma^+ = \mathrm{id}_W$. Hence, we have $h = \alpha^+$. □

Remark 9.3.19. Note that for a finitary set functor F, all elements of $\mu F \cong W$ in Theorem 9.3.18 are finite coalgebras. This follows from Proposition 9.2.19.

We conclude this section with a number of examples of concrete descriptions of initial algebras and terminal coalgebras we obtain from Theorems 9.3.10 and 9.3.18.

Example 9.3.20. (1) For the functor $FX = \{0, 1\} \times X^\Sigma$ the terminal coalgebra T is carried by the set of (isomorphism classes of) all minimal deterministic automata with initial states. The coalgebra structure $\tau\colon T \to FT$, interpreted as an automaton on T, has as accepting states those minimal automata A whose initial state is accepting in A, and the next-state function maps a pair consisting of a minimal automaton (A, δ, a_0, F) and an input symbol $s \in \Sigma$ to the minimization of the automaton $(A, \delta, \delta(a_0, s), F)$: one shifts the initial state along its s-transition and then minimizes the result.

The unique coalgebra homomorphism $\alpha^+ \colon A \to T$ assigns to a state a of the automaton A the minimization of A with that initial state.

As there are no well-founded coalgebras for F, the initial algebra W in Theorem 9.3.18 is empty.

(2) For the power-set functor $\mathscr{P}$, T is the collection of pointed, reachable, and simple graphs (cf. Example 9.3.3(2)), which is *not* a set. Indeed, for every cardinal κ the following graph is simple and reachable: the vertices are the ordinals $i < \kappa$, the edges are (i, j) with $i < j$, and the chosen point is 0. However, for the finite power-set functor $\mathscr{P}_\mathsf{f}$ the well-pointed coalgebras are the pointed, reachable and simple finitely branching graphs. Every such graph is countable because it is finitely branching and every vertex is reachable from the chosen vertex. Thus all such pointed graphs form a set up to isomorphism, and we obtain

$$\nu\mathscr{P}_\mathsf{f} \cong \text{all pointed, reachable, and simple finitely branching graphs,}$$
$$\mu\mathscr{P}_\mathsf{f} \cong \text{all pointed, reachable, simple, and well-founded finitely branching graphs}$$

(up to isomorphism).

(3) We next consider the set functor $FX = X \times X + 1$. Here we view coalgebras for F (differently than earlier in this chapter) as directed graphs with edges labelled by l or r such that every node which has successors has exactly one successor node reachable by an l-labelled edge and one by an r-labelled edge, respectively.

We call *pointed* such graphs *binary term graphs*. Here are two examples:

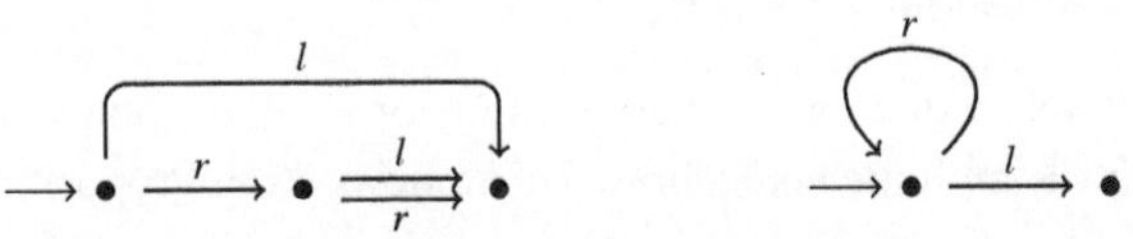

(9.8)

Recall from Remark 2.5.10 that to every node x of a binary term graph A the unique homomorphism $h : A \to \nu F$ assigns its tree expansion: the root is x and a node is either a leaf, if it is a terminating state, or has the two next states as children (left hand for input l, right hand for input r).

It follows from Remark 9.1.7 that a binary term graph is a simple coalgebra iff different nodes have different tree expansions. Furthermore, a well-pointed F-coalgebra is a *minimal* binary term graph, that is, one which is reachable and simple. Note that tree expansion of the given point yields a bijection between well-pointed coalgebras and binary trees. Its inverse is given by identifying in a given binary tree systematically all shared subtrees of the nodes. This explains why the following descriptions of μF and νF are isomorphic to those in Examples 3.2.5 and 2.5.11(4). For example, the two binary trees

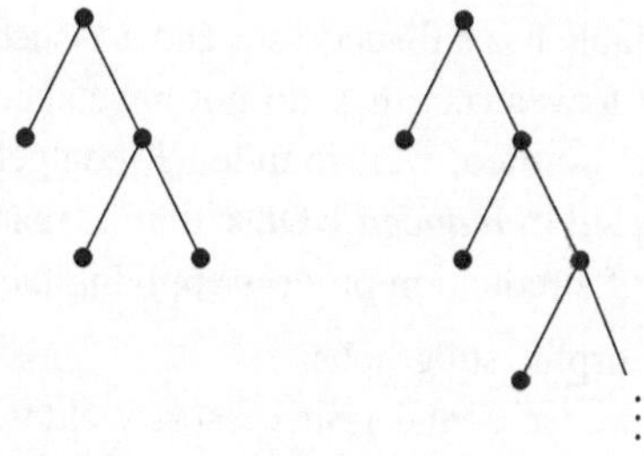

yield the two minimal binary term graphs in (9.8).

From Theorem 9.3.10 and Theorem 9.3.18 we obtain:

$$\nu F \cong \text{all minimal binary term graphs,}$$

$$\mu F \cong \text{all well-founded, minimal binary term graphs}$$

(up to isomorphism). Indeed, the well-founded coalgebras are precisely those yielding a finite tree expansion.

(4) **Σ-term graphs.** More generally, consider a polynomial functor H_Σ (see Example 2.1.5). Analogously to the previous example, well-pointed H_Σ-coalgebras may be identified with minimal Σ-term graphs. A *Σ-term graph* is a pointed directed graph such that every node with n successors is labelled by some $\sigma \in \Sigma_n$, and moreover, the successor nodes are reachable by an edge labelled in $\{0, \dots, n-1\}$ so that each edge label appears precisely once. The previous example is the special case where Σ is a signature containing a binary operations symbol ⋆ and a constant symbol c. The two binary term graphs in (9.8) above can be presented as Σ-term graphs as follows:

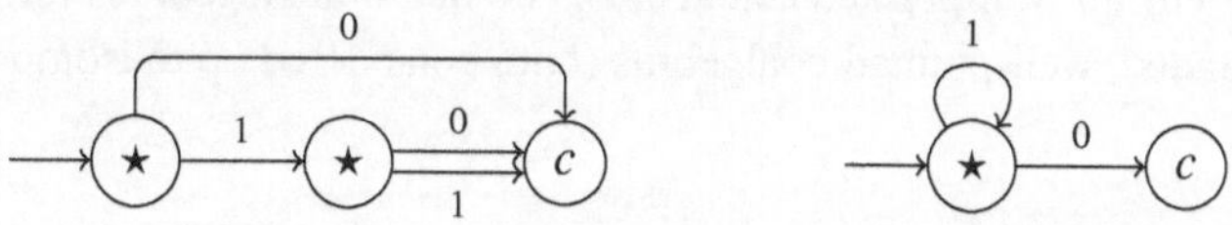

Minimality of a Σ-term graph means that every node is reachable from the given point and that the Σ-tree expansions of distinct nodes are always different. We thus obtain

$$\nu H_\Sigma \cong \text{all minimal } \Sigma\text{-term graphs,}$$

$$\mu H_\Sigma \cong \text{all well-founded minimal } \Sigma\text{-term graphs}$$

(up to isomorphism).

(5) Binary decision diagram (BDD). Given a set V (of Boolean variables), we have seen in Example 3.2.8(2) that all binary decision trees form the initial

algebra for $FX = \{0,1\} + V \times X \times X$. Now a BDD is given by a pointed directed acyclic graph whose nodes are labelled by variables or by 0 or 1, and every node labelled by a variable has a 0-successor and a 1-successor, whereas nodes labelled by 0 or 1 are leaves (i.e. they do not have successors). Thus BDDs are precisely the finite, pointed, well-founded F-coalgebras. In practice, one is usually more interested in *reduced* BDDs, that is, reachable BDDs that are obtained as the result of a reduction process applying the following two rules:

- merge any two isomorphic subgraphs;
- eliminate any node whose 0- and 1-successors yield isomorphic subgraphs.

Reduced BDDs are well-founded, well-pointed F-coalgebras, but not conversely; for example, the following well-founded, well-pointed F-coalgebra is not a reduced BDD:

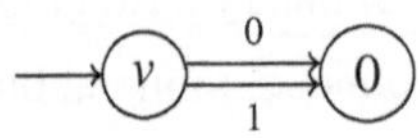

9.4 Summary

We have presented simple and reachable coalgebras and constructions of the simple quotient of a coalgebra and the reachable part of a pointed one. In addition, we have introduced well-pointed coalgebras: those which are both reachable and simple. Well-pointed coalgebras constitute a coalgebraic formulation of minimality of state-based systems.

For set functors preserving intersections, we have also seen new descriptions of the initial algebra and the terminal coalgebra. In fact, the terminal coalgebra is formed by all well-pointed coalgebras, and the initial algebra is formed by all well-founded, well-pointed coalgebras (both considered up to isomorphism).

10

Fixed Points Determined by Finite Behaviour

In this chapter we study a fixed point of an endofunctor that in general is different from its initial algebra and its terminal coalgebra. This *rational fixed point*, as it is called, arises in the study of the final semantics of coalgebras. It collects precisely the behaviours of all 'finite' coalgebras of a given endofunctor, which, in the case of state-based systems, are those with finitely many states. Thus, the rational fixed point provides a semantic domain for finite-state behaviour. Accordingly, examples of rational fixed points include regular languages, eventually periodic and rational streams, regular trees for a signature, and many other types of finite-state behaviour.

Of course, for categories more general than Set, 'finite' needs to be replaced by a suitable category-theoretic concept. We use the concept of a *finitely presentable* object and work in a category that has enough such objects. More precisely, we work in a *locally finitely presentable category*, a notion originally introduced by Gabriel and Ulmer [134]. In addition, the endofunctor under consideration needs to be determined by its action on the finitely presentable objects, which means that it is *finitary*. We briefly review these concepts in Section 10.1. We shall see that for Set and many other categories the rational fixed point is *fully abstract* with respect to final semantics; that is, it is a subcoalgebra of the terminal coalgebra.

Moreover, similar to what we have seen in the previous chapters for the initial algebra and the terminal coalgebra, the rational fixed point has two equivalent characterizations by a universal property: as a coalgebra it is the terminal locally finitely presentable coalgebra (Section 10.2), and as an algebra it is the initial iterative algebra (Section 10.3). For a rational fixed point in Set we present two explicit descriptions (Section 10.4): it is the coalgebra formed by all finite well-pointed coalgebras, and it is the quotient of the coalgebra of rational Σ-trees modulo basic equations for a given presentation of the functor. Then we treat full abstractness of the rational fixed point (Section 10.5). We conclude the chapter with a brief discussion of further work based on the rational fixed point (Section 10.6). In particular, we mention monads arising from iterative algebras which are closely related to Elgot's iterative theories [115].

10.1 Locally Finitely Presentable Categories

We need an appropriate notion of a 'finite' object for our study of the rational fixed point. We use the notion of a finitely presentable object. The name stems from general algebra, where an algebra (e.g. a group, ring, or module) is finitely presentable if it is presented by finitely many generators and finitely many relations. The categorical generalization of these concepts is formulated with the help of finitary functors (Definition 4.3.1): those preserving directed colimits. Objects whose hom-functors are finitary are called finitely presentable, and this then leads to the concepts of a locally finitely presentable category. We also recall filtered colimits, generalizing directed ones, and show that they are equivalent to directed colimits. We now briefly recall all these notions. A more comprehensive introduction may be found in [23]. Recall that a functor is finitary if it preserves directed colimits (Definition 4.3.1).

Definition 10.1.1. An object X of a category $\mathscr{A}$ is *finitely presentable* if its hom-functor

$$\mathscr{A}(X, -) : \mathscr{A} \to \mathsf{Set}$$

is finitary.

Example 10.1.2. (1) A set is finitely presentable in Set iff it is finite.

(2) Analogously, in the categories Pos (posets) and Gra (graphs) the finitely presentable objects are precisely the finite posets and graphs, respectively.

(3) In the category of K-Vec of vector spaces over the field K the finitely presentable objects are precisely the finite-dimensional vector spaces, that is, those isomorphic to K^n for some $n \in \mathbb{N}$.

(4) In the categories Met and CMet no nonempty space is finitely presentable. To see this, consider the ω-chain of spaces $B_n = \{0, 1\}$ where $d(0, 1) = 2^{-n}$ $(n < \omega)$ where the connecting maps are $\mathrm{id}_{0,1}$. The colimit $\mathrm{colim}\, B_n = 1$ is the singleton space. Given a space $A \neq \emptyset$, the morphism $f : A \to 1$ factorizes through each colimit injection, but not essentially uniquely.

Remark 10.1.3. More explicitly, an object X is finitely presentable iff every morphism f from X to a colimit $C = \mathrm{colim}_{i \in I}\, C_i$ of a directed diagram D

factorizes essentially uniquely through one of the colimit injections $c_i \colon C_i \to C$. This means that

(1) there exists an $i \in I$ and a morphism $f' \colon X \to C_i$ such that $c_i \cdot f' = f$, and

(2) given two such factorizations $c_i \cdot f' = c_i \cdot f'$ of f, there exists a connecting morphism $c_{i,j} \colon C_i \to C_j$ of D such that $c_{i,j} \cdot f' = c_{i,j} \cdot f'$.

Example 10.1.4. No nonempty cpo is finitely presentable in CPO. To see this, consider the ω-chain of inclusion maps

$$0^\top \hookrightarrow 1^\top \hookrightarrow 2^\top \hookrightarrow \cdots,$$

where $n^\top = \{0 < 1 < \cdots < n < \top\}$. Its colimit in CPO is the cpo $C = \mathbb{N} \cup \{v, \top\}$ with colimit injections given by the obvious inclusion maps, where $v = \bigvee_{n \in \mathbb{N}} n$ and $v < \top$. Then for every nonempty cpo X the continuous map $f \colon X \to C$ which is constantly v does not factorize through any of the colimit injections $n^\top \hookrightarrow C$.

Definition 10.1.5. A category $\mathscr{A}$ is called *locally finitely presentable* if

(1) it is cocomplete,

(2) it has only a set of finitely presentable objects (up to isomorphism), and

(3) every object in $\mathscr{A}$ is a directed colimit of finitely presentable objects.

Example 10.1.6. We list a number of locally finitely presentable categories:

(1) Set: every set X is a directed colimit of its finite subsets. More precisely, X is the colimit of the diagram by all its finite subsets and all inclusion maps between them. Note that this diagram is clearly directed.

(2) Pos: every poset is a directed colimit of its finite subposets. Similarly for graphs.

(3) K-Vec: every vector space X is a directed colimit of its finite-dimensional subspaces. This follows from the fact that every finite subset of X generates a finite-dimensional subspace.

(4) More generally, every finitary variety of algebras (i.e. a category of algebras specified by operations of finite arity and equations) is locally finitely presentable. The finitely presentable objects are precisely those algebras which can be presented by finitely many generators and relations [23, Theorem 3.12]. In other words, these are precisely the quotients of free algebras on finitely many generators modulo a congruence generated by finitely many pairs.

Concrete examples are the categories of monoids, groups, rings, vector spaces, modules over a (semi-)ring, and many others.

(5) Interesting special cases are *locally finite* varieties, that is, varieties where

finitely presentable algebras are precisely the finite ones. Concrete examples are the categories of Boolean algebras, join-semilattices, distributive lattices, and vector spaces over a finite field.

(6) The category Nom of nominal sets (Example 2.1.10) is locally finitely presentable (see e.g. Pitts [250, Rem. 5.17]). Petrişan [249, Prop. 2.3.7] has shown that the finitely presentable nominal sets are precisely the orbit-finite ones, where an *orbit* of an element x of a nominal set X is the set $\{\pi \cdot x : \pi \in \mathsf{Perm}(\mathbb{A})\}$, and X is *orbit-finite* if it has finitely many orbits.

(7) Presheaf categories $\mathsf{Set}^{\mathscr{C}^{\mathsf{op}}}$, where $\mathscr{C}$ is a small category, are locally finitely presentable. One particular example of interest is the category of sets in context, that is, the presheaf category $\mathsf{Set}^{\mathscr{F}}$, where $\mathscr{F}$ is the category of all finite sets and all maps between them. That this category has been used by Fiore et al. [124] to capture structural induction on terms with variable binding operators (e.g. λ-terms) by initial algebras (see Examples 2.1.8 and 2.2.18(4)). The finitely presentable objects in $\mathsf{Set}^{\mathscr{F}}$ are precisely the *super-finitary* presheaves [44]: presheaves $P\colon \mathscr{F} \to \mathsf{Set}$ where $P(\Gamma)$ is finite for every context Γ, and there is a context Γ_0 generating P in the sense that for every nonempty context Γ we have

$$P(\Gamma) = \bigcup_{\gamma\colon \Gamma_0 \to \Gamma} P\gamma[P(\Gamma_0)].$$

(8) The category CPO is not locally finitely presentable since no nonempty cpo is finitely presentable (Example 10.1.4).

(9) Neither Met nor CMet is locally finitely presentable: we have seen in Example 10.1.2(4) that no nonempty space is finitely presentable.

Directed colimits are a special case of filtered colimits. We define them next, and we also mention that the two concepts are equivalent: we could define locally finitely presentable categories using filtered colimits. Below we often apply this equivalence and use filtered colimits.

Definition 10.1.7. (1) A category $\mathscr{D}$ is *filtered* if every finite diagram in $\mathscr{D}$ has a cocone. Equivalently, the following three conditions hold:

(a) $\mathscr{D}$ is nonempty,

(b) for every pair A_1, A_2 of objects of $\mathscr{D}$ there exists a cocone $A_1 \to B \leftarrow A_2$, and

(c) for every parallel pair of morphisms $f, g\colon A_1 \rightrightarrows A_2$ in $\mathscr{D}$ there exists a morphism $h\colon A_2 \to A_3$ with $h \cdot f = h \cdot g$.

(2) A *filtered diagram* in the category $\mathscr{A}$ is a diagram $\mathscr{D} \to \mathscr{A}$ whose scheme $\mathscr{D}$ is a filtered category, and a *filtered colimit* is a colimit of a filtered diagram.

Example 10.1.8. (1) A poset is filtered iff it is directed.

(2) Finitely cocomplete categories are filtered: for (a), use the initial object, for (b), put $B = A_1 + A_2$, and for (c), choose the coequalizer of f and g as h.

Theorem 10.1.9 [23, Cor. 1.5]. *Filtered colimits are equivalent to directed ones:*

(1) *A category has filtered colimits iff it has directed ones, and*

(2) *a functor between such categories preserves filtered colimits iff it preserves directed ones.*

Remark 10.1.10. Analogously to Theorem 10.1.9, colimits of nonempty chains are also equivalent to directed colimits (in the analogous sense) [23, Cor. 1.7]. This is proved for posets in Theorem B.2.2.

Example 10.1.11. A finitary functor F on a locally finitely presentable category $\mathscr{A}$ has the following properties:

(1) the category $\mathsf{Alg}\, F$ is locally finitely presentable (Theorem 11.3.2), but $\mathsf{Coalg}\, F$ may fail to be locally finitely presentable (Corollary 11.3.11).

(2) every coalgebra (A, α) with A finitely presentable in $\mathscr{A}$ is a finitely presentable object of $\mathsf{Coalg}\, F$ (Proposition 11.3.4) (this is not true for algebras in general); but in general these are not finitely presentable objects of $\mathsf{Coalg}\, F$ (Example 11.3.9(2)).

Remark 10.1.12. (1) It is well known that colimits commute with other colimits. In a locally finitely presentable category, filtered colimits also commute with finite products. This means that for any pair of filtered diagrams $D_i\colon \mathscr{D}_i \to \mathscr{A}$, $i = 1, 2$, the colimit of the diagram $\mathscr{D}_1 \times \mathscr{D}_2 \to \mathscr{A}$ taking $(d_1, d_1) \in \mathscr{D}_1 \times \mathscr{D}_2$ to $D_1 d_1 \times D_2 d_2$ is given by $\operatorname{colim} D_1 \times \operatorname{colim} D_2$. Similarly, filtered colimits commute with all finite limits.

(2) Consequently, given a pair of finitary functors $F, G\colon \mathscr{A} \to \mathscr{B}$ where $\mathscr{B}$ is locally finitely presentable, their product $F \times G$ is finitary, too.

(3) Finitary functors between arbitrary categories are closed under taking colimits (in particular, coproducts) and composition.

Example 10.1.13. (1) As already mentioned for $\mathscr{A} = \mathsf{Set}$ in Example 4.3.5, constant functors and the identity functor are clearly finitary.

(2) It follows that polynomial set functors H_Σ for (finitary) signatures Σ are finitary.

(3) The finite power-set functor $\mathscr{P}_\mathsf{f}$ is finitary, but the full power-set functor $\mathscr{P}$ is not (cf. Example 4.3.5).

(4) On the category Nom of nominal sets and equivariant maps, the algebraic

type functor $FX = \mathbb{A} + X \times X + [\mathbb{A}]X$ (see Example 2.1.8) is finitary. Indeed, this follows from Remark 10.1.12 and the fact $[\mathbb{A}](-)$, being a left adjoint [250, Thm. 4.13], is finitary.

Similarly, every functor on Nom arising from a binding signature (Remark 2.1.11) is finitary.

(5) In Examples 10.5.11 and 12.3.17 further below we shall use the following endofunctors on Nom: $FX = \{0,1\} \times X^{\mathbb{A}}$ and $FX = \{0,1\} \times X^{\mathbb{A}} \times [\mathbb{A}]X$, where $X^{\mathbb{A}}$ is the exponential in Nom which is a cartesian closed category; see, for example, Pitts [250, Thm. 2.19]. We now explain this briefly. Recall from Example 2.1.10 the notion of support of an element of a nominal set X. For two nominal sets X and Y we have an action on the set of all functions $f\colon X \to Y$ given by $(\pi \star f)(x) = \pi \cdot f(\pi^{-1} \cdot x)$. A function $f\colon X \to Y$ is *finitely supported* if it has a finite support with respect to this action. Equivalently there exists a finite subset $A \subseteq \mathbb{A}$ such that $\pi \cdot f(x) = f(\pi \cdot x)$ for all $x \in X$ and all $\pi \in \mathsf{Perm}(\mathbb{A})$ which fix all elements of A. The nominal set X^Y of all finitely supported maps from X to Y is the exponential in Nom. Consequently, equivariant maps $Z \times Y \to X$ are in bijective correspondence with equivariant maps $Z \to X^Y$ via (un)currying.

The functors above will be used to obtain various types of nominal automata as coalgebras. In particular, the first one is the type functor of deterministic nominal automata, and the coalgebras for the one functor are deterministic orbit-finite automata with variable binding transitions considered by Kozen et al. [203]. Both functors are finitary. Indeed, $[\mathbb{A}](-)$ being a quotient functor of the finitary functor $\mathbb{A} \times (-)$ is finitary, and that $(-)^{\mathbb{A}}$ is finitary follows from [234, Cor. 6.16].

(6) The finitely supported power-set functor $\mathscr{P}_{\mathsf{fs}}$ on Nom is not finitary. To define this functor, consider the action on the set of all subsets of a nominal set X given by $\pi \cdot Y = \{\pi \cdot y : y \in Y\}$ for $Y \subseteq X$. A subset is *finitely supported* if it has a finite support with respect to this action. The nominal set $\mathscr{P}_{\mathsf{fs}}X$ consists of all finitely supported subsets of X. This yields an endofunctor on Nom which acts on morphisms like $\mathscr{P}$. (In fact, Nom is a topos, and $\mathscr{P}_{\mathsf{fs}}X$ is the power object of X.)

(7) On the category $\mathsf{Set}^{\mathscr{F}}$ of sets in context the functor $FX = V + X \times X + \delta X$ (see Example 2.1.8) is finitary. Similarly, every functor on $\mathsf{Set}^{\mathscr{F}}$ arising from a binding signature is finitary.

Remark 10.1.14. The category of finitary set functors is equivalent to the category $\mathsf{Set}^{\mathscr{F}}$ of sets in context. In fact, the equivalence restricts a given set functor $F\colon \mathsf{Set} \to \mathsf{Set}$ by pre-composing it with the inclusion functor

$J \colon \mathscr{F} \hookrightarrow \mathsf{Set}$, and in the opposite direction the equivalence assigns to a given presheaf $P \colon \mathscr{F} \to \mathsf{Set}$ its left Kan extension along J.

Under this equivalence, the super finitary presheaves in Example 10.1.6(7) correspond precisely to the quotients of polynomial functors H_Σ, where Σ contains only finitely many operation symbols [52, Cor. 3.34].

Notation 10.1.15. For a locally finitely presentable category $\mathscr{A}$ we denote by

$$\mathscr{A}_{\mathsf{fp}}$$

the (essentially small) full subcategory given by all finitely presentable objects.

Remark 10.1.16. Let us collect a few useful properties of a locally finitely presentable category $\mathscr{A}$ that we shall subsequently use.

(1) The category $\mathscr{A}$ is complete [23, Rem. 1.56].

(2) A finite colimit of finitely presentable objects is finitely presentable again. Thus, the category $\mathscr{A}_{\mathsf{fp}}$ from Notation 10.1.15 is finitely cocomplete.

(3) By [23, Prop. 1.22], every object X of $\mathscr{A}$ is the filtered colimit of the *canonical filtered diagram* D_X given by all morphisms with codomain X having a finitely presentable domain. In other words, the colimit of the following diagram

$$\mathscr{A}_{\mathsf{fp}}/X \xrightarrow{D_X} \mathscr{A}, \qquad D_X(p \colon P \to X) = P$$

is X with the colimit injections given by all the morphisms p.

(4) For example, the terminal object of $\mathscr{A}$ is the colimit of the embedding $\mathscr{A}_{\mathsf{fp}} \hookrightarrow \mathscr{A}$.

(5) Let $D \colon \mathscr{D} \to \mathscr{A}$ be a filtered diagram. A cocone $c_i \colon Di \to C$, $i \in \mathscr{D}$, is a colimit cocone iff every morphism $p \colon P \to C$ with P finitely presentable factorizes essentially uniquely through one of the colimit injections $c_i \colon C_i \to C$. This means that

(a) there exist $i \in \mathscr{D}$ and p' such that $p = c_i \cdot p'$, and

(b) given two factorizations $p = c_i \cdot p_t$ $(t = 1, 2)$ there exists a morphism $h \colon i \to j$ in $\mathscr{D}$ such that $Dh \cdot p_1 = Dh \cdot p_2$.

Indeed, this states precisely that the hom-functors $\mathscr{A}(P, -)$ of finitely presentable objects P collectively reflect filtered colimits [52, Lemma 2.5].

(6) The category $\mathscr{A}$ has (strong epi, mono)-factorizations [23, Prop. 1.61] (cf. Remark 7.2.1).

Furthermore, let F be an endofunctor on $\mathscr{A}$ preserving monomorphisms. By Proposition 7.2.2, $\mathsf{Coalg}\, F$ inherits these factorizations. Note that for a set functor F we may drop the assumption that F preserves monomorphisms (see Remark 7.2.5(1)).

Lemma 10.1.17. *Let $\mathscr{A}$ be a category with finite colimits. Then every diagram in $D\colon \mathscr{D} \to \mathscr{A}$ can be extended to a filtered diagram with the same colimit (in fact, with essentially the same cocones).*

Proof Let $J\colon \mathscr{D} \to \hat{\mathscr{D}}$ be the free completion of $\mathscr{D}$ under finite colimits. This means that pre-composition with J yields an equivalence between the functor category $[\mathscr{D}, \mathscr{A}]$ and the category of all functors in $[\hat{\mathscr{D}}, \mathscr{A}]$ preserving finite colimits. A construction of $\hat{\mathscr{D}}$ can be found, for example, in work by Bunge and Carboni [95] based on unblished work by Pitts.

Thus, D extends to a functor $\hat{D}\colon \hat{\mathscr{D}} \to \mathscr{A}$. (Note that this extension is given by left Kan extension along J, i.e. $\hat{D} = \mathsf{Lan}_J\, D$.) To verify that D and $\hat{D}$ have the same cocones, note that a cocone of D with vertex X is precisely a natural transformation from D to $\Delta_X\colon \mathscr{D} \to \mathscr{A}$, the constant functor with value X. Thus all cocones form the set $\mathsf{Nat}(D, \Delta_X)$ of all natural transformations $D \to \Delta_X$. Analogously for $\hat{D}$. We obtain a bijection

$$\mathsf{Nat}(\hat{D}, \Delta_X) \cong \mathsf{Nat}(D, \Delta_X \cdot J) \tag{10.1}$$

that maps a cocone $d\colon \hat{D} \to \Delta_X$ to its restriction $dJ\colon D \cong \hat{D} \cdot J \to \Delta_X \cdot J$. On the right $\Delta_X \cdot J$ is also the constant functor with value X with domain $\mathscr{D}$. Thus the right-hand side is precisely the cocones of D with vertex X. This implies that D and $\hat{D}$ have the same colimit; indeed, a cocone $c\colon \hat{D} \to \Delta_C$ is a colimit of $\hat{D}$ iff we have a bijection $\mathscr{C}(C, X) \cong \mathsf{Nat}(\hat{D}, \Delta_X)$ given by pre-composition with (the components of) c. Now compose that last bijection with the one in (10.1) to see that C with the restricted cocone cJ is, equivalently, a colimit of D. □

10.2 The Rational Fixed Point

We now turn to the coalgebraic description of the rational fixed point as the terminal locally finitely presentable coalgebra.

Assumption 10.2.1. Throughout the rest of this chapter we assume that $\mathscr{A}$ is a locally finitely presentable category and that $F\colon \mathscr{A} \to \mathscr{A}$ is a finitary functor.

Remark 10.2.2. Note that these assumptions ensure that F has both an initial algebra μF (by Corollary 6.1.12) and a terminal coalgebra νF (this follows from Theorem 11.2.18).

Notation 10.2.3. We denote by $\mathsf{Coalg}_\mathsf{f}\, F$ the full subcategory of $\mathsf{Coalg}\, F$ formed by all coalgebras on finitely presentable carriers.

Remark 10.2.4. The category $\mathsf{Coalg_f}\,F$ is finitely cocomplete. This follows from the fact that $\mathscr{A}_{\mathsf{fp}}$ is finitely cocomplete (see Remark 10.1.16(2)) and that the forgetful functor $U\colon \mathsf{Coalg_f}\,F \to \mathscr{A}$ creates colimits (Proposition 4.1.1). Thus $\mathsf{Coalg_f}\,F$ is an essentially small filtered category (see Example 10.1.8(2)). Indeed, for essential smallness note that there is (up to isomorphism) only a set of finitely presentable objects A, each having only a set of coalgebra structures $\alpha\colon A \to FA$.

Definition 10.2.5. A coalgebra $\alpha\colon A \to FA$ is called *locally finitely presentable* (or *lfp*, for short) if it is a colimit of a diagram of coalgebras in $\mathsf{Coalg_f}\,F$.

Remark 10.2.6. (1) Every lfp coalgebra is in fact a *filtered* colimit of coalgebras in $\mathsf{Coalg_f}\,F$. This follows from the fact that $\mathsf{Coalg_f}\,F$ is finitely cocomplete (see Remark 10.2.4).

(2) Note that a coalgebra (A,α) is lfp if it is the colimit of the filtered diagram of all homomorphisms from coalgebras of $\mathsf{Coalg_f}\,F$ to (A,α). More precisely, (A,α) is the canonical colimit of the following diagram

$$D\colon \mathsf{Coalg_f}\,F/(A,\alpha) \to \mathsf{Coalg}\,F, \qquad D(m\colon (B,\beta) \to (A,\alpha)) = (B,\beta).$$

Equivalently, the following canonical forgetful functor is final (Remark 10.3.9):

$$\mathsf{Coalg_f}\,F/(A,\alpha) \to \mathscr{A}_{\mathsf{fp}}/A.$$

This is how the notion was originally introduced [227, Def. 3.7].

(3) Suppose that F is a finitary set functor. Given a coalgebra (A,α), every element $x \in A$ *generates* a subcoalgebra $s\colon (A',\alpha') \hookrightarrow (A,\alpha)$. This means that (A',α') is the smallest subcoalgebra of (A,α) containing x. Indeed, by Theorem 4.4.3, F preserves all nonempty intersections. Therefore, the intersection of all subcoalgebras containing x is itself a subcoalgebra by Proposition 4.1.5.

(4) Note that Proposition 7.2.2 can be slightly sharpened for set functors: the image of a coalgebra homomorphism $f\colon (A,\alpha) \to (B,\beta)$ may be taken to be a subcoalgebra $(C,\gamma) \hookrightarrow (B,\beta)$. Indeed, in the proof of Proposition 7.2.2 choose $C = f[A]$.

Recall (e.g. Silva et al. [276]) that a coalgebra for a set functor is called *locally finite* if each of its elements generates a finite subcoalgebra.

Proposition 10.2.7. *Let F be a finitary set functor. Then a coalgebra is lfp iff it is locally finite.*

We will obtain this as an easy consequence of a much more general categorical result in Section 10.5. Here we provide a set-theoretic proof.

Proof For the 'only if' direction, let (A,α) be an lfp coalgebra. Then it is a filtered colimit of finite coalgebra with colimit injections $c_i\colon (A_i,\alpha_i) \to (A,\alpha)$, say. Every element $x \in A$ lies in the finite subset $c_i[A_i]$ for some i, and we know that $c_i[A_i]$ carries a subcoalgebra of (A,α) by Remark 10.2.6(4).

For the 'if' direction, form the filtered diagram of all finite subcoalgebras $s_i\colon (A_i,\alpha_i) \hookrightarrow (A,\alpha)$. By Proposition 4.1.1 A is the union of all the A_i. Therefore, for the underlying diagram in Set, the cocone $s_i\colon A_i \hookrightarrow A$ is a filtered colimit. By Remark 4.1.7, this cocone is a filtered colimit in $\mathsf{Coalg}\,F$, which proves that (A,α) is an lfp coalgebra. □

Corollary 10.2.8. *For a finitary set functor a coalgebra is lfp iff it is the union of all of its finite subcoalgebras.*

Example 10.2.9. A deterministic automaton, considered as a coalgebra for $FX = \{0,1\} \times X^{\Sigma}$ (see Example 2.4.2(4)), is lfp iff from every of its states only finitely many states are reachable (by inputs from Σ^*). Indeed, the subcoalgebra generated by a state x consists precisely of those states reachable from x.

The terminal coalgebra νF formed by all formal languages on Σ (see Example 2.5.5) is not lfp. Indeed, the states reachable from a language L in νF are its left derivatives $w^{-1}L$ for $w \in \Sigma^*$. It is well known that a language has finitely many left derivatives iff it is regular. Thus, every non-regular language demonstrates that νF is not lfp.

In contrast, the set of all regular languages over Σ carries a locally finite subcoalgebra of νF.

Notation 10.2.10. We denote the full subcategory of $\mathsf{Coalg}\,F$ given by all lfp coalgebras by $\mathsf{Coalg}_{\mathsf{lfp}}\,F$.

Remark 10.2.11. (1) Every category $\mathscr{C}$ has a free completion $\mathsf{Ind}\,\mathscr{C}$ under filtered colimits (called the Ind-*completion* of $\mathscr{C}$). This means that $\mathsf{Ind}\,\mathscr{C}$ contains $\mathscr{C}$ as a full subcategory, it has filtered colimits, and it has the following universal property: every functor $F\colon \mathscr{C} \to \mathscr{A}$ where $\mathscr{A}$ has filtered colimits has a finitary extension $F^*\colon \mathsf{Ind}\,\mathscr{C} \to \mathscr{A}$ which is unique up to natural equivalence.

(2) The Ind-completion of a category $\mathscr{C}$ with finite colimits is the category of all functors from $\mathscr{C}^{\mathsf{op}}$ to Set preserving finite limits [134, Satz 5.5]. The category $\mathscr{C}$ is considered as a full subcategory of $\mathsf{Ind}\,\mathscr{C}$ via the contravariant hom-functors $\mathscr{C}(-,C)\colon \mathscr{C}^{\mathsf{op}} \to \mathsf{Set}$. In this case $\mathsf{Ind}\,\mathscr{C}$ is locally finitely presentable.

Theorem 10.2.12 [233, Thm. 2.7]**.** *The category* $\mathsf{Coalg}_{\mathsf{lfp}}\,F$ *is the* Ind-*completion of* $\mathsf{Coalg}_{\mathsf{f}}\,F$.

Proof We will make use of the following result [178, Sect. VI.1.8]: Let $\mathscr{C}$ be a category with finite colimits and $I\colon \mathscr{C} \to \mathcal{E}$ be a full embedding into a cocomplete category $\mathcal{E}$ whose image consists of finitely presentable objects in $\mathcal{E}$. Then the unique finitary extension $I^*\colon \mathsf{Ind}\,\mathscr{C} \to \mathcal{E}$ is also a full embedding.

We apply this to $\mathcal{E} = \mathsf{Coalg}\,F$ and $\mathscr{C} = \mathsf{Coalg}_\mathsf{f}\,F$. The former is cocomplete since so is $\mathscr{A}$ and the forgetful functor $U\colon \mathsf{Coalg}\,F \to \mathscr{A}$ creates colimits (Proposition 4.1.1). Similary, $\mathscr{C}$ is finitely cocomplete (see Remark 10.2.4).

Furthermore, we know from Example 10.1.11(2) that for every finitary functor F on a locally finitely presentable category, every coalgebra with a finitely presentable carrier is a finitely presentable object in $\mathsf{Coalg}(F)$.

Therefore we can apply the above result: the unique (finitary) extension of the full embedding $\mathsf{Coalg}_\mathsf{f}\,F \hookrightarrow \mathsf{Coalg}\,F$ is itself a full embedding $\mathsf{Ind}(\mathsf{Coalg}_\mathsf{f}\,F) \to \mathsf{Coalg}\,F$. The definition of this extension is that it takes formal filtered diagrams of objects in $\mathsf{Coalg}_\mathsf{f}\,F$ and constructs their colimit in $\mathsf{Coalg}\,F$. Therefore, according to Definition 10.2.5, its image is precisely $\mathsf{Coalg}_\mathsf{lfp}(F)$. So by restricting we obtain an equivalence $\mathsf{Ind}(\mathsf{Coalg}_\mathsf{f}\,F) \cong \mathsf{Coalg}_\mathsf{lfp}\,F$. □

Corollary 10.2.13. *The category* $\mathsf{Coalg}_\mathsf{lfp}\,F$ *is locally finitely presentable.*

In particular, it follows that $\mathsf{Coalg}_\mathsf{lfp}\,F$ is complete (see Remark 10.1.16(1)) and therefore has a terminal object.

Definition 10.2.14. The *rational fixed point* of F is the terminal lfp coalgebra. We denote it by ϱF.

We shall see in Theorem 10.2.25 that the coalgebra structure of ϱF is indeed an isomorphism.

As an immediate consequence we obtain the following coalgebraic construction of the rational fixed point due to [33, Section 3]:

Corollary 10.2.15. *The rational fixed point is the colimit of all coalgebras on finitely presentable carriers:*

$$\varrho F = \mathsf{colim}(\mathsf{Coalg}_\mathsf{f}\,F \hookrightarrow \mathsf{Coalg}\,F).$$

Indeed, apply Remark 10.1.16(4).

For a finitary set functor, we obtain the following construction of the rational fixed point. It shows that the rational fixed point is fully abstract with respect to final semantics: every of its elements is the semantics of a state in a finite coalgebra ('no junk') and no two behaviourally equivalent states are separated ('no confusion'). Shortly, ϱF is a subcoalgebra of νF. We shall see a more general result in Proposition 10.5.6 where we present sufficient conditions on a general locally finitely presentable category $\mathscr{A}$ and a functor F ensuring full abstractness. Here we present a proof for the special case of set functors for the convenience of the reader.

Definition 10.2.16. We say that the rational fixed point is *fully abstract* if ϱF is a subcoalgebra of νF: the unique homomorphism $\varrho F \to \nu F$ is monic.

Remark 10.2.17. The notion of full abstraction originates in the denotational semantics of programming languages [251] where it is a statement about the relation of operational and denotational semantics stating that observational equivalence coincides with equality in denotational semantics. Here we use the term to state that behavioural equivalence for elements of finite coalgebras provided by the rational fixed point coincides with equivalence in the final semantics, which is given by equality in the terminal coalgebra.

Proposition 10.2.18. *The rational fixed point ϱF of a finitary set functor F is fully abstract.*

More precisely, it is the subcoalgebra of νF on all elements generating a finite subcoalgebra of νF.

Proof It suffices to prove that ϱF is (isomorphic to) the union of all finite subcoalgebras of νF. Let us denote this union by R and denote by $m\colon \varrho F \to \nu F$ the unique coalgebra homomorphism.

We first prove that m is injective. We know from Corollary 10.2.8 that ϱF is the union of all its finite subcoalgebras. Thus, given a pair $x, y \in \varrho F$ with $m(x) = m(y)$, we can choose a subcoalgebra (A, α) of ϱF containing x and y. The unique coalgebra homomorphism $h\colon (A, \alpha) \to (\nu F, \tau)$ merges x and y. We factorize h into a strong epimorphism $e\colon (A, \alpha) \twoheadrightarrow (C, \gamma)$ followed by an inclusion $s\colon (C, \gamma) \hookrightarrow (\nu F, \tau)$. Thus we have $e(x) = e(y)$. We also have a unique coalgebra homomorphism g from (C, γ) to ϱF. By the universal property of ϱF, we obtain $g \cdot e = i$, where i is the inclusion of the subcoalgebra (A, α) in ϱF. We thus have

$$x = i(x) = g \cdot e(x) = g \cdot e(y) = i(y) = y.$$

We proceed to prove that $R = \varrho F$. We clearly have an injection $\varrho F \rightarrowtail R$, since ϱF is the union of its finite subcoalgebras, and these may be identified with finite subcoalgebras of νF via extending the inclusions by m.

Conversely, let $s\colon (A, \alpha) \hookrightarrow (\nu F, \tau)$ be any finite subcoalgebra. Then $s = m \cdot h$ where $h\colon A \to \varrho F$ is the unique coalgebra homomorphism into the terminal lfp coalgebra. Since s is injective, it follows that h is injective, too. Thus we obtain an injection $R \rightarrowtail \varrho F$, which completes the proof. □

For a polynonial endofunctor H_Σ on Set, an initial algebra is carried by the set of all finite Σ-trees (see Example 2.2.11) and a terminal coalgebra by the set of all (finite and infinite) Σ-trees (see Theorem 2.5.9). The unique coalgebra homomorphism from a coalgebra (A, α) to the terminal coalgebra assigns to every element its tree expansion (see Remark 2.5.10). We now describe the rational fixed point.

Definition 10.2.19. A Σ-tree is *rational* if it has, up to isomorphism, only finitely many subtrees.

Ginali [139] has shown that these are precisely Courcelle's regular Σ-trees [107]. We are going to show that ϱH_Σ is formed by all rational Σ-trees.

Proposition 10.2.20. *For every finitary signature Σ the rational fixed point ϱH_Σ is the subcoalgebra of νH_Σ given by all rational Σ-trees.*

Proof Indeed, use Proposition 10.2.7 and recall from Example 2.5.8 that the coalgebra structure of νH_Σ takes a Σ-tree t whose root is labelled by an n-ary operation symbol σ to the pair $(\sigma, (t_0, \dots, t_{n-1}))$ where the t_i are the maximum proper subtrees of t. Consequently, every tree t generates the subcoalgebra of all subtrees of t up to isomorphism. □

Example 10.2.21. We present examples of rational fixed points of finitary set functors.

(1) Let $FX = \{0, 1\} \times X^\Sigma$ be the functor whose coalgebras are deterministic automata and whose terminal coalgebra is formed by all formal languages over the alphabet Σ (see Example 2.5.5). By Proposition 10.2.18, the rational fixed point is thus the subcoalgebra formed by all regular languages over Σ.

This can also be derived from Proposition 10.2.20 as follows. The functor F is (naturally isomorphic to) the polynominal functor for the signature of two operation symbols of arity $|\Sigma|$. Trees for that signature are clearly in bijective correspondence with functions $\Sigma^* \to \{0, 1\}$, that is, formal languages over Σ (cf. Remark 2.2.13). Rational trees then correspond precisely to the regular languages over Σ.

(2) As an instance of Proposition 10.2.20, consider $FX = X \times X + 1$ on Set, which is the polynomial functor for the signature Σ with one binary operation symbol and one constant. We have seen in Example 2.5.11(4) that its terminal coalgebra consists of all binary trees. For example, all finite binary trees and the following infinite tree

are rational, but the tree represented by

$$(t_0 * (t_1 * (t_2 * (\cdots) \cdots))),$$

where t_i is the complete binary tree of height i, is not rational.

(3) Similarly, all rational finitely branching strongly extensional trees form the rational fixed point of $\mathscr{P}_\mathrm{f}$ (cf. Example 2.2.7(4) and Theorem 4.5.15).

(4) For the bag functor $\mathscr{B}$: Set $\to$ Set the initial algebra consists of all finite unordered trees (see Example 3.2.11) and the terminal coalgebra consists of all unordered trees (see Example 4.3.31(4)). Thus the rational fixed point is formed by all rational unordered trees.

(5) Let $FX = \Sigma \times X$ for a set Σ, then we have $\nu F = \Sigma^\omega$, the coalgebra of all streams in Σ (see Example 2.5.3(3)). Since the coalgebra structure on νF is given by the head and tail functions, we see that the subcoalgebra generated by a stream $(\sigma_n)_{n\in\mathbb{N}}$ consists of all suffixes $(\sigma_{n+i})_{n\in\mathbb{N}}$. Clearly, a stream has finitely many different suffixes iff it is *eventually periodic*: it is of the form $uv^\omega = uvvv\cdots$ for finite words $u, v \in \Sigma^*$. Thus, the rational fixed point ϱF consists precisely of all eventually periodic streams in Σ.

(6) We recall that automatic sequences [58] are computed by Moore automata with binary input. More precisely, a Moore automaton with the next state function $\delta\colon Q\times\{0,1\} \to Q$, output function $\lambda\colon Q \to \Gamma$ and initial state $q_0 \in Q$ computes the stream $(s_n)_{n\in\mathbb{N}}$ with $s_n = \lambda(\delta^*(q_0, (n)_2))$, where $(n)_2$ denotes the binary representation of $n \in \mathbb{N}$, and δ^* is the extension of the next state function to words. A stream $s \in \Gamma^\omega$ is called 2-*automatic* if it is computed by some Moore automaton.

We have seen that Moore automata are coalgebras for $FX = \Gamma \times X^{\{0,1\}} \cong \Gamma \times X \times X$ (Example 2.4.2(6)) and that the set of streams Γ^ω carries a terminal F-coalgebra (Example 2.5.11(8)). Grabmeyer et al. [148, Thm. 52] proved that a stream s is 2-automatic if and only if the smallest subcoalgebra of Γ^ω containing it is finite. Hence, it follows from Proposition 10.2.18 that ϱF consists precisely of the 2-automatic streams.

Analogously, k-automatic sequences in A^ω are computed by Moore automata with input alphabet $\Sigma = \{1, \ldots, k\}$, which are coalgebras for the functor $FX = A \times X^k$, $k \in \mathbb{N}$. We have mentioned in Example 2.5.11(8) that Γ^ω carries a terminal F-coalgebra, and it follows from the result in op. cit. that ϱF consists precisely of the k-automatic sequences.

Remark 10.2.22. For the set functor $FX = \mathbb{N} \times X$ we shall see, in addition to the description of νF and ϱF as (eventually periodic) streams, an interesting description of these two fixed points in Section 15.5: νF is carried by the set $\mathbb{R}_{\geq 0}$ of non-negative real numbers, and ϱF is carried by the subset of all *quadratic irrationals*, that is, the real roots of irreducible polynomials with rational coefficients.

The key ingredient for proving that ϱF is a fixed point of F is Proposition 10.2.24 further below. In its proof we use a diagram that we first describe:

Construction 10.2.23. For every coalgebra (A, α) with A finitely presentable, we define a diagram $D\colon \mathscr{D} \to \mathsf{Coalg}\, F$ as follows. The diagram scheme $\mathscr{D}$ has as objects morphisms $p\colon P \to FA$ with P finitely presentable, to which D assigns the following coalgebra in $\mathsf{Coalg}_{\mathsf{f}}\, F$:

$$\bar{p} = (P + A \xrightarrow{[p,\alpha]} FA \xrightarrow{F\mathsf{inr}} F(P + A)).$$

A morphism in $\mathscr{D}$ from $p\colon P \to FA$ to $q\colon Q \to FA$ is a coalgebra homomorphism $h\colon (P + A, \bar{p}) \to (Q + A, \bar{q})$ which fulfils $[p, \alpha] = [q, \alpha] \cdot h$:

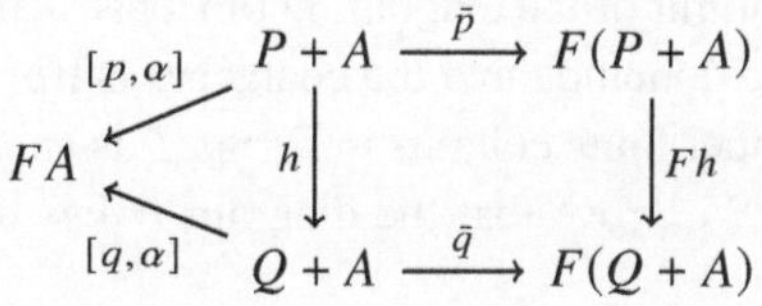

(10.2)

We make two observations:

(1) The unique morphism $u\colon 0 \to FA$, where 0 is the initial object, yields $\bar{u} = (\alpha\colon A \to FA)$. Moreover, $\alpha\colon A \to FA$ itself yields the object $(A + A, \bar{\alpha})$ for which the codiagonal $\nabla\colon A + A \to A$ is a connecting morphism of D from $(A + A, \bar{\alpha})$ to $(A, \bar{u})$ as shown by the commutative diagram below:

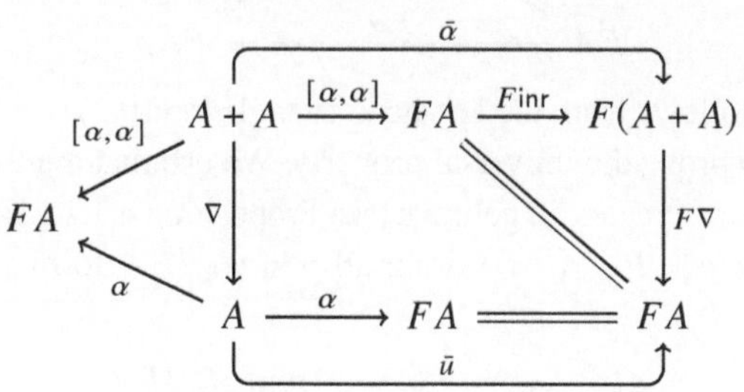

(2) Every morphism m from $p\colon P \to FA$ to $q\colon Q \to FA$ in the slice category $\mathscr{A}_{\mathsf{fp}}/FA$, that is, with $q \cdot m = p$, yields the connecting morphism $m + A$ of D;

this follows from the commutative diagram below:

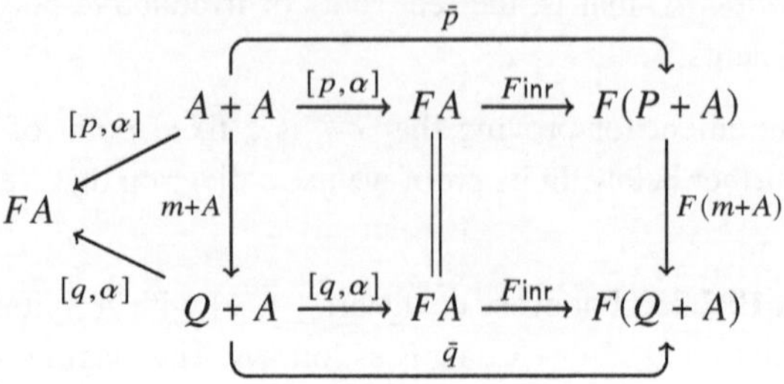

Proposition 10.2.24. *If* (A, α) *is an lfp coalgebra, then so is* $(FA, F\alpha)$.

Proof It is sufficient to prove the result for coalgebras with a finitely presentable carrier. The general case then follows: given (A, α) in $\mathsf{Coalg}_{\mathsf{lfp}}\, F$, expressed as a filtered colimit of coalgebras $\alpha_i \colon A_i \to FA_i$ $(i \in I)$ with A_i finitely presentable, then $F\alpha \colon FA \to FFA$ is a filtered colimit of the coalgebras $(FA_i, F\alpha_i)$ since F is finitary. Thus $(FA, F\alpha)$ is lfp since each $(FA_i, F\alpha_i)$ is lfp and $\mathsf{Coalg}_{\mathsf{lfp}}\, F$ is closed under filtered colimits.

Now suppose that (A, α) is a coalgebra with A finitely presentable. We prove that $(FA, F\alpha)$ is a colimit of the diagram D of Construction 10.2.23.

By proving this we conclude that the coalgebra is lfp: while D is itself not filtered, its closure under finite colimits in $\mathsf{Coalg}_{\mathsf{f}}\, F$ is and has the same colimit by Lemma 10.1.17. We prove that the diagram D has the following colimit cocone in $\mathsf{Coalg}\, F$:

$$[p, \alpha] \colon (P + A, \bar{p}) \to (FA, F\alpha),$$

where p ranges over $\mathscr{A}_{\mathsf{fp}}/FA$. Indeed, the fact that each $[p, \alpha]$ is a homomorphism is clear:

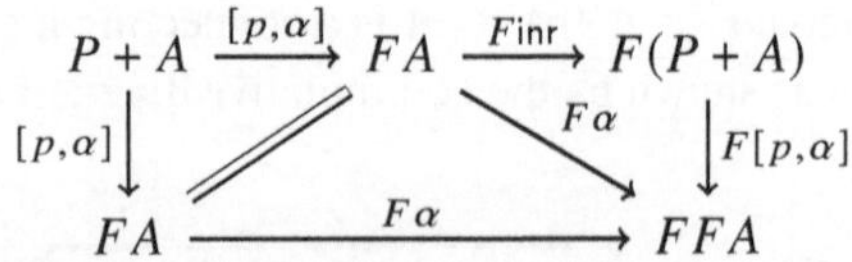

Compatibility follows from the left-hand triangle in (10.2).

It remains to prove the universal property. We use that the forgetful functor $U \colon \mathsf{Coalg}\, F \to \mathscr{A}$ creates all colimits (see Proposition 4.1.1). Hence, it suffices to prove that $[p, \alpha] \colon P + A \to FA$, for all p in $\mathscr{A}_{\mathsf{fp}}/FA$, form a colimit of UD. Suppose that

$$c_p \colon P + A \to C \qquad (p \text{ in } \mathscr{A}_{\mathsf{fp}}/FA)$$

is a cocone of UD in $\mathscr{A}$. We shall show that there exists a unique $c \colon FA \to C$ such that $c_p = c \cdot [p, \alpha]$ for all p.

First, it is easy to see that the morphisms $c_p \cdot \mathsf{inl}\colon P \to C$ form a cocone of the canonical diagram $\mathscr{A}_{\mathsf{fp}}/FA \to \mathscr{A}$ whose colimit is FA (Remark 10.1.16(3)): in fact, given a morphism m as in Construction 10.2.23(2), the given cocone fulfils $c_q \cdot (m + A) = c_p$, hence, $c_q \cdot \mathsf{inl} \cdot m = c_p \cdot \mathsf{inl}$. Therefore there exists a unique $c\colon FA \to C$ such that $c_p \cdot \mathsf{inl} = c \cdot p$ for all p in $\mathscr{A}_{\mathsf{fp}}/FA$. Thus, to conclude $c_p = c \cdot [p, \alpha]$, it only remains to prove that $c_p \cdot \mathsf{inr} = c \cdot \alpha$ holds for all p. The connecting morphism ∇ in Construction 10.2.23(1) yields

$$c_u \cdot \nabla = c_\alpha\colon A + A \to C.$$

This implies that

$$c_u = c_\alpha \cdot \mathsf{inl} = c \cdot \alpha \tag{10.3}$$

by the definition of c for $p = \alpha$. Moreover, the unique morphism $u\colon 0 \to P$ is a morphism in $\mathscr{A}_{\mathsf{fp}}/FA$ from the unique $0 \to FA$ to $p\colon P \to FA$ and therefore yields the connecting morphism $u + A = \mathsf{inr}\colon (A, \bar{u}) \to (P + A, \bar{p})$ using Construction 10.2.23(2). Thus, we conclude

$$c_p \cdot (u + A) = c_p \cdot \mathsf{inr} = c_u = c \cdot \alpha,$$

where the last step uses (10.3). □

The following result was first proved in [33, Thm. 3.3]. We now obtain a short proof by virtue of Proposition 10.2.24.

Theorem 10.2.25. *The coalgebra structure of ϱF is an isomorphism, i.e. ϱF is a fixed point of F.*

Proof We argue as in Lambek's Lemma: denote by $\omega\colon \varrho F \to F(\varrho F)$ the coalgebra structure. By Proposition 10.2.24 we know $(F(\varrho F), F\omega)$ is an lfp coalgebra. Therefore we obtain a homomorphism $s\colon (F(\varrho F), F\omega) \to (\varrho F, \omega)$. Since $\omega\colon (\varrho F, \omega) \to (F(\varrho F), F\omega)$ is clearly a homomorphism, we have that $s \cdot \omega$ is a homomorphism from $(\varrho F, \omega)$ to itself. Thus, $s \cdot \omega = \mathsf{id}$ by the universal property of $(\varrho F, \omega)$. It follows that

$$\omega \cdot s = Fs \cdot F\omega = F(s \cdot \omega) = F\mathsf{id} = \mathsf{id}.$$ □

Remark 10.2.26. We end this section by remarking that a rational fixed point can also be considered for non-finitary functors F. Indeed, in lieu of F one considers its *finitary coreflection* (cf. Section C.9 for set functors). This is the finitary functor $F_{\mathsf{f}}\colon \mathscr{A} \to \mathscr{A}$ given by the left Kan extension of FJ along J where $J\colon \mathscr{A}_{\mathsf{fp}} \hookrightarrow \mathscr{A}$ is the inclusion functor. Since F_{f} concides with F on $\mathscr{A}_{\mathsf{fp}}$, the two functors have the same coalgebras on finitely presentable carriers.

10.3 Iterative Algebras

Recall from Theorem 7.3.14 that the terminal coalgebra νF is the initial cia for F. A very similar characterization holds for the rational fixed point ϱF: it is the initial *iterative algebra* for F. This notion is weaker than that of a completely iterative algebra (see Definition 7.3.2) in that one only requires unique solutions of flat equation morphisms whose domain is finitely presentable. (In Set this means that only finite systems (7.3) are required to have unique solutions.)

Iterative algebras for polynomial set functors were introduced by Nelson [242] (see Tiuryn [289] for a related concept) in connection with Elgot's *iterative theories* [115]. Much later this was generalized from Set to arbitrary locally finitely presentable categories [30, 33]. In this section we only mention the most important results on iterative algebras.

Recall from Definition 7.3.2 the notions of a flat equation morphism and its solution in an F-algebra.

Definition 10.3.1 [33, Def. 2.5]. A flat equation morphism $e\colon X \to FX + A$ is called *finitary* if X is a finitely presentable object of $\mathscr{A}$.

An algebra (A, α) is called *iterative* if every finitary flat equation morphism has a unique solution $e^\dagger$ in A.

Example 10.3.2. (1) Every cia is, of course, an iterative algebra.

For example, we saw in Example 7.3.10(2) that the interval $[0, 1]$ is a cia for the functor $FX = X + X$ on Set, and so it is iterative. It is interesting to note that the solutions of the finitary flat equation morphisms are precisely the rational numbers in the unit interval.

(2) Unary algebras $\alpha\colon A \to A$ in Set are the algebras for the identity functor. Iterativity of (A, α) means that α has a unique cycle, and this is a fixed point (see [33, Example 2.13] and cf. Example 7.3.7(4)).

An example of an iterative algebra that is not a cia is the unary algebra $A = \mathbb{N}$ with the algebra structure given by $\alpha(0) = 0$ and $\alpha(n + 1) = n$. Indeed, its only cycle is the fixed point 0. However, for $X = \{x_n : n \in \mathbb{N}\}$ the flat equation morphism $e\colon X \to X + A$ with $e(x_n) = x_{n+1}$ has two solutions in A: the constant function with value 0 and the function $x_n \mapsto n$.

(3) The algebra of addition on the extended real numbers $\widetilde{\mathbb{R}} = \mathbb{R} \cup \{\infty\}$ is an iterative algebra for the functor $FX = X \times X$ [33]. Note that this is not a cia: the system of equations $x_0 \approx x_1 + 1, x_1 \approx x_2 + 1, \ldots$ has more than one solution (e.g. ${x_n}^\dagger = \infty$ or ${x_n}^\dagger = -n$).

Remark 10.3.3. Note that in an iterative algebra (A, α) we have a unique solution for every lfp coalgebra $e\colon X \to FX + A$. Indeed, write (X, e) as a filtered colimit of a diagram D of coalgebras (X_i, e_i) in $\mathsf{Coalg_f}(F(-)+A)$. Then the latter coalgebras are finitary flat equation morphisms. Their unique solutions $e_i^\dagger : X_i \to A$ form a cocone of D and therefore induce a unique factorizing morphism $e^\dagger : X \to A$. A routine verification then shows that $e^\dagger$ is the unique solution of e, as desired.

Iterative algebras constitute a full subcategory of $\mathsf{Alg}\,F$. The proof is similar to the corresponding proof for cias (cf. Proposition 7.3.5), but it is a bit more involved. Since the reasoning in the second part of the proof is quite typical for an argument in a locally finitely presentable category, we fully spell out the details:

Proposition 10.3.4 [33, Prop. 2.18]. *Let (A, α) and (B, β) be iterative algebras. Then a morphism $h\colon A \to B$ is an algebra homomorphism if and only if it preserves solutions.*

Proof For the 'only if' direction ones proceeds as in the proofs of Propositions 7.3.5 and 7.1.5: given a homomorphism $h\colon (A, \alpha) \to (B, \beta)$, one shows for every finitary flat equation morphism $e: X \to FX + A$ that $h \cdot e^\dagger$ is a solution of $(FX + h) \cdot e$ and concludes using uniqueness of solutions.

For the 'if' direction, suppose that $h\colon A \to B$ preserves solutions. By Remark 10.1.16(3), A is the colimit of the canonical diagram D_A of all morphisms $p: P \to A$ with P finitely presentable. Similarly, since F is finitary, we have $FA = \operatorname{colim} D_{FA}$. Hence, in order to prove that h is a homomorphism, $h \cdot \alpha = \beta \cdot Fh$, it suffices to prove

$$h \cdot \alpha \cdot p = \beta \cdot Fh \cdot p \quad \text{for all } p: P \to FA \text{ with } P \text{ finitely presentable.} \tag{10.4}$$

As P is finitely presentable, we know that the hom-functor $\mathscr{A}(P, -)$ preserves that colimit. Therefore, $p: P \to FA$ factorizes through a colimit injection: we obtain morphisms q in $\mathscr{A}_{\mathsf{fp}}/A$ and $p': P \to FQ$ such that the triangle below commutes:

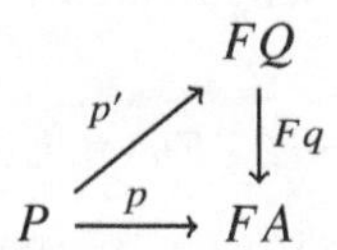

We then form the following finitary flat equation morphism

$$e = (P + Q \xrightarrow{p'+Q} FQ + Q \xrightarrow{F\mathsf{inr}+q} F(P + Q) + A).$$

The solution $e^\dagger$ yields the commutative square below:

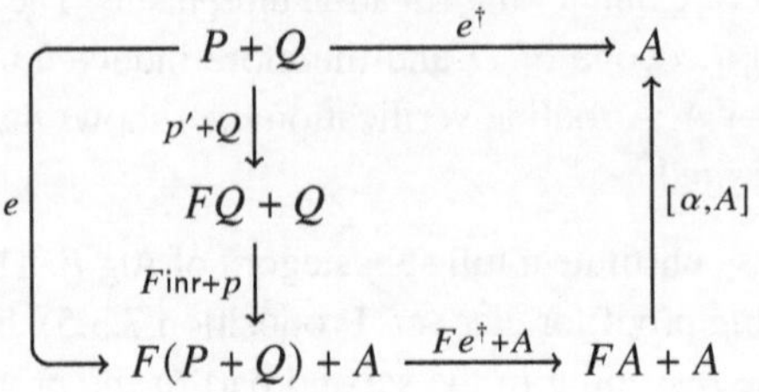

This shows that $e^\dagger \cdot \mathsf{inr} = q$ and therefore

$$e^\dagger \cdot \mathsf{inl} = \alpha \cdot F(e^\dagger \cdot \mathsf{inr}) \cdot p' = \alpha \cdot Fq \cdot p' = \alpha \cdot p.$$

Since h preserves the solution of e, we have $h \cdot e^\dagger = f^\dagger$ for

$$f = (X \xrightarrow{e} F(P+Q) + A \xrightarrow{F(P+Q)+h} F(P+Q) + B).$$

Thus, we obtain the following equations:

$$f^\dagger = h \cdot e^\dagger = h \cdot [\alpha \cdot p, q] = [h \cdot \alpha \cdot p, h \cdot q]. \tag{10.5}$$

We now obtain the following commutative diagram:

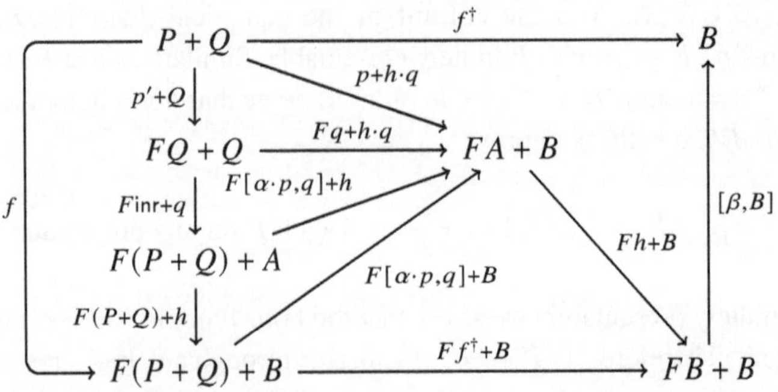

The outside commutes because $f^\dagger$ is a solution of f, for the lowest triangle use (10.5), and the remaining triangles are trivial. Thus the upper right-hand part commutes:

$$f^\dagger = [\beta \cdot Fh \cdot p, h \cdot q]. \tag{10.6}$$

The left-hand components of (10.5) and (10.6) establish the desired equality (10.4). □

Lemma 10.3.5. *If (A, α) is an iterative algebra, then so is $(FA, F\alpha)$.*

The proof is the same as for cias in Lemma 7.3.11(1).

Proposition 10.3.6 [33, Thm. 2.20]. *Iterative algebras are closed under filtered colimits in the category* $\mathsf{Alg}\,F$ *of* F*-algebras.*

Proof Suppose we have a filtered diagram D of iterative algebras (A_i, α_i), $i \in I$, and let (A, α) be its colimit in $\mathsf{Alg}\,F$ with colimit injections $c_i\colon (A_i, \alpha_i) \to (A, \alpha)$. Since F is finitary, filtered colimits of F-algebras are formed on the level of $\mathscr{A}$; more precisely, the forgetful functor $\mathsf{Alg}\,F \to \mathscr{A}$ creates filtered colimits (cf. Remark 4.1.7). Let $e\colon X \to FX + A$ be a finitary flat equation morphism. Since X is finitely presentable and $FX + A = \mathsf{colim}_{i \in I}(FX + A_i)$, there exist an i and a morphism $e_i\colon X \to FX + A_i$ such that the triangle below commutes:

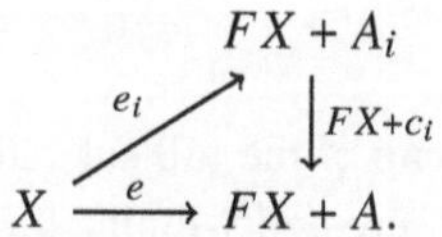

By Proposition 10.3.4, we know that $c_i \cdot e_i^\dagger\colon X \to A$ is a solution of e in A.

In order to prove uniqueness, we show for any solution $s\colon X \to A$ of e in A that $s = c_i \cdot e_i^\dagger$. First, use that X is finitely presentable to obtain some $j \in I$ and s' such that the following triangle commutes:

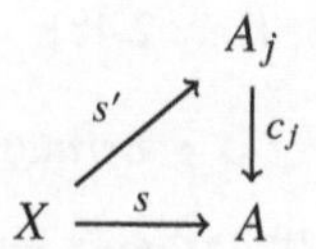

Since D is filtered, we can assume that for j above we have a homomorphism $h\colon (A_i, \alpha_i) \to (A_j, \alpha_j)$ in the diagram, that is, $c_j \cdot h = c_i$. Furthermore, we show that j can be chosen such that s' is a solution of $e_j = (FX + h) \cdot e_i$. To see this we consider the following diagram:

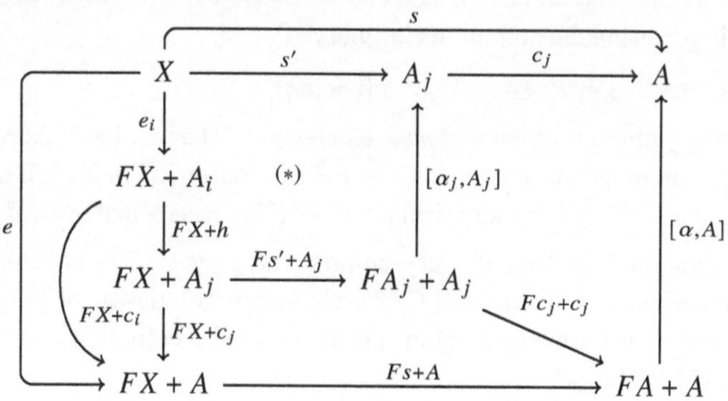

Note that all its inner parts except perhaps the part marked (∗) commute. Thus, this part commutes when post-composed by c_j. By filteredness we can therefore assume that $(*)$ commutes (for otherwise we can use Condition (c) in Definition 10.1.7(1): there exists a connecting morphism $g\colon (A_j, \alpha_j) \to (A_k, \alpha_k)$ in the diagram D equating the sides of $(*)$; then we work with k in lieu of j). Thus, by the uniqueness of solutions in A_j, we have $s' = e_j^\dagger$. Moreover, by Proposition 10.3.4, we obtain $e_j^\dagger = h \cdot e_i^\dagger$ and we conclude

$$s = c_j \cdot s' = c_j \cdot e_j^\dagger = c_j \cdot h \cdot e_i^\dagger = c_i \cdot e_i^\dagger. \qquad \square$$

Corollary 10.3.7. *The full subcategory of* $\mathsf{Alg}\, F$ *given by iterative algebras is reflective and has limits and filtered colimits formed at the level of the base category* $\mathscr{A}$.

Proof We have proved closure of the full subcategory under filtered colimits in Proposition 10.3.6. For closure under limits, the argument is identical to that in Lemma 7.3.11(2) but using the 'only if' part of Proposition 10.3.4 in lieu of Proposition 7.3.5.

Reflectivity now follows from the fact that the category $\mathsf{Alg}\, F$ of all algebras for F is locally finitely presentable (cf. [23, Cor. 2.75]). Moreover, every full subcategory of a locally finitely presentable category closed under limits and filtered colimits is reflective [23, Thm. 2.48]. □

Corollary 10.3.8. *Every object of* $\mathscr{A}$ *generates a free iterative algebra.*

Indeed, it follows from Proposition 2.2.24 and Theorem 3.1.8 that on every object of $\mathscr{A}$ a free algebra (A, α) exists. The reflection of (A, α) in the category of iterative algebras is the free iterative algebra on A. (This follows from the fact that left adjoint functors compose.)

In particular, the reflection of the initial algebra μF in the category of iterative algebras is the initial iterative F-algebra. Furthermore, we next prove that this algebra is ϱF using the colimit in Corollary 10.2.15.

Remark 10.3.9. Let $P\colon \mathscr{D} \to \mathscr{C}$ be a functor.

(1) For every object C of $\mathscr{C}$ the *comma category* C/P has as objects pairs (X, f) where X is an object of $\mathscr{D}$ and $f\colon C \to PX$ is a morphism in $\mathscr{C}$. A morphism from (X, f) to (Y, g) is a morphism $d\colon X \to Y$ of $\mathscr{D}$ such that $g = Pd \cdot f$.

(2) The functor P is *final* if every comma category C/P is nonempty and connected (every pair of objects of C/P is connected by a zig-zag of morphisms).

It follows that for every diagram $D\colon \mathscr{C} \to \mathscr{A}$ the colimits of D and DP coincide [217, Thm. IX.3.1].

(3) If $\mathscr{D}$ is filtered, then P is final iff the following two conditions hold:

(a) for every object C of $\mathscr{C}$ there exists a morphism $C \to PX$ for some object X of $\mathscr{D}$,

(b) for every parallel pair $f, g \colon C \rightrightarrows PX$ in $\mathscr{C}$ there exists $d \colon X \to X'$ in $\mathscr{D}$ with Pd merging f and g, that is, $Pd \cdot f = Pd \cdot g$.

For the proof that ϱF is the initial iterative algebra in Theorem 10.3.12 further below we use the following filtered category:

Construction 10.3.10. Let $e \colon X \to FX + \varrho F$ be a finitary flat equation morphism. For every coalgebra (A, α) in $\mathsf{Coalg_f}\, F$ we denote by $\alpha^\sharp \colon A \to \varrho F$ the unique homomorphism (cf. Definition 10.2.14).

We define a category $\mathscr{D}$ whose objects are pairs consisting of a coalgebra (A, α) in $\mathsf{Coalg_f}\, F$ and a morphism $f \colon X \to FX + A$ (where X is the above domain of e) such that for the unique homomorphism $\alpha^\sharp \colon (A, \alpha) \to (\varrho F, \omega)$ the following triangle commutes:

$$\begin{array}{ccc} & & FX + A \\ & \overset{f}{\nearrow} & \Big\downarrow {\scriptstyle FX+\alpha^\sharp} \\ X & \xrightarrow{\quad e \quad} & FX + \varrho F \end{array} \tag{10.7}$$

A morphism in $\mathscr{D}$ from $((A, \alpha), f)$ to $((B, \beta), g)$ is a homomorphism $h \colon (A, \alpha) \to (B, \beta)$ such that $(FX + h) \cdot f = g$:

$$\begin{array}{cccc} & A & \xrightarrow{\ \alpha\ } & FA \\ {\scriptstyle f}\nearrow & \Big\downarrow {\scriptstyle h} & & \Big\downarrow {\scriptstyle Fh} \\ X \quad \underset{g}{\searrow} & B & \xrightarrow[\ \beta\ ]{} & FB \end{array}$$

The category $\mathscr{D}$ is clearly essentially small, and we now prove that it is filtered by verifying (a)–(c) in Definition 10.1.7(1).

(a) To see that the category $\mathscr{D}$ is nonempty observe first that, since the forgetful functor $U \colon \mathsf{Coalg}\, F \to \mathscr{A}$ creates colimits (see Proposition 4.1.1), the unique homomorphisms $\alpha^\sharp \colon (A, \alpha) \to (\varrho F, \omega)$, for (A, α) in $\mathsf{Coalg_f}\, F$, form a colimit cocone in $\mathscr{A}$ (cf. Corollary 10.2.15). Furthermore, since the functor $FX + (-) \colon \mathscr{A} \to \mathscr{A}$ clearly is finitary (see Example 10.1.13(1) and Remark 10.1.12(3)), we see that the morphisms $FX + \alpha^\sharp \colon FX + A \to FX + \varrho F$ form a filtered colimit, too. Since X is finitely presentable, the morphism $e \colon X \to FX + \varrho F$ admits a factorization as in (10.7), which proves that $\mathscr{D}$ is nonempty. Moreover, the factorization is essentially unique (see Definition 10.1.7(1)), which we will use in the next item.

(b) Suppose we are given two objects $((A, \alpha), f)$ and $((B, \beta), f')$ of $\mathscr{D}$. Then we first use that $\mathsf{Coalg_f}\, F$ is filtered to obtain homomorphisms $h \colon (A, \alpha) \to (C, \gamma)$

and $h' \colon (B, \beta) \to (C, \gamma)$. This implies that

$$\gamma^\sharp \cdot h = \alpha^\sharp \qquad \text{and} \qquad \gamma^\sharp \cdot h' = \beta^\sharp.$$

Furthermore, we see that $FX + \gamma^\sharp$ merges $(FX + h) \cdot f$ and $(FX + h') \cdot g$:

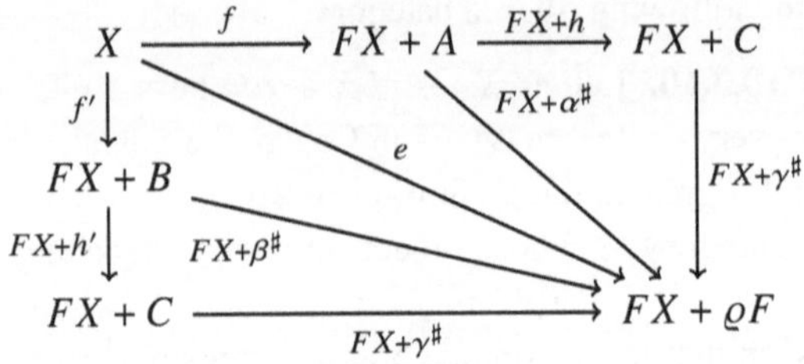

Since $FX + \gamma^\sharp$ is the colimit injection of a filtered colimit, we thus use the essential uniqueness of the above factorization to obtain a homomorphism $k \colon (C, \gamma) \to (D, \delta)$ in $\mathsf{Coalg_f}\, F$ such that $FX + k$ merges $(FX + h) \cdot f$ and $(FX + h') \cdot f'$. Defining

$$f' = (FX + k \cdot h) \cdot f = (FX + k \cdot h') \cdot f',$$

we see that $k \cdot h \colon ((A, \alpha), f) \to ((D, \delta), f')$ and $k \cdot h' \colon ((B, \beta), f') \to ((D, \delta), f')$ are morphisms in $\mathscr{D}$, as desired.

(c) First recall that $\mathsf{Coalg_f}\, F$ has coequalizers (Remark 10.2.4). Hence, given a parallel pair of morphisms $h, k \colon ((A, \alpha), f) \rightrightarrows ((B, \beta), f')$ in $\mathscr{D}$, their coequalizer $c \colon (B, \beta) \to (C, \gamma)$ in $\mathsf{Coalg_f}\, F$ yields a morphism from $((B, \beta), f')$ to $((C, \gamma), f')$ in $\mathscr{D}$, where $f' = (FX + c) \cdot f'$. Indeed, we have

$$(FX + \gamma^\sharp) \cdot f' = (FX + \gamma^\sharp) \cdot (FX + c) \cdot f' = (FX + \beta^\sharp) \cdot f' = e.$$

Remark 10.3.11. We will also use the fact that the forgetful functor

$$P \colon \mathscr{D} \to \mathsf{Coalg_f}\, F \qquad \text{given by} \qquad ((A, \alpha), f) \mapsto (A, \alpha)$$

is final. Indeed, for condition (a) in Remark 10.3.9(3), suppose we are given (B, β) in $\mathsf{Coalg_f}\, F$. We choose some $((A, \alpha), f)$ in $\mathscr{D}$ (using the nonemptiness of $\mathscr{D}$). Since $\mathsf{Coalg_f}\, F$ is filtered we obtain (C, γ) in $\mathsf{Coalg_f}\, F$ and homomorphisms $h \colon (A, \alpha) \to (C, \gamma)$ and $k \colon (B, \beta) \to (C, \gamma)$. Then the desired object of $\mathscr{D}$ is $((C, \gamma), g)$ where

$$g = (X \xrightarrow{f} FX + A \xrightarrow{FX+h} FX + C)$$

because the diagram below clearly commutes:

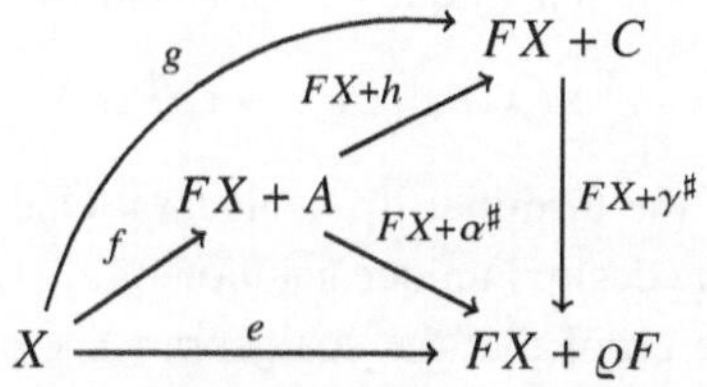

The desired morphism of $\mathsf{Coalg_f}\, F$ is $k\colon (B,\beta) \to P((C,\gamma),g)$.

Condition (b) is verified similarly. Given a parallel pair

$$h, k\colon (B,\beta) \rightrightarrows P((C,\gamma),g)$$

of homomorphisms, their coequalizer $c\colon (C,\gamma) \to (C',\gamma')$ in $\mathsf{Coalg_f}\, F$ is a morphism in $\mathscr{D}$ whose image under P merges h and k; this is shown by a similar argument as in Construction 10.3.10(c) above.

Theorem 10.3.12 [33, Thm. 3.3]**.** *The rational fixed point ϱF is the initial iterative algebra.*

Proof Denote by $\omega\colon \varrho F \to F(\varrho F)$ the coalgebra structure. We prove in part (2) below that the algebra $(\varrho F, \omega^{-1})$ is iterative. Its initiality is then easy to establish:

(1) Initiality of ϱF. Let (A,α) be an iterative algebra for F. First observe that for every coalgebra $e\colon X \to FX$ in $\mathsf{Coalg_f}\, F$ we have a unique coalgebra-to-algebra morphism $e^\dagger\colon X \to A$. The proof is identical to what we saw in Remark 7.3.3(1). It follows that for every lfp coalgebra $e\colon X \to FX$ we have a unique coalgebra-to-algebra morphism $e^\dagger\colon X \to A$. Indeed, write (X,e) as a colimit of coalgebras (X_i, e_i), $i \in I$, in $\mathsf{Coalg_f}\, F$. Then the $e_i^\dagger\colon X_i \to A$ form a cocone, and the induced morphism $e^\dagger$ with $e^\dagger \cdot e_i^\sharp = e_i^\dagger$ is the desired unique coalgebra-to-algebra morphism. Applying this to the lfp coalgebra $(\varrho F, \omega)$ we see that there exists a unique coalgebra-to-algebra morphism from $(\varrho F, \omega)$ to (A,α), i.e. a unique algebra homomorphism from $(\varrho F, \omega^{-1})$ to (A,α), as desired.

(2) Iterativity of ϱF. The proof follows the same reasoning as that in Proposition 7.3.6. Here we need to establish that the construction performed there yields an lfp coalgebra. More precisely, given a finitary flat equation morphism $e\colon X \to FX + \varrho F$, we form the coalgebra

$$\bar{e} = \big(X + \varrho F \xrightarrow{[e,\mathsf{inr}]} FX + \varrho F \xrightarrow{FX+\omega} FX + F(\varrho F) \xrightarrow{\mathsf{can}} F(X + \varrho F)\big),$$

where $\mathsf{can} = [F\mathsf{inl}, F\mathsf{inr}]$. We prove below that this is an lfp coalgebra.

The remainder of the proof then proceeds precisely as the one of Proposition 7.3.6. In fact, we obtain a unique homomorphism

$$\bar{e}^\sharp \colon (X + \varrho F, \bar{e}) \to (\varrho F, \omega)$$

using that $(\varrho F, \omega)$ is the terminal lfp coalgebra. One then proves that $e^\dagger = \bar{e}^\sharp \cdot \mathsf{inl} \colon X \to \varrho F$ is the desired unique solution of e.

We now turn to the proof that the coalgebra $(X + \varrho F, \bar{e})$ is lfp. Thus, our proof will be finished once we present a filtered diagram $D \colon \mathscr{D} \to \mathsf{Coalg}_\mathsf{f}\, F$ whose colimit is $(X + \varrho F, \bar{e})$.

We use the filtered category $\mathscr{D}$ from Construction 10.3.10 and define D as follows:

For every object $((A, \alpha), f)$ of $\mathscr{D}$ we have a coalgebra as follows:

$$\bar{f} = \big(X + A \xrightarrow{[f,\mathsf{inr}]} FX + A \xrightarrow{FX+\alpha} FX + FA \xrightarrow{\mathsf{can}} F(X + A)\big).$$

Since X and A are finitely presentable, so is $X + A$, and therefore $(X + A, \bar{f})$ lies in $\mathsf{Coalg}_\mathsf{f}\, F$. This defines the diagram D on objects: $D((A, \alpha), f) = (X + A, \bar{f})$. For a morphism $h \colon ((A, \alpha), f) \to ((B, \beta), g)$ of $\mathscr{D}$ we define $Dh = X + h$. We need to verify that Dh is a homomorphism. This follows from the commutativity of the following diagram:

$$
\begin{array}{ccccccc}
 & & & \bar{f} & & & \\
\ulcorner & \text{—} & \text{—} & \text{—} & \text{—} & \text{—} & \downarrow \\
X + A & \xrightarrow{[f,\mathsf{inr}]} & FX + A & \xrightarrow{FX+\alpha} & FX + FA & \xrightarrow{\mathsf{can}} & F(X + A) \\
{\scriptstyle X+h}\downarrow & & \downarrow{\scriptstyle FX+h} & & \downarrow{\scriptstyle FX+Fh} & & \downarrow{\scriptstyle F(X+h)} \\
X + B & \xrightarrow{[g,\mathsf{inr}]} & FX + B & \xrightarrow{FX+\beta} & FX + FB & \xrightarrow{\mathsf{can}} & F(X + B) \\
\llcorner & \text{—} & \text{—} & \text{—} & \text{—} & \text{—} & \uparrow \\
 & & & g & & &
\end{array}
\tag{10.8}
$$

We are ready to prove that $(X + \varrho F, \bar{e})$ is the colimit of D. For every object $((A, \alpha), f)$ in $\mathscr{D}$ we have a homomorphism

$$X + \alpha^\sharp \colon (X + A, \bar{f}) \to (X + \varrho F, \bar{e}),$$

which follows from the condition (10.7) using a diagram similar to (10.8). Given a morphism $h \colon ((A, \alpha), f) \to ((B, \beta), g)$ of $\mathscr{D}$ we obtain from $\alpha^\sharp \cdot h = \beta^\sharp$ the following commutative triangle

$$
\begin{array}{ccc}
X + A & \xrightarrow{X+h} & X + B \\
 & {\scriptstyle X+\alpha^\sharp}\searrow \quad \swarrow{\scriptstyle X+\beta^\sharp} & \\
 & X + \varrho F &
\end{array}
$$

Thus, the homomorphisms $X + \alpha^\sharp$ form a cocone on D.

In order to prove that this is a colimit cocone, it suffices to show that the morphisms $X + \alpha^\sharp \colon X + A \to X + \varrho F$ form a colimit of UD in $\mathscr{A}$, since the forgetful functor $U\colon \mathsf{Coalg}\, F \to \mathscr{A}$ creates colimits (Proposition 4.1.1). Observe that UD can be decomposed as follows:

$$UD = (\mathscr{D} \xrightarrow{P} \mathsf{Coalg_f}\, F \xhookrightarrow{I} \mathsf{Coalg}\, F \xrightarrow{U} \mathscr{A} \xrightarrow{X+(-)} \mathscr{A}),$$

where P is the functor from Remark 10.3.11 and I is the inclusion. Since $X+(-)$ is finitary (see Example 10.1.13(1) and Remark 10.1.12(3)) and $\varrho F = \mathsf{colim}\, UI$ we are done because P is final. □

Corollary 10.3.13. *For every finitary functor on a locally finitely presentable category, the initial iterative algebra is precisely the terminal lfp coalgebra.*

Remark 10.3.14. Similar to the alternative definition of cias in Remark 7.3.21, an iterative algebra can, equivalently, be defined as an algebra (A, α) such that for every coalgebra $e\colon X \to F(X + A)$ with X finitely presentable there exists a unique morphism $e^\dagger\colon X \to A$ such that the square below commutes:

$$\begin{array}{ccc} X & \xrightarrow{e^\dagger} & A \\ {\scriptstyle e}\downarrow & & \uparrow{\scriptstyle \alpha} \\ F(X+A) & \xrightarrow{F[e^\dagger, \mathrm{id}_A]} & FA \end{array} \tag{10.9}$$

As before for cias, the theory of iterative algebras could be developed based on this alternative definition. The proof of the equivalence is considerably more involved than for cias.

10.4 The Rational Fixed Point of a Set Functor

In this section we exhibit two descriptions of the rational fixed point for $\mathscr{A} = \mathsf{Set}$, in addition to the one in Proposition 10.2.18. The first one is as the coalgebra of all finite well-pointed coalgebras (see Section 9.3), and the latter is related to a presentation of the functor F by operations and basic equations (see Section 4.3), which means that F is the quotient of a polynomial functor H_Σ. Recall our assumption that F is a finitary functor. Hence, it has a terminal coalgebra carried by a set T representing all well-pointed coalgebras (Corollary 9.3.11). Recall also the coalgebra structure $\tau\colon T \to FT$ from (9.6).

Notation 10.4.1. Let R be a set representing all finite well-pointed coalgebras (see Definition 9.3.2). Thus, R can be chosen as a subset of T.

Proposition 10.4.2. *For every finitary set functor F the set R of all finite well-pointed coalgebras is the rational fixed point ϱF.*

In particular, R is a subcoalgebra of the terminal coalgebra νF by Proposition 10.2.18.

Proof Given an element of R, that is, a finite well-pointed coalgebra (A, α, x), the well-founded modification $\alpha^+(x)$ (Notation 9.3.4) is also finite, and so it can be chosen to lie in R; in symbols: $\tau[R] \subseteq FR$. Let $\omega\colon R \to FR$ be the restriction of τ. For every finite coalgebra (A, α) the unique homomorphism $\alpha^+\colon (A, \alpha) \to (T, \tau)$ yields a finite subcoalgebra $\alpha^+[A]$. (This follows from Proposition 7.2.2 if F preserves monomorphisms. If it does not, then use the Trnková hull $\bar{F}$: due to $x \in A$ we have $A \neq \emptyset$, thus the image $\alpha^+[A]$ is the same with respect to F or $\bar{F}$.) Hence we can assume that $\alpha^+[A]$ lies in R. It is clear that the corresponding restriction of α^+ is the unique homomorphism from (A, α) to (R, ω). □

Example 10.4.3. For $FX = \{0, 1\} \times X^\Sigma$ we get that ϱF consists of all finite minimal deterministic automata. This coalgebra is isomorphic to that formed by all regular languages on Σ.

Our second description of ϱF is based on the notion of a presentation (cf. Definition 4.3.7).

Recall from Proposition 4.3.19 that every finitary functor F can be presented by a finitary signature Σ and a set $\mathcal{E}$ of basic equations: $F \cong H_\Sigma/\mathcal{E}$. As before we denote by $\varepsilon\colon H_\Sigma \twoheadrightarrow F$ the corresponding natural epi-transformation.

We have seen in Proposition 4.3.22 and Theorem 4.3.26 that the initial algebra and the terminal coalgebra for F arise as the quotients of the initial algebra and terminal coalgebra for H_Σ (carried by the sets of all finite Σ-trees and all Σ-trees), respectively modulo the equivalence given by application of the basic equations in $\mathcal{E}$:

$$\mu F = \mu H_\Sigma/{\sim} \qquad \text{and} \qquad \nu F = \nu H_\Sigma/{\approx}.$$

Here the equivalence $\sim$ is given by finitely many applications of the basic equations in $\mathcal{E}$ (see (4.2)) and $\approx$ by finite and infinite applications (see (4.3)). It turns out that the equivalence $\approx$ restricts to the rational fixed point ϱH_Σ, which is carried by the set of all rational Σ-trees (see Proposition 10.2.20). In fact, we consider ϱH_Σ as an F-coalgebra

$$\varrho H_\Sigma \xrightarrow{\ \omega\ } H_\Sigma(\varrho H_\Sigma) \xrightarrow{\ \varepsilon_{\varrho H_\Sigma}\ } F(\varrho H_\Sigma)$$

and obtain the following result:

Theorem 10.4.4 [19, Thm. 5.7]. *Given a presentation of a finitary set functor F, the rational fixed point of F is a quotient of the F-coalgebra ϱH_Σ of all rational Σ-trees modulo the equivalence $\approx$; in symbols:*

$$\varrho F = \varrho H_\Sigma / \approx .$$

Corollary 10.4.5. *Given a presentation of a finitary set functor F, we have the following commutative square of F-coalgebra homomorphisms:*

$$\begin{array}{ccc} \varrho H_\Sigma & \twoheadrightarrow & \varrho F \\ \downarrow & & \downarrow \\ \nu H_\Sigma & \twoheadrightarrow & \nu F \end{array}$$

Each of the above homomorphisms is a uniquely determined and represents a canonical subcoalgebra or quotient coalgebra, respectively.

Example 10.4.6. (1) For the finite power-set functor $\mathscr{P}_\mathsf{f}$ we have the presentation of Example 4.3.27 with $H_\Sigma X = X^*$. Hence we obtain $\varrho \mathscr{P}_\mathsf{f} = R_\Sigma / \approx$, where R_Σ is the coalgebra of all rational finitely branching ordered trees and $\approx$ the equivalence of Example 4.2.10.

(2) Analogously, for every analytic functor $FX = \coprod_{\sigma \in \Sigma} X^k / G_\sigma$, where k is the arity of σ (cf. Example 4.3.31(3)), the rational fixed point is the quotient $\varrho F = R_\Sigma / \approx$ of all rational Σ-trees modulo permutations of children of any σ-labelled node (using elements of the permutation group G_σ).

(3) In particular, for the bag functor $\mathscr{B}$ we have the presentation with $H_\Sigma X = X^*$ (Example 4.3.14(4)), and we obtain the rational fixed point $\varrho \mathscr{B}$ as the quotient of the coalgebra of all rational finitely branching trees modulo arbitrary permutations of children of nodes. We thus recover the description of $\varrho \mathscr{B}$ as the coalgebra of all rational unordered trees from Example 10.2.21(4).

10.5 Full Abstractness and Finitely Generated Objects

We have seen in Proposition 10.2.18 that the rational fixed point of a finitary set functor is fully abstract, that is, it is a subcoalgebra of the terminal coalgebra. We shall see that for a finitary functor F on a general locally finitely presentable category $\mathscr{A}$ this is not always the case. In this section we will study sufficient conditions on $\mathscr{A}$ and F that ensure full abstractness of the rational fixed point. Those conditions are closely tied to an alternative notion of 'finite' object one may consider in a locally finitely presentable category. We have so far considered finitely presentable objects, which are a very useful formalization of 'finiteness'.

In this section we use also the notion of a *finitely generated* object, which is a broader concept. We will see that ϱF is fully abstract whenever the classes of finitely presentable and finitely generated objects coincide in the base category. Besides being our key for the full abstractness of ϱF, we shall see that finitely generated objects lead to another interesting fixed point of F. In the special case where $\mathscr{A}$ is a finitary variety, we will also consider the more restrictive concept of free finitely generated algebras (aka. ffg objects) in lieu of finitely presentable objects modelling 'finiteness'. This yields a third fixed point for finite behaviour.

Definition 10.5.1. An object X is *finitely generated* if $\mathscr{A}(X, -)$ preserves filtered colimits of monomorphisms, that is, colimits of directed (equivalently, filtered) diagrams $D\colon \mathscr{D} \to \mathscr{A}$ such that Dh is a monomorphism for every morphism h in $\mathscr{D}$.

Like finite presentability, this concept stems from general algebra, where it designates the algebras which are generated by a finite subset (a simplification of Example 10.1.6(4)).

Remark 10.5.2. (1) Every finitely presentable object is finitely generated. But the converse may fail (see e.g. Example 10.5.4(6) and (7)).

(2) An object X is finitely generated iff it is a strong quotient (cf. Remark 7.2.1(1)) of a finitely presentable object [23, Prop. 1.69]. In more detail, there exists a finitely presentable object X_0 and a strong epimorphism $X_0 \twoheadrightarrow X$. It follows that finitely generated objects are closed under strong quotients and that $\mathscr{A}_{\mathsf{fp}} = \mathscr{A}_{\mathsf{fg}}$ iff finitely presentable objects are closed under strong quotients.

(3) Since $\mathscr{A}$ is locally finitely presentable, it is co-well-powered (see [23, Thm. 1.58]). Moreover, since it has only a set of finitely presentable objects, it also has only a set of finitely generated objects (up to isomorphism).

Notation 10.5.3. We denote by

$$\mathscr{A}_{\mathsf{fg}}$$

the (essentially small) full subcategory given by all finitely generated objects of $\mathscr{A}$.

Example 10.5.4. (1) In the categories Set, Pos, Gra, and K-Vec every finitely generated object is finitely presentable: the finitely generated objects are the finite sets, posets, graphs, and finite-dimensional vector spaces, respectively. Indeed, we used only filtered diagrams of monomorphisms in Example 10.1.6(1)–(3).

(2) In the category Nom of nominal sets, we have that $\mathscr{A}_{\mathsf{fp}} = \mathscr{A}_{\mathsf{fg}}$ are the orbit-finite sets (cf. Example 10.1.6(6)). This follows from Remark 10.5.2(2).

(3) In the category $\mathsf{Set}^{\mathscr{F}}$ of sets in context, we have that $\mathscr{A}_{\mathsf{fp}} = \mathscr{A}_{\mathsf{fg}}$ [52, Cor. 3.34], and this is the class of super-finitary presheaves (see Example 10.1.6(7)).

(4) In a finitary variety of algebras (see Example 10.1.6(4)), the finitely generated objects are precisely the algebras X which are generated by a finite subset: that is, there exists a finite set $S \subseteq X$ such that any subalgebra containing S is all of X [23, Theorem 3.12]. In other words, finitely generated algebras are precisely the quotients of free algebras on finitely many generators.

(5) Every locally finite variety (see Example 10.1.6(5)) clearly satisfies $\mathscr{A}_{\mathsf{fp}} = \mathscr{A}_{\mathsf{fg}}$ by Remark 10.5.2(2).

(6) For a simple example where $\mathscr{A}_{\mathsf{fp}} \neq \mathscr{A}_{\mathsf{fg}}$, consider the signature Σ consisting of contant symbols c_n $(n \in \mathbb{N})$. Then a Σ-algebra is a set A with chosen elements a_n $(n \in \mathbb{N})$. It is finitely generated iff the set $A \setminus \{a_n\}$ is finite. In contrast, it is finitely presentable, iff, moreover, there exists $k \in \mathbb{N}$ such that all a_n with $k \geq n$ are pairwise distinct. For example, the terminal (one-element) algebra is finitely generated but not finitely presentable.

(7) In the categories of groups we have $\mathscr{A}_{\mathsf{fp}} \neq \mathscr{A}_{\mathsf{fg}}$. Examples demonstrating this are non-trivial. However, there are only countably many finitely presented groups (up to isomorphism), while there are already $2^{\aleph_0}$ groups on 2 generators. A simple example of a finitely generated group which is not finitely presented is the standard wreath product $\mathbb{Z} \mathbin{\mathsf{wr}} \mathbb{Z}$ [261, §14.1.4].

(8) Also in the category of monoids, we have $\mathscr{A}_{\mathsf{fp}} \neq \mathscr{A}_{\mathsf{fg}}$, see Campbell et al. [97, Example 4.5].

In contrast, for the full subcategory $\mathscr{A}$ of commutative monoids we have $\mathscr{A}_{\mathsf{fp}} = \mathscr{A}_{\mathsf{fg}}$. This is known as Redei's theorem [259]; see Freyd [127] for a very short proof.

(9) In the category $\mathscr{A} = \mathsf{Set}^{\mathscr{F}}$ of sets in context, we have $\mathscr{A}_{\mathsf{fp}} = \mathscr{A}_{\mathsf{fg}}$ [52, Cor. 3.34].

We now proceed to investigate the case where $\mathscr{A}_{\mathsf{fp}} = \mathscr{A}_{\mathsf{fg}}$ holds. It turns out that the results proved in Section 10.2 for finitary set functors smoothly generalize to finitary functors on $\mathscr{A}$. In the following we say that a subobject of A is *finitely presentable* if it is represented by a monomorphism $s\colon S \rightarrowtail A$ such that S is a finitely presentable object.

Proposition 10.5.5. *Suppose that $\mathscr{A}_{\mathsf{fp}} = \mathscr{A}_{\mathsf{fg}}$ and that F preserves monomorphisms. Then a coalgebra (A, α) is lfp iff every finitely presentable subobject of A is contained in a subcoalgebra from* $\mathsf{Coalg}_{\mathsf{f}}\, F$.

More precisely, (A, α) is lfp iff for every finitely presentable subobject $s\colon S \rightarrowtail A$

there exists a subcoalgebra $m\colon (B,\beta) \rightarrowtail (A,\alpha)$ with (B,β) in $\mathsf{Coalg_f}\, F$ and $s \le m$ (Remark 2.1.18).

Proof It follows from Remark 10.1.16(5) that a coalgebra (A,α) is lfp iff the following two conditions hold:

(1) For every morphism $p\colon P \to A$ where P is a finitely presentable object, there exist a coalgebra (B,β) in $\mathsf{Coalg_f}\, F$, a morphism $p'\colon P \to B$ and a coalgebra homomorphism $h\colon (A,\alpha) \to (B,\beta)$ such that $p = h \cdot p'$:

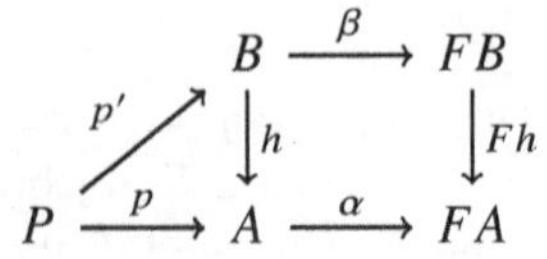

(2) Given two such factorizations $p_t\colon P \to B$ $(t = 1,2)$ with $h \cdot p_1 = h \cdot p_2$, there exist a coalgebra homomorphisms $h'\colon (B',\beta') \to (A,\alpha)$ and $g\colon (B,\beta) \to (B',\beta')$ such that $h' \cdot g = h$ and $g \cdot p_1 = g \cdot p_2$.

In order to prove the proposition we proceed in two steps:

(i) we show that under our current assumptions (1) implies (2), and

(ii) we verify that (1) is equivalent to the condition in the statement of the proposition.

Ad (i). Suppose that (A,α) satisfies condition (1). Let p_t and h be as in (2). Take the coequalizer $c\colon B \to C$ of p_1, p_2 in $\mathscr{A}$. Since P and B are finitely presentable so is C (cf. Remark 10.1.16(2)). By the universal property of c we obtain a morphism $g\colon C \to A$ such that $g \cdot c = h$. We now apply condition (1) to g, and we obtain a coalgebra (B',β') in $\mathsf{Coalg_f}\, F$, a morphism $g'\colon C \to B'$ and a homomorphism $h'\colon (B',\beta') \to (A,\alpha)$ such that $h' \cdot g' = g$. Next we factorize the homomorphism h' into a homomorphism $e\colon (B',\beta') \twoheadrightarrow (B',\beta')$ carried by a strong epimorphism followed by $m\colon (B',\beta') \rightarrowtail (A,\alpha)$ carried by a monomorphism (see Remark 10.1.16(6)). In summary we have the following commutative diagram:

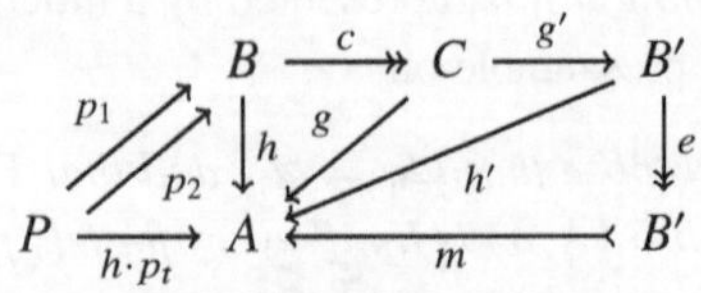

(10.10)

Clearly, $e \cdot g' \cdot c$ merges p_1 and p_2, since so does c. To see that $e \cdot g' \cdot c$ is a homomorphism (in $\mathsf{Coalg_f}\, F$) consider the diagram below:

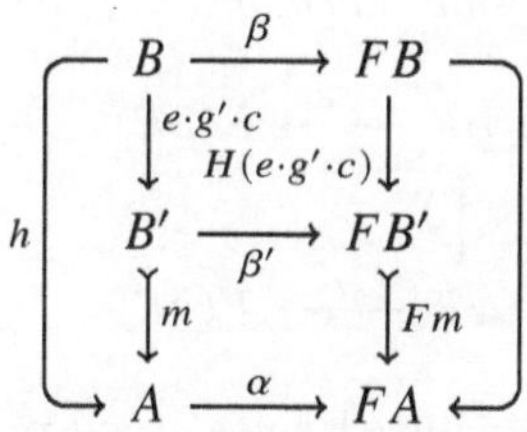

Its outside commutes since h is a homomorphism, the left- and right-hand parts commute by (10.10), and the lower square commutes since m is a homomorphism. Thus the desired upper square commutes when extended by Fm. Since Fm is a monomorphism by assumption, we see that the upper innner square commutes.

Ad (ii). Suppose that (1) holds and let $s\colon S \rightarrowtail A$ be a finitely presentable subobject. We obtain a homomorphism $h\colon (B_0, \beta_0) \to (A, \alpha)$ and a morphism $s_0'\colon S \to B_0$ with $h \cdot s_0' = s$. Then the image $m\colon (B, \beta) \rightarrowtail (A, \alpha)$ of h (cf. Remark 7.2.3) yields the desired subcoalgebra.

Conversely, assume that the condition in the statement of the lemma holds, and suppose towards a proof of (1) that $p\colon P \to A$ is a morphism with P finitely presentable. Take the (strong epi, mono)-factorization $p = s \cdot e$ and apply the condition to s to obtain s' and a subcoalgebra $m\colon (B, \beta) \rightarrowtail (A, \alpha)$ with $m \cdot s' = s$. With $p' = s' \cdot e$ this yields the desired factorization of p. □

The next result presents sufficient conditions on $\mathscr{A}$ and F for the rational fixed point to be fully abstract (Definition 10.2.16). It also entails the concrete descriptions of ϱF in the subsequent examples.

Proposition 10.5.6. *Suppose that $\mathscr{A}_{\mathsf{fp}} = \mathscr{A}_{\mathsf{fg}}$ and that F preserves monomorphisms. Then ϱF is the subcoalgebra of νF given by the union of images of all F-coalgebra morphisms $(A, \alpha) \to (\nu F, \tau)$ where (A, α) ranges over* $\mathsf{Coalg_f}\, F$.

Proof Let $\mathscr{D}$ be the full subcategory formed by all subcoalgebras of νF in $\mathsf{Coalg_f}\, F$, that is, all coalgebras (A, α) with A finitely presentable and such that the unique coalgebra homomorphism $(A, \alpha) \to (\nu F, \tau)$ is a monomorphism. By Corollary 10.2.15, it is sufficient to show that the inclusion $\mathscr{D} \hookrightarrow \mathsf{Coalg_f}\, F$ is final (Remark 10.3.9). Since $\mathscr{D}$ is a full subcategory and $\mathsf{Coalg_f}\, F$ is filtered (Remark 10.2.4) it suffices to verify that for every (A, α) in $\mathsf{Coalg_f}\, F$ we have a homomorphism $(A, \alpha) \to (A', \alpha')$ with (A', α') in $\mathscr{D}$. Indeed, given (A, α) let (A', α') be its image under the unique homomorphism $(A, \alpha) \to (\nu F, \tau)$ (see Remark 7.2.3). Then we have a homomorphism $e\colon (A, \alpha) \twoheadrightarrow (A', \alpha')$ where e is a strong epimorphism in $\mathscr{A}$. Using Remark 10.5.2(2) and that $\mathscr{A}_{\mathsf{fp}} = \mathscr{A}_{\mathsf{fg}}$, we conclude that A' is finitely presentable, whence (A', α') lies in $\mathscr{D}$. □

Corollary 10.5.7. *Under the assumptions in Proposition 10.5.6, the rational fixed point $(\varrho F, \omega)$ is a cartesian subcoalgebra of the terminal coalgebra:*

$$\begin{array}{ccc} \varrho F & \xrightarrow{\ \omega\ } & F(\varrho F) \\ \big\downarrow & \lrcorner & \big\downarrow \\ \nu F & \xrightarrow{\ \tau\ } & F(\nu F) \end{array}$$

Indeed, the above square is a pullback since ω and τ are isomorphisms.

Before presenting further examples of rational fixed points, we are going to prove that whenever $\mathscr{A}$ is a variety of algebras (see Example 10.1.6(4)), the union in Proposition 10.5.6 may be restricted to range only over F-coalgebra whose carrier is a *free finitely generated* algebra in $\mathscr{A}$, that is, an algebra which is free on a finite set of generators.

Remark 10.5.8. Let $\mathscr{A}$ be a variety of algebras. Then every free algebra A is *projective*, that is, for every surjective morphism $e\colon B \twoheadrightarrow C$ and every morphism $f\colon A \to C$ there exists a morphism $f'\colon A \to B$ with $e \cdot f' = f$:

$$\begin{array}{ccc} A & \overset{f'}{\dashrightarrow} & B \\ & {\scriptstyle f}\searrow & \big\downarrow {\scriptstyle e} \\ & & C \end{array}$$

Indeed, let A be free on the set X of generators with the universal map $\eta\colon X \to A$. Since e is surjective, we have a map $m\colon C \rightarrowtail A$ such that $e \cdot m = \mathrm{id}_C$ (in Set). Now extend the map $m \cdot f \cdot \eta\colon X \to B$ to a morphism $f'\colon A \to B$ in $\mathscr{A}$ using the freeness of A. Then $e \cdot f' = f$ because this holds when we restrict to the generators of A (i.e. we pre-compose with η):

$$e \cdot f' \cdot \eta = e \cdot m \cdot f \cdot \eta = f \cdot \eta.$$

Corollary 10.5.9. *Let $\mathscr{A}$ be a variety such that $\mathscr{A}_{\mathsf{fp}} = \mathscr{A}_{\mathsf{fg}}$, and let $F\colon \mathscr{A} \to \mathscr{A}$ be finitary and preserve injective as well as surjective morphisms. Then ϱF is the subcoalgebra of νF given by the union of all coalgebra homomorphisms $(A, \alpha) \to (\nu F, \tau)$ where A is a free finitely generated algebra.*

Proof By Proposition 10.5.6 we know that ϱF is the union of images of homomorphisms $h\colon (B, \beta) \to (\nu F, \tau)$, where B ranges over finitely generated algebras in $\mathscr{A}$. Given any such coalgebra (B, β) we know from Example 10.1.6(4) that there exists a free finitely generated algebra A and a surjective morphism $e\colon A \twoheadrightarrow B$. Using projectivity (see Remark 10.5.8) and that F preserves surjective morphisms, we obtain some coalgebra structure α on A such that e is a coalgebra homomorphism:

$$\begin{array}{ccc} A & \overset{\alpha}{\dashrightarrow} & FA \\ {\scriptstyle e}\downarrow & & \downarrow{\scriptstyle Fe} \\ B & \xrightarrow{\beta} & FB \end{array}$$

Thus the image of B in νF is the same that of A, which completes the proof. □

We proceed to presenting further examples of rational fixed points, in addition to the rational fixed point for set functors we have seen in Example 10.2.21. In all of the following examples, ϱF is fully abstract by Proposition 10.5.6.

Example 10.5.10. (1) For the set functor $FX = \mathbb{R} \times X$ the rational fixed point consists of eventually periodic streams of real numbers (cf. Example 10.2.21(5)). Let us consider F as an endofunctor on the category of real vector spaces. The following interesting description of ϱF then follows from the work of Rutten [267]. The *convolution product* of two streams in $\mathbb{R}^\omega$ is given by

$$(\sigma \times \tau)(n) = \sum_{i=0}^{n} \sigma(i) \cdot \tau(n-i).$$

For every stream σ with $\sigma(0) \neq 0$ there exists an inverse σ^{-1}, that is, $\sigma \times \sigma^{-1} = (1, 0, 0, 0, \ldots)$. A stream is called *rational* if it has the form $\sigma \times \tau^{-1}$, where σ and τ have finitely many non-zero entries and $\tau(0) \neq 0$. For example, the powers of 2 form a rational stream which is not eventually periodic. Rutten [267, Thm. 8.1] characterized the rational streams as those representable by linear systems of finite dimension, which means that a stream is rational iff it lies in the image of some coalgebra homomorphism $A \to \nu F$ where (A, α) is in $\mathsf{Coalg}_\mathsf{f}\, F$. Using Proposition 10.5.6 we thus obtain that the rational fixed point for F consists of all rational streams.

(2) The previous example can be generalized from real streams to formal power series (also called *weighted languages*) over a given semiring $\mathbb{S}$. Recall from Example 2.4.6 that for weighted automata one considers the functor $FX = \mathbb{S} \times X^\Sigma$ on the category $\mathbb{S}$-Mod of $\mathbb{S}$-semimodules and that the F-coalgebras carried by the free finitely generated $\mathbb{S}$-semimodules $\mathbb{S}^n$ can be identified with weighted automata. We have also seen in Example 2.5.12 that the terminal coalgebra for F is carried by the set $\mathbb{S}^{\Sigma^*}$ of formal power series. A formal power series is *recognizable* if it is recognized by a finite weighted automaton (see e.g. [113]). By the Kleene–Schützenberger theorem [271] (see also [78, Prop. 6.1 & Thm. 7.1]) it follows that recognizable power series are, equivalently, the *rational* formal power series. For those semirings for which finitely generated semimodules coincide with finitely presentable ones, $\mathbb{S}\text{-}\mathsf{Mod}_\mathsf{fp} = \mathbb{S}\text{-}\mathsf{Mod}_\mathsf{fg}$, Corollary 10.5.9

yields that the rational fixed point ϱF is the subcoalgebra of $\nu F = \mathbb{S}^{\Sigma^*}$ given by the rational formal power series. This coincidence holds, for example, for

(a) every field,
(b) every finite semiring,
(c) every principal ideal domain (e.g. the ring $\mathbb{Z}$ of integers), and therefore
(d) every finitely generated commutative ring (by Hilbert's basis theorem).

More generally, whenever $\mathbb{S}$ is a *Noetherian* semiring in the sense of Ésik and Maletti [121] (i.e. every subsemimodule of a finitely generated semimodule is itself finitely generated) then the classes of finitely generated semimodules and finitely presented ones coincide [89, Prop. 2.6]. The tropical semiring $(\mathbb{N} \cup \{\infty\}, \min, +, \infty, 0)$ is not Noetherian [120].

Furthermore, the condition of being Noetherian is not necessary for the coincidence of finitely generated and finitely presentable semimodules. For example, the usual semiring of natural numbers is also not Noetherian: the $\mathbb{N}$-semimodule $\mathbb{N} \times \mathbb{N}$ is finitely generated but its subsemimodule generated by the infinite set $\{(n, n+1) \mid n \geq 1\}$ is not. However, $\mathbb{N}$-semimodules are precisely the commutative monoids, and finitely generated monoids are finitely presentable (Example 10.5.4(8)).

(3) Let $F = V + X \times X + \delta X$ on the category $\mathsf{Set}^{\mathscr{F}}$ of sets in context (see Example 10.1.6(7)). As we have seen in Example 2.2.18(4), the initial algebra μF is the presheaf of λ-terms (or finite λ-trees) up to α-equivalence. Similarly, νF is the presheaf of finite and infinite λ-trees up to α-equivalence (see Example 2.5.15). The finitely generated and finitely presentable objects coincide in $\mathsf{Set}^{\mathscr{F}}$ [52, Cor. 3.34]. Using Example 10.5.4(9), the rational fixed point is fully abstract. Moreover, ϱF is the subcoalgebra of νF given by *rational* λ-trees, that is, those α-equivalence classes of λ-trees which contain a tree having only finitely many subtrees (up to isomorphism) [44, Thm. 2.17].

Example 10.5.11. We now present examples of rational fixed points in the category Nom of nominal sets.

(1) For the functor $FX = \mathbb{A} + X \times X + [\mathbb{A}]X$ on the category Nom, we have seen that μF is formed by α-equivalence classes of λ-terms (see Example 2.2.18(1)), and νF is formed by α-equivalence classes of all λ-trees with finitely many free variables (see Example 2.5.14). Using Example 10.5.4(2), the rational fixed point of F is fully abstract. In fact, ϱF is the nominal set of all rational λ-trees modulo α-equivalence [230, Thm. 3.4]. More generally, for every functor on nominal sets arising from a binding signature (Remark 2.1.11) a similar result holds [234].

(2) For the functor $FX = \{0, 1\} \times X^{\mathbb{A}}$ (see Example 10.1.13(4)) the category

$\mathsf{Coalg}_{\mathsf{f}}\, F$ consists of precisely the orbit-finite deterministic nominal automata (cf. Example 12.3.17). Its terminal coalgebra is carried by the set of all finitely supported languages over $\mathbb{A}$. That is, one takes the nominal set $\mathscr{P}_{\mathsf{fs}}(\mathbb{A}^*)$ and equips it with the following automaton structure

$$\langle o, t\rangle : \mathscr{P}_{\mathsf{fs}}(\mathbb{A}^*) \to \{0,1\} \times \mathscr{P}_{\mathsf{fs}}(\mathbb{A}^*)^{\mathbb{A}}$$

where $o(L) = 1$ if L contains the empty word and t is given by $t(L)(a) = a^{-1}L = \{w : aw \in L\}$ (similarly as for ordinary deterministic automata). It is easy to verify that this coalgebra structure is equivariant and indeed forms the terminal coalgebra.

It follows from Proposition 10.5.6 that ϱF is the subcoalgebra of all finitely supported languages over $\mathbb{A}$ accepted by orbit-finite deterministic nominal automata.

(3) For $FX = \{0,1\} \times X^{\mathbb{A}} \times [\mathbb{A}]X$ (Example 10.1.13(5)) the terminal coalgebra has been described by Kozen et al. [203, Thm. 4.10]. We shall see in Example 11.5.8(2) that an equivalent description is obtained by *bar languages* [270, Def. 3.3] that we are now going to describe. Put $\overline{\mathbb{A}} = \mathbb{A} + \{|a : a \in \mathbb{A}\}$ and then take *bar strings* which are words over this set modulo α-equivalence. More precisely, the α-equivalence $=_\alpha$ on $\overline{\mathbb{A}}^*$ is generated by $w|au =_\alpha w|bv$ if $\langle a\rangle u = \langle b\rangle v$ in the abstraction set $[\mathbb{A}](\overline{\mathbb{A}}^*)$ (see Example 2.1.10). We write $[w]_\alpha$ for the α-equivalence class of a word $w \in \overline{\mathbb{A}}^*$. A *bar language* is a finitely supported set of bar strings. The terminal coalgebra for F is carried by the nominal set of all bar languages:

$$\nu F = \mathscr{P}_{\mathsf{fs}}(\overline{M}), \qquad \text{where } \overline{M} = \mathscr{P}_{\mathsf{fs}}(\overline{\mathbb{A}}^*/{=_\alpha}).$$

The coalgebra structure is given by

$$\tau = \langle \tau_1, \tau_2, \tau_3\rangle : \mathscr{P}_{\mathsf{fs}}\overline{M} \to \{0,1\} \times (\mathscr{P}_{\mathsf{fs}}\overline{M})^{\mathbb{A}} \times [\mathbb{A}](\mathscr{P}_{\mathsf{fs}}\overline{M}),$$

where $\tau_1(L) = 1$ iff L contains $[\varepsilon]_\alpha$, $\tau_2(L)(a) = \{[w]_\alpha : [aw]_\alpha \in L\}$ and $\tau_3(L) = \langle a\rangle\{[w]_\alpha : [|aw]_\alpha \in L\}$ for any a which is fresh for L (i.e. not contained in its support). For a full proof, see Frank et al. [126, Prop. 4.18].

Again, it follows by Proposition 10.5.6 that ϱF consists of precisely all bar languages accepted by the deterministic RNNA, that is, deterministic orbit-finite automata with name binding transitions (cf. Remark 12.5.18).

All categories considered in Examples 10.5.10 and 10.5.11 satisfy $\mathscr{A}_{\mathsf{fp}} = \mathscr{A}_{\mathsf{fg}}$. We now turn to categories $\mathscr{A}$ in which $\mathscr{A}_{\mathsf{fp}}$ and $\mathscr{A}_{\mathsf{fg}}$ do not coincide. This is known to be the case for a number of categories relevant for the coalgebraic modelling of systems. For example, certain varieties of algebras (e.g. for the categories of groups, monoids, or semimodules for the semiring $\mathbb{S} = (\mathbb{Z}_2)^{\mathbb{N}}$), the target categories of the generalized power-set construction (Construction 12.5.7). Another example is the category of finitary monads on sets [51, Cor. 4.13] used in the categorical study of Courcelle's algebraic trees [43].

There are also categories for which it is not known whether $\mathscr{A}_{\mathsf{fp}} = \mathscr{A}_{\mathsf{fg}}$, for example, the category of idempotent semirings (used in the treatment of context-free grammars [311]) or the category of algebras for the stack monad (used for modelling configurations of stack machines [144, 145]).

In such categories the rational fixed point may fail to be fully abstract, that is, ϱF is not a canonical subcoalgebra of νF (cf. [89, Ex. 3.15] for a related example):

Example 10.5.12 [228, Ex. 2.18(1)]. Let $\mathscr{A}$ be the category of algebras for the signature Σ containing two unary operation symbols u and v. The set of natural numbers $\mathbb{N}$ equipped with the successor function as both operations $u^{\mathbb{N}}$ and $v^{\mathbb{N}}$ forms an object of $\mathscr{A}$. We consider the functor $FX = \mathbb{N} \times X$ on $\mathscr{A}$. Coalgebras for F are automata carried by an algebra A in $\mathscr{A}$ equipped with two Σ-algebra homomorphisms, an output morphism $A \to \mathbb{N}$ and a next state morphism $A \to A$. The terminal coalgebra is carried by the set $\mathbb{N}^{\omega}$ of streams of natural numbers with the coordinatewise algebra operations and with the coalgebra structure given by the usual head and tail functions.

Note that the free Σ-algebra on a set X of generators is $TX = \{u, v\}^* \times X$; we denote its elements by $w(x)$ for $w \in \{u, v\}^*$ and $x \in X$. The operations are given by prefixing words by the letters u and v, respectively: $s^{TX}: w(x) \mapsto sw(x)$ for $s = u$ or v.

Now consider the F-coalgebra $\alpha: A \to FA$, where $A = T\{x\}$ is free Σ-algebra on one generator x, and α is determined by $\alpha(x) = (0, u(x))$. We write $\dagger\alpha: A \to \nu F$ for the unique coalgebra homomorphism. Clearly, $\dagger\alpha(x)$ is the stream $(0, 1, 2, 3, \cdots)$ of all natural numbers, and since $\dagger\alpha$ is a Σ-algebra morphism we have

$$\dagger\alpha(u(x)) = \dagger\alpha(v(x)) = (1, 2, 3, 4, \cdots). \tag{10.11}$$

Since A is (free) finitely generated, it is of course, finitely presentable. Thus, (A, α) is a coalgebra in $\mathsf{Coalg}_{\mathsf{f}}\, F$.

However, one can prove that the (unique) F-coalgebra morphism $\alpha^{\sharp}: A \to \varrho F$ satisfies $\alpha^{\sharp}(u(x)) \neq \alpha^{\sharp}(v(x))$. For details see [228]. Therefore the unique homomorphism $\varrho F \to \nu F$ is not monic since it merges $u(x)$ and $v(x)$ by (10.11).

The Locally Finite Fixed Point. One problem we have encountered with the rational fixed point is that, in general, it need not be fully abstract (see Example 10.5.12). However, in all applications, the regular behaviour of systems is defined by taking the image of all the behaviours of finite(ly presented) systems in the semantic domain, i.e. the terminal coalgebra. Hence, in applications one wants the rational fixed point to be a subcoalgebra of νF. Unfortunately, the condition in Proposition 10.5.6 that finitely generated objects be finitely presentable is, if true at all, often difficult to verify in cases where it is unknown.

The a priori choice of taking finitely presentable objects as the right abstraction of 'finite set' is not completely canonical. One may ask what happens if one chooses finitely generated objects instead. This has been investigated by Milius, Pattinson, and Wißmann [235]. In fact, one can rework much of the theory we have seen by systematically replacing finitely presentable by finitely generated objects. We now briefly mention the key definitions and results; for details see op. cit.

Assumption 10.5.13. In addition to Assumption 10.2.1, we now assume that $F\colon \mathscr{A} \to \mathscr{A}$ preserves nonempty monomorphisms (see Remark 7.2.5(2)).

First one replaces lfp coalgebras by the corresponding concept based on finitely generated objects. The definition given in op. cit. is equivalent to the following

Definition 10.5.14 [235, Cor. 3.6]**.** An F-coalgebra (A, α) is called *locally finitely generated* (or *lfg*, for short) if it is the colimit of a diagram of coalgebras with a finitely generated carrier.

Theorem 10.5.15 [235, Thm. 3.8]**.** *Suppose that $F\colon \mathscr{A} \to \mathscr{A}$ preserves nonempty monomorphisms. The full subcategory of* Coalg F *given by lfg coalgebras has a terminal object which is given by the colimit of all coalgebras with finitely generated carrier.*

Definition 10.5.16. The terminal lfg coalgebra is called the *locally finite fixed point* of F and denoted by ϑF.

That the name is justified follows from the first part of the following result. Furthermore, an advantage of ϑF is that, unlike the rational fixed point, it is always fully abstract with respect to behavioural equivalence.

Theorem 10.5.17 [235, Thm. 3.12 & 3.14]**.** (1) *The coalgebra structure of ϑF is an isomorphism.*

(2) *The locally finite fixed point ϑF is a (canonical) subcoalgebra of νF.*

Corollary 10.5.18 [235, Cor. 3.10]**.** *If $\mathscr{A}_{\mathsf{fp}} = \mathscr{A}_{\mathsf{fg}}$, then the locally finite fixed point coincides with the rational fixed point: $\vartheta F \cong \varrho F$.*

Indeed, the colimits in Corollary 10.2.15 and Theorem 10.5.15 are the same.

Under additional assumptions on $\mathscr{A}$ (which hold e.g. for all varieties $\mathscr{A}$), there is a strong connection of ϑF to the rational fixed point: ϑF is the image of ϱF in the terminal coalgebra [235, Thm. 5.4]; in symbols we have

$$\varrho F \twoheadrightarrow \vartheta F \rightarrowtail \nu F. \tag{10.12}$$

Example 10.5.19. (1) By Corollary 10.5.18, all instances of rational fixed points in Examples 10.2.21 and 10.5.10 are examples of locally finite fixed points, too. Hence, all examples of rational fixed points which are of widespread interest may be studied as locally finite fixed points.

(2) In addition, we also obtain several examples of finite behaviour domains that could not so far be obtained as instances of the rational fixed point, for example, context-free languages (and their real-time deterministic and non-deterministic variants, resp.), constructively $\mathbb{S}$-algebraic formal power series (and any other instance of the generalized power-set construction in Construction 12.5.7), and the monad of Courcelle's algebraic trees (see [107, Sec. 5]). That description of algebraic trees as a locally finite fixed point yields the first characterization of those trees by a universal property, solving an open problem from [43]. For details, see [235, Sec. 6].

Finally, the locally finite fixed point ϑF is also characterized by a universal property as an algebra in a similar way as ϱF: define *fg-iterative* algebras for F by replacing the finitely presentable object X by a finitely generated one in Definition 10.3.1. Then one obtains the following result.

Theorem 10.5.20 [235, Cor. 4.9]**.** *The locally finite fixed point is the initial fg-iterative algebra.*

The Locally ffg Fixed Point. Whenever the base category $\mathscr{A}$ is a finitary variety of algebras, one has yet another choice of the notion of a 'finite object', namely the class of all free algebras on a finite set of generators; we call them *ffg objects*, for short. They form a proper subclass of the class of all finitely presentable objects. This class is interesting because it consists precisely of the carriers of coalgebras arising as the targets of the generalized power-set construction (Construction 12.5.7). Again one could attempt to rework the theory of the rational fixed point, this time replacing finitely presentable objects by ffg ones.

To do that reworking, a slightly stronger condition needs to be required of F, namely that it is *strongly finitary* which means finitary and preserving reflexive coequalizers. These are coequalizers of parallel pairs $f, g\colon X \rightrightarrows Y$ having a joint spitting $s\colon Y \to X$ (with $f \cdot s = g \cdot s = \mathrm{id}_Y$). Strong finitarity is a fairly mild restriction; for example, every finitary set functor is strongly finitary (this follows from [41, Cor. 6.30]). Furthermore, every functor F which is a lifting of a finitary set functor to the variety $\mathscr{A}$ is strongly finitary. This follows from the fact [41, Cor. 11.9] that the forgetful functor $\mathscr{A} \to \mathsf{Set}$ preserves and reflects filtered colimits and reflexive coequalizers. (However, in general, not every finitary functor is strongly finitary [41, Ex. 7.11].)

Definition 10.5.21 [228]**.** Let F be a strongly finitary endofunctor on a variety. We define the coalgebra φF as the colimit of all coalgebras carried by ffg objects. It is called the *locally ffg fixed point* of F.

Theorem 10.5.22 [297, Lemma 4.5]**.** *The coalgebra structure of φF is an isomorphism.*

Somewhat surprisingly, the coalgebra φF fails to have the finality property with respect to coalgebras with ffg carriers [55, Section 2.5]. This also shows that φF cannot have a universal property as some kind of iterative algebra (i.e. where solutions are unique). However, one can adjust the notion of an Elgot algebra (see Section 10.6 below), and prove that φF is the initial ffg-Elgot algebra [55, Thm. 4.11]. In addition, in op. cit. it is shown that free ffg-Elgot algebras exist, and for an ffg object X the free ffg-Elgot algebra is given by $\varphi(F(-)+X)$.

Relations Between Four Fixed Points. As in the setting of the previous paragraph, let $\mathscr{A}$ be a finitary variety and $F\colon \mathscr{A} \to \mathscr{A}$ be finitary and preserve surjective as well as nonempty injective morphisms.[1] Then we already know from (10.12) that the subcoalgebra ϑF of νF is a quotient of ϱF. In addition, ϑF is a quotient of φF [228, Prop. 3.9]. Hence, (10.12) extends to the following picture:

$$\varphi F \twoheadrightarrow \varrho F \twoheadrightarrow \vartheta F \rightarrowtail \nu F.$$

Whenever $\mathscr{A}_{\mathsf{fp}} = \mathscr{A}_{\mathsf{fg}}$, we have additionally that $\varrho F \cong \vartheta F$ by Corollary 10.5.18, thus ϱF is fully abstract. Moreover, if all finitely generated algebras are free (which implies that $\mathscr{A}_{\mathsf{fp}} = \mathscr{A}_{\mathsf{fg}}$), then ϱF and ϑF coincide with φF as well. However, this only happens in very rare cases, for example, in Set and K-Vec. Milius [228] introduced the notion of a *proper* functor (generalizing the notion of a proper semiring of Ésik and Maletti [120]) and proved that a functor F is proper if and only if the first three fixed points coincide, that is, the picture above collapses to $\varphi F \cong \varrho F \cong \vartheta F \hookrightarrow \nu F$. Finally, there are instances where $\varrho F \cong \vartheta F \cong \nu F \cong 1$ are trivial but φF is non-trivial [55, Section 2.5].

[1] These are mild assumptions; for example, if $\mathscr{A}$ is single-sorted and F a lifting of a finitary set functor, then these conditions are fulfilled.

Unified Fixed Points. The technical development for the four fixed points above shows a number of parallels. Urbat [297] gives a uniform account of several results on fixed points that are based on coalgebras. His theory is parametric in a class $\mathbb{I}$ of diagram schemes and a class $\mathcal{M}$ of morphisms coming from a factorization system $(\mathcal{E}, \mathcal{M})$ on the base category $\mathscr{A}$, where $\mathcal{E}$ is some class of epimorphisms. For an endofunctor $F\colon \mathscr{A} \to \mathscr{A}$ he considers those coalgebras that have an $(\mathbb{I}, \mathcal{M})$-presentable carrier, that is, a carrier X such that $\mathscr{A}(X, -)$ preserves colimits in $\mathscr{A}$ that have a diagram scheme in $\mathbb{I}$ and whose colimit injections lie in $\mathcal{M}$.

Under suitable conditions on $\mathbb{I}$, $\mathcal{M}$, and a given endofunctor $F\colon \mathscr{A} \to \mathscr{A}$, Urbat proves that the colimit T_F of all coalgebras with an $(\mathbb{I}, \mathcal{M})$-presentable carrier is a fixed point of F and that this is characterized by a universal property both as a coalgebra and as an algebra (the proofs of these characterizations use that all diagrams in $\mathbb{I}$ are filtered). As instances one obtains νF, ϱF, ϑF, and φF as shown in the following table:

$\mathbb{I}$	$(\mathcal{E}, \mathcal{M})$	T_F
categories with a terminal object	(iso, all)	νF
small filtered categories	(iso, all)	ϱF
small filtered categories	(strong epis, monos)	ϑF
small sifted categories	(iso, all)	φF

Note that in the last case $\mathbb{I}$ does not consist of filtered diagrams, and hence, the characterization of φF by a universal property does not follow from the uniform proofs.

10.6 Beyond the Rational Fixed Point

In this section we briefly discuss further work which is based on the ideas and results on the rational fixed point we have seen in Sections 10.4 and 10.5.

Iterative Monads. For a finitary functor F on a locally finitely presentable category $\mathscr{A}$ free iterative algebras RX exist on every every object X of $\mathscr{A}$ (Corollary 10.3.8). They are, equivalently, the rational fixed points for the functors $F(-) + X$ (see [33, Prop. 2.24]). More precisely, given the rational fixed point $\omega_X\colon RX \to FRX + X$, then pre-composing ω_X^{-1} by the two coproduct injections yields the structure $FRX \to RX$ and the universal morphism $X \to RX$ of a free iterative algebra on X. It follows that R is the object assignment of monad on $\mathscr{A}$. This monad is called the *rational monad* of F, and it is characterized by a universal property: it is the free iterative monad on F [33, Thm. 5.12].

For example, for a polynomial set functor H_Σ, the rational monad R assigns to a set X the set of all rational Σ-trees over X. These are the rational Σ-trees (cf. Proposition 10.2.20) whose leaves are labelled by constants from Σ or variables from X. The monad structure arises from substitution of rational Σ-trees for variables in rational Σ-trees. This generalizes and extends classical work on iterative theories by Elgot [115] and on iterative algebras for a signature by Nelson [242] and Tiuryn [289].

Elgot Algebras. In a similar way as we have discussed in Remark 7.3.23(4), one is led to the notion of an *Elgot algebra* $(A, \alpha, (\_)^\dagger)$, which is simply the variation of the notion of a complete Elgot algebras where the solution operation $(\_)^\dagger$ is restricted to finitary flat equation morphisms $e\colon X \to FX + A$. Every iterative algebra is an Elgot algebra. Continuous algebras in $\mathscr{A} = \mathsf{CPO}$ are examples of Elgot algebras not admitting unique solutions $e^\dagger$ in general.

Theorem 10.3.12 can be augmented to state that the rational fixed point is the initial Elgot algebra. More generally, the rational fixed point of $F(-) + X$ is the free Elgot algebra on X [32, Prop. 4.7]. The main result [32, Thm. 4.8] states that the Elgot algebras form precisely the category of Eilenberg–Moore algebras for the rational monad R above.

Elgot Monads and Iteration Theories. Going one level up from the locally finitely presentable category $\mathscr{A}$ to the category $[\mathscr{A}, \mathscr{A}]_{\mathsf{fin}}$ of finitary endofunctors on $\mathscr{A}$, one sees that assigning a free iterative monad R_F to a given finitary functor $F\colon \mathscr{A} \to \mathscr{A}$ is the object assignment of a monad $\mathcal{R}$ on $[\mathscr{A}, \mathscr{A}]_{\mathsf{fin}}$. It is natural to ask what the Eilenberg–Moore algebras for this monad are: these turn out to be closely related to Bloom and Ésik's iteration theories [84]. In fact, this relationship can be clearly stated for $\mathscr{A} = \mathsf{Set}$: assigning to a signature Σ the rational monad $\mathcal{R}(H_\Sigma + \{\bot\})$ (of rational trees over Σ augmented with a new constant symbol $\bot$) yields the free iteration theory on Σ. Morever this assignment yields a monad on the category of signatures whose Eilenberg–Moore algebras are precisely Bloom and Ésik's iteration theories [35].

The move from Elgot's iterative theories (featuring unique solutions of recursive equations) to Bloom and Ésik's iteration theories (featuring solution operators $(\_)^\dagger$ obeying natural equational laws) can also be performed in a more general category-theoretic setting. For a locally finitely presentable category $\mathscr{A}$ having well-behaved coproducts (see [36] for a precise definition), one considers the assignment that maps a finitary functor $F\colon \mathscr{A} \to \mathscr{A}$ to the free iterative monad on $F + C_1$, where C_1 is the constant functor on 1: $F \mapsto \mathcal{R}(F + C_1)$. This yields a monad on $[\mathscr{A}, \mathscr{A}]_{\mathsf{fin}}$ whose Eilenberg–Moore algebras are *Elgot monads*, the appropriate categorical generalization of iteration theories [39, 42].

10.7 Summary

We have seen that besides the initial algebra and the terminal coalgebra a number of interesting fixed points of a finitary functor arise from 'finite' coalgebras. We have looked in detail at the rational fixed point for a finitary functor F on a locally finitely presentable category $\mathscr{A}$. In this setting the role of 'finite' objects is played by the finitely presentable objects (a notion suitably generalizing that of finite sets or of algebras presented by finitely many generators and relations). We have considered all those coalgebras whose carrier is finitely presentable and have obtained lfp coalgebras as their (filtered) colimits. The rational fixed point ϱF is then defined to be the terminal lfp coalgebra for F. We proved that it is indeed a fixed point of F. Furthermore, it can be constructed as the colimit of *all* coalgebras with finitely presentable carrier. In this sense ϱF is determined by 'finite behaviour' as the title of the chapter suggests. In addition, we have shown that whenever $\mathscr{A}_{\mathsf{fp}} = \mathscr{A}_{\mathsf{fg}}$ and F preserves monomorphisms, the rational fixed point is fully abstract with respect to behavioural equivalence, that is, it is a subcoalgebra of the terminal coalgebra νF. This result entails concrete descriptions of rational fixed points for endofunctors on sets and many other categories of interest in the coalgebraic semantics of systems.

We then turned to iterative algebras and proved that free iterative algebras exist and that the rational fixed point is an initial iterative algebra for F. Thus, the initial iterative algebra is precisely the terminal lfp coalgebra.

Finally, we have seen two explicit descriptions of the rational fixed point for a finitary endofunctor F on sets. First, ϱF consists of all finite well-pointed coalgebras for F. Second, using a presentation of F by a finitary signature Σ and a set $\mathcal{E}$ of basic equations, ϱF is the quotient of the rational fixed point ϱH_Σ of all rational Σ-trees modulo finite and infinite applications of the basic equations in $\mathcal{E}$.

We concluded the chapter with a discussion of further work inspired by results and ideas arising from the theory of rational fixed points.

11

Sufficient Conditions for Initial Algebras and Terminal Coalgebras

This chapter presents a number of results which allow us to conclude that an endofunctor has an initial algebra or a terminal coalgebra. These results are diverse, and they do not have close connections. Section 11.1 gives a sufficient condition using the Adjoint Functor Theorem. Following this, we generalize the result that accessible set functors have terminal coalgebras, Theorem 6.5.11, to the setting of accessible functors on general categories. In Section 11.3 we prove that the category of coalgebras for an accessible endofunctor is locally presentable whenever the base category is. We continue in Section 11.4 with a variation on this same theme: we generalize accessible functors to preaccessible functors, and we prove results on the existence of μF and νF for such functors. Section 11.5 is yet a different application of adjoints to the theme of this chapter.

11.1 Bounded Functors

We begin with a short section generalizing the notion of *bounded set functor* by Kawahara and Mori [186].

Remark 11.1.1. A *generating set* in a category is a set $\mathbb{G}$ of objects such that whenever parallel morphisms $p, q: X \to Y$ are distinct, then there exists a morphism $g: G \to X$ with $G \in \mathbb{G}$ and $p \cdot g \neq q \cdot g$.

Definition 11.1.2. A functor $F: \mathscr{A} \to \mathscr{A}$ is called *bounded* by a set $\mathbb{G}$ of objects if for every coalgebra $\alpha: A \to FA$ and every morphism $g: G \to A$ with $G \in \mathbb{G}$, there exists a coalgebra $\alpha': A' \to FA'$ with $A' \in \mathbb{G}$ and a coalgebra homomorphism $h\colon (A', \alpha') \to (A, \alpha)$ through which g factorizes:

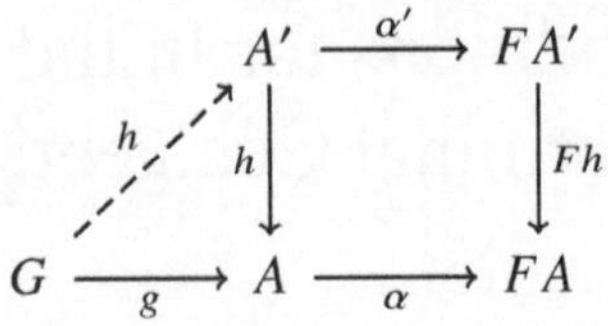

This was defined by Kawahara and Mori [186] for $\mathscr{A}$ = Set and $\mathbb{G}$ the (essentially small) collection of all sets of cardinality at most λ [186]. Then F is called *bounded by* λ. Observe that in this case it is sufficient to request that for every element $x \in A$ there exists a subcoalgebra $A' \hookrightarrow A$ containing x and with $|A'| \le \lambda$. In fact, the generalization to maps $g\colon G \to A$ in lieu of elements follows from the observation that a union of subcoalgebras is always a subcoalgebra.

Remark 11.1.3. For set functors the concept of being finitely bounded (equivalently, finitary) in Definition 4.3.3 is different from being bounded by finite sets:

Example 11.1.4. The identity functor $\mathsf{Id}\colon \mathsf{Set} \to \mathsf{Set}$ is bounded by countable sets but not by finite ones. For example, the coalgebra $\alpha\colon \mathbb{N} \to \mathbb{N}$ given by the successor function $\alpha(n) = n + 1$ has no finite subcoalgebra containing 0.

The following is an easy generalization of the result for set functors proved by Kawahara and Mori [186].

Theorem 11.1.5. *Let $\mathscr{A}$ be a cocomplete and co-well-powered category with a terminal object. Every endofunctor bounded by a generating set has a terminal coalgebra.*

Proof This uses Freyd's Special Adjoint Functor Theorem: we prove that $U\colon \mathsf{Coalg}\, F \to \mathscr{A}$ has a right adjoint $R\colon \mathscr{A} \to \mathsf{Coalg}\, F$ (then $R1$ is terminal in $\mathsf{Coalg}\, F$). For that we need, since $\mathsf{Coalg}\, F$ is cocomplete and co-well-powered (by Corollary 4.1.3) and U preserves colimits (by Proposition 4.1.1), just a generating set in $\mathsf{Coalg}\, F$. Let $\mathbb{G}$ be a generating set in $\mathscr{A}$ such that F is bounded by it. Then the set of all coalgebras $\alpha\colon A \to FA$ with $A \in \mathbb{G}$ is generating. (This is a set by our standing assumption in this book that all categories are locally small.) In fact, given distinct homomorphisms $p, q\colon (B, \beta) \to (\bar{B}, \bar{\beta})$, there exists $g\colon G \to B$ with $G \in \mathbb{G}$ and $p \cdot g \neq q \cdot g$. Let $h\colon (B', \beta') \to (B, \beta)$ be a homomorphism with $B' \in \mathbb{G}$ such that g factorizes through h. Then clearly $p \cdot h \neq q \cdot h$. □

Corollary 11.1.6. *Let $\mathscr{A}$ a cocomplete, well-, and co-well-powered category with a terminal object and a smooth class $\mathcal{M}$ of monomorphisms. Every endofunctor* *bounded by a generating set and preserving $\mathcal{M}$-monomorphisms has an initial algebra.*

This follows from Theorem 6.3.4 since the terminal coalgebra, which exists by Theorem 11.1.5, is an $\mathcal{M}$-pre-fixed point.

11.2 Accessible Functors

We proved in Corollary 4.3.12 that every finitary set functor has a terminal coalgebra. We now generalize this to accessible endofunctors on more general categories.

Assumption 11.2.1. Throughout this section λ denotes a regular infinite cardinal. Recall that λ is the linearly ordered set of all ordinals $i < \lambda$.

Definition 11.2.2. We generalize the notion of a filtered category $\mathscr{D}$ in Definition 10.1.7 to a *λ-filtered category* by strengthening the definition to require that every set of morphisms of size less than λ has a cocone. Equivalently:

(1) every set of less than λ objects has a cocone, and

(2) every family of less than λ parallel morphisms $f_i \colon A \to B$ ($i \in I$ with $|I| < \lambda$) has a coequalizing morphism $g \colon B \to C$ (i.e. $g \cdot f_i \colon A \to C$ is independent of $i \in I$).

A *λ-filtered diagram* in a category $\mathscr{A}$ is a diagram $\mathscr{D} \to \mathscr{A}$ whose scheme $\mathscr{D}$ is a λ-filtered category, and a *λ-filtered colimit* is a colimit of a λ-filtered diagram.

Example 11.2.3. (1) An important example of a filtered colimit is a colimit of an ω-chain (cf. Definition 10.1.7). Analogously, an important example of a λ-filtered colimit is a colimit of a λ-chain (see Remark 6.0.5), which is a diagram whose scheme is the poset λ.

(2) For every set X, the collection of all subsets of cardinality less than λ ordered by inclusion forms a λ-filtered diagram. Its λ-filtered colimit is X, with colimit injections given by the inclusion maps.

Remark 11.2.4. Every λ-filtered diagram is clearly filtered. More generally, if $\lambda' < \lambda$, then every λ-filtered diagram is λ'-filtered. That is, the larger λ is, the smaller is the collection of λ-filtered diagrams.

The following concept generalizes finitely presentable objects (see Definition 10.1.1).

Definition 11.2.5. An object X of a category $\mathscr{A}$ is called *λ-presentable* if its hom-functor $\mathscr{A}(X, -) : \mathscr{A} \to \mathsf{Set}$ preserves λ-filtered colimits.

Remark 11.2.6. More explicitly, an object X is λ-presentable iff every morphism f from X to a colimit $C = \mathrm{colim}_{i \in I} C_i$ of a λ-filtered diagram D *factorizes essentially uniquely* through one of the colimit injections $c_i : C_i \to C$. This means that

(1) there exists an $i \in I$ and a morphism $f' : X \to C_i$ such that $c_i \cdot f' = f$, and

(2) given less than λ factorizations f'_k ($k \in I$ with $|I| < \lambda$) of f through c_i (so that $f = f'_k \cdot c_i$ for all k) there exists a connecting morphism $c_{i,j} : C_i \to C_j$ of D such that $c_{i,j}$ merges all f'_k, $k < \lambda$.

Example 11.2.7. (1) For the case $\lambda = \omega$ of finitely presentable objects, see Examples 10.1.2 and 10.1.6. In particular, recall that in a variety of algebras, such as monoids, groups, or vector spaces, finitely presentable objects are precisely the algebras presented by finitely many generators and relations.

(2) In Set the λ-presentable objects are precisely the sets of cardinality less than λ, and similarly in the categories of posets or graphs.

(3) In the category K-Vec of vector spaces over the field K the λ-presentable objects are precisely the vector spaces of dimension less than λ.

(4) More generally, λ-presentable objects in a variety are precisely the algebras presented by less than λ generators and relations [23, Thm. 3.12].

(5) Let F be a finitary set functor. For an uncountable λ, we prove in Corollary 11.3.8 that a coalgebra is λ-presentable in $\mathsf{Coalg}\, F$ iff it has cardinality less than λ. This result does not extend to $\lambda = \omega$. For a counterexample consider $\mathsf{Coalg}\,\mathsf{Id}$, which is equal to the category $\mathsf{Alg}\,\mathsf{Id}$ of algebras having one unary operation. The infinite (co)algebra $(\mathbb{N}, \mathsf{succ})$ is finitely presentable.

(6) In contrast to item (5), for every cardinal λ we can find a finitary set functor F such that there are finite F-algebras which are not λ-presentable. For example, let F be the constant functor with value λ. An F-algebra is a set X equipped with constants $c_i \in X$, $i < \lambda$. It is λ-presentable iff the number of pairs $i < j < \lambda$ with $x_i = x_j$ is less than λ (cf. item (4)). It follows that the terminal (singleton) algebra is not λ-presentable.

(7) In the category of coalgebras for $\mathscr{P}_\mathsf{f}$, the finitely presentable objects are precisely the finite graphs (see Theorem 11.3.10).

(8) In the category Met of metric spaces, no nonempty space is finitely presentable. The same is true for CMet, the full subcategory of complete metric spaces [49, Rem. 2.10]. For an uncountable cardinal λ, a space is λ-presentable in Met iff it has cardinality less than λ. In CMet a space is

$\aleph_1$-presentable iff is it *separable*, which means that it has a countable dense subset [49, Cor. 2.8]. For all $\lambda > \aleph_1$, the λ-presentable spaces are precisely those of cardinality less than λ.

(9) Analogously, in CPO no nonempty cpo is finitely presentable (see Examples 10.1.4). A cpo is $\aleph_1$-presentable iff it contains a countable dense subset X, that is, no proper subcpo contains X. For all $\lambda > \aleph_1$, the λ-presentable cpos are precisely those of cardinality less than λ.

(10) In the category DCPO of directed complete posets, for every λ the only λ-presentable object is the empty dcpo. The proof is analogous to Examples 10.1.4.

The following concept is due to Gabriel and Ulmer [134].

Definition 11.2.8. A category is *locally λ-presentable* if it is cocomplete, every object is a λ-filtered colimit of λ-presentable objects, and there is only a set of λ-presentable objects up to isomorphism.

A category is *locally presentable* if it is locally λ-presentable for some λ.

Notation 11.2.9. If $\mathscr{A}$ is a locally λ-presentable category, we denote by $\mathscr{A}_\lambda$ a (small) full subcategory representing all λ-presentable objects up to isomorphism.

Example 11.2.10. (1) We have seen examples of locally finitely presentable categories in Example 10.1.6. In particular, every (many-sorted) finitary variety is locally finitely presentable.

(2) Varieties of algebras with operations of arities less than λ (cf. Example 6.1.16) are locally λ-presentable categories [23, Thm. 3.28] (cf. item (1)). For example, the category of σ-Boolean algebras (those having countable joins) and σ-homomorphisms is locally $\aleph_1$-presentable.

(3) Every Grothendieck topos is a locally presentable category [92, Prop. 3.4.16].

(4) The categories Met and CMet are locally $\aleph_1$-presentable [49, Cor. 2.8]. Analogously, the category CPO is locally $\aleph_1$-presentable.

(5) A Banach space is a normed vector space (over the real or complex numbers) with a complete norm. The category of Banach spaces and linear contractions is locally $\aleph_1$-presentable [23, Ex. 1.48].

(6) In contrast, DCPO is not locally λ-presentable for any λ (cf. Example 11.2.7(10)).

Remark 11.2.11. Every locally λ-presentable category is locally λ'-presentable for all $\lambda' \geq \lambda$ [23, Thm. 1.20].

Remark 11.2.12. Analogously to Remark 10.1.16 we collect a few useful properties of locally presentable categories that we shall subsequently use. Let $\mathscr{A}$ be a locally λ-presentable category.

(1) The category $\mathscr{A}$ is complete, well-powered, and co-well-powered [23, Rem. 1.56 and Thm. 1.58].

(2) A colimit of a diagram is λ-presentable whenever the diagram has less than λ objects which are all λ-presentable, and less than λ morphisms [23, Prop. 1.16]. It follows that $\mathscr{A}_\lambda$ is closed under such colimits, whence it is a λ-filtered category.

(3) By [23, Prop. 1.22], every object X of a locally λ-presentable category $\mathscr{A}$ is the λ-filtered colimit of the *canonical λ-filtered diagram* given by all morphisms with codomain X having a λ-presentable domain. More precisely, let $\mathscr{A}_\lambda/X$ be the (slice) category of all morphisms $p\colon P \to X$ with $P \in \mathscr{A}_\lambda$ (as objects) and the obvious commutative triangles as morphism. Then X is a colimit of the following diagram

$$\mathscr{A}_\lambda/X \xrightarrow{D_X} \mathscr{A}, \qquad D_X(p\colon P \to X) = P$$

with the colimit injections given by all the morphisms p. For example, the terminal object of $\mathscr{A}$ is the colimit of the embedding $\mathscr{A}_\lambda \hookrightarrow \mathscr{A}$.

(4) From item (3) and the fact that colimit cocones are collectively epic, it follows that $\mathscr{A}_\lambda$ is a generating set (Remark 11.1.1).

(5) Let $e\colon A \twoheadrightarrow B$ be a split epimorphism with $e \cdot m = \mathsf{id}$ for some $m\colon B \rightarrowtail A$. If A is λ-presentable, so is B. Indeed, use item (2): e is the coequalizer of id_A and $m \cdot e$.

(6) Remark 10.1.16(5) generalizes in the obvious way from $\lambda = \omega$ to all λ, giving an important characterization of colimit cocones in λ-presentable categories.

Definition 11.2.13. A functor F is *λ-accessible* if it preserves λ-filtered colimits. (When $\lambda = \omega$, we call F *finitary*. This matches our usage in Chapters 4 and 10.)

A functor F is called *accessible*if it is λ-accessible for some λ.

Example 11.2.14. (1) We have seen examples of accessible set functors in Example 6.5.9. We also discuss this in more detail in Section C.8.

(2) The Hausdorff functor $\mathcal{H}$ on CMet (Example 5.2.6(4)) is finitary [49], which improves an earlier result by van Breugel [302] that $\mathcal{H}$ is accessible.

(3) Another interesting finitary functor on CMet is the *Kantorovich functor*. Given a metric space (X, d) the *Borel subsets* of X form the smallest σ-algebra $\mathscr{B}(X, d)$ containing all open sets. A *probability measure* is a σ-additive function $\mu\colon \mathscr{B}(X, d) \to [0, 1]$ such that $\mu(X) = 1$. It is called *tight* if for every $\varepsilon > 0$ there exists a compact set $C \subseteq X$ such that $\mu(X \setminus C) < \varepsilon$. The Kantorovich

functor maps (X, d) to the space $\mathscr{K}X$ formed by all tight probability measures on X with the metric defined by

$$\bar{d}(\mu, \nu) = \sup_f \left(\int_X f \, d\mu - \int_X f \, d\nu \right),$$

where f ranges over all non-expanding maps $X \to [0, 1]$ (cf. Kantorovich [183]). For a non-expanding map $f\colon (X, d) \to (X', d')$, the non-expanding map $\mathscr{K}f\colon \mathscr{K}X \to \mathscr{K}X'$ assings to a measure μ on X the measure $B \mapsto \mu(f^{-1}(B))$ on X'. It was shown by van Breugel and Worrell [303] that this is a well-defined endofunctor on CMet, and like the Hausdorff functor, it is finitary [49], improving the result by van Breugel et al. [305] that $\mathscr{K}$ is accessible.

Remark 11.2.15. (1) Every finitary functor is λ-accessible. More generally, if $\lambda' \geq \lambda$, then every λ-accessible functor is λ'-accessible. This follows from Remark 11.2.4.

(2) A composite $F_1 F_2$ of accessible functors is accessible: if F_i is λ_i-accessible, the composite is λ-accessible for $\lambda = \max\{\lambda_1, \lambda_2\}$.

Analogously, a coproduct $\coprod_{i \in I} F_i$ of accessible functors is accessible: choose λ larger than all λ_i, where F_i is λ_i-accessible.

(3) Every left adjoint is of course finitary since it preserves all colimits. Given an adjoint situation $F \dashv U\colon \mathscr{A} \to \mathscr{B}$ (Remark 2.2.28(2)) with U accessible, we thus see that the induced monad $T = UF$ on $\mathscr{B}$ is accessible.

We now present a proof that every accessible endofunctor on a locally presentable category has an initial algebra and a terminal coalgebra. Later we use our proof to derive other properties of Coalg F, for example, characterizing the λ-presentable objects. The result in Theorem 11.2.18 is a consequence of Makkai and Paré's Limit Theorem [218, Thm. 5.1.6], which moreover implies that Coalg F (being the inserter of Id and F) is an accessible category; we recall that the concept of a λ-accessible category is the weakening of a locally λ-presentable one where, in lieu of cocompleteness, only existence of λ-directed colimits is assumed. Since we know that Coalg F is cocomplete (Proposition 4.1.1), it is locally presentable and therefore also complete (Remark 11.2.12(1)). In particular, it has an initial and a terminal object.

Our main result will follow from Proposition 11.2.17 showing that for uncountable λ every λ-accessible functor is bounded by λ-presentable objects (Definition 11.1.2). However, for $\lambda = \aleph_0$ this result does not hold:

Example 11.2.16. The coalgebra A for $\mathscr{P}_\mathsf{f}$ given by the graph consisting of an infinite path has no homomorphism $G \to A$ from any finite graph G (considered as a coalgebra for $\mathscr{P}_\mathsf{f}$). Thus, $\mathscr{P}_\mathsf{f}$ is not bounded by any set of finite coalgebras.

Proposition 11.2.17. *Let λ be an uncountable regular cardinal. Every λ-accessible endofunctor on a locally λ-accessible category $\mathscr{A}$ is bounded by $\mathscr{A}_\lambda$.*

Proof (1) Let $\beta\colon B \to FB$ be a coalgebra and $g\colon A \to B$ a morphism, where $A \in \mathscr{A}_\lambda$. Then it is our task to define a coalgebra $\alpha^+\colon A^+ \to FA^+$ with A^+ λ-presentable, and a homomorphism $g^+\colon (A^+, \alpha^+) \to (B, \beta)$ through which g factorizes in $\mathscr{A}$. In part (2), we shall obtain A^+ as a colimit of an ω-chain $A_0 \xrightarrow{a_0} A_1 \xrightarrow{a_1} \cdots$ of objects in $\mathscr{A}_\lambda$ with a colimit cocone $c_n\colon A_n \to C$. We shall also define morphisms p_n so that all of the squares below commute:

$$\begin{CD}
A_0 @>{a_0}>> A_1 @>{a_1}>> A_2 @>{a_2}>> \cdots \\
@V{p_0}VV @V{p_1}VV @V{p_2}VV \\
FA_1 @>{Fa_1}>> FA_2 @>{Fa_2}>> FA_3 @>{Fa_3}>> \cdots
\end{CD} \tag{11.1}$$

Finally, we shall also define a cocone $g_n\colon A_n \to B$ of our ω-chain. After defining all of these morphisms, we prove several important properties about them. We finish this proof by defining a coalgebra structure on A^+ using the morphisms p_n and $g^+\colon A^+ \to B$ as the factorizing morphism of the cocone g_n. Finally we verify that g^+ is a homomorphism.

(2) Our construction proceeds inductively. At step 0, we let A_0 be the given object A, and g_0 the given morphism g. The inductive step begins with A_n and g_n. We then define a_n, p_n, A_{n+1}, and g_{n+1} and finally arrange that the diagrams in (11.2) below commute for n; the square on the right is considered for $n \geq 1$.

$$\begin{array}{ccc}
A_n & \xrightarrow{a_n} & A_{n+1} \\
& {\scriptstyle g_n}\searrow \quad \swarrow{\scriptstyle g_{n+1}} & \\
& B &
\end{array}
\qquad
\begin{CD}
A_n @>{p_n}>> FA_{n+1} \\
@V{g_n}VV @VV{Fg_{n+1}}V \\
B @>{\beta}>> FB
\end{CD}
\qquad
\begin{CD}
A_{n-1} @>{a_{n-1}}>> A_n \\
@V{p_{n-1}}VV @VV{p_n}V \\
FA_n @>{Fa_n}>> FA_{n+1}
\end{CD} \tag{11.2}$$

We now describe the inductive step in detail. Express B as the λ-filtered colimit $B = \operatorname{colim} D_B$ of its canonical λ-filtered diagram; see Remark 11.2.12(3). The diagram has as colimit injections all morphisms $g'\colon A' \to B$ with $A' \in \mathscr{A}_\lambda$. Since F is λ-accessible, we have $FB = \operatorname{colim} FD_B$ with colimit injections Fg'. For the morphism $\beta \cdot g_n\colon A_n \to FB$ we choose a colimit injection $h\colon C \to B$ of $\mathscr{A}_\lambda/B$ such that $\beta \cdot g_n$ factorizes through Fh. We thus obtain the following commutative square

$$\begin{CD}
A_n @>{p'_n}>> FC \\
@V{g_n}VV @VV{Fh}V \\
B @>{\beta}>> FB
\end{CD}$$

For the objects g_n and h of the filtered category $\mathscr{A}_\lambda/B$, we choose an object $k\colon E \to B$ and morphisms u and v from g_n and h to it, respectively:

$$\begin{array}{ccccc} A_n & \xrightarrow{u} & E & \xleftarrow{v} & C \\ & {\scriptstyle g_n}\searrow & \downarrow{\scriptstyle k} & \swarrow{\scriptstyle h} & \\ & & B & & \end{array} \tag{11.3}$$

We next define the following morphism: $\gamma = (A_n \xrightarrow{p'_n} FC \xrightarrow{Fv} FE)$. We observe that the following square commutes:

$$\begin{array}{ccc} A_n & \xrightarrow{\gamma} & FE \\ {\scriptstyle g_n}\downarrow & & \downarrow{\scriptstyle Fk} \\ B & \xrightarrow{\beta} & FB \end{array} \tag{11.4}$$

(2a) The first inductive step is completed by taking $a_0 = u$, $p_0 = \gamma$, $A_1 = E$, and $g_1 = k$. The first two diagrams in (11.2) commute for $n = 0$ by definition, and the third one is not defined.

(2b) For the inductive steps with $n > 0$, consider the diagram below:

$$\begin{array}{ccccc} A_{n-1} & \xrightarrow{a_{n-1}} & A_n & \xrightarrow{g_n} & B \\ {\scriptstyle p_{n-1}}\downarrow & & \downarrow{\scriptstyle \gamma} & & \downarrow{\scriptstyle \beta} \\ FA_n & \xrightarrow{Fu} & FE & \xrightarrow{Fk} & FB \end{array} \qquad \text{with } g_{n-1}\colon A_{n-1} \to B \text{ (top)},\ Fg_n\colon FA_n \to FB \text{ (bottom)} \tag{11.5}$$

The lower part and the right-hand square commute by (11.4). The outside and the upper part commute by the induction hypothesis, more precisely by the left and middle diagram in (11.2) for $n-1$. The left-hand square does not necessarily commute, and this is the source of much of the rest of our work below. However, the morphism Fk merges the two sides of the left-hand square:

$$Fk \cdot (\gamma \cdot a_{n-1}) = Fk \cdot (F\gamma \cdot p_{n-1}).$$

Recall that A_{n-1} is λ-presentable and that k is a colimit injection for the canonical λ-filtered diagram $D_B\colon \mathscr{A}_\lambda/B \to \mathscr{A}$. Therefore Fk is a colimit injection for the diagram FD_B, since F is λ-accessible. By the essential uniqueness in Remark 11.2.6, there is some morphism w in A_λ/B so that Fw merges the two sides of (11.5). In more detail, we can choose $g_{n+1}\colon A_{n+1} \to B$ and $w\colon E \to A_{n+1}$ such that $k = g_{n+1} \cdot w$ and

$$Fw \cdot (\gamma \cdot a_{n-1}) = Fw \cdot (Fu \cdot p_{n-1}). \tag{11.6}$$

We now can supply the data needed in this step of the construction. We take A_{n+1} and g_{n+1} as just above, and a_n and p_n to be given by

$$a_n = (A_n \xrightarrow{u} E \xrightarrow{w} A_{n+1}) \quad \text{and} \quad p_n = (A_n \xrightarrow{\gamma} FE \xrightarrow{Fw} FA_{n+1}). \tag{11.7}$$

Then all three diagrams of (11.2) commute for n (the last one provided $n \geq 1$):

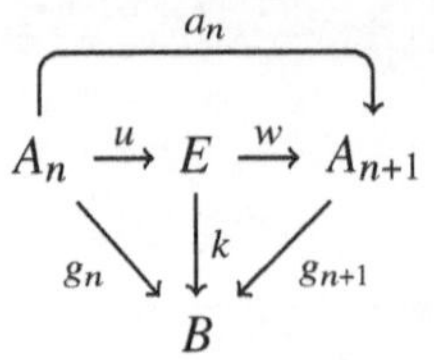

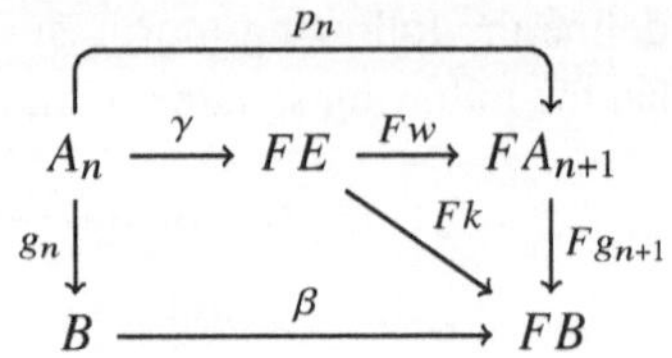

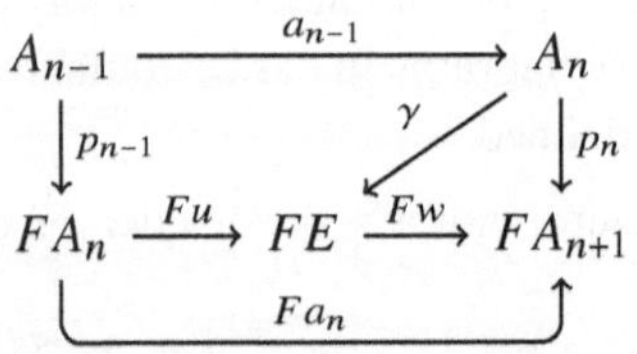

We have used (11.3), (11.4), (11.5), and (11.7). For the third diagram note that by (11.6) the left-hand part commutes when post-composed by Fw.

This completes our construction. At this point, we have objects A_n and morphisms $p_n \colon A_n \to FA_{n+1}$ making (11.1) commute.

(3) Let A^+ be the colimit of the chain (A_n) in (11.1) with colimit injections $c_n \colon A_n \to A^+$. Since $\lambda > \omega$ and all A_n are λ-presentable, the object A^+ is λ-presentable as well by Remark 11.2.12(3). From the commutativity of the squares in (11.1) it follows that the morphisms $Fc_{n+1} \cdot p_n \colon A_n \to FA^+$ form a compatible cocone. Let $\alpha^+ \colon A \to FA^+$ be the unique morphism factorizing that cocone:

$$\begin{array}{ccc} A_n & \xrightarrow{p_n} & FA_{n+1} \\ {\scriptstyle c_n}\downarrow & & \downarrow{\scriptstyle Fc_{n+1}} \\ A^+ & \xrightarrow{\alpha^+} & FA^+ \end{array} \tag{11.8}$$

(4) We are ready to present the desired factorization of g. The morphisms $g_n \colon A_n \to B$ form a compatible cocone by the left triangle in (11.2). Let $g^+ \colon A^+ \to B$ be the unique morphism with $g_n = g^+ \cdot c_n$ for all n. We check that g^+ is a homomorphism; this concludes the proof, since g factorizes as

$g = g^+ \cdot c_0$. For each n, consider the diagram below:

$$\begin{array}{ccccc} & A_n & \xrightarrow{p_n} & FA_{n+1} & \\ & {\scriptstyle c_n}\downarrow & & \downarrow{\scriptstyle Fc_{n+1}} & \\ g_n & A^+ & \xrightarrow{\alpha^+} & FA^+ & Fg_{n+1} \\ & {\scriptstyle g^+}\downarrow & & \downarrow{\scriptstyle Fg^+} & \\ & B & \xrightarrow{\beta} & FB & \end{array} \tag{11.9}$$

Its upper square is (11.8), the outside is the middle of (11.2), and the sides follow from the definition of g^+. Thus, the lower square commutes, since it does when pre-composed by each c_n, and the colimit cocone $(c_n)_{n<\omega}$ is collectively epic. □

Theorem 11.2.18. *Every accessible endofunctor on a locally presentable category has an initial algebra and a terminal coalgebra.*

Proof The existence of an initial algebra follows from Corollary 6.1.12: given a λ-accessible functor F, the initial-algebra chain converges in λ steps because the first λ steps of the initial-algebra chain form a λ-filtered diagram.

We prove that νF exists. Since $\mathscr{A}$ is locally presentable, there exists $\lambda > \omega$ for which $\mathscr{A}$ is locally λ-presentable and F is λ-accessible (Remark 11.2.15(1)). Moreover, $\mathscr{A}$ is cocomplete and co-well-powered, and the set $\mathscr{A}_\lambda$ is generating (see Remark 11.2.12(4)). From Proposition 11.2.17 we obtain that F is bounded by $\mathscr{A}_\lambda$, so that we may apply Theorem 11.1.5. □

Accessible endofunctors on general categories $\mathscr{A}$ need not have terminal coalgebras. For example, on the category $\mathsf{Pos}_{0,1}$ of posets with distinct least and greatest elements and monotone maps preserving them, the identity functor is finitary, but has no terminal coalgebra [184, Ex. 2.4(6) & Rem. 6.4]. When $\mathscr{A}$ is accessible, conditions on an accessible endofunctor to have a terminal coalgebra are presented by Karazeris et al. [184].

Remark 11.2.19. In the proof of Proposition 11.2.17 we worked with a coalgebra (B, β) and a morphism $g \colon A \to B$ with A in $\mathscr{A}_\lambda$. We shall need the following variation on this result in Theorem 11.3.7. Suppose that, in addition to the assumptions in Proposition 11.2.17, we are given another morphism $w \colon C \to A$ with C in $\mathscr{A}_\lambda$ together with a coalgebra $\gamma \colon C \to FC$ such that $g \cdot w \colon (C, \gamma) \to (B, \beta)$ is a homomorphism. Then our factorization $g = g^+ \cdot c_0$ in that proof can be chosen so that $c_0 \cdot w \colon (C, \gamma) \to (A^+, \alpha^+)$ is also a homomorphism:

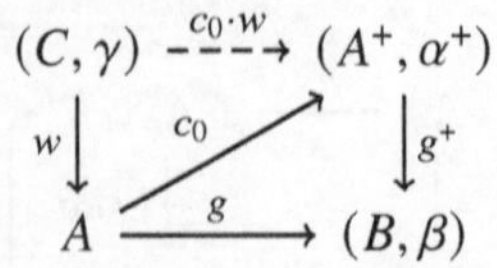

To see this, we only need to modify step 0 to arrange that the following square commutes:

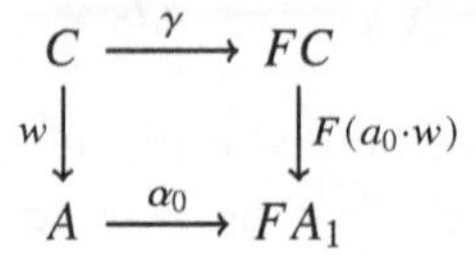

(11.10)

The following diagram then shows that $c_0 \cdot w$ is indeed a homomorphism:

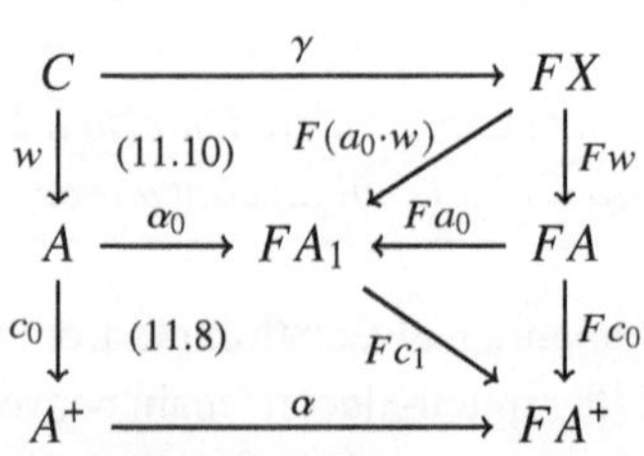

The modification above is possible since C is λ-presentable and the colimit injection $Fg_1 \colon FA_1 \to FB$ merges the two sides of the square in (11.10): since $g \cdot w \colon (C, \gamma) \to (B, \beta)$ is a homomorphism, we obtain

$$Fg_1 \cdot (Fa_0 \cdot Fw) \cdot \gamma = Fg \cdot Fw \cdot \gamma = F(g \cdot w) \cdot \gamma = \beta \cdot (g \cdot w) = Fg_1 \cdot \alpha_0 \cdot w,$$

where the last step is due to the middle square in (11.2). Thus, if $k \colon C \to FA_1$ denotes the composite in (11.10), both sides of the desired square are factorizations of k through the colimit morphism Fg_1 of $FB = \mathsf{colim}\, FD_B$. Here we use the essential uniqueness of such factorizations (Remark 11.2.6) once again.

Example 11.2.20. (1) We have already seen that every Kripke polynomial set functor (Definition 4.4.13) has a terminal coalgebra (Proposition 4.4.16) and an initial algebra (Example 6.3.11). We could also obtain these results by an application of Theorem 11.2.18. Indeed, an easy induction over the structure of Kripke polynomial functors shows that every such functor F is λ-accessible, where λ is the cardinality of the largest index set I of a product in the definition of F. So in addition to the previous results, we see here that the initial-algebra chain for F converges in λ steps.

(2) All endofunctors on CMet formed by products, coproducts, and compositions starting from the Hausdorff functor $\mathcal{H}$ (Example 5.2.6(4)) and constant

functors have an initial algebra and a terminal coalgebra. In other words, these are the functors on CMet formed according to the following grammar

$$F ::= \mathcal{H} \mid A \mid \mathsf{Id} \mid \textstyle\prod_{i\in I} F_i \mid \coprod_{i\in I} F_i \mid FF,$$

where A ranges over all complete metric spaces and I is an arbitrary index set. Indeed, this follows from Theorem 11.2.18 since CMet is locally $\aleph_1$-presentable. In Section 15.2, we shall see a similar terminal coalgebra result for the (non-accessible) category of topological spaces, where the Hausdorff functor is replaced by the Vietoris functor mapping a topological space to the space of all its compact subspaces.

(3) Recall the Kantorovich functor $\mathscr{K}$ from Example 11.2.14(3). We can extend the grammar in item (2) by $\mathscr{K}$: and every endofunctor on CMet according to the extended grammar (in particular the Kantorovich functor itself) has an initial algebra and a terminal coalgebra.

Remark 11.2.21. The Hausdorff functor can also be defined on the larger category Met. However, on this category it fails to be λ-accessible for any λ. To see this note first that the colimit of a diagram of metric spaces (X_i, d_i) $(i \in I)$ in Met is formed as follows: first one takes the colimit $X = \mathsf{colim}_{i\in I} X_i$ in Set and equips it with the pseudometric d given by

$$d(x, y) = \inf\{d_i(x', y') : i \in I, \mathsf{in}_i(x') = x, \mathsf{in}_i(y') = y\},$$

where $\mathsf{in}_i\colon X_i \to X$ denotes the colimit injection. Since this infimum may be zero, one then takes the quotient $X/{\sim}$ where $x \sim y$ iff $d(x, y) = 0$. With the induced metric on the quotient, this is the colimit of the given diagram with injections given by $q \cdot \mathsf{in}_i\colon X_i \to X/{\sim}$, where $q\colon X \to X/{\sim}$ is the (non-expanding) quotient map. We thus see that colimit injections in Met form jointly surjective families [24, Lemma 2.4]; in contrast, epimorphisms are just dense maps.

Now for a counterexample to λ-accessibility of $\mathcal{H}$, consider the compact metric space $[0, 1]^\lambda$ as a λ-filtered colimit of all its subspaces $\mathsf{in}_S\colon S \hookrightarrow [0, 1]^\lambda$ with $|S| < \lambda$. Then the cocone $(\mathcal{H}\mathsf{in}_S)$ is not a colimit in Met because it is not jointly surjective: the element $[0, 1]^\lambda$ of $\mathcal{H}([0, 1]^\lambda)$ is not contained in $\mathcal{H}S$ for any subspace S in the above diagram.

We also have the following result on the terminal-coalgebra chain (see Definition 6.5.1).

Theorem 11.2.22 [27, Thm. 2.6]**.** *Let F be an accessible endofunctor on a locally presentable category. If F preserves monomorphisms, then its terminal-*

coalgebra chain converges:

$$\nu F = V_j \qquad \text{for some ordinal number } j.$$

We will later present the proof of a more general result (Theorem 11.4.15).

Open Problem 11.2.23. Does Theorem 11.2.22 hold without the preservation of monomorphisms?

11.3 Presentability of Algebras and Coalgebras

For an accessible endofunctor F on a locally presentable category, the categories $\mathsf{Alg}\,F$ and $\mathsf{Coalg}\,F$ are also locally presentable. As mentioned in Section 11.2, this follows from Makkai and Paré's Limit Theorem [218, Thm. 5.1.6]; one uses that $\mathsf{Alg}\,F$ and $\mathsf{Coalg}\,F$ are inserters of (F, Id) and (Id, F), respectively. We give direct full proofs for the sharpened results that fix the cardinal (Theorems 11.3.2 and 11.3.8). We again denote by λ an arbitrary infinite regular cardinal.

Remark 11.3.1. (1) In the proof of Theorem 11.3.2 we apply Beck's monadicity theorem [217, Thm. VI.7.1]. This makes use of the notion of a *split coequalizer*; a parallel pair $f, g\colon A \rightrightarrows B$ has such a coequalizer if there are morphisms c, s, and t as follows:

$$A \underset{g}{\overset{f}{\rightrightarrows}} B \xrightarrow{c} C \quad (t\colon B \to A,\ s\colon C \to B) \qquad \text{satisfying} \qquad \begin{array}{ll} c \cdot f = c \cdot g, & c \cdot s = \mathsf{id}_C, \\ f \cdot t = \mathsf{id}_B, & g \cdot t = s \cdot c. \end{array} \tag{11.11}$$

(Note that this implies that c is a coequalizer of f and g.)

Beck's monadicity theorem states that for a right adjoint functor $U\colon \mathscr{C} \to \mathscr{A}$ with the corresponding monad T, the category $\mathscr{C}$ is isomorphic to $\mathscr{A}^T$ provided that U creates coequalizers of *U-split pairs*; these are parallel pairs $f, g\colon A \to B$ in $\mathscr{C}$ such that the pair Uf, Ug has a split coequalizer in $\mathscr{A}$, and creation of the coequalizer means that there exists a unique morphism $c'\colon B \to C$ in $\mathscr{C}$ such that $Uc' = c$ and, moreover, c' is a coequalizer of the pair f, g in $\mathscr{C}$.

(2) The above isomorphism is given by the canonical *comparison functor* $K\colon \mathscr{C} \to \mathscr{A}^T$ defined by $KC = (UC, U\varepsilon_C)$ for the counit $\varepsilon\colon FU \to \mathsf{Id}$. The functor U is then said to be *strictly monadic*.

Theorem 11.3.2. *Let F be a λ-accessible endofunctor on a locally λ-presentable category $\mathscr{A}$. Then the category* $\mathsf{Alg}\,F$ *is locally λ-presentable.*

Proof (1) The functor F is a varietor (Definition 2.2.26). Indeed, for every object X of $\mathscr{A}$ we know that the functor $F(-) + X$ is λ-accessible (Remark 11.2.15(1)). Therefore it has an initial algebra by Theorem 11.2.18, equivalently a free F-algebra on X (see Proposition 2.2.24). Thus, the forgetful functor $U\colon \mathsf{Alg}\, F \to \mathscr{A}$ has a left adjoint (Remark 2.2.28). Furthermore, we know from Remark 4.1.7 that U is λ-accessible. Thus, we obtain a λ-accessible monad T obtained by composing U with its left adjoint.

(2) The category $\mathscr{A}^T$ of Eilenberg–Moore algebras for the monad T is locally λ-presentable [134, Satz 10.3] (see also [81, Thm. 6.9]). We will show that $\mathsf{Alg}\, F$ and $\mathscr{A}^T$ are isomorphic categories. We already know that the free algebras $(A^\sharp, \varphi_A)$ for F yield the free algebras $(A^\sharp, \mu_A)$ for T (Remark 2.2.28). Therefore, it is sufficient to verify that U is a (strictly) monadic functor. It then follows that the comparison functor $\mathsf{Alg}\, F \to \mathscr{A}^T$ is an isomorphism of categories.

(3) We now use Beck's monadicity theorem (Remark 11.3.1) to prove that U is monadic. Since it is a right adjoint, it remains to prove that it creates coequalizers of U-split pairs. Let $f, g\colon (A, \alpha) \rightrightarrows (B, \beta)$ be a parallel pair, and let a diagram (11.11) be given in $\mathscr{A}$. It is our task to prove that there exists a unique algebra structure $\gamma\colon FC \to C$ such that c is a homomorphism and that, furthermore, c is a coequalizer of f and g in $\mathsf{Alg}\, F$. For the existence put

$$\gamma = \big(FC \xrightarrow{Fs} FB \xrightarrow{\beta} B \xrightarrow{c} C\big).$$

Then c is easily seen to be a homomorphism:

$$\begin{aligned}
\gamma \cdot Fc &= c \cdot \beta \cdot F(s \cdot c) && \text{def. of } \gamma\\
&= c \cdot \beta \cdot F(g \cdot t) && s \cdot c = g \cdot t\\
&= c \cdot g \cdot \alpha \cdot Ft && g \text{ a homomorphism}\\
&= c \cdot f \cdot \alpha \cdot Ft && c \cdot g = c \cdot f\\
&= c \cdot \beta \cdot F(f \cdot t) && f \text{ a homomorphism}\\
&= c \cdot \beta && f \cdot t = \mathsf{id}_B.
\end{aligned}$$

The uniqueness is also easy: if γ' makes c a homomorphism, then we have

$$\begin{aligned}
\gamma' &= \gamma' \cdot Fc \cdot Fs && c \cdot s = \mathsf{id}_C\\
&= c \cdot \beta \cdot Fs && c \text{ a homomorphism.}
\end{aligned}$$

Finally, c is a coequalizer in $\mathsf{Alg}\, F$. Suppose we have a homomorphism $d\colon (B, \beta) \to (D, \delta)$ with $d \cdot f = d \cdot g$. Then the unique morphism $h\colon C \to D$

such that $d = h \cdot c$ satisfies $h = d \cdot s$. It follows that it is a homomorphism:

$$\begin{aligned} h \cdot \gamma &= h \cdot c \cdot \beta \cdot Fs && \text{def. of } \gamma \\ &= d \cdot \beta \cdot Fs && h \cdot c = d \\ &= \delta \cdot F(d \cdot s) && d \text{ a homomorphism} \\ &= \delta \cdot Fh && \text{def. of } h. \end{aligned}$$

This completes the proof. □

Remark 11.3.3. If $\mathscr{A}$ has regular factorizations and F preserves regular epimorphisms, a characterization of the λ-presentable objects of $\mathsf{Alg}\, F$ follows from recent results by Adámek et al. [51, Thm. 3.7 and Rem. 4.14]: An F-algebra A is finitely presentable if and only if it can be presented as a coequalizer of a parallel pair of homomorphisms between free algebras on λ-presentable objects. Equivalently, A is a quotient of a free algebra on a λ-presentable object modulo a λ-generated congruence (see op. cit. for the definition of the latter notion).

In the remainder of this section we prove that for a λ-accessible endofunctor F

(1) every coalgebra (A, α) with A λ-presentable is a λ-presentable object of $\mathsf{Coalg}\, F$,

(2) the converse holds whenever λ is uncountable.

Moreover, for uncountable λ we show that $\mathsf{Coalg}\, F$ is a locally λ-presentable category whenever $\mathscr{A}$ is (Corollary 11.3.8). This does not hold for ω: we show that the category of coalgebras for $\mathscr{P}_\mathsf{f}$ is not locally finitely presentable (Corollary 11.3.11).

Proposition 11.3.4. *Let F be a λ-accessible endofunctor on a category $\mathscr{A}$. Every coalgebra with a λ-presentable underlying object is λ-presentable in* $\mathsf{Coalg}\, F$.

Proof We can restrict ourselves to a coalgebra $\alpha\colon A \to FA$ with A in $\mathscr{A}_\lambda$. Let a λ-filtered diagram D in $\mathsf{Coalg}\, F$ be given with objects (B_i, β_i) and a colimit cocone $c_i\colon (B_i, \beta_i) \to (C, \gamma)$ for $i \in I$. It is our task to prove that every homomorphism $f\colon (A, \alpha) \to (C, \gamma)$ factorizes essentially uniquely (Remark 11.2.6) through some c_i via a homomorphism $(A, \alpha) \to (B_i, \beta_i)$.

By Proposition 4.1.1, the underlying diagram UD, where $U\colon \mathsf{Coalg}\, F \to \mathscr{A}$ is the forgetful functor has the colimit cocone $c_i\colon B_i \to C$ ($i \in I$), and UD is of course λ-filtered. Since A is λ-presentable in $\mathscr{A}$, there exists $j \in I$ and a factorization $f = c_j \cdot g$ for some $g\colon A \to B_j$. Because f and c_j are homomorphisms, in the following diagram the outside and all inner parts except the upper square commute:

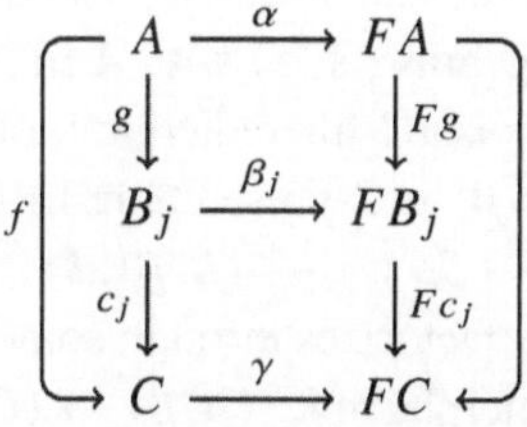

Since F is λ-accessible, all Fc_j form the colimit cocone of the λ-filtered diagram FUD. The object A is λ-presentable and the colimit injection Fc_j merges the two sides of the upper square. Therefore the composite morphism from A to FC has two factorizations through Fc_j given by the two sides of the upper square. By the essential uniqueness of such factorizations (Remark 11.2.6), there exists a connecting morphism of FUD merging the two sides, too. In symbols, we have a homomorphism $u\colon (B_j,\beta_j) \to (B_i,\beta_i)$ with $c_i \cdot u = c_j$ such that

$$Fu \cdot (Fg \cdot \alpha) = Fu \cdot (\beta_j \cdot g).$$

Then we see that $u \cdot g\colon (A,\alpha) \to (B_i,\beta_i)$ is a homomorphism:

$$F(u \cdot g) \cdot \alpha = Fu \cdot (\beta_j \cdot g) = \beta_i \cdot (u \cdot g),$$

where the second equation uses that u is a homomorphism. Moreover, $u \cdot g$ is the desired factorization of f:

$$f = c_j \cdot g = c_i \cdot (u \cdot g).$$

The essential uniqueness of this factorization is clear since this holds in $\mathscr{A}$. □

Example 11.3.5. (1) If F is a finitary endofunctor, then every coalgebra with a λ-presentable carrier is λ-presentable. This follows from the fact that F is λ-accessible for all λ (cf. Remark 11.2.15(1)).

(2) The analogous result for algebras does not holds (Example 11.2.7(6)).

Remark 11.3.6. Recall from Remark 10.3.9 the notion of a final functor $P\colon \mathscr{D} \to \mathscr{C}$ and the two equivalent conditions for finality in the case where $\mathscr{C}$ and $\mathscr{D}$ are filtered. By Remark 11.2.4, these conditions entail that for every diagram $D\colon \mathscr{C} \to \mathscr{A}$ the colimits of D and DP coincide when $\mathscr{C}$ and $\mathscr{D}$ are λ-filtered.

Theorem 11.3.7 [21, Thm. IV.2]. *Let λ be an uncountable cardinal. For every λ-accessible endofunctor F on a locally λ-presentable category, a coalgebra is λ-presentable in* Coalg F *iff its underlying object is λ-presentable.*

Proof (1) By Proposition 11.3.4, we only need to prove that every λ-presentable coalgebra has a λ-presentable carrier. Let us denote by $\mathsf{Coalg}_\lambda F$ the full subcategory of all coalgebras (A, α) with A in $\mathscr{A}_\lambda$. In item (2) below we prove that every F-coalgebra is a λ-filtered colimit of a diagram of coalgebras from $\mathsf{Coalg}_\lambda F$. Thus if (B, β) is λ-presentable in $\mathsf{Coalg}\, F$, we can express it as a λ-filtered colimit $c_i\colon (C_i, \gamma_i) \to (B, \beta)$, $i \in I$, of coalgebras with C_i in $\mathscr{A}_\lambda$. It follows that $\mathsf{id}_{(B,\beta)}$ factorizes through some colimit injection, which means that we have a homomorphism $h\colon (B, \beta) \to (C_i, \gamma_i)$ for some $i \in I$ with $\mathsf{id}_{(B,\beta)} = c_i \cdot h$. Thus c_i is a split epimorphism in $\mathscr{A}$. Since C_i lies in $\mathscr{A}_\lambda$ we conclude that B is λ-presentable (Remark 11.2.12(5)).

(2) Let (B, β) be a coalgebra. We show in item (3) below that the following forgetful functor is final:

$$E\colon \mathsf{Coalg}_\lambda F/(B, \beta) \to \mathscr{A}_\lambda/B$$

assigning to every homomorphism $g\colon (A, \alpha) \to (B, \beta)$, where A is in $\mathscr{A}_\lambda$, the morphism $g\colon A \to B$. We then obtain that (B, β) is the colimit of the canonical functor $P\colon \mathsf{Coalg}_\lambda F/(B, \beta) \to \mathsf{Coalg}\, F$ mapping g to its domain (A, α). Indeed, for the forgetful functor $U\colon \mathsf{Coalg}\, F \to \mathscr{A}$ and the canonical diagram $D_B\colon \mathscr{A}_\lambda/B \to \mathscr{A}$ (see Remark 11.2.12(3)), we have a commutative square

$$\begin{array}{ccc} \mathsf{Coalg}_\lambda F/(B,\beta) & \xrightarrow{\ P\ } & \mathsf{Coalg}\, F \\ {\scriptstyle E}\downarrow & & \downarrow{\scriptstyle U} \\ \mathscr{A}_\lambda/B & \xrightarrow{\ D_B\ } & \mathscr{A} \end{array}$$

Since U creates colimits (Proposition 4.1.1), E is final, and $B = \mathsf{colim}\, D_B$, we conclude that $(B, \beta) = \mathsf{colim}\, P$.

(3) We now verify the two conditions mentioned in Remark 10.3.9(3) for E (and recall Remark 11.3.6). Condition (a) states that for every A in $\mathscr{A}_\lambda$ and every morphism $g\colon A \to B$ there exists a coalgebra homomorphism $g^+\colon (A^+, \alpha^+) \to (B, \beta)$ with A^+ in $\mathscr{A}_\lambda$, through which g factorizes. This holds by Remark 11.2.19.

Condition (b) concerns the situation of two factorizations of a given morphism $h_0\colon C_0 \to B$ in $\mathscr{A}_\lambda/B$ through a homomorphism $h\colon (C, \gamma) \to (B, \beta)$ in $\mathsf{Coalg}_\lambda F/(B, \beta)$: we are given $u_i\colon A \to C$ with $h_0 = h \cdot u_i$ for $i = 1, 2$. The claim is that there exists a homomorphism $g^+\colon (A^+, \alpha^+) \to (B, \beta)$ and a morphism v from h to g^+ in $\mathsf{Coalg}_\lambda F/(B, \beta)$ such that $v \cdot u_1 = v \cdot u_2$.

Since $\mathscr{A}_\lambda/B$ is filtered, we have a morphism $g\colon A \to B$ and a morphism w from (C, h) to (A, g) in $\mathscr{A}_\lambda/B$ such that $w \cdot u_1 = w \cdot u_2$. By condition (a) we can find a homomorphism $g^+\colon (A^+, \gamma^+) \to (B, \beta)$ and a morphism $c_0\colon A \to A^+$

with $g^+ \cdot c_0 = g$. By Remark 11.2.19 these data can be chosen such that $c_0 \cdot w$ is a homomorphism from (C, γ) to (A^+, α^+). Hence, $c_0 \cdot g$ is a morphism from h to g^+ in $\mathsf{Coalg}_\lambda\, F/(B, \beta)$:

$$g^+ \cdot (c_0 \cdot w) = g \cdot w = h.$$

Since w merges u_1 and u_2, so does $c_0 \cdot w$, and we are done. □

Corollary 11.3.8 [21, Thm. IV.2]. *Let λ be an uncountable cardinal. For every λ-accessible endofunctor F on a locally λ-presentable category,* $\mathsf{Coalg}\, F$ *is also locally λ-presentable.*

Proof Since $\mathscr{A}$ is cocomplete, so is $\mathsf{Coalg}\, F$ by Proposition 4.1.1. From the proof of Theorem 11.3.7 recall the full subcategory $\mathsf{Coalg}_\lambda\, F$ of all coalgebras (A, α) with A in $\mathscr{A}_\lambda$. This is clearly a small category, and by Proposition 11.3.4 it consists of λ-presentable coalgebras. Moreover, we have shown in item (2) of that proof that every coalgebra is a λ-filtered colimit of a diagram of coalgebras from $\mathsf{Coalg}_\lambda\, F$. □

Example 11.3.9. (1) Let F be a finitary set functor. If $\lambda > \omega$, then a coalgebra is λ-presentable in $\mathsf{Coalg}\, F$ iff its cardinality is less than λ. Indeed, this follows from Theorem 11.3.7 since F is λ-accessible. For $\lambda = \omega$, we know from Proposition 11.3.4 that every finite coalgebra is finitely presentable in $\mathsf{Coalg}\, F$ but the converse may fail as we now demonstrate.

(2) Consider a polynomial set functor H_Σ and recall that its terminal coalgebra is formed by all Σ-trees (Theorem 3.3.10). Every Σ-tree t yields a coalgebra where the carrier set A_t is the set of all nodes of t, and the coalgebra structure $\alpha_t : A_t \to H_\Sigma A_t$ is given by the structure of t: for every node x of t labelled by σ of arity n and having the child nodes $x_1, \ldots, x_n$ we let $\alpha_t(x) = \sigma(x_1, \ldots x_n) \in H_\Sigma A_t$. It is known that these *tree coalgebras* (A_t, α_t) are finitely presentable in $\mathsf{Coalg}\, F_\Sigma$ [21, Prop. II.9]. However, they are not finite in general if Σ has any operation symbols of non-zero arity or infinitely many constants. For example, for a signature Σ with a unary operation u symbol, consider the infinite Σ-tree, namely an infinite path of nodes labelled by u.

Surprisingly, for the finite power-set we have the following result:

Theorem 11.3.10 [21, Thm. III.3]. *A coalgebra for the finite power-set functor is finitely presentable iff it is finite.*

Proof By Proposition 11.3.4 we know that every finite $\mathscr{P}_\mathsf{f}$-coalgebra is finitely presentable. We now prove the converse.

(1) We first show that a graph G with an infinite path (x_n) with no repetitions

is not a finitely presentable $\mathscr{P}_\mathsf{f}$-coalgebra. Let G' be the graph obtained from G by adding new nodes y_n and new edges for every $n \in \mathbb{N}$ as follows:

(a) $x_n \to y_{n+1}$, and

(b) $y_n \to z$ for every edge $x_n \to z$ in G.

For example, for G being the infinite path $x_0 \to x_1 \to x_2 \to \cdots$ we obtain the following graph G':

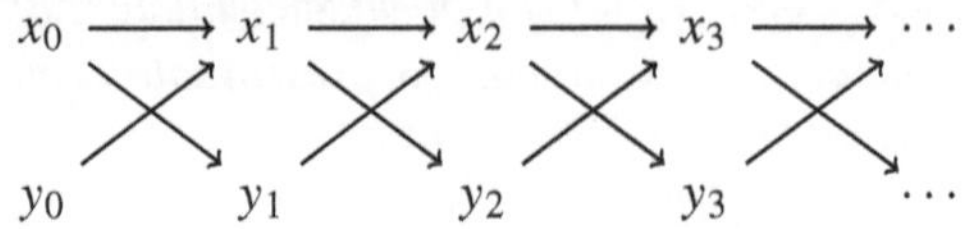

Now consider the ω-chain of quotient coalgebras $q_n\colon G' \twoheadrightarrow G'_n$, where q_n merges all x_k and y_k with $k < n$. Then all q_n and the obvious quotient maps $G'_n \twoheadrightarrow G'_{n+1}$ are coalgebra homomorphisms. Moreover, the colimit of the latter ω-chain is G. However, id_G does not factorize through any of the colimit injections $c_n\colon G'_n \to G$: for if we have a coalgebra homomorphism $h\colon G \to G'_n$ for some n, then the edge $h(x_n) = x_n \to x_{n+1}$ in G' implies that $h(x_{n+1}) = x_{n+1}$ and the edge $h(x_n) = x_n \to y_{n+1}$ implies $h(x_{n+1}) = y_{n+1}$, a contradiction. Thus, G is not finitely presentable.

(2) Suppose now that G is a graph which is a finitely presentable $\mathscr{P}_\mathsf{f}$-coalgebra. Every finite set $S \subseteq G$ of nodes generates the subcoalgebra $\bar{S}$ of G given by all nodes reachable from nodes in S by a path. This yields the filtered diagram $\mathcal{D}$ of all subcoalgebras $\bar{S} \hookrightarrow G$ where $S \subseteq G$ ranges over all finite subsets of G. Moreover G is clearly the union of all $\bar{S}$. Thus G is the colimit of $\mathcal{D}$ in $\mathsf{Coalg}\,\mathscr{P}_\mathsf{f}$. Since G is finitely presentable, the identity id_G factorizes through one of the inclusion maps $\bar{S} \hookrightarrow G$, which implies that $G \cong \bar{S}$ for some finite subset $S \subseteq G$. By item (1) we know that G does not have any infinite path without repetitions, and G is also finitely branching and has finitely many connected components. Hence, each of these components is a finitely branching connected graph. By Kőnig's Lemma [191] (cf. Proposition 9.2.19) we conclude that each component of G is finite. Therefore, so is G, as desired. □

Corollary 11.3.11. *The category of coalgebras for $\mathscr{P}_\mathsf{f}$ is not locally finitely presentable (while it is locally $\aleph_1$-presentable according to Corollary 11.3.8).*

Proof If $\mathsf{Coalg}\,\mathscr{P}_\mathsf{f}$ were locally finitely presentable, then the colimit of all finite(ly presentable) $\mathscr{P}_\mathsf{f}$-coalgebras would be the terminal coalgebra $\nu\mathscr{P}_\mathsf{f}$ (Remark 11.2.12(3)). Indeed, in every locally λ-presentable category $\mathscr{A}$, since $\mathscr{A}_\lambda$ is colimit dense, the colimit of the embedding $\mathscr{A}_\lambda \hookrightarrow \mathscr{A}$ is the terminal object. Now $\nu\mathscr{P}_\mathsf{f}$ is formed by all strongly extensional finitely branching trees (see

Theorem 4.5.15). However, the colimit of all finite coalgebras is the rational fixed point $\varrho\mathscr{P}_f$ formed by all *rational* such trees (see Example 10.2.21(3)). □

Example 11.3.12. (1) There are non-finitary set functors whose category of coalgebras is locally finitely presentable. For instance, consider the functor $FX = X^{\mathbb{N}}$. Using (un)currying, we see that its coalgebras $A \to A^{\mathbb{N}}$ are, equivalently, algebras $\mathbb{N} \times A \to A$. Moreover, $\mathsf{Coalg}\, F$ is (isomorphic to) the category of algebras for the finitary set functor $FX = \mathbb{N} \times X$, which is locally finitely presentable by Theorem 11.3.2 (applied to $\lambda = \omega$).

(2) For every, not necessarily finitary, signature Σ, the category $\mathsf{Coalg}\, H_\Sigma$ is locally finitely presentable [21, Cor. II.7].

Remark 11.3.13. (1) Generalizing from $FX = X^{\mathbb{N}}$, for every right adjoint endofunctor F on a locally λ-presentable category, the category $\mathsf{Coalg}\, F$ is locally λ-presentable, too. Indeed, coalgebras $A \to FA$ are in bijective correspondence with algebras $LA \to A$, where L is the left adjoint of F. Since L preserves colimits, it is finitary, and so $\mathsf{Alg}\, L \cong \mathsf{Coalg}\, F$ is locally finitely presentable by Theorem 11.3.2 (see also Bird's thesis [81, Thm. 2.14]).

(2) Every finitary functor F on a locally finitely presentable category preserving limits is a right adjoint by the Adjoint Functor Theorem for locally presentable categories [23, Thm. 1.66]. Therefore $\mathsf{Coalg}\, F$ is locally finitely presentable.

In Chapter 10 we have studied, for a finitary endofunctor F, the rational fixed point, which is the colimit of all coalgebras carried by finitely presentable objects. In contrast, the analogous fixed point for λ-accessible functors with λ uncountable is just νF:

Corollary 11.3.14 [33, Prop. 5.16]**.** *For every λ-accessible endofunctor on a locally λ-presentable category, where λ is uncountable, the terminal coalgebra is the colimit of all coalgebras carried by λ-presentable objects.*

Proof Since we know that $\mathsf{Coalg}\, F$ is locally λ-presentable and $\mathsf{Coalg}_\lambda\, F$ represents all its locally λ-presentable objects (up to isomorphism), we conclude that the colimit of the embedding $\mathsf{Coalg}_\lambda\, F \hookrightarrow \mathsf{Coalg}\, F$ is the terminal coalgebra (see Remark 11.2.12(3)). □

11.4 Preaccessible Functors

We know that every λ-accessible endofunctor has an initial algebra obtained in λ steps of the initial-algebra chain. We now introduce *λ-preaccessible* functors as those preserving colimits of λ-chains of strong subobjects. We prove that every λ-preaccessible endofunctor has an initial algebra, provided the base category is cocomplete and co-well-powered. The initial-algebra chain is proved to converge, but not necessarily in λ steps. This result stems from a fundamental article of Koubek and Reiterman [200] in which a number of other properties are also proved about the category of algebras for a preaccessible endofunctor F, for example, that $\mathsf{Alg}\, F$ is cocomplete, and every partial F-algebra has a free cocompletion. Our presentation is influenced by the work of Kelly [187] who used and extended Koubek and Reiterman's ideas. Following this, we prove that every λ-preaccessible endofunctor has a terminal coalgebra obtained by iterating the terminal-coalgebra chain. For this, we need to assume that the base category is locally presentable and that the functor preserves monomorphisms.

Assumption 11.4.1. Throughout this section $\mathscr{A}$ denotes a cocomplete and co-well-powered category.

Analogously to Remark 7.2.1(3), $\mathscr{A}$ has factorizations of morphisms as epimorphisms followed by *strong monomorphisms* [90, Prop. 4.4.3].

Definition 11.4.2. A monomorphism $m\colon C \rightarrowtail D$ is *strong* if for every epimorphism $e\colon A \to B$ and every commutative square

$$\begin{array}{ccc} A & \xrightarrow{e} & B \\ {\scriptstyle f}\downarrow & \overset{d}{\swarrow} & \downarrow{\scriptstyle g} \\ C & \xrightarrow[m]{} & D \end{array}$$

there is a unique diagonal morphism d as above making the triangles commute. Thus, this concept is dual to that of a strong epimorphism which we have seen in Chapter 7.

Unlike Chapter 7, in this section we denote by $\twoheadrightarrow$ epimorphisms and by $\rightarrowtail$ strong monomorphisms.

Example 11.4.3. (1) In Set and K-Vec, every monomorphism is strong.

(2) In the category Pos of posets, a monomorphism $m\colon P \to Q$ is strong iff it is order-reflecting: $m(p) \le m(q)$ implies $p \le q$. Thus, m represents the subobject given by the subset $m[P] \subseteq Q$ with the order induced by that of Q. The same holds for the categories of cpos, groups, or unary algebras.

(3) Analogously, in the category Met be the category of metric spaces, strong subobjects of a metric space Q represent precisely the *closed* metric subspaces $P \subseteq Q$: for every sequence in P with a limit $x \in Q$ we have $x \in P$.

(4) Let Top be the category of topological spaces and continuous functions. A strong monomorphism $m\colon C \to D$ represents the same subobject as the image $m[C]$ with the induced topology.

Remark 11.4.4. (1) Strong monomorphisms are closed under composition.

(2) If a composite $g \cdot f$ is a strong monomorphism, so is f.

Definition 11.4.5. For an infinite regular cardinal λ, a functor F is *λ-preaccessible* if it preserves colimits of λ-chains of strong monomorphisms. The functor F is *preaccessible* if F is λ-preaccessible for some λ.

Observe that preaccessible functors are not required to preserve strong monomorphisms.

Example 11.4.6. (1) Since λ-chains are λ-directed diagrams, every λ-accessible functor is λ-preaccessible.

(2) In Proposition C.8.12 we will present an ω-preaccessible set functor which is not accessible. This depends on a set-theoretical assumption; we do not know of an example which is set-theory independent.

(3) The only λ-presentable topological spaces are the discrete ones [134, Sec. 6.2]. Thus, for a non-discrete space A we have that $\mathsf{Top}(A, -)\colon \mathsf{Top} \to \mathsf{Set}$ is not accessible. But this functor $\mathsf{Top}(A, -)$ is λ-preaccessible whenever $|A| < \lambda$. Indeed, as mentioned in Example 11.4.3(4), strong monomorphisms represent subspaces with the induced topology, and colimits of chains of them are formed by taking unions. It is easy to see that $\mathsf{Top}(A, -)$ preserves these colimits.

Remark 11.4.7. Koubek and Reiterman [200] work with a general factorization system $(\mathcal{E}, \mathcal{M})$ and with functors preserving colimits of λ-chains of morphisms in $\mathcal{M}$. This does not gain much generality:

(1) If $\mathcal{E}$ consists of epimorphisms, then every morphism in $\mathcal{M}$ is a strong monomorphism [38, Prop. 14.10]. Thus, λ-preaccessibility is weaker than the condition above.

(2) If $\mathcal{M}$ consists of monomorphisms, then every morphism in $\mathcal{M}$ is a strong monomorphisms again: since our category is complete and co-well-powered, $\mathcal{E}$ then consists of epimorphisms [8, Prop. IV.1].

We now turn to the proof that preaccessible functors have initial algebras. First, the following technical fact is useful.

Lemma 11.4.8. *Every monotone function from the linearly ordered class of ordinals to a poset is eventually constant.*

Proof Given a poset P and a monotone function $h\colon \mathsf{Ord} \to P$, there must be some $p \in P$ such that $h^{-1}(p)$ is unbounded: if not, then $\mathsf{Ord} = \bigcup_{p \in P} h^{-1}(p)$ would be a bounded set of ordinals – a contradiction. Suppose that both $h^{-1}(p)$ and $h^{-1}(q)$ were unbounded. Let $i \le j \le k$ be such that $h(i) = p$, $h(j) = q$, and $h(k) = p$. Then $h(i) \le h(j) \le h(k)$. So $p \le q \le p$. Thus, $p = q$. It follows that there is a unique p such that $h^{-1}(p)$ is unbounded. Then $\bigcup_{q \neq p} h^{-1}(q)$ is bounded, say by the ordinal i. If $j > i$, then $h(j) = p$. □

Remark 11.4.9. (1) Recall from Remark 4.2.2 that quotients of an object A are represented by epimorphisms $e\colon A \twoheadrightarrow B$ and that we speak about the quotient e in lieu of the quotient object represented by e. Since $\mathscr{A}$ is co-well-powered, we have a poset $\mathsf{Quot}(A)$ of quotients of A, ordered as follows: $e \leq f$ iff e factorizes through f.

(2) Consider a collection (e_j) ($j \in \mathsf{Ord}$, $j \geq j_0$) of quotients of an object A with the property that for $j_0 \leq j \leq k$, e_k factorizes through e_j. Such a collection defines a monotone function f from the ordinals to $\mathsf{Quot}(A)^{\mathsf{op}}$ which is given by $f(i) = e_k$ for the ordinal $k = j_0 + i$. By Lemma 11.4.8, there is some ordinal $j_1 \leq j_0$ so that all e_k with $k \geq j_1$ represent the same quotient of A.

Remark 11.4.10. Let $(A_i)_{i<\lambda}$ and $(B_i)_{i<\lambda}$ be two chains in $\mathscr{A}$ for some limit ordinal λ, and suppose that we have natural transformations $f_i\colon A_i \to B_i$ and $g_i\colon B_i \to A_{i+1}$. Then the cocones of these chains are in a bijective correspondence as follows: Every cocone $b_i\colon B_i \to C$ of (B_i) yields a cocone $b_i \cdot f_i\colon A_i \to C$ of (A_i). Every cocone $a_i\colon A_i \to C$ of (A_i) yields a cocone $a_{i+1} \cdot g_i\colon B_i \to C$ of (B_i). Consequently, $\mathsf{colim}_{i<\lambda} A_i = \mathsf{colim}_{i<\lambda} B_i$ with the colimit cocones related as above.

Theorem 11.4.11. *For every preaccessible endofunctor on a cocomplete and co-well-powered category, the initial-algebra chain converges.*

Proof Recall from Definition 6.1.4 our notation W_i for the objects in the initial-algebra chain for an endofunctor F and $w_{i,j}\colon W_i \to W_j$ for the connecting morphisms. We start with statements that do not use the assumption that the given functor F is preaccessible.

(1) We shall define a function f on the class of ordinals by recursion. Our definition ensures that f is monotone and continuous, that is, $f(\ell) = \bigvee_{i<\ell} f(i)$ for limit ordinals ℓ. It also takes at every step i the following (epi, strong mono)-factorizations of the morphisms

$$w_{f(i),j} = \big(W_{f(i)} \xrightarrow{e_i^j} E_i^j \xrightarrow{m_i^j} W_j\big) \qquad \text{for } j \geq f(i). \tag{11.12}$$

Moreover f is defined so that e_i^j is the same quotient of $W_{f(i)}$ for every $j \geq f(i+1)$.

To simplify our notation we are going to write Z_i for $W_{f(i)}$ and shorten the connecting maps, too:

$$z_{i,j} = \big(W_{f(i)} \xrightarrow{w_{f(i),f(j)}} W_{f(j)}\big) \qquad \text{for } j \geq i.$$

We set $f(0) = 0$, and for limit ordinals ℓ, we define $f(\ell) = \sup_{i<\ell} f(i)$. Thus, we need only define $f(i+1)$ from $f(i)$.

For $k \geq j \geq f(i)$, we show for the factorizations defined in (11.12) that e_i^k factorizes through e_i^j. To see this, consider the square below:

$$\begin{array}{ccc} W_{f(i)} = Z_i & \xrightarrow{\;e_i^j\;} & E_i^j \\ {\scriptstyle e_i^k}\downarrow & \overset{d}{\swarrow} & \downarrow{\scriptstyle w_{j,k}\cdot m_i^j} \\ E_i^k & \xrightarrow{\;m_i^k\;} & W_k \end{array} \tag{11.13}$$

It commutes because $m_i^k \cdot e_i^k = w_{f(i),k} = w_{j,k} \cdot w_{f(i),j}$, whence the morphism d exists by diagonal fill-in (cf. Remark 7.2.1(1)).

This gives a collection $(e_i^j : Z_i \to E_i^j)$ of quotients of Z_i for $j \geq f(i)$. By Remark 11.4.9(2), applied to $j_0 = f(i)$ we can choose an ordinal $f(i+1)$ such that all epimorphisms e_i^j with $j \geq f(i+1)$ represent the same quotient of Z_i. We may also take $f(i+1) > f(i)$. This concludes the transfinite recursion defining f and the factorizations (11.12).

(2) To further simplify our notation, put

$$D_i = E_i^{f(i+1)}, \quad e_i = \big(Z_i \xrightarrow{\;e_i^{f(i+1)}\;} D_i\big), \quad \text{and} \quad m_i = \big(D_i \xrightarrow{\;m_i^{f(i+1)}\;} Z_{i+1}\big).$$

We next verify that the composites of m_i with the connecting morphisms $W_{f(i+1)} \to W_j$ are strong monomorphisms. Let $j \geq f(i+1)$, and consider the following diagonal fill-in:

$$\begin{array}{ccc} Z_i & \xrightarrow{\;e_i\;} & D_i \\ & & \downarrow{\scriptstyle m_i} \\ {\scriptstyle e_i^j}\downarrow & \swarrow & Z_{i+1} = W_{f(i+1)} \\ & & \downarrow{\scriptstyle w_{f(i+1),j}} \\ E_i^j & \xrightarrow{\;m_i^j\;} & W_j \end{array} \tag{11.14}$$

In this case, d in (11.13) is an isomorphism since the quotients $e_i = e_i^{f(i+1)}$ and e_i^j are the same. Hence $w_{f(i+1),j} \cdot m_i$ is a strong monomorphism, since m_i^j is one.

(3) We now show that the objects D_i form a λ-chain $(D_i)_{i<\lambda}$ whose connecting morphisms $d_{i,j}$ are strong monomorphisms. They are defined as the following

diagonal fill-ins for $i \le j < \lambda$:

$$\begin{array}{ccc} Z_i & \xrightarrow{e_i} & D_i \\ {\scriptstyle z_{i,j}}\downarrow & & \downarrow{\scriptstyle m_i} \\ Z_j & {\scriptstyle d_{i,j}} \swarrow & Z_{i+1} \\ {\scriptstyle e_j}\downarrow & & \downarrow{\scriptstyle z_{i+1,j+1}} \\ D_j & \xrightarrow{m_j} & Z_{j+1} \end{array} \tag{11.15}$$

The right-hand composite is strongly monic as we have seen at the end of item (2) (taking $f(j+1)$ in lieu of j). So $m_j \cdot d_{i,j}$ is strongly monic, too. It follows from Remark 11.4.4 that $d_{i,j}$ is strongly monic.

(4) Fix λ so that F is λ-preaccessible. We set $\mu = f(\lambda)$. We are going to prove our theorem by showing that F preserves the colimit of the μ-chain $(W_i)_{i<\mu}$. The proof involves considering the chains $(Z_j)_{j<\lambda}$ and $(D_j)_{j<\lambda}$.

We claim that for every cocone $a_j \colon Z_j \to A$ $(j < \lambda)$ of the chain $Z_j = W_{f(j)}$ there is a unique cocone $\overline{a}_i \colon W_i \to A$ $(i < \mu)$ such that (a_j) is obtained from $(\overline{a}_i)$ by *thinning* via f: $a_j = \overline{a}_{f(j)}$. Indeed, given $i < \mu$, choose j with $i \le f(j)$, and put $\overline{a}_i = a_j \cdot w_{i,f(j)} \colon W_i \to A$. We show that this is independent of the choice of j. Indeed, for $k < \lambda$ such that $i \le j \le k$, we have $a_j = a_k \cdot z_{j,k} = a_k \cdot w_{f(j),f(k)}$. Therefore

$$a_k \cdot w_{i,f(k)} = a_k \cdot w_{f(j),f(k)} \cdot w_{i,f(j)} = a_j \cdot w_{i,f(j)}$$

due to the compatibility of the cocone a_j. It is easy to see, using this independence, that $(\overline{a}_i)$ is a cocone obtained by thinning (a_j). The uniqueness is also clear.

Thus, the passage from $(a_j)_{j<\lambda}$ to $(\bar{a}_i)_{i<\mu}$ is a bijection between the cocones of $(Z_j)_{j<\lambda}$ and $(W_i)_{i<\mu}$; the inverse takes a cocone $b_i \colon W_i \to A$ $(i < \mu)$ to $a_j := b_{f(j)}$ $(j < \lambda)$. In addition, the colimits of the two chains are the same (with colimit cocones related by the above bijection). The same argument shows that the two chains $(FZ_j)_{j<\lambda}$ and $(FW_i)_{i<\mu}$ have the same colimit. Hence, it suffices to prove that F preserves the colimit of $(Z_j)_{j<\lambda}$.

(5) Compare the λ-chain $(D_j)_{j<\lambda}$ of strong monomorphisms with the λ-chain $(Z_j)_{j<\lambda}$. Diagram (11.15) yields natural transformations

$$e_j \colon Z_j \to D_j \qquad \text{and} \qquad m_j \colon D_j \to Z_{j+1}.$$

By Remark 11.4.10, the two chains have the same colimit. Applying that remark to the natural transformations $(Fe_j)_{j<\lambda}$ and $(Fm_j)_{j<\lambda}$, we see that the chains $(FD_j)_{j<\lambda}$ and $(FZ_j)_{j<\lambda}$ also have the same colimit. Since F is λ-preaccessible, it preserves the colimit of the former chain, whence it also preserves the colimit of the latter one. Thus, the initial-algebra chain of F converges in μ steps. □

Using this result and Corollary 6.1.13, we also have:

Corollary 11.4.12. *Every preaccessible functor is a varietor (Definition 2.2.26).*

We now turn to terminal coalgebras. Here we assume that our base category is locally presentable (Definition 11.2.8), which is stronger than merely being cocomplete and co-well-powered (by [23, Thm. 1.58]). We have already proved that every accessible endofunctor on a locally presentable category has a terminal coalgebra (Theorem 11.2.18). That proof does not guarantee that the terminal-coalgebra chain converges. However, as we have mentioned in Theorem 11.2.22, that chain converges for accessible endofunctors preserving monomorphisms. We now generalize this result to preaccessible functors preserving monomorphisms.

We first recall some further facts about locally presentable categories that we will use.

Remark 11.4.13. Let $\mathscr{A}$ be a locally λ-presentable category. From Notation 11.2.9 we recall that $\mathscr{A}_\lambda$ denotes a (small) full subcategory representing all λ-presentable objects up to isomorphism.

(1) Every object is λ'-presentable for some regular infinite cardinal λ' [23, Prop. 1.22].

(2) The category $\mathscr{A}$ is locally λ'-presentable for every regular infinite cardinal $\lambda' \geq \lambda$ (Remark 11.2.11).

(3) There exists a small category $\mathscr{C}$ and a full subcategory of $\mathsf{Set}^{\mathscr{C}}$ (the category of funcctors from $\mathscr{C}$ to Set) which is equivalent to $\mathscr{A}$ and closed under limits and λ-filtered colimits in $\mathsf{Set}^{\mathscr{C}}$ [23, Thm. 1.46]. Note that monomorphisms in $\mathsf{Set}^{\mathscr{C}}$ are the natural transformations all of whose components are injective functions and that limits in $\mathsf{Set}^{\mathscr{C}}$ are formed objectwise.

(4) Every λ-presentable object A of $\mathsf{Set}^{\mathscr{C}}$, for $\mathscr{C}$ small, is a colimit of less than λ hom-functors [134, Par. 6.3]. If $\mathscr{C}$ has less than λ morphisms, it follows from the Yoneda Lemma that each set $A(C)$, for C in $\mathscr{C}$, has cardinality less than λ.

(5) In Definition 11.2.8, λ-filtered colimits can, equivalently, be replaced by λ-directed ones [23, Rem. 1.21].

Lemma 11.4.14. *Every locally λ-presentable category $\mathscr{A}$ has a set $\mathscr{A}_0$ of objects such that every object of $\mathscr{A}$ is a λ-directed colimit of a diagram formed by strong monomorphisms between object of $\mathscr{A}_0$. Moreover, the colimit injections are strong monomorphisms.*

Proof By Remark 11.4.13(5), every object X of $\mathscr{A}$ is the colimit of some λ-

directed diagram D of objects D_i in $\mathscr{A}_\lambda$ ($i \in I$) and morphisms $d_{i,j}\colon D_i \to D_j$ for $i \le j$ in I. We denote the colimit injections by $c_i\colon D_i \to X$. Since $\mathscr{A}$ is co-well-powered, the collection of all quotients of all objects of $\mathscr{A}_\lambda$ forms a set, which we denote by $\mathscr{A}_0$. We factorize every c_i as an epimorphism $e_i\colon D_i \twoheadrightarrow E_i$ with $E_i \in \mathscr{A}_0$ followed by a strong monomorphism $m_i\colon E_i \to X$. This yields a λ-directed diagram E of objects E_i ($i \in I$) and connecting morphisms $e_{i,j}\colon E_i \to E_j$ (for $i \le j$ in I) obtained via diagonal fill-in as shown below:

$$\begin{array}{ccc} D_i & \xrightarrow{\;e_i\;} & E_j \\ {\scriptstyle e_j\cdot d_{i,j}}\downarrow & \overset{e_{i,j}}{\swarrow} & \downarrow{\scriptstyle m_i} \\ E_j & \xrightarrow{\;m_j\;} & X \end{array}$$

Each e_{ij} is a strong monomorphism because $m_j \cdot e_{i,j}$ is (Remark 11.4.4(2)). It is easy to verify that X is a colimit of E with colimit injections $m_i\colon E_i \rightarrowtail X$. □

Theorem 11.4.15. *Suppose that $\mathscr{A}$ be locally presentable and $F\colon \mathscr{A} \to \mathscr{A}$ is preaccessible and preserves monomorphisms. Then its terminal-coalgebra chain converges.*

Proof (1) Let $\mathscr{A}$ be locally λ-presentable and $\mathscr{A}_0$ be a set as in Lemma 11.4.14. Then there exists a regular cardinal $\mu \ge \lambda$ such that the objects of $\mathscr{A}_0$ are μ-presentable (Remark 11.4.13(1)). Further, we may assume that F is μ-preaccessible (Remark 11.4.13(2)).

Finally, using Remark 11.4.13(3) we can find a small subcategory $\mathscr{C}$ such that $\mathscr{A}$ is (equivalent to) a full subcategory of $\mathsf{Set}^{\mathscr{C}}$ which closed under limits and λ-filtered colimits. The cardinal μ can be chosen arbitrarily large. Thus, we can assume that $\lambda \le \mu$ and that $\mathscr{C}$ has less than μ morphisms.

(2) In the terminal-coalgebra chain, express V_μ as a λ-filtered colimit of a diagram with objects D_t in $\mathscr{A}_0$ ($t \in T$) and strong monomorphisms between them with the colimit cocone $d_t\colon D_t \to V_\mu$ formed by strong monomorphisms (Lemma 11.4.14). We now prove that for every $t \in T$ there exists some $i < \mu$ such that $v_{\mu,i} \cdot d_t\colon D_t \to V_i$ is monic. In other words: for every object C of $\mathscr{C}$, the component $(v_{\mu,i} \cdot d_t)_C\colon D_t(C) \to V_i(C)$ is an injective function. To this end, fix an object C of $\mathscr{C}$ and consider $x \ne y$ in $D_t(C)$. Since d_t is monic, we have $(d_t)_C(x) \ne (d_t)_C(y)$ in V_μ. The limit cone $v_{\mu,i}\colon V_\mu \to V_i$ is collectively monic. It is a limit cone in $\mathscr{A}$ but also in $\mathsf{Set}^{\mathscr{C}}$, since $\mathscr{A}$ is closed under limits in $\mathsf{Set}^{\mathscr{C}}$. It follows that the functions $(v_{\mu,i})_C$ ($i < \mu$) are collectively monic in Set. Thus, there exists some $i < \mu$ such that $(v_{\mu,i} \cdot d_t)_C(x) \ne (v_{\mu,i} \cdot d_t)_C(y)$.

In the previous paragraph, i depends on C and on $x, y \in D_t(C)$. However, a larger i' would also work. It remains to verify that we can choose i independently of (C, x, y). Recall from Remark 11.4.13(4) that $\mathscr{C}$ has less than μ objects and that $D_t(C)$ has less than μ elements, since every object D_t in $\mathscr{A}_0$ is μ-presentable. Thus, there are less than $\mu^3 = \mu$ triples (C, x, y) as above. Consequently, we can choose $i < \mu$ such that $(v_{\mu,i} \cdot d_t)_C$ is injective for all C in $\mathscr{C}$, as claimed.

(3) We next prove that the connecting morphism $v_{\mu+1,\mu} \colon FV_\mu \to V_\mu$ in the terminal-coalgebra chain for F is monic. Note that $FV_\mu = \mathsf{colim}_{t \in T} FD_i$ since F is μ-preaccessible (due to $\mu \geq \lambda$). Fix an object C of $\mathscr{C}$ and suppose that $x_1 \neq x_2$ in $FV_\mu(C)$. We will prove that $(v_{\mu+1,\mu})_C$ does not merge x_1 and x_2. Here we use the fact that $\mathscr{A}$ is closed under λ-directed colimits in $\mathsf{Set}^{\mathscr{C}}$. Therefore, $FV_\mu = \mathsf{colim}_{t \in T} FD_t$ in $\mathscr{A}$ implies that $FV_\mu(C) = \mathsf{colim}_{t \in T} FD_t(C)$ in Set. Consequently, for $x_1, x_2 \in FV_\mu(C)$ there exists $t \in T$ and $y_1, y_2 \in FD_t(C)$ with $x_k = (Fd_t)_C(y_k)$, $k = 1, 2$. Moreover, $y_1 \neq y_2$ due to $x_1 \neq x_2$. Use item (2) to find $i < \mu$ with $v_{\mu,i} \cdot d_t$ monic. Then $F(v_{\mu,i} \cdot d_t)$ is also monic since F preserves monomorphisms. From $y_1 \neq y_2$ we obtain $F(v_{\mu,i})_C(x_1) \neq F(v_{\mu,i})_C(x_2)$. Finally, we have

$$F(v_{\mu,i})_C = v_{\mu+1,i+1} = v_{\mu,i+1} \cdot v_{\mu+1,\mu}.$$

Since $F(v_{\mu,i})_C$ does not merge x_1, x_2 neither does $(v_{\mu+1,\mu})_C$, as desired.

(4) By an easy transfinite induction we see that all connecting morphisms $v_{i,\mu} \colon V_i \to V_\mu$ $(i \geq \mu)$ are monic. Indeed, for $i = \mu + 1$ use item (2). For the isolated step, use

$$v_{i+1,\mu} = v_{\mu+1,\mu} \cdot v_{i+1,\mu+1} = v_{\mu+1,\mu} \cdot Fv_{i,\mu}$$

and also recall that $v_{i,\mu}$ is monic by induction hypothesis and that F preserves monomorphisms. The limit step is trivial: a limit cone of a chain of monomorphisms is formed by monomorphisms.

(5) Since $\mathscr{A}$ is well-powered (Remark 11.2.12(1)), the collection $\mathsf{Sub}(V_\mu)$ of subobjects of V_μ is a poset. Apply Lemma 11.4.8 to the monotone function $i \mapsto v_{\mu+i,\mu}$ from Ord to $\mathsf{Sub}(V_\mu)$ to conclude that the terminal-coalgebra chain of F converges. □

Open Problem 11.4.16. Find conditions on a category $\mathscr{A}$ such that for every preaccessible endofunctor on $\mathscr{A}$ the terminal coalgebra exists (and the terminal-coalgebra chain converges).

11.5 Adjunctions

The main idea in using adjunctions to obtain initial algebra is quite simple. If $G \colon \mathscr{A} \to \mathcal{B}$ is left adjoint, then since G preserves colimits, it preserves initial objects. So if one has an adjunction between the categories of algebras for two functors, then the left adjoint takes an initial algebra to an initial algebra. What is needed is a sufficient condition for the adjunction between the categories of algebras. Here is one due to Hermida and Jacobs [162]. (Hermida and Jacobs credit part of the result to unpublished work of Backhouse and Bijsterveld.) Recall the notion of an adjunction from Remark 2.2.28(2).

Remark 11.5.1. Recall (e.g. from MacLane [217]) that a functor $L\colon \mathscr{A} \to \mathscr{B}$ is a *left adjoint* to a functor $R\colon \mathscr{B} \to \mathscr{A}$ if there are natural transformations $\eta\colon \mathsf{id}_{\mathscr{B}} \to RL$ and $\varepsilon\colon LR \to \mathsf{Id}_{\mathscr{A}}$ such that the following *triangular identities* hold:

$$\varepsilon L \cdot L\eta = \mathsf{id}_L \qquad \text{and} \qquad R\varepsilon \cdot \eta R = \mathsf{id}_R.$$

We say that R is the *right adjoint* to L, and we write $L \dashv R$.

Theorem 11.5.2 [162, Thm. 2.14]**.** *Suppose that we are given functors F and G and an adjunction L ⊣ R as shown below:*

$$G \circlearrowright \mathscr{B} \underset{R}{\overset{L}{\underset{\longleftarrow}{\overset{\longrightarrow}{\bot}}}} \mathscr{A} \circlearrowleft F \tag{11.16}$$

For every natural isomorphism $\varrho\colon FL \to LG$, *let* $\widehat{L}\colon \mathsf{Alg}\,G \to \mathsf{Alg}\,F$ *be defined by*

$$(GB \xrightarrow{\beta} B) \mapsto (FLB \xrightarrow{\varrho_B} LGB \xrightarrow{L\beta} LB)$$

on objects, and by $f \mapsto Lf$ *on morphisms. Then* $\widehat{L}$ *is a left adjoint functor.*

Proof We omit the straightforward verification that $\widehat{L}$ is a functor. We form the following adjoint transpose σ:

$$\frac{LGR \xrightarrow{\varrho^{-1}R} FLR \xrightarrow{F\varepsilon} F}{GR \xrightarrow{\sigma} RF}$$

Define a functor $\widehat{R}$ on objects by

$$\widehat{R}\colon \mathsf{Alg}\,F \to \mathsf{Alg}\,G \quad \text{by} \quad (FA \xrightarrow{\alpha} A) \mapsto (GRA \xrightarrow{\sigma_A} RFA \xrightarrow{R\alpha} RA)$$

and on morphisms by $g \mapsto Rg$. The verification that $\widehat{R}$ is a functor is again easy. We verify that we have an adjunction:

$$\mathsf{Alg}\,F \underset{\widehat{R}}{\overset{\widehat{L}}{\underset{\longleftarrow}{\overset{\longrightarrow}{\bot}}}} \mathsf{Alg}\,G$$

We define $\widehat{\eta}\colon \mathsf{id}_{\mathsf{Alg}\,F} \to \widehat{R}\widehat{L}$ and $\widehat{\varepsilon}\colon \widehat{L}\widehat{R} \to \mathsf{Id}_{\mathsf{Alg}\,G}$ as follows: for a G-algebra (B,β), $\widehat{\eta}_{(B,\beta)} = \eta_B$; and for an F-algebra (A,α), $\widehat{\varepsilon}_{(A,\alpha)} = \varepsilon_A$.

The triangular identities (Remark 2.2.28(2)) for the original adjunction imply those for the new one. Let us verify that $\widehat{\eta}_{(B,\beta)}$ is a homomorphism for every algebra $\beta\colon GB \to B$. Denote by $\gamma\colon GRLB \to RLB$ the structure of the algebra $\widehat{R}\widehat{L}(B,\beta)$, shown on the bottom below:

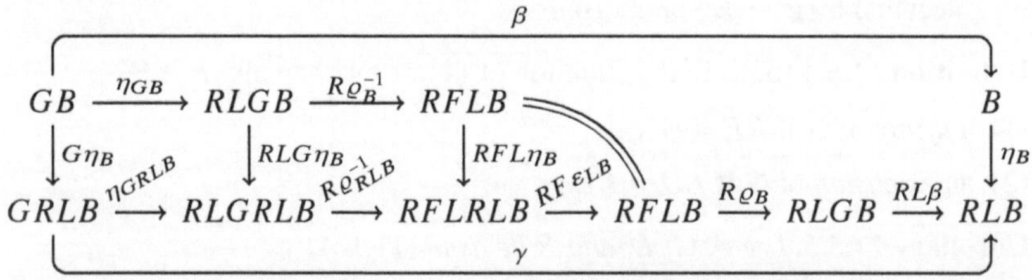

The triangle commutes since $\varepsilon L \cdot L\eta = \mathsf{id}_L$, and the remaining inner parts commute by naturality (for the upper part also use that $\varrho \cdot \varrho^{-1} = \mathsf{id}$). So the outside commutes, which shows that $\widehat{\eta}_{(B,\beta)}\colon (B,\beta) \to \widehat{R}\widehat{L}(B,\beta)$ is an algebra homomorphism.

The proof that each morphism $\widehat{\varepsilon}_{(A,\alpha)}$ is a homomorphism is similar. Denote $\delta\colon FLRA \to LRA$ the structure of the algebra $\widehat{L}\widehat{R}(A,\alpha)$, shown at the top below:

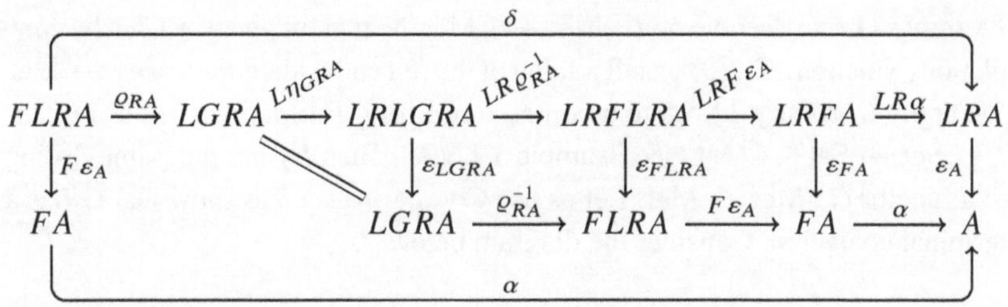

The triangle commutes since $\varepsilon L \cdot L\eta = \mathsf{id}_L$, and the other inner parts commute by naturality of ε (for the lower left-hand part also use that $\varrho^{-1} \cdot \varrho = \mathsf{id}$). So the outside commutes, as desired. □

Dually, we have the following result:

Corollary 11.5.3 [162, Cor. 2.15]. *Suppose that we are given functors F and G and an adjunction $L \dashv R$ as in* (11.16). *For every natural isomorphism $\varrho\colon RF \to GR$, let $\widehat{R}\colon$ $\mathsf{Coalg}\,F \to \mathsf{Coalg}\,G$ be defined by*

$$(A \xrightarrow{\alpha} FA) \mapsto (RA \xrightarrow{R\alpha} RFA \xrightarrow{\varrho_A} GRA)$$

on objects, and by $f \mapsto Rf$ on morphisms. Then $\widehat{R}$ is a right adjoint functor.

In fact, a left adjoint $\widehat{L}$ to $\widehat{R}$ is obtained by taking the adjoint transpose $\sigma\colon LG \to FL$ of $G \xrightarrow{G\eta} GRL \xrightarrow{\varrho^{-1}L} RFL$ and defining the functor

$$\widehat{L}\colon \mathsf{Coalg}\, G \to \mathsf{Coalg}\, F \quad \text{by} \quad (B \xrightarrow{\beta} GB) \mapsto (LB \xrightarrow{L\beta} LGB \xrightarrow{\sigma_B} FLB)$$

on object and by $g \mapsto Lg$ on morphisms.

Definition 11.5.4 [63]. In the situation of (11.16) we say that F is

(1) a *lifting* of G if $RF = GR$.

(2) an *extension* of G if $LG = FL$.

Corollary 11.5.5. *Let F, G, L, and R be as in* (11.16).

(1) *If F is an extension of G and $g\colon GX \to X$ is an initial G-algebra, then $Lg\colon FLX \to LX$ is an initial F-algebra: $\mu F = L(\mu G)$.*

(2) *If F is a lifting of G and $f\colon X \to FX$ is a terminal F-coalgebra, then $Rf\colon RX \to GRX$ is a terminal G-coalgebra: $\nu G = R(\nu F)$.*

Item (1) follows from Theorem 11.5.2 and the fact that left adjoints preserve (colimits and thus) initial objects. Item (2) uses a dual formulation of Theorem 11.5.2.

Example 11.5.6. Denote by $L\colon \mathsf{Met} \to \mathsf{CMet}$ the functor given by Cauchy completion (Notation 3.2.2(3)), a left adjoint of the full embedding $R\colon \mathsf{CMet} \hookrightarrow \mathsf{Met}$. We apply Corollary 11.5.5 to a contracting lifting F of a polynomial functor $H_\Sigma\colon \mathsf{Set} \to \mathsf{Set}$ to CMet (see Example 5.2.6(3)). Then H_Σ has the 'same' lifting to a functor $G\colon \mathsf{Met} \to \mathsf{Met}$. Let us use Corollary 11.5.5 to show that G has a terminal coalgebra. Consider the diagram below:

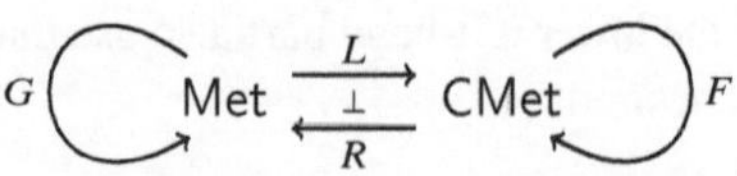

By Theorem 5.2.20, F has a terminal coalgebra. Then by Theorem 11.5.2 and the preservation of terminal objects under right adjoints, G has the terminal coalgebra carried by $R(\nu F)$. Moreover, whenever G has an initial algebra $(\mu G, \iota)$, its Cauchy completion $(L(\mu G), L\iota)$ is an initial F-algebra.

Remark 11.5.7. A common instance of Theorem 11.5.2 is where $L \dashv R$ is a dual adjunction:

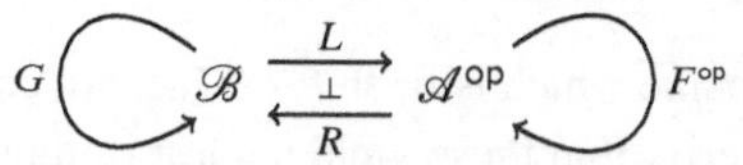

For a given natural isomorphism $\varrho\colon F^{\mathrm{op}}L \to LG$ one then obtains a left adjoint

$$\widehat{L}\colon \mathsf{Alg}\, G \to (\mathsf{Alg}\, F^{\mathrm{op}}) = (\mathsf{Coalg}\, F)^{\mathrm{op}}.$$

Since the initial object of $(\mathsf{Coalg}\, F)^{\mathrm{op}}$ is the terminal F-algebra, we thus have

$$\widehat{L}(\mu G) = \nu F.$$

Example 11.5.8. (1) Consider $\mathscr{A} = \mathscr{B} = \mathsf{Set}$ and let us abuse notation and write $\mathscr{P} \cong \mathsf{Set}(-, 2)\colon \mathsf{Set} \to \mathsf{Set}^{\mathrm{op}}$ for the *contravariant* power-set functor, which is self-adjoint on the right, that is, left adjoint to $\mathscr{P}^{\mathrm{op}}\colon \mathsf{Set}^{\mathrm{op}} \to \mathsf{Set}$. So we can take $L = R = \mathscr{P}$. For the set functor $GX = 1 + \Sigma \times X$, the initial algebra is carried by the set Σ^* of all words over the alphabet Σ (Example 2.2.15(2)). We take $FX = 2 \times X^{\Sigma}$, and we have an obvious natural isomorphism with components

$$\varrho_X\colon 2 \times (\mathscr{P}X)^{\Sigma} \to \mathscr{P}(1 + \Sigma \times X).$$

Thus, the terminal coalgebra for F is given by $\mathscr{P}\Sigma^*$, the set of all formal languages over Σ (cf. Example 2.5.5). The structure of the initial algebra is $[\varepsilon, c]\colon 1 + \Sigma \times \Sigma^* \to \Sigma^*$, given by the empty word and the concatenation map c. It is easy to check that $\widehat{\mathscr{P}}$ maps this structure to the coalgebra structure $\langle o, \partial\rangle\colon \mathscr{P}(\Sigma^*) \to 2 \times \mathscr{P}(\Sigma^*)^{\Sigma}$ given by the predicate o of all languages containing the empty word (i.e. $o(L) = 1$ iff $\varepsilon \in L$) and the map ∂ taking left derivatives (i.e. $\partial(L)(a) = a^{-1}L = \{w : aw \in L\}$). This corresponds to Example 2.5.5.

(2) Let us now take $\mathscr{A} = \mathscr{B} = \mathsf{Nom}$, the category of nominal sets and equivariant maps (Example 2.1.10). Consider the endofunctor $GX = 1 + \mathbb{A} \times X + [\mathbb{A}]X$. We write $P\colon \mathsf{Nom} \to \mathsf{Nom}^{\mathrm{op}}$ for the contravariant functor mapping a nominal set X to the exponential 2^X; this is isomorphic to the set $\mathscr{P}_{\mathsf{fs}}X$ of all finitely supported subsets of X. An equivariant function $f\colon X \to Y$ is then mapped to $Pf\colon \mathscr{P}_{\mathsf{fs}}Y \to \mathscr{P}_{\mathsf{fs}}X$ taking inverse images under f: $Pf(S) = f^{-1}[S]$. So P is the nominal version of the contravariant power-set functor in item (1). Similarly as in the previous item, P is self-adjoint on the right. For every nominal set we have a canonical isomorphism $\mathscr{P}_{\mathsf{fs}}(\mathbb{A} \times S) \cong (\mathscr{P}_{\mathsf{fs}}S)^{\mathbb{A}}$. It follows from Pitts [250,

Prop. 4.14] that there is a canonical isomorphism $\mathscr{P}_{\mathsf{fs}}([\mathbb{A}]X) \cong [\mathbb{A}](\mathscr{P}_{\mathsf{fs}}X)$ which is given by

$$S \mapsto \langle a \rangle \{x : \langle a \rangle x \in S\},$$

where $a \in \mathbb{A}$ is any name which is fresh for S (i.e. not contained in its support). It is not difficult to verify that these yield natural isomorphisms

$$(P-)^{\mathbb{A}} \cong P(\mathbb{A} \times (-)) \qquad \text{and} \qquad [\mathbb{A}]P(-) \cong P([\mathbb{A}]-).$$

We clearly have a natural isomorphism

$$PX \times PY \cong P(X + Y),$$

and since $P1 \cong 2$, we obtain for the functor $FX = 2 \times X^{\mathbb{A}} \times [\mathbb{A}]X$ on Nom a natural isomorphism $\varrho \colon F^{\mathsf{op}}P \to PG$ with components

$$2 \times (\mathscr{P}_{\mathsf{fs}}X)^{\mathbb{A}} \times [\mathbb{A}](\mathscr{P}_{\mathsf{fs}}X) \cong \mathscr{P}_{\mathsf{fs}}(1 + \mathbb{A} \times X + [\mathbb{A}]X).$$

By Remark 11.5.7 we thus have a left adjoint $\widehat{P} \colon \mathsf{Alg}\, G \to (\mathsf{Coalg}\, F)^{\mathsf{op}}$ so that $\widehat{P}(\mu G) = \nu F$.

It follows from the work of Gabbay and Pitts [132] (see also Pitts' book [250, Thm. 8.15] and cf. Example 2.5.14) that the initial algebra μG is carried by all words over the alphabet $\overline{\mathbb{A}} = \mathbb{A} + \{|a : a \in \mathbb{A}\}$ modulo α-equivalence (as defined in Example 10.5.11(3)). Consequently, νF is carried by the nominal set $\mathscr{P}_{\mathsf{fs}}(\mathscr{P}_{\mathsf{fs}}(\overline{\mathbb{A}}^* /{=_\alpha}))$ of all bar languages, as already announced in that example.

(3) In a similar but easier way one sees that the terminal coalgebra for $FX = 2 \times X^{\mathbb{A}}$ on Nom from Example 10.5.11(2) is carried by the nominal set $\mathscr{P}_{\mathsf{fs}}(\overline{\mathbb{A}}^*)$.

11.6 Summary

We began with a result in Section 11.1 giving a sufficient condition for the existence of a terminal coalgebra using the Adjoint Functor Theorem. The centrepiece of the chapter is a collection of results related to locally presentable categories. We have seen that an accessible endofunctor F on such a category has an initial algebra and a terminal coalgebra. Furthermore, the categories $\mathsf{Alg}\, F$ and $\mathsf{Coalg}\, F$ are also locally presentable, and we obtained characterizations of the presentable objects in $\mathsf{Coalg}\, F$.

We then introduced preaccessible functors in Section 11.4. On a cocomplete and co-well-powered category, the initial-algebra chain of a preaccessible functor F converges, and so μF exists. If the base category is locally presentable and F preserves monomorphisms, the same holds for νF. This is the first set of results on preaccessible functors, and much remains to be done; in particular, we do not yet have an application of our results which shows that either μF or νF exists and which does not have a proof using the results we mentioned above on accessible endofunctors. On the other hand, our results in Section 11.5 pertaining to adjoint situations have many applications, as we shall see in Chapter 12. Further, we have provided a sufficient condition to upgrade an adjunction $L \dashv R$ from one given category $\mathscr{B}$ to another category $\mathscr{A}$ to an adjunction from F-algebras to G-algebras, where $G\colon \mathscr{B} \to \mathscr{B}$ and $F\colon \mathscr{A} \to \mathscr{A}$.

12
Liftings and Extensions from Set

In this chapter we will present some results that show that terminal coalgebras and initial algebras of endofunctors on Set lift to terminal coalgebras and initial algebras on categories of sets with structure such as complete partial orders, complete metric spaces, and (free) algebras for a monad.

To obtain the existence and constructions of a terminal coalgebra for an endofunctor on those more complicated categories, it is often sufficient to call on the existence and construction of the terminal coalgebra for an endofunctor on Set underlying the endofunctor of interest. More concretely in almost all of the following discussions (except the one for Kleisli categories) we will consider a category $\mathscr{A}$ with a forgetful functor $U\colon \mathscr{A} \to \mathsf{Set}$ and a set functor F having a *lifting* to an endofunctor on $\mathscr{A}$, that is, an endofunctor $\bar{F}$ such that the following square commutes:

$$\begin{array}{ccc} \mathscr{A} & \xrightarrow{\bar{F}} & \mathscr{A} \\ {\scriptstyle U}\downarrow & & \downarrow{\scriptstyle U} \\ \mathsf{Set} & \xrightarrow{F} & \mathsf{Set} \end{array} \tag{12.1}$$

In this case an $\bar{F}$-coalgebra (A, α) is a called a *lifting* of the F-coalgebra UA with the structure $U\alpha\colon UA \to UFA = FUA$.

Theorem 12.0.1. *Suppose that $\mathscr{A}$ is complete and well-powered, and $U\colon \mathscr{A} \to$ Set preserves limits and is faithful. If a set functor F has a terminal coalgebra, then every lifting $\bar{F}$ has one too, and it is a lifting of the terminal F-coalgebra.*

Proof Since F has a terminal coalgebra νF, we know from Theorem 6.5.15 that the terminal-coalgebra chain V_k converges at some ordinal λ. It follows that for every $i \geq \lambda$ we have $v_{i+1,i}$ invertible, thus $v_{i+1,i}^{-1}\colon V_i \to FV_i$ is a terminal F-coalgebra. Now consider the terminal-coalgebra chain $(\bar{V}_k)$ of $\bar{F}$. Since $U\colon \mathscr{A} \to \mathsf{Set}$ preserves limits, one easily shows by transfinite induction that U

preserves the terminal-coalgebra chain of $\bar{F}$. More precisely, we can choose that chain so that

$$U\bar{v}_{k,j} = (U\bar{V}_k = V_k \xrightarrow{v_{k,j}} V_j = U\bar{V}_j)$$

holds. Since $v_{\lambda+1,\lambda}$ (and hence every $v_{k,j}$ with $k, j \geq \lambda$) is an isomorphism in Set and U is faithful, we see that all $\bar{v}_{k,\lambda}$, $k \geq \lambda$ are monomorphisms in $\mathscr{A}$. For the fact that $\mathscr{A}$ is well-powered, it follows that there exists an ordinal $i \geq \lambda$ such that $\bar{v}_{i+1,i}$ is an isomorphism in $\mathscr{A}$. Thus, by (the dual of) Theorem 6.1.11, $\bar{V}_i$ is a terminal coalgebra for $\bar{F}$ with the structure $\bar{v}_{i+1,i}^{-1}$. □

12.1 Complete Partial Orders

Here we consider the category $\mathscr{A}$ = CPO of complete partial orders (not necessarily with a least element) and continuous maps, and the forgetful functor U: CPO → Set. Note first that Theorem 5.1.24 does not help us to obtain terminal coalgebras because of the 'missing' least elements, and in addition, some endofunctors F of interest are not locally continuous (see Example 12.1.2).

Recall from Example 2.1.7(1) that in CPO products are cartesian products ordered pointwise, and coproducts are disjoint unions with elements of different components incomparable.

Let us consider set functors which lift to CPO.

Example 12.1.1. (1) Every constant functor on Set has a lifting, as does the identity functor.

(2) Products, coproducts, and composites of endofunctors having a lifting have a lifting.

(3) Consequently, every polynomial endofunctor H_Σ on Set has a lifting given by the formula we have seen in (2.2):

$$H_\Sigma X = \coprod_{n \in \mathbb{N}} \Sigma_n \times X^n,$$

where each Σ_n is equipped with the discrete order.

Notice that in each of the cases of the previous example, the lifting is always locally continuous. But this need not always be the case:

Example 12.1.2. The power-set functor $\mathscr{P}$ lifts to CPO by taking $\bar{\mathscr{P}}(X, \sqsubseteq)$ to be the power-set with joins given by union of subsets (this disregards the order on X completely). This lifting is not locally continuous, in fact not even locally monotone: the two maps $0: 1 \to 2$ and $1: 1 \to 2$, where 2 is the two-element chain, satisfy $0 \leq 1$ but not $\mathscr{P}0 \sqsubseteq \mathscr{P}1$.

Since CPO is complete and well-powered and U preserves limits, we obtain from Theorem 12.0.1 the following

Corollary 12.1.3. *Let a set functor F have a lifting $\bar{F}\colon \mathsf{CPO} \to \mathsf{CPO}$ and terminal coalgebra. Then there is a cpo-order on νF making it the terminal $\bar{F}$-coalgebra with the same coalgebra structure.*

Example 12.1.4. (1) In the case of a polynomial functor F the terminal coalgebra $\nu H_\Sigma = T_\Sigma$ consists of all Σ-trees (see Example 3.3.9). This becomes a terminal coalgebra for the lifting $\bar{H}_\Sigma$ when equipped with the discrete cpo structure.

(2) The cpo structure becomes more interesting if we consider the coalgebra $T_\Sigma X$ of Σ-trees over a set X of generators, that is, Σ-trees where leaves are either labelled by constant symbols from Σ_0 or generators from X. This is the terminal coalgebra for the set functor $H_\Sigma(-) + X$; the initial algebra consists of finite Σ-trees over X (cf. Example 2.2.25(2)). For a cpo $(X, \sqsubseteq)$ of generators, the lifting has the terminal coalgebra $T_\Sigma X$ equipped with the following cpo structure: we have $t \sqsubseteq s$ for two Σ-trees over X iff they have the same shape, and for each leaf in t labelled by a generator $x \in X$, the corresponding leaf in s is labelled by a generator $y \in X$ with $x \sqsubseteq y$.

One may ask whether a lifting result similar to Corollary 12.1.3 also holds for initial algebras. However, for CPO this is not the case even if the lifting is locally continuous (Definition 5.1.16). A monad is locally continuous if so is its underlying functor.

Example 12.1.5. The functor $FX = X + 1$ on Set has the locally continuous lifting $\bar{F}X = X_\perp$ (see Example 5.1.17(4)). But the initial algebra for F is carried by the set of natural numbers (Example 2.2.7(3)) whereas the initial algebra for the lifting is $\mathbb{N}^\top$ (with the structure fixing $\top$).

Remark 12.1.6. Example 12.1.4(2) can be generalized to parametrized terminal coalgebras. Let F be a set functor which has a free cia TX on every set X; equivalently, a terminal coalgebra for $F(-) + X$ exists (Theorem 7.3.16). As we mentioned in Remark 7.3.23(2), $X \mapsto TX$ is the object part of a monad on Set (in fact, the free completely iterative monad on F). For every cpo $(X, \leq)$ the functor $\bar{F}(-) + (X, \leq)$ is a locally continuous lifting of $F(-) + X$.

Recall from Definition 7.3.17 that an endofunctor is iteratable if it has all free cias.

Theorem 12.1.7 [37, Thm. 4.2]**.** *Suppose that F is an iteratable set functor that has a locally continuous lifting $\bar{F}$ on* CPO*. Then the free cia monad T lifts to a locally continuous monad on* CPO*.*

Note that the fact that T has a lifting follows from Corollary 12.1.3; it also appears in previous work [229, Ex. 2.8]. Theorem 12.1.7 yields local continuity in addition.

Proof Since $\overline{F}$ is locally continuous, Corollary 12.1.3 implies that TX carries a cpo turning it into the terminal coalgebra for the lifting of $F(-) + X$ mentioned in Remark 12.1.6. To see that T lifts to a monad on CPO, recall from Remark 7.3.23(3) how T is endowed with the structure of a Kleisli triple: the morphisms $\eta_X : X \to TX$ are obtained by composing the inverse of the terminal coalgebra structure $TX \cong F(TX) + X$ with the right-hand coproduct injection. This yields continuous maps η_X. Furthermore, recall that the Kleisli extension mapping a morphism $s : X \to TY$ to $\hat{s} : TX \to TY$ is defined using primitive corecursion with respect to $F(-) + Y$ (Theorem 2.6.4). Thus, for a continuous map s we obtain a continuous $\hat{s}$.

It remains to show that the lifting $\overline{T}$ of T is locally continuous. For this we apply Theorem 12.0.1 to the functors $F(-) + X$ for every object X and use the fact that limits are computed objectwise in functor categories. This shows that $\overline{T}$ is a large limit of the endofunctors $\overline{V}_i : \mathsf{CPO} \to \mathsf{CPO}$ ($i \in \mathsf{Ord}$), where $\overline{V}_0$ is the constant functor of value 1, $\overline{V}_i = F\overline{V}_i + \mathsf{Id}$, and on limit ordinals one forms a limit. Since F is locally continuous, it is easy to verify that $\overline{V}_i$ is locally continuous using transfinite induction on i. Thus, $\overline{T} = \lim \overline{V}_i$ is also locally continuous. □

12.2 Complete Metric Spaces and Pseudometric Spaces

We will now consider the category $\mathscr{A} = \mathsf{CMet}$ of complete metric spaces and non-expanding maps (Notation 3.2.2(2)). Let $U : \mathsf{CMet} \to \mathsf{Set}$ denote the forgetful functor. We have seen in Corollary 5.2.18 that (almost) every locally contracting endofunctor on CMet has a terminal coalgebra. Furthermore, it follows from Theorem 11.2.18 that every accessible endofunctor on CMet has a terminal coalgebra.

For endofunctors that have a lifting on CMet, the terminal coalgebra lifts, too:

Corollary 12.2.1. *Let a set functor F have a terminal coalgebra νF and a lifting $\overline{F}$ on* CMet. *Then there is a complete metric space on νF making it a terminal $\overline{F}$-coalgebra with the same coalgebra structure.*

Indeed, this follows from Theorem 12.0.1 since CMet is complete and well-powered and U preserves limits.

Example 12.2.2. (1) Every polynomial functor H_Σ on Set has a lifting $\overline{H}_\Sigma$ on

CMet where on each summand $\Sigma_n \times X^n$ we take the maximum metric (with Σ_n equipped with the discrete metric): the distance of $\sigma(x_i)$ and $\sigma(y_i)$ is the maximum of $d(x_i, y_i)$, where $i = 1, \ldots, n$ for an n-ary operation symbol σ, and all other distances are 1. This is different from the contracting lifting H'_Σ considered in Example 5.2.6 – the lifting $\overline{H}_\Sigma$ is *not* locally contracting.

The terminal coalgebra for $\overline{H}_\Sigma$ is the set T_Σ of all Σ-trees with the discrete metric. More interestingly, for a complete metric space (X, d) the functor $\overline{H}_\Sigma(-) + (X, d)$ has the terminal coalgebra given by the set $T_\Sigma X$ of all Σ-trees on X with the metric determined by considering the metric d on leaves: trees t and s of different shape have distance 1, and if t and s have the same shape, then their distance is the supremum of the distances $d(x, y)$, where x ranges over the leaves of t in X and for every such x, y is the corresponding leaf in s.

(2) The finite power-set functor $\mathscr{P}_\mathsf{f}$ has a lifting $\overline{\mathscr{P}}_\mathsf{f}$ on CMet: given a metric space X we equip $\mathscr{P}_\mathsf{f} X$ with the Hausdorff metric (Example 5.2.6(4)). The terminal coalgebra-chain of $\overline{\mathscr{P}}_\mathsf{f}$ is given by the sets V_i from the terminal-coalgebra chain for $\mathscr{P}_\mathsf{f}$ with the discrete metric. Indeed, this follows by an easy transfinite induction using that for a discrete metric space X the metric space $\overline{\mathscr{P}}_\mathsf{f} X$ is discrete again and that discrete spaces are closed under limits in CMet. Thus, the terminal-coalgebra chain for $\overline{\mathscr{P}}_\mathsf{f}$ converges in exactly $\omega + \omega$ steps (cf. Theorem 4.4.10), and $\nu\overline{\mathscr{P}}_\mathsf{f} = V_{\omega+\omega}$ is the set of all finitely branching strongly extensional trees (Theorem 4.5.15) equipped with the discrete metric.

(3) Every Kripke polynomial set functor F (Definition 4.4.13) has a lifting on CMet. This is easy to see using item (2) and how products (Example 3.2.6) and coproducts (Example 3.2.3) are formed in CMet: the existence of a lifting of F follows by induction on the structure of F since a composition FG of functors F, G having liftings $\bar{F}, \bar{G}$ has the lifting $\bar{F}\bar{G}$.

For every Kripke polynomial set functor F the terminal-coalgebra chain of the lifting $\bar{F}$ is given by that of F with the metric on each set of the chain induced in an obvious way by the constant functors from which F is built. The terminal coalgebra for $\bar{F}$ is $\nu F = V_{\omega+\omega}$ (Proposition 4.4.16) with that metric.

Like for CPO, initial algebras of a lifted functor $\bar{F}$ need not be lifted from Set:

Example 12.2.3. The functor $FX = X + 1$ has the lifting $\bar{F}(X, d) = (X, \frac{1}{2}d) + 1$ (see Example 5.2.6). But whereas the initial algebra for F is the set of natural numbers, the initial algebra of $\bar{F}$ is $\mathbb{N} \cup \{\infty\}$, the natural numbers with an extra element ∞ serving as the limit of the Cauchy sequence $0, 1, 2, 3, \ldots$.

Remark 12.2.4. (1) Let us come back to parametrized terminal coalgebras as in Remark 12.1.6. Recall from Definition 5.2.5 the notion of a locally contracting endofunctor. Suppose that F is a set functor with $F1$ nonempty and which has all free cias and a locally contracting lifting $\bar{F}$ to CMet (such as the lifting of polynomial functors in Example 5.2.6). Then the functor $\bar{F}(-) + (X, \varepsilon d)$ is a locally contracting lifting of $F(-) + X$, where ε is the contraction constant of $\bar{F}$, and εd is the scaling of the metric d by ε (Example 5.2.6(1)).

(2) Note that the limit of a diagram of complete metric spaces (X_i, d_i) $(i \in I)$ is given by the limit X of the sets X_i in Set equipped with the sup-metric d on X: for every $x, y \in X$ we have

$$d(x, y) = \sup_{i \in I} d_i\big(p_i(x), p_i(y)\big),$$

where $p_i : X \to X_i$ is the limit projection. In fact, this gives the smallest distance such that every limit projection is a non-expanding map.

(3) It follows that a limit of locally contracting functors is locally contracting, provided that there is a uniform bound on the contraction coefficients. For suppose $F\colon \mathsf{CMet} \to \mathsf{CMet}$ is the limit of a diagram of locally contracting functors F_i $(i \in I)$, all with contraction coefficient less than ε. Then for every pair of non-expanding maps $f, g\colon (X, d) \to (Y, d')$ we have

$$\begin{aligned} d_{FX,FY}(Ff, Fg) &= \sup_{i \in I} d_{FX,FY}(F_i f, F_i g) \\ &= \sup_{i \in I}(\varepsilon \cdot d_{X,Y}(f, g)) \\ &\leq \varepsilon \cdot d_{X,Y}(f, g). \end{aligned}$$

By a *locally contracting monad* below is meant a monad whose underlying functor is locally contracting.

Theorem 12.2.5. *Suppose that F is an iteratable set functor which lifts to a locally contracting endofunctor $\bar{F}$ on* CMet. *Then the free cia monad T on F lifts to a locally contracting monad on* CMet.

Proof Suppose that $\bar{F}$ is locally contracting with a contraction factor ε. By Corollary 12.2.1, TX carries a complete metric space turning it into the terminal coalgebra for the above-mentioned lifting of $F(-) + X$. To see that T lifts to a monad on CMet, one proceeds analogously as in the proof of Theorem 12.1.7 for non-expanding maps in lieu of continuous ones.

It remains to show that T is locally contracting. In fact, Theorem 5.2.20 implies that $\bar{T}$ is the limit of an ω^{op}-chain of the following locally contracting endofunctors H^n on CMet:

$$H^0 = C_1 \qquad \text{and} \qquad H^{n+1} = \bar{F}H^n + \mathsf{Id}_{\varepsilon},$$

where C_1 is the constant functor on the trivial one-point space and $\mathsf{Id}_\varepsilon(X,d) = (X, \varepsilon d)$; the connecting natural transformations are obvious.

To see that T is indeed the limit of this chain, notice that for each complete metric space (X, d), the space TX is the limit of $H^n(X, d)$ (cf. Example 5.2.10); in order to apply Theorem 5.2.20, one uses that $\mathsf{CMet}(1, H(X, d))$ is nonempty since $F1 \neq \emptyset$. By an easy induction argument, each H^n is locally contracting (with the factor ε). Since limits of locally contracting functors are locally contracting (Remark 12.2.4(3)), we are done. □

Besides the simple liftings of Kripke polynomial functors we have seen in Example 12.2.2, Baldan et al. [65] have shown how to lift set functors to metric spaces using two different approaches. These may be viewed as generalizations of the Kantorovich and Wasserstein distances on probability distributions, as explained below (see e.g. [310]). Both approaches work with pseudometric spaces. A *pseudometric* is defined like a metric d (see Notation 4.4.5) except that one omits the requirement that $d(x, y) = 0$ implies $x = y$ (but $d(x, x) = 0$ still holds). We will briefly review one of the two liftings, present some examples and mention the induced pseudometrics on the lifted terminal coalgebras. Note that op. cit. works with distances in $[0, \top]$ where $\top \in (0, \infty]$.

Notation 12.2.6. Let us denote by PMet the category of pseudometric spaces with distances in $[0, 1]$ and non-expanding maps.

We write $[0, 1]$ for the (pseudo)metric space given by the interval with its usual Euclidean metric $d_e(x, y) = |x - y|$. Liftings of a set functor F are induced by *evaluation functions* as defined below. We discuss the Kantorovich lifting and subsequently briefly mention the Wasserstein lifting.

Definition 12.2.7 [65]**.** (1) An *evaluation function* for a set functor F is a function

$$\mathsf{ev}_F \colon F[0, 1] \to [0, 1].$$

It induces the endofunctor $\widetilde{F}$ on the slice category $\mathsf{Set}/[0, 1]$ (whose objects are all maps $g \colon X \to [0, 1]$) defined by $F(g) = \mathsf{ev}_F \cdot g$ on objects and $\widetilde{F}m = Fm$ on morphisms.

(2) Suppose that F is a set functor equipped with an evaluation function ev_F. The *Kantorovich distance* d^K on FX assigns to a pair $x, y \in FX$ the supremum of the numbers $|\widetilde{F}f(x) - \widetilde{F}f(y)|$ where f ranges over all non-expanding maps from (X, d) to $[0, 1]$; in symbols, $d^K \colon FX \times FX \to [0, 1]$ is defined by

$$d^K(x, y) = \sup\left\{d_e(\widetilde{F}f(x), \widetilde{F}f(y)) \mid f \colon (X, d) \to [0, 1] \text{ non-expanding}\right\}$$

for all $x, y \in FX$.

(3) The *Kantorovich lifting* of F is the functor $\bar{F}\colon \mathsf{PMet} \to \mathsf{PMet}$ defined by

$$\bar{F}(X, d) = (FX, d^K) \qquad \text{and} \qquad \bar{F}f = Ff.$$

It is easy to see that the Kantorovich distance is a pseudometric on FX and that Ff is non-expanding if so is $f\colon (X, d) \to (Y, d')$ [65, Thms. 5.5 & 5.7]. Functoriality of $\bar{F}$ is then immediate.

Example 12.2.8 [65]. (1) The leading example is the distribution functor $\mathscr{D}$. A *discrete probability distribution* on a set X is a function $\mu\colon X \to [0, 1]$ with the property that $\sum_{x\in X} \mu(x) = 1$. (An infinite sum of non-negative real numbers is the supremum of the sums of its finite subsets.) It follows that the set of $x \in X$ such that $\mu(x) \neq 0$ is countable. We call this set $\mathsf{supp}(\mu)$; it is the *support* of μ. Moreover, μ extends to an operation $\mathscr{P}(X) \to [0, 1]$, by $\mu(A) = \sum_{x\in A} \mu(x)$. The set functor $\mathscr{D}$ assigns to a set X the set of all discrete probability distributions on it. For a function $f\colon X \to Y$, $\mathscr{D}f$ assigns to $\mu \in \mathscr{D}X$ the measure $\mathscr{D}f(\mu)$ defined by $\mathscr{D}f(\mu)(y) = \mu(f^{-1}(y))$ for $y \in Y$ (cf. Example 2.4.3(4)). Notice that $\mathsf{supp}(\mathscr{D}f(\mu)) = f[\mathsf{supp}\,\mu]$.

The evaluation function $\mathsf{ev}_{\mathscr{D}}\colon \mathscr{D}[0, 1] \to [0, 1]$ takes a discrete probability distribution to the expected value of the identity function on $[0, 1]$:

$$\mathsf{ev}_{\mathscr{D}}(\mu) = \sum_{x\in[0,1]} x \cdot \mu(x).$$

Consequently, for every $g\colon X \to [0, 1]$ and $\mu \in \mathscr{D}X$ we see that

$$\tilde{\mathscr{D}}g(\mu) = \sum_{x\in[0,1]} g(x) \cdot \mu(x).$$

The ensuing Kantorovich distance d^K is then the usual Kantorovich pseudometric on discrete probability distributions:

$$d^K(\mu_1, \mu_2) = \sup \left\{ \left| \sum_{x\in X} f(x) \cdot (\mu_1(x) - \mu_2(x)) \right| \mid f\colon (X, d) \to [0, 1] \right\}.$$

Thus, we obtain the lifting with $\bar{\mathscr{D}}(X, d) = (\mathscr{D}X, d^K)$ as expected.

(2) For the power-set functor $\mathscr{P}$ with the evaluation function

$$\sup\colon \mathscr{P}[0, 1] \to [0, 1]$$

the Kantorovich distance yields the Hausdorff metric on $\mathscr{P}X$ (cf. Example 5.2.6(4) and Example 7.3.10(3)): for a pseudometric space (X, d) we have

$$d^K(S, T) = \max(\sup_{s\in S} d(s, T)\ \sup_{t\in T} d(t, S)) \qquad \text{for } S, T \subseteq X,$$

where $d(s, T) = \inf_{t\in T} d(s, t)$.

Remark 12.2.9. Note that the Kantorovich distance can be characterized by a universal property, namely, it is the smallest pseudometric on $F[0,1]$ making the evaluation function $\mathsf{ev}_F\colon F[0,1] \to [0,1]$ non-expanding.

Baldan et al. [65] also provide a different way to produce a lifting of a set functor to PMet. It is obtained by generalizing the following Wasserstein pseudometric on probability measures:

Example 12.2.10. Consider the distribution functor $\mathscr{D}$ (Example 12.2.8(1)). A *coupling* of probability distributions $\mu_1, \mu_2 \in \mathscr{D}X$ is an arbitrary probability distribution μ on $X \times X$ whose marginal distributions are μ_1, μ_2, that is $\mathscr{D}\pi_i(\mu) = \mu_i$ where $\pi_i\colon X \times X \to X$ are the product projections for $i = 1, 2$. The *Wasserstein*[1] *pseudometric* on $\mathscr{D}X$ is defined by

$$d^{\mathrm{W}}(\mu_1, \mu_2) = \inf\left\{ \textstyle\sum_{x,y \in X} d(x,y) \cdot \mu(x,y) \mid \mu \text{ a coupling of } \mu_1, \mu_2 \right\}.$$

A result known as Kantorovich–Rubinstein duality [310] states that this distance coincides with the Kantorovich pseudometric of Example 12.2.8(1). So this yields the same lifting of $\mathscr{D}$ to PMet.

Remark 12.2.11. The generalized Wasserstein distance [65, Def. 5.12] does not always coincide with the Kantorovich distance. Given a set functor F equipped with an evaluation map ev_F, one assumes that F preserves weak pullbacks and that ev_F satisfies three conditions (which we do not mention). Under these assumptions the Wasserstein distance yields a pseudometric on FX yielding a lifting of F [65, 5.20–5.22].

Finally, an explicit description of the pseudometric on a terminal coalgebra for a lifting of a set functor is obtained by the following

Construction 12.2.12. Let $\bar{F}\colon \mathsf{PMet} \to \mathsf{PMet}$ be a lifting of a set functor F and let (A, α) be an F-coalgebra. Given a pseudometric d on A we write d^F for the pseudometric of $\bar{F}(A, d)$. The collection of all pseudometrics on A forms a complete lattice, and the following endofunction is monotone:

$$d \mapsto d^F \cdot (\alpha \times \alpha).$$

It therefore has a least fixed point

$$d_\alpha = \inf\{d \mid d\colon A \times A \to [0,1] \text{ is a pseudometric with } d^F \cdot (\alpha \times \alpha) \le d\}.$$

Theorem 12.2.13 [65, Thm. 6.2]**.** *Let F be a set functor which has a terminal coalgebra and a lifting $\bar{F}\colon$ PMet $\to$ PMet. Then $(\nu F, \tau)$ lifts to a terminal*

[1] This name goes back to Dobrushin who learned about this distance in a paper by Wasserstein [306]; the distance was first introduced by Kantorovich [182].

coalgebra for $\bar{F}$:

$$\tau\colon (\nu F, d_\tau) \to (F(\mu F), d^F_\tau).$$

Example 12.2.14 [65]. For the functor $FX = \{0,1\} \times X^\Sigma$ whose coalgebras are deterministic automata, we take the evaluation function $\mathsf{ev}_F\colon \{0,1\} \times [0,1]^\Sigma \to [0,1]$ given by

$$\mathsf{ev}_F(o,t) = c \cdot \max_{a\in\Sigma} t(a) \qquad \text{for some } 0 < c < 1.$$

For a pseudometric space (X,d) the induced Wasserstein pseudometric on FX is

$$d^W\big((o_1,t_1),(o_2,t_2)\big) = \max\{\delta(o_1,o_2), c \cdot \max_{a\in\Sigma} d(t_1(a),t_2(a))\}.$$

Furthermore, recall from Example 2.5.5 that the terminal coalgebra for F is formed by all formal languages; here considered as predicates $L\colon \Sigma^* \to \{0,1\}$. The pseudometric on formal languages induced by Construction 12.2.12 is

$$d_\tau(L_1,L_2) = c^k, \qquad \text{where } k = \min\{n \in \mathbb{N} : \exists w \in \Sigma^*.\, |w| = n \text{ and } L_1(w) \neq L_2(w)\}.$$

This means that the distance between L_1 and L_2 is c^k, where k is the length of the shortest word w that distinguishes L_1 from L_2.

Remark 12.2.15. (1) Construction 12.2.12 and the ensuing pseudometrics on lifted terminal coalgebras are used to study *behavioural distances*, which replace behavioural equivalence by a notion that takes into account the quantitative information about elements of (terminal) coalgebras available in the pseudometric setting. For probabilistic systems considered as coalgebras for certain functors on PMet, this was first investigated by van Breugel and Worrell [303] based on work by other authors who worked in non-coalgebraic settings.

The coalgebraic understanding of probabilistic bisimilarity as coalgebraic behavioural equivalence goes back to de Vink and Rutten [269].

We further refer to Baldan et al., especially to their discussion [65, Sec. 10.1] of related work.

(2) The Kantorovich lifting of a set functor also appears as a special instance in more general frameworks where set functors are lifted to the category of small $\mathscr{V}$-categories for a commutative unital quantale $\mathscr{V}$. For more on this we refer to work by Balan et al. [64] and by Katsumata and Sato [185].

12.3 Kleisli Categories

We now investigate conditions under which initial algebras and terminal coalgebras of endofunctors on a category $\mathscr{A}$ extend to the Kleisli category (Definition 12.3.1) of a monad on $\mathscr{A}$. Kleisli categories have applications in the study of the semantics of computation with side effects, started by Moggi's seminal paper [238] and also in the coalgebraic trace semantics first studied by Hasuo, Jacobs, and Sokolova [158] for $\mathscr{A} = \mathsf{Set}$. In the latter one considers coalgebras $\alpha\colon A \to MFA$ where M is a monad and F an endofunctor on the same category. The monad is supposed to model a type of branching, for example, non-deterministic or probabilistic, and F models the type of transitions of the coalgebra (A, α). For example, labelled transitions systems are of this form (see Example 2.4.2(7)), and we shall also consider probabilistic transition systems. The main result in coalgebraic trace semantics is an extension theorem (Theorem 12.3.5): under certain conditions, including that a given endofunctor F has an extension $\bar{F}$ to the Kleisli category of M, the initial algebra for F yields a canonical fixed point of $\bar{F}$. In op. cit. the existence of a canonical extension $\bar{F}$ was proven for polynomial set functors F provided that M is a commutative set monad (Remark 12.3.11). This was later extended to analytic functors [231].

The salient point of these results is that in concrete instances, the unique homomorphism from an MF-coalgebra (considered as a coalgebra for the extension $\bar{F}$) to the canonical fixed point assigns to a state its finite traces. Hence, behavioural equivalence for $\bar{F}$ is trace equivalence, while behavioural equivalence for F is bisimilarity.

We will first discuss the main results on canonical fixed points of Kleisli extensions and then mention a few instances of coalgebraic trace semantics.

Definition 12.3.1. Let (M, η, μ) be a monad on a category $\mathscr{A}$. Its Kleisli category $\mathscr{A}_M$ has the same objects as $\mathscr{A}$; a morphism from an object X to an object Y is a morphism $f\colon X \to MY$ of $\mathscr{A}$. The composition of f with $g\colon Y \to MZ$ (denoted by $g \circ f$ for distinction) is defined by

$$g \circ f = (X \xrightarrow{f} MY \xrightarrow{Mg} MMZ \xrightarrow{\mu_Z} MZ).$$

We denote the morphisms in $\mathscr{A}_M$ by $X \twoheadrightarrow Y$. There is a canonical (identity-on-objects) functor $J\colon \mathscr{A} \to \mathscr{A}_M$ given on morphisms by

$$J(X \xrightarrow{f} Y) = (X \xrightarrow{f} Y \xrightarrow{\eta_Y} MY).$$

This functor is a left adjoint to the functor $K\colon \mathscr{A}_M \to \mathscr{A}$ given by

$$K\big(X \xrightarrow{f} Y\big) = \big(MX \xrightarrow{Mf} MMY \xrightarrow{\mu_Y} MY\big).$$

Example 12.3.2. We list some examples of set monads and their Kleisli categories.

(1) The *maybe monad* is given by $MX = X+1$ with the unit $\mathsf{inl}\colon X \to X+1$ and the multiplication $\mathsf{id}_X + \nabla_1 \colon X+1+1 \to X+1$ (where $\nabla_1 = [\mathsf{inl}, \mathsf{inr}] \colon 1+1 \to 1$ is the codiagonal). Its Kleisli category is isomorphic to the category of sets and partial functions.

(2) The *power-set monad* $M = \mathcal{P}$ has as unit the 'singleton' map $\eta_X \colon X \to \mathcal{P}X$ defined by $x \mapsto \{x\}$ and as multiplication the 'union' map $\mu_X \colon \mathcal{P}\mathcal{P}X \to \mathcal{P}X$ defined by $\{S_i : i \in I\} \mapsto \bigcup_{i\in I} S_i$. Its Kleisli category is isomorphic to the category of sets and relations.

(3) The *subdistribution monad* $M = \mathcal{D}_{\leq}$ assigns to a set X the set $\mathcal{D}_{\leq}X$ of functions $d\colon X \to [0,1]$ with $\sum_{x\in X} d(x) \leq 1$. On functions it acts in the same way as the distribution functor in Example 12.2.8(1). The unit $\eta_X \colon X \to \mathcal{D}_{\leq}X$ assigns to $x \in X$ the Dirac distribution δ_x with $\delta_x(x) = 1$ and $\delta_x(y) = 0$ for $y \neq x$. Moreover for every map $s\colon X \to \mathcal{D}_{\leq}Y$ we obtain the map $\widehat{s}\colon \mathcal{D}_{\leq}X \to \mathcal{D}_{\leq}Y$ given by

$$\widehat{s}(\mu)(y) = \sum_{x\in X} s(x)(y) \cdot \mu(x).$$

These data are easily seen to satisfy the axioms of a Kleisli triple (see e.g. Moggi [238]); it follows that $\mathcal{D}_{\leq}$ is a set monad [219, Thm. 3.18 and Exer. 12]. Its Kleisli category is isomorphic to the category of substochastic relations, that is functions $R\colon X \times Y \to [0,1]$ such that $\sum_{y\in Y} R(x,y) \leq 1$. (This implies that for every $x \in X$ only countably many $R(x,y)$ are non-zero.) For finite sets X and Y these are precisely the substochastic matrices, and the composition of morphisms in the Kleisli category then corresponds to matrix multiplication.

(4) The *environment monad* is given by $MX = X^E$, where E is a fixed set. Its unit $\eta_X \colon X \to X^E$ assigns to an element $x \in X$ the constant map of value x, and the multiplication μ_X assigns to an element of $(X^E)^E$ (i.e. an $|E| \times |E|$ matrix with values in X) its diagonal (considered as a map from E to X). The morphisms of the Kleisli category are functions from X to Y which take, in addition to an input from X, an input from the environment E to compute a value in Y.

(5) The *bag monad* $M = \mathcal{B}$ assigns to a set X the free commutative monoid on X or, equivalently, the set of all bags (i.e. finite multisets) on X. Its Kleisli category consists of weighted relations $R\colon X \times Y \to \mathbb{N}$. These are the functions with the property that for every $x \in X$ the sum $\sum_{y\in Y} R(x,y)$ is finite. For finite

sets X and Y these are $|X|\times|Y|$ matrices of natural numbers, and the composition of morphisms is given by matrix multiplication.

Definition 12.3.3. We say that an endofunctor $\bar{F}\colon \mathscr{A}_M \to \mathscr{A}_M$ *extends* an endofunctor $F\colon \mathscr{A} \to \mathscr{A}$ if the following square commutes:

$$\begin{array}{ccc} \mathscr{A}_M & \xrightarrow{\bar{F}} & \mathscr{A}_M \\ {\scriptstyle J}\uparrow & & \uparrow{\scriptstyle J} \\ \mathscr{A} & \xrightarrow{F} & \mathscr{A} \end{array} \tag{12.2}$$

Note that this is the special case of Definition 11.5.4(2) for the adjunction $J \dashv K$. Thus, Corollary 11.5.5(2) gives the following result as a special case.

Proposition 12.3.4 [158]. *Let M be a monad on $\mathscr{A}$, let $F\colon \mathscr{A} \to \mathscr{A}$ have an extension $\bar{F}$ on $\mathscr{A}_M$. If $(\mu F, \iota)$ is an initial F-algebra, then μF is an initial $\bar{F}$-algebra with the structure $J\iota = \eta_{\mu F}\cdot\iota\colon \bar{F}(\mu F) \twoheadrightarrow \mu F$.*

For various monads of interest, it turns out that the Kleisli category is left-strict CPO-enriched (Definitions 5.1.1 and 5.1.30). Recall that a canonical fixed point of an endofunctor is an initial algebra which coincides with the terminal coalgebra (Definitions 5.0.1).

Theorem 12.3.5. *Let M be a monad on $\mathscr{A}$, let $F\colon \mathscr{A} \to \mathscr{A}$ have an extension $\bar{F}$ to $\mathscr{A}_M$. Assume that $\mathscr{A}$ is left-strict CPO-enriched, that $\bar{F}$ is locally continuous and that F has an initial algebra $(\mu F, \iota)$. Then $(\mu F, J\iota^{-1})$ is a terminal $\bar{F}$-coalgebra, whence a canonical fixed point.*

Proof Immediate from Propositions 12.3.4 and 5.1.33. □

Remark 12.3.6. (1) An analogous result holds when $\mathscr{A}_M$ is left-strictly enriched in the category DCPO of directed-complete partial orders and $\bar{F}$ is locally monotone [126, Thm. 3.5]. Indeed for the proof, combine Proposition 12.3.4 with the result mentioned in Remark 5.1.34.

(2) The first extension result of this kind appears in work by Hasuo, Jacobs, and Sokolova [158, Thm. 3.3] for $\mathscr{A}$ = Set. While assuming that $\mathscr{A}_M$ is left-strict CPO-enriched, they only assume local monotonicity of $\bar{F}$. But in addition, they require that $\mathscr{A}_M$ has a zero object and that the initial-algebra chain (Definition 3.1.4) converges in ω steps (Definition 6.1.9). Their proof is then more involved than ours, making use of the classical limit-colimit coincidence as we have seen in Theorems 5.1.10 and 5.1.24. A related result is in Jacobs' book [173, Thm. 5.3.4]: It admits a general base category $\mathscr{A}$ and assumes that $\mathscr{A}_M$ is enriched in DCPO, but retains the requirement of a zero object in Set_M and the convergence of the initial-algebra chain in ω steps.

Remark 12.3.7. Note that an MF-coalgebra $A \longrightarrow MFA$ is, equivalently, a coalgebra $A \twoheadrightarrow \bar{F}A$ for the extension $\bar{F}$. As observed by Jacobs et al. [174, Lemma 1], this correspondence yields the object part of a functor $\mathsf{Coalg}(MF) \rightarrow \mathsf{Coalg}\,\bar{F}$ which lifts the functor $J\colon \mathscr{A} \longrightarrow \mathscr{A}_M$ to coalgebras (the vertical arrows are the forgetful functors):

$$
\begin{array}{ccc}
\mathsf{Coalg}(MF) & \xrightarrow{\hat{J}} & \mathsf{Coalg}\,\bar{F} \\
\downarrow & & \downarrow \\
\mathscr{A} & \xrightarrow{J} & \mathscr{A}_M
\end{array}
$$

Definition 12.3.8 (Coalgebraic Trace Semantics). Let F and M on the category $\mathscr{A}$ satisfy the assumptions in Theorem 12.3.5. Given a coalgebra $\alpha\colon A \longrightarrow MFA$, the *coalgebraic trace map* is the unique coalgebra homomorphism tr_α from (A, α) to $(\mu F, J\iota^{-1})$ in $\mathscr{A}_M$:

$$
\begin{array}{ccc}
A & \xrightarrow{\alpha} & \bar{F}A \\
{\scriptstyle \mathsf{tr}_\alpha}\downarrow & & \downarrow{\scriptstyle \bar{F}\,\mathsf{tr}_\alpha} \\
\mu F & \xrightarrow{J\iota^{-1}} & \bar{F}(\mu F)
\end{array}
$$

There remains the question of how to obtain extensions of endofunctors F to $\mathscr{A}_M$. We shall see in Example 12.3.13(2) that every polynominal set functor has a canonical extension to the Kleisli categories for each of the monads mentioned in Example 12.3.2. Moreover, Theorem 12.3.14 states that this holds, more generally, for every analytic set functor (see Example 3.2.13).

It is well known and easy to prove (see Mulry [241]) that extensions of F to $\mathscr{A}_M$ are in bijective correspondence with *distributive laws* of F over M. The latter are natural transformations $\lambda\colon FM \longrightarrow MF$ compatible with the monad structure of M:

$$
\begin{array}{ccc}
F & \xrightarrow{F\eta} & FM \\
& {\scriptstyle \eta F}\searrow & \downarrow{\scriptstyle \lambda} \\
& & MF
\end{array}
\qquad\qquad
\begin{array}{ccccc}
FMM & \xrightarrow{\lambda M} & MFM & \xrightarrow{M\lambda} & MMF \\
{\scriptstyle F\mu}\downarrow & & & & \downarrow{\scriptstyle \mu F} \\
FM & & \xrightarrow{\quad\lambda\quad} & & MF
\end{array}
\tag{12.3}
$$

Remark 12.3.9. Let M be a monad on the category $\mathscr{A}$ and let $F\colon \mathscr{A} \longrightarrow \mathscr{A}$ have an extension to $\mathscr{A}_M$ via the distributive law $\lambda\colon FM \longrightarrow MF$. Suppose we are given a natural transformation $q\colon F \twoheadrightarrow G$ with epimorphic components. Then an object-indexed family of morphisms $\varrho_X\colon GMX \longrightarrow MGX$ is a distributive law of G over M provided that the following squares commute

$$\begin{array}{ccc} FMX & \xrightarrow{\lambda_X} & MFX \\ {\scriptstyle q_{MX}}\downarrow & & \downarrow{\scriptstyle Mq_X} \\ GMX & \xrightarrow{\varrho_X} & MGX \end{array} \qquad \text{for every object } X \text{ of } \mathscr{A}.$$

The proof of this fact is a straightforward diagram chase: for naturality of ϱ pre-compose the naturality square by q_X and use naturality of λ. Similarly, the instances for G and ϱ of the two diagrams in (12.3) are proven commutative by pre-composing with $q\colon F \twoheadrightarrow G$ and $q_{MM}\colon FMM \twoheadrightarrow GMM$, respectively, and using naturality of λ, η, and μ as well as the laws for λ in (12.3). The details are in [231, Lemma A.2] and [126, Details for Rem. 3.1(3) in appendix].

Example 12.3.10. Let M be a monad on the category $\mathscr{A}$.

(1) Constant functors and the identity functor on $\mathscr{A}$ obviously extend to $\mathscr{A}_M$.

(2) Suppose that $\mathscr{A}$ has coproducts. Then $\overline{F+G} = \bar{F} + \bar{G}$ for a pair F, G of endofunctors which extend to $\mathscr{A}_M$. Indeed, for a coproduct $F + G$ one uses that $J\colon \mathscr{A} \to \mathscr{A}_M$, being a left adjoint, preserves coproducts. Given extensions $\bar{F}$ and $\bar{G}$, it is then clear that $\bar{F}+\bar{G}$ extends $F+G$: for every morphism $f\colon X \to TY$ in $\mathscr{A}_M$ one has

$$\overline{F+G}(f) = \big(FX + FY \xrightarrow{\bar{F}f+\bar{G}f} MFY + MGY \xrightarrow{[M\mathsf{inl}, M\mathsf{inr}]} M(FY+GY)\big),$$

where $FY \xrightarrow{\mathsf{inl}} FY + GY \xleftarrow{\mathsf{inr}} GY$ are the coproduct injections. This works similarly for arbitrary coproducts.

(3) Furthermore, $\overline{GF} = \bar{G}\bar{F}$ for a pair F, G of endofunctors which extend to $\mathscr{A}_M$.

We will see in Theorem 12.3.14 below that for finite products a similar canonical construction can be performed when M is a commutative monad.

Remark 12.3.11. (1) The notion of a commutative monad was first introduced by Kock [196, Def. 3.1]. It is based on the notion of a *strong* monad: a monad M equipped with a natural transformation $s_{X,Y}\colon X \times MY \to M(X \times Y)$ (called *strength*) satisfying four natural equational laws (two with respect to 1 and $\times$ on $\mathscr{A}$ and two with respect to the monad structure), which we do not recall since they are not needed for our exposition. A strength gives rise to a *costrength* $t_{X,Y}\colon MX \times Y \to M(X \times Y)$ defined by

$$t_{X,Y} = \big(X \times MY \cong MY \times X \xrightarrow{s_{X,Y}} M(Y \times X) \xrightarrow{M(\cong)} M(X \times Y)\big).$$

A strong monad M is *commutative* if the following diagram commutes:

$$\begin{array}{ccccccc} & & M(MX\times Y) & \xrightarrow{Mt_{X,Y}} & MM(X\times Y) & & \\ & \nearrow^{s_{MX,Y}} & & & & \searrow^{\mu_{X\times Y}} & \\ MX\times MY & & & \xrightarrow{\quad d_{X,Y}\quad} & & & M(X\times Y) \\ & \searrow_{t_{X,MY}} & & & & \nearrow_{\mu_{X\times Y}} & \\ & & M(X\times MY) & \xrightarrow{Ms_{X,Y}} & MM(X\times Y) & & \end{array}$$

The ensuing natural transformation d in the middle is called a *double strength*, and it exhibits M as a symmetric monoidal monad. In fact, Kock considered commutative monads for arbitrary symmetric monoidal products, and he proved [196, 197] that giving the structure of a commutative monad is equivalent to giving the structure of a symmetric monoidal monad.

(2) Every set monad is strong via a canonical strength; this follows, for example, from Moggi's result [238, Thm. 3.4]: one defines $s_{X,Y}$ by the commutativity of the following diagrams for every element $t \in X$ (considered as a map $t\colon 1 \to X$):

$$\begin{array}{ccc} X\times MY & \xrightarrow{s_{X,Y}} & M(X\times Y) \\ {\scriptstyle t\times MY}\uparrow & & \uparrow{\scriptstyle M(t\times Y)} \\ 1\times MY \cong & MY & \cong M(1\times Y) \end{array}$$

However, not every set monad is commutative. The commutative set monads are those in whose corresponding algebraic theory every operation is a homomorphism.

Example 12.3.12. (1) All the monads mentioned in Example 12.3.2 are commutative. This is easy to prove.

(2) In contrast, the exception monad on Set given by $MX = X+E$ (with unit and multiplication analogous to the ones in Example 12.3.2(1)) is not commutative whenever $|E| > 1$.

(3) The monad on Set given by $MX = X^*$, the free monoid on X, is not commutative, while the bag monad (Example 12.3.2(5)) is.

Example 12.3.13. Suppose that $\mathscr{A}$ has finite products and M is a commutative monad on $\mathscr{A}$ with the ensuing double strength d.

(1) For a product of functors F and G having extensions $\bar{F}$ and $\bar{G}$ to $\mathscr{A}_M$, respectively, one then obtains an extension of $F \times G$: for every $f\colon X \to MY$ in $\mathscr{A}_M$ put

$$\overline{F\times G}(f) = \big(FX\times GX \xrightarrow{\bar{F}f\times\bar{G}f} MFY\times MGY \xrightarrow{d_{FY,GY}} M(FY\times GY)\big).$$

(2) Combining this with Example 12.3.10, we see that for a commutative set

monad M every polynomial set functor H_Σ has a canonical extension to $\mathscr{A}_M$. More generally, this holds for every base category with coproducts and finite products.

(3) Recall from Example 3.2.13 that an analytic functor is of the form

$$FX = \coprod_{\sigma \in \Sigma} X^k / G_\sigma,$$

where Σ is a finitary signature, k the arity of σ and G_σ a group of permutations on k. As proved by Joyal [179, 180], a finitary set functor is analytic iff it weakly preserves wide pullbacks. As previously mentioned, every polynomial set functor H_Σ is analytic, and so is the bag functor $\mathscr{B}$, but the finite power-set functor $\mathscr{P}_\mathsf{f}$ is not analytic.

Theorem 12.3.14 [231, Thm. 2.9]. *Let M be a commutative set monad. Then every analytic set functor has a canonical extension to* Set_M.

We sketch the proof; full details may be found in the cited work [231, Appendix A]. Note that the proof makes no explicit use of sets at all. In fact, the result holds for every commutative monad on a cocomplete category with finite products. There one defines the functors $X \mapsto X^n/G$ for a group G of permutation on n by using the coequalizers $q_X \colon X^n \twoheadrightarrow X^n/G$ exhibited in the proof sketch below.

Proof sketch By Example 12.3.10(2) it suffices to exhibit an extension of the set functor $H(X) = X^n/G$, where G is a group of permutations on n. This is done by exhibiting a distributive law $\varrho \colon HM \to MH$. First note that for the functor $Q_n X = X^n$ we have a canonical distributive law $\lambda \colon Q_n M \to M Q_n$ corresponding to the extension obtained as in Example 12.3.13(1). For each permutation $\pi \in G$ we let $\gamma_X^\pi \colon X^n \to X^n$ be the map which permutes the components of a given tuple according to π. Clearly, $\gamma^\pi \colon Q_n \to Q_n$ is a natural transformation.

Observe that H is a quotient functor of Q_n, where the corresponding epi-transformation $q \colon Q_n \twoheadrightarrow H$ is the coequalizer of all γ^π for $\pi \in G$. One then proves that the following squares commute

$$\begin{array}{ccc} Q_n M & \xrightarrow{\lambda} & M Q_n \\ {\scriptstyle \gamma^\pi M}\downarrow & & \downarrow{\scriptstyle M\gamma^\pi} \\ Q_n M & \xrightarrow{\lambda} & M Q_n \end{array} \qquad \text{for all } \pi \in G.$$

This is done using that every permutation in G is the composite of transpositions of the form $i \leftrightarrow i+1$. Hence it suffices to show the desired commutativity for all natural transformation $\gamma^\pi \colon Q_n \to Q_n$ where π is a transposition $i \leftrightarrow i+1$. In this case $\gamma_X^\pi \colon X^n \to X^n$ swaps the ith and $(i+1)$st components of every n-tuple of elements of X. One proceeds by induction on n; we omit these straightforward details here.

To complete the proof, one considers the following diagram:

$$\begin{array}{ccc} Q_n M & \xrightarrow{\lambda} & M Q_n \\ {\scriptstyle \gamma^\pi M}\downarrow & & \downarrow{\scriptstyle M\gamma^\pi} \\ Q_n M & \xrightarrow{\lambda} & M Q_n \\ {\scriptstyle qM}\downarrow & & \downarrow{\scriptstyle Mq} \\ HM & \overset{\varrho}{\dashrightarrow} & MH \end{array}$$

Using that $qM\colon Q_n M \to Q_n M$ is a coequalizer of all $\gamma^\pi M$ and that the upper square commutes for every γ^π, we obtain a unique $\varrho\colon HM \to MH$ such that the lower square commutes. This is a distributive law of H over M by an application of Remark 12.3.9. □

Coming back to the assumptions in Theorem 12.3.5, we still need to see for which monads of interest the Kleisli category Set_M is left-strict CPO-enriched. Let M be one of the monads $(-)+1$, $\mathscr{P}$ and $\mathscr{D}_\leq$ (see Example 12.3.2). Then we have obvious cpo structures $\sqsubseteq$ on MX for every set X: for $(-)+1$ take the flat cpo (with $\bot$ the element of 1), for $\mathscr{P}$ take subset inclusion, and for $\mathscr{D}_\leq$ take the pointwise order: $d \sqsubseteq d'$ iff $d(x) \leq d'(x)$ for all $x \in X$. This gives the pointwise cpo structure on $\mathsf{Set}_M(X,Y)$, and it is easy to see that composition is continuous and left-strict. For the canonical extension $\bar{F}$ of an analytic set functor F according to Theorem 12.3.14, it is not difficult to prove that $\bar{F}$ is locally continuous [231, Ex. 4.3(2)]. From Theorem 12.3.5 we obtain

Corollary 12.3.15. *Let F be an analytic set functor, let M be one of the monads $(-)+1$, $\mathscr{P}$, or $\mathscr{D}_\leq$. Then an initial F-algebra yields a canonical fixed point of the canonical extension $\bar{F}$ to Set_M.*

Example 12.3.16. We present two leading examples of combinations of analytic functors F and commutative monads M and the ensuing coalgebraic trace semantics provided by the canonical fixed point of the extended functor $\bar{F}$.

(1) Recall from Example 2.4.2(7) that labelled transition systems may be regarded as coalgebras. Here we slightly change the type functor by considering $M = \mathscr{P}$ and $FX = 1 + \Sigma \times X$. Coalgebras for MF are of the form $\alpha\colon A \to \mathscr{P}(1 + \Sigma \times A)$; they are labelled transition systems with explicit terminating states, where a state $x \in A$ is *terminating* if $\alpha(x)$ contains the element of 1. From Example 2.2.15(2) we know that the initial algebra for F is carried by the set Σ^* of all words on Σ. It is not hard to work out that the unique $\bar{F}$-coalgebra homomorphism from $(A,\alpha) \twoheadrightarrow (J\Sigma^*, J\iota)$ is the map $t\colon A \to \mathscr{P}(\Sigma^*)$ which

assigns to a state $x \in A$ the set of all *terminating traces*: words $s_0 s_2 \cdots s_n$ over Σ such that there is a sequence of transitions labelled by the letters s_i which starts in x and ends in a terminating state.

Note that this is slightly different from the usual trace semantics of labelled transition systems. In fact, given a labelled transition system $A \to \mathcal{P}(\Sigma \times A)$, the set of traces of a state $x \in A$ consists of all words $w \in \Sigma^*$ with some sequence of transitions in A labelled by the letters of w. Unfortunately, this example cannot be obtained as an instance by simply replacing F above by $F'X = \Sigma \times X$ because the initial algebra for the latter functor is empty. So the unique $\bar{F}'$-coalgebra homomorphism is the unique morphism $A \to \mathcal{P}(\emptyset) \cong 1$.

(2) As explained in Hasuo's thesis [156, Ch. 4], the trace semantics of probabilistic labelled transitions systems can be handled by replacing the power-set monad by the subdistribution monad $M = \mathcal{D}_{\leq}$ (Example 12.3.2(3)). We take $FX = 1 + \Sigma \times X$ once again. A probabilistic labelled transition system is then a coalgebra $\alpha \colon A \to \mathcal{D}_{\leq}(1 + \Sigma \times A)$ (cf. Example 2.4.3(4)). The unique $\bar{F}$-coalgebra homomorphism $(A, \alpha) \twoheadrightarrow (J\Sigma^*, J\iota)$ is the map $t \colon A \to \mathcal{D}_{\leq}(\Sigma^*)$ assigning to a state $x \in A$ the sum of the probabilities of all its terminating traces in (A, α); using $*$ to denote the element of 1 we thus have

$$t(x)(\varepsilon) = \alpha(x)(*),$$
$$t(x)(s_0 s_1 \cdots s_n) = \sum_{x, x_1, \ldots, x_n \in X} P(x \xrightarrow{s_0} x_1 \xrightarrow{s_1} \cdots \xrightarrow{s_{n-1}} x_n \downarrow),$$

where the summands on the right are the probabilities of a single terminating trace:

$$\begin{aligned} &P(x \xrightarrow{s_0} x_1 \xrightarrow{s_1} \cdots \xrightarrow{s_{n-1}} x_n \downarrow) \\ &\quad = \alpha(x)(s_0, x_1) \cdot \alpha(x_1)(s_1, x_2) \cdot \cdots \cdot \alpha(x_{n-1})(s_{n-1}, x_n) \cdot \alpha(x_n)(*). \end{aligned}$$

Languages Accepted by Nominal Automata We now present an extended example of coalgebraic trace semantics taken from Frank et al. [126]. Data structures over infinite alphabets have received increasing attention. Infinite alphabets model the communication of values from infinite data types such as nonces [206], channel names [161], process identifiers [88], URLs [80], or data values in XML documents [243]. There are numerous automata models for infinite alphabets; partial surveys are found in [86, 274]. *Data languages* are sets of words over an infinite alphabet where one can only observe letters up to a symmetry (e.g. the equality of a pair of letters) rather than the letters themselves. One principled way of handling infinite alphabets uses nominal sets (Example 2.1.10). For example, Kaminski and Francez' classical register automata (with non-deterministic update) [181] are equivalent to *nominal orbit-finite automata* [87], which in turn can be understood as coalgebras for a functor on Nom, the category of nominal sets and equivariant functions (Example 2.1.10). To see this, recall the notions of an orbit-finite nominal set (Example 10.1.6(6)) and the finitely supported power-set functor $\mathscr{P}_{\mathsf{fs}}$ (Example 10.1.13(6)).

Example 12.3.17. (1) A non-deterministic nominal orbit-finite automaton (nofa) over $\mathbb{A}$ consists of an orbit-finite nominal set S of states equipped with equivariant subsets I and O of initial and final states, respectively, and an equivariant next-state relation $R \subseteq S \times \mathbb{A} \times S$; we write $s \xrightarrow{a} s'$ for $(s, a, s') \in R$. Equivalently, the relation R can be described as an equivariant map $S \to \mathscr{P}_{\mathsf{fs}}(\mathbb{A} \times S)$. Moreover, the equivariant subset $O \subseteq S$ is equivalently described as an equivariant function $o\colon S \to \{0, 1\} = \mathscr{P}_{\mathsf{fs}}1$, where the codomain is equipped with the discrete nominal structure: $\pi \cdot i = i$ for $i = 0, 1$. Disregarding initial states, a nofa is thus a coalgebra $S \to \mathscr{P}_{\mathsf{fs}}1 \times \mathscr{P}_{\mathsf{fs}}(\mathbb{A} \times S) \cong \mathscr{P}_{\mathsf{fs}}(1 + \mathbb{A} \times S)$. So it is a coalgebra for the functor MF on Nom, where $FX = 1 + \mathbb{A} \times X$ and $M = \mathscr{P}_{\mathsf{fs}}$.

(2) Deterministic orbit-finite nominal automata can also be described as coalgebras. They are the special case of nofas where the next-state relation is actually an equivariant function $S \times \mathbb{A} \to S$. Using currying (cf. Example 10.1.13(5)) and disregarding initial states once again, a deterministic orbit-finite nominal automaton on the orbit-finite nominal set S is described by two equivariant functions $o\colon S \to \{0, 1\}$ and $t\colon S \to S^{\mathbb{A}}$. These functions form a coalgebra for the functor $FX = \{0, 1\} \times X^{\mathbb{A}}$ on Nom.

Remark 12.3.18. The *data language* accepted by a state $s \in S$ of a nofa is defined in the same way as for ordinary non-deterministic automata. It is the set of words $a_1 a_2 \cdots a_n$ over $\mathbb{A}^*$ for which there is a sequence of transitions

$$s \xrightarrow{a_1} s_1 \xrightarrow{a_2} s_2 \xrightarrow{a_3} \cdots \xrightarrow{a_n} s_n,$$

where s_n is a final state. Note that nofas are expressively equivalent to *finite memory automata* [181].

In contrast to ordinary automata, deterministic orbit-finite nominal automata are strictly weaker than nofas (i.e. the class of data languages accepted by them is strictly contained in that accepted by nofas). The intuitive reason is that the set $\mathscr{P}_{\mathsf{fs}}S$ of finitely supported subsets of an orbit-finite (state) set S may fail to be orbit-finite again, and therefore determinization of nofas fails. Bojanczyk et al. [87, Lemma 5.4] have shown that complementation of nofas fails whereas deterministic nominal automata are easy to complement. In addition,

Veldhoven [307, Ch. 5] provides a nominal language which is accepted by a nofa but not by any deterministic orbit-finite nominal automaton.

We now sketch how the language semantics of nofa is an instance of coalgebraic trace semantics. More details and proofs can be found in the work of Frank et al. [126].

First note that the initial algebra for $FX = 1 + \mathbb{A} \times X$ on Nom is the nominal set $\mathbb{A}^*$; this is completely analogous to Example 2.2.15(2). Moreover, observe that $\mathscr{P}_{\mathsf{fs}}$ is a strong monad (cf. Remark 12.3.11) on Nom, with the monad structure and strength inherited from the power-set monad:

$$\begin{array}{lll} \eta_X \colon X \to \mathscr{P}_{\mathsf{fs}}X & \mu_X \colon \mathscr{P}_{\mathsf{fs}}\mathscr{P}_{\mathsf{fs}}X \to \mathscr{P}_{\mathsf{fs}}X & s_{X,Y} \colon X \times \mathscr{P}_{\mathsf{fs}}Y \to \mathscr{P}_{\mathsf{fs}}(X \times Y) \\ \quad x \mapsto \{x\} & \quad \mathcal{S} \mapsto \bigcup \mathcal{S} & \quad (x, S) \mapsto \{(x, s) : s \in S\}. \end{array}$$

Thus, the monad $\mathscr{P}_{\mathsf{fs}}$ also inherits commutativity from the power-set monad.

Remark 12.3.19. (1) Further below we shall also need uniformly finitely supported sets introduced by Turner and Winskel [294]. A nominal set X is *uniformly finitely supported* (ufs) if there exists a finite set of names which supports all elements of X; equivalently, $\bigcup_{x \in X} \mathsf{supp}(x)$ is finite. We write $\mathscr{P}_{\mathsf{ufs}}X$ for the set of all uniformly finitely supported subsets of a nominal set X. It is easy to see that the unit, multiplication, and strength of $\mathscr{P}_{\mathsf{fs}}$ all restrict to ufs subsets. Thus $\mathscr{P}_{\mathsf{ufs}}$ is a commutative submonad of $\mathscr{P}_{\mathsf{fs}}$.

(2) We show that the ufs subsets of an orbit-finite nominal set X are precisely the finite subsets, in symbols: $\mathscr{P}_{\mathsf{ufs}}X = \mathscr{P}_{\mathsf{f}}X$. To see this, observe first every finite subset $S \subseteq X$ is clearly ufs since $\bigcup_{s \in S} \mathsf{supp}(s)$ is a finite union of finite sets (and clearly supports S). Conversely, let $S \subseteq X$ ufs. Observe that permutations act by exchanging elements of the support

$$\mathsf{supp}(\pi \cdot x) = \pi \cdot \mathsf{supp}(x),$$

in other words, $\mathsf{supp} \colon X \to \mathscr{P}_{\mathsf{f}}X$ is an equivariant map. Thus, for every finite set $T \subseteq \mathbb{A}$ of names, there are only finitely many members of every orbit of X whose support is contained in T. Since the support of every element $s \in S$ is contained in the finite set $T = \mathsf{supp}(S)$, we conclude that S must be finite.

An obstacle seems to be that the nominal sets $\mathscr{P}_{\mathsf{fs}}X$ and $\mathscr{P}_{\mathsf{ufs}}X$ are not in general complete lattices (not even cpos) since the union of a chain of (uniformly) finitely supported subsets of X may fail to be (uniformly) finitely supported again. The following result may thus come as a mild surprise.

Proposition 12.3.20 [126, Prop. 3.7]**.** *For every pair X, Y of nominal sets, the sets $\mathsf{Nom}_{\mathscr{P}_{\mathsf{fs}}}(X, Y)$ and $\mathsf{Nom}_{\mathscr{P}_{\mathsf{ufs}}}(X, Y)$ form complete lattices (whence dcpos with bottom).*

Moreover, composition in those Kleisli categories is easily seen to be continuous and left-strict, and therefore these categories are clearly left-strictly enriched in DCPO, the category of directed-complete partial orders (cf. Remark 12.3.6(1)).

Using Example 12.3.10, we see that the functor $FX = 1 + \mathbb{A} \times X$ has an extension $\bar{F}$ to $\mathsf{Nom}_{\mathscr{P}_{\mathsf{fs}}}$, and it is easy to see that $\bar{F}$ is locally monotone. Thus, using Remark 12.3.6(1), the initial F-algebra $\overline{\mathbb{A}}^*$ lifts to a terminal coalgebra for $\bar{F}$ on $\mathsf{Nom}_{\mathscr{P}_{\mathsf{fs}}}$. It is then not difficult to work out the following result.

Theorem 12.3.21 [126, Thm. 3.19]. *For every nofa, i.e. a coalgebra $S \to \mathscr{P}_{\mathsf{fs}}(1 + \mathbb{A} \times S)$, its coalgebraic trace map $S \to \mathscr{P}_{\mathsf{fs}}(\overline{\mathbb{A}}^*)$ maps a state to the data language it accepts.*

Regular Non-deterministic Nominal Automata and Bar Languages One of the hallmarks of automata for data languages, for example, Kaminski and Francez' classical register automata [181], is their capability of explicitly storing an input letter for comparing it with a letter read later. In nominal automata this is realized by 'storing' letters in the support of a state, and the operation of storing is performed by *binding transitions*. Regular non-deterministic nominal automata (RNNA) [270] are a model with binding transitions. They can be equipped with either *global* or *local* freshness semantics, that is, names in a binding transition can either be viewed as fresh with respect to the input word read so far or as fresh with respect to boundedly many names currently stored in memory. Under global freshness semantics, RNNA are essentially equivalent to Bollig et al.'s session automata [88]. Under local freshness semantics, RNNA embed into the standard register automata model [181]. They retain a reasonable level of expressivity, in particular they not only allow any number of 'registers' but can express languages that are not acceptable by deterministic or unambiguous register automata such as 'some letter occurs twice'. Crucially, RNNA, unlike unrestricted register automata, nevertheless have a decidable inclusion problem and in fact of elementary complexity (exponential space, or more precisely parametrized polynomial space); see [270] for details. RNNA have also been used for model checking a nominal fixed point logic over finite data words [160], and they have been studied as acceptors of languages of infinite data words using a Büchi condition [299].

Here we will explain how the language semantics of RNNA is an instance of coalgebraic trace semantics. The coalgebraic description of RNNA involves the binding functor $\mathbb{A}[X]$ (see Example 2.1.10) and the uniformly finitely supported power-set functor monad $\mathscr{P}_{\mathsf{ufs}}$ (Remark 12.3.19).

Definition 12.3.22 [270]. A *regular non-deterministic nominal automaton*

(RNNA) is an orbit-finite coalgebra for the functor

$$GX = \{0,1\} \times \mathscr{P}_{\mathsf{ufs}}(\mathbb{A} \times X) \times \mathscr{P}_{\mathsf{ufs}}([\mathbb{A}]X).$$

Remark 12.3.23. (1) We thus see that RNNA are equivalently coalgebras for the functor $\mathscr{P}_{\mathsf{ufs}}F$ where $FX = 1 + \mathbb{A} \times X + [\mathbb{A}]X$. Indeed, use that we have natural isomorphisms

$$\mathscr{P}_{\mathsf{ufs}}X \times \mathscr{P}_{\mathsf{ufs}}Y \cong \mathscr{P}_{\mathsf{ufs}}(X + Y).$$

(2) Op. cit. defines RNNA more explicitly as tuples consisting of an orbit-finite state set S, an initial state, an equivariant subset of final states, and a transition relation $R \subseteq S \times \overline{\mathbb{A}} \times S$, where one uses the infinite alphabet

$$\overline{\mathbb{A}} = \mathbb{A} \cup \{|a : a \in \mathbb{A}\}.$$

(Recall bar strings and languages from Example 10.5.11(3).) We write $q \xrightarrow{\sigma} q'$ for transitions $(q, \sigma, q') \in R$. Transitions with input letter $|a$ are the above-mentioned binding transitions. The idea here is that along such a transition a (barred) input letter is read and its name bound to a for possible comparison with (unbarred) letters read later (see Example 12.3.24 for a concrete example). The relation R has to satisfy two technical conditions which precisely state that it yields a coalgebra structure for the functor G in Definition 12.3.22; we omit mentioning these conditions as they are not needed for the following development. As usual, we also omit initial states in the coalgebraic description.

(3) A word in $\overline{\mathbb{A}}^*$ is a *bar string*. As already explained in Example 10.5.11(3), one defines α-equivalence on bar strings to be the least equivalence relation $=_\alpha$ such that

$$x|av =_\alpha x|bw \qquad \text{for all } a, b \in \mathbb{A} \text{ and } x, v, w \in \overline{\mathbb{A}}^* \\ \text{such that } \langle a\rangle v = \langle b\rangle w \text{ in } [\mathbb{A}]\overline{\mathbb{A}}^*.$$

We denote by $\overline{\mathbb{A}}^*/{=_\alpha}$ the set of α-equivalence classes of finite bar strings, and we write $[w]_\alpha$ for the α-equivalence class of $w \in \overline{\mathbb{A}}^*$. The equivalence relation $=_\alpha$ is equivariant, whence $\overline{\mathbb{A}}^*/{=_\alpha}$ forms a nominal nominal set with the group action $\pi \cdot [w]_\alpha = [\pi \cdot w]_\alpha$ for $\pi \in \mathsf{Perm}(\mathbb{A})$ and $w \in \overline{\mathbb{A}}^*$.

(4) A state $s \in S$ of an RNNA *accepts* a bar string $w = \sigma_1\sigma_2 \cdots \sigma_n \in \overline{\mathbb{A}}^*$ if there is a sequence of transitions

$$s \xrightarrow{\sigma_1} s_1 \xrightarrow{\sigma_2} \cdots \xrightarrow{\sigma_n} s_n,$$

where s_n is a final state. The *bar language* accepted by the state s is defined by

$$L_\alpha = \{[w]_\alpha : w \in \overline{\mathbb{A}}^*, s \text{ accepts } w\}.$$

Example 12.3.24. Here is an easy example of an RNNA, presented graphically:

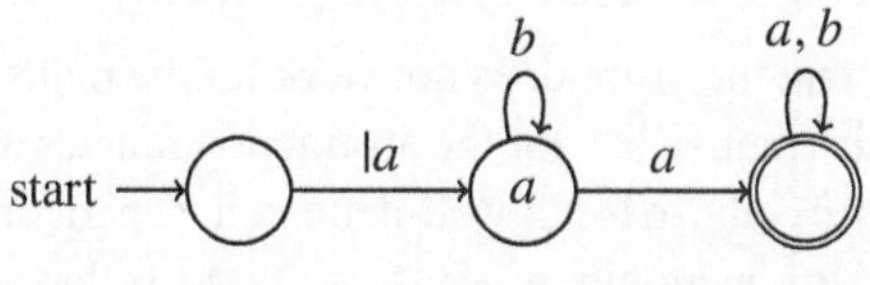

Each of the three nodes in the picture represents an orbit of the state space $1 + \mathbb{A} + 1$. The label a of the second node indicates the support of (that particular representative) state. The loop labelled b at the second node stands for any letter which is different from the one in the support of the state. An accepting run of this automaton, started in the left-hand state, reads the first letter of a word w, binds it on the name a, and then proceeds to the second state (orbit). The automaton then keeps reading letters, and if the first letter occurs a second time, then the automaton proceeds to the accepting right-hand state. Hence, the bar language of the left-hand state consists of all α-equivalence classes of words $|ab_1b_2 \cdots b_nac_1c_2 \cdots c_k$, that is words whose first barred letter occurs again.

We have already seen that the Kleisli category of the monad $\mathscr{P}_{\mathsf{ufs}}$ is left-strictly enriched in directed-complete partial orders (Proposition 12.3.20).

Furthermore, we have mentioned in Example 11.5.8(2) that the initial algebra for $FX = 1 + \mathbb{A} \times X + [\mathbb{A}]X$ is carried by the nominal set $\overline{\mathbb{A}}^* /{=_\alpha}$ of all bar strings modulo α-equivalence. We have also mentioned that there are canonical natural isomorphisms $\mathscr{P}_{\mathsf{fs}}([\mathbb{A}]X) \cong [\mathbb{A}](\mathscr{P}_{\mathsf{fs}}X)$.

Proposition 12.3.25 [126, Prop. 3.11]**.** *The above isomorphisms restrict to* $\mathscr{P}_{\mathsf{ufs}}$ *and yield a distributive law* $[\mathbb{A}]\mathscr{P}_{\mathsf{ufs}}X \to \mathscr{P}_{\mathsf{ufs}}([\mathbb{A}]X)$ *of the abstraction functor over the monad* $\mathscr{P}_{\mathsf{ufs}}$*. Moreover, the ensuing extension of* $[\mathbb{A}](-)$ *to* $\mathsf{Nom}_{\mathscr{P}_{\mathsf{ufs}}}$ *is locally monotone.*

Remark 12.3.26. (1) The proof makes use of the distributive law λ of the functor $\mathbb{A} \times (-)$ over the monad $\mathscr{P}_{\mathsf{ufs}}$ given by the strength; that is, we have $\lambda_X \colon \mathbb{A} \times \mathscr{P}_{\mathsf{ufs}}X \to \mathscr{P}_{\mathsf{ufs}}(\mathbb{A} \times X)$ given by $(a, S) \mapsto \{(a, s) : s \in S\}$. One then defines a family of equivariant maps $\varrho_X \colon [\mathbb{A}]\mathscr{P}_{\mathsf{ufs}}X \to \mathscr{P}_{\mathsf{ufs}}[\mathbb{A}]X$ by $\varrho_X(\langle a \rangle S) = \{\langle a \rangle s : s \in S\}$ and verifies that the following diagram commutes:

$$\begin{array}{ccc} \mathbb{A} \times \mathscr{P}_{\mathsf{ufs}}X & \xrightarrow{\lambda_X} & \mathscr{P}_{\mathsf{ufs}}(\mathbb{A} \times X) \\ {\scriptstyle q_{\mathscr{P}_{\mathsf{ufs}}X}} \big\downarrow & & \big\downarrow {\scriptstyle \mathscr{P}_{\mathsf{ufs}} q_X} \\ [\mathbb{A}]\mathscr{P}_{\mathsf{ufs}}X & \xrightarrow{\varrho_X} & \mathscr{P}_{\mathsf{ufs}}[\mathbb{A}]X \end{array}$$

(Furthermore, ϱ_X is the inverse of the restricted isomorphism $\mathscr{P}_{\mathsf{ufs}}([\mathbb{A}]X) \to [\mathbb{A}](\mathscr{P}_{\mathsf{ufs}}X)$ from Example 11.5.8(2).) By Remark 12.3.9, we thus see that ϱ is a distributive law of the abstraction functor over the monad $\mathscr{P}_{\mathsf{ufs}}$. One readily shows that the ensuing extension on $\mathsf{Nom}_{\mathscr{P}_{\mathsf{ufs}}}$ is locally monotone.

(2) However, note that the proof does not work for the monad $\mathscr{P}_{\mathsf{fs}}$, whence one does not obtain a distributive law for the abstraction functor over $\mathscr{P}_{\mathsf{fs}}$ in this way. The problem is that ϱ_X above is not well-defined in general if S is not uniformly finitely supported. For example, given $\mathbb{A} \in \mathscr{P}_{\mathsf{fs}}\mathbb{A}$ we have $\langle a\rangle\mathbb{A} = \langle b\rangle\mathbb{A}$ for every pair a, b of names. However, if $a \neq b$, then the sets $\{\langle a\rangle c : c \in \mathbb{A}\}$ and $\{\langle b\rangle c : c \in \mathbb{A}\}$ differ: $\langle a\rangle b$ is contained in the former but not in the latter set. In fact, since $a \neq b$, $\langle a\rangle b = \langle b\rangle c$ can hold only if a is fresh for $\{b, c\}$ and $b = (a\,b) \cdot c$ (see Pitts [250, Lemma 4.3]). The latter means that $c = a$, contradicting freshness of a for it.

Finally, using Remark 12.3.6(1) we obtain that $\bar{\mathbb{A}}^*/{=_\alpha}$ extends to a terminal coalgebra for the extension $\bar{F}$.

Theorem 12.3.27 [126, Thm. 3.21]**.** *For an RNNA, regarded as a coalgebra $S \to \mathscr{P}_{\mathsf{ufs}}(1 + \mathbb{A} \times A + [\mathbb{A}]X)$, the coalgebraic trace map $S \to \mathscr{P}_{\mathsf{ufs}}(\bar{\mathbb{A}}^*/{=_\alpha})$ assigns to each state the bar language it accepts.*

This theorem and Theorem 12.3.21 can be seen as instances of a more general result for a class of functors that we now briefly mention.

Remark 12.3.28. (1) The class of *binding polynomial functors* on Nom is the smallest class of functors containing the constant functors, the identity and abstraction functors, and being closed under coproducts, finite products, and composition. In other words, binding polynomial functors are formed according to the grammar:

$$F ::= C \mid \mathsf{Id} \mid [\mathbb{A}](-) \mid F \times F \mid \textstyle\coprod_{i \in I} F_i \mid FF, \tag{12.4}$$

where C ranges over all constant functors on Nom and I is an arbitrary index set.

Recall from Example 10.1.6(6) that Nom is locally finitely presentable. Moreover every binding polynomial functor is finitary; this holds since $[\mathbb{A}](-)$ is a left adjoint [250, Thm. 4.13] and finitary functors are closed under coproducts, finite products, and composition (cf. Example 10.1.13(1)). By Theorem 11.2.18, every binding polynomial functor on Nom has an initial algebra and a terminal coalgebra.

(2) A special instance are functors F_Σ arising from a binding signature (Remark 2.1.11). The initial algebra for F_Σ is carried by the nominal set T_Σ of α-equivalence classes of terms of the binding signature (see Gabbay and Pitts [133, Rem. 6.13] and cf. Pitts [250, Thm. 8.15]).

Moreover, for every countable binding signature Σ, the terminal coalgebra for F_Σ is carried by the nominal set of α-equivalence classes of all finite and infinite trees over Σ with finitely many free variables (see Kurz et al. [209, Cor. 5.35]).

(3) Similarly, endofunctors on Nom obtained from nominal algebraic signatures in the sense of Pitts [250, Def. 8.12] are instances of binding polynomial functors.

(4) The fact that the abstraction functor admits a distributive law over $\mathscr{P}_{\mathsf{ufs}}$ can be combined with what we have seen in Examples 12.3.10 and 12.3.13 to obtain that every binding polynomial functor F on Nom has a canonical extension to the Kleisli category of $\mathscr{P}_{\mathsf{ufs}}$. Moreover, it is easy to show that the canonical extension $\bar{F}$ on $\mathsf{Nom}_{\mathscr{P}_{\mathsf{ufs}}}$ is locally monotone.

Similarly, every polynomial functor on Nom, that is, one obtained from the grammar in (12.4) without the abstraction functor $[\mathbb{A}](-)$, has a canonical locally monotone extension to the Kleisli category of $\mathscr{P}_{\mathsf{fs}}$.

Thus from Remark 12.3.6(1) we obtain that every (binding) polynominal functor on Nom has a canonical fixed point on $\mathscr{A}_{\mathscr{P}_{\mathsf{fs}}}$ (and $\mathscr{A}_{\mathscr{P}_{\mathsf{ufs}}}$, respectively) and an ensuing coalgebraic trace semantics.

Corollary 12.3.29. (1) *For every polynomial functor F on* Nom*, its canonical extension $\bar{F}$ on $\mathscr{A}_{\mathscr{P}_{\mathsf{fs}}}$ has a canonical fixed point given by μF.*

(2) *For every binding polynomial functor F on* Nom*, its canonical extension $\bar{F}$ on $\mathscr{A}_{\mathscr{P}_{\mathsf{ufs}}}$ has a canonical fixed point given by μF.*

Further Results for Kleisli Categories of Set Monads We proceed to show that without the cpo-structure on hom-sets of Set_M in Theorem 12.3.5, an extended endofunctor need not have a canonical fixed point. To see this let $MX = X^E$ be the environment monad (Example 12.3.2(4)). Then every endofunctor on Set (and, more generally, on every category with powers) has a canonical distributive law over M given by:

$$\lambda_X \colon F(X^E) \to (FX)^E \qquad \text{with } \pi_e \cdot \lambda_X = F\pi_e \text{ for every } e \in E.$$

So according to Proposition 12.3.4 an initial F-algebra extends to an initial $\bar{F}$-algebra. Moreover, from [231, Corollary 5.6] we obtain the following result:

Proposition 12.3.30. *Let $MX = X^E$ be the environment monad, let F be a set functor and $\bar{F}$ be its canonical extension to* Set_M*. Then a terminal F-coalgebra $(\nu F, \tau)$ yields a terminal $\bar{F}$-coalgebra $(\nu F, J\tau)$.*

So here we see an example of an endofunctor F on Set (namely X^E) for which the initial algebra and terminal coalgebra do not coincide, and the same is true for the extended functor $\bar{F}$.

Now consider the power-set monad $M = \mathcal{P}$: here is a result which allows one to extend terminal coalgebras for polynomial set functors due to Jacobs:

Theorem 12.3.31 [171, Thm. 6.5]. *Let $M = \mathcal{P}$ be the power-set monad. For every polynomial set functor F the terminal coalgebra (F, τ) yields a weakly terminal coalgebra $(\nu F, J\tau)$ for the canonical extension $\bar{F}$.*

Moreover, note that the hom-sets of $\mathsf{Set}_{\mathcal{P}}$ (the category of relations) form complete lattices under the pointwise order structure. Given a coalgebra (A, α) for $\bar{F}$, every coalgebra morphism h from (A, α) to the weakly terminal coalgebra $(\nu F, J\tau)$ is a fixed point of the monotone function

$$h \mapsto J\tau^{-1} \cdot \bar{F}h \cdot \alpha. \tag{12.5}$$

By the Knaster–Tarski Theorem B.1.7, these homomorphisms form a complete lattice. As shown by Jacobs [171], the greatest homomorphism plays a special role because it yields infinite trace semantics.

Example 12.3.32. Consider the polynomial set functor $FX = \Sigma \times X + 1$. Its terminal coalgebra is formed by the set $\Sigma^* + \Sigma^\infty$ of all finite and infinite sequences of elements of the label alphabet Σ (Example 2.5.3(4)). Given a labelled transition system $\alpha\colon A \to \mathcal{P}(\Sigma \times A + 1)$, the greatest coalgebra homomorphism $h\colon A \to \mathcal{P}(\Sigma^* + \Sigma^\infty)$ assigns to a state $x \in A$ the set of all its finite terminating traces and all of its infinite ones.

Remark 12.3.33. Theorem 12.3.31 has inspired a number of subsequent studies on (infinite) trace semantics. We mention a few of them explicitly:

(1) Urabe and Hasuo [295, Prop. 4.2] give a list of conditions on a monad M and a functor F on a category $\mathcal{A}$ under which the terminal F-coalgebra is weakly terminal for the extension $\bar{F}$. They show that this applies to the sub-Giry monad (a variation of the well-known Giry monad [140]) and polynomial functors on the category of measurable spaces.

(2) Cîrstea [104, 105] provides a related but slightly different setting where a terminal F-coalgebra lifts to a weakly terminal coalgebra for $\bar{F}$. Technically this is based on extending the terminal-coalgebra ω^{op}-chain to Set_M, with its limit in Set becoming a weak limit in Set_M.

Again, as a consequence, the terminal F-coalgebra extends to a weakly terminal coalgebra on Set_M [105, Prop. 2]. That setting applies to the Giry monad on the category of standard Borel spaces, but it does not handle the sub-Giry monad or the power-set monad. However, the *nonempty* power-set monad is handled in op. cit. For that monad, a trace maps turns out to be an *op-lax* coalgebra homomorphism, this means that it is a pre-fixed point of the function in (12.5).

(3) Cîrstea [106] provides a setting for possibly infinite trace semantics for commutative partially additive monads M based on generalized relation liftings (cf. Section 12.4) of the functor F; the involved relations are maps $X \times Y \to M1$, where the assumptions on M ensure that $M1$ is a partial commutative semiring. Trace semantics for a coalgebra $\alpha\colon A \to MFA$ in this setting is provided by taking the greatest fixed point of an endofunction given by α on the set of generalized relations from A to νF; this is reminiscent of what we saw above for homomorphisms being fixed points.

(4) Kerstan and König [189, Thm. 3.33] provide trace semantics for continuous probabilistic systems by giving explicit descriptions of the terminal coalgebras for the extensions of $FX = \Sigma \times X$ and $GX = \Sigma \times X + 1$ on the category of measurable spaces to the Kleisli categories of the Giry and sub-Giry monads: $\nu\bar{F} = \emptyset$, $\nu\bar{G} = \Sigma^*$ for the sub-Giry monad, and $\nu\bar{F} = \Sigma^\omega$ and $\nu\bar{G} = \Sigma^\infty$ for the Giry monad, each equipped with suitable measurable space structures

(5) Urabe et al. [296] provide a coalgebraic view on the semantics of Büchi automata and parity automata based on solving alternating fixed point equations in the complete lattice $\mathsf{Set}_{\mathscr{P}}(A, \nu F)$. Here is the idea: a non-deterministic Büchi automaton can be understood as a coalgebra $\alpha\colon A \to \mathscr{P}FA$, where $FX = \Sigma \times X + 1$. One decomposes the carrier set as a coproduct $A = A_n + A_f$ where A_f consists of the final (or terminating) states and A_n contains all non-final ones; the restrictions of α to the components of the coproduct (in $\mathsf{Set}_{\mathscr{P}}$) are denoted as follows

$$\alpha_n\colon A_n \rightarrow \bar{F}A \qquad \text{and} \qquad \alpha_f\colon A_f \rightarrow \bar{F}A.$$

The language accepted by the Büchi automaton is then the solution $t = [t_n, t_f]\colon A \to \nu F$ of the alternating fixed point equation given by the following squares

$$\begin{array}{ccc} A_n & \xrightarrow{\alpha_n} & \bar{F}A \\ {\scriptstyle t_n}\big\downarrow & =_\mu & \big\downarrow{\scriptstyle \bar{F}[t_n, t_e]} \\ \nu F & \xrightarrow{J\tau} & \bar{F}(\nu F) \end{array} \qquad \begin{array}{ccc} A_f & \xrightarrow{\alpha_f} & \bar{F}A \\ {\scriptstyle t_f}\big\downarrow & =_\nu & \big\downarrow{\scriptstyle \bar{F}[t_n, t_f]} \\ \nu F & \xrightarrow{J\tau} & \bar{F}(\nu F) \end{array}$$

where $=_\mu$ means taking least and $=_\nu$ taking greatest fixed points. More precisely, in order to obtain t, one solves the following mutally recursive system of alternating fixed point equations in the complete lattice $\mathsf{Set}_{\mathscr{P}}(A, \nu F)$:

$$t_n =_\mu J\tau^{-1} \cdot \bar{F}[t_n, t_f] \cdot \alpha_n \qquad t_f =_\nu J\tau^{-1} \cdot \bar{F}[t_n, t_f] \cdot \alpha_f.$$

Urabe et al. [296, Appendix B] gives an explanation how the solution is obtained by taking least and greatest fixed points of appropriate monotone maps on complete lattices.

(6) For parity (tree) automata, the treatment is similar by decomposing $A = A_1 + \cdots + A_n$ where A_i consists of all states with parity i (and replacing F by an arbitrary polynomial set functor). As shown by Urabe et al. [296, Thm. 4.6], the solution of the ensuing system of n alternating fixed point equations provides the language map $t\colon A \to \Sigma^* + \Sigma^\infty$ of the given automaton.

12.4 Relation Liftings

In the setting of sets and relations (the Kleisli category $\mathsf{Set}_{\mathscr{P}}$) there is a different way to obtain an extension of a set functor F via spans. To see this, note that a relation $R \subseteq X \times Y$ can be written as a span of maps r_0, r_1 by restricting the product projections of $A \times B$ to R, as shown on the left below:

As shown on the right above, one can apply the functor F to the span and then take the image in $FA \times FB$ to obtain the relation

$$\bar{F}(R) = \{(Fr_0(u), Fr_1(u)) \mid u \in FR\} \subseteq FX \times FY.$$

This is the *relation lifting* of F. The relation $\bar{F}$ extends F in the sense that (12.2) clearly commutes. As for the functoriality of the relation lifting, observe that $\bar{F}$ clearly preserves identity relations. For relation composition we have the following result (see Section C.2 for weak pullback preservation):

Theorem 12.4.1. *The relation lifting of a set functor F satisfies $\bar{F}(R \cdot S) \subseteq \bar{F}R \cdot \bar{F}S$. Moreover, $\bar{F}(R \cdot S) = \bar{F}R \cdot \bar{F}S$ if and only if F preserves weak pullbacks.*

The definition of relation lifting as well as the above theorem appear in Barr's paper [70] on relational algebras, the only exception being the 'only if' part of the theorem, which was proved by Trnková [293].

Example 12.4.2. (1) For the power-set functor $\mathscr{P}$, the relation lifting $\bar{\mathscr{P}}$ maps a relation $R \subseteq X \times Y$ to the relation defined for all $S \subseteq X$ and $T \subseteq Y$ by

$$S\ (\bar{\mathscr{P}}R)\ T \qquad \text{iff} \qquad \forall x \in S.\, \exists y \in T.\, x\, R\, y \text{ and } \forall y \in T.\, \exists x \in S.\, x\, R\, y.$$

(2) For the functor R from Example 7.4.6 the relation lifting is not a functor since R does not preserve weak pullbacks (this follows from Lemma C.3.17 and Example C.3.8(6)).

Remark 12.4.3. (1) As explained by Kurz and Velebil [208], the relation lifting $\bar{F}$ is locally monotone. Moreover, $\bar{F}$ is the unique locally monotone extension of the set functor F to Rel. In fact, it has been shown by Carboni et al. [102] that this is part of a bijection between locally monotone endofunctors on Rel and set functors preserving weak pullbacks.

(2) Being an extension, $\bar{F}$ can of course be obtained by a distributive law of F over $\mathscr{P}$. Kurz and Velebil ask whether there is an extension of a set functor on Rel which is not a relation lifting. One such example is the functor underlying the canonical comonad given by the standard adjunction $J \dashv K \colon \mathsf{Rel} = \mathsf{Set}_{\mathscr{P}} \to \mathsf{Set}$: the functor $JK \colon \mathsf{Rel} \to \mathsf{Rel}$ clearly is an extension of $\mathscr{P} \colon \mathsf{Set} \to \mathsf{Set}$ which is not the relation lifting.

Remark 12.4.4. Relation liftings are of interest because the notion of a bisimulation (Definition 2.6.6) can be equivalently formulated with their help. This goes back to Rutten [263, Cor. 3.1]. More recently, Marti and Venema [223] have shown how to achieve this for set functors not preserving weak pullbacks by using lax extensions to Rel. Other authors have studied extensions and generalization of the notion of relation liftings in order to capture other notions of process equivalences such as (metric) simulations in a coalgebraic setting. In these works Set and Rel are replaced by a different base category and appropriate category of 'relations' over it. Hughes and Jacobs [166] present a detailed study in which preordered sets form the base category, whereas Levy [215] works over the category of posets. The conceptual difference between those two settings is that in the former case one captures similarity (via the order on the terminal coalgebra) and bisimilarity (via equality) simultaneously. This reflects the fact that mutual similarity is weaker than bisimilarity, a difference not captured using posets.

Worrel [317] works over $\mathscr{V}$-categories in lieu of sets, where $\mathscr{V}$ is a commutative unital quantale. For example, preordered sets and generalized (ultra-)metric spaces are instances of $\mathscr{V}$-categories. Relations are replaced by $\mathscr{V}$-bimodules, and Worrell considers lax extensions of a functor F on the category of all small $\mathscr{V}$-categories to the category of $\mathscr{V}$-bimodules. The ensuing notion of an F-simulation yields the $\mathscr{V}$-category structure on the terminal coalgebra, which instantiates to similarity or a metric similarity in the leading examples.

The above works are complemented by Balan and Kurz [63] who show how to extend set functors to the categories of preorders and posets and, jointly with Velebil [64], to $\mathscr{V}$-categories, respectively.

We refer to Kurz and Velebil's survey [208] for further details on relation liftings and pointers to the literature.

12.5 Eilenberg–Moore Algebras

Next we turn to coalgebras with a carrier having an extra algebraic structure given again by a monad. Recall that every monad (M, η, μ) on a category $\mathscr{A}$ yields its *Eilenberg–Moore* category $\mathscr{A}^M$: the objects are the Eilenberg–Moore algebras for M which are M-algebras $\alpha\colon MA \to A$ such that

$$\alpha \cdot \eta_A = \mathsf{id}_A \qquad \text{and} \qquad \alpha \cdot \mu_A = \alpha \cdot T\alpha,$$

and the morphisms are the usual M-algebra homomorphisms.

Leading examples of Eilenberg–Moore categories are *varieties of algebras*. A variety is specified by a pair (Σ, E), where Σ is a signature of operation symbols and E a set of equations (i.e. pairs of Σ-terms). It is the category of Σ-algebras satisfying the equations in E and Σ-algebra homomorphisms. Varieties where the arities of operations in Σ are bounded above by a cardinal λ are precisely the Eilenberg–Moore algebras for λ-accessible monads on Set. In fact, for a variety specified by (Σ, E), the corresponding monad M assigns to a set X the free algebra of the variety on X; this can be described as the algebra $T_\Sigma X/{\sim_E}$ of Σ-terms on X modulo the (least congruence given by the) equations in E. Typical examples of varieties are the categories of (semi)groups, monoids, (semi)rings, modules for a fixed (semi)ring, vector spaces over a fixed field, and many others. Observe that all these examples are *finitary* varieties (that is, they can be specified by (Σ, E) where every operation in Σ has finite arity).

Note that every object X of $\mathscr{A}$ generates a *free* Eilenberg–Moore algebra MX with structure and universal morphism given as follows:

$$\mu_X\colon MMX \to MX \qquad \text{and} \qquad \eta_X\colon X \to MX.$$

The Kleisli category of M in Section 12.3 can then be regarded as the full subcategory of all free Eilenberg–Moore algebras via the functor $K\colon \mathscr{A}_M \to \mathscr{A}^M$ given by

$$K\big(X \xrightarrow{f} Y\big) = \big(MX \xrightarrow{Mf} MMY \xrightarrow{\mu_Y} MY\big).$$

Indeed, one easily verifies that $\mu_Y \cdot Tf$ is an M-algebra homomorphism from (MX, μ_X) to (MY, μ_Y).

Thus, Theorem 12.3.5 provides a terminal coalgebra for an extended functor on the category of free Eilenberg–Moore algebras for a monad. Now we will ask when the terminal coalgebra for a given endofunctor F on $\mathscr{A}$ lifts to the category $\mathscr{A}^M$. Not surprisingly, this is true whenever the functor F has a lifting to $\mathscr{A}^M$, analogously as for initial algebras in Kleisli categories (see Proposition 12.3.4).

Remark 12.5.1. (1) Analogously to (12.1) and instantiating Definition 11.5.4(1) we say that an endofunctor $\bar{F}\colon \mathscr{A}^M \to \mathscr{A}^M$ is a *lifting* of an endofunctor $F\colon \mathscr{A} \to \mathscr{A}$ if the following square commutes:

$$\begin{array}{ccc} \mathscr{A}^M & \xrightarrow{\bar{F}} & \mathscr{A}^M \\ {\scriptstyle U}\downarrow & & \downarrow{\scriptstyle U} \\ \mathscr{A} & \xrightarrow{F} & \mathscr{A} \end{array}$$

Here $U\colon \mathscr{A}^M \to \mathscr{A}$ is the forgetful functor mapping an Eilenberg–Moore algebra to its underlying object.

(2) As shown by Applegate [60] (see also Johnstone [177]), liftings of F to $\mathscr{A}^M$ are in bijective correspondence with *distributive laws of M over F*: this is a natural transformation $\lambda\colon MF \to FM$ satisfying

$$\lambda \cdot \eta F = F\eta \qquad \text{and} \qquad \lambda \cdot \mu F = F\mu \cdot \lambda M \cdot M\lambda.$$

The following theorem follows from the work of Turi and Plotkin [252] (see also Bartels [73, Theorem 3.2.3]).

Theorem 12.5.2. *Let F have a lifting $\bar{F}$ on the Eilenberg–Moore category of the monad M. If F has a terminal coalgebra, then it lifts to a terminal $\bar{F}$-coalgebra.* More concretely, suppose that F has a terminal coalgebra. Then there exists a unique F-coalgebra homomorphism $\alpha\colon M(\nu F) \to \nu F$ such that the following diagram commutes:

$$\begin{array}{ccccc} M(\nu F) & \xrightarrow{M\tau} & MF(\nu F) & \xrightarrow{\lambda_{\nu F}} & FM(\nu F) \\ {\scriptstyle \alpha}\downarrow & & & & \downarrow{\scriptstyle F\alpha} \\ \nu F & & \xrightarrow{\qquad\quad\tau\qquad\quad} & & F(\nu F) \end{array}$$

It is then easy to prove that $(\nu F, \alpha)$ is an Eilenberg–Moore algebra for M such that the coalgebra structure $\tau\colon \nu F \to \bar{F}(\nu F)$ is an M-algebra homomorphism (in fact, this is precisely what the commutativity of the above diagram expresses). Moreover, $(\nu F, \tau)$ is a terminal $\bar{F}$-coalgebra.

Remark 12.5.3. The forgetful functor $U\colon \mathscr{A}^M \to \mathscr{A}$ has the left adjoint L given by $LX = (MX, \mu_X)$, where $M = UL$ is the monad induced by the adjunction $L \dashv U$. Note that by Theorem 11.5.2, $\bar{F}$ being a lifting of F entails that this adjunction lifts to the level of coalgebras as displayed in the diagram below (where the vertical arrows are the forgetful functors):

$$\begin{array}{ccc} \mathsf{Coalg}\,F & \underset{\hat{U}}{\overset{\hat{L}}{\underset{\longleftarrow}{\overset{\longrightarrow}{\scriptstyle\perp}}}} & \mathsf{Coalg}\,\bar{F} \\ \downarrow & & \downarrow \\ \mathscr{A} & \underset{U}{\overset{L}{\underset{\longleftarrow}{\overset{\longrightarrow}{\scriptstyle\perp}}}} & \mathscr{A}^M \end{array}$$

However, Theorem 12.5.2 is *not* just a consequence of the corresponding lifting result Corollary 11.5.5(1); that result just states that $U(\nu\bar{F}) = \nu F$, which does not imply that $\nu\bar{F}$ exists if νF does nor what the algebra structure on νF is.

The following result is helpful in obtaining liftings of functors F to $\mathscr{A}^M$. Given an adjunction $L \dashv R$, the *adjoint transpose* of a morphism $f\colon X \to RY$ is its image under the inverse of the natural isomorphism $\mathscr{C}(LX, Y) \cong \mathscr{C}(X, RY)$,

Proposition 12.5.4. *Let $M\colon \mathscr{A} \to \mathscr{A}$ be a monad and $L \dashv R\colon \mathscr{A} \to \mathscr{A}$ an adjunction with counit $\varepsilon\colon LR \to \mathsf{Id}$. If $\lambda\colon LM \to ML$ is a distributive law of L over M, then we obtain a distributive law $\varrho\colon MR \to RM$ of M over R as the adjoint transpose of*

$$LMR \xrightarrow{\lambda R} MLR \xrightarrow{M\varepsilon} M.$$

Proof The adjoint transpose in the statement is given by

$$LX \xrightarrow{Lf} LRY \xrightarrow{\varepsilon_Y} Y.$$

This implies that a unique natural transformation $\varrho\colon MR \to RM$ is defined by the commutativity of the square below:

$$\begin{array}{ccc} LMR & \xrightarrow{L\varrho} & LRM \\ {\scriptstyle \lambda R}\downarrow & & \downarrow{\scriptstyle \varepsilon M} \\ MLR & \xrightarrow{M\varepsilon} & M \end{array}$$

We verify that the two properties of a distributive law for ϱ follow from those of λ by taking adjoint transposes. That is, it suffices to show that the desired diagrams commute when we apply L to them and post-compose with εM. For the unit law $\varrho \cdot \eta R = R\eta$ we obtain the commutative diagram below:

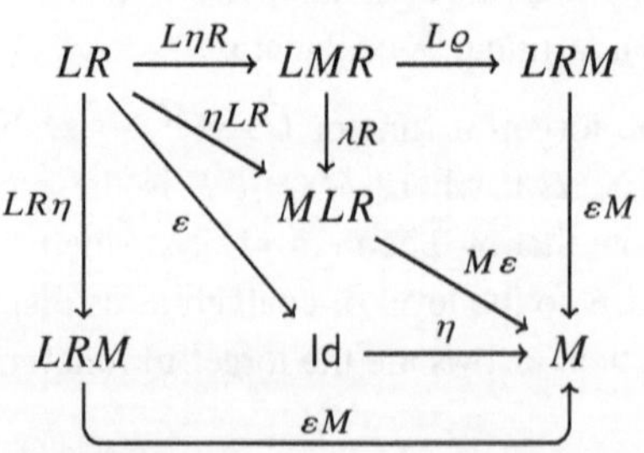

The right-hand part commutes by the definition of ϱ, the left-hand triangle by the unit law for λ, the middle triangle by the naturality of η and the remaining lower part by the naturality of ε.

For the multiplication law $\varrho \cdot \mu R = R\mu \cdot \varrho M \cdot M\varrho$ we obtain the diagram below:

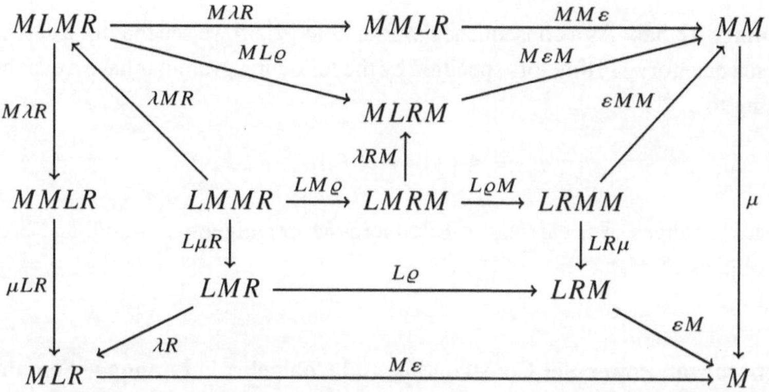

To see that the outside commutes remove $M\lambda R$ at the beginning of both paths and use the naturality of μ. The upper triangle commutes by the definition of ϱ, the left-hand part below it by the naturality of λ, the part to its right commutes by the definition of ϱ, the left-hand part commutes by the multiplication law for λ, the right-hand part commutes by the naturality of ε, and the lower part by the definition of ϱ. Thus, the middle rectangle commutes when post-composed by εM, as required. □

Example 12.5.5. (1) The identity functor on $\mathscr{A}$ obviously lifts to $\mathscr{A}^M$, and so does a constant functor on the carrier object of an Eilenberg–Moore algebra for M.

(2) Suppose that $\mathscr{A}$ has products. Then for a product $F \times G$ of endofunctors one uses that $U \colon \mathscr{A}^M \to \mathscr{A}$ preserves products. Given liftings $\bar{F}$ and $\bar{G}$, it is then clear that $\bar{F} \times \bar{G}$ is a lifting of $F \times G$. This works similarly for arbitrary products.

(3) Liftings compose: if F and G have liftings $\bar{F}$ and $\bar{G}$, then GF has a lifting defined by $\overline{GF} = \bar{G}\bar{F}$.

(4) If $\mathscr{A}$ is cartesian closed and the monad M is strong (Remark 12.3.11), then the exponentiation functor $(-)^A$ lifts to $\mathscr{A}^M$ for every object A of $\mathscr{A}$. In fact, we apply Proposition 12.5.4 to the adjunction $A \times (-) \dashv (-)^A$ and use that two of the axioms of the strength $A \times MX \to M(A \times X)$ state that it is a distributive law of $A \times (-)$ over M.

(5) As a concrete example, suppose that (A, α) is any Eilenberg–Moore algebra for a set monad M, and let Σ be a set (input alphabet). Then the functor $FX = A \times X^\Sigma$ has a lifting on $\mathscr{A}^M$. It assigns to an Eilenberg–Moore algebra

(B,β) the product algebra on $A \times B^{\Sigma}$ carrying the canonical coordinatewise algebra structure.

Remark 12.5.6. As a consequence of Example 12.5.5 we see that for a cartesian closed category $\mathscr{A}$, functors specified by the following grammar have a canonical lifting to $\mathscr{A}^M$:

$$F ::= A \mid \mathsf{Id} \mid F \times F \mid F^{\Sigma} \mid FF,$$

where A ranges over carriers of Eilenberg–Moore algebras (A,α), and Σ over all objects of $\mathscr{A}$.

Generalized Power-Set Construction and Coalgebraic Language Semantics
We will now present an important application of lifted terminal coalgebras. This concerns a generalization of the power-set (or subset) construction from automata theory. This construction, due to Rabin and Scott [258], turns a given non-deterministic automaton into an equivalent deterministic one. The branching type of non-determinism can be replaced by an arbitrary monad M and the transition type of deterministic automata by a coalgebraic type functor F. This idea goes back to Bartels' thesis [73, Sec. 4.4.3] and was elaborated and fully worked out by Silva et al. [278].

Construction 12.5.7 (Generalized Power-Set Construction). Let (M,η,μ) be a monad and F an endofunctor on the category $\mathscr{A}$. Suppose that $\alpha\colon A \to FMA$ is a coalgebra such that its codomain carries an Eilenberg–Moore algebra structure $\beta\colon M(FMA) \to FMA$. Using that (MA,μ_A) is the free Eilenberg–Moore algebra on A we obtain a unique homomorphism of M-algebras

$$\alpha^{\sharp}\colon (MA,\mu_A) \to (FMA,\beta)$$

such that $\alpha^{\sharp} \cdot \eta_A = \alpha$. The F-coalgebra $(MA,\alpha^{\sharp})$ is called the *(generalized) determinization* of (A,α).

Remark 12.5.8. If the functor F has a lifting $\bar{F}\colon \mathscr{A}^M \to \mathscr{A}^M$, then the Eilenberg–Moore algebra structure β can be taken as the structure of $\bar{F}(MA,\mu_A)$. So a determinization exists for every coalgebra $\alpha\colon A \to MFA$, and $(MA,\alpha^{\sharp})$ is then an $\bar{F}$-coalgebra.

Definition 12.5.9 (Coalgebraic Language Semantics). Given (M,η,μ), F and (A,α) as in Construction 12.5.7, the *coalgebraic language morphism* $\ddagger\alpha\colon A \to \nu F$ is obtained by pre-composing the unique F-coalgebra homomorphism h from the determinization of (A,α) to νF with η_A, which is summarized in the diagram below:

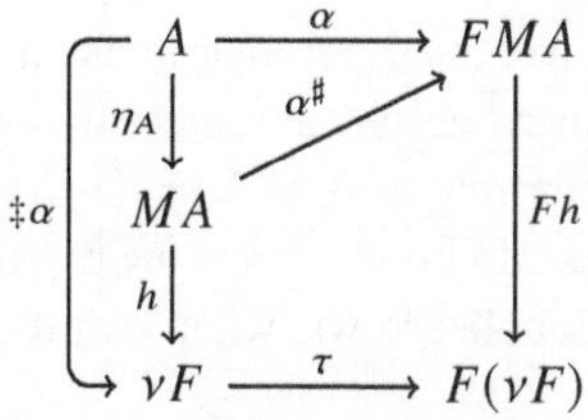

For $\mathscr{A}$ = Set, there is an ensuing notion of *coalgebraic language equivalence* of a pair x, y of states in FM-coalgebras; it is defined to be behavioural equivalence (with respect to F) of $\eta(x)$ and $\eta(y)$ in the determinizations of the given FM-coalgebras. This equivalence was first studied by Silva et al. [278]; the name goes back to [89].

Example 12.5.10. (1) The language semantics of non-deterministic automata is our first instance. Recall from Example 4.5.23 that non-deterministic automata with input alphabet Σ are coalgebras $\alpha\colon A \to \{0,1\} \times (\mathscr{P}_{\mathsf{f}}A)^{\Sigma}$. That is, we consider $M = \mathscr{P}_{\mathsf{f}}$ and $FX = \{0,1\} \times X^{\Sigma}$ on Set. By Example 12.5.5(5), F lifts to the Eilenberg–Moore category $\mathsf{Set}^{\mathscr{P}_{\mathsf{f}}}$, which is well known to be (isomorphic to) the category of join-semilattices with bottom.

Given a non-deterministic automaton as an FM-coalgebra, one readily verifies that its determinization $\alpha^{\sharp}\colon \mathscr{P}_{\mathsf{f}}A \to \{0,1\} \times (\mathscr{P}_{\mathsf{f}}A)^{\Sigma}$ is the deterministic automaton obtained by the well-known power-set construction. Now recall from Example 2.5.5 that νF is carried by the set $\mathscr{P}\Sigma^*$ of all formal languages on Σ. The coalgebraic language map $\ddagger\alpha\colon A \to \mathscr{P}(\Sigma^*)$ therefore assigns to each state of A the formal language accepted by that state.

(2) Languages accepted by weighted automata. We have already seen in Example 2.4.6 that automata with weights in a semiring $\mathbb{S}$ can be understood as coalgebras for the functor $\bar{F}X = \mathbb{S} \times X^{\Sigma}$ on the category $\mathbb{S}$-Mod of $\mathbb{S}$-semimodules. According to Example 12.5.5(5) this functor is the lifting of the set functor F given by the same formula. It is well known that for the additive monoid of a semiring $\mathbb{S}$ the ensuing functor $\mathbb{S}^{(-)}$ (Example 2.4.3(1)) is a monad on Set. It assigns to a set X the free $\mathbb{S}$-semimodule on X. For example, if X is a finite set, say $|X| = n$, then we have $\mathbb{S}^{(X)} = \mathbb{S}^n$. It follows that weighted automata are precisely the determinizations of finite coalgebras for $F\mathbb{S}^{(-)}$. To see this, recall that a weighted automaton is a coalgebra $\beta\colon \mathbb{S}^n \to \mathbb{S} \times (\mathbb{S}^n)^{\Sigma}$ for $\bar{F}$. Replacing $\mathbb{S}^n$ by $\mathbb{S}^{(X)}$ for some n-element set X and pre-composing by the unit $\eta_X\colon X \to \mathbb{S}^{(X)}$ we obtain an $F\mathbb{S}^{(-)}$-coalgebra

$$\alpha\colon X \xrightarrow{\eta_X} \mathbb{S}^{(X)} \xrightarrow{\beta} \mathbb{S} \times (\mathbb{S}^{(X)})^{\Sigma}.$$

Then we have $\beta = \alpha^{\#}$, since β is a morphism of $\mathbb{S}$-semimodules which restricts to the map α and $\alpha^{\#}$ is the unique $\mathbb{S}$-semimodule morphism with that property. Note that the presentation of a weighted automaton as an $F\mathbb{S}^{(-)}$-coalgebra (X, α) corresponds nicely to the usual graphical representation. For example, consider the following weighted automaton over $\Sigma = \{a, b\}$ with weights from the semiring of natural numbers, so $\mathbb{N}^{(-)} = \mathscr{B}$ is the bag monad (output weights in states are represented by double arrows; when omitted they are 0)

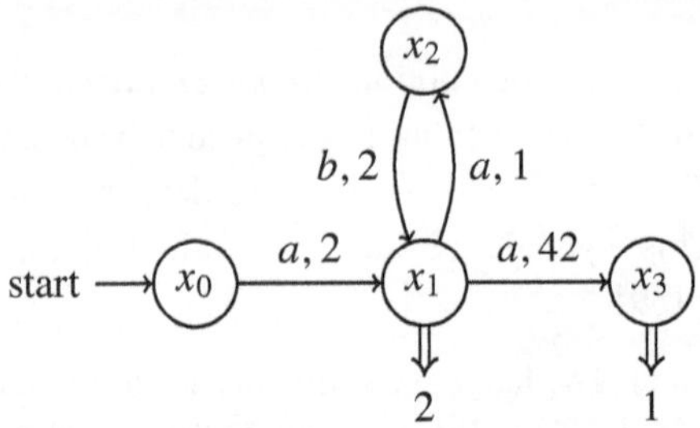

It represents a coalgebra on $X = \{x_0, x_1, x_2, x_3\}$ with structure $\alpha = \langle o, t\rangle : X \to \mathbb{N} \times (\mathscr{B}X)^{\Sigma}$ given as follows (where we use an obvious set-like notation for multisets):

$$
\begin{aligned}
&o(x_0) = 0, && t(x_0)(a) = \{2 \cdot x_1\}, && t(x_0)(b) = \emptyset,\\
&o(x_1) = 2, && t(x_1)(a) = \{1 \cdot x_2, 42 \cdot x_3\}, && t(x_1)(b) = \emptyset,\\
&o(x_2) = 0, && t(x_2)(a) = \emptyset, && t(x_2)(b) = \{2 \cdot b\},\\
&o(x_3) = 1, && t(x_3)(a) = \emptyset, && t(x_3)(b) = \emptyset.
\end{aligned}
$$

Coming back to general weighted automata, the terminal coalgebra νF is carried by the set $\mathbb{S}^{\Sigma^*}$ of all formal power series (Example 2.5.7(1)), and according to Theorem 12.5.2 this lifts to a terminal coalgebra for $\bar{F}$ (cf. Example 2.5.12). It follows that the coalgebraic language map $\ddagger\alpha : X \to \mathbb{S}^{\Sigma^*}$ assigns to each state $x \in X$ the language accepted by that state in the weighted automaton. So coalgebraic language equivalence is the usual weighted language equivalence.

In contrast, the behavioural equivalence of $F\mathbb{S}^{(-)}$-coalgebras in Set is precisely the weighted bisimilarity introduced by Buchholz [94] (cf. Example 2.4.3(1)). This was proved by Silva et al. [277, Prop. 3.4].

(3) Silva et al. [278, Sec. 3.1] discuss probabilistic Moore automata. Here we put $M = \mathscr{D}_{\mathsf{f}}$ (cf. Example 2.4.3(4)), with the monad structure completely analogous to that of $\mathscr{D}_{\leq}$ in Example 12.3.2(3), and $FX = [0, 1] \times X^{\Sigma}$. Note that $[0, 1] = \mathscr{D}_{\mathsf{f}}\{0, 1\}$, and so F lifts to the category $\mathsf{Set}^{\mathscr{D}_{\mathsf{f}}}$ (Example 12.5.5(5)). Further, this category is isomorphic to the category of positively convex

algebras (aka convex sets); this goes back to Pumplün [255] (see also [256] and Jacobs [172] for a proof).

As we have seen in Example 2.5.7(1), the terminal F-coalgebra is carried by the set $[0, 1]^{\Sigma^*}$ of formal power series.

The coalgebras for FM are (Rabin) probabilistic automata [257]. For every probabilistic automaton considered as a coalgebra $\alpha\colon A \to [0, 1] \times (\mathscr{D}_f A)^{\Sigma}$, its determinization is a Moore automaton in $\mathsf{Set}^{\mathscr{D}_f}$ with outputs in $[0, 1]$ with the algebra $\mathscr{D}_f A$ of states. The coalgebraic language map $\ddagger\alpha\colon A \to [0, 1]^{\Sigma^*}$ assigns to a state $x \in A$ the formal power series accepted by it.

(4) Markov decision processes are systems with mixed non-deterministic and probabilistic branching. We consider one particular type called *simple Segala systems*, which are coalgebras of the form $\alpha\colon A \to \mathscr{P}(\Sigma \times \mathscr{D}X)$. Here we put $M = \mathscr{D}$, and $FX = \mathscr{P}(\Sigma \times X)$ is the coalgebraic type functor of labelled transition systems (LTS), see Example 2.4.2(7). As shown by Jacobs et al. [174, Sec. 5.2], the determinzation of an FM-coalgebra (A, α) has the shape $\alpha^\sharp\colon \mathscr{D}A \to \mathscr{P}(\Sigma \times \mathscr{D}A)$; this is called the distribution LTS of the given simple Segala system [108].

Context-Free Languages and Beyond The languages generated by a context-free grammar are an instance of coalgebraic language semantics. The main idea for this goes back to Winter, Bonsangue, and Rutten [311], but it was only fully realized as an instance later by Milius, Pattinson, and Wissmann [235, Sec. 6.4] (cf. Remark 12.5.11(1)).

Here one takes the monad M given by $MX = \mathscr{P}_f((X + \Sigma)^*)$. This is actually the composition of three monads: the exception monad $(-) + \Sigma$, the free-monoid monad $(-)^*$ and the finite power-set monad. Even though, in general, the composite of two monads need not be a monad, in this case the composition happens to be a monad again. Its Eilenberg–Moore algebras are precisely the idempotent semirings S *pointed* by Σ, which means that S comes equipped with a map $\Sigma \to S$ providing additional constants in S for every letter $a \in \Sigma$. Algebra homomorphisms preserve these additional constants.

The functor is again $FX = \{0, 1\} \times X^{\Sigma}$ as used for deterministic automata with $\nu F = \mathscr{P}(\Sigma^*)$. For every Σ-pointed idempotent semiring S one can equip FS with the structure of a Σ-pointed idempotent semiring so that F has a lifting $\bar{F}$ on Set^M [235, Lemma 6.10]. This structure makes a non-trivial use of the Σ-point S, and one should note that this is *not* the canonical coordinatewise structure that we saw in Example 12.5.5(5), which is the reason why we consider Σ-points (cf. Remark 12.5.11 below).

A context-free grammar G in Greibach normal over the terminal alphabet Σ gives rise to an FM-coalgebra $\alpha = \langle e, p\rangle\colon A \to \{0, 1\} \times (\mathscr{P}_f(A + \Sigma)^*)^{\Sigma}$, where

A is the set of non-terminals. For a non-terminal symbol $V \in A$ we have $o(V) = 1$ iff $V \to \varepsilon$ is a rule in the grammar, and for every $a \in \Sigma$ we define $t(V) \subseteq (A + \Sigma)^*$ as follows:

$$t(V)(a) = \{V_1 V_2 \cdots V_n \mid V \to aV_1 V_2 \cdots V_n \text{ is a rule of the grammar } G\}.$$

Actually, FM-coalgebras allow grammar rules to be of the form $V \to aw$ for $a \in \Sigma$ and $w \in (A + \Sigma)^*$, which is slightly more general than the rule format of Greibach normal form grammars.

The generalized power-set construction turns an FM-coalgebra into a deterministic automaton in the category of Σ-pointed idempotent semirings with the state space $\mathcal{P}_{\mathrm{f}}((A + \Sigma)^*)$. The coalgebraic language map $\ddagger\alpha\colon A \to \mathcal{P}(\Sigma^*)$ assigns to each non-terminal symbol $V \in A$ the language generated from V using the given grammar G.

Remark 12.5.11. (1) Winter et al. [311] use the monad $MX = \mathcal{P}_{\mathrm{f}}(X^*)$, where Set^M is (isomorphic to) the category of idempotent semirings. An FM-coalgebra is then precisely a grammar with rules in Greibach normal form. However, in that case FMA does not carry the structure of an idempotent semiring in such a way that the determinization $\alpha^\sharp$ arises as the extension of $\alpha\colon A \to FMA$ to the free Eilenberg–Moore algebra (MA, μ_A). Instead, they first extend α to $\bar{\alpha}\colon A^* \to FMA$ by a special-purpose inductive definition on words involving a kind of product rule for concatenation and union of finite languages. Then $\bar{\alpha}$ is extended to the free join-semilattice $\mathcal{P}_{\mathrm{f}}(A^*) = MA$ on A^* noting that $FMA = \{0, 1\} \times (\mathcal{P}_{\mathrm{f}} A^*)^\Sigma$ does carry the obvious coordinatewise structure of a join-semilattice.

(2) In subsequent work, Winter et al. [312] generalize this approach to constructively $\mathbb{S}$-algebraic languages for a semiring $\mathbb{S}$. These were originally introduced by Fliess [125], see also Droste et al. [113], and they are the 'context-free' formal power series (in contrast to recognizable formal power series mentioned in Example 10.5.10(2)). Once again, by adding a Σ-point to the Eilenberg–Moore algebras for the respective monad M used by Winter et al., which means pre-composing M by the exception monad $(-) + \Sigma$, one obtains the constructive $\mathbb{S}$-algebraic formal power series as a proper instance of the coalgebraic language semantics [235, Sec. 6.4].

(3) With the help of the coalgebraic language semantics it is also possible to capture the class of languages accepted by machines with extra memory, such as (deterministic) pushdown automata or Turing machines [145]. That paper considers M-*automata*, which are coalgebras for FM, where $FX = \Gamma \times X^\Sigma$ is the type functor of Moore automata from Example 2.4.2(6). In each instance the monad M models the computational side effects arising from manipulating

the extra memory such as a stack or a Turing tape. The details are non-trivial and the language semantics is less directly obtained than in Example 12.5.10; we therefore refer our readers to op. cit.

Languages accepted by Nominal Automata Following Frank et al. [126] we come back to nofas (Example 12.3.17(1)) and RNNAs (Definition 12.3.22), and we show that their language semantics is an instance of coalgebraic language semantics in Definition 12.5.9.

Example 12.5.12. First we consider the monad $\mathscr{P}_{\mathsf{fs}}\colon \mathsf{Nom} \to \mathsf{Nom}$. For every nominal set X we have a canonical isomorphism $\mathscr{P}_{\mathsf{fs}}(\mathbb{A} \times X) \cong (\mathscr{P}_{\mathsf{fs}}X)^{\mathbb{A}}$ since $\mathscr{P}_{\mathsf{fs}}X$ acts as the power object in the topos Nom. Therefore nofas are coalgebras in Nom of the form

$$\alpha\colon S \to \mathscr{P}_{\mathsf{fs}}1 \times \mathscr{P}(\mathbb{A} \times S) \cong \{0,1\} \times (\mathscr{P}_{\mathsf{fs}}S)^{\mathbb{A}}.$$

Thus, they are coalgebras for FM on Nom where $M = \mathscr{P}_{\mathsf{fs}}$ and $FX = \{0,1\} \times X^{\mathbb{A}}$. The Eilenberg–Moore algebras for $\mathscr{P}_{\mathsf{fs}}$ are internal complete join-semilattices in Nom, that is, nominal sets equipped with an equivariant partial order in which every finitely supported subset has a join. Clearly, $\{0,1\}$ with $0 < 1$ carries such a structure, whence the functor F lifts to $\mathsf{Nom}^{\mathscr{P}_{\mathsf{fs}}}$ according to Remark 12.5.6.

Recall from Example 10.5.11(2) that the terminal coalgebra for $FX = \{0,1\} \times X^{\mathbb{A}}$ is carried by the set $\mathscr{P}_{\mathsf{fs}}(\mathbb{A}^*)$ of all finitely supported languages over $\mathbb{A}$ [126, Cor. 4.16]. One now readily works out that for a given nofa $\alpha\colon S \to \{0,1\} \times (\mathscr{P}_{\mathsf{fs}}S)^{\mathbb{A}}$ its coalgebraic language map $\ddagger\alpha\colon S \to \mathscr{P}_{\mathsf{fs}}(\mathbb{A}^*)$ assigns to every state $s \in S$ its data language.

We now turn to RNNA, where the situation is a bit more delicate.

Remark 12.5.13. Recall from Definition 12.3.22 that RNNA are coalgebras of the form $S \to \{0,1\} \times \mathscr{P}_{\mathsf{ufs}}(\mathbb{A} \times S) \times \mathscr{P}_{\mathsf{ufs}}([\mathbb{A}]S)$.

(1) As we have seen in Proposition 12.3.25 there are canonical isomorphisms

$$\mathscr{P}_{\mathsf{ufs}}([\mathbb{A}]X) \cong [\mathbb{A}](\mathscr{P}_{\mathsf{ufs}}X),$$

which are restrictions of the ones for $\mathscr{P}_{\mathsf{fs}}$ mentioned in Example 11.5.8(2).

(2) We also have canonical injections $\mathscr{P}_{\mathsf{ufs}}(\mathbb{A}\times X) \rightarrowtail (\mathscr{P}_{\mathsf{ufs}}X)^{\mathbb{A}}$ restricting one direction of the canonical isomorphisms $\mathscr{P}_{\mathsf{fs}}(\mathbb{A} \times X) \cong (\mathscr{P}_{\mathsf{fs}}X)^{\mathbb{A}}$ mentioned in Example 11.5.8(2). However, note that the inverse does not restrict to uniformly finitely supported subsets.

Using the above two items we see that RNNAs may be regarded as coalgebras

$$S \to \{0,1\} \times (\mathscr{P}_{\mathsf{ufs}}S)^{\mathbb{A}} \times [\mathbb{A}](\mathscr{P}_{\mathsf{ufs}}S).$$

That is, RNNAs are coalgebras for FM where $FX = \{0,1\} \times X^{\mathbb{A}} \times [\mathbb{A}]X$ (cf. Example 10.1.13(5)) and $M = \mathscr{P}_{\mathsf{ufs}}$.

The key to seeing that the functor F lifts to the Eilenberg–Moore category of $\mathscr{P}_{\mathsf{ufs}}$ is the following result.

Proposition 12.5.14 [126, Prop. 4.10]. *The abstraction functor* $[\mathbb{A}](-)$ *has a canonical lifting to the Eilenberg–Moore category of* $\mathscr{P}_{\mathsf{ufs}}$.

Proof The abstraction functor has a left adjoint $\mathbb{A} * (-)$, where $*$ denotes the *fresh product* defined for two nominal sets X and Y by

$$X * Y = \{(x,y) : x \in X,\ y \in Y,\ \mathsf{supp}(x) \cap \mathsf{supp}(y) = \emptyset\},$$

see [250, Thm. 4.12]. The strength of $\mathscr{P}_{\mathsf{ufs}}$ restricts to the fresh product; we have

$$s_{X,Y} \colon X * \mathscr{P}_{\mathsf{ufs}}Y \to \mathscr{P}_{\mathsf{ufs}}(X * Y) \qquad (x,S) \mapsto \{(x,s) : s \in S\}.$$

Indeed, if $\mathsf{supp}\, x \cap \mathsf{supp}\, S = \emptyset$, then $\mathsf{supp}\, x \cap \mathsf{supp}\, s = \emptyset$ for every $s \in S$ because S is uniformly finitely supported and thus $\mathsf{supp}\, s \subseteq \mathsf{supp}\, S$. It follows that $s_{\mathbb{A},X} \colon \mathbb{A} * \mathscr{P}_{\mathsf{ufs}}X \to \mathscr{P}_{\mathsf{ufs}}(\mathbb{A} * X)$ yields a distributive law of $\mathbb{A} * (-)$ over $\mathscr{P}_{\mathsf{ufs}}$ because the respectice equational laws already hold with respect to the cartesian product and the power-set monad $\mathscr{P}$ on Set. By Proposition 12.5.4 we thus obtain a distributive law of $\mathscr{P}_{\mathsf{ufs}}$ over $[\mathbb{A}](-)$. □

Remark 12.5.15. The proof of Proposition 12.5.14 does not work for $\mathscr{P}_{\mathsf{fs}}$, similarly to Proposition 12.3.25. The problem is that the strength does not restrict to the fresh product for all finitely supported subsets. Indeed, even if $\mathsf{supp}(x)$ and $\mathsf{supp}(S)$ are disjoint, the support of x may not be disjoint from that of every element $s \in S$, whence (x,s) does not lie in $X * Y$. For example, take $X = Y = \mathbb{A}$ and $S = \mathbb{A} \setminus \{a\}$ for some $a \in \mathbb{A}$. Clearly, $\mathsf{supp}(S) = \{a\}$. Thus, (b,S) lies in $\mathbb{A} * \mathscr{P}_{\mathsf{fs}}\mathbb{A}$ for every $b \neq a$. However, while $b \in S$ we do not have that $(b,b) \in \mathbb{A} * \mathbb{A} = \{(a,a') : a, a' \in \mathbb{A}, a \neq a'\}$, which means that $s_{\mathbb{A},\mathbb{A}}(b,S)$ does not lie in $\mathscr{P}_{\mathsf{fs}}(\mathbb{A} * \mathbb{A})$.

Remark 12.5.16. (1) Combining Proposition 12.5.14 with what we have seen in Example 12.5.5, we conclude that the functor $FX = \{0,1\} \times X^{\mathbb{A}} \times [\mathbb{A}]X$ lifts to $\mathsf{Nom}^{\mathscr{P}_{\mathsf{ufs}}}$ using that $\{0,1\}$ carries the obvious structure of an Eilenberg–Moore algebra for $\mathscr{P}_{\mathsf{ufs}}$.

(2) More generally, we see that every functor formed according to the following grammar has a canonical lifting to $\mathsf{Nom}^{\mathscr{P}_{\mathsf{ufs}}}$:

$$F ::= A \mid \mathsf{Id} \mid [\mathbb{A}](-) \mid \textstyle\prod_{i \in I} F_i \mid F^N,$$

where A ranges over all nominal sets equipped with the structure $\alpha \colon \mathscr{P}_{\mathsf{ufs}}A \to A$

of an Eilenberg–Moore algebra for $\mathscr{P}_{\mathsf{ufs}}$, I is an arbitrary index set, and N ranges over all nominal sets. That is, we can essentially extend the grammar in Remark 12.5.6 by the abstraction functor (and obviously by infinite products, since they exist in Nom).

Now recall from Example 10.5.11(3) that the terminal coalgebra for the functor $FX = \{0,1\} \times X^{\mathbb{A}} \times [\mathbb{A}]X$ is formed by the set $\mathscr{P}_{\mathsf{fs}}(\bar{A}^*/{\equiv_\alpha})$ of all bar languages. Further recall from Remark 12.3.23(4) that every state of an RNNA accepts a bar language.

Theorem 12.5.17 [126, Cor. 4.22]. *Given an RNNA regarded as a coalgebra $\alpha\colon S \to \{0,1\} \times (\mathscr{P}_{\mathsf{f}}S)^{\mathbb{A}} \times [\mathbb{A}](\mathscr{P}_{\mathsf{f}}S)$, its coalgebraic language map $\ddagger\alpha\colon S \to \mathscr{P}_{\mathsf{fs}}(\bar{\mathbb{A}}/{\equiv_\alpha})$ assigns to every state $s \in S$ the bar language it accepts.*

Remark 12.5.18. A deterministic variant of RNNA was considered by Kozen et al. [203]. These are orbit-finite coalgebras for the functor

$$FX = \{0,1\} \times X^{\mathbb{A}} \times [\mathbb{A}]X$$

on Nom. Their language semantics is simply given by taking the image of the unique coalgebra homomorphism in the terminal coalgebra. Moreover, the languages accepted by these automata form precisely the rational fixed point of F (Example 10.5.11(3)).

12.6 Connecting Coalgebraic Trace and Language Semantics

Jacobs et al. [174] show how the coalgebraic trace semantics (Definition 12.3.8) and the coalgebraic language semantics (Definition 12.5.9) are connected in cases where both are applicable. We present this connection in Theorem 12.6.10.

Assumption 12.6.1. We assume that M is a monad and F, G are endofunctors, all on the category $\mathscr{A}$, such that F extends to $\bar{F}\colon \mathscr{A}_M \to \mathscr{A}_M$ via the distributive law $\lambda\colon FM \to MF$ and G lifts to $\widehat{G}\colon A^M \to \mathscr{A}^M$ via the distributive law $\varrho\colon MG \to GM$.

The source of the connection of the two semantics is the existence of an extension natural transformation connecting λ and ϱ in the following way.

Definition 12.6.2. An *extension* is a natural transformation $\varepsilon\colon MF \to GM$ compatible with the two distributive laws:

$$\begin{array}{ccc} MFM \xrightarrow{M\lambda} MMF \xrightarrow{\mu F} MF & \quad & MMF \xrightarrow{\quad\mu F\quad} MF \\ {\scriptstyle \varepsilon M}\downarrow \qquad\qquad \downarrow{\scriptstyle \varepsilon} & & {\scriptstyle M\varepsilon}\downarrow \qquad\qquad \downarrow{\scriptstyle \varepsilon} \\ GMM \xrightarrow{\quad G\mu \quad} GM & & MGM \xrightarrow{\varrho M} GMM \xrightarrow{G\mu} GM \end{array} \tag{12.6}$$

Example 12.6.3. (1) The leading example is that of non-deterministic automata with input alphabet Σ. Here we have $\mathscr{A}$ = Set, $FX = 1+\Sigma\times X$, $GX = \{0,1\}\times X^\Sigma$ and $M = \mathscr{P}$ (cf. Example 12.3.16(1) and 12.5.10(1)). The extension is given by the canonical isomorphisms

$$MF = \mathscr{P}(1+\Sigma\times X) \cong \mathscr{P}1 \times \mathscr{P}(\Sigma\times X) \cong \{0,1\}\times(\mathscr{P}X)^\Sigma = GM$$

defined by $\varepsilon_X(S) = (b, f)$, where $b = 1$ iff $* \in S$ and $f(s) = \{x : (s,x) \in S\}$.

(2) The case of nofas is completely analogous. This time one takes the functors on Nom given by $FX = 1 + \mathbb{A}\times X$ and $GX = \{0,1\}\times X^{\mathbb{A}}$ and the monad $M = \mathscr{P}_{\mathsf{fs}}$ (cf. Examples 12.3.17(1) and 12.5.12). As we have mentioned before, the isomorphisms in the previous item restrict to $\mathscr{P}_{\mathsf{fs}}$, yielding isomorphisms in Nom. Thus, they yield an extension since the validity of the equational laws described by (12.6) is inherited from the extension in the previous item.

(3) For RNNAs the situation is more interesting. Here we take

$$FX = 1 + \mathbb{A}\times X + [\mathbb{A}]X, \quad GX = \{0,1\}\times X^{\mathbb{A}} \times [\mathbb{A}]X, \quad \text{and} \quad M = \mathscr{P}_{\mathsf{ufs}}$$

on Nom (cf. Theorem 12.3.27 and 12.5.17). We obtain an extension by composing the canonical isomorphisms

$$\mathscr{P}_{\mathsf{ufs}}(1 + \mathbb{A}\times X + [\mathbb{A}]X) \cong \{0,1\}\times\mathscr{P}_{\mathsf{ufs}}(\mathbb{A}\times X)\times\mathscr{P}_{\mathsf{ufs}}([\mathbb{A}]X)$$

with the injections

$$\{0,1\}\times\mathscr{P}_{\mathsf{ufs}}(\mathbb{A}\times X)\times\mathscr{P}_{\mathsf{ufs}}([\mathbb{A}]X) \rightarrowtail \{0,1\}\times(\mathscr{P}_{\mathsf{ufs}}X)^{\mathbb{A}}\times[\mathbb{A}](\mathscr{P}_{\mathsf{ufs}}X)$$

obtained using the injection $\mathscr{P}_{\mathsf{ufs}}(\mathbb{A}\times X) \rightarrowtail (\mathscr{P}_{\mathsf{ufs}}X)^{\mathbb{A}}$ in Remark 12.5.13(2). See Frank et al. [126, Lemma 4.20] for the proof that this yields an extension.

(4) Probabilistic transition systems are FM-coalgebras, where we take

$$FX = 1 + \Sigma\times X \quad \text{and} \quad M = \mathscr{D}_\leq$$

on Set (Example 12.3.16(2)). We also consider $GX = [0,1]\times X^\Sigma$. Coalgebras for GM are very similar to the probabilistic Moore automata we have seen in Example 12.5.10(3), except that we now allow subdistributions. There are canonical injections

$$\mathscr{D}_\leq(X+Y) \rightarrowtail \mathscr{D}_\leq X\times\mathscr{D}_\leq Y \quad \text{and} \quad \mathscr{D}_\leq(A\times X) \rightarrowtail (\mathscr{D}_\leq X)^A.$$

We also see that $\mathscr{D}_\leq 1 \cong [0,1]$. It is easy to verify that combining these we obtain an extension $\varepsilon\colon MF \to GM$ with injective components

$$\varepsilon_X\colon \mathscr{D}_\leq(1+\Sigma\times X) \rightarrowtail [0,1]\times(\mathscr{D}_\leq)^\Sigma \quad \text{given by} \quad \varepsilon_X(f) = (f(*), f'),$$

where $f'(s)(x) = f(s,x)$ for every $s \in S$ and $x \in X$.

(5) Finally, we consider weighted labelled transition systems with outputs. Once again we take $\mathscr{A} = \mathsf{Set}$, $FX = 1 + \Sigma \times X$, and for a semiring $\mathbb{S}$, we take $GX = \mathbb{S} \times X^\Sigma$ and $M = \mathbb{S}^{(-)}$, the monoid-valued functor for the additive monoid of $\mathbb{S}$ (cf. Example 2.4.3(1)). Coalgebras for MF are finitely branching transition systems whose states have an output value in $\mathbb{S}$ and whose transitions carry a label from Σ and a weight from $\mathbb{S}$. Moreover, coalgebras for GM are weighted automata as we have seen in Example 12.5.10(2). There are canonical isomorphisms $\mathbb{S}^{(X+Y)} \cong \mathbb{S}^{(X)} \times \mathbb{S}^{(Y)}$, and we have canonical injections $\mathbb{S}^{(A \times X)} \rightarrowtail (\mathbb{S}^{(X)})^A$. It is easy to verifiy that combining these with the obvious isomorphism $\mathbb{S}^{(1)} \cong \mathbb{S}$, we obtain an extension $\varepsilon\colon MF \to GM$ with injective components

$$\varepsilon_X\colon \mathbb{S}^{(1+\Sigma\times X)} \rightarrowtail \mathbb{S} \times (\mathbb{S}^{(X)})^\Sigma \quad \text{given by} \quad \varepsilon_X(f) = (f(*), f'),$$

where $f'(s)(x) = f(s,x)$ for every $s \in S$ and $x \in X$.

Remark 12.6.4. (1) For every object X of $\mathscr{A}$ the morphism ε_X is a homomorphism of Eilenberg–Moore algebras for M from (MFX, μ_{FX}) to $\widehat{G}(MX, \mu_X)$. Indeed, this is precisely what the commutativity of the diagram on the right in (12.6) expresses.

(2) The canonical *comparison functor* $K\colon \mathscr{A}_M \to \mathscr{A}^M$ maps an object X of $\mathscr{A}_M$ to the free Eilenberg–Moore algebra (MX, μ_X) and a morphism $f\colon X \to MY$ to its homomorphic extension, that is the unique M-algebra homomorphism

$$Kf = f^\sharp\colon (MX, \mu_X) \to (MY, \mu_Y) \qquad \text{such that } f^\sharp \cdot \eta_X = f.$$

Establishing functoriality of K is straightforward using the Kleisli laws (12.10).

(3) For every coalgebra $\gamma\colon X \to MFX$ the extension natural transformation ε yields a coalgebra $\varepsilon_X \cdot \gamma\colon X \to GMX$, and we take its determinization $E(X,\gamma) = (MX, (\varepsilon_X \cdot \gamma)^\sharp)$. For an $\bar{F}$-coalgebra homomorphism $h\colon (X,\gamma) \to (Y,\delta)$ we put $Eh = h^\sharp\colon MX \to MY$, the homomorphic extension of $h\colon X \to MY$ (in $\mathscr{A}$). Thus, E is defined to be a lifting of the comparison functor K in item (2) above.

The following result is proved as part of Jacobs et al. [174, Thm. 2], and we include an explicit proof here.

Proposition 12.6.5. *The above assignment yields a functor*

$$E\colon \mathsf{Coalg}\,\bar{F} \to \mathsf{Coalg}\,\widehat{G}.$$

Remark 12.6.6. We first collect a few facts needed for the proof below.

(1) The action of $\bar{F}\colon \mathscr{A}_M \to \mathscr{A}_M$ on a morphism $f\colon X \to MY$ is given using the corresponding distributive law $\lambda\colon FM \to MF$ as follows:

$$\bar{F}(f) = \big(FX \xrightarrow{Ff} FMY \xrightarrow{\lambda_Y} MFY\big). \tag{12.7}$$

(2) Given $f\colon X \to MY$ in $\mathscr{A}$ the homomorphic extension fulfils

$$f^\sharp = (MX \xrightarrow{Mf} MMY \xrightarrow{\mu_Y} MY). \tag{12.8}$$

(3) Given an Eilenberg–Moore algebra (A, α), a morphism $f\colon X \to A$, and a morphism $h\colon (A, \alpha) \to (B, \beta)$ in $\mathscr{A}^M$ we have

$$\begin{array}{ccc} MX & \xrightarrow{f^\sharp} & A \\ & \searrow^{(h\cdot f)^\sharp} & \downarrow h \\ & & B \end{array} \tag{12.9}$$

Indeed, both paths are morphisms in $\mathscr{A}^M$ which agree when pre-composed with the universal morphism $\eta_X\colon X \to MX$. The universal property of the free algebra (MX, μ_X) thus yields the commutativity of the above triangle.

(4) Restricting homomorphic extensions to free algebras yields the extension operation from the presentation of M as a Kleisli triple (cf. Remark 7.3.23(3)). This means that for morphisms $f\colon X \to MY$ and $g\colon Y \to MZ$ we have

$$f^\sharp \cdot \eta_X = f, \qquad \eta_X^\sharp = \mathsf{id}_{MX}, \qquad g^\sharp \cdot f^\sharp = (g^\sharp \cdot f)^\sharp. \tag{12.10}$$

Proof We will prove that $h^\sharp\colon E(X, \gamma) \to E(Y, \delta)$ is a $\widehat{G}$-coalgebra homomorphism; it is then clearly functorial being a lifting of the comparison functor K (Remark 12.6.4(2)). First we prove that the following square commutes in $\mathscr{A}^M$:

$$\begin{array}{ccc} MX & \xrightarrow{h^\sharp} & MY \\ {\scriptstyle (\varepsilon_X\cdot\gamma)^\sharp}\downarrow & & \downarrow{\scriptstyle (\varepsilon_Y\cdot\delta)^\sharp} \\ GMX & \xrightarrow{Gh^\sharp} & GMY \end{array}$$

It suffices to show that the diagram commutes (in $\mathscr{A}$) when pre-composed by the universal morphism $\eta_X\colon X \to MX$ of the free Eilenberg–Moore algebra (MX, μ_X). By using $h^\sharp \cdot \eta_X = h$ and similarly for $\varepsilon_X \cdot \gamma$ it is thus our task to show that the outside of the following diagram commutes:

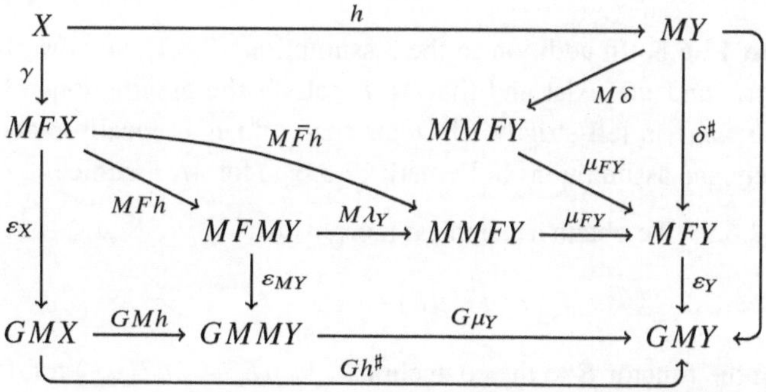

(12.11)

Indeed, the upper inner part commutes since h is a homomorphism of coalgebras for $\bar{F}$ (unfolding the definition of the composition in $\mathscr{A}_M$; see Definition 12.3.1). The triangle below it commutes due to (12.7), and the triangle on the right as well as the lowest part by (12.8). The right-hand part commutes by (12.9) using that ε_Y is a morphism in $\mathscr{A}^M$ (Remark 12.6.4(1)). Finally, the lower left-hand part commutes by the naturality of ε, and the lower right-hand one due to the left-hand law in (12.6). □

Remark 12.6.7. For a coalgebra $\gamma\colon X \to MFX$, the coalgebra $E(X, \gamma)$ can also be described as

$$MX \xrightarrow{M\gamma} MMFX \xrightarrow{\mu_{FX}} MFX \xrightarrow{\varepsilon_X} GMX.$$

This follows by combining (12.8) and (12.9); cf. the right-hand part of diagram (12.11).

In addition, one obtains the following commutative cube of functors [174, Thm. 3], where the diagonal functors are forgetful and the vertical arrows are (the liftings of) the standard left adjoints (the functor E_0 acts by post-composing an MF-coalgebra by ε and as the identity on morphisms):

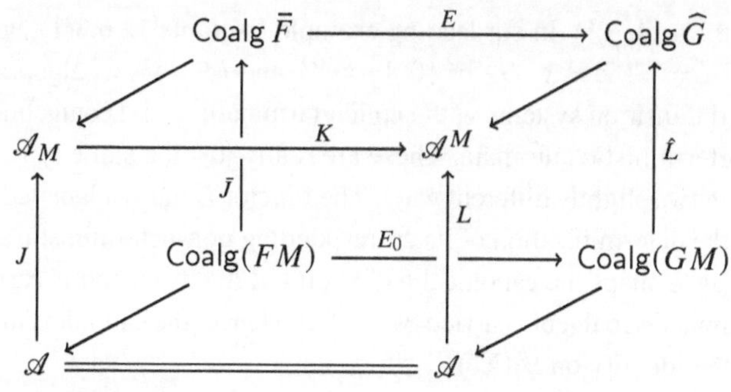

Note that the left and right faces of the cube are the squares in Remarks 12.3.7 and 12.5.3, respectively.

Assumption 12.6.8. In addition to the Assumption 12.6.1, we now also assume that μF and νG exist and that M, F satisfy the assumptions of Theorem 12.3.5: $\mathscr{A}_M$ is left-strictly CPO-enriched and $\overline{F}$ is locally monotone. (Alternatively, the assumptions in Remark 12.3.6(1) for M, F suffice.)

Notation 12.6.9. We obtain a canonical morphism

$$e\colon M(\mu F) \to \nu G$$

by applying the functor E to the coalgebra $J\iota^{-1}\colon \mu F \to MF(\mu F)$ (cf. Proposition 12.3.4) and taking the unique coalgebra homomorphism from it to the terminal $\widehat{G}$-coalgebra (Theorem 12.5.2).

The coalgebraic trace and coalgebraic languages semantics are connected via the canonical morphism e as follows.

Theorem 12.6.10. *For every coalgebra $\gamma\colon X \to TFX$ we have*

$$\ddagger(\varepsilon_X \cdot \gamma) = \big(X \xrightarrow{\mathsf{tr}_\gamma} M(\mu F) \xrightarrow{e} \nu G\big).$$

Proof By Remark 12.6.4(3), we know that $E\,\mathsf{tr}_\gamma = \mathsf{tr}_\gamma^\sharp\colon TX \to T(\mu F)$ is a homomorphism of $\widehat{G}$-coalgebras from the determinization $E(X,\gamma)$ of $(X, \varepsilon_X{\cdot}\gamma)$ to $E(\mu F, J\iota^{-1})$. Since νG carries the terminal $\widehat{G}$ coalgebra, we have $e \cdot \mathsf{tr}_\gamma^\sharp = h$ where $h\colon (MX, \varepsilon_X \cdot \gamma) \to \nu G$ is the unique $\widehat{G}$-coalgebra homomorphism. Pre-composing by η_X we obtain the desired equation $\ddagger(\varepsilon_X \cdot \gamma) = h \cdot \eta_X = e \cdot \mathsf{tr}_\gamma^\sharp \cdot \eta_X = e \cdot \mathsf{tr}_\gamma$. □

If the canonical morphism $e\colon M(\mu F) \to \nu G$ is an embedding, then Theorem 12.6.10 states that the coalgebraic trace semantics and coalgebraic language semantics coincide, or rather that one can be obtained from the other. We show that this is indeed the case in the instances from Example 12.6.3.

Example 12.6.11. (1) In our leading example Example 12.6.3(1), we have the functors $FX = 1 + \Sigma \times X$, $GX = \{0,1\} \times X^\Sigma$ and $M = \mathscr{P}$. Coalgebras for MF are labelled transition systems with explicit termination, and coalgebras for GM are non-deterministic automata. These are really just the same type of system presented in two slightly different ways. The functor E maps a labelled transition system to the determinization of its corresponding non-deterministic automaton. In particular, E maps the canonical fixed point of the extension $\overline{F}$ carried by Σ^* to the terminal $\widehat{G}$-coalgebra carried by $\mathscr{P}(\Sigma^*)$. Hence, the canonical morphism e is simply the identity on $\mathscr{P}(\Sigma^*)$.

(2) Similarly in the case of nofas we have $\mu F = \mathbb{A}^*$, and $\nu G = \mathscr{P}_{\mathsf{fs}}(\mathbb{A}^*)$ and the canonical morphism is again the identity on $\mathscr{P}_{\mathsf{fs}}(\mathbb{A}^*)$.

(3) In the case of RNNAs we have $\nu F = \overline{\mathbb{A}}^*/{=_\alpha}$, the nominal set of all bar strings and $\nu G = \mathscr{P}_{\mathsf{fs}}(\overline{\mathbb{A}}^*/{=_\alpha})$, the nominal set of all bar languages. The canonical morphism is the inclusion map $e\colon \mathscr{P}_{\mathsf{ufs}}(\overline{\mathbb{A}}^*/{=_\alpha}) \hookrightarrow \mathscr{P}_{\mathsf{fs}}(\overline{\mathbb{A}}^*/{=_\alpha})$ [126, Lemma 4.21].

(4) For probabilistic transitions systems we have $\mu F = \Sigma^*$ and $\nu G = [0,1]^{\Sigma^*}$. The canonical morphism is the inclusion map $e\colon \mathscr{D}_{\leq}(\Sigma^*) \hookrightarrow [0,1]^{\Sigma^*}$.

Remark 12.6.12. Even if μF does not exist or if M, F do not satisfy the assumptions of Theorem 12.3.5, we may still use an extension $\varepsilon\colon MF \to GM$ to obtain a semantics for MF-coalgebras. Indeed, given $\gamma\colon X \to MFX$ one forms the GM-coalgebra $(X, \varepsilon_X \cdot \gamma)$ and takes its coalgebraic language morphism $\ddagger(\varepsilon_X \cdot \gamma)\colon X \to \nu G$.

For example, for a weighted labelled transition systems (Example 12.6.3(5)) the Kleisli category for the monad $M = \mathbb{S}^{(-)}$ fails to admit the desired enrichment. But we have $\nu G = \mathbb{S}^{\Sigma^*}$, the set of all formal power series (see Example 12.5.10(2)). So we can still define a trace semantics as follows: given a weighted transition system $\gamma\colon X \to \mathbb{S}^{(1+\Sigma\times X)}$, one takes the weighted automaton $\varepsilon_X \cdot \gamma\colon X \to \mathbb{S} \times (\mathbb{S}^{(X)})^\Sigma$, and $\ddagger(\varepsilon_X \cdot \gamma)\colon X \to \mathbb{S}^{\Sigma^*}$ provides the trace semantics of the given weighted labelled transition system.

12.7 Coalgebraic Language Semantics as a Quotient

In connection with the generalized power-set construction, there is an interesting relation between $\nu(FM)$ and νF (if both exist). It turns out that the latter is a canonical quotient $\bar{F}$-coalgebra of the former. This gives a coalgebraic account of the well-known fact that in special instances such as non-deterministic automata, behaviour modulo language equivalence is a canonical quotient of behaviour modulo bisimilarity.

Assumption 12.7.1. In the following we assume that M is a monad on $\mathscr{A}$ and that F is a functor on $\mathscr{A}$ with a lifting $\bar{F}\colon \mathscr{A}^M \to \mathscr{A}^M$ given by a distributive law $\lambda\colon MF \to FM$. We also assume that the terminal coalgebras for FM and F exist and denote their structures by

$$\tau\colon \nu F \to F(\nu F) \qquad \text{and} \qquad \tau_0\colon \nu(FM) \to FM(\nu(FM)).$$

Remark 12.7.2. (1) The lifting $\bar{F}$ maps an Eilenberg–Moore algebra (A, α) to the M-algebra

$$MFA \xrightarrow{\lambda_A} FMA \xrightarrow{F\alpha} FA.$$

It is easy to see that this structure satisfies the two laws of an Eilenberg–Moore algebra.

(2) Given an FM-coalgebra (C, γ), the coalgebra structure of its determinization is

$$\gamma^\sharp = (MC \xrightarrow{M\gamma} MFMC \xrightarrow{\lambda_{MC}} FMMC \xrightarrow{F\mu_C} FMC).$$

Indeed, one proves that the above morphism is an M-algebra homomorphism from (MC, μ_C) to $\bar{F}(MC, \mu_C)$ which extends γ. By the uniqueness, this is therefore equal to $\gamma^\sharp$. To see this note that the above structure clearly is an M-algebra homomorphism: $M\gamma$ is one since γ is so by one of the laws of Eilenberg–Moore algebras, and so is $F\mu_C \cdot \lambda_{MC}$ being the algebra structure on FMC (see item (1)). Moreover, we have an extension of γ:

$$\begin{aligned} F\mu_C \cdot \lambda_{MC} \cdot M\gamma \cdot \eta_C &= F\mu_C \cdot \lambda_{MC} \cdot \eta_{MC} \cdot \gamma && \text{naturality of } \eta \\ &= F\mu_C \cdot F\eta_{MC}\gamma && \lambda \text{ is a distributive law} \\ &= \gamma && \text{since } \mu \cdot \eta M = \text{id}. \end{aligned}$$

Lemma 12.7.3. *Every fixed point (C, γ) of FM carries a unique algebra structure $\alpha\colon MC \to C$ such that γ is an M-algebra homomorphism. Furthermore,*

$$\alpha\colon (MC, \gamma^\sharp) \to (C, F\alpha \cdot \gamma)$$

is an $\bar{F}$-coalgebra homomorphism.

Proof Let $\gamma\colon C \to FMC$ be a fixed point. The forgetful functor $U\colon \mathscr{A}^M \to \mathscr{A}$ creates isomorphisms, which implies that the Eilenberg–Moore algebra structure on FMC yields the following Eilenberg–Moore algebra structure on C:

$$\alpha = \big(MC \xrightarrow{M\gamma} MFMC \xrightarrow{(F\mu\cdot\lambda M)_C} FMC \xrightarrow{\gamma^{-1}} C\big). \tag{12.12}$$

This is the unique structure of an Eilenberg–Moore algebra on C such that γ is an M-algebra homomorphism. In order to show that α is an F-coalgebra homomorphism, consider the diagram below:

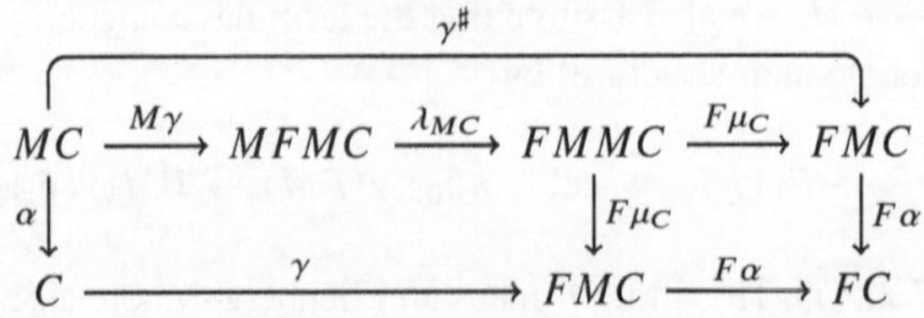

The upper part commutes by Remark 12.7.2(2), the left-hand part does by (12.12) using that γ is an isomorphism, and the right-hand square trivially commutes.

We already know that $(C, \gamma^{\sharp})$ is an $\bar{F}$-coalgebra. We now show that the lower row is also an $\bar{F}$-coalgebra. Recall that γ is an M-algebra homomorphism and so is α (by one of the axioms of Eilenberg–Moore algebras). Hence, $F\alpha = \bar{F}\alpha$ is an M-algebra homomorphism. This demonstrates that $(C, F\alpha \cdot \gamma)$ is an $\bar{F}$-coalgebra, which implies that α is an $\bar{F}$-coalgebra homomorphism, as desired. □

Theorem 12.7.4. *The terminal $\bar{F}$-coalgebra is a canonical quotient coalgebra of the terminal FM-coalgebra.*

More precisely, we know from Theorem 12.5.2 that νF carries the Eilenberg–Moore algebra structure $\alpha\colon M(\nu F) \to \nu F$, turning it into the terminal $\bar{F}$-coalgebra. From Lemma 12.7.3 we have the Eilenberg–Moore algebra structure $\beta\colon M(\nu(FM)) \to \nu(FM)$ such that the following is an $\bar{F}$-coalgebra:

$$\nu(FM) \xrightarrow{\tau_0} FM(\nu(FM)) \xrightarrow{F\beta} F(\nu(FM)).$$

The theorem states that the unique $\bar{F}$-coalgebra homomorphism below represents a strong quotient in $\mathscr{A}^M$:

$$q\colon (\nu(FM), F\beta \cdot \tau_0) \twoheadrightarrow (\nu F, \tau).$$

Proof Just within this proof let us abuse notation and write F in lieu of $\bar{F}$ (since all the diagrams below can be taken to be in $\mathscr{A}$), Z for νF and Z_0 for $\nu(FM)$.

It is our task to show that the following unique coalgebra homomorphism is a strong epimorphism in $\mathscr{A}^M$:

$$\begin{array}{ccccc} Z_0 & \xrightarrow{\tau_0} & FMZ_0 & \xrightarrow{F\beta} & FZ_0 \\ {\scriptstyle q}\downarrow & & & & \downarrow{\scriptstyle Fq} \\ Z & & \xrightarrow{\qquad\tau\qquad} & & FZ \end{array} \tag{12.13}$$

This follows from the fact that q has a splitting $s\colon Z \rightarrowtail Z_0$ in $\mathscr{A}$. To obtain s, we use that (Z_0, τ_0) is the terminal FM-coalgebra in $\mathscr{A}$; there exists a unique FM-coalgebra homomorphism as shown in the diagram below:

$$\begin{array}{ccccc} Z & \xrightarrow{\tau} & FZ & \xrightarrow{F\eta_Z} & FMZ \\ {\scriptstyle s}\downarrow & & & & \downarrow{\scriptstyle FMs} \\ Z_0 & & \xrightarrow{\qquad\tau_0\qquad} & & FMZ_0 \end{array} \tag{12.14}$$

It fulfills $q \cdot s = \mathrm{id}_Z$ in $\mathscr{A}$ since the following diagram commutes; the universal property of $\nu F = Z$ then yields the desired result:

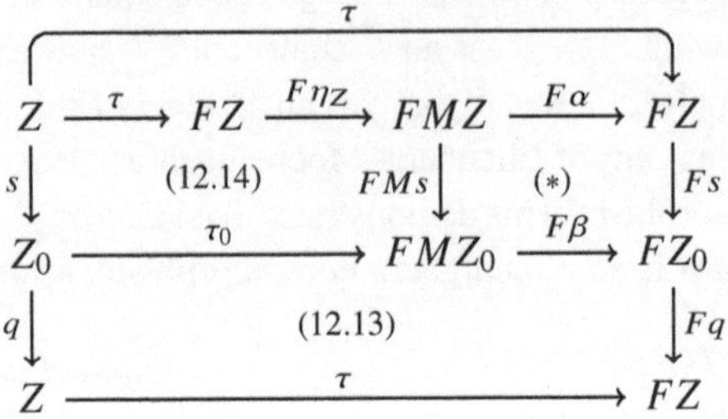

The upper part commutes since $\alpha \cdot \eta_Z = \mathsf{id}_Z$, and the upper left-hand and lower rectangles commute as indicated. We do not claim that the upper right-hand square $(*)$ commutes, but it does when pre-composed with $F\eta_Z$. To see this, remove F and consider the diagram below:

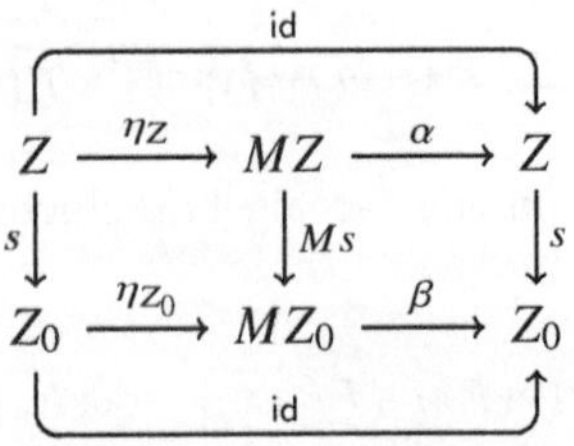

Here, the left-hand square commutes by the naturality of η, the upper and lower parts by the unit law of Eilenberg–Moore algebras, and the outside trivially commutes. Therefore the right-hand square pre-composed by η_Z commutes, as desired. □

Theorem 12.7.4 already provides a coalgebraic version of the desired result that language equivalence is a quotient of bisimilarity. In the following we will work this out in a more explicit form.

Remark 12.7.5. Note that the generalized power-set construction is part of a functor $\hat{L}\colon \mathsf{Coalg}(FM) \to \mathsf{Coalg}(\bar{F})$. In fact, as observed by Jacobs et al. [174, Lemma 1], this functor is a lifting of the left adjoint L of the forgetful functor $U\colon \mathscr{A}^M \to \mathscr{A}$ to coalgebras (the vertical arrows in the diagram below are the forgetful functors):

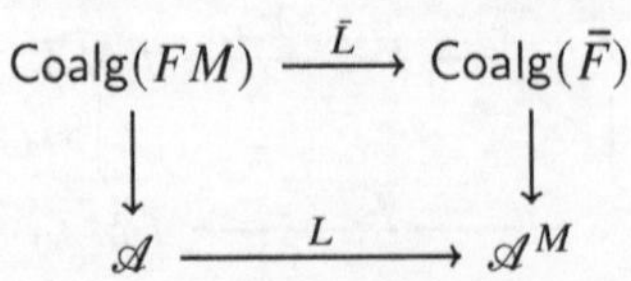

Indeed, every FM-coalgebra homomorphism $h\colon (C, \gamma) \to (D, \delta)$ yields an

$\bar{F}$-coalgebra homomorphism $Mh\colon (MC, \gamma^\sharp) \to (MD, \delta^\sharp)$ due to the following diagram

$$\begin{array}{ccccc} & & \gamma^\sharp & & \\ MC & \xleftarrow{\eta_C} & C & \xrightarrow{\gamma} & FMC \\ {\scriptstyle Mh}\downarrow & & \downarrow{\scriptstyle h} & & \downarrow{\scriptstyle FMh} \\ MD & \xleftarrow{\eta_D} & D & \xrightarrow{\delta} & FMD \\ & & \delta^\sharp & & \end{array}$$

The upper and lower parts commute by the definition of the determinization, and the left-hand square commutes by the naturality of η. Furthermore, note that the outside consists of M-algebra homomorphisms. Thus, using that η_C is a universal morphism, we see that the right-hand inner square commutes in $\mathscr{A}$ if and only if the outside commutes in $\mathscr{A}^M$.

That $\hat{L}$ is functorial and the claimed lifting is immediate.

Lemma 12.7.6. *For every FM-coalgebra homomorphism $h\colon (C, \gamma) \to (D, \delta)$ we have*

$$\ddagger\gamma = \big(C \xrightarrow{h} D \xrightarrow{\ddagger\delta} \nu F\big).$$

Proof Given h, we have an $\bar{F}$-coalgebra homomorphism

$$Mh\colon (C, \gamma^\sharp) \to (D, \delta^\sharp)$$

by Remark 12.7.5. Let $k\colon (D, \delta^\sharp) \to (\nu F, \tau)$ be the unique $\bar{F}$-coalgebra homomorphism. Then $k \cdot Mh$ is the unique $\bar{F}$-coalgebra homomorphism from $(C, \gamma^\sharp)$ to $(\nu F, \tau)$. Composing with η and using the definition of the coalgebraic language morphism (Definition 12.5.9), we obtain the desired result:

$$\begin{aligned} \ddagger\delta \cdot h &= k \cdot \eta_D \cdot h && \text{definition of } \ddagger\delta \\ &= k \cdot Mh \cdot \eta_C && \text{naturality of } \eta \\ &= \ddagger\gamma && \text{definition of } \ddagger\gamma. \end{aligned}$$

□

Notation 12.7.7. In the following two results, we denote the unique coalgebra homomorphisms obtained by the universal property of $\nu(FM)$ by

$$\dagger\gamma\colon (C, \gamma) \to (\nu(FM), \tau_0).$$

Proposition 12.7.8. *For every FM-coalgebra (C, γ) the coalgebraic language morphism is given by*

$$\ddagger\gamma = \big(C \xrightarrow{\dagger\gamma} \nu(FM) \xrightarrow{q} \nu F\big).$$

Proof (1) We first prove that for the terminal coalgebra $(\nu(FM), \tau_0)$ we have

$$\ddagger\tau_0 = q\colon \nu(FM) \twoheadrightarrow \nu F.$$

Let $\beta\colon M(\nu(FM)) \to \nu(FM)$ is the Eilenberg–Moore algebra from Lemma 12.7.3. We observe first that $q \cdot \beta\colon M(\nu(FM)) \to \nu F$ is an $\bar{F}$-coalgebra homomorphism:

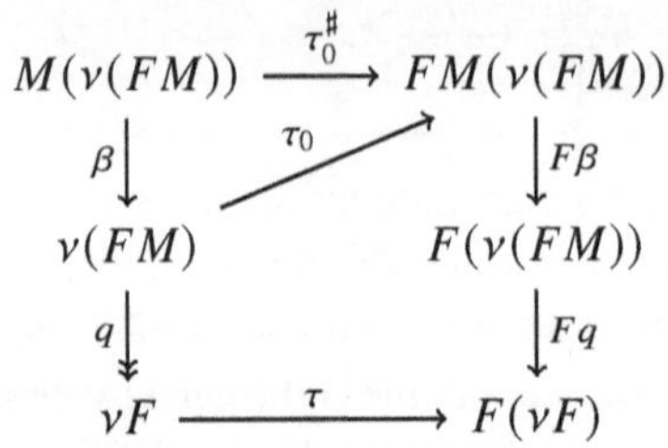

The lower part commutes by the definition of q in (12.13), and for the upper triangle one uses the freeness of the Eilenberg–Moore algebra $M(\nu(FM))$: both $\tau_0^\sharp$ and $\tau_0 \cdot \beta$ are M-algebra homomorphisms, and they are equal when pre-composed by η:

$$\tau_0^\sharp \cdot \eta_{\nu(FM)} = \tau_0 = \tau_0 \cdot \beta \cdot \eta_{\nu(FM)}.$$

Using the definition of the coalgebraic language morphism (Definition 12.5.9), we obtain $\ddagger\tau_0$ by pre-composing the coalgebra homomorphism $q \cdot \beta$ by η; thus $\ddagger\tau_0 = q \cdot \beta \cdot \eta_{\nu(FM)} = q$.

(2) The desired statement follows by combining item (1) and Lemma 12.7.6: Given an FM-coalgebra (C, γ) we conclude

$$q \cdot \dagger\gamma = \ddagger\tau_0 \cdot \dagger\gamma = \ddagger\gamma. \qquad \square$$

As a corollary, we obtain the main result by Silva et al. [278] stating that behavioural equivalence (with respect to FM) implies coalgebraic language equivalence (viz. behavioural equivalence with respect to F). We formulate this for $\mathscr{A}$ = Set, as in op. cit.

Corollary 12.7.9. *Let (D, γ) and (D, δ) be FM-coalgebras. For every $x \in C$ and $y \in D$ we have the following*

$$\dagger\gamma(x) = \dagger\delta(y) \qquad \Longrightarrow \qquad \ddagger\gamma(x) = \ddagger\delta(y).$$

Rational Fixed Points in Eilenberg–Moore Categories We now connect coalgebraic language semantics and the topic of this section with rational fixed points of functors, which we treated in Section 10.2.

First, in contrast to Theorem 12.5.2, the rational fixed point of a finitary functor on a locally presentable category $\mathscr{A}$ does not lift to $\mathscr{A}^M$ in general:

Example 12.7.10. Consider $\mathscr{A}$ = Set and $M = \mathbb{R}^{(-)}$ for the semiring (actually field) $\mathbb{R}$ of real numbers. Then $\mathsf{Set}^M = \mathbb{R}\text{-}\mathsf{Vec}$ is the category of real vector spaces. The rational fixed point for the set functor $FX = \mathbb{R} \times X$ is formed by eventually periodic streams (Example 10.2.21(5)), while that of the lifting $\bar{F}$ on $\mathbb{R}$-Vec consists of all rational streams (Example 10.5.10(2)).

However, when Set^M is a locally finite variety (Example 10.1.6(5)), the rational fixed point does lift, analogously as in Theorem 12.5.2. More generally, this holds for finitary monads on a locally finitely presentable category $\mathscr{A}$, where the forgetful functor $U\colon \mathscr{A}^M \to \mathscr{A}$ preserves finitely presentable objects.

Remark 12.7.11. Recall from Remark 10.3.9 the notion of a final functor $\mathscr{D} \to \mathscr{C}$ and the two conditions for finality when $\mathscr{C}$ and $\mathscr{D}$ are filtered. We note that in this case every right adjoint $P\colon \mathscr{D} \to \mathscr{C}$ is final. Indeed suppose that P has the left adjoint L. Ad condition (a) take the unit $C \to PLC$ of the adjunction. Ad condition (b), given a parallel pair $f, g\colon C \rightrightarrows PX$ in $\mathscr{C}$ take their adjoint transposes $f^\sharp, g^\sharp\colon LC \rightrightarrows X$ in $\mathscr{D}$. Since $\mathscr{D}$ is filtered, there exists some $d\colon X \to X'$ in $\mathscr{D}$ with $d \cdot f^\sharp = d \cdot g^\sharp$. Transposing back, we obtain $Pd \cdot f = Pd \cdot g$.

Theorem 12.7.12. *Suppose that M is a finitary monad on a locally finitely presentable category $\mathscr{A}$ such that $U\colon \mathscr{A}^M \to \mathscr{A}$ preserves finitely presentable objects. Let $F\colon \mathscr{A} \to \mathscr{A}$ be finitary and have the lifting $\bar{F}\colon \mathscr{A}^M \to \mathscr{A}^M$. Then the rational fixed point for F lifts to the rational fixed point for $\bar{F}$.*

More concretely, let $\omega\colon \varrho F \to F(\varrho F)$ be the rational fixed point. There exists a unique F-coalgebra homomorphism $\alpha\colon M(\varrho F) \to \varrho F$ such that the following diagram commutes:

$$\begin{array}{ccccc} M(\varrho F) & \xrightarrow{M\omega} & MF(\varrho F) & \xrightarrow{\lambda_{\varrho F}} & FM(\varrho F) \\ {\scriptstyle \alpha}\downarrow & & & & \downarrow{\scriptstyle F\alpha} \\ \varrho F & & \xrightarrow{\quad\omega\quad} & & F(\varrho F) \end{array} \tag{12.15}$$

We will prove that $(\varrho F, \alpha)$ is an Eilenberg–Moore algebra for M such that the coalgebra structure $\omega\colon \varrho F \to F(\varrho F)$ is an M-algebra homomorphism. In addition, $(\varrho F, \omega)$ is then the rational fixed point of $\bar{F}$.

Proof The lifting $\bar{F}$ on $\mathscr{A}^M$ is given by a distributive law $\lambda\colon MF \to FM$. It is well known that λ also entails that M lifts to a monad $\overline{M}$ on Coalg F; see e.g. Turi and Plotkin [252]. In fact, we obtain from Theorem 11.5.2 that the standard adjunction $L \dashv U : \mathscr{A}^M \to \mathscr{A}$ lifts to an adjunction $\hat{L} \dashv \hat{U} : \mathsf{Coalg}\,\bar{F} \to \mathsf{Coalg}\,F$. The ensuing monad $\overline{M}$ on Coalg F maps a coalgebra (C, γ) to the coalgebra

$$MC \xrightarrow{M\gamma} MFC \xrightarrow{\lambda_C} FMC$$

and a coalgebra homomorphism h to Mh. We use the easily established fact that U is finitary; thus, its left adjoint L preserves finitely presentable objects [51, Lemma 2.4], and U does so by assumption, the adjunction $\hat{L} \dashv \hat{U}$ further restricts to an adjunction $\hat{L}_{\mathsf{f}} \dashv \hat{U}_{\mathsf{f}} : \mathsf{Coalg}_{\mathsf{f}}\,\bar{F} \to \mathsf{Coalg}\,F$ so that we obtain the following commutative diagram (with the unlabelled arrows being the forgetful functors):

$$\begin{array}{ccccc}
\mathsf{Coalg}_{\mathsf{f}}\,\bar{F} & \xhookrightarrow{\bar{I}} & \mathsf{Coalg}\,\bar{F} & \longrightarrow & \mathscr{A}^M \\
{\scriptstyle \hat{L}_{\mathsf{f}}}\uparrow \dashv \downarrow{\scriptstyle \hat{U}_{\mathsf{f}}} & & {\scriptstyle \hat{L}}\uparrow \dashv \downarrow{\scriptstyle \hat{U}} & & {\scriptstyle L}\uparrow \dashv \downarrow{\scriptstyle U} \\
\mathsf{Coalg}_{\mathsf{f}}\,F & \xhookrightarrow{I} & \mathsf{Coalg}\,F & \longrightarrow & \mathscr{A}
\end{array}$$

(1) We show that the rational fixed point for $\bar{F}$ exists and that its underlying object is ϱF. Since M is finitary, we know that $\mathscr{A}^M$ is locally finitely presentable and the forgetful functor $U : \mathscr{A}^M \to \mathscr{A}$ creates filtered colimits [134, Satz 10.3] (see also [81, Thm. 6.9]). This implies that $\bar{F}$ is finitary. Indeed, for every filtered diagram $D : \mathscr{D} \to \mathscr{A}^M$ we have

$$\begin{aligned}
U\bar{F}(\mathsf{colim}\,D) &= FU(\mathsf{colim}\,D) = \mathsf{colim}(FUD) = \mathsf{colim}(U\bar{F}D) \\
&= U(\mathsf{colim}\,\bar{F}D).
\end{aligned}$$

Since U creates filtered colimits, we obtain $\bar{F}(\mathsf{colim}\,D) = \mathsf{colim}\,\bar{F}D$, as desired. Similarly, $\overline{M}$ is finitary.

(2) Next we show that the underlying object of $\varrho\bar{F}$ is ϱF. Indeed, since $\hat{U}_{\mathsf{f}}$ is a right adjoint we know that it is a final functor (Remark 12.7.11). This implies $\hat{U}(\varrho\bar{F}) = \varrho F$ as follows:

$$\begin{aligned}
\hat{U}(\varrho\bar{F}) &= \hat{U}(\mathsf{colim}\,\bar{I}) && \text{def. of } \varrho\bar{F} \\
&= \mathsf{colim}(\hat{U}\bar{I}) && \text{since } \hat{U} \text{ is finitary} \\
&= \mathsf{colim}(I\hat{U}_{\mathsf{f}}) && \text{since } \hat{U}\bar{I} = I\hat{U}_{\mathsf{f}} \\
&= \mathsf{colim}\,I && \text{since } \hat{U}_{\mathsf{f}} \text{ is final} \\
&= \varrho F && \text{def. of } \varrho F.
\end{aligned}$$

It remains to show that the Eilenberg–Moore algebra structure of $\varrho\bar{F}$ is as claimed.

(3) The monad $\overline{M}$ preserves lfp coalgebras (Definition 10.2.5): Given an lfp coalgebra (C, γ) for F, it is the colimit of the canonical diagram

$D\colon \mathsf{Coalg_f}\, F/(C,\gamma) \to \mathsf{Coalg}\, F$ (cf. Remark 10.2.6(2)). Since $M = UL$ preserves finitely presentable objects, we conclude that $\overline{M}$ restricts to $\mathsf{Coalg_f}\, F$. Thus $\overline{M}D$ is a diagram of F-coalgebras with finitely presentable carriers. Its colimit is

$$\overline{M}(C,\gamma) = \overline{M}(\mathsf{colim}\, D) = \mathsf{colim}\, \overline{M}D.$$

Hence, this is an lfp coalgebra, as desired.

(4) The previous item implies that the coalgebra $\overline{M}(\varrho F, \omega)$ in the top row of (12.15) is an lfp coalgebra. Since ϱF is the terminal lfp coalgebra for F, we obtain a unique morphism $\alpha\colon M(\varrho F) \to \varrho F$ such that (12.15) commutes.

(5) The proof that α is the structure of an Eilenberg–Moore algebra is completely analogous to the proof for νF in Theorem 12.5.2 (using that $\overline{M}$ preserves lfp coalgebras). Since (12.15) commutes, $\omega\colon \varrho F \to F(\varrho F)$ is an M-algebra homomorphism. □

We finish this part with a version of Theorem 12.7.4 for rational fixed points. This result is relevant for the axiomatization of regular system behaviour. For $\mathscr{A} = \mathsf{Set}$, the rational fixed point of FM may be understood as the domain of behaviours for finite FM-coalgebras modulo bisimilarity, and the rational fixed point of $\bar{F}$ as the domain of behaviour of the same systems but modulo (coalgebraic) language equivalence. That the latter appears as a quotient of the former entails that it is possible to add to a sound and complete expression calculus for bisimilarity some axioms to arrive at a sound and complete calculus for language equivalence. This has been demonstrated in previous work [89] on the instance of weighted automata. First we cite a technical lemma:

Lemma 12.7.13 [89, Lemma 3.16]. *Suppose that F is a finitary functor preserving monomorphisms on a locally finitely presentable category $\mathscr{A}$ in which $\mathscr{A}_{\mathsf{fp}} = \mathscr{A}_{\mathsf{fg}}$. Then lfp coalgebras are closed under strong quotients.*

Remark 12.7.14. In the following theorem we assume that $\mathscr{A} = \mathsf{Set}$ and that the lifting $\bar{F}\colon \mathsf{Set}^M \to \mathsf{Set}^M$ is a *proper* functor [228, Def. 5.2] (recall that we have mentioned that notion on page 359). We do not recall the definition because it is not needed in the proof below. We will only need [228, Cor. 5.7] that for a proper functor $\bar{F}$ the rational fixed point $\varrho\bar{F}$ is isomorphic to the locally ffg fixed point $\varphi\bar{F}$ (cf. Definition 10.5.21 and recall from the discussion before it that liftings of finitary set functors are strongly finitary).

More technically, this means that $\varrho\bar{F}$ is the colimit of the full embedding $\mathscr{D} \hookrightarrow \mathsf{Coalg}\,\bar{F}$, where $\mathscr{D}$ is formed by all determinizations $(MC, \gamma^\sharp)$ of FM-coalgebras (C,γ). We denote the colimit cocone by $i_\gamma\colon MC \to \varrho\bar{F}$.

Theorem 12.7.15. *Suppose M is a finitary set monad such that in* Set^M *finitely* *generated and finitely presentable algebras form the same class. Let* $F: \mathscr{A} \to \mathscr{A}$ *be a finitary functor preserving monomorphisms which lifts to a proper functor* $\bar{F}: \mathscr{A}^M \to \mathscr{A}^M$*. Then* $\varrho\bar{F}$ *is a canonical quotient strong coalgebra of* $\varrho(FM)$*.*

More precisely, for the rational fixed point $(\varrho(FM), \omega_0)$ we obtain from Lemma 12.7.3 an Eilenberg–Moore algebra structure

$$\beta: M(\varrho(FM)) \to \varrho(FM)$$

such that the following is a coalgebra for $\bar{F}$:

$$\varrho(FM) \xrightarrow{\omega_0} FM(\varrho(FM)) \xrightarrow{F\beta} F(\varrho(FM)).$$

Moreover, this is an lfp coalgebra. The theorem states that the unique $\bar{F}$-coalgebra homomorphism p to the terminal lfp coalgebra represents a strong quotient in Set^M:

$$p: (\varrho(FM), F\beta \cdot \omega_0) \twoheadrightarrow (\varrho\bar{F}, \omega).$$

In addition, we know that both rational fixed points are subcoalgebras of the respective terminal coalgebra so that we obtain the following commutative square in $\mathsf{Coalg}\,\bar{F}$:

$$\begin{array}{ccc} \varrho(FM) & \rightarrowtail & \nu(FM) \\ {\scriptstyle p}\downarrow & & \downarrow{\scriptstyle q} \\ \varrho\bar{F} & \rightarrowtail & \nu\bar{F} \end{array}$$

This means that $\varrho\bar{F}$ is the image of $\varrho(FM)$ in $\nu\bar{F}$ under the canonical strong quotient q. Again, this states that regular behaviour modulo coalgebraic language equivalence is the quotient of regular behaviour modulo bisimilarity.

Proof We know that the rational fixed point $\varrho(FM)$ is the colimit of the inclusion functor $I: \mathsf{Coalg}_\mathsf{f}(FM) \hookrightarrow \mathsf{Coalg}(FM)$ (Corollary 10.2.15). Recall from Remark 12.7.5 the functor $\bar{L}: \mathsf{Coalg}(FM) \to \mathsf{Coalg}\,\bar{F}$. It is finitary since it is a lifting of the left adjoint (whence finitary) functor $L: \mathscr{A} \to \mathscr{A}^M$ and the forgetful functors $\mathsf{Coalg}(FM) \to \mathscr{A}$ and $\mathsf{Coalg}\,\bar{F} \to \mathscr{A}^M$ create colimits.

Therefore, $\bar{L}$ preserves the colimit $\varrho(FM) = \mathsf{colim}\, I$. Hence, the determinization

$$M(\varrho(FM)) \xrightarrow{\omega_0^\sharp} FM(\varrho(FM))$$

is a colimit of the diagram $\bar{L}I: \mathsf{Coalg}_\mathsf{f}(FM) \to \mathsf{Coalg}\,\bar{F}$ formed by all determinizations $(MC, \gamma^\sharp)$ of FM-coalgebras (C, γ) with C in $\mathscr{A}_\mathsf{fp}$. Its connecting morphisms are of the form Mh for an FM-coalgebra homomorphism $h\colon (C,\gamma) \to (D,\delta)$ in $\mathsf{Coalg}_\mathsf{f}(FM)$. We denote the colimit cocone by $j_\gamma\colon MC \to M(\varrho(FM))$.

Note that the Eilenberg–Moore algebra structure $\beta\colon M(\varrho(FM)) \to \varrho(FM)$ is a strong epimorphism in $\mathscr{A}^M$ since it has the splitting $\eta_{\varrho(FM)}$ in $\mathscr{A}$. Moreover, by Lemma 12.7.3 it is an $\bar{F}$-coalgebra homomorphism from $(M(\varrho(FM)), \omega_0^\sharp)$ to $(\varrho(FM), F\beta \cdot \omega_0)$. We apply Lemma 12.7.13 to $\bar{F}\colon \mathscr{A}^M \to \mathscr{A}^M$ to see that the latter coalgebra is therefore lfp. Thus we obtain a unique $\bar{F}$-coalgebra homomorphism $p\colon (\varrho(FM), F\beta \cdot \omega_0) \to (\varrho\bar{F}, \omega)$ as claimed.

It remains to prove that p is a strong epimorphism in $\mathscr{A}^M$. Since $\bar{F}$ is proper we know that $\varrho\bar{F}$ is the colimit of the inclusion functor $\mathscr{D} \hookrightarrow \mathsf{Coalg}\,\bar{F}$ (Remark 12.7.14). From the fact that $\varrho\bar{F}$ is the terminal lfp coalgebra, we obtain the following description of the colimit cocone of the diagram D from Remark 12.7.14:

$$i_\gamma = \big(MX \xrightarrow{j_\gamma} M(\varrho(FM)) \xrightarrow{\beta} \varrho(FM) \xrightarrow{p} \varrho\bar{F}\big) \text{ for } (C,\gamma) \text{ in } \mathsf{Coalg}_\mathsf{f}(FM).$$

Since the colimit injections i_γ form a strongly epic family, it follows that p is strongly epic. This completes the proof. □

Remark 12.7.16. Let us further discuss the assumptions of Theorem 12.7.15.
(1) Note that $\bar{F}$ is proper whenever finitely generated algebras in Set^M are closed under kernel pairs and F preserves weak pullbacks [228, Prop. 5.10]. The closure under kernel pairs has the following equivalent formulation in terms of standard general algebra: every congruence R on a finitely generated algebra A is finitely generated as a subalgebra $R \hookrightarrow A \times A$. (Observe that this is *not* equivalent to stating that R is a finitely generated congruence.) This property implies that the classes of finitely generated and finitely presentable algebras in Set^M coincide [89, Lemma 3.19].

For example, these assumptions hold for $\bar{F}X = \mathbb{S} \times X^\Sigma$ on the category $\mathbb{S}$-Mod of semimodules for a Noetherian semiring $\mathbb{S}$ (cf. Example 10.5.10(2)). The monad here is $\mathbb{S}^{(-)}$ for the additive monoid of the semiring $\mathbb{S}$.

(2) Under the above assumptions, $(\varrho(FM), F\beta \cdot \omega_0)$ is a weakly terminal lfp coalgebra for $\bar{F}$ [89, Lemma 3.39]. The key to proving this is to show that every coalgebra in $\mathsf{Coalg}_\mathsf{f}\,\bar{F}$ is the coequalizer of a pair of parallel homomorphism in the full subcategory $\mathscr{D}$ of all determinizations.

(3) The original proof of Theorem 12.7.15 [89, Thm. 3.41] used the assumptions in item (1). That properness of $\bar{F}$ is a strictly weaker assumption is demonstrated by the functor $\bar{F}X = \mathbb{N} \times X^\Sigma$ on the category of commutative monoids (equivalently, $\mathbb{N}$-semimodules for the usual semiring of natural numbers). In this category there are congruences on finitely generated objects which are not

finitely generated as algebras, for example, the congruence on $\mathbb{N} \times \mathbb{N}$ generated by the set $\{(n, n+1) \mid n \in \mathbb{N}\}$ is easily seen not to be a finitely generated algebra.

However, $\bar{F}$ above is proper [228, Ex. 5.3] and strongly finitary, and the classes of finitely generated and finitely presentable commutative monoids coincide (again cf. Example 10.5.10(2)). Moreover, for every proper semiring in the sense of Ésik and Maletti [120], the functor $\bar{F}X = \mathbb{S} \times X^{\Sigma}$ on the category $\mathbb{S}$-Mod of $\mathbb{S}$-semimodules is proper.

(4) Sokolova and Woracek [282] have proved that the non-negative rationals $\mathbb{Q}_{\geq 0}$ and non-negative reals $\mathbb{R}_{\geq 0}$ form proper semirings, thus the corresponding type functors for weighted automata are proper. In addition, they have proved that on the category $\mathsf{Set}^{\mathscr{D}_\mathsf{f}}$ the functor $\bar{F}X = [0, 1] \times X^{\Sigma}$ of positively convex algebras (cf. Example 12.5.10(3)) is proper. Moreover, in $\mathsf{Set}^{\mathscr{D}_\mathsf{f}}$ the classes of finitely generated and finitely presentable algebras coincide [281].

Initial Algebras We conclude this section with a few remarks on initial algebras and further work. In contrast to Theorem 12.5.2, an initial algebra does not lift to an initial algebra for the lifting $\bar{F}$ in general:

Example 12.7.17. Consider the monad $MX = X + 1$ on Set with unit $\eta_X = \mathsf{inl}\colon X \to X + 1$ and multiplication given by the codiagonal

$$\mu_X = (MMX = X + 1 + 1 \xrightarrow{X+\nabla} X + 1 = MX).$$

Eilenberg–Moore algebras for M are *pointed sets*: sets with a chosen element $\bot$; homomorphisms are maps preserving $\bot$. The identity functor $FX = X$ has the empty initial algebra, but its lifting to Set^M has the initial algebra $\{\bot\}$.

Remark 12.7.18. (1) Suppose that $\mathscr{A}$ is a locally λ-presentable category (Definition 11.2.8) and that M is λ-accessible (Definition 11.2.13). It is well known that $\mathscr{A}^M$ is locally λ-presentable, too [134, Satz 10.3] (see also [81, Thm. 6.9]). Thus, by Corollary 6.1.12, for every λ-accessible functor G on $\mathscr{A}^M$ the initial-algebra chain converges in λ steps. In particular, for every λ-accessible endofunctor $F\colon \mathscr{A} \to \mathscr{A}$ every lifting is λ-accessible, too, since the forgetful functor $U\colon \mathscr{A}^M \to \mathscr{A}$ creates λ-filtered colimits.

Notice, however, that the initial-algebra chain for a lifting $\bar{F}$ looks different from that for F; it starts with the initial object $(M0, \mu_0)$ of $\mathscr{A}^M$, and after ω-steps the colimit $\mathsf{colim}_{n<\omega} \bar{F}^n(M0, \mu_0)$ may not be carried by $\mathsf{colim}_{n<\omega} F^n M0$ in $\mathscr{A}$ (this can happen when $\lambda > \omega$).

(2) The above considerations imply that in certain special cases the initial algebra does indeed lift. Suppose that $\mathscr{A}$ is locally finitely presentable, that F

and M are finitary, F has a lifting $\bar{F}$ on $\mathscr{A}^M$ and $M0 \cong 0$. Let us write I for the initial object $(M0, \mu_0)$ of $\mathscr{A}^M$. Then, the initial-algebra chain for $\bar{F}$ is a lifting of the initial-algebra chain for F; in symbols: $U(\bar{F}^n I) \cong F^n 0$. Since the forgetful functor U creates filtered colimits, we obtain

$$\begin{aligned} U(\mu\bar{F}) &= U(\operatorname*{colim}_{n<\omega} \bar{F}^n I) \\ &= \operatorname*{colim}_{n<\omega} U(\bar{F}^n I) \\ &\cong \operatorname*{colim}_{n<\omega} F^n 0 \\ &= \mu F. \end{aligned}$$

Example 12.7.19. As we have pointed out at the beginning of Section 12.5, finitary monads on $\mathscr{A}$ = Set are precisely those that can be specified by a finitary signature Σ and a set of equations. In this case, the condition $M\emptyset = \emptyset$ means that there are no constants in the signature Σ. This condition may seem quite restrictive. However, there are a number of monads that do satisfy it; for example, the monad of (free) semigroups $MX = X^+$, the finite nonempty power-set monad, the nonempty bag monad, the monad M for the variety of algebras with one (commutative) binary operation. Besides, for every monad M on Set there is the monad M^0 which agrees with M on nonempty sets and functions and with $M^0\emptyset = \emptyset$.

Remark 12.7.20. There is also a 'dual' result to Theorem 12.5.2. We have already mentioned that from a distributive law λ of the monad M over the functor F one obtains a lifting of the monad M to the category of F-coalgebras:

$$\begin{array}{ccc} \mathsf{Coalg}\,F & \xrightarrow{\ \overline{M}\ } & \mathsf{Coalg}\,F \\ {\scriptstyle U}\downarrow & & \downarrow{\scriptstyle U} \\ \mathscr{A} & \xrightarrow{\quad M \quad} & \mathscr{A} \end{array}$$

It is not difficult to verify that the unit and multiplication of M are F-coalgebra homomorphisms. Hence, the initial Eilenberg–Moore algebra for $\overline{M}$ is a lifting of the initial Eilenberg–Moore algebra $(M0, \mu_0)$ for M: $\overline{M}(0, u)$, where $u\colon 0 \to F0$ is the trivial initial F-coalgebra, carries the initial Eilenberg–Moore algebra for the lifting $\overline{M}$.

Remark 12.7.21. (1) There is an equivalent way to study coalgebras for a lifted functor $\bar{F}\colon \mathscr{A}^M \to \mathscr{A}^M$. In fact, the category of $\bar{F}$-coalgebras is equivalent to the category of *λ-bialgebras*, which are objects of $\mathscr{A}$ equipped with an Eilenberg–Moore algebra for M as well as an F-coalgebra structure subject to a natural compatibility condition; see Turi and Plotkin [252] and Klin [192] for an excellent introduction to λ-bialgebras.

(2) One motivation of distributive laws is that they capture (sets of) operational rules defining algebraic operations over system behaviour. One concrete example is the well-known GSOS format of Bloom, Istrail, and Meyer [83] in process algebra, which ensures that bisimilarity is a congruence. This result was generalized by Turi and Plotkin [252] to the level of distributive laws. The monad M here represents the syntax of the program constructs while the functor F presents the type of system behaviour. The distributive law $\lambda\colon MF \to FM$ can be understood as by (generalized) operational rules specifying transitions on program terms, thus defining the behaviour of the syntactic program constructs. The λ-bialgebras are then the models of the specification. Two models of special importance are the *operational model* and the *denotational model*. The operational model is given by the initial λ-bialgebra (which is carried by the initial Eilenberg–Moore algebra $(M0, \mu_0)$ for M, see Remark 12.7.20) – it consists of all programs (i.e. terms of operations) with a behaviour (namely, the coalgebra $\gamma\colon M0 \to FM0$) induced by the distributive law λ. Furthermore, the denotational model is given by the terminal λ-bialgebra (which is carried by the terminal F-coalgebra; see Theorem 12.5.2) – it consists of process behaviours (modulo bisimilarity) on which the program constructs specified by the distributive law operate via the Eilenberg–Moore algebra $\alpha\colon M(\nu F) \to \nu F$.

(3) We have only briefly considered one form of distributive law (of a monad over a functor). There are other forms of distributive laws which are considered in connection with their λ-bialgebras (e.g. of a functor over another one, or of a monad over a comonad, abstract GSOS rules of Turi and Plotkin etc.). The beautiful, and in various applications highly relevant, topic of mathematical operational semantics proposed by Turi and Plotkin [252] and further worked out in work of Bartels [73] and others goes much beyond the scope of our book. So we refrain from providing more details. Again, a good source to start reading is Klin [192].

12.8 Summary

We have seen a number results which imply that initial algebras and terminal coalgebras for set functors yield initial algebras and terminal coalgebras for related endofunctors on categories of sets with extra structure.

For the categories CPO (complete partial orders) and CMet (complete metric spaces) we have shown that, given a set functor F having a terminal coalgebra νF and a lifting on these categories, νF can be equipped with a canonical order or metric, respectively, so that this yields the terminal coalgebra for the lifting (Corollaries 12.1.3 and 12.2.1). These are instances of a general lifting result for complete and well-powered categories (Theorem 12.0.1). Initial algebras, however, need not lift from Set to CPO or CMet, respectively.

In the metric case, besides these simple generic lifting results, we have also looked at more specific liftings of a set functor F to the category of pseudometric spaces. They arise from evaluation functions which are algebras for F on the unit interval. We have discussed one particular way an evaluation function yields a lifting, the Kantorovich lifting (Definition 12.2.7), and we briefly mentioned a second way, the Wasserstein lifting. We have also shown that the pseudometric on the terminal coalgebra νF that makes it the terminal coalgebra of the lifted functor is obtained as a least fixed point (Theorem 12.2.13). Finally, we have mentioned that this can be used to study behavioural distances, which allows a more informative analysis of the difference between the behaviour of states of coalgebras than standard behavioural equivalence.

A large portion of this section is devoted to extensions of functors to Kleisli categories and liftings to Eilenberg–Moore categories, as well as the semantics that arises there from initial algebras and terminal coalgebras. First, we have seen that for a functor F on a category $\mathscr{A}$ which has an initial algebra μF and an extension $\bar{F}$ to the Kleisli category of a monad M, μF yields the initial algebra for the extended functor on $\mathscr{A}_M$ (Proposition 12.3.4). This arises as a simple application of the adjoint lifting results we have seen in Section 11.5. Moreover, if the Kleisli category $\mathscr{A}_M$ is left-strictly CPO-enriched and the extension is locally continuous, then μF yields a canonical fixed point of $\bar{F}$ (Theorem 12.3.5). We have seen that for a commutative monad M, every polynomial functor has a canonical extension to $\mathscr{A}_M$, and for $\mathscr{A}$ = Set, every analytic set functor has a locally monotone extension (Theorem 12.3.14). Coalgebraic trace semantics is the semantics induced by the finality of the canonical fixed point of $\bar{F}$. Instances are the classical trace semantics of (probabilistic) labelled transition systems (for $\mathscr{A}$ = Set) and languages accepted by nominal automata (for $\mathscr{A}$ = Nom).

Concerning Eilenberg–Moore categories, we have seen that if an endofunctor F on $\mathscr{A}$ has a terminal coalgebra νF and a lifting $\bar{F}$ on $\mathscr{A}^M$, then the structure of νF is a homomorphism of Eilenberg–Moore algebras making it the terminal coalgebra for $\bar{F}$ (Theorem 12.5.2). We have provided generalized determinization of FM-coalgebras (Construction 12.5.7) and the coalgebraic language semantics which arises from $\nu\bar{F}$ (Definition 12.5.9). We have presented many instances of this semantics: the language semantics of non-deterministic automata, languages accepted by weighted and probabilistic automata, context-free languages, their weighted versions, and languages accepted by nominal automata.

We have also seen that in instances where both the coalgebraic trace semantics and the coalgebraic language semantics are applicable, the two semantics are connected (Theorem 12.6.10).

The fact that νF is a quotient coalgebra of $\nu(FM)$ for the lifted functor $\bar{F}$ (Theorem 12.7.4) corresponds to the well-known phenomenon that linear-type (trace or language) semantics are canonical quotients of branching-type semantics, here coalgebraic behavioural equivalence or bisimilarity (Proposition 12.7.8 and Corollary 12.7.9).

We have also discussed rational fixed points in connection with generalized determinization. We have shown that the rational fixed point of an endofunctor F does not yield the rational fixed point of its lifting $\bar{F}$ on $\mathscr{A}^M$ in general, and we have formulated conditions under which it does (Theorem 12.7.12). We have also seen conditions under which the above canonical quotient $\nu(FM) \twoheadrightarrow \nu F$ restricts to the respective rational fixed points (Theorem 12.7.15).

13

Interaction between Initial Algebras and Terminal Coalgebras

The theme of this chapter is that for many set functors F the terminal coalgebra carries

(1) a complete metric which is the Cauchy completion of the initial algebra, and

(2) a cpo ordering which is the ideal completion of the initial algebra.

Example 13.0.1. Let E be a nonempty set. The terminal coalgebra for the polynomial functor $FX = X \times X + E$ consists of all binary trees with leaves labelled in E (cf. Example 2.5.11(4)). We assume that $E \neq \emptyset$ and choose an element $b \in E$. By the *cutting* of a tree t at level n, we mean the tree obtained by first removing all nodes of height greater than n, and then labelling all nodes of height n by b. (Note that this is different than the cutting we have previously used, e.g. in Example 3.3.9.)

The metric on νF we consider is

$$d(s,t) = 2^{-n}, \quad \text{for the largest } n \text{ such that the cuttings of } s \text{ and } t \text{ at level } n \text{ are equal trees,} \tag{13.1}$$

for all trees $s \neq t$. This is a complete metric space which is, whenever $E \neq \emptyset$, a Cauchy completion of μF, the subalgebra of νF formed by all finite binary trees with leaves labelled in E (cf. Example 2.2.15(1)). Moreover, νF is ordered as follows:

$$s \sqsubseteq t \qquad \text{if } s = t \text{ or } s \text{ is obtained from } t \text{ by cutting it at some level } n. \tag{13.2}$$

This is a continuous algebra (see Example 2.1.7). The finite trees form a subposet which is dense in νF: every tree t in νF is the join of an increasing ω-sequence in μF. Namely: let t_n be the cutting of t at level n. Then

$$t_0 \sqsubseteq t_1 \sqsubseteq t_2 \sqsubseteq \cdots \qquad \text{with} \qquad \bigsqcup_{n<\omega} t_n = t.$$

Indeed, νF is an ideal completion of the poset μF. (We discuss ideal completions in Remark 13.3.5 below.)

So the themes of this chapter are metric and order structures on νF for set functors, and then the relation of μF to those structures. The root of the matter is the way ω^{op}-limits and ω-colimits are defined in Set. We need hypotheses on F, of course, to ensure that μF and νF are related to ω^{op}-limits and ω-colimits, respectively. Under weaker hypotheses, we obtain the same results for μF and $V_\omega = \lim_{n<\omega} F^n 1$, the limit which we saw in Chapter 4. Indeed, the additional hypotheses are needed only to show that $\nu F = V_\omega$. So the work of this chapter could just as well be described as about the relationship between μF and V_ω.

We also introduce the concept of a precontinuous set functor and show that for such functors Example 13.0.1 has a generalization: the terminal coalgebra carries a canonical metric (and a canonical partial order) such that when the initial algebra is nonempty, the two have the same Cauchy completion (and ideal completion). For set functors preserving ω^{op}-limits, we shall see that νF is the Cauchy completion (and ideal completion) of μF.

We study the Cauchy completion in Section 13.2 and the ideal completion in Section 13.3. Before that, we show that for *every* set functor F with a terminal coalgebra, the initial algebra exists and is a subcoalgebra of νF. More generally, this holds for endofunctors preserving monomorphisms whenever νF has smooth subobjects (see Definition 6.2.2). We also establish some general results on the initial-algebra and the terminal-coalgebra chains.

13.1 Canonical Morphism

Recall from Remark 2.5.16 that for every endofunctor F the initial algebra and terminal coalgebra, when they exist, are connected by the canonical morphism $k\colon \mu F \to \nu F$:

$$\begin{array}{ccc} \mu F & \xleftarrow{\iota} & F(\mu F) \\ {\scriptstyle k}\downarrow & & \downarrow{\scriptstyle Fk} \\ \nu F & \xrightarrow{\tau} & F(\nu F) \end{array} \tag{13.3}$$

We have mentioned *canonical fixed points* in Chapter 5: this means that k is an isomorphism. For set functors this is relatively rare (but see Proposition 5.0.2). For every set functor F, if νF exists, then so does μF (Corollary 6.3.9), and it follows from Corollary 6.3.10 that k is a monomorphism. This same fact often holds in other familiar categories.

Notation 13.1.1. Suppose that for $F\colon \mathscr{A} \to \mathscr{A}$ both the initial-algebra chain (W_i) and the terminal-coalgebra chain (V_i) exist. We know that for every ordinal i the coalgebra $w_{i,i+1}\colon W_i \to FW_i$ is recursive (Example 7.4.3(5)), and that the algebra $v_{i+1,i}\colon FV_i \to V_i$ is corecursive (Example 7.1.4(5)). So for every $i \in \mathsf{Ord}$ we have a unique coalgebra-to-algebra morphism k_i as shown below:

$$\begin{array}{ccc} W_i & \xrightarrow{w_{i,i+1}} & FW_i \\ {\scriptstyle k_i}\downarrow & & \downarrow{\scriptstyle Fk_i} \\ V_i & \xleftarrow{v_{i+1,i}} & FV_i \end{array} \tag{13.4}$$

Recall that both μF and νF are often obtained by convergence of the initial-algebra and terminal-coalgebra chains, respectively (Theorems 6.1.11 and 6.5.4).

Corollary 13.1.2. *If a functor F has both an initial algebra and a terminal coalgebra obtained via convergence of the corresponding chains, then the canonical morphism $k\colon \mu F \to \nu F$ has the form $k = k_i\colon W_i \to V_i$ for some ordinal i.*

Indeed, we have an ordinal i with such that the initial algebra for F is $(W_i, w_{i,i+1}^{-1})$ and its terminal coalgebra is $(V_i, v_{i+1,i}^{-1})$. This means that (13.3) and (13.4) are the same square.

Lemma 13.1.3. (1) *For every ordinal i, we have $k_{i+1} = Fk_i$.*
(2) *For every pair $i \leq j$ of ordinals, the following square commutes:*

$$\begin{array}{ccc} W_i & \xrightarrow{w_{i,j}} & W_j \\ {\scriptstyle k_i}\downarrow & & \downarrow{\scriptstyle k_j} \\ V_i & \xleftarrow{v_{j,i}} & V_j \end{array} \tag{13.5}$$

Proof (1) Applying F to the square (13.4), we see that Fk_i is a coalgebra-to-algebra morphism from

$$w_{i+1,i+2}\colon W_{i+1} = FW_i \xrightarrow{Fw_{i,i+1}} FFW_i = FW_{i+1}$$

to $v_{i+2,i+1}\colon FV_{i+1} = FFV_i \xrightarrow{Fv_{i+1,i}} FV_i = V_{i+1}$. By uniqueness of such a morphism, we obtain the desired result.

(2) We know that $w_{i,j}$ is a coalgebra homomorphism from $(W_i, w_{i,i+1})$ to $(W_j, w_{j,j+1})$, that k_j is a coalgebra-to-algebra morphism as defined in (13.4), and that $v_{j,i}$ is an algebra homomorphism from $(V_j, v_{j+1,j})$ to $(V_i, v_{i+1,i})$. Therefore their composite $v_{j,i} \cdot k_j \cdot w_j$ is a coalgebra-to-algebra morphism. Thus, the desired square commutes by the uniqueness of k_i. □

Remark 13.1.4. Recall from [217, Section IX.4] that for a pair of mixed variant functors $F, G\colon \mathscr{C}^{\mathsf{op}} \times \mathscr{C} \to \mathscr{D}$ a *dinatural transformation from F to G* is a family of morphisms $d_C\colon F(C,C) \to G(C,C)$ indexed by the objects of $\mathscr{C}$ such that for every pair of morphisms $f\colon C' \to C$ and $g\colon C \to C'$ in $\mathscr{C}$ an obvious hexagon commutes. The collection $k_i\colon W_i \to V_i$ (i in Ord) may be regarded as a dinatural transformation. To see this, we pre-compose $W\colon \mathsf{Ord} \to \mathscr{A}$ and $V\colon \mathsf{Ord}^{\mathsf{op}} \to \mathscr{A}$ by the projection functors from the category $\mathsf{Ord}^{\mathsf{op}} \times \mathsf{Ord}$ to Ord and $\mathsf{Ord}^{\mathsf{op}}$, respectively, to obtain a pair of functors $\bar{W}, \bar{V}\colon \mathsf{Ord}^{\mathsf{op}} \times \mathsf{Ord} \to \mathscr{A}$. Then a dinatural transformation from $\bar{W}$ to $\bar{V}$ is precisely a collection of morphisms $k_i\colon W_i \to V_i$ such that the squares (13.5) commute for every pair i, j of ordinals. We will abuse terminology and say that (k_i) is a dinatural transformation from W to V.

Corollary 13.1.5. *The collection (k_i) is the unique dinatural transformation from W to V such that $k_{i+1} = Fk_i$ for all i.*

Indeed, instantiating (13.5) with $j = i+1$ and using $k_{i+1} = Fk_i$ the commutativity of that square states precisely that k_i is a coalgebra-to-algebra morphism and therefore is uniquely determined.

13.2 Cauchy Completion

Returning to the theme of this chapter, we are ready to endow V_ω with the structure of a complete metric space. The goal is to show that V_ω is a completion (in some sense) of the structure on μF which is inherited via k. It is not surprising that we are going to request $F\emptyset \neq \emptyset$; otherwise $\mu F = \emptyset$, and so our result has no chance.

Recall that throughout this book, we take metrics to be bounded by 1. Let us consider each $F^n 1$, $n < \omega$, as a discrete metric space: the distance of two distinct elements is always 1. We shall endow the limit $V_\omega = \lim_{n<\omega} F^n 1$ (with the limit cone $v_{\omega,n}\colon V_\omega \to F^n 1$) with a natural non-discrete metric.

Notation 13.2.1. Define a metric d on V_ω by

$$d(x,y) = \begin{cases} 0 & \text{if } x = y, \\ 2^{-n} & \text{if } n \text{ is largest such that } v_{\omega,n}(x) = v_{\omega,n}(y), \text{ otherwise.} \end{cases}$$

Proposition 13.2.2 [13, Thm. 3.3]. *The metric space (V_ω, d) is complete.*

Proof The triangle inequality is the only interesting point in the verification that d is a metric. Indeed, it is an ultrametric:

$$d(x,z) \leq \max\{d(x,y), d(y,z)\}.$$

If the maximum on the right is 2^{-n}, then $v_{\omega,n}$ merges x, y as well as y, z. Thus it merges x, z.

We verify that this metric is complete. Fix a Cauchy sequence $(x_n)_{n<\omega}$. For every $\varepsilon = 2^{-k}$, we then have a natural number n_k so that for all $m \geq n_k$, $d(x_{n_k}, x_m) < 2^{-k}$. Thus:

$$v_{\omega,n_k}(x_{n,k}) = v_{\omega,n_k}(x_m) \qquad \text{for all } m \geq n_k. \tag{13.6}$$

We choose an increasing sequence (n_k) with this property. Consider the ω-sequence of sets $S_k = V_{n_k}$ with connecting maps $s_k = v_{n_{k+1},n_k} : S_{k+1} \to S_k$. It has the same limit V_ω with the limit cone $v_{\omega,n_k} : V_\omega \to S_k$. The elements $\overline{x}_k = v_{\omega,n_k}(x_{n_k})$ of S_k are compatible: due to (13.6) we have $s_{k+1}(\overline{x}_{k+1}) = \overline{x}_k$ for all k.

Consequently, there exists $\overline{x} \in V_\omega$ such that for all k, $v_{\omega,n_k}(\overline{x}) = \overline{x}_k$. This implies that $d(\overline{x}, \overline{x}_k) < 2^{-k}$, since $v_{\omega,n_k}(\overline{x}) = v_{\omega,n_k}(x_{n_k})$. We thus see that $\overline{x}$ is a limit of the sequence $(x_{n_k})_{k<\omega}$, hence a limit of $(x_n)_{n<\omega}$. □

Definition 13.2.3. An endofunctor is *continuous* if it preserves ω^{op}-limits.

Example 13.2.4. (1) For every, possibly infinitary, signature, the polynomial functor H_Σ is continuous.

(2) The finite power-set functor $\mathscr{P}_\mathrm{f}$ is not continuous. We have seen in Example 4.4.6 that it does not preserve the limit of its terminal-coalgebra ω^{op}-chain.

(3) In contrast, for every $n \in \mathbb{N}$, the subfunctor $\mathscr{P}_n$ of $\mathscr{P}$ of all subsets of cardinality less than n is continuous.

Definition 13.2.5. (1) A *prelimit* of a diagram is a cone which is collectively monic. (That is, every cone has at most one factorization through a prelimit.)

(2) A set functor F is *precontinuous* if it preserves nonempty prelimits of ω^{op}-chains. That is, if $p_i : P \to X_n$ is a prelimit of the ω^{op}-chain (X_n) and P is nonempty, then Fp_n $(n < \omega)$ is a collectively monic cone.

Remark 13.2.6. Let D be a diagram with a limit L. The prelimits of D are precisely those cones of D which have a monic factorization through L. Consequently, prelimits correspond bijectively to subobjects of L.

Proposition 13.2.7. *Every finitary set functor is precontinuous.*

Proof Let (X_n) $(n < \omega)$ be an ω^{op}-chain, and let $p_n : P \to X_n$ be a nonempty prelimit. Let $x, y \in FP$, and suppose that for all k, $Fp_k(x) = Fp_k(y)$. Since F is finitary, it is finitely bounded (Proposition 4.3.4), thus there is a finite subset $m : P_0 \hookrightarrow P$ and elements $\overline{x}$ and $\overline{y}$ such that $Fm(\overline{x}) = x$ and $Fm(\overline{y}) = y$. Because P is nonempty, we may assume that $P_0 \neq \emptyset$ without loss of generality.

Since P_0 is finite, there is a fixed n so that $p_n \cdot m$ is monic. Then $P_0 \neq \emptyset$ implies that $p_n \cdot m$ is a split monomorphism. Thus, $F(p_n \cdot m)$ is also monic. Now $F(p_n \cdot m)(\overline{x}) = Fp_n(x) = Fp_n(y) = F(p_n \cdot m)(\overline{y})$ implies that $\overline{x} = \overline{y}$, whence $x = y$. □

Example 13.2.8. (1) If our definition of precontinuity had required preservation of *all* prelimits, finitary functors would no longer be precontinuous. For example, consider the functor $C_{2,1}$ which takes the empty set to 2 and all other sets to 1. This functor does not preserve any empty prelimit of an ω^{op}-chain of nonempty sets. For example, consider the ω^{op}-chain formed by the sets $X_n = \mathbb{N} \setminus \{i : i < n\}$ and the inclusion maps between them. Its limit (viz. the intersection of all X_n) is empty, but the limit cone is mapped by $C_{2,1}$ to the cone formed by the non-injective map $2 \to 1$.

(2) Every continuous set functor is precontinuous. To see this, take an ω^{op}-chain $(X_n)_{n<\omega}$ and a nonempty prelimit, say $p_n \colon C \to X_n$. Let $\ell_n \colon L \to X_n$ be the limit. By Remark 13.2.6, the factorizing morphism $k \colon C \to L$ is monic. By continuity the limit of $(FX_n)_{n<\omega}$ is $F\ell_n \colon FL \to FX_n$. The factorization morphism of $Fp_n \colon FC \to FX_n$ through this limit is Fk. Since C is nonempty, k is a split monomorphism, and so Fk is monic. We conclude from Remark 13.2.6 that $Fp_n \colon FP \to FX_n$ is a prelimit.

It follows that for every signature Σ, the associated polynomial functor H_Σ is precontinuous.

(3) The class of precontinuous functors is closed under composition, products, and coproducts.

(4) Every Kripke polynomial functor (Definition 4.4.13) is precontinuous.

We now show that precontinuous functors form a strictly larger class than the classes of finitary or continuous functors. Recall the distribution functor from Example 12.2.8(1).

Proposition 13.2.9 [316]. *The distribution functor $\mathscr{D}$ is precontinuous. It is neither finitary nor continuous.*

Proof (1) Consider an ω^{op}-chain:

$$A_0 \xleftarrow{a_0} A_1 \xleftarrow{a_1} A_2 \xleftarrow{a_2} \cdots$$

Let $p_n \colon P \to A_n$ be a nonempty prelimit. We shall prove that $\mathscr{D}p_n \colon \mathscr{D}P \to \mathscr{D}A_n$ is a prelimit, too. Let $\mu, \mu' \in \mathscr{D}P$ satisfy $\mathscr{D}p_n(\mu) = \mathscr{D}p_n(\mu')$ for all $n < \omega$; we show that $\mu = \mu'$. For this, let $x \in P$ and put $S_n = p_n^{-1}(p_n(x)) \subseteq P$. Now $\mathscr{D}p_n(\mu)(p_n(x)) = \mu(p_n^{-1}(p_n(x))) = \mu(S_n)$, and similarly for μ'. So $\mu(S_n) = \mu'(S_n)$. Since C is a prelimit, $\{x\} = \bigcap_{n\in\omega} S_n$.

Also, $S_0 \supseteq S_1 \supseteq \cdots$ and $\mu(S_i) < \infty$. By continuity of measures (a standard result), $\mu(x) = \mu(\{x\}) = \inf_n \mu(S_n)$. It follows that $\mu(x) = \mu'(x)$, since both numbers are the infima of the same sequence of real numbers. This for all $x \in C$ shows that $\mu = \mu'$.

(2) $\mathscr{D}$ is not finitary. Let $\mu \in \mathscr{D}(\omega)$ be defined by $\mu(n) = 2^{-(n+1)}$. For each $n = \{0, 1, \ldots, n-1\}$, let $i_n : n \to \omega$ be the inclusion. Then $\mu \notin \mathscr{D}i[\mathscr{D}n]$, since every measure in $\mathscr{D}i[\mathscr{D}n]$ has finite support. So $\mathscr{D}$ is not finitely bounded and hence not finitary; see Proposition 4.3.4.

(3) $\mathscr{D}$ does not preserve limits of ω^{op}-chains. Consider $a_n : 2^{n+1} \to 2^n$ mapping each binary sequence of length $n+1$ to its first n values. The limit of this chain is 2^ω, with evident projections $p_n : 2^\omega \to 2^n$. For each n, let $\mu_n \in \mathscr{D}(2^n)$ be the uniform distribution. Then the family $(\mu_n)_{n<\omega}$ is compatible with the chain. Fix $\mu \in \mathscr{D}(2^\omega)$. For all $x \in 2^\omega$, we again use that μ is continuous to see that $\mu(x) = \inf_n p_n(x) = 0$. This for all x shows that $\mu(2^\omega) = 0$, contradicting μ being a probability measure. □

Remark 13.2.10. The functor $\mathscr{D}$ also preserves all intersections. We know of no precontinuous set functor that does not preserve intersections.

Theorem 13.2.11. *Let F be a precontinuous set functor. Then*

(1) *The morphism $v_{\omega+1,\omega} : FV_\omega \to V_\omega$ is monic.*

(2) *The functor F has an initial algebra and a terminal coalgebra, and both μF and νF are subalgebras of $(V_\omega, v_{\omega+1,\omega})$.*

Proof (1) This is trivial if F is the constant functor with value $\emptyset$. Otherwise, $V_\omega \neq \emptyset$ (as in the proof of Corollary 4.4.9). Thus, F takes the limit $v_{\omega,n} : V_\omega \to F^n 1$ to the collectively monic cone $Fv_{\omega,n}$ $(n < \omega)$. Since $v_{\omega+1,\omega}$ is the factorization of that cone through the limit, it is monic by Remark 13.2.6.

(2) We have seen in item (1) that $(V_\omega, v_{\omega+1,\omega})$ is a pre-fixed point of F. Moreover, we have shown in Example 7.1.4(5) that this is a corecursive algebra. Now our result follows from Theorem 6.3.4 and Proposition 7.2.8. □

Remark 13.2.12. (1) This result generalizes. On any complete and well-powered category, every functor which preserves monomorphisms and ω^{op}-prelimits has a terminal coalgebra which is a subalgebra of V_ω.

(2) The morphism $k : \mu F \to \nu F$ from (13.3) is monic by Corollary 6.3.10.

(3) We consider νF as a subspace of the metric space V_ω (Notation 13.2.1). In addition, μF is a metric subspace of νF via $k : \mu F \to \nu F$ from (13.3).

Definition 13.2.13. A set functor is *grounded* if $F\emptyset \neq \emptyset$.

The first item of the following theorem was proved by Barr [72].

Theorem 13.2.14. *Let F be a grounded set functor.*

(1) *If F is continuous, then νF is the Cauchy completion of μF.*

(2) *If F is precontinuous, then μF and νF both have V_ω as their Cauchy completion.*

(Note that in both items, μF and νF exist by Theorem 13.2.11(2).)

Proof The first item follows from the second one since for a continuous functor $\nu F = V_\omega$.

In the proof of the second item, we abbreviate $v_{1,0} \colon F1 \to 1$ as v and $v_{\omega,n} \colon V_\omega \to F^n 1$ as $\bar{v}_n$. Analogously, $w_{n,\omega} \colon F^n \emptyset \to W_\omega$ is abbreviated as $\bar{w}_n$. We prove some auxiliary statements first.

(1) Since F is grounded, we can choose an element $b \colon 1 \to F\emptyset$. Form the maps

$$c_n = (F^n 1 \xrightarrow{F^n b} F^{n+1}\emptyset \xrightarrow{\bar{w}_{n+1}} W_\omega \xrightarrow{k_\omega} V_\omega). \tag{13.7}$$

We prove that for all $x \in V_\omega$, $\lim_{n\to\infty} d(c_n \cdot \bar{v}_n(x), x) = 0$. It suffices to verify the following claim

$$\bar{v}_k = \left(V_\omega \xrightarrow{\bar{v}_n} F^n 1 \xrightarrow{c_n} V_\omega \xrightarrow{\bar{v}_k} F^k 1\right) \qquad \text{for all } n > k. \tag{13.8}$$

Here is why this implies our result: fix x and n, and suppose that k is such that $\bar{v}_k(c_n \bar{v}_n(x)) \neq \bar{v}_k(x)$. By our claim, we must have $k > n$. Thus

$$d(c_n \bar{v}_n(x), x) \leq \frac{1}{2^n}.$$

To prove (13.8) consider the following diagram:

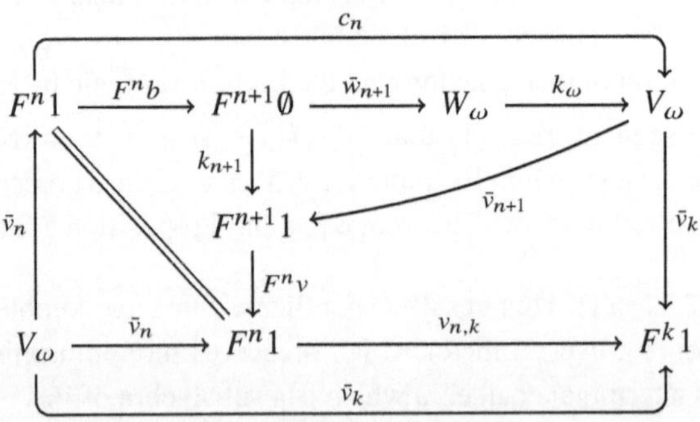

The top is the definition of c_n, and the bottom comes from the fact that V is a chain. The lower-left triangle is obvious. For the upper-right triangle, we use the obvious equality $v \cdot k_1 \cdot b = \mathrm{id}_1$ and the fact that $k_{n+1} = F^{n+1} k_0$ to see that

$$F^n v \cdot F^{n+1} k_0 \cdot F^n b = F^n v \cdot F^n k_1 \cdot F^n b = F^n (v \cdot k_1 \cdot b) = \mathrm{id}_{F^n 1}.$$

A special case of (13.5) is that $k_{n+1} = \bar{v}_{n+1} \cdot k_\omega \cdot \bar{w}_{n+1}$, and this is the remaining interior region. In the bottom, $v_{n,k}$ is the map from the terminal-coalgebra chain. All the inside regions commute. The outside yields the desired equation (13.8).

(2) The image $k_\omega[W_\omega]$ is a dense subset of V_ω. Indeed, fix $x \in V_\omega$. Then for all n, put $x_n = \bar{w}_{n+1} \cdot F^n b \cdot \bar{v}_n(x) \in W_\omega$. The distance from x to $k_\omega(x_n) = k_\omega \cdot \bar{w}_{n+1} \cdot F^n b \cdot \bar{v}_n(x) = c_n \cdot \bar{v}_n(x)$ tends to 0 as $n \to \infty$ by item (1).

(3) Since V_ω is complete, it is the Cauchy completion of each of its dense subspaces. Thus the desired result follows from the previous item. □

Remark 13.2.15. (1) The morphisms $\bar{v}_n = v_{\omega,n}$ and $\bar{w}_n = w_{n,\omega}$ used in the proof of Theorem 13.2.14 bear an analogy to those used in Theorem 5.2.11.

(2) For a finitary functor F, V_ω is the Cauchy completion of μF by the second item in Theorem 13.2.14. However, the first item does not hold for a finitary functor in general, see Example 13.2.16(2) below.

Example 13.2.16. (1) For $FX = X \times X + E$, the metric of Example 13.0.1 is precisely the metric on νF coming from Theorem 13.2.14. This example immediately generalizes to all polynomial functors on Set (see Example 6.1.16). The signature need not be finitary: for all Σ, this functor H_Σ is continuous; thus νH_Σ carries the metric of (13.1). Assume $H_\Sigma \emptyset \neq \emptyset$; this means that there is at least one constant in Σ. When we choose a constant $b \in \Sigma$, we can consider the cuttings in the formula (13.1) to be Σ-trees again: the new leaves obtained by the cutting are labelled by b. All these cuttings have finite height and therefore lie in μH_Σ, indeed even in $H_\Sigma^\omega \emptyset = \bigcup_{n<\omega} H_\Sigma^n \emptyset$. This explains why $H_\Sigma^\omega \emptyset$ is dense in νH_Σ.

(2) For the finite power-set functor we have described $\mu \mathscr{P}_\mathsf{f}$ as the set of all finite extensional trees (Example 3.2.9), and $\nu \mathscr{P}_\mathsf{f}$ as the set of all finitely branching, strongly extensional trees (Theorem 4.5.15). The metric $\bar{d}$ of Notation 13.2.1 is just the restriction of the metric d_τ on $\mathbb{F}$ (the set of all saturated, strongly extensional trees) of Remark 4.5.20. Thus, in this case the Cauchy completion of $\mu \mathscr{P}_\mathsf{f}$ is not $\nu \mathscr{P}_\mathsf{f}$.

Our final topic in this section concerns approximations to the unique coalgebra homomorphisms to νF for grounded precontinuous functors. For all coalgebras A, we express the homomorphism $h\colon A \to \nu F$ as the limit of a canonical sequence of morphisms $h_n\colon A \to \mu F$. More precisely, $h = \lim(k \cdot h_n)$ in the function space $(\nu F)^A$ (equipped with the usual supremum metric: the distance of $f, g\colon A \to \mu F$ is the supremum of $d(f(x), g(x))$ for $x \in A$). We call these maps *approximations* to the unique coalgebra morphism. Recall our choice of a point $b\colon 1 \to F\emptyset$.

Definition 13.2.17. Let (A, α) be a coalgebra for a grounded set functor F. We define morphisms $h_n\colon A \to \mu F$ for $n \in \mathbb{N}$ by the following induction:

$$\begin{array}{ccc} A & \xrightarrow{h_0} & \mu F \\ {\scriptstyle !}\downarrow & & \uparrow{\scriptstyle w_1} \\ 1 & \xrightarrow[b]{} & F\emptyset \end{array} \qquad\qquad \begin{array}{ccc} A & \xrightarrow{h_{n+1}} & \mu F \\ {\scriptstyle \alpha}\downarrow & & \uparrow{\scriptstyle \iota} \\ FA & \xrightarrow[Fh_n]{} & F(\mu F) \end{array}$$

Thus, h_0 is a constant map obtained from $b \in F\emptyset$, and h_{n+1} is obtained canonically from h_n using the coalgebra structure α. We now prove that the maps h_n form approximations of the unique coalgebra homomorphism $h\colon A \to \nu F$.

Proposition 13.2.18. *For every coalgebra A of a grounded precontinuous set functor F, the unique coalgebra homomorphism $h\colon A \to \nu F$ satisfies $h = \lim_n (k \cdot h_n)$ in the metric space $(\nu F)^A$.*

Proof The terminal coalgebra structure is $v_{\delta+1,\delta}^{-1}$ for some ordinal δ (see Theorem 6.5.15). The coalgebra homomorphism h satisfies $h = v_{\delta+1,\delta} \cdot Fh \cdot \alpha$.

We are going to show that

$$v_{\delta,n} \cdot h = v_{\delta,n} \cdot k \cdot h_n \qquad \text{for all } n < \omega. \tag{13.9}$$

This implies that the distance in $(\mu F)^A$ between h and $k \cdot h_n$ is at most 2^{-n}. The proof is by induction on n. The case $n = 0$ is trivial since $v_{i,0}$ has codomain 1. Assuming (13.9) for n, consider the diagram below:

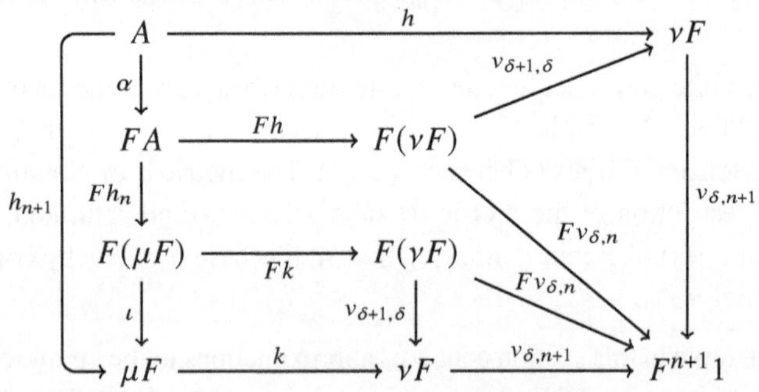

The upper part commutes since h is a homomorphism, and the two inner triangles commute because $\nu F = V_\delta$. The left-hand part is the definition of h_{n+1}, and the lower left-hand square is (13.3). The remaining inner region is the induction hypothesis with F applied to it. Thus, the outside commutes, which is the desired (13.9) for $n + 1$. □

Example 13.2.19. (1) Let us first consider deterministic systems with output in an alphabet Σ, allowing of terminating states. The functor is $FX = \Sigma \times X + 1$ (Example 2.5.3(4)). We have $\nu F = \Sigma^\infty$, and $\mu F = \Sigma^*$ (Example 2.2.15(2)). For a coalgebra (A, α), the homomorphism $h\colon A \to \nu F$ takes a state $a \in A$ to the (possibly infinite) word $h(a)$ of outputs when we regard a as the initial state. For each natural number n, $h_n(a)$ takes a to the prefix of length n of $h(a)$, or to $h(a)$ itself if $h(a)$ has length at most n. Indeed, this holds for 0 since $F\emptyset = 1$, represented by $\{\varepsilon\}$; so h_0 is the constant function with value the empty word. If our description holds for n, then it holds for $n + 1$ because $h_{n+1} = \iota \cdot F h_n \cdot \alpha$. In more detail,

$$\iota = [c, \varepsilon]\colon \Sigma \times \Sigma^* + 1 \to \Sigma^*,$$

where $c(s, w) = sw$ is concatenation. Thus, if $\alpha(a) = (s, b) \in \Sigma \times A$, we see that $h_{n+1}(a) = s h_n(b)$.

(2) Let H_Σ be a polynomial functor. The results in this section only apply when Σ has a nullary (constant) symbol, since this is what groundedness comes to. Choose such a symbol $b \in H_\Sigma 0$. The terminal coalgebra is νH_Σ is carried by the set T_Σ of all finite and infinite Σ-trees (Theorem 2.5.9). Given a coalgebra A, the homomorphism $h\colon A \to T_\Sigma$ takes $a \in A$ to its tree expansion. The initial algebra μH_Σ is carried by the set of well-founded trees (Example 6.1.16). The approximation $h_n\colon A \to \mu H_\Sigma$ takes a to $\partial_n a$, the cutting of $h(a)$ at height n. This means that all nodes of height larger than n are deleted, and all leaves at height n in the resulting tree are labelled with the chosen symbol b.

(3) Recall that coalgebras for $\mathscr{P}_\mathsf{f}$ are precisely the finitely branching graphs (cf. Example 2.4.2(4)). Here $\nu\mathscr{P}_\mathsf{f}$ is the coalgebra of all finitely branching strongly extensional trees (Theorem 4.5.15). Given a finitely branching graph A, $h\colon A \to \nu F$ takes a node $a \in A$ to the tree expansion t_a of a. We have $\mathscr{P}_\mathsf{f}\emptyset = \{\emptyset\}$. The initial algebra $\mu\mathscr{P}_\mathsf{f}$ consists of finite extensional trees (Example 3.2.9(2)). The approximation $h_n\colon A \to \mu\mathscr{P}_\mathsf{f}$ takes a node $a \in A$ to the extensional quotient of the cutting $\partial_n t_a$ of t_a at level n.

Remark 13.2.20. If F is continuous, then $\nu F = V_\omega$ is a complete metric space which is also a corecursive algebra for F. For each coalgebra (A, α), $(\nu F)^A$ has the same properties. In this case, the approximations h_n for $n > 0$ are exactly what is obtained from the Contracting Mapping Theorem B.3.2 to obtain the coalgebra-to-algebra morphism, starting with the map h_0 as we did. (As with all applications of the Contracting Mapping Theorem, the starting point is arbitrary.) In the general case, when F is precontinuous but not continuous, this line of reasoning does not work and we need the more elaborate argument of Proposition 13.2.18.

Remark 13.2.21. The results of this section hold not only for set functors, but also for endofunctors of a wide variety of categories called *strictly locally finitely presentable* [14]. These include the categories of posets, vector spaces, Boolean algebras, and graphs. Given a grounded finitary endofunctor F, there is a canonical metric space associated with νF, and the subspace associated with μF has the same Cauchy completion. Homomorphisms into νF are approximated by sequences of morphisms into μF.

13.3 Ideal Completion

In lieu of complete metrics we can also use directed-complete partial orders (dcpos) to show how the terminal coalgebra is determined by the initial algebra. We again denote by F a precontinuous set functor. The role of the Cauchy completion is now played by the ideal completion recalled in Remark 13.3.5.

Notation 13.3.1. Given a grounded set functor F, we choose an element $b\colon 1 \to F\emptyset$ and use $k_\omega\colon W_\omega \to V_\omega$ from Notation 13.1.1. Let us denote by $e_n\colon F^n 1 \to V_\omega$ the following composite:

$$e_n = \left(F^n 1 \xrightarrow{Fb} F^{n+1} 0 \xrightarrow{w_{n+1,\omega}} W_\omega \xrightarrow{k_\omega} V_\omega\right) \qquad \text{for } n < \omega.$$

We also put

$$e = \left(1 \xrightarrow{b} F\emptyset \xrightarrow{F!} F1\right).$$

Definition 13.3.2. We define a partial order on V_ω by putting

$$x \sqsubseteq y \qquad \text{iff} \qquad x = y \text{ or } x = e_n \cdot v_{\omega,n}(y) \text{ for some } n < \omega.$$

Analogously, each $F^n 1$ is a poset under the order given by

$$x \sqsubseteq y \qquad \text{iff} \qquad x = y \text{ or } e_n(x) \sqsubseteq e_n(y).$$

Recall from Remark 5.1.14 the coincidence of limits and colimits of chains of embedding-projection pairs.

Theorem 13.3.3 [12, Thm. 3.3 and 3.7]**.** *For every grounded precontinuous set functor, the set V_ω is a dcpo under the order in Definition 13.3.2. Moreover, for every $k < \omega$ we have the following embedding projection pairs between dcpos:*

$$F^k 1 \underset{v_{\omega,k}}{\overset{e_k}{\rightleftarrows}} V_\omega \qquad \text{and} \qquad F^k 1 \underset{v_{k+1,k}}{\overset{F^k e}{\rightleftarrows}} F^{k+1} 1$$

such that

$$\bigsqcup_{k<\omega} e_k \cdot v_{\omega,k} = \mathrm{id}_{V_\omega}. \tag{13.10}$$

Proof sketch (1) The relation $\sqsubseteq$ is a partial order on V_ω in which every strictly increasing ω-chain has a unique upper bound ([12, Thm. 3.3]).

(2) The set V_ω is a dcpo: let $M \subseteq V_\omega$ be a directed poset without a largest element (otherwise the join of M would be that element). Then there exists a strictly increasing ω-chain (x_n) in M. Its unique upper bound x is the join of M. Indeed, for every upper bound of M is also an upper bound of (x_n) and therefore equal to x, and we verify for every $y \in M$ that $y \sqsubseteq x$: Since M is directed, there exists an ω-chain (y_n) such that $y_0 = y$, and for every n we have $x_n \sqsubseteq y_n$. The unique upper bound of (y_n) is also an upper bound of (x_n) and therefore equal to x. Thus $y = y_0 \sqsubseteq x$ as desired.

(3) For every $k < \omega$ we have an embedding $F^k e \colon F^k 1 \to F^{k+1} 1$ with projections $v_{k+1,k}$ [12, Thm. 2.7]. We thus obtain an ω-chain $(F^k e)$ of embeddings. Furthermore, *loc. cit.* states that e_k and $v_{\omega,k}$ form an embedding-projection pair and that the cocone (e_k) is a colimit of the ω-chain $(F^k e)$. By Theorem 5.1.10 we obtain equation (13.10). □

Example 13.3.4. (1) For $FX = X \times X + E$ the above order on νF is precisely that of Example 13.0.1.

(2) More generally, let H_Σ be a grounded polynomial functor (not necessarily finitary). Then νH_Σ is the dcpo with order given by $t < s$ iff the tree s has the cutting t at some level $n < \omega$, where all leaves at that level are (re)labelled by $b \in \Sigma_0$. Shortly, we have $t = \partial_n s$ (Example 13.2.19(2)).

(3) Analogously for the finite power-set functor $\mathscr{P}_\mathrm{f}$: recall that $\mu\mathscr{P}_\mathrm{f}$ is described as the set of all finite extensional trees (Example 3.2.9). It is partially ordered by

$$t \sqsubseteq s \quad \text{iff} \quad t = s \text{ or } t = \partial_n s \text{ for some } n < \omega.$$

Let us remark that although $\mu\mathscr{P}_\mathrm{f}$ is also described as the set of all hereditarily finite sets, this partial order is *not* set inclusion. For example, for the trees

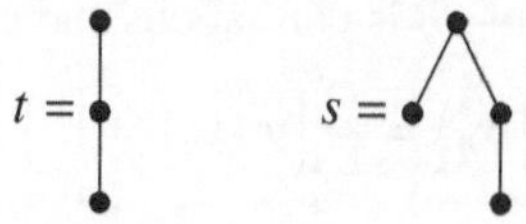

we see that t and s are incompatible. However t represents $\{\{\emptyset\}\}$ which is a subset of $\{\emptyset, \{\emptyset\}\}$ (represented by s).

The ideal completion of the above poset $(\mu\mathscr{P}_\mathrm{f}, \sqsubseteq)$ is not $\nu\mathscr{P}_\mathrm{f}$ but the limit $\mathbb{F} = \lim_{n<\omega} V_n$. This is a special case of the following:

Remark 13.3.5. The *ideal completion* of a poset A is a dcpo A^+ containing A as a subposet such that every monotone map $f \colon A \to B$, where B is a dcpo, has a unique continuous extension $f^+ : A^+ \to B$. The ideal completion A^+ is unique up to isomorphism.

The ideal completion can be constructed as the poset of all *ideals* of A; an ideal is a subset $I \subseteq A$ which is downwards closed and upwards directed. This poset is ordered by inclusion, and A appears as a subposet via the identification of every $x \in A$ with the ideal $\downarrow x = \{y \in A : y \leq x\}$.

Lemma 13.3.6. *Let A^+ be a dcpo containing A. Then A^+ is the ideal completion of A whenever the following conditions hold:*

(1) *For every element $x \in A^+$ an ω-chain (x_n) is chosen in A such that $x = \bigsqcup_{n<\omega} x_n$,*

(2) *if $x \in A$ then $x = x_n$ for some $n < \omega$,*

(3) *every ideal $I \subseteq A$ has the form $I = \{x_n : n < \omega\}$ for some $x \in A^+$, and*

(4) *every strictly increasing ω-chain in A^+ has a unique upper bound.*

Proof Let B be a dcpo. For every monotone map $f : A \to B$ define $f^+ : A^+ \to B$ by

$$f^+(x) = \bigsqcup_{n<\omega} f(x_n).$$

This extends f due to (2). To see that f^+ is monotone, suppose that $x \leq y$ in A^+. Then y is an upper bound of the ω-chain (x_n). By (4) we either have $x = y$, if (x_n) is strictly increasing, or $x = x_n$ for some $n < \omega$, otherwise. In the former case we are done, and in the latter case $x = x_n$ lies in the ideal $\downarrow y$, which has the form $\{y_m : m < \omega\}$ due to (3). Thus $x = x_n = y_m$ for some $m < \omega$ so that

$$f^+(x) \sqsubseteq f^+(y_m) = f(y_m) \sqsubseteq \bigsqcup_{m<\omega} f(y_m) = f^+(y),$$

where in the equation in the second step we use (2) once more.

That f^+ is continuous now follows from (3). Finally, the uniqueness of f^+ is easy to establish: if $g : A^+ \to B$ is continuous and extends f, then we have

$$g(x) = g\Big(\bigsqcup_{n<\omega} x_n\Big) = \bigsqcup_{n<\omega} g(x_n) = \bigsqcup_{n<\omega} f(x_n) = f(x). \qquad \square$$

Remark 13.3.7. The conditions in Lemma 13.3.6 are rather special. They imply that A is closed under existing joins in A^+. In other words, A^+ is also the free *conservative* dcpo-completion of A. This implies that the conditions are sufficient but not in general necessary for A^+ to be the ideal completion of A.

We have also seen in the proof that we could work with cpos (i.e. joins of ω-chains in lieu of directed ones) and obtain the same result: a cpo A^+ is the free cpo-completion of A if it satisfies the four conditions.

Theorem 13.3.8. *Let F be a grounded set functor.*

(1) *If F is continuous, then νF is the ideal completion of μF.*

(2) *If F is precontinuous, then μF and νF both have V_ω as their ideal completion.*

(Note that in both items, μF and νF exist by Theorem 13.2.11(2).)

Proof The first item follows from the second one as $\nu F = V_\omega$ for every continuous functor (Theorem 3.3.4).

For the second item, let $i \geq \omega$ be an ordinal with $\nu F = V_i$. Denote by A the image of the following map:

$$W_\omega \xrightarrow{w_{\omega,i}} \mu F \xrightarrow{k} \nu F = V_i \xrightarrow{v_{i,w}} V_\omega .$$

This map is monic because $v_{i,\omega}$ is, due to precontinuity (Remark 13.2.6), k is monic by Corollary 6.3.10, and $w_{\omega,i}$ is due to Corollary C.6.14.

We apply Lemma 13.3.6 to $A^+ = V_\omega$ which is a dcpo by Theorem 13.3.3. This will prove that V_ω is the ideal completion of W_ω, and it will also imply that the four conditions hold for $A = v_{i,\omega}[\nu F]$ and $A = v_{i,\omega} \cdot k[\mu F]$, yielding the desired result.

Ad (1). For $x \in V_\omega$ put $x_n = e_n \cdot v_{\omega,n}(x)$ $(n < \omega)$. These elements of A form an ω-chain with join x [12, Proof of Prop. 4.1].

Ad (2). For every $n < \omega$ the following square commutes [12, Proof of Prop. 4.4]:

$$\begin{array}{ccc} F^n\emptyset & \xrightarrow{w_{n,\omega}} & W_\omega \\ {\scriptstyle k_n}\downarrow & & \downarrow{\scriptstyle k_\omega} \\ F^n 1 & \xrightarrow{e_n} & V_\omega \end{array}$$

Given $x \in A$, there exists $n < \omega$ and $y \in F^n\emptyset$ such that $x = k_\omega \cdot w_{n,\omega}(y)$. Thus, we obtain $x = e_n \cdot k_n(y)$, and therefore $v_{\omega,n} \cdot e_n = \mathrm{id}$ implies

$$x_n = e_n \cdot v_{\omega,n} \cdot e_n \cdot k_n(y) = e_n \cdot k_n(y) = k_\omega \cdot w_{n,\omega}(y) = x.$$

Ad (3), see [12, Proof of Prop. 4.1].

Ad (4), see [12, Item (f) in the proof of Thm. 3.3]. □

Remark 13.3.9. If F is finitary, there is a result analogous to Proposition 13.2.18: for every F-coalgebra (A, α) the unique homomorphism $h\colon A \to \nu F$ is the join of the ω-chain of approximations [15, Cor. 56]:

$$h = \bigsqcup_{n<\omega} k \cdot h_n \qquad \text{in } (\nu F)^A.$$

13.4 Summary

We have seen a close connection between initial algebras and terminal coalgebras. Besides the canonical homomorphism $k\colon \mu F \to \nu F$ which is often monic (e.g. for all set functors with a terminal coalgebra), we have also presented a canonical collection of morphisms $k_i\colon W_i \to V_i$ forming a dinatural transformation between the initial-algebra chain the terminal-coalgebra chain.

We introduced precontinuous set functors, a collection containing all finitary and all continuous functors. We proved that if a precontinuous functor F is grounded, that is $F\emptyset \neq \emptyset$, then νF carries a metric whose Cauchy completion is the same as that of its subspace μF. Moreover, coalgebra homomorphisms into νF can be approximated by canonical Cauchy sequences of morphisms into μF.

Analogous results also hold for a partial order defined on νF for a finitary set functor F: its ideal completion is the same as that of the subposet μF, and approximations of homomorphisms are obtained by ω-joins of morphisms into μF.

14

Derived Functors

Given an endofunctor F we can form various *derived endofunctors* whose initial algebras and terminal coalgebras are related to those for F. The most prominent example of derived functors are coproducts of a functor with constant functors – this yields free F-algebras, cofree F-coalgebras, and free completely iterative algebras, and we discuss those in Section 14.1. Section 14.2 is devoted to compositions of functors, while in Section 14.3 we study functors on slice categories. Then in Section 14.4 we turn to functors on product categories, and Section 14.6 covers coproducts of functors with a treatment of *double-algebras*. Finally, in Section 14.7 we discuss an application of initial double-algebras: coproducts of monads.

14.1 Coproducts and Products with a Constant Functor

We begin with coproducts of F and constant functors. Recall from Proposition 2.2.24 that the initial algebra $\mu X.\,(FX + A)$ is precisely a free F-algebra $A^\sharp$ on A.

We have also mentioned varietors (Definition 2.2.26), which are endofunctors F for which every free algebra $A^\sharp$ exists. We have seen that free algebras define a monad M on $\mathscr{A}$, the free monad on F (Remark 2.2.28). Moreover on a category with powers, an endofunctor is a varietor iff it has a free monad (Theorem 2.2.29). In fact, if the monad M is free on F via the universal natural transformation $\kappa\colon F \to M$, then MX is a free F-algebra with structure and universal morphism given by

$$FMX \xrightarrow{\kappa_{MX}} MMX \xrightarrow{\mu_X} MX \qquad \text{and} \qquad X \xrightarrow{\eta_X} MX.$$

For example, the free monad on H_Σ for a finitary signature Σ is given by

$$MA = \text{all finite } (\Sigma + A)\text{-trees.}$$

For infinitary signatures cf. Example 6.1.16.

We now turn to iterative constructions of free algebras. They can be constructed using the initial-algebra (ω-)chain (see Definitions 3.1.4 and 6.1.4) for $F(-) + A$ which looks as follows:

$$0 \xrightarrow{w_0} F0 + A \xrightarrow{w_1} F(F0 + A) + A \xrightarrow{w_2} \cdots,$$

where w_0 is unique and $w_{n+1} = Fw_n + A$. However, a more direct construction is possible, as we now explain. It starts with the following ω-chain

$$A \xrightarrow{\mathsf{inr}} FA + A \xrightarrow{F\mathsf{inr}+\mathsf{id}} F(FA + A) + A \xrightarrow{F(F\mathsf{inr}+\mathsf{id})+\mathsf{id}} \cdots$$

Construction 14.1.1 [7]. Let F be an endofunctor on a cocomplete category. Given an object A, we define the *free-algebra chain* $\mathsf{Ord} \to \mathscr{A}$ of A for F. Its objects are denoted by $A^\sharp_i$ and its connecting morphisms by $a^\sharp_{i,j}$, $i \le j \in \mathsf{Ord}$. They are defined by transfinite recursion:

$A^\sharp_0$	$= A,$	
$A^\sharp_{j+1}$	$= FA^\sharp_j + A$	for all ordinals j,
$A^\sharp_j$	$= \mathsf{colim}_{i<j} A^\sharp_i$	for all limit ordinals j, and
$a^\sharp_{0,1}$	$= (A^\sharp_0 = A \xrightarrow{\mathsf{inr}} FA^\sharp_0 + A = A^\sharp_1)$	is the right-hand coproduct injection,

$$a^\sharp_{j+1,k+1} = (A^\sharp_{j+1} = FA^\sharp_j + A \xrightarrow{Fa^\sharp_{j,k} + \mathsf{id}_A} FA^\sharp_k + A = A^\sharp_{k+1}),$$

$a^\sharp_{i,j}$ $(i < j)$ is the colimit cocone for limit ordinals j.

Convergence of the free-algebra chain in λ steps means that $a^\sharp_{\lambda,\lambda+1}$ is an isomorphism.

Remark 14.1.2. From the ordinal ω onwards, the free-algebra chain for A coincides with the initial-algebra chain (W_i) for $F(-) + A$. To see this, it suffices to prove that the ω-colimits of W_n $(n < \omega)$ and $A^\sharp_n$ $(n < \omega)$ coincide. The subsequent steps of the two chains are then the same: we obtain $W_i = A^\sharp_i$ for $i \ge \omega$.

We use the following natural transformations for $n < \omega$

$s_n : W_n \to A^\sharp_n$	defined by	$s_0 = \,!$ and $s_{n+1} = Fs_n + A$, and
$t_n : A^\sharp_n \to W_{n+1}$	defined by	$t_0 = \mathsf{inr}$ and $t_{n+1} = Ft_n + A$,

where inr is the right-hand coproduct injection:

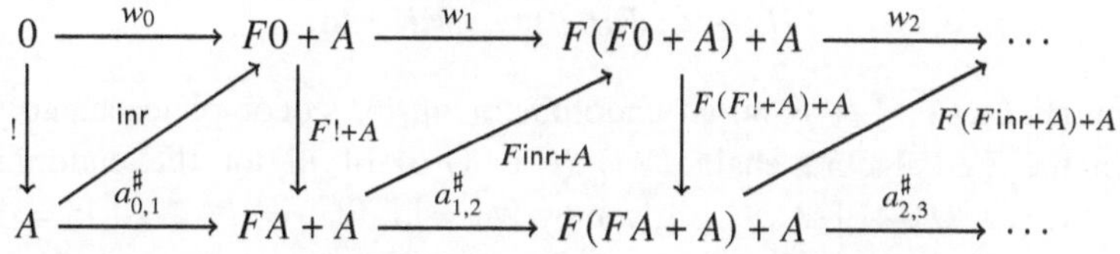

It is easy to prove by induction that for all $n < \omega$ we have $t_n \cdot s_n = w_n$ and $s_{n+1} \cdot t_n = a^\sharp_{n,n+1}$. This implies that the unique induced maps

$$s_\omega = \operatorname*{colim}_{n<\omega} s_n : W_\omega \to A^\sharp_\omega \qquad \text{and} \qquad t_\omega = \operatorname*{colim}_{n<\omega} t_n : A^\sharp_\omega$$

are mutually inverse. Since colimits are unique up to isomorphism, a colimit of W_n $(n < \omega)$ is given by $A^\sharp_\omega$ with colimit cocone $t_\omega \cdot a^\sharp_{n,\omega}$ $(n < \omega)$. This states that the two ω-chains have the same colimit.

Analogously to the respective results for initial algebras (Theorem 6.1.11 and Corollary 6.1.12) we obtain:

Corollary 14.1.3 [9, 2nd Prop.]. *Let F be an endofunctor on a cocomplete category. If the free-algebra chain of an object A converges in j steps, then $A^\sharp_j$ is the free F-algebra on A with structure and universal morphism given by the coproduct components of*

$$(a^\sharp_{j,j+1})^{-1} : FA^\sharp_j + A \to A^\sharp_j.$$

Corollary 14.1.4. *Let $\mathscr{A}$ be a cocomplete category, and let $F : \mathscr{A} \to \mathscr{A}$ preserve colimits of λ-chains for some λ. Then the free-algebra chain of every object A converges in λ steps. Therefore, $A^\sharp = A^\sharp_\lambda$.*

We can view the free-algebra chains of all objects A together at once by switching to the category $[\mathscr{A}, \mathscr{A}]$ of all endofunctors on $\mathscr{A}$ and natural transformations between them. (Strictly speaking, we should say 'large' category since the natural transformations between a given pair of functor may form a proper class. However, this does not play any role in what follows.)

Suppose that $\mathscr{A}$ has finite coproducts. Then so does $[\mathscr{A}, \mathscr{A}]$: the coproduct $F + G$ of two endofunctors is formed objectwise.

Definition 14.1.5. For every endofunctor F on $\mathscr{A}$ we define an endofunctor $\mathscr{F}$ on the functor category $[\mathscr{A}, \mathscr{A}]$ by

$$\mathscr{F} = F \cdot (-) + \mathsf{Id} : [\mathscr{A}, \mathscr{A}] \to [\mathscr{A}, \mathscr{A}].$$

It assigns to a every functor $H : \mathscr{A} \to \mathscr{A}$ the functor $FH + \mathsf{Id}$ and to every natural transformation $\tau\colon H \to K$ the natural transformation

$$H\tau + \mathsf{Id}\colon FH + \mathsf{Id} \to FK + \mathsf{Id}.$$

Remark 14.1.6. Let F be an endofunctor on the cocomplete category $\mathscr{A}$. Then the free-algebra chain (W_i) ($i \in \mathsf{Ord}$) of Id for the endofunctor $F \cdot (-)\colon [\mathscr{A}, \mathscr{A}] \to [\mathscr{A}, \mathscr{A}]$ is given by $W_0 = \mathsf{Id}$, $W_{i+1} = F \cdot W_i + \mathsf{Id} = \mathscr{F} W_i$ and $W_j = \lim_{i<j} W_i$ for limit ordinals j.

When applying these endofunctors to an object A of $\mathscr{A}$, we obtain the free-algebra chain of A for F:

$$W_i A = A_i^\sharp \qquad \text{for all } i \in \mathsf{Ord}.$$

Recall from Definition 2.2.26 that a varietor is an endofunctor such that a free algebra exists on every object.

Theorem 14.1.7. *Let $\mathscr{A}$ have finite coproducts. Given a varietor F on $\mathscr{A}$, an initial algebra for $\mathscr{F}$ exists in $[\mathscr{A}, \mathscr{A}]$.*

Proof (1) For every object A of $\mathscr{A}$ let $\varphi_A\colon FMA \to MA$ denote a free algebra on A and $\eta_A\colon A \to MA$ the universal morphism. This states that the forgetful functor $U\colon \mathsf{Alg}\,F \to \mathscr{A}$ has a left adjoint and the ensuing monad in denoted by (M, η, μ) (cf. Remark 2.2.28). Recall that for every morphism $f\colon A \to B$ the morphism Mf is given as a homomorphism of F-algebras:

$$\begin{array}{ccc} FMA & \xrightarrow{\varphi_A} & MA \\ {\scriptstyle FMf}\big\downarrow & & \big\downarrow{\scriptstyle f} \\ FMB & \xrightarrow{\varphi_B} & B \end{array}$$

Thus we have natural transformations $\varphi\colon FM \to M$ and $\eta\colon \mathsf{Id} \to M$. We shall show these form the structure and universal morphism of a free algebra on Id for the functor $F \cdot (-)$ on $[\mathscr{A}, \mathscr{A}]$. The desired result then follows by Proposition 2.2.24.

(2) Suppose that H is an endofunctor equipped with two natural transformations

$$\psi\colon FN \to N \qquad \text{and} \qquad \varepsilon\colon \mathsf{Id} \to N.$$

It is our task to prove that there is a unique homomorphism $\bar{\varepsilon}\colon (M, \varphi) \to (N, \psi)$ of algebras for $F \cdot (-)$ such that $\bar{\varepsilon} \cdot \eta = \varepsilon$.

Uniqueness is clear: given any such natural transformation $\bar{\varepsilon}\colon M \to N$, its components are clearly F-algebra homomorphisms from (MA, φ_A) to (NA, ψ_A) such that $\bar{\varepsilon}_A \cdot \eta_A = \varepsilon_A$. Thus, $\bar{\varepsilon}$ is uniquely determined by the freenees of the F-algebras MA.

For the existence, given an object A we take the unique F-algebra homomorphism $\bar{\varepsilon}_A\colon (MA, \varphi_A) \to (NA, \psi_A)$ such that $\bar{\varepsilon}_A \cdot \eta_A = \varepsilon_A$. Then it is easy to show that these morphisms form the components of a natural transformation: for every morphism $f\colon A \to B$ both paths in the desired naturality square

$$\begin{array}{ccc} MA & \xrightarrow{\bar{\varepsilon}_A} & NA \\ {\scriptstyle Mf}\downarrow & & \downarrow{\scriptstyle Nf} \\ MB & \xrightarrow{\bar{\varepsilon}_B} & NB \end{array}$$

are F-algebra homomorphisms from (MA, φ_A) to (NB, ψ_B) which are equal when pre-composed by η_A. By definition, $\bar{\varepsilon}\colon M \to N$ is a homomorphism of algebra for $F \cdot (-)$ such that $\bar{\varepsilon} \cdot \eta = \varepsilon$, and so we are done. □

Corollary 14.1.8. *A set functor F is a varietor if and only if $\mathscr{F}$ has an initial algebra.*

Proof For 'only if' see Theorem 14.1.7. For 'if' suppose let $\alpha\colon FM + \mathsf{Id} \to M$ be the initial algebra for $\mathscr{F}$. By Lambek's Lemma 2.2.5, each component $\alpha_A\colon FMA + A \to MA$ is an isomorphism. Thus, MA is a fixed point of the set functor $F(-) + A$. By Proposition 2.2.24 and Corollary 6.3.9, F has a free algebra on every set A. □

Remark 14.1.9. For endofunctors F preserving monomorphisms, Corollary 14.1.8 generalizes to cocomplete, well-powered categories $\mathscr{A}$ with smooth monomorphisms (Definition 6.2.2) assuming that, moreover, monomorphisms are closed under binary coproducts. To see this, observe that the last condition on $\mathscr{A}$ implies that for every object A the functor $F(-) + A$ preserves monomorphisms. Moreover, MA is a fixed point. Hence, we can apply Corollary 6.3.9.

Open Problem 14.1.10. Does the equivalence

$$F \text{ varietor} \quad\Longleftrightarrow\quad \mathscr{F} \text{ has an initial algebra}$$

hold for all endofunctor on a category $\mathscr{A}$ with finite coproducts?

Observe that if the initial algebra for $\mathscr{F}$ is obtained by convergence of the initial-algebra chain, then F is a varietor:

Corollary 14.1.11. *Let F be an endofunctor on a cocomplete category. If the initial-algebra chain for $\mathscr{F}$ converges in λ-steps, the free-algebra chains for all objects converge in λ steps, too.*

Indeed, this follows from Remark 14.1.6.

Example 14.1.12. We present a varietor F on Set such that the initial-algebra chain for $F \cdot (-) + \mathsf{Id}$ does not converge in $[\mathsf{Set}, \mathsf{Set}]$. Our example is a variation on Example 6.5.24. Recall that a *strong limit cardinal* is a cardinal closed under cardinal exponentiation. No cardinal 2^κ is a strong limit cardinal.

Let Card be the class of cardinals, and let $f\colon \mathsf{Ord} \to \mathsf{Card}$ be defined as follows:

$$\begin{array}{lll} f(0) & = & \aleph_0, \\ f(i+1) & = & \text{the smallest strong limit cardinal of cofinality} > f(i), \\ f(j) & = & \bigcup_{i<j} f(i) \quad \text{for } j \text{ a limit ordinal.} \end{array}$$

In the successor step, we use the fact that for every limit ordinal j there are arbitrarily large strong limit cardinals of cofinality larger than j. Notice that $f(i+1) > 2^{f(i)}$, since $2^{f(i)}$ is not closed under exponentiation.

Let C be the class of cardinals κ such that for some i, $2^{f(i)} < \kappa \leq f(i+1)$. Define a set functor $\mathscr{P}_C$ by

$$\mathscr{P}_C X = \{\emptyset\} \cup \{M \subseteq X \colon |M| \in C\} \text{ and } \mathscr{P}_C f(M) = \begin{cases} f[M] & \text{if } f|_M \text{ is monic,} \\ \emptyset & \text{else.} \end{cases}$$

(1) We prove that $\mathscr{P}_C$ is a varietor using Corollary 6.4.1: we find 'enough' pre-fixed points. For every ordinal i let X be a set of size $2^{f(i+1)}$. The largest element of C below $|X|$ is $f(i)$. Therefore

$$|\mathscr{P}_C X| \leq \sum_{j \leq 2^{f(i+1)}, j \in C} |X|^j \leq \sum_{j \leq f(i)} |X|^j \leq f(i) \cdot (2^{f(i+1)})^{f(i)} = 2^{f(i+1)} = |X|.$$

Thus, $\mathscr{P}_C$ has arbitrarily large pre-fixed points.

(2) Next we prove that there are arbitrarily large cardinals λ such that the initial-algebra chain for $\mathscr{P}_C \cdot (-) + \mathsf{Id}$ does not converge in λ steps. This proves that it does not converge, as stated. The cardinals that we use are of the form

$$\lambda = 2^{2^{f(j)}} \qquad \text{for any ordinal number } j.$$

Since $f(j+1)$ is closed under exponentiation, $\lambda < f(j+1)$. We show that if X is a set of power λ, then the free-algebra chain $X_i^\sharp$ for $\mathscr{P}_C$ does not converge in less than $f(j)$ steps. Indeed, we prove that for $0 \leq i < k \leq f(j)$,

$$\lambda \leq |X_i^\sharp| < |X_k^\sharp| < f(j+1).$$

The proof is by induction on the ordinal $k < f(j)$. The case $k = 0$ is trivial. For $k+1 < j$, assume our result for k. Let $\kappa = |X_k|$. By induction hypothesis, $f(j) < \lambda < \kappa < f(j+1)$. Hence $\kappa \in C$. We need only show that $\kappa < |X_{k+1}^\sharp| < f(j+1)$. Notice that $\mathscr{P}_C(X_k)$ contains all subsets of X_k size κ. Thus $|X_{k+1}^\sharp| = |\mathscr{P}_C(X_k)| = 2^\kappa$. Moreover $\lambda < 2^\kappa < f(j+1)$, since 2^κ is not closed under exponentiation and $f(j+1)$ is closed.

We conclude with the limit step for $k \leq f(j)$. Let $\kappa = \sup\{|\mathscr{P}_C(X_i)| : i < k\}$.

Notice that $\mathrm{cf}(\kappa) \leq \mathrm{cf}(k) \leq k \leq f(j)$. By definition, $\mathrm{cf}(f(j+1)) > f(j)$. Hence $\kappa < f(j+1)$. The rest of this step is easy.

Corollary 14.1.11 implies that the initial-algebra chain for $\mathscr{P}_C \cdot (-) + \mathsf{Id}$ does not converge in λ steps. Since λ can be arbitrarily large, we are done.

Remark 14.1.13. (1) As we have seen in Theorem 7.3.14, the terminal coalgebras for $F(-) + A$ are also interesting: they are precisely the *free completely iterative* algebras on A. The ensuing monad T assigning to every object A the free completely iterative algebra on A is the free completely iterative monad on F (cf. Remark 7.3.23(2)).

(2) Dually to what we have seen for $F(-) + A$, the terminal coalgebra for $F(-) \times A$ is precisely the cofree F-coalgebra on A (see Proposition 2.6.15). Moreover, by the dual of Theorem 2.2.29, a functor F on a category with copowers has all cofree coalgebras iff it admits a cofree comonad $F_\sharp$.

By dualizing Construction 14.1.1, we obtain an iterative construction of cofree coalgebras using the *cofree-coalgebra chain* which starts as follows (where π_r denotes the right-hand product projection):

$$A \xleftarrow{\pi_r} FA \times A \xleftarrow{F\pi_r \times \mathsf{id}} F(FA \times A) \times A \xleftarrow{F(F\pi_r \times \mathsf{id}) \times \mathsf{id}} \cdots .$$

If it converges, then it yields the cofree F-coalgebra (dual of Corollary 14.1.3). For a functor F preserving limits of λ^{op}-chains for some limit ordinal λ, we know that it does converge at λ (dual of Corollary 14.1.4).

(3) For every λ-accessible set functor F the cofree-coalgebra chain converges in $\lambda + \lambda$ steps. This follows from Theorem 6.5.11 applied to the λ-accessible functors $F(-) \times A$.

(4) Finally, the initial algebra for $F(-) \times A$ is precisely the same as a cofree parametrically recursive coalgebra on A (Remark 7.4.5). This yields a comonad C assigning to every object A the cofree parametrically recursive coalgebra on A. It is characterized as the cofree recursive comonad [301], dual to what we mentioned in item (1) above.

Another functor which is derived from $F\colon \mathscr{A} \to \mathscr{A}$ is the functor $F \cdot (-)$ on the category $[\mathscr{A}, \mathscr{A}]$ of endofunctors on $\mathscr{A}$. The following is a consequence of a result by Kelly [187, Thm. 23.3]:

Theorem 14.1.14. *Let F be an endofunctor on a category $\mathscr{A}$ with finite coproducts and colimits of chains. If the free-algebra chain for $F \cdot (-)$ and Id converges, then F has a free monad.*

Remark 14.1.15. (1) There is a close connection between the free-algebra chain of Id for $F \cdot (-)$ in $[\mathscr{A}, \mathscr{A}]$ and the free-algebra chains for F. In fact, it is easy to see that for every object A and every ordinal i we have $\mathsf{Id}_i^\sharp(A) \cong A_i^\sharp$. In other words, the free algebra chain of Id for $F \cdot (-)$ starting with

$$\mathsf{Id} \xrightarrow{\mathsf{inr}} F + \mathsf{Id} \xrightarrow{F\mathsf{inr}+\mathsf{Id}} F(F+\mathsf{Id}) + \mathsf{Id} \xrightarrow{F(F\mathsf{inr}+\mathsf{Id})+\mathsf{Id}} \cdots$$

yields an object-free version of the free-algebra chains for F. It converges at an ordinal λ iff so do all the free-algebra chains for F, and then it yields the free monad on F as $\mathsf{Id}_\lambda^\sharp$.

(2) If F preserves colimits of λ-chains, then so does $F \cdot (-)$ on $[\mathscr{A}, \mathscr{A}]$. Using Proposition 2.2.24, we see that the free monad on F is obtained as the initial algebra for the functor $\mathscr{F}$ on $[\mathscr{A}, \mathscr{A}]$. Indeed, the initial-algebra chain for that functor without the first step C_0, the constant functor on the initial object of $\mathscr{A}$, is the same as the free-algebra chain for $F \cdot (-)$ and Id.

(3) Dually, suppose that $\mathscr{A}$ is a category with finite products and limits of chains and that $F \colon \mathscr{A} \to \mathscr{A}$ preserves limits of λ^{op}-chains. Then the terminal coalgebra for the functor $F \cdot (-) \times \mathsf{Id}$ on $[\mathscr{A}, \mathscr{A}]$ is, equivalently, the cofree comonad on F. Moreover, the cofree-coalgebra chain for $F \cdot (-)$ and Id is related to the cofree-coalgebra chains for F mentioned in Remark 14.1.13(2) dually as in Remark 14.1.15.

14.2 Composite Functors

In this section we derive a few results which pertain to initial algebras for composites of functors. In particular we are going to present a new simple proof of Freyd's Iterated Square Theorem [128]. We also discuss the converse of that theorem; this seems to be new. We begin with the following result of Eppendahl:

Lemma 14.2.1 [117, Lemma 3.1]. *Given functors $F \colon \mathscr{A} \to \mathscr{B}$ and $G \colon \mathscr{B} \to \mathscr{A}$, consider an algebra $\alpha \colon GFA \to A$ and a coalgebra $\beta \colon B \to FGB$. Then there is a bijective correspondence between coalgebra-to-algebra morphisms $f \colon (GB, G\beta) \to (A, \alpha)$ and coalgebra-to-algebra morphisms $g \colon (B, \beta) \to (FA, F\alpha)$:*

$$\begin{array}{ccc} GB & \xrightarrow{G\beta} & GFGB \\ {\scriptstyle f}\downarrow & & \downarrow{\scriptstyle GFf} \\ A & \xleftarrow{\alpha} & GFA \end{array} \qquad\qquad \begin{array}{ccc} B & \xrightarrow{\beta} & FGB \\ {\scriptstyle g}\downarrow & & \downarrow{\scriptstyle Fg} \\ FA & \xleftarrow{F\alpha} & FGFA \end{array} \tag{14.1}$$

Proof The direction from f to g is given by

$$g = (B \xrightarrow{\beta} FGB \xrightarrow{Ff} FGFA).$$

If the left-hand square in (14.1) commutes, so does the right-hand one:

$$g = Ff \cdot \beta = F\alpha \cdot FGFf \cdot FG\beta \cdot \beta = F\alpha \cdot FGg \cdot \beta.$$

The opposite direction from g to f is given by

$$f = (GB \xrightarrow{Gg} GFA \xrightarrow{\alpha} A).$$

Again, it is easy to see that g is a coalgebra-to-algebra morphism. Furthermore, the two passages are inverse to each other. For given f, compute g and then get f back using (14.1):

$$\alpha \cdot Gg = \alpha \cdot G(Ff \cdot \beta) = f.$$

Analogously for the reverse composite. □

Recall the notion of a recursive coalgebra from Definition 7.4.1.

Corollary 14.2.2 [117, Cor. 4.3]. *Let $F\colon \mathscr{A} \to \mathscr{B}$ and $G\colon \mathscr{B} \to \mathscr{A}$ be functors. If (B, β) is a recursive coalgebra for FG, then $(GB, G\beta)$ is a recursive coalgebra for GF.*

Corollary 14.2.3. *Let $F\colon \mathscr{A} \to \mathscr{B}$ and $G\colon \mathscr{B} \to \mathscr{A}$ be functors. If FG has an initial algebra, then so does GF, and we have*

$$\mu(GF) = G\mu(FG).$$

Proof Let $i\colon FGI \to I$ be initial. Then by Lambek's Lemma 2.2.5, i is invertible, and by Corollary 7.4.4, (I, i^{-1}) is a recursive coalgebra for FG. By Lemma 14.2.1, the coalgebra-to-algebra morphisms from (GI, Gi^{-1}) into a GF-algebra (A, a) bijectively correspond to those from (I, i^{-1}) to (FA, Fa). Since the latter exists uniquely, so does the former, and we conclude that (GI, Gi) is an initial algebra for GF. □

Theorem 14.2.4 (Freyd's Iterated Square Theorem [128]). *An endofunctor F has an initial algebra whenever FF does.*

Proof We write F^2 for FF. Let $\iota\colon F^2 I \to I$ be an initial algebra for F^2. Then there is a unique F^2-algebra morphism $t\colon (I, \iota) \to (FI, F\iota)$. Recalling the proof of Lambek's Lemma 2.2.5, we observe first that

$$\iota^{-1} = (I \xrightarrow{t} FI \xrightarrow{Ft} F^2 I), \tag{14.2}$$

and ι^{-1} is an F^2-algebra morphism as shown in the following diagram:

$$\begin{array}{ccc} F^2 I & \xrightarrow{\iota} & I \\ {\scriptstyle F^2 t}\downarrow & & \downarrow{\scriptstyle t} \\ F^3 I & \xrightarrow{F\iota} & FI \\ {\scriptstyle F^3 t}\downarrow & & \downarrow{\scriptstyle Ft} \\ F^4 I & \xrightarrow{F^2\iota} & F^2 I \end{array}$$

We will prove that $\iota \cdot Ft \colon FI \to I$ is an initial F-algebra. Given an algebra $\alpha \colon FA \to A$, we prove that a unique F-algebra homomorphism exists.

(1) Existence: we know that there exists a unique F^2-algebra homomorphism f from (I, ι) to $(A, \alpha \cdot F\alpha)$. To prove that f is an F-algebra homomorphism from $(I, \iota \cdot Ft)$ to (A, α), we first show that

$$f = \alpha \cdot Ff \cdot t.$$

We use initiality of (I, ι): we prove that $\alpha \cdot Ff \cdot t$ also is an F^2-algebra homomorphism:

$$\begin{array}{ccccc} F^2 I & & \xrightarrow{\quad\quad\iota\quad\quad} & & I \\ {\scriptstyle F^2 t}\downarrow & & & & \downarrow{\scriptstyle t} \\ F^3 I & & \xrightarrow{\quad\quad F\iota\quad\quad} & & FI \\ {\scriptstyle F^3 f}\downarrow & & & & \downarrow{\scriptstyle Ff} \\ F^3 A & \xrightarrow{F^2\alpha} & F^2 A & \xrightarrow{F\alpha} & FA \\ {\scriptstyle F^3 f}\downarrow & & & & \downarrow{\scriptstyle \alpha} \\ F^2 A & \xrightarrow{F\alpha} & FA & \xrightarrow{\alpha} & A \end{array}$$

The top square commutes by definition of t. The square in the middle commutes since f is an F^2-algebra homomorphism. The square on the bottom obviously commutes. Now we obtain that f is an F-algebra morphism, as desired:

$$\alpha \cdot Ff = \alpha \cdot F(\alpha \cdot Ff \cdot t) = \alpha \cdot F\alpha \cdot F^2 f \cdot Ft = f \cdot \iota \cdot Ft.$$

(2) For the uniqueness, given any F-algebra morphism $f \colon (I, \iota \cdot Ft) \to (A, \alpha)$, we have

$$\alpha \cdot Ff \cdot t = f \cdot \iota \cdot Ft \cdot t \overset{(14.2)}{=} f \cdot \iota \cdot \iota^{-1} = f,$$

and therefore

$$\alpha \cdot F\alpha \cdot F^2 f \cdot Ft = \alpha \cdot F(\alpha \cdot Ff \cdot t) = \alpha \cdot Ff = f \cdot \iota \cdot Ft.$$

Now Ft is an epimorphism since $Ft \cdot t \cdot \iota = \mathrm{id}$. Thus, we see that f is an F^2-algebra morphism from (I, ι) to $(A, \alpha \cdot F\alpha)$. This completes the proof. □

We next prove that given an initial algebra $\alpha\colon FA \to A$, then FF has the initial algebra $\alpha \cdot F\alpha\colon FFA \to A$. This result is useful for deriving the existence of canonical fixed points (cf. Chapter 5) of functors of several variables with mixed variance [128].

Theorem 14.2.5 (Converse of the Iterated Square Theorem). *Let $\mathscr{A}$ have finite products. An endofunctor F on $\mathscr{A}$ has an initial algebra iff FF does. Moreover*

$$\mu F = \mu(FF).$$

Proof Suppose that $i\colon FI \to I$ is an initial F-algebra. Given an F^2-algebra $\alpha\colon F^2A \to A$, form the F-algebra $\bar{\alpha}\colon F(FA \times A) \to FA \times A$ to have the components

$$F\pi_r\colon F(FA \times A) \to FA \qquad \text{and} \qquad \alpha \cdot F\pi_\ell\colon F(FA \times A) \to A,$$

where π_ℓ and π_r denote the product projections of $FA \times A$. Then there exists a unique F-algebra homomorphism $\langle f, g\rangle\colon I \to FA \times A$; that is, f and g are the unique morphisms such that the following equations hold:

$$g \cdot i = \alpha \cdot Ff \qquad \text{and} \qquad f \cdot i = Fg. \tag{14.3}$$

It follows that g is an F^2-algebra homomorphism from $(I, i \cdot Fi)$ to (A, α):

$$g \cdot (i \cdot Fi) = \alpha \cdot Ff \cdot Fi = \alpha \cdot F^2 g.$$

For the uniqueness, suppose that $g\colon (I, i \cdot Fi) \to (A, \alpha)$ is any F^2-algebra homomorphism. Let $f = Fg \cdot i^{-1}$. Then these f and g satisfy the two equations in (14.3); we obviously have $f \cdot i = Fg$ and we compute

$$g \cdot i \cdot Fi = \alpha \cdot F^2 g = \alpha \cdot Ff \cdot Fi.$$

The first equation in (14.3) follows since Fi is an isomorphism. Consequently, the morphism g is unique. □

Corollary 14.2.6. *Let $\mathscr{A}$ have finite coproducts. An endofunctor F on $\mathscr{A}$ has a terminal coalgebra iff FF does.*

Example 14.2.7. The converse of the Iterated Square Theorem does not hold without finite products. For consider the discrete category with three objects A, B, and C (i.e. a three-element set) and let F be the functor (map) with $A \mapsto A$, $B \mapsto C$, and $C \mapsto B$. Then A is an initial F-algebra but FF does not have an initial algebra (as A, B, and C are all fixed points of FF).

14.3 Relatively Terminal Coalgebras

Given an algebra $\alpha\colon FA \to A$, the functor F lifts to the slice category $\mathscr{A}/A$ of all objects (X,x) where $x\colon X \to A$: define $F^\alpha\colon \mathscr{A}/A \to \mathscr{A}/A$ by

$$F^a(X \xrightarrow{x} A) = FX \xrightarrow{Fx} FA \xrightarrow{\alpha} A$$

on objects and for every $f\colon (X,x) \to (X',x')$ in $\mathscr{A}/A$ put $F^\alpha f = Ff$. Terminal coalgebras of F^α are investigated in [27], they are called *relatively terminal coalgebras* for the given algebra. To explain this, recall from Definition 7.4.1 the notion of a coalgebra-to-algebra morphism.

Definition 14.3.1 [27]. Let $\alpha\colon FA \to A$ be an F-algebra. Its *relatively terminal coalgebra* is a coalgebra $t\colon T \to FT$ with a universal coalgebra-to-algebra homomorphism into (A,α). This is a coalgebra-to-algebra morphism $\eta\colon (T,t) \to (A,\alpha)$ such that for every coalgebra-to-algebra morphism $f\colon (C,c) \to (A,\alpha)$ there exists a unique F-coalgebra homomorphism $h\colon (C,c) \to (T,t)$ such that $f = \eta \cdot h$.

Remark 14.3.2. The relatively terminal coalgebra of an F-algebra (A,α) is precisely a terminal coalgebra for F^α. Indeed, a coalgebra for F^α consists of an object $f\colon C \to A$ of $\mathscr{A}/A$ and a morphism $c\colon C \to FC$ such that

$$c = \alpha \cdot Ff \cdot c.$$

This means precisely that f is a coalgebra-to-algebra morphism from (C,c) to (A,α). Thus, the terminal coalgebra for F^α has the universal property of relative terminal coalgebras, and vice versa.

Example 14.3.3 [27]. (1) If F preserves limits of ω^{op}-chains, then the relatively terminal coalgebra is the limit of the following ω^{op}-chain:

$$A \xleftarrow{\alpha} FA \xleftarrow{F\alpha} FFA \xleftarrow{FF\alpha} \cdots.$$

(2) Every set functor has a relatively terminal coalgebra for every pre-fixed point $\alpha\colon FA \rightarrowtail A$. More generally, let $\mathscr{A}$ be a complete and well-powered category. If an endofunctor F preserves monomorphisms, then it has relatively terminal coalgebras for all pre-fixed points the relative terminal coalgebra is obtained via a limit of the $\mathsf{Ord}^{\mathsf{op}}$-chain generalizing the previous item.

(3) For a signature Σ, let A be a Σ-algebra. For the polynomial set functor H_Σ, the relatively terminal coalgebra can be described as the coalgebra of all 'unrestricted computations' in A; its elements can be represented by ordered

trees with nodes labelled in $\Sigma \times A$ in such a way that for every node

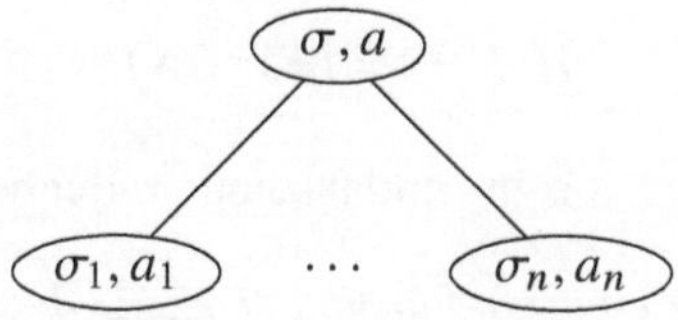

we have: σ has arity n and $a = \sigma^A(a_1, \ldots, a_n)$.

Theorem 14.3.4 [27]. *Let $\mathscr{A}$ be a complete and well-powered category, and let F be an endofunctor which preserves monomorphisms. Every pre-fixed point of F has a relatively terminal coalgebra.*

Proof We are given a monomorphism $\alpha\colon FA \rightarrowtail A$. By Remark 8.2.2, $\mathsf{Sub}(A)$ is a complete lattice. We are going to follow some of the steps in Proposition 7.2.8. We shall omit details that are found in the earlier proof. We have a monotone operator $\Phi\colon \mathsf{Sub}(A) \to \mathsf{Sub}(A)$ which maps a given subobject $u\colon X \rightarrowtail A$ to

$$\Phi(u) = \big(FX \overset{Fu}{\rightarrowtail} FA \overset{\alpha}{\rightarrowtail} A\big).$$

There is a greatest fixed point $\nu\Phi$ by the Knaster–Tarski Theorem B.1.7, and again

$$\nu\Phi = \bigvee \{m \in \mathsf{Sub}(A) \mid m \leq \Phi(m)\}. \tag{14.4}$$

Write $m^*\colon X^* \rightarrowtail A$ for $\nu\Phi$. Since this is a fixed point of Φ, there is an isomorphism $i\colon X^* \to FX^*$. We claim that (X^*, i) is a relatively terminal coalgebra for (A, m). For this, let $\gamma\colon Y \to FY$ have a coalgebra-to-algebra morphism $e^\dagger\colon Y \to A$. As in Proposition 7.2.8, (stong-epi, mono) factorizations in the category $\mathscr{A}$ imply that $e^\dagger$ factorizes through m^*, and uniquely so. It is then routine to check that this factorization of e^* gives a (unique) coalgebra morphism $(Y, e^\dagger) \to (X^*, i)$. □

14.4 Mutual Recursion

Given an endofunctor F the recursive equation

$$X \cong FX$$

has the initial solution whenever μF exists: this is Lambek's Lemma 2.2.5. Now assuming two endofunctors F and G on $\mathscr{A}$ are given, what can we say about the mutual recursive equations

$$X \cong FY \qquad \text{and} \qquad Y \cong GX? \tag{14.5}$$

If we form the endofunctor H on $\mathscr{A} \times \mathscr{A}$ by

$$H(X,Y) = (FY, GX),$$

then whenever μH exists, it is the initial solution. But here is a simpler answer:

Proposition 14.4.1. *For a pair of functors* $\mathscr{A} \underset{G}{\overset{F}{\rightleftarrows}} \mathscr{B}$, *an initial solution of the equations* (14.5) *exists iff GF has an initial algebra. Moreover the components of the initial solutions are $X = \mu(FG)$ and $Y = \mu(GF)$.*

Remark 14.4.2. We have seen in Corollary 14.2.3 that whenever GF has an initial algebra $y\colon GFY \to Y$ then $(X,x) = (FY, Fy)$ is an initial algebra for FG. We shall prove that for the endofunctor $H(X,Y) = (FY, GX)$ on $\mathscr{A} \times \mathscr{A}$ we have

$$\mu H = (X,Y) \qquad \text{with structure} \qquad (\mathsf{id}, y) = (FY, GX) \to (X,Y).$$

Proof (1) Suppose $\mu(GF)$ exists. Given an H-algebra with components

$$\alpha\colon FB \to A \qquad \text{and} \qquad \beta\colon GA \to B,$$

we prove that there exists a unique homomorphism from $((X,Y), (\mathsf{id}, y))$ into it. Indeed, to give an H-algebra morphism from (X,Y) to (A,B) means to give morphisms $a\colon X \to A$ and $b\colon Y \to B$ such that the following squares commute:

$$\begin{array}{ccc} FY & = & X \\ {\scriptstyle Fb}\downarrow & & \downarrow{\scriptstyle a} \\ FB & \xrightarrow{\alpha} & A \end{array} \qquad\qquad \begin{array}{ccc} GX & \xrightarrow{Gx} & Y \\ {\scriptstyle Ga}\downarrow & & \downarrow{\scriptstyle b} \\ GA & \xrightarrow{\beta} & B \end{array}$$

Since the left-hand square states simply that $a = \alpha \cdot Fb$, this is the same as to give a morphism b such that the following square commutes:

$$\begin{array}{ccccc} GFY & & \xrightarrow{Gx} & & Y \\ {\scriptstyle GFb}\downarrow & & & & \downarrow{\scriptstyle b} \\ GFB & \xrightarrow{G\alpha} & GA & \xrightarrow{\beta} & B \end{array}$$

Moreover, b is unique: Corollary 14.2.3 shows that (Y, Gx) is an initial GF-algebra.

(2) Conversely, let $(X,Y) = \mu H$ with the structure morphism given by $x\colon FY \to X$ and $y\colon GX \to Y$. Then $y \cdot Gx\colon GFY \to Y$ is an initial GF-algebra. Indeed, given a GF-algebra $\beta\colon GFB \to B$, consider the following H-algebra with $A = FB$ and structure given by $\mathsf{id}\colon FB \to A$ and $\beta\colon GA \to B$. There

exists a unique homomorphism from μH with components a and b:

$$\begin{array}{ccc} FY & \xrightarrow{x} & X \\ {\scriptstyle Fb}\downarrow & & \downarrow{\scriptstyle a} \\ FB & = & A \end{array} \qquad \begin{array}{ccc} GX & \xrightarrow{y} & Y \\ {\scriptstyle Ga}\downarrow & & \downarrow{\scriptstyle b} \\ GA & \xrightarrow{\beta} & B \end{array} \tag{14.6}$$

Then b is a homomorphism of GF-algebras:

$$\begin{array}{ccccc} GFY & \xrightarrow{Gx} & GX & \xrightarrow{y} & Y \\ {\scriptstyle GFb}\downarrow & & {\scriptstyle Ga}\downarrow & & \downarrow{\scriptstyle b} \\ GFB & = & GA & \xrightarrow{\beta} & B \end{array}$$

with the outer bottom arrow $\beta\colon GFB \to B$.

We turn to the uniqueness. Lambek's Lemma 2.2.5 implies that x and y are invertible, and if the outside of the last diagram commutes, then $a = Fb \cdot x^{-1}$ makes (a, b) a pair of morphisms for which the two squares (14.6) commute. □

14.5 Parametric Fixed Points

In many applications one is interested in mixed inductive/coinductive definitions, that is, a mix of taking initial algebras and terminal coalgebras. For example, given a functor $F\colon \mathscr{C}^3 \to \mathscr{C}$, one would like to form

$$\nu X_1.\, \mu X_2.\, \nu X_3.\, F(X_1, X_2, X_3). \tag{14.7}$$

More generally, one would like to iteratively take, for an endofunctor with several arguments, the initial algebra in some arguments and the terminal coalgebra in others.

In this section we shall prove that such mixed fixed points exist provided that $\mathscr{C}$ is a locally presentable category and F an accessible functor. More precisely, fixing all but one argument and taking an initial algebra or terminal coalgebra gives a functor which is accessible (see Theorems 14.5.5 and 14.5.9). We then iterate this procedure.

Example 14.5.1. The functor $F\colon \mathsf{Set} \times \mathsf{Set} \to \mathsf{Set}$ defined by $F(A, X) = X + A$ and similarly on maps yields

$$\nu A.\, \mu X.\, F(A, X) = \mathbb{N}^\infty.$$

Indeed, we have the set functor $T\colon \mathsf{Set} \to \mathsf{Set}$ given by $TA = \mu X.\, (X + A) =$

$\mathbb{N} \times A$ (cf. Example 2.2.9), and we then compute $\nu A.\, TA = \mathbb{N}^\omega$ (see Example 2.5.3(3)). In the reverse order we have

$$\mu A.\, \nu X.\, F(A, X) = \mathbb{N}^*.$$

In this case we consider the set functor $SA = \nu X.\, (X + A) = \mathbb{N} \times A + 1$ (use Example 2.5.8 for a signature with one unary operation symbol and a constant symbols indexed by A; cf. Example 2.5.11(1)). Thus, we have $\mu A.\, SA = \mathbb{N}^*$ by Example 2.2.15(2).

In lieu of concentrating on functors from $\mathscr{C}^n$ to $\mathscr{C}$, we work with a general parameter category $\mathscr{A}$. The desired special case of interest is $\mathscr{A} = \mathscr{C}^{n-1}$.

Assumption 14.5.2. For the rest of this section we assume that $\mathscr{A}$ and $\mathscr{C}$ are locally λ-presentable categories and that

$$F \colon \mathscr{A} \times \mathscr{C} \to \mathscr{C}$$

is a λ-accessible functor.

First, it is clear that for every object A, the functor $F(A, -)$ is λ-accessible again and therefore has an initial algebra and a terminal coalgebra (Theorem 11.2.18). Our aim is to prove that the assigments $A \mapsto \mu X.\, F(A, X)$ and $A \mapsto \nu A.\, F(A, X)$, given by taking parametrized initial algebras or terminal-coalgebras, respectively, yield λ-accessible functors $\mathscr{A} \to \mathscr{C}$.

Notation 14.5.3. We denote by $T \colon \mathscr{A} \to \mathscr{C}$ the functor defined on objects by

$$TA = \mu F(A, -);$$

that is, TA is the initial algebra for the endofunctor $F(A, -) \colon \mathscr{C} \to \mathscr{C}$. These initial algebras exist by Theorem 11.2.18, and we denote their structures by

$$\iota_A \colon F(A, TA) \to TA.$$

Furthermore, T assigns to every $f \colon A \to B$ the unique $F(A, -)$-algebra homomorphism Tf displayed below:

$$\begin{array}{ccccc} F(A,TA) & & \xrightarrow{\quad\iota_A\quad} & & TA \\ {\scriptstyle F(A,Tf)}\downarrow & & & & \downarrow{\scriptstyle Tf} \\ F(A,TB) & \xrightarrow{F(f,TB)} & F(B,TB) & \xrightarrow{\iota_B} & TB \end{array} \tag{14.8}$$

That this assignment is functorial is then easy to verify using the universal property of initial algebras.

We next prove that the functor T is itself an initial algebra for an endofunctor on the category $[\mathscr{A}, \mathscr{C}]$ of all functors from $\mathscr{A}$ to $\mathscr{C}$. Indeed, let $\mathcal{F}$ be the endofunctor on $[\mathscr{A}, \mathscr{C}]$ that assigns to a functor $H\colon \mathscr{A} \to \mathscr{C}$ the functor $\mathcal{F}(H)\colon \mathscr{A} \to \mathscr{C}$ given by $A \mapsto F(A, HA)$ and $f \mapsto F(f, Hf)$.

Proposition 14.5.4. *The functor T is the initial algebra for $\mathcal{F}$.*

Proof First, the commutativity of (14.8) states that the initial algebra structures $\iota_A\colon F(A,TA) \to TA$ form a natural transformation from $\mathcal{F}(T)$ to T.

Given a functor $H\colon \mathscr{A} \to \mathscr{C}$ and an $\mathcal{F}$-algebra, which is a natural transformation $\alpha_A\colon F(A,HA) \to HA$, we define morphisms $\varphi_A\colon TA \to HA$ by initiality of TA for every object A:

$$\begin{array}{ccc} F(A,TA) & \xrightarrow{\iota_A} & TA \\ {\scriptstyle F(A,\varphi_A)}\downarrow & & \downarrow{\scriptstyle \varphi_A} \\ F(A,HA) & \xrightarrow{\alpha_A} & HA \end{array} \tag{14.9}$$

We only need to prove that the morphisms φ_A form a natural transformation. That φ is then a unique $\mathcal{F}$-algebra homomorphism follows from the fact that each φ_A is the unique morphism such that the above square (14.9) commutes. To verify the naturality of φ we prove for every morphism $f\colon A \to B$ of $\mathscr{A}$ that $\varphi_B \cdot Tf = Hf \cdot \varphi_A$ using the initiality of TA. Indeed, $\varphi_B \cdot Tf$ is a homomorphism due to the following diagram

$$\begin{array}{ccccc} F(A,TA) & & \xrightarrow{\qquad\qquad \iota_A \qquad\qquad} & & TA \\ {\scriptstyle F(A,Tf)}\downarrow & & & & \downarrow{\scriptstyle Tf} \\ F(A,TB) & \xrightarrow{F(f,TB)} & F(B,TB) & \xrightarrow{\iota_B} & TB \\ {\scriptstyle F(A,\varphi_B)}\downarrow & & {\scriptstyle F(B,\varphi_B)}\downarrow & & \downarrow{\scriptstyle \varphi_B} \\ F(A,HB) & \xrightarrow{F(f,HB)} & F(B,HB) & \xrightarrow{\alpha_B} & HB \end{array}$$

Its upper rectangle commutes by the definition (14.8) of Tf, the lower right-hand part commutes by the definition (14.9) of φ, and the lower left-hand one trivially does.

Furthermore, $Hf \cdot \varphi_A$ is a homomorphism due to the diagram below:

$$\begin{array}{ccccc} F(A,TA) & & \xrightarrow{\qquad\qquad \iota_A \qquad\qquad} & & TA \\ {\scriptstyle F(A,\varphi_A)}\downarrow & & & & \downarrow{\scriptstyle \varphi_A} \\ F(A,HA) & & \xrightarrow{\qquad\qquad \alpha_A \qquad\qquad} & & HA \\ {\scriptstyle F(A,Hf)}\downarrow & \searrow^{F(f,Hf)} & & & \downarrow{\scriptstyle Hf} \\ F(A,HB) & \xrightarrow{F(A,Hf)} & F(B,HB) & \xrightarrow{\alpha_B} & HB \end{array}$$

Its upper rectangle commutes by the definition (14.9) of φ, the lower right-hand part by the naturality of α, and the remaining triangle trivially commutes.

The desired equation thus follows from the initiality of (TA, ι_A). □

Theorem 14.5.5. *The functor $T\colon \mathscr{A} \to \mathscr{C}$ is λ-accessible.*

Proof For every object A of $\mathscr{A}$, let us denote the components of the initial-algebra chain for $F(A,-)$ by $T_i A$. Recall from Corollary 6.1.12 that this chain converges in λ steps, whence $TA = T_\lambda A$. By an easy transfinite induction one shows that every T_i is a functor: for any given $f\colon A \to B$ define $T_i f$ by transfinite recursion as follows:

$$\begin{aligned} T_0 f &= \mathsf{id}_0, \\ T_{i+1} f &= (T_{i+1}A = F(A, T_i A) \xrightarrow{F(f, T_i f)} F(B, T_i B) = T_{i+1} B), \end{aligned} \tag{14.10}$$

and on limit ordinals j, $T_j f$ is uniquely determined such that all connecting morphisms $w_{i,j}\colon T_i A \to T_j A$ form natural transformations. We thus obtain a chain in the category $[\mathscr{A}, \mathscr{C}]$ of all functors from $\mathscr{A} \to \mathscr{C}$.

It remains to prove that each T_i is a λ-accessible functor. We proceed by transfinite induction.

Base case: The functor T_0 is the constant functor of value 0, the initial object of $\mathscr{C}$. It preserves all colimits.

Isolated step: The projection $\Pi\colon \mathscr{A} \times \mathscr{C} \to \mathscr{A}$ preserves colimits since those are computed coordinate-wise in $\mathscr{A} \times \mathscr{C}$. Since T_i is λ-accessible, so is the functor $\langle \Pi, T_i \rangle\colon \mathscr{A} \times \mathscr{C} \to \mathscr{A} \times \mathscr{C}$. The functor T_{i+1} is the following composite

$$T_{i+1} = \big(\mathscr{A} \times \mathscr{C} \xrightarrow{\langle \Pi, T_i \rangle} \mathscr{A} \times \mathscr{C} \xrightarrow{F} \mathscr{C}\big).$$

Since both $\langle \Pi, T_i \rangle$ and F are λ-accessible, so is T_{i+1}.

Limit step: all λ-accessible functors form a full subcategory of $[\mathscr{A}, \mathscr{C}]$ closed under colimits (computed object-wise). Thus $T_i = \mathsf{colim}_{j<i} T_j$ is λ-accessible if so is each T_j. □

Dually to the functor T given by initial algebras for $F(A,-)$ we obtain a functor S from the terminal coalgebras. They exist since $F(A,-)$ is λ-accessible (Theorem 11.2.18).

Notation 14.5.6. We denote by $S\colon \mathscr{A} \to \mathscr{C}$ the functor defined on objects by $SA = \nu F(A,-)$ and on morphisms dually to that in Notation 14.5.3.

Dually to Proposition 14.5.4 we have the following result:

Corollary 14.5.7. *The functor S is the terminal coalgebra for $\mathfrak{F}$.*

The analogous result to Theorem 14.5.5 for terminal coalgebras is more subtle: we need to assume that λ is uncountable. The key to this is the following construction of the terminal coalgebras for $F(A,-)$.

Construction 14.5.8. Let λ be an uncountable cardinal. Recall from Notation 11.2.9 that the locally λ-presentable category $\mathscr{C}$ has a small full subcategory $\mathscr{C}_\lambda$ representing all λ-presentable objects up to isomorphism. For an object A of $\mathscr{A}$ denote by

$$\mathsf{Coalg}_\lambda F(A,-)$$

the full subcategory of the category of $\mathsf{Coalg}\, F(A,-)$ given by all coalgebras $X \to F(A,X)$ with a carrier X in $\mathscr{C}_\lambda$.

The category $\mathsf{Coalg}\, F(A,-)$ is cocomplete, and $\mathsf{Coalg}_\lambda F(A,-)$ is closed in it under colimits of diagrams with less than λ objects and morphisms. This follows from Proposition 4.1.1 and the fact that λ-presentable objects are closed under such colimits (in every locally λ-presentable category).

Consequently, $\mathsf{Coalg}_\lambda F(A,-)$ is a λ-filtered category, whence the inclusion functor

$$I_A\colon \mathsf{Coalg}_\lambda F(A,-) \hookrightarrow \mathsf{Coalg}\, F(A,-)$$

is a (small) λ-filtered diagram. Since λ is uncountable, we know from Corollary 11.3.14 that the terminal coalgebra $\tau_A\colon SA \to F(A,SA)$ is the colimit of I_A:

$$SA = \mathsf{colim}\, I_A \quad \text{with colimit injections } \alpha^\#\colon (X,\alpha) \to (SA,\tau_A).$$

Note that by Proposition 4.1.1, the underlying object of the coalgebra SA is the colimit of the diagram UI_A of the underlying objects X of all coalgebras (X,α) in $\mathsf{Coalg}_\lambda F(A,-)$. Moreover, the coalgebra structure $\tau_A\colon SA \to F(A,SA)$ is unique such that all colimit injections $\alpha^\#$ are coalgebra homomorphisms.

Theorem 14.5.9. *If λ is uncountable, then the functor $S\colon \mathscr{A} \to \mathscr{C}$ is λ-accessible.*

Proof Suppose that $D\colon \mathscr{D} \to \mathscr{A}$ is a λ-filtered diagram and $a_k\colon A_k \to A$, $k \in \mathscr{D}$, form its colimit. Let us consider the category $\mathcal{E}$ whose objects are morphisms $\alpha\colon X \to F(A_k,X)$ with X in $\mathscr{C}_\lambda$ and whose morphisms from α to $\beta\colon Y \to F(A_\ell,Y)$ are pairs (u,h) where $h\colon X \to Y$ is a morphism of $\mathscr{C}$ and $u\colon A_k \to A_\ell$ is a connecting morphism in the diagram D such that the following square commutes:

$$\begin{array}{ccc} X & \xrightarrow{\alpha} & F(A_k,X) \\ {\scriptstyle h}\big\downarrow & & \big\downarrow{\scriptstyle F(u,h)} \\ Y & \xrightarrow{\beta} & F(A_\ell,Y) \end{array}$$

We have the forgetful functor $E\colon \mathcal{E} \to \mathscr{C}$ mapping objects $\alpha\colon X \to F(A_k, X)$ to X. Observe that the colimit

$$\operatorname*{colim}_{k\in\mathscr{D}} SA_k = \operatorname*{colim}_{k\in\mathscr{D}} \operatorname{colim} I_{A_k}$$

is precisely the colimit of E. Further, we have a functor $J\colon \mathcal{E} \to \mathsf{Coalg}_\lambda F(A,-)$ assigning to $\alpha\colon X \to F(A_k, X)$ in $\mathcal{E}$ the following coalgebra

$$X \xrightarrow{\alpha} F(A_k, X) \xrightarrow{F(a_k,X)} F(A, X)$$

and defined by $J(u,h) = h$ on morphisms. The desired proof of $SA = \operatorname{colim}_{k\in\mathscr{D}} SA_k$ will be complete once we prove that J is final (Remark 10.3.9). Since both $\mathcal{E}$ and $\mathsf{Coalg}_\lambda F(A,-)$ are λ-filtered diagrams, this requires that we verify the following two properties for every object $\alpha\colon X \to F(A, X)$ of $\mathsf{Coalg}_\lambda F(A,-)$:

(1) there is some $k \in \mathscr{D}$ and $\beta\colon Y \to F(A_k, Y)$ such that (X,α) has a morphism to $J(Y,\beta)$ in $\mathsf{Coalg}_\lambda F(A,-)$,

(2) for every parallel pair of morphisms

$$f, g\colon (X,\alpha) \to J(Y,\beta)$$

in $\mathsf{Coalg}_\lambda F(A,-)$, where $\beta\colon Y \to F(A_k, Y)$ in $\mathcal{E}$, there exists a morphism $(u,h)\colon (Y,\beta) \to (Z,\gamma)$ in $\mathcal{E}$ for some object $\gamma\colon Z \to F(A_\ell, Z)$ ($\ell \in \mathscr{D}$) such that $h \cdot f = h \cdot g$.

Ad (1). Given α, we know that $F(A, X) = \operatorname{colim}_{k\in\mathscr{D}} F(A_k, X)$ because $F(-, X)$ is λ-accessible. Since X is λ-presentable, we thus obtain a factorization of α through some colimit injection $F(a_k, X)$, that is, there exists $k \in \mathscr{D}$ and $\beta\colon X \to F(A_k, X)$ such that $\alpha = F(a_k, X)\cdot\beta$. This means that $J(X,\beta) = (X,\alpha)$. Thus id_X is the desired morphism.

Ad (2). Given α, f, and g, since $\mathsf{Coalg}_\lambda F(A,-)$ is λ-filtered, there exists $\zeta\colon Z \to F(A, Z)$ with Z in $\mathscr{C}_\lambda$ and an $F(A,-)$-coalgebra homomorphism $h\colon J(Y,\beta) \to (Z,\zeta)$ such that $h \cdot f = h \cdot g$:

$$\begin{array}{ccccc}
X & & \xrightarrow{\quad\alpha\quad} & & F(A,X) \\
f\downarrow\downarrow g & & & & F(A,f)\downarrow\downarrow F(A,g) \\
Y & \xrightarrow{\beta} & F(A_k,Y) & \xrightarrow{F(a_k,Y)} & F(A,Y) \\
h\,\downarrow & & & & \downarrow F(A,h) \\
Z & & \xrightarrow{\quad\zeta\quad} & & F(A,Z)
\end{array}$$

Since Z is λ-presentable we can choose some $m \in \mathscr{D}$ and $\zeta'\colon Z \to F(A_m, Z)$ such that $\zeta = F(a_m, Z)\cdot\zeta'$. Further, since the A_k form a λ-filtered diagram we may assume without loss of generality that we have a connecting morphism $v\colon A_k \to A_m$ of the diagram $\mathscr{D}$.

Now in the following diagram

$$\begin{array}{ccccc}
Y & \xrightarrow{\beta} & F(A_k, Y) & \xrightarrow{F(a_k,Y)} & F(A,Y) \\
{\scriptstyle h}\downarrow & & \downarrow{\scriptstyle F(v,h)} & & \downarrow{\scriptstyle F(A,h)} \\
Z & \xrightarrow{\zeta'} & F(A_m, Z) & \xrightarrow{F(a_m,Z)} & F(A,Z)
\end{array}
\qquad \zeta\colon Z \to F(A,Z)$$

the outside and the right-hand square commute. Hence the left-hand square commutes when extended by $F(a_m, Z)$. Thus by λ-filteredness, we can choose a connecting morphism $w\colon A_m \to A_\ell$ of $\mathscr{D}$ such that $F(w, Z)$ merges the left-hand square above. This means that we have a commutative diagram

$$\begin{array}{ccccc}
Y & & \xrightarrow{\beta} & & F(A_k, Y) \\
{\scriptstyle h}\downarrow & & \swarrow{\scriptstyle F(v,h)} & & \downarrow{\scriptstyle F(w\cdot v,h)} \\
Z & \xrightarrow{\zeta'} & F(A_m, Z) & \xrightarrow{F(w,Z)} & F(A_\ell, Z)
\end{array}$$

Thus, $(w \cdot v, h)$ is a morphism from (Y, β) to $(Z, F(w, Z) \cdot \zeta')$ in $\mathcal{E}$. Consequently, putting $\gamma = F(w, Z) \cdot \zeta'$ and $u = w \cdot v$ we have the desired object (Z, γ) and morphism (u, h) of $\mathcal{E}$ such that $h \cdot f = h \cdot g$. □

Example 14.5.10. Coming back to our motivating example in (14.7), let us consider a λ-accessible functor $F\colon \mathscr{C}^n \to \mathscr{C}$. Then every mixed fixed point

$$\chi_1 X_1.\, \chi_2 X_2.\, \cdots \chi X_n.\, F(X_1, \ldots, X_n), \tag{14.11}$$

where each χ_i is μ or ν, exists. Indeed, take $\mathscr{A} = \mathscr{C}^{n-1}$ and apply Theorem 14.5.5 to obtain the accessible functor $T = \mu X_n.F(-, X_n)\colon \mathscr{C}^{n-1} \to \mathscr{C}$ if $\chi_n = \mu$ or Theorem 14.5.9 to obtain the accessible functor $S = \nu X_n.F(-, X_n)\colon \mathscr{C}^{n-1} \to \mathscr{C}$ if $\chi_n = \nu$, respectively.

And so on for another $n - 1$ steps.

Example 14.5.11. Theorem 14.5.9 does not hold for $\lambda = \omega$. For a counterexample consider the finitary functor $F\colon \mathsf{Set} \times \mathsf{Set} \to \mathsf{Set}$ with $F(A, X) = X \times X + A$. Then SA is the set of all binary trees with leaves labelled in A (cf. Example 2.5.11(4)). The functor S is not finitary. To see this, take $A = \mathbb{N}$ and write it as the directed union of the chain of finite ordinals $A_n = \{0, \ldots, n - 1\}$. Then

the following tree

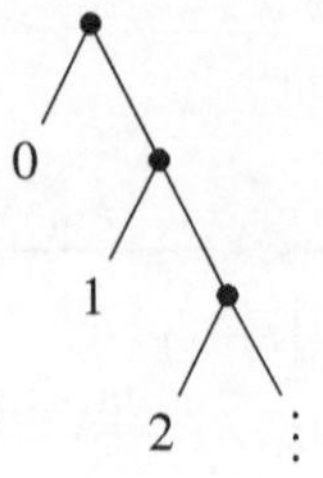

lies in $S\mathbb{N}$ but not in any SA_n.

However, we have the following result.

Corollary 14.5.12. *If* $\mathscr{A}$ *and* $\mathscr{C}$ *are locally* $\aleph_1$*-presentable and* $F\colon \mathscr{A} \times \mathscr{C} \to \mathscr{C}$ *is finitary, then* T *is* $\aleph_1$*-accessible.*

Indeed, F is also $\aleph_1$-accessible, and so one can apply Theorem 14.5.9 with $\lambda = \aleph_1$.

Example 14.5.13. Let $\mathscr{C} = \mathscr{A} = \mathsf{Set}$. By the *$\lambda$-ary polynomial functors* in $[\mathsf{Set}^2, \mathsf{Set}]$ we mean the smallest class of functors $F\colon \mathsf{Set}^2 \to \mathsf{Set}$ containing all constant functors and the projections $P_i(X_1, X_2) = X_i$, $i = 1, 2$, and which is closed under products of size less than λ and arbitrary coproducts. For every polynomial functor F, the functor S of Notation 14.5.6 is a polynomial functor on Set (associated to a λ-ary signature if λ is uncountable, and to an $\aleph_1$-ary signature if $\lambda = \omega$). This follows from Example 6.5.6.

For example, for $F(A, X) = A \times X$, which is finitary, $SA = A^\omega$ is the polynomial functor associated to a signature with one ω-ary operation symbol.

Example 14.5.14. (1) Mixed fixed points were considered by several authors. For example, Ghani et al. [137] consider the functor $F(X, Y) = B \times X + Y^A$ on Set and form $\nu X.\, \mu Y.\, F(X, Y)$. By Example 6.1.16, we know that $\mu Y.\, F(X, Y)$ consists of all A-branching well-founded trees with leaves labelled in $B \times X$. According to Example 14.5.13, this is the polynomial functor for the signature whose n-ary operation symbols are all A-branching trees with n leaves labelled in B. It follows from Example 6.5.6 that the desired mixed fixed point $\nu X.\mu Y.F(X, Y)$ consists of all partial labellings of the nodes of the complete A-branching tree by elements of B such that on every path from the root one finds an infinite sequence of labels in B. Indeed, every path of a tree in $\mu Y.\, F(X, Y)$ ends with a node labelled by an element of B and an element of X. Replace the latter by such tree and continue on a path to a leaf of that tree, etc. ad infinitum.

(2) Note that the order of taking fixed points in mixed inductive/coinductive definitions matters, as we have seen in Example 14.5.1. This point is also

demonstrated by the following example, which is interesting in its own right. Let us consider the functor $F(X,Y) = A \times X + B \times Y$ on Set. Then the two mixed fixed points $\mu X.\, \nu Y.\, F(X,Y)$ and $\nu Y.\, \mu X.\, F(X,Y)$ are non-isomorphic. Indeed, on the one hand we have

$$\nu Y.\, F(X,Y) = (X \mapsto B^* \times A \times X + B^\omega)$$

and therefore

$$\mu X.\, \nu Y.\, F(X,Y) = (B^* \times A)^* \times B^\omega \cong (A + B)^* \times B^\omega$$

is the set of all streams on $A + B$ that contain only finitely many elements of A.

On the other hand, we have

$$\mu X.\, F(X,Y) = (Y \mapsto A^* \times B \times Y)$$

and therefore

$$\nu Y.\, \mu X.\, F(X,Y) = (A^* \times B)^\omega$$

is the set of streams on $A + B$ that contain no substream of only elements of A. So the two fixed points are clearly not the same, for example, the stream $[a, b, a, b, a, b, \ldots]$, for any $a \in A$, $b \in B$, is contained in $\nu Y.\, \mu X.\, F(X,Y)$ but not in $\mu X.\, \nu Y.\, F(X,Y)$.

We end this section with a result which states that in the mixed fixed-point expression (14.11) one can always strictly alternate between μ and ν.

Theorem 14.5.15. *For every functor $F\colon \mathscr{C} \times \mathscr{C} \to \mathscr{C}$ we have*

$$\mu X.\, F(X,X) \cong \mu X.\, \mu Y.\, F(X,Y).$$

More precisely, suppose that $TX = \mu F(X,-)$ exists for every object X and that, moreover, μT exists, too. Then μT carries an initial algebra for the endofunctor

$$\mathscr{C} \xrightarrow{\Delta} \mathscr{C} \times \mathscr{C} \xrightarrow{F} \mathscr{C},$$

where $\Delta\colon \mathscr{C} \to \mathscr{C} \times \mathscr{C}$ is the diagonal functor given by $\Delta(X) = (X, X)$.

Proof As before we denote by $\iota_X\colon F(X,TX) \to TX$ the structures of the initial algebras for $F(X,-)$ for every object X. Furthermore, let $I = \mu T$ with the algebra structure $\iota\colon TI \to I$. By Lambek's Lemma 2.2.5 we know that ι is an isomorphism, and so we have the following structure of an algebra for $F\Delta$ on I:

$$F(I,I) \xrightarrow{F(I,\iota^{-1})} F(I,TI) \xrightarrow{\iota_I} TI \xrightarrow{\iota} I. \tag{14.12}$$

We shall prove that this is an initial algebra for $F\Delta$. Suppose that $\alpha\colon F(A,A) \to A$ is any algebra.

(1) Existence of a homomorphism. Using the initiality of TA we obtain a unique T-algebra structure $\overline{\alpha}$ on A such that the diagram below commutes:

$$\begin{array}{ccc} F(A,TA) & \xrightarrow{\iota_A} & TA \\ {\scriptstyle F(A,\overline{\alpha})}\big\downarrow & & \big\downarrow{\scriptstyle \overline{\alpha}} \\ F(A,A) & \xrightarrow{\alpha} & A \end{array} \tag{14.13}$$

Furthermore, using initiality of I we obtain a unique T-algebra homomorphism h from (I,ι) to $(A,\overline{\alpha})$:

$$\begin{array}{ccc} TI & \xrightarrow{\iota} & I \\ {\scriptstyle Th}\big\downarrow & & \big\downarrow{\scriptstyle h} \\ TA & \xrightarrow{\overline{\alpha}} & A \end{array} \tag{14.14}$$

This morphism is an $F\Delta$-algebra homomorphism from the algebra in (14.12) to (A,α):

$$\begin{array}{ccccccc} F(I,I) & \xrightarrow{F(I,\iota^{-1})} & F(I,TI) & \xrightarrow{\iota_I} & TI & \xrightarrow{\iota} & I \\ & & {\scriptstyle F(h,Th)}\big\downarrow & & \big\downarrow{\scriptstyle Th} & & \\ {\scriptstyle F(h,h)}\big\downarrow & & F(A,TA) & \xrightarrow{\iota_A} & TA & & \big\downarrow{\scriptstyle h} \\ & \swarrow{\scriptstyle F(A,\overline{\alpha})} & & & & \searrow{\scriptstyle \overline{\alpha}} & \\ F(A,A) & & & \xrightarrow{\alpha} & & & A \end{array} \tag{14.15}$$

The right-hand part commutes by (14.14), the upper middle square commutes by the definition of Th (cf. (14.8)), and the lower part does by (14.13). For the remaining left-hand part consider the two arguments of F separately: the left-hand argument trivially commutes, and the right-hand one follows from (14.14) using that ι is invertible.

(2) Uniqueness. Suppose we are given a morphism h such that the outside of (14.15) commutes. It suffices to prove that the square (14.14) commutes. This follows from the initiality of TI: we prove that both compositions are homomorphisms from (TI,ι_I) to $(A,\alpha\cdot F(h,A))$. Indeed, to see that $h\cdot\iota$ is a homomorphism we use the following diagram:

$$\begin{array}{ccccccc} F(I,TI) & & & \xrightarrow{\iota_I} & & & TI \\ {\scriptstyle F(I,\iota)}\big\downarrow & \diagdown\!\!\diagdown & & & & & \big\downarrow{\scriptstyle \iota} \\ F(I,I) & \xrightarrow{F(I,\iota^{-1})} & F(I,TI) & \xrightarrow{\iota_I} & TI & \xrightarrow{\iota} & I \\ {\scriptstyle F(I,h)}\big\downarrow & \searrow{\scriptstyle F(h,h)} & & & & & \big\downarrow{\scriptstyle h} \\ F(I,A) & \xrightarrow{F(h,A)} & F(A,A) & & \xrightarrow{\alpha} & & A \end{array}$$

That the lower right-hand part commutes is given, and the remaining parts trivially commute.

Moreover, that $\overline{\alpha} \cdot Th$ is also a homomorphism is shown using the diagram below:

$$\begin{array}{ccccc}
F(I,TI) & & \xrightarrow{\quad\quad\quad\iota_I\quad\quad\quad} & & TI \\
{\scriptstyle F(I,Th)}\big\downarrow & & & & \big\downarrow{\scriptstyle Th} \\
F(I,TA) & \xrightarrow{F(h,TA)} & F(A,TA) & \xrightarrow{\iota_A} & TA \\
{\scriptstyle F(I,\overline{\alpha})}\big\downarrow & & {\scriptstyle F(A,\overline{\alpha})}\big\downarrow & & \big\downarrow{\scriptstyle \overline{\alpha}} \\
F(I,A) & \xrightarrow{F(h,A)} & F(A,A) & \xrightarrow{\alpha} & A
\end{array}$$

Its upper rectangle commutes by the definition of Th (cf. (14.8)), the lower right-hand square commutes by (14.13) and the lower left-hand one trivially does. □

Remark 14.5.16. Theorem 14.5.15 essentially follows from a result of Bloom and Ésik [85, Thm. 6.1]. They consider a cartesian 2-category $\mathscr{A}$ and the operations

$$(-)^\dagger : \mathscr{A}(D \times C, C) \to \mathscr{A}(D, C)$$

that assigns to every 1-cell $f: D \times C \to C$ the initial algebra $f^\dagger: D \to C$ for the endofunctor $\mathfrak{F}$ on $\mathscr{A}(D,C)$ defined on objects by $\mathfrak{F}(h) = f \cdot \langle \mathrm{id}_D, h \rangle$ and similarly on morphisms (cf. Proposition 14.5.4). They prove that this operation satisfies the so-called Conway identities (up to canonical isomorphism). The property in Theorem 14.5.15 is a form of Bloom and Ésik's double dagger law:

$$f^{\dagger\dagger} = (D{\times}C \xrightarrow{D\times\Delta} D{\times}C{\times}C \xrightarrow{f} C)^\dagger : D \to C \text{ for every } f: D \times C \times C \to C,$$

which is well known to follow from the Conway identities in settings where those identities hold on the nose [84]. We nevertheless decided to include an explicit proof for the convenience of the reader.

14.6 Initial Double-Algebras

In this section we study initial algebras of endofunctors of the form $F + G$. It turns out that this is closely related mutual recursion as we saw it in Section 14.4: $F + G$ has an initial algebra iff FG has one.

Observe first that an algebra for $F + G$ is nothing else than a *double-algebra* for (F, G): a triple (A, α, β) consisting of an F-algebra $\varphi: FA \to A$ and a

G-algebra $\psi\colon GA \to A$ on the same carrier object A. *Homomorphisms* of double-algebras are those morphisms of the base category that are homomorphisms with respect to F as well as G. That is, given another double-algebra (B, γ, δ), a homomorphism is a morphism $h\colon A \to B$ such that the diagram below commutes:

$$\begin{array}{ccccc} FA & \xrightarrow{\alpha} & A & \xleftarrow{\beta} & GA \\ {\scriptstyle Fh}\downarrow & & \downarrow{\scriptstyle h} & & \downarrow{\scriptstyle Gh} \\ FB & \xrightarrow{\gamma} & B & \xleftarrow{\delta} & GB \end{array}$$

It is clear that the resulting category is isomorphic to the category of algebras for $F + G$.

At this point, we remind the reader of the notion of a *smooth class* of monomorphisms (see Definition 6.2.2).

Remark 14.6.1. (1) Let F and G be endofunctors of $\mathscr{A}$. The initial chain of $F + G$ can be described as the following *initial double-algebra chain* $X_i + Y_i$ ($i \in \mathsf{Ord}$) which has $X_0 = 0 = Y_0$, the limit steps are given by chain colimits, and the isolated steps by

$$X_{i+1} = F(X_i + Y_i) \qquad \text{and} \qquad Y_{i+1} = G(X_i + Y_i).$$

(2) We say that monomorphisms are *closed under coproducts* if

$$m_1 + m_2\colon A_1 + A_2 \to B_1 + B_2$$

is monic whenever both m_1 and m_2 are.

Lemma 14.6.2. *Let $\mathscr{A}$ be a well-powered category with finite coproducts and smooth monomorphisms closed under coproducts. For two endofunctors F and G preserving monomorphisms,*

(1) *an initial double-algebra exists iff the initial double-algebra chain converges.*

(2) *if so, all the three functors F, G, and GF have initial algebras.*

Proof Since monomorphisms are closed under coproduct, $F + G$ preserves monomorphisms. Thus, (1) is an application of Theorem 6.3.4 to this functor. For (2), let $X \cong FX + GX$ be an initial double-algebra. Then X is a pre-fixed point of F: the coproduct injection $\mathsf{inl}\colon FX \to X$ is monic; indeed, since monomorphisms are smooth, the unique morphism $u\colon 0 \to GX$ is monic. Thus so is $\mathsf{inl} = \mathsf{id}_{FX} + u\colon FX + 0 \to FX + GX \cong X$. Analogously, X is a pre-fixed point of G. Finally, due to the monomorphism $\mathsf{inr} \cdot G\mathsf{inl}\colon GFX \to X$, it is also a pre-fixed point of GF. In all cases, we again use Theorem 6.3.4. □

Remark 14.6.3. In the case where F and G are finitary set functors, the initial double-algebra is the colimit of the ω-chain

$$0 \to F0 + G0 \to F(F0 + G0) + G(F0 + G0) \to \cdots$$

Indeed, this is just the initial-algebra ω-chain for $F + G$.

Example 14.6.4. The converse of Lemma 14.6.2(2) above does not hold in general: from the existence of μF, μG, and $\mu(GF)$ we cannot derive the existence of an initial double-algebra. Consider $F = \mathscr{P}_0$, the nonempty power-set functor (so we have $\mathscr{P}_0\emptyset = \emptyset$), and $GX = 1$ (the constant functor). The initial double-algebra chain

$$\emptyset \to 1 \to \mathscr{P}1 + 1 \to \mathscr{P}(\mathscr{P}1 + 1) + 1 \to \cdots$$

does not converge, although $\mathscr{P}_0$ and 1 (= $1 \cdot \mathscr{P}_0$) have initial algebras.

Remark 14.6.5. In the rest of this section we study double-algebras for a pair of monads. It turns out that double-algebras are closely related to initial algebras of certain derived functors. The reason for introducing double-algebras for pairs of monads **S** and **T** will be made clear in the next subsection, where we study coproducts in the category $\mathsf{Mnd}\,\mathscr{A}$ of all monads on a given category $\mathscr{A}$. We will present a formula for coproducts based on initial double-algebras. All results below on these monadic initial double-algebras are taken from [45].

Recall from Section 12.5 the notion of an Eilenberg–Moore algebra for a monad $\mathbf{S} = (S, \eta, \mu)$. In the following we will frequently shorten the name to **S**-*algebra*.

Definition 14.6.6. Let **S** and **T** be monads on a given category $\mathscr{A}$. Then *double-algebras* for $(\mathbf{S}, \mathbf{T})$ are triples (A, σ, τ) where $\sigma\colon SA \to A$ is an **S**-algebra and $\tau\colon TA \to A$ a **T**-algebra. Morphisms are those morphisms of $\mathscr{A}$ which are double-algebra homomorphisms for the functors (S, T). We denote the category of double-algebras for $(\mathbf{S}, \mathbf{T})$ by $\mathscr{A}^{\mathbf{S},\mathbf{T}}$. The forgetful functor $U^{\mathbf{S},\mathbf{T}}\colon \mathscr{A}^{\mathbf{S},\mathbf{T}} \to \mathscr{A}$ maps a double-algebra (A, σ, τ) to its underlying object A.

Example 14.6.7. (1) As shown by Hyland, Plotkin, and Power [167, Thm. 4], if **T** is a free monad on an endofunctor F (Remark 2.2.28), then the initial bialgebra for $(\mathbf{S}, \mathbf{T})$ is $\mu(SF)$, whenever the composite SF has an initial algebra.

(2) In particular, let $\mathbf{M}_E$ be the *exception monad* for an object E: here $M_E X = X + E$ with the unit $\eta_X = \mathsf{inl}\colon X \to X + E$ being the left-hand coproduct injections and the multiplication $\mu_X = X + \nabla\colon X + E + E \to X + E$ is given by the codiagonal. This is the free monad on the constant functor with value E. An initial double-algebra for $(\mathbf{S}, \mathbf{M}_E)$ is carried by the free **S**-algebra SE.

(3) For the *terminal monad* $\mathbf{1}$ given by the constant functor of value 1, all algebras are isomorphic to $(1, \mu_1)$. Thus an initial double-algebra for $(\mathbf{S}, \mathbf{1})$ is carried by 1.

Analogously, the following *modified terminal monad* $\mathbf{1}^0$ on Set is given by $\emptyset \mapsto \emptyset$ and $X \mapsto 1$ for X nonempty. It has, up to isomorphism, only two algebras, carried by $\emptyset$ and 1, respectively. An initial double-algebra for $(\mathbf{S}, \mathbf{1})$ is carried by $\emptyset$ if $S\emptyset = \emptyset$, or by 1 else.

For the remainder of this section we restrict ourselves to monads on Set. We need the following concept:

Definition 14.6.8. A monad on Set is called *consistent* if it is neither isomorphic to $\mathbf{1}$ nor to $\mathbf{1}_0$.

Consequently, consistent monads are precisely those which have algebras with more than one element. In fact, they admit arbitrarily large algebras: for consistent monads $\mathbf{S}$ the unit $\eta\colon \mathsf{Id} \to \mathbf{S}$ has monic components. Moreover, as Theorem 14.6.9 shows, the complements $SX \setminus \eta_X[X]$ of these components form a subfunctor of S when S is restricted to the category Set_m of all sets and all injective maps between them. The following result follows from [45, Lemma IV.1, Prop. IV.5, Thm. VII.12].

Theorem 14.6.9. *For a consistent monad* $\mathbf{S}$ *the functor* S *preserves monomorphisms. Its restriction to* Set_m *has the form*

$$S = \bar{S} + \mathsf{Id}$$

for an endofunctor $\bar{S}$ *on* Set_m*, where the right-hand coproduct injection is the unit* η*. Moreover, there exists a cardinal* λ *such that* $|\bar{S}X| \geq |X|$ *for all sets* X *with* $|X| \geq \lambda$.

Example 14.6.10. (1) For the power-set monad $\mathscr{P}$ the functor $\bar{\mathscr{P}}$ assigns to X the collection of all sets $M \subseteq X$ with $|M| \neq 1$. This is not a set functor but an endofunctor on Set_m, where it is a subfunctor of $\mathscr{P}$.

(2) For $\mathbf{M}_E$ the functor $\overline{M}_E$ is constant with value E.

Remark 14.6.11. Let $\mathbf{S}$ and $\mathbf{T}$ be consistent monads. Consider the following mutually recursive 'equation' system in Set_m:

$$X \cong \bar{S}Y \quad \text{and} \quad Y \cong \bar{T}X.$$

This has an initial solution given by $X = \mu(\bar{S}\bar{T})$ and $Y = \mu(\bar{T}\bar{S})$ whenever the initial algebra exists (Proposition 14.4.1). For this initial solution a *canonical double-algebra* is given on $X + Y$ as follows: $X + Y$ is both the free $\mathbf{S}$-algebra on Y due to $X + Y \cong \bar{S}Y + Y = SY$ and a free **T**-algebra on X due to $X + Y \cong X + \bar{T}X = TX$.

Theorem 14.6.12 [45, Prop. VI.3]. *Let* **S** *and* **T** *be consistent monads on* Set. *They have an initial double-algebra iff* $\bar{S}\bar{T}$ *has an initial algebra in* Set_m. *The initial double-algebra is then the canonical one on the set* $\mu(\overline{S}\overline{T}) + \mu(\overline{T}\overline{S})$.

Example 14.6.13. (1) An initial double-algebra for **S** and $\mathbf{M}_E$ is SE, the free **S**-algebra on E.

(2) An initial double-algebra for **S** and $\mathscr{P}$ exists for a consistent monad **S** iff $|S1| \leq 2$. Indeed, then $\bar{S}1 \leq 1$ and $\bar{S}\overline{\mathscr{P}}1 = \mu(\bar{S}\overline{\mathscr{P}})$. Conversely, from $|S1| > 2$ we obtain that $\bar{S}\overline{\mathscr{P}}$ has no fixed points.

14.7 Coproducts of Monads on Set

The formation of coproducts of monads on Set is an important construction both in general algebra and in computer science. From the algebraic view, given two algebraic theories, their disjoint union (both of the sets of operation symbols and of equations) forms an algebraic theory – and the corresponding monad is just the coproduct. Baader and Tinelli [62] prove that coproduct injections are monic; we present a more general result in Corollary 14.7.9 below. In computer science where monads are used for a description of computational effects, coproducts mean combination of effects without taking any common features of the effects into account, see e.g. Hyland et al. [167]. A complete description of coproducts of monads in the category of sets was presented [45] as an application of the Initial Double-Algebra Theorem 14.6.12. Recall from Definition 14.6.8 that a consistent monad on Set is non-isomorphic to **1** and $\mathbf{1}^0$. Coproducts with **1** or $\mathbf{1}^0$ are trivial, see Example 14.7.4 below; for all other monads we present a necessary and sufficient condition for the existence of a coproduct, as well as a coproduct formula. We also prove that coproduct injections are monic.

Notation 14.7.1. Throughout this section we work with monads on Set only. We denote by $\mathbf{S} \oplus \mathbf{T}$ the coproduct of monads in the category of all monads on Set (and monad morphisms between them).

Lemma 14.7.2 [69, §6]. *A coproduct* $\mathbf{S} \oplus \mathbf{T}$ *exists iff every set generates a free double-algebra. This means that the forgetful functor* $U^{\mathbf{S},\mathbf{T}}$ *(Definition 14.6.6) has left adjoint. The ensuing monad is* $\mathbf{S} \oplus \mathbf{T}$.

The proof is an easy consequence of the fact that monad morphisms from **M** to **P** bijectively correspond to concrete functors from $\mathscr{A}^{\mathbf{P}}$ to $\mathscr{A}^{\mathbf{M}}$.

Remark 14.7.3. (1) The unit of $\mathbf{S} \oplus \mathbf{T}$ is given by the universal arrows $\eta_A^{\mathbf{S}\oplus\mathbf{T}} : A \to \bar{A}$ where $(\bar{A}, s_A, t_A)$ is a free double-algebra on A.

(2) The coproduct injections

$$\sigma : \mathbf{S} \to \mathbf{S} \oplus \mathbf{T} \qquad \text{and} \qquad \tau : \mathbf{T} \to \mathbf{S} \oplus \mathbf{T}$$

are easy to describe: for every object A there exists a unique $\mathbf{S}$-algebra homomorphism σ_A from SA (the free $\mathbf{S}$-algebra) to $(\bar{A}, s_A)$ such that the triangle below commutes:

$$\begin{array}{ccc} A & \xrightarrow{\eta_A^{\mathbf{S}}} & SA \\ & \searrow^{\eta_A^{\mathbf{S}\oplus\mathbf{T}}} & \downarrow \sigma_A \\ & & \bar{A} = (\mathbf{S} \oplus \mathbf{T})A \end{array}$$

This is the A-component of σ, analogously for τ.

Example 14.7.4. (1) The terminal monad $\mathbf{1}$ (Example 14.6.7(3)) given by $X \mapsto 1$ has (trivial) coproducts with all monads:

$$\mathbf{S} \oplus \mathbf{1} = \mathbf{1}.$$

This follows from Lemma 14.7.2 since $\mathbf{1}$ has only one algebra (up to isomorphism) with underlying object 1.

(2) The modified terminal monad $\mathbf{1}^0$ also has (trivial) coproducts:

$$\mathbf{S} \oplus \mathbf{1}^0 = \begin{cases} \mathbf{1} & \text{if } S0 \text{ is non-initial,} \\ \mathbf{1}^0 & \text{else.} \end{cases}$$

(3) The exception monad $\mathbf{M}_E$ (Example 14.6.7(2)) has coproducts with all monads:

$$\mathbf{S} \oplus \mathbf{M}_E : X \mapsto S(X + E).$$

This also follows from Lemma 14.7.2 because to give a double-algebra on A for $\mathbf{S}$ and $\mathbf{M}_E$ is the same as to give an Eilenberg–Moore algebra for $\mathbf{S}$ plus a morphism from E to A. And this is the same as to give an Eilenberg–Moore algebra for the monad $X \mapsto S(X + E)$.

(4) The *modified exception monad* $\mathbf{M}_E^0$ given by $\emptyset \mapsto \emptyset$ and $X \mapsto X + E$ for all nonempty sets X also has coproducts with all monads: if $S\emptyset \neq \emptyset$, then $\mathbf{S} \oplus \mathbf{M}_E^0 = \mathbf{S} \oplus \mathbf{M}_E$. If $S\emptyset = \emptyset$, then $\mathbf{S} \oplus \mathbf{M}_E^0$ is the modification of $\mathbf{S} \oplus \mathbf{M}_E$ given by $\emptyset \mapsto \emptyset$.

Notation 14.7.5. For a consistent monad $\mathbf{S}$ and a set A we denote by $\bar{S}_A$ the following endofunctor on Set_m:

$$\bar{S}_A X = \bar{S}(X + A).$$

Here + means coproduct in Set. (Although + is not a coproduct on Set_m, it defines an endofunctor $(-) + A$ on Set_m.)

Theorem 14.7.6 [45, Thm. VI.4]. *A coproduct of consistent set monads* **S** *and* **T** *exists iff for every set A an initial algebra of $\bar{S}_A\bar{T}_A$ exists in* Set_m. *The coproduct formula is then*

$$(\mathbf{S} \oplus \mathbf{T})A = \mu\bar{S}_A\bar{T}_A + \mu\bar{T}_A\bar{S}_A + A. \tag{14.16}$$

Corollary 14.7.7. *If consistent monads on* Set *have a coproduct, then the coproduct injections are monic.*

This can be derived from the coproduct formula (see [45, Lemma VI.7]).

Remark 14.7.8. The right-hand side of (14.16) is a double-algebra canonically. This is completely analogous to Remark 14.6.11: we know from Proposition 14.4.1 that the objects $X = \mu\bar{S}_A\bar{T}_A$ and $Y = \mu\bar{T}_A\bar{S}_A$ fulfil

$$X \cong \bar{S}(Y + A) \qquad \text{and} \qquad Y \cong \bar{T}(X + A).$$

Thus $X + Y + A$ is a free **S**-algebra on $Y + A$ via the isomorphism

$$S(Y + A) = \bar{S}(Y + A) + (Y + A) \cong X + Y + A.$$

Analogously, it is a free **T**-algebra on $X + A$ via the isomorphism

$$T(X + A) = \bar{T}(X + A) + (X + A) \cong Y + X + A \cong X + Y + A.$$

Corollary 14.7.9 [45, Thm. VI.4]. *Two monads* **S** *and* **T** *on* Set *have a coproduct iff one of the following holds:*

(1) *both are consistent and the formula* (14.16) *is well-defined*
(2) *one of the monads is isomorphic to* $\mathbf{1}$ *or* $\mathbf{1}^0$.

Example 14.7.10. (1) The power-set monad $\mathscr{P}$ has 'almost' no coproducts: whenever $\mathscr{P} \oplus \mathbf{T}$ exists, then **T** is isomorphic to one of the monads $\mathbf{M}_E$, $\mathbf{M}_E^0$, $\mathbf{1}$, or $\mathbf{1}^0$.

Indeed, if **T** is not isomorphic to one of them, then it is consistent, and there exists a cardinal λ such that $|X| \geq \lambda$ implies $|\bar{T}X| \geq |X|$ (Theorem 14.6.9). Let A be an infinite set of cardinality at least λ, then $\bar{\mathscr{P}}_A\bar{T}_A$ does not have an initial algebra (thus, $\mathscr{P} \oplus \mathbf{T}$ does not exist). Indeed, $\bar{\mathscr{P}}_A\bar{T}_A$ has no fixed points: for every set X of cardinality κ we have $|\bar{\mathscr{P}}_A\bar{T}_A(X)| \geq 2^\kappa$. To see this note that $\bar{T}_A X = \bar{T}(X + A)$ has, since $|X + A| \geq \lambda$, cardinality at least $|X + A|$. Since this cardinal is infinite and $|\bar{\mathscr{P}}Z| = |\mathscr{P}Z| = 2^{|Z|}$ for all infinite sets Z, we conclude that the set

$$\bar{\mathscr{P}}_A \bar{T}_A(X) = \bar{\mathscr{P}}(\bar{T}(X + A) + A)$$

has cardinaltiy at least $2^{|X+A|}$.

(2) For the finite power-set monad $\mathscr{P}_{\mathsf{f}}$, given a consistent monad $\mathbf{S}$, we have that $\mathbf{S} \oplus \mathscr{P}_{\mathsf{f}}$ exists iff S is a varietor. Indeed, if S is a varietor, it has arbitrarily large infinite pre-fixed points (Corollary 6.4.1), thus $\mathscr{P}_{\mathsf{f}}$ and S have arbitrarily large joint fixed points. Thus $\mathbf{S} \oplus \mathscr{P}_{\mathsf{f}}$ exists.

Conversely, if $\mathbf{S} \oplus \mathscr{P}_{\mathsf{f}}$ exists, S has arbitrarily large fixed points. Thus, it is a varietor.

Corollary 14.7.11 [45, Cor. VIII.10]. *For a monad* $\mathbf{S}$ *on* Set *the following conditions are equivalent:*

(1) $\mathbf{S}$ *has coproducts with all monads,*

(2) $\mathbf{S}$ *has a coproduct with the power-set monad* $\mathscr{P}$*,*

(3) $\mathbf{S}$ *is isomorphic to* $\mathbf{M}_E$, $\mathbf{M}_E^0$, $\mathbf{1}$, *or* $\mathbf{1}^0$.

Example 14.7.12. A coproduct of λ-accessible consistent monads $\mathbf{S}$ and $\mathbf{T}$ always exists and is given by the formula

$$(\mathbf{S} \oplus \mathbf{T})A = X_\lambda + Y_\lambda + A,$$

where X_λ and Y_λ are the colimits of the following λ-chains defined by

$$X_0 = 0 = Y_0, \qquad X_{i+1} = \bar{S}(\bar{T}(X_i + A) + A) \quad \text{and} \quad Y_{i+1} = \bar{T}(\bar{S}(Y_i + A) + A),$$

and for limit ordinals i we use colimits of the previous i-chains.

Remark 14.7.13. The fact that the category of all accessible monads has coproducts (indeed, all colimits) was already established by Kelly [187]. But the formula for coproducts which he, rather implicitly, gives is much more involved than the formula above. The latter essentially stems from Ghani and Uustalu [135] who proved it for so-called ideal monads (which are rather rare).

14.8 Summary

This chapter was devoted to the study of initial algebra and terminal coalgebras for endofunctors derived from some given ones.

We started with a brief treatment of (co)products of functors with a constact functor. We have seen that a set functor F is a varietor (equivalently, the derived functors $F(-)+A$ have initial algebras) iff the ensuing endofunctor $F\cdot(-)+\mathsf{Id}$ on

[Set, Set] has an initial algebra. Whether this further holds for all endofunctors on every 'reasonable' category is an open problem.

For a composite functor FG, an initial algebra exists iff it does for GF. Moreover, these initial algebras yield the initial solutions to mutually resursive equations $X \cong FY$ and $Y \cong GX$. We also presented Freyd's Iterated Square Theorem and its converse stating that a functor F on category with finite coproducts has an initial algebra if so does FF.

This was followed by a brief treatment of relatively terminal coalgebras, which are terminal coalgebras on a slice $\mathscr{A}/A$ for an algebra $\alpha\colon FA \to A$.

A core topic of this chapter was the study of parametric fixed points. We have seen that for accessible functors $\mathscr{A}^n \to \mathscr{A}$ on a locally presentable category, mixed fixed points exist. This follows from the fact that taking initial algebras or terminal coalgebras in one argument yields an accessible functor again. So the process can be iterated.

The chapter concluded with a brief discussion of double-algebras and coproducts of monads on Set as an application of our methods developed in earlier sections. The presentation there is an extended reference (without proofs) of an article on the topic [45].

15
Special Topics

Our concluding chapter aims at connections of our previous topics with areas of mathematics. We shall see material from many of our previous chapters connected to topics like the unit interval and other subsets of the real line, and ordered sets.

15.1 Variations on Cantor's Theorem

This section collects a few results which imply that certain endofunctors have no terminal coalgebras, or even no fixed points. The beginning of this line of work was Cantor's Theorem [99] from which it follows that the power-set functor $\mathscr{P}\colon \mathsf{Set} \to \mathsf{Set}$ has no fixed points (Example 2.2.7(1)).

There are a number of generalizations of this result. Here is one for the category Met of metric spaces. Let $\mathscr{P}_{\mathsf{cl}}\colon \mathsf{Met} \to \mathsf{Met}$ be the functor taking a space (X, d) to the set of its closed subsets, with the Hausdorff metric d: the distance of $S, T \in \mathscr{P}_{\mathsf{cl}}X$ is

$$\bar{d}(S,T) = \max(\sup_{x\in S} d(x,T), \sup_{y\in T} d(y,S)), \tag{15.1}$$

where $d(x,T) = \inf_{y\in T} d(x,y)$. For a non-expanding map $f\colon X \to Y$, the non-expanding map $\mathscr{P}_{\mathsf{cl}}f\colon \mathscr{P}_{\mathsf{cl}}X \to \mathscr{P}_{\mathsf{cl}}Y$ sends a closed subset S of X to the closure of $f[S]$. We also consider the subfunctor $\mathscr{P}'_{\mathsf{cl}}\colon \mathsf{Met} \to \mathsf{Met}$ of $\mathscr{P}_{\mathsf{cl}}$ taking a space to the sets of its nonempty closed subsets, with the Hausdorff metric.

The functor $\mathscr{P}_{\mathsf{cl}}$ has no terminal coalgebra, as proved by van Breugel [302, Prop. 8]. In fact, we shall see below that it has no fixed point.

The functor $\mathscr{P}'_{\mathsf{cl}}$ has an initial algebra given by the empty metric space and a terminal coalgebra carried by any singleton metric space. But it has no other fixed points (see van Breugel et al. [304, Cor. 5]). We provide below a different, shorter proof.

A subset X of a metric space is *δ-discrete* if whenever $x \neq y$ are elements of X, $d(x, y) \geq \delta$. Every subset of a δ-discrete set is δ-discrete, and every such set is closed. Moreover, if C and D are different subsets of a δ-discrete set, then $\bar{d}(C, D) \geq \delta$.

Remark 15.1.1. A subset S of an ordinal i is *cofinal* if for all $j < i$ there is some $k \in S$ with $j \leq k < i$. If S is not cofinal, then its complement $i \setminus S$ must be so. (But it is possible that both S and $i \setminus S$ are cofinal in i.)

Recall (Example 6.2.5(3)) that an isometric embedding is a map preserving distances.

Theorem 15.1.2. *If $|M| \geq 2$, then there is no isometric embedding $\mathscr{P}'_{\mathsf{cl}}M \to M$.*

Proof Suppose towards a contradiction that $\iota \colon \mathscr{P}_{\mathsf{cl}}M \to M$ were an isometric embedding where $|M| \geq 2$. Fix distinct points $a, b \in M$ and put $\delta = d(a, b)/2$. Let $A = \{x \in M : d(x, a) \leq \delta\}$ and $B = M \setminus A$.

By transfinite recursion, we define an ordinal-indexed sequence of elements $x_i \in M$, and along the way we also prove that each set $\{x_j : j \leq i\}$ is δ-discrete. We obtain the desired contradiction since (x_i) is an ordinal-indexed sequence of pairwise distinct elements of M.

For $i = 0$, put $x_0 = \iota(\{a, b\})$. Given an ordinal $i > 0$, we have $\{x_j : j < i\}$ and we assume that this set is δ-discrete. We put

$$X_i = \begin{cases} A & \text{if } \{j < i : x_j \in A\} \text{ is cofinal in } i, \\ B & \text{else,} \end{cases}$$

$$S_i = \{x_j : j < i\},$$

$$x_i = \iota(X_i \cap S_i).$$

Being nonempty and δ-discrete, $X_i \cap S_i$ lies in $\mathscr{P}'_{\mathsf{cl}}M$.

The remainder of our proof consists of showing that $\{x_j : j \leq i\}$ is δ-discrete. We only need to verify that $d(x_j, x_i) \geq \delta$ when $0 \leq j < i$. We argue the case $X_i = A$; when $X_i = B$, the argument is similar, mutatis mutandis. For $j = 0$, recall that $x_0 = \iota(\{a, b\})$ and $x_i = \iota(A \cap S_i)$. Since b has distance at least δ from every element of A, we obtain $\bar{d}(\{a, b\}, A \cap S_i) \geq \delta$. As ι is an isometric embedding, this distance is also $d(x_0, x_i)$. Now let $j > 0$. Since we have $X_i = A$, let k be such that $j \leq k < i$ and $x_k \in A$. Either $x_j = \iota(A \cap S_j)$ or else $x_j = \iota(B \cap S_j)$.

In the first case, note that $x_k \in A \cap S_i$ since $k < i$, and $x_k \notin S_j$ by the definition of S_j since $k \geq j$. So $A \cap S_j$ and $A \cap S_i$ are distinct nonempty subsets of the δ-discrete set S_i. Hence, the distance between these sets is at least δ, and therefore we have $d(x_j, x_i) \geq \delta$.

In the second case, $B \cap S_j$ is a nonempty subset of B, and thus again it is not equal to $A \cap S_i$. So again we see that $d(x_j, x_i) = \bar{d}(B \cap S_j, A \cap S_i) \geq \delta$. □

Corollary 15.1.3. (1) *The functor* $\mathscr{P}'_{\mathsf{cl}}$: Met → Met *has no fixed points except the empty set and the singletons.*

(2) *The functor* $\mathscr{P}_{\mathsf{cl}}$: Met → Met *admits no isometric embedding* $\mathscr{P}_{\mathsf{cl}}M \to M$, *whence has no fixed point.*

Proof The first item is immediate from Theorem 15.1.2. For the second one, observe that the inclusion map $e\colon \mathscr{P}'_{\mathsf{cl}}M \to \mathscr{P}_{\mathsf{cl}}M$ is an isometric embedding. Assuming that there were an isometric embedding $\iota\colon \mathscr{P}_{\mathsf{cl}}M \to M$, we see that M cannot be empty (since $\mathscr{P}_{\mathsf{cl}}M$ is nonempty) or a singleton (since then $|\mathscr{P}_{\mathsf{cl}}M| = 2$). Hence $|M| \geq 2$. Moreover, we obtain an isometric embedding $\iota \cdot e\colon \mathscr{P}'_{\mathsf{cl}}M \to M$, contradicting Theorem 15.1.2. □

Remark 15.1.4. Corollary 15.1.3(2) has a simple direct proof: one constructs 1-discrete sets D_i and elements $x_i \in M$ by transfinite recursion: $D_i = \{x_j : j < i\}$ and $x_i = \iota(D_i)$ (note that this starts with $D_0 = \emptyset$ and $x_0 = \iota(\emptyset)$). The contradiction is exhibitted by proving that the D_i form a strictly increasing chain of 1-discrete sets. This uses a transfinite induction which is an easier variation of what we have seen in Theorem 15.1.2.

We now turn to an analog of Cantor's Theorem for posets. Note that monomorphisms in Pos are the monotone injections, and epimorphisms are the monotone surjections.

Notation 15.1.5. We denote by D: Pos → Pos the endofunctor taking a poset P to the poset

$$DP = \text{all down-closed subsets } S \subseteq P, \text{ including } \emptyset,$$

ordered by inclusion. It maps a monotone map $f\colon P \to Q$ to the function Df which sends a down-closed set $S \subseteq P$ to the down-closure of the image $f[S]$:

$$S \mapsto \{q \in Q : q \leq f(s) \text{ for some } s \in S\}.$$

Theorem 15.1.6. *Let P be a poset.*

(1) *There is no monomorphism $DP \to P$.*

(2) *There is no epimorphism $P \to DP$.*

Item (1) is due to Gleason and Dilworth [141], and our statement of item (2) is a reformulation of a result in op. cit. noted by Davies, Hayes, and Rousseau [110]; our proof shows the similarity to the argument for Cantor's Theorem.

Proof (1) Suppose towards a contradiction that a monomorphism $\iota\colon DP \to P$ exists. It is easy to see that DP is a dcpo with bottom. We have a monotone function $f\colon DP \to DP$ which is given by

$$f(C) = \{q \in P : q \leq \iota(C)\}. \tag{15.2}$$

Let A be the least fixed point of f; see Theorem 6.1.1. Then $\iota(A) \in f(A) = A$. Now let $\mathcal{S} \subseteq DP$ consist of all proper post-fixed points B of f not containing $\iota(B)$:

$$\mathcal{S} = \{B : \iota(B) \notin B \subsetneq f(B)\}.$$

We show that $\mathcal{S}$ is closed under f and under joins of chains. For closure under f, let B be in $\mathcal{S}$. Then we have $f(B) \subsetneq f(f(B))$ since f is clearly strictly monotone. Moreover, $B \subsetneq f(B)$ implies $\iota(B) < \iota(f(B))$ which yields that $\iota(f(B))$ does not lie in $\downarrow\iota(B) = f(B)$.

Now let (B_i) be a chain in $\mathcal{S}$ and let $B = \bigcup B_i$ be its join in DP. Assuming $\iota(B) \in B$ yields a contradiction: we have $\iota(B) \in B_i$ for some i, and then $\iota(B_i) \leq \iota(B)$ implies that $\iota(B_i) \in B_i$ since B_i is down-closed. Thus, we have that $\iota(B) \notin B_i$ for every i. Therefore we must have $B \subsetneq f(B)$, since clearly we have $\iota(B) \in f(B)$ by the definition of f.

By Corollary 7.5.1, we conclude that $A = \mu f$ lies in $\mathcal{S}$, which implies that $\iota(A) \notin A$, a contradiction.

(2) Again, suppose towards a contradiction that an epimorphism $\tau\colon P \to DP$ exists. Let

$$I = \{p \in P : \text{for all } q \leq p, q \notin \tau(q)\}.$$

Clearly I is down-closed. Since τ is surjective, there exists a $p \in P$ such that $\tau(p) = I$. If $p \in I$, then $p \in \tau(p)$. This would mean that $p \notin I$, a contradiction. Hence, we have $p \notin I$. So let $q \leq p$ be such that $q \in \tau(q)$. By monotonicity, $\tau(q) \subseteq \tau(p)$. So $q \in \tau(p) = I$. But then by definition of I, $q \notin \tau(q)$. This is a contradiction. □

Corollary 15.1.7. *The functor $D\colon \mathsf{Pos} \to \mathsf{Pos}$ has no fixed points.*

For our next result, recall from Example 5.1.18 that an *ideal* in a poset P is a directed down-set $X \subseteq P$ with the additional property that whenever $x, y \in X$, there is some $z \in X$ such that $x, y \leq z$.

We denote by Idl^+ subfunctor of $D\colon \mathsf{Pos} \to \mathsf{Pos}$ assigning to a poset the poset of all ideals of P together with the empty set as a least element.

Theorem 15.1.8 [77]. *Let P be a poset. There is no monomorphism from $\mathsf{Idl}^+(P)$ to P.*

Proof The proof is analogous to that of Theorem 15.1.6(1), with two modifications. First, in lieu of the function in (15.2) we use its restriction $f\colon \mathsf{Idl}^+(P) \to \mathsf{Idl}^+(P)$. Second, the set $\mathcal{S}$ is taken as a subset of $\mathsf{Idl}^+(P)$. □

Corollary 15.1.9. *The functor* $\mathsf{Idl}^+\colon \mathsf{Pos} \to \mathsf{Pos}$ *has no fixed points.*

Remark 15.1.10. Another negative result of the type we have seen in this section was proved by Hofmann and Nora [164, Thm. 3.16]. They consider a non-symmetric version of 'Hausdorff' functor on the category of small $\mathscr{V}$-categories for a quantale $\mathscr{V}$, and they show that it has no fixed points, provided that $\mathscr{V}$ is non-trivial. One instance of $\mathscr{V}$-categories are preorders; here their result states that the functor D' assigning to a preordered set the preordered set of all up-closed subsets has no fixed point (cf. Corollary 15.1.7).

15.2 Vietoris Polynomial Functors on Topological Spaces

This section contains a terminal coalgebra result in the topological setting. We study the category Top of topological spaces and continuous functions, and the endofunctor is a version of the finite power-set functor relevant for that category.

Recall that a *base* of a topology is a collection $\mathcal{B}$ of open sets such that every open sets is a union of members of $\mathcal{B}$. A *subbase* is a collection of open sets whose finite intersections form a base. For every collection $\mathcal{B}$ of subsets of the space, there is a smallest topology for which $\mathcal{B}$ is a (sub)base, the family of unions of finite intersections from $\mathcal{B}$.

Definition 15.2.1. (1) Let X be a topological space. Let $\mathscr{V}X$ be the space of compact subsets of X. It is equipped with the 'hit-and-miss' topology which has as a subbase all sets of the following forms:

$$\begin{aligned} U^{\diamond} &= \{R \in VX : R \cap U \neq \emptyset\} && (R \text{ hits } U), \\ U^{\square} &= \{R \in VX : R \subseteq U\} && (R \text{ misses } X \setminus U), \end{aligned} \tag{15.3}$$

where U ranges over the opens of X. This means that every open set in $\mathscr{V}X$ is a union of finite intersections of the sets in (15.3). We call $\mathscr{V}X$ the *Vietoris space of* X, also known as the *hyperspace* of X.

(2) For a continuous function $f\colon X \to Y$ we put $\mathscr{V}f(A) = f[A]$ for every compact subset A of X.

Remark 15.2.2. (1) For a compact Hausdorff space X, Vietoris [309] defined $\mathscr{V}X$ to consist of all closed subsets of X, which are the same as the compact subsets in this case. In the coalgebraic literature, $\mathscr{V}X$ has also mostly been studied for spaces X which are compact and Hausdorff. However, note that the 'classic Vietoris space' (using closed subsets) does not yield a functor on Top (see Hofmann et al. [165, Rem. 2.28]. Hofmann et al. [165, Def. 2.27] call the functor $\mathscr{V}$ in Definition 15.2.1 the compact Vietoris functor.

(2) Michael [225, Thm. 4.9.8] proved that X is Hausdorff iff so is $\mathscr{V}X$. Indeed, for 'only if', given $K_i \in \mathscr{V}X$, $i = 1, 2$, satisfying $K_1 \neq K_2$, assume without loss of generality that $K_1 \setminus K_2 \neq \emptyset$ and let $x \in K_1 \setminus K_2$. Since X is Hausdorff there are disjoint open sets U_i, $i = 1, 2$, such that $x \in U_1$ and $K_2 \subseteq U_2$. Hence, $K_1 \in U_1^\diamond$ and $K_2 \in U_2^\square$. Moreover, $U_1^\diamond$ and $U_2^\square$ are disjoint, since for $R \subseteq U_2$ we have $R \cap U_1 = \emptyset$. We omit the proof of 'if'.

(3) In addition, Vietoris [309] originally proved that for a compact Hausdorff space X (the classic Vietoris space) $\mathscr{V}X$ is compact Hausdorff, too.

Proposition 15.2.3. *For every continuous function $f : X \to Y$ and every open $U \subseteq Y$, $(f^{-1}(U))^\diamond = (\mathscr{V}f)^{-1}(U^\diamond)$, and $(f^{-1}(U))^\square = (\mathscr{V}f)^{-1}(U^\square)$.*

Proof Let $R \in \mathscr{V}X$. Recall that the image of a compact set under a continuous function is also compact. Observe that

$$R \cap f^{-1}(U) \neq \emptyset \Leftrightarrow f[R] \cap U \neq \emptyset \Leftrightarrow f[R] \in U^\diamond \Leftrightarrow R \in (\mathscr{V}f)^{-1}(U^\diamond).$$

This proves our first assertion for all R. For the second assertion, we have

$$R \subseteq f^{-1}(U) \Leftrightarrow f[R] \subseteq U \Leftrightarrow f[R] \in U^\square \Leftrightarrow R \in (\mathscr{V}f)^{-1}(U^\square). \qquad \square$$

Corollary 15.2.4. *The mapping $X \mapsto \mathscr{V}X$ is the object assignment of a functor $\mathscr{V}$ on* Top*, where for a continuous function $f : X \to Y$, we put $\mathscr{V}f(A) = f[A]$.*

Indeed, Proposition 15.2.3 shows that for every subbasic open set of $\mathscr{V}Y$ its inverse image under $\mathscr{V}f$ is open in $\mathscr{V}X$, which establishes continuity of $\mathscr{V}f$.

Notation 15.2.5. We denote by Haus and KHaus the full subcategories of Top given by all Hausdorff and all compact Hausdorff spaces, respectively. By Remark 15.2.2(2) and (3), $\mathscr{V}$ restricts to both of these full subcategories, and we denote the restrictions by $\mathscr{V}$ as well.

Remark 15.2.6. (1) Assume that

$$X_0 \xleftarrow{f_0} X_1 \xleftarrow{f_1} X_2 \xleftarrow{f_2} \cdots \tag{15.4}$$

is an ω^{op}-chain in Top. Its limit $L = \lim X_n$ in Top is carried by the limit in Set. This means that the set L consists of all sequences $(x_n)_{n<\omega}$, $x_n \in X_n$ that are compatible: $f_n(x_{n+1}) = x_n$ for every $n < \omega$. The limit cone consist of the continuous maps $\ell_n : L \to X_n$ given by $\ell_n((x_i)) = x_n$. The topology on L has as a base the sets $\ell_n^{-1}(U)$, for U open in X_n. This collection forms a base since it is closed under intersections: if $n \le m$, then $\ell_n^{-1}(U_1) \cap \ell_m^{-1}(U_2) = \ell_m^{-1}(U_2 \cap W)$, where $W = f_{m,n}^{-1}(U_1)$ and $f_{m,n} \colon X_m \to X_n$ is the evident composition.

(2) The full subcategories Haus and KHaus are closed under limits in Top. In particular, the inclusion functors $\mathsf{Haus} \hookrightarrow \mathsf{Top}$ and $\mathsf{KHaus} \hookrightarrow \mathsf{Top}$ preserve and reflect limits.

(3) If the ω^{op}-chain in (15.4) consists of surjective continuous maps between compact Hausdorff spaces, then each projection map $\ell_n \colon L \to X_n$ is surjective, too (see Ribes and Zalesskii [260, Prop. 1.1.10]).

(4) Note that if X has a base $\mathcal{B}$ which is closed under finite unions, then the sets $U^\Diamond$ and $U^\Box$ for $U \in \mathcal{B}$ already form a subbase of $\mathscr{V}X$. Indeed, given a set $\mathcal{S}$ of open subsets of X we have $(\bigcup \mathcal{S})^\Diamond = \bigcup\{U^\Diamond : U \in \mathcal{S}\}$. Moreover, it is easy to see that

$$\left(\bigcup \mathcal{S}\right)^\Box = \bigcup\left\{\left(\bigcup \mathcal{F}\right)^\Box : \mathcal{F} \subseteq \mathcal{S} \text{ finite}\right\};$$

'$\supseteq$' is trivial, and for '$\subseteq$' use compactness of $R \in \mathscr{V}X$. Hence, if $\mathcal{S}$ consists of basic open sets from $\mathcal{B}$, then $\bigcup F \in \mathcal{B}$ due to its closure under finite unions. Thus, $(\bigcup \mathcal{S})^\Box$ is a union of sets of the form $U^\Box$ for $U \in \mathcal{B}$.

The following result was stated by Hofmann et al. [165, Thm. 3.39], and we supply a full proof for the convenience of the reader.

Proposition 15.2.7. *The functor* $\mathscr{V} \colon \mathsf{Haus} \to \mathsf{Haus}$ *preserves limits of* ω^{op}-*chains.*

Proof Consider an ω^{op}-chain as in (15.4). Let $M = \lim \mathscr{V}X_n$, with limit cone $r_n \colon M \to \mathscr{V}X_n$. Let $m \colon \mathscr{V}L \to M$ be the unique continuous map such that $\mathscr{V}\ell_n = r_n \cdot m$ for all $n < \omega$. We shall prove that m is a bijection and then that its inverse is continuous, which proves that m is an isomorphism.

(1) Injectivity of m follows from the fact that $\mathscr{V}\ell_n$ ($n < \omega$) forms a jointly monic family, as we will now prove. Suppose that $A, B \in \mathscr{V}L$ satisfy $\ell_n[A] = \ell_n[B]$ for every $n < \omega$. We prove that $A \subseteq B$; by symmetry $A = B$ follows. Given $a \in A$, we show that every open neighbourhood of a has nonempty intersection with B. Since B is closed, we then have $a \in B$ (otherwise $L \setminus B$ would be an open neighbourhood of a disjoint from B). It suffices to prove the desired property for the basic open neighbourhoods $\ell_n(U)$ of a, for U open in X_n (see Remark 15.2.6(1)). Since $\ell_n[A] = \ell_n[B]$ we have some $b \in B$ which satisfies $\ell_n(a) = \ell_n(b)$. Then $b \in \ell_n^{-1}(U) \cap B$.

(2) Surjectivity of m. Suppose we are given any element of M, which means that we have a sequence $(K_n)_{n<\omega}$ of compact subsets $K_n \subseteq X_n$ such that $f_n[K_{n+1}] = K_n$ for every $n < \omega$. We need to find a compact set $K \subseteq L$ such that $\ell_n[K] = K_n$ for every $n < \omega$. Observe that the connecting maps

$f_n\colon X_{n+1} \to X_n$ restrict to surjective maps $K_{n+1} \twoheadrightarrow K_n$. Thus, the K_n form an ω^{op}-chain of surjections in KHaus. Let K be the limit of this ω^{op}-chain with projections $p_n\colon K \twoheadrightarrow K_n$; these are surjective by Remark 15.2.6(3). Clearly, p_n is a restriction of ℓ_n, and so we have $K \in \mathscr{V}L$ such that $\ell_n[K] = p_n[K] = K_n$ as desired.

(3) Finally, we prove that the inverse $k\colon M \to \mathscr{V}L$, say, of m is continuous. We know that the sets $\ell_n^{-1}(U)$, for U open in X_n, form a base of L. Moreover, this base is closed under finite unions. By Remark 15.2.6(4) and using Proposition 15.2.3 we obtain that $\mathscr{V}L$ has a subbase given by the sets

$$(\mathscr{V}\ell_n)^{-1}(U^\diamond) = (\ell_n^{-1}(U))^\diamond \text{ and } (\mathscr{V}\ell_n)^{-1}(U^\square) = (\ell_n^{-1}(U))^\square \text{ for } U \text{ open in } X_n.$$

It suffices to show that the inverse images of these subbasic open sets of $\mathscr{V}L$ are open in M. For $\mathscr{V}\ell_n^{-1}(U^\diamond)$ with U open in X_n we use that $\mathscr{V}\ell_n \cdot k = r_n$ clearly holds to obtain

$$k^{-1}(\mathscr{V}\ell_n^{-1}(U^\diamond) = r_n^{-1}(U^\diamond),$$

which is a basic open set of M by Remark 15.2.6(1). For the subbasic open sets $\mathscr{V}\ell_n^{-1}(U^\square)$ the proof is similar. □

Corollary 15.2.8. *The restriction of $\mathscr{V}$ to* KHaus *preserves limits of ω^{op}-chains.*

Remark 15.2.9. (1) Proposition 15.2.7 and Corollary 15.2.8 hold more generally for codirected limits, which are limits of diagrams whose diagram scheme is P^{op} for a directed poset P. In fact, the proof is the same.

(2) The terminal coalgebra for $\mathscr{V}$ is the topological space of all compactly branching strongly extensional trees. We have seen in Theorem 4.5.19 that for the finite power-set functor $\mathscr{P}_{\mathsf{f}}$ these trees form the limit $V_\omega = \lim \mathscr{P}_{\mathsf{f}}^n 1$. Since the forgetful functor $\mathsf{Top} \to \mathsf{Set}$ preserves limits, we know that this also yields the limit of the terminal-coalgebra ω^{op}-chain for $\mathscr{V}$, which is $\nu\mathscr{V}$ by Theorem 15.2.12. Note that $\mathscr{V}^n 1$ is the discrete space on the set $\mathscr{P}_{\mathsf{f}}^n 1$.

We proceed to describe the topology on $\nu\mathscr{V} = V_\omega$: it has a base given by the open sets U_s, where s is a finite extensional tree, and a tree t lies in U_s if there exists some $n < \omega$ such that s is the extensional quotient $\partial_n t$ of the cutting of t at level n (see Example 4.2.10). Indeed, in Set, the limit projections $v_{\omega,n}\colon V_\omega \to \mathscr{P}_{\mathsf{f}}^n 1$ are given by $v_{\omega,n}(t) = \partial_n t$. By Remark 15.2.6(1), the sets $v_{\omega,n}^{-1}(U)$ with U ranging over all subsets of $\mathscr{P}_{\mathsf{f}}^n 1$ form a base. Moreover, it clearly suffices to consider singleton sets $U = \{s\}$ to get a (smaller) base, and a tree t lies in $v_{\omega,n}^{-1}(\{s\})$ iff $s = \partial_n t$.

One should compare Proposition 15.2.7 with the situation in Set regarding the finite power-set functor $\mathscr{P}_{\mathsf{f}}$. As we saw in Example 3.3.15, the limit of the terminal ω^{op}-chain of $\mathscr{P}_{\mathsf{f}}$ is *not* preserved by the functor, and thus the terminal coalgebra is not this limit.

The following definition is due to Kupke et al. [205] for Stone spaces; Hofmann et al. [165, Def. 2.29] use arbitrary topological spaces but they later essentially restrict constants to be (compact) Hausdorff, stably compact or spectral spaces.

Definition 15.2.10. The *Vietoris polynomial functors* are the endofunctors on Top built from the Vietoris functor $\mathscr{V}$, the constant functors, and the identity functor, using product, coproduct, and composition.

Remark 15.2.11. Thus, the Vietoris polynomial functors can be defined by the following grammar

$$F ::= \mathscr{V} \mid A \mid \mathsf{Id} \mid \prod_{i \in I} F_i \mid \coprod_{i \in I} F_i \mid FF,$$

where A ranges over all topological spaces and I is an arbitrary index set.

The following theorem essentially appears in work by Hofmann et al. [165, Thm. 3.42] (except for the convergence ordinal). There is an earlier variant by Kurz et al. [205] for Stone spaces. But our proof is different then the previous ones using the convergence of the terminal-coalgebra chain.

Theorem 15.2.12. *Every Vietoris polynomial functor $F\colon \mathsf{Top} \to \mathsf{Top}$ in which all constants A are Hausdorff spaces has a terminal coalgebra obtained by finitary iteration: $\nu F = V_\omega$. Moreover, νF is a Hausdorff space.*

Proof An easy induction on Vietoris polynomial functors F as in the statement shows that each such functor has the following properties:

(1) it has a restriction $F_0\colon \mathsf{Haus} \to \mathsf{Haus}$,

(2) the functor F_0 preserves limits of ω^{op}-chains; the most important step is done in Proposition 15.2.7.

We obtain νF as the limit of the terminal-coalgebra ω^{op}-chain $F_0^n 1$ $(n < \omega)$. Since Haus is closed under limits in Top and $F_0^n 1 = F^n 1$, the functor F has the same terminal coalgebra $\nu F = \operatorname{colim} F^n 1$. □

Corollary 15.2.13. *The restriction of a Vietoris polynomial functor to* Haus *is a covarietor.*

Indeed, if F is a Vietoris polynomial functor, then so is the functor $F(-) \times A$ for every Hausdorff space A. Now apply Proposition 2.2.24 and Theorem 15.2.12.

In the following corollary, a Vietoris polynomial functor is *compact* if it is built from the grammar in Remark 15.2.11 restricted to constant functors for compact Hausdorff spaces and finite coproducts.

Corollary 15.2.14. *For every compact Vietoris polynomial functor* $F\colon \mathsf{Top} \to \mathsf{Top}$ *its terminal coalgebra* $\nu F = V_\omega$ *is a compact Hausdorff space.*

Proof Every compact Vietories functor F has a restriction to an endofunctor on KHaus. Hence, the terminal-coalgebra ω^{op}-chain $F^n 1$ lies in KHaus. Moreover, KHaus is closed under limits in Top (because it is a full reflective subcategory: the reflection of a space is its Stone–Čech compactification). Thus, $\nu F = \lim_{n<\omega} F^n 1$ is compact Hausdorff. □

Proposition 15.2.15. *Every Vietoris polynomial functor on* Top *has an initial algebra.*

Proof It is easy to see that Top has smooth monomorphisms (Definition 6.2.2). Indeed, monomorphisms are precisely the injective continuous functions. Moreover, the forgetful functor $U\colon \mathsf{Top} \to \mathsf{Set}$ preserves colimits; more precisely, the colimit of a diagram in Top is its colimit Set equipped with the final topology induced by the maps in the colimit cocone. Now the Vietoris functor $\mathscr{V}$ clearly preserves monomorphisms. Hence, an easy induction on the structure of Vietoris polynomial functors shows that every such functor F does, too. Now apply Theorem 6.3.4 using the pre-fixed point $\tau^{-1}\colon F(\nu F) \to \nu F$. □

Corollary 15.2.16. *The restriction of a Vietoris polynomial functor to* Haus *is a varietor.*

Indeed, note that if F is a Vietoris polynomial functor, then so is every functor $F(-) + A$ for every Hausdorff space A. Now apply Proposition 2.2.24 and Proposition 15.2.15.

Open Problem 15.2.17. Are Vietoris polynomial functors on Top varietors or covarietors?

This is equivalent to asking whether the result that μF or νF exist for every Vietoris polynomial functor would remain true if we allowed non-Hausdorff constants.

Remark 15.2.18. Note that Theorem 15.2.12 also holds for Vietoris polynomial functors when we take Haus as our base category. Other full subcategories of Top have been studied by Hofmann et al. [165], and they also study the completeness of Coalg F for Vietoris polynomial functors F (however, they restrict to using finite products and finite coproducts in their definition of Vietoris polynomial functors).

(1) For a Vietoris polynomial functor F restricted to Haus, the category Coalg F is complete [165, Cor. 3.41]. In addition, every subfunctor of F has a terminal coalgebra [165, Cor. 4.6].

(2) A *Stone space* is a compact Hausdorff space having a basis of clopen sets. Let Stone be the full subcategory of KHaus determined by the Stone spaces. It was shown by Michael [225, Thm. 4.9.9] that if X is a Stone space, so is $\mathcal{V}X$. Thus $\mathcal{V}$ restricts to an endofunctor on Stone. The special case of Corollary 15.2.14 for Stone was proved by Kupke et al. [205]. If a compact Vietoris polynomial functor on Top is built from constant functors for Stone spaces and the other constructors (the identity, $\mathcal{V}$, finite product, finite coproduct, and composition), then its terminal coalgebra is a Stone space.

Remark 15.2.19. Hofmann et al. [165, Ex. 2.27(2)] also consider a related construction called the *lower Vietoris space* of X. It is the set of all closed subsets of X with the topology generated by all sets $U^{\diamond}$, cf. (15.3). This again yields a functor on Top: a given continuous function is mapped to $A \mapsto \overline{f[A]}$, where $\overline{f[A]}$ denotes the closure of $f[A]$. Furthermore, one has a corresponding notion of lower Vietoris polynomial functors. They prove that for such functors F on the category of stably compact spaces, Coalg F is complete [165, Thm. 3.35]. Furthermore, if a lower Vietoris polynomial functor F on Top can be restricted to that category, then it has a terminal coalgebra obtained by finite iteration: $\nu F = V_{\omega}$ [165, Thm. 3.36]. Similar results hold for the category of spectral spaces and spectral maps.

15.3 Hausdorff Polynomial Functors on Metric Spaces

We come back to the Hausdorff functor (Example 5.2.6(4)), which maps a (complete) metric space to the (complete) metric space of all its nonempty compact subspaces equipped with the Hausdorff metric. Hence, it performs a similar construction on metric spaces as the Vietoris functor on topological spaces. In fact, a metric space X is compact iff the topology induced by the Hausdorff metric on $\mathcal{H}X$ is the same as the Vietoris topology [262, Cor. of Thm. 6].

We have already seen in Example 11.2.20(2) that Hausdorff polynomial functors on CMet are accessible, and therefore they have an initial algebra and a terminal coalgebra. In contrast, the Hausdorff functor on Met is not accessible (Remark 11.2.21). Nevertheless, we prove here that Hausdorff polynomial functors on Met have a terminal coalgebra, and that their terminal-coalgebra chain converges in $\omega + \omega$ steps. As a consquence, we obtain that every Hausdorff polynomial functor has an initial algebra using Theorem 6.3.4.

Remark 15.3.1. (1) The Hausdorff functor $\mathcal{H}$: Met $\rightarrow$ Met (Example 5.2.6(3)) maps a metric space X to the space $\mathcal{H}X$ of all nonempty compact subsets of X equipped with the Hausdorff metric given by

$$\bar{d}(S,T) = \max\left(\sup_{x\in S} d(x,T), \sup_{y\in T} d(y,S)\right), \qquad \text{for } S,T \subseteq X \text{ compact,}$$

where $d(x,S) = \inf_{y\in T} d(x,y)$. For a non-expanding map $f\colon X \to Y$ we have $\mathcal{H}f\colon S \mapsto f[S]$. By Example 5.2.6(4), this functor restricts to a functor on CMet, which we denote by the same symbol $\mathcal{H}$.

(2) For a discrete metric space X (where all distances are 0 or 1), $\mathcal{H}X$ is the discrete space formed by all nonempty finite subsets of X.

(3) For an arbitrary metric space X, the nonempty finite subsets of X form a dense set in $\mathcal{H}X$. Indeed, given a compact set $S \subseteq X$, for every $\varepsilon > 0$, there exists a nonempty finite set $T \subseteq S$ such S is covered by ε-balls around the points in T. Therefore $d(x,T) \leq \varepsilon$ for all $x \in S$, and we have $d(y,S) = 0$ for all $y \in T$. This implies that $\bar{d}(S,T) \leq \varepsilon$.

Example 15.3.2. (1) The terminal coalgebra for $\mathcal{H}$ is 1, the terminal one-point space, because we have $\mathcal{H}1 \cong 1$.

(2) For the functor $\mathcal{H}(-) + \{\emptyset\}$, a terminal coalgebra is carried by the space of all finitely branching strongly extensional trees (Theorem 4.5.15) equipped with the discrete metric. This follows from the finite power-set functor $\mathcal{P}_\mathsf{f}$ having its terminal coalgebra formed by those trees. Indeed, the terminal-coalgebra chain V_i ($i \in \mathsf{Ord}$) for $\mathcal{H}(-) + \{\emptyset\}$ is obtained by equipping the sets in the chain for $\mathcal{P}_\mathsf{f}$ with the discrete metric. Since furthermore, limits in Met (or CMet) are formed as in Set, we see that both chains converge in exactly $\omega + \omega$ steps so that $\nu(\mathcal{H}(-) + \{\emptyset\}) = V_{\omega+\omega}$.

Remark 15.3.3. (1) Let $f_n\colon X_{n+1} \to X_n$ ($n < \omega$) be an ω^{op}-chain in Met. Its limit $L = \lim X_n$ in Met is carried by the limit in Set. This means that the set L consists of all sequences $(x_n)_{n<\omega}$, $x_n \in X_n$ that are compatible: $f_n(x_{n+1}) = x_n$ for every $n < \omega$. The limit cone consists of the non-expanding maps $\ell_n\colon L \to X_n$ given by $\ell_n((x_i)) = x_n$. The metric on L is defined by $d((x_n),(y_n)) = \sup_{n<\omega} d(x_n, y_n)$.

(2) Unlike the Vietoris functor, the Hausdorff functor does not preserve limits of ω^{op}-chains. We have seen that the terminal-coalgebras chain for $\mathcal{H}(-) + \{\emptyset\}$ converges in exactly $\omega + \omega$ steps (see Example 15.3.2(2)). Thus this functor does not preserve the limit $V_\omega = \lim_{n<\omega} V_n$. Hence, $\mathcal{H}$ does not preserve the limit of the terminal-coalgebra ω^{op}-chain for $\mathcal{H}(-) + \{\emptyset\}$ either.

Recall from Example 6.2.5(3) that isometric embeddings (functions which are distance preserving) are monomorphisms representing precisely the subspaces of a metric space.

Definition 15.3.4. Let $(X_n)_{n<\omega}$ be an ω^{op}-chain in Met. A cone $r_n \colon M \to X_n$ is *isometric* if for all $x, y \in M$ we have $d(x, y) = \sup_{n\in\mathbb{N}} d(r_n(x), r_n(y))$.

The property established in the following proposition is reminiscent of precontinuity (Definition 13.2.5).

Proposition 15.3.5. *The Hausdorff functor preserves isometric cones of ω^{op}-chains.*

Proof Let $(X_n)_{n<\omega}$ be an ω^{op}-chain with connecting maps $f_n \colon X_{n+1} \to X_n$. Given an isometric cone $\ell_n \colon M \to X_n$ $(n < \omega)$, we prove that the cone formed by $\mathcal{H}\ell_n \colon \mathcal{H}M \to \mathcal{H}X_n$ is also isometric:

$$\bar{d}(S,T) = \sup_{n<\omega} \bar{d}(\ell_n(S), \ell_n(T)) \qquad \text{for all compact subset } S, T \subseteq M.$$

We can assume that S and T are finite: since finite sets are dense in $\mathcal{H}M$ by Remark 15.3.1(3) and the maps ℓ_n are (non-expanding whence) continuous, the desired equality then holds for all pairs in $\mathcal{H}M$.

Since every ℓ_n is non-expanding, we only need to prove that $\bar{d}(S,T) \le c$ holds for $c = \sup_{n<\omega} \bar{d}(\ell_n[S], \ell_n[T])$.

We proceed to show that for every $\varepsilon > 0$ we have $\bar{d}(S,T) \le c + \varepsilon$, from which the desired inequality follows. By the definition $\bar{d}$ it suffices to prove that for every $x \in S$ we have $d(x,T) \le c + \varepsilon$. By symmetry, $d(y, S) \le c + \varepsilon$ for every $y \in T$.

Given $y \in T$ we have $d(x, y) = \sup_{n<\omega} d(\ell_n(x), \ell_n(y))$. Thus, there exists some $k < \omega$ such that

$$d(x, y) \le d(\ell_k(x), \ell_k(y)) + \varepsilon.$$

Since T is finite, we can chose k such that this inequality holds for all $y \in T$. By definition we have

$$d(\ell_k(x), \ell_k[T]) = \inf_{y\in T} d(\ell_k(x), \ell_k(y)) \qquad \text{in } X_k.$$

Since T is finite, we can pick some $y \in T$ such that

$$d(\ell_k(x), \ell_k[T]) = d(\ell_k(x), \ell_k(y)).$$

We conclude that

$$\begin{aligned} d(x,T) &\le d(x, y) \le d(\ell_k(x), \ell_k(y)) + \varepsilon = d(\ell_k(x), \ell_k[T]) + \varepsilon \\ &\le \bar{d}(\ell_k[S], \ell_k[T]) + \varepsilon \\ &\le c + \varepsilon. \end{aligned}$$

□

Remark 15.3.6. The Hausdorff functor preserves isometric embeddings and their intersections. Indeed, for every subspace X of a metric space Y, a set $S \subseteq X$

is compact in X iff it is compact in Y. Moreover, given $S, T \in \mathcal{H}X$, their distances in $\mathcal{H}X$ and $\mathcal{H}Y$ are the same. Thus, $\mathcal{H}$ preserves isometric embeddings.

Given a collection $X_i \subseteq Y$ ($i \in I$) of subspaces, a set $S \subseteq \bigcap_{i \in I} X_i$ is compact iff it is so in Y (and therefore in every X_i). Thus $\mathcal{H}$ preserves that intersection.

The following definition extends the class of functors on CMet that we have seen in Example 11.2.20(2) to the larger category Met.

Definition 15.3.7. The *Hausdorff polynomial functors* are the endofunctors on Met built from the Hausdorff functor, the constant functors, and the identity functor, using product, coproduct, and composition.

Remark 15.3.8. Thus, the Hausdorff polynomial functors can be defined by the following grammar

$$F ::= \mathcal{H} \mid A \mid \mathsf{Id} \mid \textstyle\prod_{i \in I} F_i \mid \coprod_{i \in I} F_i \mid FF,$$

where A ranges over all metric spaces and I is an arbitrary index set.

Example 15.3.9. Let Σ be a metric space (of actions). The Hausdorff polynomial functor $FX = \mathcal{H}(\Sigma \times X) + \{\emptyset\}$ has as coalgebras compactly branching labelled transition systems (cf. Example 4.4.15(2)). If $\mathrm{b}\Sigma = \{0, 1\}$ with $d(0, 1) = \delta < 1$, then we may describe the terminal coalgebra as follows [6]. We work with trees whose edges are labelled in Σ. The terminal coalgebra is the set of all finitely branching, strongly extensional labelled trees t, with the metric

$$d(t, u) = \begin{cases} 0 & \text{if } t = s, \\ \delta & \text{if } t \neq s \text{ but } |t| = |s|, \\ 1 & \text{otherwise.} \end{cases}$$

Here $|t|$ denotes the unlabelled tree which underlies t.

Theorem 15.3.10. *For every Hausdorff polynomial functor $F\colon \mathsf{Met} \to \mathsf{Met}$ its terminal-coalgebra chain converges in $\omega + \omega$ steps: $\nu F = V_{\omega+\omega}$.*

Proof An easy induction over the structure of Hausdorff polynomial functors shows that each such functor F preserves:

(1) isometric cones of ω^{op}-chains, and

(2) isometric embeddings and their intersections.

The most important step in done in Proposition 15.3.5 and Remark 15.3.6.

We conclude that in the terminal-coalgebra chain for F the map $v_{\omega+1,\omega}\colon V_{\omega+1} \to V_\omega$ is an isometric embedding by item (1), thus $v_{\omega+2,\omega+1} = Fv_{\omega+1,\omega}$ is one by item (2), whence so is $v_{\omega+2,\omega}$, etc. We thus see that all maps $v_{\omega+n,\omega}\colon V_{\omega+n} \to V_\omega$ ($n < \omega$) are isometric embeddings. By item (2), F

preserves the intersection of the ensuing subspaces of V_ω, namely the limit $V_{\omega+\omega} = \lim_{n<\omega} V_{\omega+n}$. Consequently, we have $\nu F = V_{\omega+\omega}$. □

Remark 15.3.11. Note that if a Hausdorff polynomial functor F uses only contants given by complete metric spaces A, then its terminal coalgebra νF is complete, too. Indeed, by an easy induction on the structure of F one shows that FX is complete whenever X is complete. Moreover, complete metric spaces are closed unter limits in Met. Hence, by an easy transfinite induction we see that every metric space V_i in the terminal-coalgebra chain for F is complete, whence $\nu F = V_{\omega+\omega}$ is complete.

Corollary 15.3.12. *Every Hausdorff polynomial functor F on* Met *or* CMet *is a covarietor.*

Indeed, if F is a Hausdorff polynomial functor, then so is every functor $F(-)\times A$. Now apply Proposition 2.6.15 and Theorem 15.3.10.

Corollary 15.3.13. *Every Hausdorff polynomial functor F on* Met *or* CMet *has an initial algebra. Moreover, F is a varietor.*

We have seen in Propositions 6.2.8 and 6.2.10 that isometric embeddings form a smooth class of monomorphisms in both categories. Since F preserves isometric embeddings, we can apply Theorem 6.3.4 using the pre-fixed point $\tau^{-1}\colon F(\nu F) \to \nu F$. For the second statement note that if F is a Hausdorff polynomial functor, then so is every functor $F(-) + A$. Hence, we can apply Corollary 2.2.27.

15.4 The Interval $[0, 1]$ as a Terminal Coalgebra

We mentioned the category BiP of *bipointed sets* in Example 3.2.14 and the endofunctor $FX = X \oplus X$ on that category, for which we checked that the initial algebra is the set D of dyadic rational numbers in $[0, 1]$ with $\bot = 0$ and $\top = 1$. In this section, we shall prove that the canonical fixed point (see Definition 5.0.1) of this functor is the unit interval $\mathbb{I} = [0, 1]$, again with $\bot = 0$ and $\top = 1$. Moreover, the usual metric structure of $\mathbb{I}$ can be included: we introduce the category BiMet of bipointed metric spaces and use an endofunctor which is a variation of $X \oplus X$. Then we prove that $\mathbb{I}$ is the canonical fixed point.

Definition 15.4.1. A *bipointed metric space* is a triple $(X, \bot, \top)$ consisting of a metric space X and two points $\bot, \top$ of X of distance 1.

The category BiMet of bipointed metric spaces has as morphisms the non-expanding functions which preserve the distinguished points.

In the following construction, we denote by $\frac{1}{2}X$ the metric space obtained from X by halving all distances.

Construction 15.4.2. Given bipointed metric spaces X and Y, we form their coproduct (disjoint union) $X + Y$ in Met. We denote by

$$X \oplus Y$$

the bipointed metric space which is the quotient of $\frac{1}{2}(X + Y)$ merging $\top_X$ and $\bot_Y$.

In more detail: the underlying set of $X \oplus Y$ is $(X + Y)/\sim$ where the only non-trivial equivalence class of $\sim$ is $\top_X \sim \bot_Y$; this element of $X \oplus Y$ is called the *midpoint* m. The metric space $\frac{1}{2}X$ is a subspace of $X \oplus Y$ (called the *left subspace*), and $\frac{1}{2}Y$ is also a subspace (the *right subspace*). The distance between $x \in X$ and $y \in Y$ in $X \oplus Y$ is

$$\frac{1}{2}(d(x, \top_X) + d(\bot_Y, y)).$$

It is easy to check that the triangle inequality holds.

Notation 15.4.3. (1) The endofunctor $FX = X \oplus X$ on BiMet assigns to every bipointed space X the space $X \oplus X$. To every morphism f of BiMet the corresponding morphism $f \oplus f$ acting as f on the left subspace as well as on the right one. Hence, $f \oplus f$ preserves the midpoint.

(2) The full subcategory BiCMet of BiMet on complete bipointed spaces has the property that F restricts to an endofunctor on it. We write $\bar{F}$ for this functor.

Example 15.4.4. The unit interval $\mathbb{I}$ is an algebra for $X \mapsto X \oplus X$ on BiMet (and BiCMet) with the structure $\iota \colon \mathbb{I} \oplus \mathbb{I} \to \mathbb{I}$ given by $x \mapsto \frac{x}{2}$ for x in the left sub-interval, and $x \mapsto \frac{x+1}{2}$ for x in the right one.

Notation 15.4.5. We write $\mathbb{D}$ for the subalgebra of $\mathbb{I}$ given by the diadic rationals (see Example 3.2.14).

Proposition 15.4.6. (1) *The space* $\mathbb{D}$ *is the initial algebra for* $FX = X \oplus X$ *on* BiMet.

(2) *The space* $\mathbb{I}$ *is the initial algebra for* $\bar{F}X = X \oplus X$ *on* BiCMet.

Proof (1) This is a variation of the argument which we saw in Example 3.2.14. The bipointed sets in that discussion are indeed metric spaces in the evident way. The maps in the initial-algebra ω-chain are isometric embeddings (Example 6.2.5(3)), and thus the colimit in BiP is the colimit in BiMet.

(2) Let $C\colon$ BiMet $\to$ BiCMet be the Cauchy-completion functor and

$U\colon \mathsf{BiCMet} \to \mathsf{BiMet}$ the forgetful functor. We have the commuting diagram:

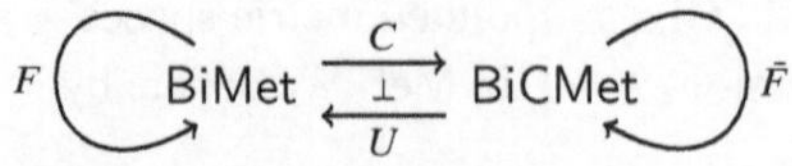

It is not hard to check that for a complete bipointed space X, $C(UX \oplus UX) = U(X \oplus X)$ (see Bhattacharya et al. [79] for a similar proof). Analogously as in Example 11.5.6, the functor C takes initial algebras to initial algebras. Thus the initial $\bar{F}$-algebra is carried by the Cauchy completion of $\mathbb{D}$; in symbols, $C(\mathbb{D}) = \mathbb{I}$. Moreover, it is easy to check that C takes the structure $\iota\colon \mathbb{D} + \mathbb{D} \to \mathbb{D}$ from Example 3.2.14 to the structure $\iota\colon \mathbb{I} \oplus \mathbb{I} \to \mathbb{I}$ which we saw in Example 15.4.4. □

The category BiCMet has no terminal object, so Theorem 5.2.20 does not apply. There are, however, several results concerning the unit interval as a terminal coalgebra. The original result in this direction is due to Freyd [130], who proved that as a set, $\mathbb{I}$ is the terminal coalgebra of $X \oplus X$ on BiP; see Corollary 15.4.9. Second, Leinster [213] showed that as a topological space with the Euclidean topology, $\mathbb{I}$ is terminal for the same functor on bipointed topological spaces. We are not going to prove this, but instead present the following 'metric variant'.

Remark 15.4.7. (1) Note that for metric spaces X, Y the hom-set $\mathsf{Met}(X, Y)$ is the metric space of all non-expanding maps $f\colon X \to Y$ equipped with the supremum metric $d_{X,Y}(f, g) = \sup_{x \in X} d_Y(f(x), g(x))$ (cf. Example 5.2.3(1)).
(2) If Y is complete, then so is $\mathsf{Met}(X, Y)$.

Theorem 15.4.8 [79]. *The space $\mathbb{I}$ is the canonical fixed point of $X \mapsto X \oplus X$ on both* BiMet *and* BiCMet.

Proof By Proposition 15.4.6 and Lambek's Lemma 2.2.5, the structure $\iota\colon \mathbb{I} \oplus \mathbb{I} \to \mathbb{I}$ (Example 15.4.4) is an isomorphism in BiCMet.

Let $e\colon X \to X \oplus X$ be a coalgebra. The subspace S of the complete metric space $\mathsf{Met}(X, \mathbb{I})$ given by all morphisms from X to $\mathbb{I}$ in BiCMet is closed. Thus, S is itself a complete metric space. We verify that S contains the function

$$d(-, \bot)\colon X \to \mathbb{I}.$$

This is a morphism of bipointed spaces, since it takes $\top$ to 1 and $\bot$ to 0. Let us check that this function is non-expanding. Fix points x and y. By the triangle inequality, $d(x, \bot) - d(y, \bot) \le d(x, y)$ and $d(y, \bot) - d(x, \bot) \le d(x, y)$. Thus

$$d_{\mathbb{I}}(d(x, \bot), d(y, \bot)) = |d(x, \bot) - d(y, \bot)| \le d(x, y).$$

This shows that S is nonempty.

We claim that $FX = X \oplus X$ is a contracting endofunctor on BiMet: for $f, g\colon X \to Y$,

$$d(Ff, Fg) \leq \frac{1}{2} d(f, g).$$

Here is the reason: for each $x \in X$, we have copies $\mathsf{inl}\, x$ and $\mathsf{inr}\, x$. For $\mathsf{inl}\, x$,

$$\begin{aligned} d(Ff(\mathsf{inl}\, x), Fg(\mathsf{inl}\, x)) &= \frac{1}{2} d(f(x), g(x)) \leq \frac{1}{2} \sup_{x' \in X} d(f(x'), g(x')) \\ &= \frac{1}{2} d(f, g) \end{aligned} \tag{15.5}$$

and similarly for $\mathsf{inr}\, x$. Both Ff and Fg preserve the midpoint. This establishes our inequality above.

The endofunction $\psi\colon S \to S$ given by

$$\psi(f) = \iota^{-1} \cdot Ff \cdot e$$

is a contraction; indeed, (15.5) above shows that $f \mapsto Ff$ is a contraction, and it is easy to see that pre- and post-composing by the non-expanding maps e and ι^{-1} preserves that property. Hence, by Banach's Fixed Point Theorem 5.2.4, there is a unique fixed point $f = \psi(f)$. But such an f is exactly a coalgebra morphism from (X, e) to $(\mathbb{I}, \iota^{-1})$. □

For further results in this direction and connections with fractal sets such as the Sierpinski gasket, see Bhattacharya et al. [79].

Corollary 15.4.9 [130]. *The terminal coalgebra for* $F\colon \mathsf{BiP} \to \mathsf{BiP}$ *is* $\iota^{-1}\colon \mathbb{I} \to \mathbb{I} \oplus \mathbb{I}$.

Indeed, given a bipointed set X, consider it as a discrete (and hence complete) metric space: all distances are 0 or 1.

We should mention that the work here is only partially successful. As Freyd [130, p. 218] points out:

"To this date, no one has found a functor whose terminal coalgebra is usefully the reals."

15.5 Terminal Coalgebras Related to Subsets of the Reals

We have seen in Section 15.4 that $\mathbb{I} = [0, 1]$ is a terminal coalgebra for certain endofunctors. However, that result did not use Set but rather the categories BiP and BiCMet. The theme of this section is that by working with different functors, we can return to Set and again characterize sets like the half-open interval $[0, 1)$ as terminal coalgebras.

$\mathbb{N} \times X$

carrier	inverse of structure
$\mathbb{A}$	$\alpha(n,r) = n + \frac{1}{1+r}$
$\mathbb{B}$	$\beta(n,r) = \frac{1}{1+n+r}$
$\mathbb{R}_{\geq 0}$	$\delta(n,r) = n + \frac{r}{1+r}$
$[0,1)$	$\sigma(n,r) = \frac{n+r}{1+n+r}$

$\mathbb{N} \times X + 1$

carrier	inverse of structure
$\mathbb{R}_{\geq 0}$	$[\varrho, 0]$, where $\varrho(n,r) = n + \frac{1}{1+r}$
$(0,1]$	$[\tau, 1]$, where $\tau(n,r) = \frac{n+r}{1+n+r}$

Figure 15.1 Terminal coalgebras obtained in this section.

In this section we work with the set functors $FX = \mathbb{N} \times X$ (denoted $\mathbb{N} \times X$ for short), and $GX = \mathbb{N} \times X + 1$ (denoted $\mathbb{N} \times X + 1$). We present several subsets of the reals from 'everyday mathematics' as either terminal coalgebras or corecursive algebras (see Section 7.3). The results on terminal coalgebras are inspired by Pavlović and Pratt [247]. Our proofs are different, mainly due to our use of corecursive algebras. Figure 15.1 summarizes the results for $\mathbb{N} \times X$ and $\mathbb{N} \times X + 1$, where notations are as follows.

Notation 15.5.1. (1) We denote by $\mathbb{R}_{\geq 0}$ the set non-negative real numbers. We write $\mathbb{A}$ for the subset of positive irrationals and $\mathbb{B}$ for the set of irrationals in $\mathbb{I} = [0,1]$.

(2) The functions δ, σ, α, β, and ϱ are algebra structures for F and G, respectively defined in Figure 15.1.

(3) As usual, the right-hand component of $[\varrho, 0] : \mathbb{N} \times \mathbb{R}_{\geq 0} + 1 \to \mathbb{R}_{\geq 0}$ takes the element of the singleton set to 0. Analogously, the right-hand component of $[\sigma, 1]$ takes the element of the singleton set to 1.

(4) Notice that our charts have morphisms α and ϱ with the same formula, and also σ and τ. The reason why we use different letters is simply that the codomains are different.

Lemma 15.5.2. *The morphism $\varrho\colon \mathbb{R}_{\geq 0} \to \mathbb{N} \times \mathbb{R}_{\geq 0}$ in Figure 15.1 is one-to-one, and $\sigma\colon \mathbb{N} \times [0,1) \to [0,1)$ is a bijection.*

Proof The verifications are elementary. For σ, notice that $\sigma = i \cdot g \cdot h$, where $h\colon \mathbb{N} \times [0,1) \to [1,\infty)$ is $h(n,r) = 1+n+r$, $g\colon [1,\infty) \to (0,1]$ is $g(x) = 1/x$, and $i\colon (0,1] \to [0,1)$ is $i(x) = 1 - x$. □

Notation 15.5.3. (1) We shall use some notation which is standard in the theory of continued fractions:

$$[b] = b, \qquad [b_0, \ldots, b_{k+1}] = b_0 + 1/[b_1, \ldots, b_k].$$

We mainly use this notation with lists of positive natural numbers.

(2) For all infinite sequences (a_i) of natural numbers with $a_i > 0$ for $i \geq 1$, the limit

$$\lim_{k\to\infty} [a_0, \ldots, a_k]$$

exists. For the proof, see Hardy and Wright [155, Thm. 165], Loya [216, Cor. 8.13], or Niven et al. [244, Thm. 7.6]. This limit is denoted in several ways in the literature, including by

$$a_0 + \frac{1}{a_1+} + \frac{1}{a_2+} \cdots \qquad \text{and by} \qquad a_0 + \cfrac{1}{a_1 + \cfrac{1}{a_2 + \cfrac{1}{\cdots}}}.$$

Lemma 15.5.4 [155, Thm. 154]. *If* $a_i \geq 1$ *for all* i*, then* $\lim_{k\to\infty}[a_0, \ldots, a_k] > 1$.

We also recall from Definition 7.1.1 the notation $e^\dagger$ for solutions of equation morphisms in a coalgebra.

Theorem 15.5.5. *The algebra* $\mathbb{R}_{\geq 0}$ *for* $\mathbb{N}\times X$ *with structure* $\varrho(n,r) = n+1/(1+r)$ *is corecursive.*

Proof We derive this result from several facts from the theory of continued fractions. Put $FX = \mathbb{N} \times X$. It is convenient to work with a naturally isomorphic functor, $HX = \mathbb{N}_{\geq 1} \times X$, where $\mathbb{N}_{\geq 1}$ is the set of positive natural numbers. The natural isomorphism $\eta\colon F \to H$ has components $\eta_X(n,x) = (n+1,x)$. There is a bijective correspondence between F-coalgebras $e\colon X \to FX$ and H-coalgebras $\eta_X \cdot e\colon X \to HX$. This extends to a bijective correspondence between solutions with respect to F and those with respect to H; indeed in the diagram below the upper square commutes iff so does the outside:

$$\begin{array}{ccc}
X & \xrightarrow{e^\dagger} & \mathbb{R}_{\geq 0} \\
{\scriptstyle e}\downarrow & & \uparrow{\scriptstyle \varrho} \\
FX & \xrightarrow{Fe^\dagger} & F\mathbb{R}_{\geq 0} \\
{\scriptstyle \eta_X}\downarrow & & \uparrow{\scriptstyle \eta^{-1}_{\mathbb{R}_{\geq 0}}} \\
HX & \xrightarrow{He^\dagger} & H\mathbb{R}_{\geq 0}
\end{array}$$

Furthermore, the isomorphism $x \mapsto x + 1$ between $\mathbb{R}_{\geq 0}$ and $\mathbb{R}_{\geq 1}$ in Set carries the H-algebra $(\mathbb{R}_{\geq 0}, \varrho \cdot \eta^{-1}_{\mathbb{R}_{\geq 0}})$ to $(\mathbb{R}_{\geq 1}, \varrho')$, where $\varrho'(n, r) = n + 1/r$. As a result, it is sufficient to prove that the H-algebra $(\mathbb{R}_{\geq 1}, \varrho')$ is corecursive.

Given a coalgebra $e\colon X \to \mathbb{N}_{\geq 1} \times X$, denote its components by $u\colon X \to \mathbb{N}_{\geq 1}$ and $\mathsf{next}\colon X \to X$. For each $x \in X$, we have an infinite sequence a^x_i, $i \in \mathbb{N}$, where

$$a^x_i = u(\mathsf{next}^i(x)). \tag{15.6}$$

Thus, we may use Notation 15.5.3(2) to define $e^\dagger$ by

$$e^\dagger(x) = \lim_{k\to\infty} [a^x_0, \ldots, a^x_k].$$

For a fixed $x \in X$, put $y = \mathsf{next}(x)$. Then $a^x_{k+1} = a^y_k$ for all k. Using Notation 15.5.3(2),

$$\begin{aligned} e^\dagger(x) &= \lim_{k\to\infty} [a^x_0, a^y_0, \ldots, a^y_k] \\ &= \lim_{k\to\infty} a^x_0 + 1/[a^y_0, \ldots, a^y_k] \\ &= a^x_0 + 1/(\lim_{k\to\infty} [a^y_0, \ldots, a^y_k]) \\ &= u(x) + 1/(e^\dagger(\mathsf{next}(x))) \end{aligned}$$

This confirms that $e^\dagger\colon X \to \mathbb{R}_{\geq 1}$ is a solution of e in the H-algebra $(\mathbb{R}_{\geq 1}, \varrho')$.

We conclude by checking the uniqueness of $e^\dagger$. Let $s\colon X \to \mathbb{R}_{\geq 1}$ be any solution of e. Then for all x, $s(x) = u(x) + 1/s(\mathsf{next}(x))$. By Loya [216, Thm. 8.14], $s(x) = \lim_{k\to\infty}[a_0, \ldots, a_k]$. This is also $e^\dagger(x)$. This completes the proof. □

Corollary 15.5.6. *The set functor* $\mathbb{N}\times X+1$ *has the following terminal coalgebras:*

(1) *The set* $\mathbb{R}_{\geq 0}$ *with structure inverse to* $[\varrho, 0]$.

(2) *The set* $(0, 1]$ *with structure inverse to* $[\tau, 1]$.

Proof (1) We will first argue that $(\mathbb{R}_{\geq 0}, [\varrho, 0])$ is a corecursive algebra for $GX = \mathbb{N} \times X + 1$. Let $e\colon Y \to \mathbb{N} \times Y + 1$ be a coalgebra. We say that $y \in Y$ is 0-*grounded* if $e(y)$ lies in the summand 1. For $k \in \omega$ we say that y is $(k + 1)$-*grounded* if there is a natural number n and a k-grounded $y' \in Y$ so that $e(y) = (n, y')$.

Let $Z \subseteq Y$ be the set of elements which are not k-grounded for any k. Define $f\colon Z \to \mathbb{N} \times Z$ by restricting e. This is the unique map such that the diagram below commutes, where $j\colon Z \to Y$ is the inclusion:

$$\begin{array}{ccccc} Z & \xrightarrow{f} & \mathbb{N}\times Z & \xrightarrow{\mathsf{inl}} & \mathbb{N}\times Z+1 \\ {\scriptstyle j}\downarrow & & & & \downarrow{\scriptstyle \mathbb{N}\times j+1} \\ Y & & \xrightarrow{\quad e \quad} & & \mathbb{N}\times Y+1 \end{array}$$

Hence (Z, f) is a coalgebra for $\mathbb{N} \times X$. We have a unique solution $f^\dagger : Z \to \mathbb{R}_{\geq 0}$.

If $e(y)$ is not the element of the right-hand summand, let us write $e(y)$ as $(u(y), \mathsf{next}(y))$, where $u(y) \in \mathbb{N}$, and $\mathsf{next}(y) \in Y$. Notice that if y is k-grounded and $k \geq 1$, then $\mathsf{next}(y)$ is $(k-1)$-grounded. Then the map $e^\dagger : Y \to \mathbb{R}_{\geq 0}$ below is easily seen to be the unique solution of e:

$$e^\dagger(y) = \begin{cases} f^\dagger(y) & \text{if } y \in Z, \\ 0 & \text{if } y \text{ is 0-grounded}, \\ u(x) + 1/(e^\dagger(\mathsf{next}(x)) + 1) & \text{if } y \text{ is } k\text{-grounded for } k \geq 1. \end{cases}$$

This confirms that $(\mathbb{R}_{\geq 0}, [\varrho, 0])$ is a corecursive algebra for $\mathbb{N} \times X + 1$. By Lemma 15.5.2, the algebra structure $[\varrho, 0]$ is a bijection: ϱ is injective, the only value which it does not take is 0, and this is rectified in $[\varrho, 0]$. Proposition 7.1.2(2) thus implies that $(\mathbb{R}_{\geq 0}, [\varrho, 0]^{-1})$ is a terminal coalgebra for G.

(2) It is easy to see that the diagram below commutes, where $f : \mathbb{R}_{\geq 0} \to (0, 1]$ is the bijection $f(x) = 1/(1+x)$:

$$\begin{array}{ccc} \mathbb{N} \times \mathbb{R}_{\geq 0} + 1 & \xrightarrow{[\varrho,0]} & \mathbb{R}_{\geq 0} \\ {\scriptstyle \mathbb{N}\times f+1}\big\downarrow & & \big\downarrow{\scriptstyle f} \\ \mathbb{N} \times (0,1] + 1 & \xrightarrow{[\tau,1]} & (0,1] \end{array}$$

We omit the verification that $f \cdot \varrho = \tau \cdot (\mathbb{N} \times f)$. Thus f is an isomorphism between the two coalgebras of our corollary. □

Our next result is also obtainable from standard results on continued fractions. But most presentations do not exactly fit our development, and so we present an elementary proof.

Lemma 15.5.7. *Let $e : X \to \mathbb{N} \times X$ be a coalgebra, and let $e^\dagger$ be the solution in $(\mathbb{R}_{\geq 0}, \varrho)$. For all $x \in X$, the number $e^\dagger(x)$ is irrational.*

Proof Suppose not. Let $Y \subseteq X$ be the set of $x \in X$ such that $e^\dagger(x)$ is rational. Note that each $e^\dagger(x)$ is non-zero. For each $x \in Y$, write $e^\dagger(x)$ as p_x/q_x, where $gcd(p_x, q_x) = 1$. Let $x \in Y$ minimize the sum $p_x + q_x$. Write $e(x)$ as (n, y). By the definition of $e^\dagger$,

$$e^\dagger(x) = \varrho(n, e^\dagger(y)) = n + \frac{1}{1 + e^\dagger(y)}.$$

Thus, $e^\dagger(y)$ is rational, too. That is, $y \in Y$. So we have p_y and q_y such that $e^\dagger(y) = p_y/q_y$, and $\gcd(p_y, q_y) = 1$. Our choice of x tells us that

$$p_x + q_x \leq p_y + q_y. \tag{15.7}$$

Now

$$\frac{p_x}{q_x} = n + \frac{1}{1 + \frac{p_y}{q_y}}.$$

Hence

$$\frac{p_x - nq_x}{q_x} = \frac{q_y}{p_y + q_y}.$$

The fraction on the right is in lowest terms, because

$$gcd(q_y, p_y + q_y) = gcd(q_y, p_y) = 1.$$

Therefore the denominator on the right must divide that on the left. In particular, $p_y + q_y \leq q_x$. Since $p_x > 0$, we have $p_y + q_y < p_x + q_x$. This is a contradiction to (15.7). □

As usual we denote, for every real number r, by $\lfloor r \rfloor$ the largest natural number $\leq r$.

Corollary 15.5.8 [247]. *The set functor* $\mathbb{N} \times X$ *has the following terminal coalgebras:*

(1) *The set* $\mathbb{A}$ *of positive irrational numbers, with structure inverse to* $\alpha\colon \mathbb{N} \times \mathbb{A} \to \mathbb{A}$ *given by* $\alpha(n, r) = n + 1/(1 + r)$.

(2) *The set* $\mathbb{B}$ *of irrational numbers in* $[0, 1]$, *with structure inverse to* $\beta\colon \mathbb{N} \times \mathbb{B} \to \mathbb{B}$ *given by* $\beta(n, r) = 1/(1 + n + r)$.

Proof (1) We first check that $\mathbb{A}$ is a fixed point of $\mathbb{N} \times X$. By Lemma 15.5.2, $\varrho\colon \mathbb{N} \times \mathbb{R}_{\geq 0} \to \mathbb{R}_{\geq 0}$ is injective. Hence so is its restriction to $\mathbb{N} \times \mathbb{A}$. Moreover, every irrational number x is $\varrho(n, r)$ for $n = \lfloor x \rfloor$ and $r = 1/(x - n) - 1$; indeed, observe that r is a positive irrational and that $\varrho(n, r) = x$. Thus, ϱ restricts to a bijection $\alpha\colon \mathbb{N} \times \mathbb{A} \to \mathbb{A}$. Further, the inclusion $i\colon \mathbb{A} \hookrightarrow \mathbb{R}_{\geq 0}$ is an algebra homomorphism $i\colon (\mathbb{A}, \alpha) \to (\mathbb{R}_{\geq 0}, \varrho)$. Lemma 15.5.7 tells us that every solution morphism takes irrational values. By Theorem 15.5.5 and Lemma 7.2.7, $(\mathbb{A}, \alpha^{-1})$ is a terminal coalgebra.

(2) Straightforward calculations show that for the bijective map $f\colon \mathbb{A} \to \mathbb{B}$ given by $r \mapsto 1/(1 + r)$ the square below commutes:

$$\begin{array}{ccc} \mathbb{N} \times \mathbb{A} & \xrightarrow{\alpha} & \mathbb{A} \\ {\scriptstyle \mathbb{N}\times f}\downarrow & & \downarrow{\scriptstyle f} \\ \mathbb{N} \times \mathbb{B} & \xrightarrow{\beta} & \mathbb{B} \end{array}$$

Thus, f is an isomorphism of coalgebras, and therefore $(\mathbb{B}, \beta)$ is a terminal coalgebra. □

Representations of irrational numbers by continued fractions Although it is not needed for our results in this section, we do wish to mention the main connection of our work to the theory of continued fractions. We denote by $\mathbb{A}$ the set of all positive irrational numbers, and by $\mathbb{A}_{\geq 1}$ the set of all irrationals ≥ 1.

Proposition 15.5.9. *For every irrational number $x > 0$, there is a unique infinite sequence of natural numbers $a_0, a_1, \ldots, a_k, \ldots$ such that*

$$x = \lim_{k \to \infty} [a_0, a_1, \ldots, a_k].$$

Proof Let $\beta \colon \mathbb{A}_{\geq 1} \to \mathbb{N}_{\geq 1} \times \mathbb{A}_{\geq 1}$ be given by $\beta(x) = (\lfloor x \rfloor, 1/(x - \lfloor x \rfloor))$. The terminal coalgebra for $HX = \mathbb{N}_{\geq 1} \times X$ is carried by the set $(\mathbb{N}_{\geq 1})^\omega$ of infinite sequences of positive natural numbers (Example 2.5.3(3)). Consider the unique coalgebra homomorphism $f \colon \mathbb{A}_{\geq 1} \to (\mathbb{N}_{\geq 1})^\omega$. For $x \in \mathbb{A}_{\geq 1}$, $f(x)$ is the *canonical continued fraction representation of* x. Write $f(x)$ as $(a_0, a_1, \ldots, a_k, \ldots)$. By Loya [216, Thm. 8.14], $\lim_{k \to \infty} [a_0, a_1, \ldots, a_k] = x$. The uniqueness of the canonical continued fraction representation follows from Hardy and Wright [155, Thm. 170], or Niven et al. [244, Thm. 7.10]. □

The next result provides a corecursive algebra structure on the unit interval $\mathbb{I}$. The proof uses a different technique than what we have seen in this section, linear fractional transformations. It is too long to include here, but we do draw a corollary.

Theorem 15.5.10 [240, Thm. 5.9]. *The algebra $\mathbb{I}$ for $\mathbb{N} \times X$ with structure $\tau(n, r) = (n + r)/(1 + n + r)$ is corecursive.*

Corollary 15.5.11 [247]. *The functor $\mathbb{N} \times X$ has the following terminal coalgebras:*

(1) *The half-open interval $[0, 1)$ with structure inverse to $\sigma(n, r) = (n+r)/(1+n+r)$.*

(2) *The set $\mathbb{R}_{\geq 0}$ with structure inverse to $\delta(n, r) = n + \frac{r}{1+r}$.*

Proof (1) Observe that τ does not take the value 1. Hence the unique solution $e^\dagger \colon X \to \mathbb{I}$ of every coalgebra $e \colon X \to \mathbb{N} \times X$ also does not take that value. We conclude that τ has a restriction to the bijection $\sigma \colon \mathbb{N} \times [0, 1) \to [0, 1)$, and every solution restricts to a function $X \to [0, 1)$. This restriction is clearly the unique solution of e in $([0, 1), \sigma^{-1})$. Thus, the latter coalgebra is corecursive by Lemma 7.2.7.

(2) It is easy to verify that we have the following isomorphism of algebras

$$h \colon ([0, 1), \sigma) \to (\mathbb{R}_{\geq 0}, \delta) \quad x \mapsto x/(1 + x).$$

Thus h is also an isomorphism between the two coalgebras above. □

coalgebra	coalgebra structure $x \mapsto \cdots$
$(\mathbb{A}, \alpha^{-1})$	$(\lfloor x \rfloor, \frac{1}{x \bmod 1} - 1)$
$(\mathbb{B}, \beta^{-1})$	$(\lfloor \frac{1}{x} \rfloor - 1, \frac{1}{x} \bmod 1)$
$(\mathbb{R}_{\geq 0}, \delta^{-1})$	$(\lfloor x \rfloor, \frac{x \bmod 1}{1-x \bmod 1})$
$([0,1), \sigma^{-1})$	$(\lfloor \frac{x}{1-x} \rfloor, \frac{x}{1-x} \bmod 1)$

Figure 15.2 Terminal coalgebra structures for $\mathbb{N} \times X$.

Remark 15.5.12. The terminal coalgebra structures in this section may be given explicitly [240]. For the functor $\mathbb{N} \times X$, we exhibit the structures in Figure 15.2. For $\mathbb{N} \times X + 1$, they are given as follows:

(1) The map $[\varrho, 0]^{-1} \colon \mathbb{R}_{\geq 0} \to \mathbb{N} \times \mathbb{R}_{\geq 0} + 1$ is defined by:

$$[\varrho, 0]^{-1}(x) = \begin{cases} * \in 1 & x = 0, \\ (x-1, 0) & x \text{ a positive natural number,} \\ (\lfloor x \rfloor, \frac{1}{x \bmod 1} - 1) & \text{otherwise.} \end{cases}$$

(2) The map $[\tau, 1]^{-1} \colon (0, 1] \to \mathbb{N} \times [0, 1) + 1$ is defined by

$$[\tau, 1]^{-1}(x) = \begin{cases} * \in 1 & x = 1, \\ (\frac{1}{x} - 2, 1) & \text{if } \frac{1}{x} \text{ is a natural number other than 1,} \\ (\lfloor \frac{1}{x} \rfloor - 1, \frac{1}{x} \bmod 1) & \text{otherwise.} \end{cases}$$

The initial algebra of $\mathbb{N} \times X + 1$ and the rational fixed point of $\mathbb{N} \times X$ Our work in this section has called upon results concerning continued fractions. By appealing to such results, one can go further and characterize the initial algebra for $\mathbb{N} \times X + 1$ and the rational fixed point of $\mathbb{N} \times X$. The initial algebra is the set of rational numbers, as a subalgebra of the terminal $(\mathbb{N} \times X + 1)$-coalgebra $(\mathbb{R}_{\geq 0}, [\varrho, 0])$. The rational fixed point of $\mathbb{N} \times X$ turns out to be the subalgebra of the terminal coalgebra $(\mathbb{A}, \alpha^{-1})$ consisting of all quadratic irrationals; these are the real roots of irreducible polynomials with rational coefficients. This follows from a classical result due to Lagrange (see e.g. [190, Thm. 28]) that the eventually periodic streams of natural numbers correspond uniquely to irrational numbers in $[0, 1]$ which are roots of quadratic polynomials with integer coefficients.

15.6 Summary

We began this chapter with some negative results: similarly to the power-set functor, the endofunctor on Met mapping a space to the set of all its closed subspaces does not have a fixed point, and neither do the endofunctors on Pos mapping a poset to its down-closed subsets or its ideals, respectively.

We moved on to investigate the Vietoris endofunctor on the category of topological spaces, which assigns to every space the space of all its compact subsets. This functor as well as the Vietoris polynomial functors with constant functors restricted to Hausdorff spaces, has an initial algebra and a terminal coalgebra. Moreover, the terminal-coalgebra chain for these functors converges in ω steps due to them preserving limits of ω^{op}-chains.

Analogously we investigated the Hausdorff functor on Met and CMet. This functor as well as the corresponding Hausdorff polynomial functors also have an initial algebra and a terminal coalgebra. Here the terminal-coalgebra chain converges in $\omega + \omega$ steps.

Finally, we have shown various structures on the unit interval $[0, 1]$ making it an initial algebra or terminal coalgebra of an endofunctor on certain categories obtained from Set, Met or CMet by equipping objects with a 'bi-pointing'. Moreover, various subsets of the real line we are described as a terminal coalgebra or a corecursive algebra for the set functors $X \mapsto \mathbb{N} \times X$ and $X \mapsto \mathbb{N} \times X + 1$.

Appendix A
Functors with Initial Algebras or Terminal Coalgebras

This short appendix is meant as a quick roadmap to some results which we have seen about the existence of μF and νF for various endofunctors. We are also interested in fixed points here, especially canonical fixed points. Recall that a *canonical fixed point* is an initial algebra $\alpha\colon FA \to A$ whose inverse $\alpha^{-1}\colon A \to FA$ is a terminal coalgebra (shortly: $A = \mu F = \nu F$).

Locally Presentable Categories. The great majority of the categories mentioned explicitly in this book are locally presentable (Definition 11.2.8). For example, Set, Pos, CPO, Met, PMet, CMet, K-Vec, and Nom. Other examples include all varieties of algebras whose operations are of a bounded (possibly infinite) arity, all presheaf categories, and all complete lattices (as preorder categories). Non-examples which we have seen include Top, Haus, KHaus, DCPO, and BiP.

Similarly, most of the endofunctors on the above locally presentable categories mentioned explicitly in this book are accessible (Definition 11.2.13). Whether the terminal-coalgebra chain converges for *all* accessible functors is an open problem.

We have seen in Theorem 11.2.18 that accessible endofunctors on locally presentable categories have both an initial algebra μF and a terminal coalgebra νF. Moreover, the initial-algebra chain converges. When F preserves monomorphisms, the terminal-coalgebra chain also converges (Theorem 11.2.22). Whether the terminal-coalgebra chain converges for *all* accessible functors is an open problem.

The *preaccessible* functors (Definition 11.4.5) are a wider class. For preaccessible endofunctors, μF exists and the initial-algebra chain converges, provided the base category $\mathscr{A}$ is cocomplete and co-well-powered (Theorem 11.4.11). Whenever F additionally preserves monomorphisms and $\mathscr{A}$ is locally presentable, νF exists and the terminal-coalgebra chain converges (Theorem 11.4.15).

The Category of Sets. We discussed equivalent formulations of accessibility for set functors in Section C.8. Examples may be found in Example C.9.4. The collection of accessible set functors is closed under products and coproducts, composition, subfunctors, and quotient functors. It contains all Kripke polynomial functors (Definition 4.4.13). The main non-example is the power-set functor $\mathscr{P}$, but each of its bounded versions $\mathscr{P}_\lambda$ is accessible.

Set functors that are not accessible seldom have a terminal coalgebra; for an (artificial) example, see Example 6.5.24.

The existence of μF can be characterized: it is equivalent to F having a fixed point (Corollary 6.3.9). For νF there is no such full characterization. But a fixed-point pair implies that νF exists, assuming the Generalized Continuum Hypothesis (Theorem 6.5.18).

Strict CPO-enriched categories. We turn to results on enriched categories where hom-sets carry an additional order. First, consider strict CPO-enriched categories (Definition 5.1.21) and *locally continuous endofunctors* (Definition 5.1.16): they preserve joins in hom-cpos. This class is closed under products, coproducts, and compositions (Example 5.1.17). If $\mathscr{A}$ has ω-colimits or ω^{op}-limits and F is locally continuous, then F has a canonical fixed point $\mu F = \nu F$ (Theorem 5.1.24).

We continue with *stable functors* on strict CPO-enriched categories. They have the property that for all morphisms $f\colon A \to A$, $\mathsf{id}_A \sqsubseteq f$ implies $\mathsf{id}_{FA} \sqsubseteq Ff$ (Definition 6.7.1). Equivalently, F preserves embedding-projection pairs (Observation 6.7.5), provided that $\mathscr{A}$ has split idempotents. A stable functor has a canonical fixed point whenever it has any fixed point (Corollary 6.7.8).

Dropping part of the strictness assumption, for a locally continuous endofunctor F on a *left-strict* CPO-enriched categories (Definition 5.1.30) we have seen that μF, if it exists, is a canonical fixed point (Proposition 5.1.33).

Strict CMet-enriched categories. We turn to results on enriched categories where hom-sets carry an additional metric structure. Strict CMet-enriched categories were defined in Definition 5.2.2. An endofunctor F is *locally contracting* if for some $\varepsilon < 1$, F shrinks distances on hom-spaces by a factor of ε (Definition 5.2.5). Every such functor on a strict CMet-enriched category with ω-colimits has a canonical fixed point (Theorem 5.2.17). The class of such functors is closed under finite products, all coproducts, and composition (Example 5.2.6).

As a corollary one obtains for the category CMet itself, that every locally contracting endofunctor F such that $F\emptyset \neq \emptyset$ has a canonical fixed point (Corollary 5.2.18). Furthermore, dropping the strictness assumption, consider any CMet-enriched category $\mathscr{A}$ with ω-colimits. Then every locally contractive endofunctor $F\colon \mathscr{A} \to \mathscr{A}$ with $\mathscr{A}(1, F1)$ nonempty has a terminal coalgebra (Theorem 5.2.20).

Hausdorff Polynomial Functors. This is the class of endofunctors on Met and CMet obtained from the Hausdorff functor, the identity functor, and constant functors using products, coproducts, and composition (Definition 15.3.7). As endofunctors on both Met and CMet, functors in this class have terminal coalgebras (Theorem 15.3.10 and Remark 15.3.11) and initial algebras (Corollary 15.3.13).

Vietoris Polynomial Functors. Here we work with the category Top of topological spaces and all continuous maps and consider the class of endofunctors obtained from the Vietoris functor, the identity functor, and constant functors with Hausdorff values using products, coproducts, and composition (Definition 15.2.10). As endofunctors on both Top and Haus, functors in this class have terminal coalgebras (Theorem 15.2.12) and initial algebras (Proposition 15.2.15).

Binding Polynomial Functors on Nom. Here one considers the locally finitely presentable category Nom of nominal sets and equivariant maps (Example 2.1.10) and the class of endofunctors obtained from the constant, identity, and abstraction functors using coproducts, finite products, and composition (Remark 12.3.28(1)). These functors are finitary and so have initial algebras and terminal coalgebras by Theorem 11.2.18.

Special instances are the functors F_Σ associated to a binding signature Σ in the sense of Fiore et al. [124] (Remark 2.1.11). We have mentioned in Example 2.2.18(1) that the initial algebra for F_Σ is carried by the nominal set of α-equivalence classes of terms over Σ. Moreover, in Example 2.5.14 we stated that for a countable binding signature Σ, the terminal coalgebra for F_Σ is carried by the nominal set of α-equivalence classes of all finite and infinite trees over Σ with finitely many free variables (see Kurz et al. [209, Cor. 5.35]).

Binding Polynomial Functors on $\mathsf{Set}^{\mathscr{F}}$. Similar results hold for the presheaf category $\mathsf{Set}^{\mathscr{F}}$ having functors $\mathscr{F} \to \mathsf{Set}$ as objects, where $\mathscr{F}$ is the category of all finite sets and maps between them (Example 2.5.15). Here one considers the class of endofunctors obtained from the constant, identity, and functor δ, using coproducts, finite products, and composition (recall from Example 2.1.8 that the presheaf $\delta(X)\colon \mathscr{F} \to \mathsf{Set}$ is defined by $\delta(X)(\Gamma) = \Gamma + 1$). These are finitary functors, and so they have initial algebras and terminal coalgebras by Theorem 11.2.18.

Special instances are the functors F_Σ associated to a binding signature Σ (see Remark 2.1.9). We have mentioned in Example 2.2.18(4) that the initial algebra μF_Σ assigns to a context Γ the α-equivalence classes of terms over Σ with variables from Γ (see Fiore et al. [124, Thm. 2.1]). Similarly, we stated in Example 2.5.15 that the terminal coalgebra νF_Σ assigns to Γ the set of α-equivalence classes of finite and infinite trees over Σ; this is a straightforward extension of results on λ-Σ-terms (for an ordinary signature Σ) [44, Thms. 3.11 & 4.18].

Appendix B
A Primer on Fixed Points in Ordered and Metric Structures

In this chapter we provide an overview of various fixed point theorems for functions on posets and metric spaces. On the one hand, we collect these results here to make use of them in the book. On the other hand, one of the main topics of this book is to provide generalizations of some of the following results and the underlying constructions to the level of categories and functors. In addition, there seem to be very few comprehensive accounts on fixed point theorems in the literature.

We begin with a review of classical fixed point theorems for functions on posets in Section B.1, in particular two famous results: Kleene's theorem (Theorem B.1.2) and the Knaster–Tarski Theorem (Theorem B.1.7). These results work with continuous and monotone functions, respectively. But we also mention less well-known results for inflationary functions. In Section B.2 we present Pataraia's theorem that every monotone function on a directed-complete poset has a least fixed point. Finally, we recall Banach's fixed point theorem for contracting functions on complete metric spaces in Section B.3.

Some of the results have been stated and even proved in previous chapters. We repeat them here and we also repeat the basic definitions necessary to understand them in order to make this chapter completely self-contained. It can be read without any prior knowledge of the material in the book.

B.1 Fixed Points in Posets

We begin by recalling Kleene's classical fixed point theorem. A or *cpo*[1] for short, is a poset $(P, \sqsubseteq)$ in which all ω-chains $a_0 \sqsubseteq a_1 \sqsubseteq a_2 \sqsubseteq \cdots$ have joins. A

[1] This is often called an ω-complete partial order, in contrast to dcpos in Example 5.1.2(3).

continuous function from a cpo P to a cpo Q is a monotone function $f : P \to Q$ preserving joins of ω-chains.

Example B.1.1. (1) Every set X can be regarded as a cpo with the discrete order. We can also add a least element to obtain $X_\perp = X + \{\perp\}$ with the order given by $x \leq y$ iff $x = y$ or $x = \perp$.

(2) For every set X, the power-set $\mathscr{P}X$ ordered by subset inclusion is a cpo with the least element $\emptyset$.

(3) Given two sets X and Y, the set $\mathsf{Pfn}(X, Y)$ of partial functions from X to Y is a cpo with the following order: $f \sqsubseteq g$ if g extends f which means that for every $x \in X$, $g(x)$ is defined whenever $f(x)$ is, and in this case $f(x) = g(x)$. The least element in this cpo is the nowhere defined function from X and Y.

Theorem B.1.2 (Kleene). *Let P be a cpo with a least element $\perp$. Then every continuous endofunction $f : P \to P$ has a least fixed point*

$$\mu f = \bigvee_{n<\omega} f^n(\perp).$$

Proof First, an induction on $n < \omega$ shows that $f^n(\perp) \leq f^{n+1}(\perp)$. So $\{f^n(\perp) : n < \omega\}$ is an ω-chain. Write μf for its join. By continuity, $f(\mu f) = \bigvee_n f(f^n(\perp))$. It is easy to check that $\bigvee_n f^n(\perp) = \bigvee_n f^{n+1}(\perp)$, and so $f(\mu f) = \mu f$. Thus, we have a fixed point of f. If x is another fixed point (in fact, $f(x) \leq x$ suffices), then an easy induction on n shows that $f^n(\perp) \leq x$; hence $\mu f \leq x$. □

As a consequence we obtain the following induction principle:

Corollary B.1.3. *Let P be a cpo with a least element $\perp$, and let S be a subset which contains $\perp$ and is closed under joins of ω-chains. If a continuous function $f : P \to P$ perserves S (in symbols: $f[S] \subseteq S$), then $\mu f \in S$.*

Indeed, the set S then forms a subcpo of P and one applies Theorem B.1.2 to it and the restriction of f to it.

Remark B.1.4. More generally, a *pre-fixed point* of f is an element x with $f(x) \leq x$. In every cpo with $\perp$ an element p is a least pre-fixed point iff it is a least fixed point. Indeed, for 'if' observe that at the end of the proof of Theorem B.1.2 it sufficed to have $f(x) \leq x$. For 'only if' note that a least pre-fixed point is a fixed point: given the least fixed point p we have $f(f(p)) \leq f(p)$ by monotonicity of f. Therefore $p \leq f(p)$, whence $f(p) = p$.

Dually, a *post-fixed point* of f is an element x with $x \leq f(x)$.

In another classical fixed point theorem, the Knaster–Tarski Theorem, one may drop the requirement that the function f be continuous. As a tradeoff, one assumes that the poset P is a *complete lattice*: every subset of P has a join.

Remark B.1.5. In then follows that P also has all meets. In fact, the meet of a subset $S \subseteq P$ is given by the join of all lower bounds of S:

$$\bigvee\{p : p \leq s \text{ for every } s \in S\}.$$

In particular, P has a least element $\perp$ and a greastest element $\top$.

Example B.1.6. (1) The power-set $\mathscr{P}X$ of every set X is a complete lattice. Moreover, subposets of $\mathscr{P}X$ often form complete lattices; this is the case whenever the subposet is closed under unions or intersections, for example, the open subsets of a topological space, the subgroups of a group, the ideals of a ring, or the submodules of a module, or the convex sets of a vector space.

(2) A *lattice* is a poset having joins and meets of all pairs of elements. Every finite nonempty lattice is clearly a complete lattice. However, in general a lattice need not be complete. For example, the poset $\mathbb{N} \times \{+, -\}$ in which all $(n, +)$ form an increasing chain, all $(n, -)$ a decreasing one, and $(n, +) < (m, -)$ for all $n, m \in \mathbb{N}$ is a lattice which is not complete.

(3) The unit interval $[0, 1]$ and the extended real numbers $\mathbb{R}^\infty = \mathbb{R} + \{-\infty, \infty\}$ with the usual order form complete lattices.

(4) The the set of natural numbers ordered by divisibility form a complete lattice. The least element is 1 and the greatest element is 0. For a finite subset, the join is the least common multiple, and for infinite subsets the join is 0.

(5) All transitive (or equivalence) relations on set ordered by inclusion form a complete lattice.

(6) The flat cpos in Example B.1.1(1) and the cpo of partial functions in Example B.1.1(3) are usually not even lattices.

In its most general form stated below, the following theorem is due to Tarski [284], whereas Knaster [194] previously proved it for power-set lattices.

Theorem B.1.7 (Knaster and Tarski). *Let $f\colon P \to P$ be a monotone function on a complete lattice P. Then the subposet of all fixed points of f is also a complete lattice.*

In particular, f has a least and a greatest fixed point.

Proof (1) We first prove that the least fixed point of f is the meet of all pre-fixed points and, dually, the greatest fixed point is the join of all post-fixed points:

$$\mu f = \bigwedge\{x \in P : f(x) \leq x\} \qquad \text{and} \qquad \nu f = \bigvee\{x \in P : x \leq f(x)\}.$$

Indeed, define μf as the above meet. Then by monotonicity of f we have $f(\mu f) \leq f(x) \leq x$ for every pre-fixed point x of f. Therefore $f(\mu f) \leq \mu f$, which makes μf a pre-fixed point of f, and indeed the least such. Hence, this the least fixed point of f by Remark B.1.4. Dually, νf above is the greatest fixed point of f.

(2) Let $Q = \{x \in P : x = f(x)\}$. This is a poset with the order inherited from P, and we show that Q has joins of all subsets. (Note that these joins need not coincide with those of P.) Therefore Q is a complete lattice.

Given $D \subseteq Q$, let $w = \bigvee D$ be its join in P. Then we have $x = f(x) \leq f(w)$ for every $x \in D$, since f is monotone. So $f(w)$ is an upper bound of D. Therefore $w = \bigvee D \leq f(w)$. Let $W = \{y \in P : w \leq y\}$. By monotonicity, f restricts to $g \colon W \to W$. Moreover, W with the inherited order is clearly a complete lattice, and g is monotone. By item (1), μg exists; moreover, it clearly lies in Q. We claim that μg is the join of D in Q. It is an upper bound of D since it belongs to W. It is the least one because any upper bound d of D belonging to Q is also a fixed point of g, so $\mu g \leq d$. Therefore $p = \bigvee D$ in Q. □

Example B.1.8. The corresponding statement for bounded lattices does not hold in general. For example, consider the lattice $\mathbb{N} \times \{+, -\}$ (Example B.1.6(2)) and the monotone endofunction f defined by $f(n, i) = (n + 1, i)$ for $i \in \{+, -\}$. Then f has no fixed point.

Along with this result, we have an induction principle that is useful for establishing properties of μf. (By duality, there is a similar principle for νf.)

Proposition B.1.9. *In the setting of Theorem B.1.7, let $S \subseteq P$ be a set which is closed under joins and under f, that is, $f[S] \subseteq S$. Then $\mu f \in S$.*

Proof Let $T = \{x \in S : x \leq \mu f\}$, and put $t = \bigvee T$. Then $t \in S$, since S is closed under joins, so $t \leq \mu f$. Moreover, $f(t) \in S$ since S is closed under f. By monotonicity, $f(t) \leq f(\mu f) = \mu f$. Thus $f(t) \in T$. Hence $f(t) \leq \bigvee T = t$. Therefore t is a pre-fixed point of f, and so $\mu f \leq t$ by Remark B.1.4. □

We now turn to results concerning least fixed points that cover the middle ground between Kleene's theorem and the Knaster–Tarski theorem. We still just assume monotonicity of the function but only need weaker completeness assumptions on P than in the Knaster–Tarski theorem, yet stronger ones than in the Kleene theorem. In particular, one assumes joins of chains that are longer than ω. These results make use of ordinals and transfinite induction. Readers not so familiar with those might want to consider the discussion in Chapter 6.

An *i-chain* in a poset P, where i is an ordinal, is a family of elements $x_j \in P$ indexed by all ordinals $j < i$ such that $p_j \leq p_k$ for all $j \leq k < i$.

Definition B.1.10. A poset P is *chain-complete* if every i-chain in P has a join (cf. Remark 6.0.5(2)). In particular, P has a least element $\bot$, namely the join of the empty chain.

Example B.1.11. (1) Every complete lattice is a chain-complete poset, of course.

(2) The flat cpos and the cpo of partial functions (Example B.1.6(6)) are chain-complete posets.

Let $f: P \to P$ be a monotone map on the chain-complete poset P. Then we can define an ordinal-indexed sequence $f^j(\bot)$ as follows:

$$f^0(\bot) = \bot, f^{j+1}(\bot) = f(f^j(\bot)), \text{ and}$$
$$f^j(\bot) = \bigvee_{i<j} f^i(\bot) \text{ for limit ordinals } j. \tag{B.1}$$

It is easy to verify that this is a chain in P; that is, $f^i(\bot) \le f^j(\bot)$ if $i \le j$.

Theorem B.1.12 (Zermelo)**.** *Let P be a chain-complete poset. Every monotone function $f: P \to P$ has a least fixed point μf. Moreover, for some ordinal j,*

$$\mu f = f^j(\bot).$$

Proof Take i to be an ordinal larger than $|P|$. For this i, there must be some $j < i$ such that $f^j(\bot) = f^{j+1}(\bot)$. Indeed, this follows from Hartogs' Lemma, stating that for every set P there exists an ordinal i such that there is no injection $i \rightarrowtail P$. Thus, there are $j < k < i$ such that $f^j(\bot) = f^k(\bot)$. Since $f^j(\bot) \le f^{j+1}(\bot) \le f^k(\bot)$ we have $f^j(\bot) = f^{j+1}(\bot)$. Hence $f^j(\bot)$ is a fixed point of f. Let $f(x) = x$. An easy transfinite induction shows that $x \ge f^i(\bot)$ for all i. This shows that $f^j(\bot)$ is the least fixed point of f. □

We attribute this theorem to Zermelo, since its mathematical content appears in his 1904 paper [319] proving the Wellordering Theorem.

There are also variations on Theorem B.1.12, such as the next result often called the Bourbaki–Witt Theorem [93, 315].

Definition B.1.13. Let P be a poset. A map $f: P \to P$ is *inflationary* if $x \le f(x)$ holds for every $x \in P$.

Note that an inflationary map need not be monotone.

Theorem B.1.14. *Every inflationary map on a chain-complete poset has a fixed point above every point.*

Proof Let P be a chain-complete poset, and let $f: P \to P$ be inflationary. Given $x_0 \in P$, we define an ordinal-indexed chain by transfinite recursion in almost the same way as in (B.1). The base case is the given point x_0. In the isolated step we put $x_{j+1} = f(x_j)$ so that $x_j \leq x_{j+1}$ since f is inflationary. For a limit ordinal j, we put $x_j = \bigvee_{i<j} x_i$. By Hartogs' Lemma, there exists an ordinal i such that there is no injection from i to the set P. Thus i is larger than the cardinality of P. Hence there exist ordinals $j < k < i$ such that $x_j = x_k$. It follows that $x_j = x_{j+1} = f(x_j)$ is the desired fixed point of f. □

Example B.1.15. Let $\mathbb{N}^\top$ be the poset of natural numbers with an extra element added on top, in the usual order. Let $P = \mathbb{N}^\top \oplus \mathbb{N}^{\mathrm{op}}$ be a copy of $\mathbb{N}^\top$ with a copy of $\mathbb{N}^{\mathrm{op}}$ on top of it. Every chain in P has an upper bound, and P also has a least element. Let $f(n) = n + 1$ in the copy of $\mathbb{N}^\top$, and $f(n) = n$ in the copy of $\mathbb{N}^{\mathrm{op}}$. Then f has a fixed point by the Bourbaki–Witt theorem, but it does not have a least fixed point. In fact, we see that, while f is monotone, P is not chain-complete, so we cannot apply Theorem B.1.12.

Remark B.1.16. In all of the results in this section, it is critical that P be a set. For consider the successor function on the ordered class Ord of ordinals. It has no fixed point.

B.2 Pataraia's Theorem

The proof of Theorem B.1.12 makes crucial use of ordinals and transfinite induction. In this section we present an alternative result which is ordinal-free and indeed constructive. It makes use of directed-complete posets in lieu of chain-complete ones.

A *directed poset* is a poset P in which every finite subset has an upper bound; in particular, P is nonempty, having an upper bound of the empty set. A poset P is called *directed-complete*, or a *dcpo* for short, if every directed subset of P has a join. Such joins are called *directed joins*. A dcpo with a least element if called a *dcpo with bottom*.

Being directed-complete may seem like a stronger condition than chain-completeness. However, this is illusory, as shown by Markowsky [222]. The proof of his theorem uses Iwamura's Lemma:

Lemma B.2.1 (Iwamura [169]). *Every infinite directed poset D is the union of a chain of directed subposets of strictly smaller cardinalities.*

In more detail, for $\lambda = |D|$, there exist directed sets D_i $(i < \lambda)$ with $|D_i| < \lambda$ and such that

$$D = \bigcup_{i<\lambda} D_i, \quad D_i \subseteq D_j \text{ for } i \leq j, \quad D_i = \bigcup_{j<i} D_j \text{ for every limit ordinal } i < \lambda.$$

Proof We express D as $\{d_i : i < \lambda\}$. We first define a function $S \mapsto S^*$ on the power-set of D. If S is finite, choose an upper bound $s \in D$ and put $S^* = S \cup \{s\}$. If S is infinite, first define sets S_n by $S_0 = S$ and S_{n+1} is obtained from S_n by adding an upper bound for each pair of elements from S_n; then we have $S^* = \bigcup_{n<\omega} S_n$. The desired directed sets D_i are now given by transfinite recursion as follows:

$$D_0 = \emptyset, \quad D_{i+1} = (D_i \cup \{d_i\})^*, \quad D_i = \bigcup_{j<i} D_j \text{ for every limit ordinal } i < \lambda.$$

For finite i, the set D_i is finite. An easy induction shows that for infinite i, we have $|D_i| = i$. □

In the following proof we use well-founded induction on a cardinal in the following way: a statement S_λ depending on some cardinal number λ is true whenever it holds for all finite cardinals λ, and S_μ for all $\mu < \lambda$ implies S_λ.

Theorem B.2.2 (Markowsky [222]). *A poset is a dcpo with bottom iff it is chain-complete.*

Proof The 'only if' direction is clear since nonempty chains are special directed posets. For 'if' suppose that P is chain-complete, and let $D \subseteq P$ be a directed set. We prove that D has a join in P. We proceed by well-founded induction on the cardinality $\lambda = |D|$.

If D is finite, then D has an upper bound in itself by directedness, namely a greatest element and therefore the join of D. In the induction step we suppose that D is infinite. Then we apply Lemma B.2.1 and obtain a chain (D_i) of directed subsets of D whose union is D and which satisfy $|D_i| < \lambda$ for every i. By the induction hypothesis each D_i has a join x_i, say. Since $D_i \subseteq D_j$ implies $x_i \le x_j$ for every $i \le j$, we obtain a chain (x_i) in P. Its join x is then the join of D. For x is clearly an upper bound of D, and if $d \le y$ for every $d \in D$, we clearly have $d \le y$ for every D_i and every i. Thus $x_i \le y$ for every i, and therefore we have $x \le y$. □

Remark B.2.3. In the proof of Theorem B.2.2 one may be tempted to avoid Iwamura's lemma constructing a cofinal chain within the given directed set D. However, D need not have one, see e.g. [146, Ex. 4.2.25]: Let X be an uncountable set, say with $|X| = \kappa$. Consider the complete lattice $\mathscr{P}(X)$, and let $D = \mathscr{P}_f(X)$ be the directed subset of all finite subsets of X. Its join is X. However, X is not the union of any chain of elements of D. For suppose that $(C_i)_{i<\kappa}$ is a chain in D such that $X = \bigcup_{i<\kappa} C_i$. Moreover, let $Y = \{i < \kappa : C_i \setminus \bigcup_{j<i} C_j \neq \emptyset\}$. If Y were bounded, then $\bigcup C_i$ would be finite, contradicting the uncountability of X.

Otherwise, let i be the ωth element of Y. Then C_i is infinite, and this contradicts $C_i \in D$.

It was Pataraia [246] who presented an ordinal-free version of Theorem B.1.12 using dcpos. Sadly, he never published his result in written form; it has appeared for example, in work by Escardó [118], Goubault-Larrecq [147], Bauer and Lumsdane [75] based on a preprint by Dacar [109], and Taylor [288]. We present a proof based on Martin's presentation [224].

Remark B.2.4. (1) Observe that all (not necessarily monotone) maps on a poset P form a poset with the pointwise order: $f \leq g$ if for every $x \in P$ we have $f(x) \leq g(x)$. Hence, a map $f: P \to P$ is inflationary iff $\mathsf{id}_P \leq f$.

(2) Function composition is *left-monotone*: we clearly have $f \cdot h \leq g \cdot h$ whenever $f \leq g$. Right-monotonicity requires that the fixed argument f be a monotone map: we then have $f \cdot g \leq f \cdot h$ for every $g \leq h$.

Definition B.2.5. (1) An element z of a monoid $(M, \cdot, 1)$ is a *zero* if $z \cdot m = z = m \cdot z$ for every $m \in M$.

(2) A monoid $(M, \cdot, 1)$ is *partially ordered* if M carries a partial order such that the multiplication is monotone: $a \leq b$ and $a' \leq b'$ implies $a \cdot a' \leq b \cdot b'$. It is *directed-complete* if its underlying poset is a dcpo.

Proposition B.2.6 [224, Thm. 1]**.** *Every directed complete monoid $(M, \cdot, 1)$ whose unit 1 is the least element has a top element which is a zero.*

Proof The set M itself is directed: for $m, n \in M$ we see that $m \cdot n$ is an upper bound since

$$m = m \cdot 1 \leq m \cdot n \geq n \cdot 1 = n,$$

using that 1 is the bottom and multiplication is monotone. Thus M has a top element $\top = \bigvee M$. We have $\top = \top \cdot 1 \leq \top \cdot m$ for every $m \in M$, and clearly $\top \cdot m \leq \top$. Thus, $\top \cdot m = \top$. Similarly $m \cdot \top = \top$, whence $\top$ is a zero. □

Theorem B.2.7 (Pataraia's Theorem)**.** *Let P be a dcpo with bottom. Then every monotone map on P has a least fixed point.*

Proof (1) Let M the set of all monotone inflationary maps on P. This is a monoid under function composition with the unit id_P. Furthermore, M is a dcpo with bottom under the pointwise order from Remark B.2.4: the least element is id_P, and directed joins are computed pointwise in P. Function composition is monotone (in both arguments) by Remark B.2.4(2). By Proposition B.2.6, M therefore has a top element $t: P \to P$ which is a zero.

(2) Let $f: P \to P$ be inflationary and monotone. Then $f \in M$ and therefore

$f \cdot t = t$. This means that for every $x \in P$, $f(t(x)) = t(x)$, whence $t(x)$ is a fixed point of f.

(3) Now let $f: P \to P$ be just monotone. Let $\mathcal{M}$ be the collection of all subsets S of P which contain $\bot$, and which are closed under f as well as under directed joins. (In more detail, we require that if $s \in S$, then $f(s) \in S$; and if $X \subseteq S$ is directed, $\bigvee X \in S$.) The set $\mathcal{M}$ is closed under arbitrary intersections. Let $T = \bigcap \mathcal{M}$ be the minimal element of $\mathcal{M}$.

The set of all post-fixed points $x \leq f(x)$ belongs to $\mathcal{M}$. Indeed, $\bot \leq f(\bot)$, and $f(x) \leq f(f(x))$ whenever $x \leq f(x)$. Moreover, for every directed set D of post-fixed points of f, the join $d = \bigvee D$ is a post-fixed point: d satisfies $x \leq f(x) \leq f(d)$ for every $x \in D$ due to the monotonicity of f; thus, $d \leq f(d)$. By the minimality of T, we deduce that T consists of post-fixed points of f. Thus, f restricts to a function on T, which is inflationary (and monotone, of course) and therefore has a fixed point p by item (2).

We show that p is a least fixed point of $f: P \to P$. Suppose that x is any fixed point. The set $L = \{y \in P : y \leq x\}$ belongs to $\mathcal{M}$. Therefore $T \subseteq L$, which implies $p \leq x$. □

Remark B.2.8. As in Theorem B.1.7, the fixed points of f form a sub-dcpo of P. The argument is almost identical to the last part of the proof of Theorem B.1.7. Indeed, in item (2) of the proof take a directed set D and observe that the set W of all upper bounds of D is a dcpo.

The following principle also appears in work by Escardó [118, Thm. 2.2] and Taylor [288].

Corollary B.2.9 (Pataraia Induction Principle). *Let P be a dcpo with bottom. If $f: P \to P$ is monotone, then μf belongs to every subset $S \subseteq P$ which contains $\bot$ and is closed under f and under directed joins.*

This follows from the proof of Theorem B.2.7: item (3) shows that $\mu f \in S$.

To illustrate the above principle, we apply it to prove the following result about the preservation of least fixed points. A monotone function f on a dcpo D with bottom is *continuous* if it preserves directed joins, and *strict* if $f(\bot) = \bot$.

Lemma B.2.10 (Uniformity). *Let P and Q be dcpos with bottom and let $f: P \to P$ and $g: Q \to Q$ be monotone. For every strict continuous map $h: P \to Q$ such that $g \cdot h = h \cdot f$ we have $h(\mu f) = \mu g$.*

Proof First, $h(\mu f)$ is a fixed point of g: we have $g(h(\mu f)) = h(f(\mu f)) = h(\mu f)$. Therefore $\mu g \leq h(\mu f)$. For the reverse inequality, let $S = \{x \in P : h(x) \leq \mu g\}$. Since h is strict, we see that $\bot \in S$. Moreover, S is closed under f, for if $x \in S$ we obtain $h(f(x)) = g(h(x)) \leq g(\mu g) = \mu g$ using monotonicity of g in the second step. Finally, S is closed under directed joins: if $D \subseteq S$ is directed, then we obtain $h(\bigvee D) = \bigvee_{x \in D} h(x) \leq \bigvee_{x \in D} \mu g = \mu g$, whence $\bigvee D$ lies in S. Thus, by Corollary B.2.9, $\mu f \in S$, which means that $h(\mu f) \leq \mu g$. □

B.3 Fixed Points in Complete Metric Spaces

We finish this chapter by presenting a proof of Banach's fixed point theorem. Recall metric spaces from Notation 3.2.2 and that a metric space is *complete* if every Cauchy sequence has a limit.

Definition B.3.1. Let X be a metric space. A *contracting map* f on X is a function $f: X \to X$ with the property that for some $0 < \varepsilon < 1$, we have $d(f(x), f(y)) < \varepsilon d(x, y)$ for all $x, y \in X$.

Theorem B.3.2 (Banach's Fixed Point Theorem [66]). *Every contracting map on a nonempty complete metric space has a unique fixed point.*

Proof (1) Let X be a nonempty complete metric space, and $f: X \to X$ be contracting. Choose some element $x \in X$, and define a sequence (x_n) of points in X by: $x_0 = x$, and $x_{n+1} = f(x_n)$. The fact that f is contracting implies that this is a Cauchy sequence. Indeed, for $q = d(x_0, x_1)$ we have $d(x_1, x_2) = d(f(x_0), f(x_1)) < \varepsilon q$. By induction, we obtain

$$d(x_n, x_{n+1}) < \varepsilon^n q \qquad \text{for every } n \in \mathbb{N}.$$

Using the triangle inequality we conclude that for all $k < m$ we have

$$d(x_k, x_m) \leq \sum_{n=k}^{m-1} d(x_n, x_{n+1}) \leq \sum_{n=k}^{\infty} \varepsilon^n q = \varepsilon^k \frac{q}{1-\varepsilon}.$$

The last expression has limit 0 as k tends to ∞.

(2) Put $x^* = \lim_{n \to \infty} x_n$. Then

$$x^* = \lim_n x_n = \lim_n f(x_n) = f(\lim_n x_n) = f(x^*).$$

To check uniqueness, suppose that y is a fixed point of f. Then

$$d(x^*, y) = d(f(x^*, f(y)) \leq \varepsilon d(x^*, y).$$

Thus, $d(x^*, y^*) = 0$, which implies that $x^* = y$. □

We have the following induction/coinduction principle called *Metric Coinduction*, due to Kozen [201].

Corollary B.3.3. *In the setting of Theorem B.3.2, let $S \subseteq X$ be a nonempty closed set. If S is closed under f, then the fixed point belongs to S.*

Proof S is a complete metric space since it is closed in X. The map f restricts to a (contracting) map on S. So by Theorem B.3.2, the restriction of f has a fixed point in S. This is the fixed point in X as well, by uniqueness. □

The reason that this is called a coinduction principle may be gleaned from a principle which uses greatest fixed points and which turns out to be equivalent. This principle is called *Contraction Coinduction* [122, Thm. 1].

For every partial order $(X, \leq)$, the *up-set* of $x \in X$ is $\uparrow x = \{y \in X : x \leq y\}$.

Corollary B.3.4. *In Theorem B.3.2, let $\leq$ be a partial order on X such that f is monotone and all up-sets are closed. Then the unique fixed point x^* is the greatest post-fixed point of f.*

Proof Let $x \leq f(x)$. We see that $x \leq x^*$ by applying Corollary B.3.3 to $\uparrow x$: it contains x and is closed by assumption; it is closed under f by monotonicity. Thus, it contains x^*. □

Example B.3.5 [122]. Corollary B.3.4 may be used to establish a fact about matrices. Let P be an $n \times n$ matrix with entries in the unit interval $[0, 1]$. We assume that for some $\varepsilon < 1$, the sum of each row of P is at most ε. We prove that $(I - P)^{-1}$ exists and has only non-negative entries, where I is the identity $n \times n$ matrix.

Let $M = \mathbb{R}^{n \times n}$ be the set of real $n \times n$ matrices. We order M pointwise and define a metric on it by $d(X, Y) = \|X - Y\|$, where $\|X\| = \max_{i,j} |x_{i,j}|$. This endows M with the structure of a complete metric space. All up-sets $\uparrow X$ are closed. Let $f : M \to M$ be defined by $f(X) = I + PX$, which is easily seen to be monotone. To check that f is a contracting map, first observe that for all $X \in M$, $\|PX\| \leq \varepsilon \|X\|$. Since $\varepsilon < 1$, we have a contracting map (with any contraction factor ε' such that $\varepsilon < \varepsilon' < 1$). Then

$$d(f(X), f(Y)) = \|P(X - Y)\| \leq \varepsilon \|X - Y\| = \varepsilon \cdot d(X, Y).$$

Let X^* be the fixed point of f. So $X^* = I + PX^*$, and thus

$$(I - P)X^* = X^* - PX^* = I.$$

Hence $X^* = (I - P)^{-1}$. Note that $f(0) = I \geq 0$, where 0 is the zero matrix. By Corollary B.3.4, $(I - P)^{-1} = X^* \geq 0$.

Appendix C
Set Functors

The subject of set functors is a highly developed area. It began with Vera Trnková's 1969 paper [291] and was developed largely by her and her students and colleagues, and of course with later contributions by many others. It is not our aim in this appendix to present the main lines of research in this area, even in a survey form. Instead, we aim for a self-contained presentation of most of the results which are needed in this book. In addition, we offer new proofs of known results and several new results as well.

Notation C.0.1. As previously in the book, we denote by $\rightarrowtail$ (and $\twoheadrightarrow$) injective (surjective) maps (and more generally monomorphisms and epimorphisms) throughout this chapter. In Set, we also have *inclusion maps* for subsets, which we denote by $\hookrightarrow$.

C.1 Examples of Set Functors

We have seen set functors throughout this book, and in this section we review most of our examples. Except where noted, the action on morphisms is as expected (and is explained in the referenced places in the main text).

Example C.1.1. (1) Polynomial functors of possibly infinitary signatures are defined by $H_\Sigma X = \coprod_{\sigma \in \Sigma} X^{\mathsf{ar}(\sigma)}$ (Example 6.1.16). Up to natural isomorphism, the class of all polynomial functors is the smallest class containing Id and closed under products and coproducts. It contains all constant functors: the empty product is the constant functor of value 1, and for an arbitrary set M the coproduct of M copies of it yields the constant functor c_M of value M.

(2) The power-set functor $\mathscr{P}\colon \mathsf{Set} \to \mathsf{Set}$ (Example 2.1.3(4)). It takes a set to the set of all its subsets. For every cardinal λ, we have a subfunctor $\mathscr{P}_\lambda\colon \mathsf{Set} \to \mathsf{Set}$, taking X to the set of subsets of X of size at most λ. Finally, we have the

subfunctor $\mathscr{P}^+$ of $\mathscr{P}$ assigning to a set X the set of all nonempty subsets: $\mathscr{P}^+X = \mathscr{P}X \setminus \{\emptyset\}$.

(3) The discrete probability measure functor $\mathscr{D}$ (Example 12.2.8(1)). It assigns to a set X the set of functions $\mu\colon X \to [0,1]$ with the property that $\sum_{x\in X} \mu(x) = 1$. We also have the subfunctor $\mathscr{D}_{\mathrm{f}}$ (Example 2.4.3(4)). It gives the discrete probability distributions of finite support. For $f\colon X \to Y$, $\mathscr{D}f\colon \mathscr{D}X \to \mathscr{D}Y$ is given by $\mathscr{D}f(\mu)(y) = \mu(f^{-1}(y))$; similarly for $\mathscr{D}_{\mathrm{f}}$.

(4) The bag functor $\mathscr{B}$ from Example 2.4.3(3). For all X, $\mathscr{B}(X)$ is the set of functions $b\colon X \to \mathbb{N}$ with finite support. For $f\colon X \to Y$, $\mathscr{B}f(b)(y) = \sum\{b(x)\colon f(x) = y\}$.

(5) The filter functor $\mathscr{F}$ (Example 6.5.14). It assigns to a set X the set $\mathscr{F}X$ of all filters (e.g. the filter $\mathscr{P}X$). For $f\colon X \to Y$, $\mathscr{F}f$ maps a filter $\mathbb{F}$ to

$$\mathscr{F}f(\mathbb{F}) = \{M \subseteq Y : f^{-1}[M] \in \mathbb{F}\}.$$

(6) The Aczel–Mendler functor $(-)^3_2$ (Example 4.3.14(4)) is given by

$$(X)^3_2 = \{(a,b,c) \in X \times X \times X : (a = b) \text{ or } (a = c) \text{ or } (b = c)\}.$$

(7) The functor R (Example 7.4.6): $R(X)$ is $X \times X$ with its diagonal removed, and then with a fresh element d added.

(8) We denote by C_M the constant functor with value M. For C_1 we also use two variations: $C_{0,1}$ and $C_{2,1}$. We put $C_{0,1}\emptyset = \emptyset$, and $C_{2,1}\emptyset = 2 = 1 + 1$. Both take all nonempty sets to 1.

C.2 Basic Facts

In this section, we collect some properties which all set functors enjoy. In the following section, we study a number of other properties which hold 'typically in practice'.

Proposition C.2.1. *Let F be a set functor.*

(1) *If $f\colon X \to Y$ is an epimorphism, so is Ff.*

(2) *If $f\colon X \to Y$ is a monomorphism with nonempty domain, then Ff is again a monomorphism.*

(3) *If $FX = \emptyset$ for some nonempty X, then $F = C_\emptyset$.*

(4) *If $F \neq C_\emptyset$, then every set V_i in the terminal-coalgebra chain of F is nonempty.*

Proof For (1) and (2) use that every endofunctor on any category preserves split epimorphisms and split monomorphisms; (1) comes from the fact that every epimorphism in Set splits, and (2) from the fact that every monomorphism in Set with nonempty domain splits.

For (3), given an arbitrary set Y, take a map $f\colon Y \to X$. From $Ff\colon FY \to FX = \emptyset$, we see that $FY = \emptyset$.

For (4), if $F \neq C_\emptyset$, then $F1 \neq \emptyset$. Choose a coalgebra $\alpha\colon 1 \to F1$. In Remark 6.5.2(4), we saw how α induces a canonical cone (α_i) on the terminal-coalgebra chain. For all i, $\alpha_i\colon 1 \to V_i$ shows $V_i \neq \emptyset$. □

Our next result is a classic result due to Trnková [291]. In the statement, a *nonempty intersection* is a pullback of two injective functions as on the left square in (C.1), with $P \neq \emptyset$. Section C.3 below studies a series of other preservation properties, most of which are defined in terms of pullbacks with additional features.

Theorem C.2.2. *Every set functor F preserves nonempty finite intersections.*

Proof By Proposition C.2.1(3), we may assume that F takes nonempty sets to nonempty sets.

(1) Consider a commutative square as shown on the left below, with $P \neq \emptyset$:

$$\begin{array}{ccc} P & \overset{p_2}{\rightarrowtail} & X_2 \\ {\scriptstyle p_1}\downarrow & & \downarrow{\scriptstyle f_2} \\ X_1 & \underset{f_1}{\rightarrowtail} & Y \end{array} \qquad\qquad \begin{array}{ccc} P & \overset{\bar{p}_2}{\twoheadleftarrow} & X_2 \\ {\scriptstyle p_1}\downarrow & & \downarrow{\scriptstyle f_2} \\ X_1 & \underset{\bar{f}_1}{\twoheadleftarrow} & Y \end{array} \qquad (C.1)$$

We prove that this square is a pullback if and only if there are retractions $\bar{f}_1\colon Y \to X_1$ and $\bar{p}_2\colon X_2 \to P$ of f_1 and p_2, respectively, so that the square on the right commutes.

For the 'only if' direction, let $\bar{z} \in P$. We define $\bar{p}_2$ by $\bar{p}_2(p_2(z)) = z$ for every $z \in P$, and for $x \notin p_2[P]$, $\bar{p}_2(x) = \bar{z}$. Similarly, we define $\bar{f}_1$ by $\bar{f}_1(f_1(x)) = x$ for $x \in X_1$, and for $y \notin f_1[X_1]$ we define $\bar{f}_1(y) = p_1(\bar{z})$. We check that $p_1 \cdot \bar{p}_2 = \bar{f}_1 \cdot f_2$. This is true when pre-composed by p_2:

$$p_1 \cdot \bar{p}_2 \cdot p_2 = p_1 = \bar{f}_1 \cdot f_1 \cdot p_1 = \bar{f}_1 \cdot f_2 \cdot p_2.$$

Thus, it is sufficient to prove that for $x \in X_2 \setminus p_2[P]$ that $p_1 \cdot \bar{p}_2(x) = f_1 \cdot \bar{f}_2(x)$. For such x, $\bar{p}_2(x) = \bar{z}$. For every $y \in X_1$, the equality $f_1(y) = f_2(x)$ would imply that $x \in p_2[P]$ by the pullback property. Therefore $f_2(x) \notin f_1[X_1]$. So $\bar{f}_1(f_2(x)) = p_1(\bar{z})$ by definition of $\bar{f}_1$. Thus, $\bar{f}_1(f_1(x)) = p_1(\bar{z}) = p_1\bar{p}_2(x)$.

For the 'if' direction, let $x_i \in X_i$, $i = 1, 2$ fulfil $f_1(x_1) = f_2(x_2)$. We need to show that there is a unique $z \in P$ such that $p_i(z) = x_i$ for $i = 1, 2$. Let

$z = \bar{p}_2(x_2)$. Then we have

$$f_1(p_1(z)) = f_1(p_1(\bar{p}_2(x_2)) = f_1(\bar{f}_1(f_2(x_2)) = f_1(\bar{f}_1(f_1(x_1)) = f_1(x_1)$$

Since f_1 is monic, $p_1(z) = x_1$. Further, $f_2(p_2(z)) = f_1(p_1(z)) = f_1(x_1) = f_2(x_2)$. Because f_2 is monic, $p_2(z) = x_2$. Thus, z is as desired. Finally, z is unique with these properties because p_2 is injective.

(2) Now given that the square on the left is a pullback, we apply F to the square on the right. The resulting square commutes, and $F\bar{p}_2$ and $F\bar{f}_1$ are retractions of Fp_2 and Ff_1, respectively. Moreover, the F-image of the square on the left consists of injective maps with $P \neq \emptyset$ by Proposition C.2.1(2). Thus, item (1) yields that this square is a pullback, whence an intersection, as desired. □

Remark C.2.3. We have mentioned in Proposition 8.1.13 that for every set functor F its Trnková hull $\bar{F}$ preserves *all* finite intersections (including empty ones). In Section C.6 we shall show how to construct $\bar{F}$ and prove that proposition.

Example C.2.4. The restriction to nonempty intersections in Theorem C.2.2 is important: the functor $C_{0,1}$ does not preserve the intersection of the coproduct injections of $1 + 1$; it maps the pullback on the left below to the non-pullback on the right:

$$\begin{array}{ccc} \emptyset & \longrightarrow & \{1\} \\ \downarrow & & \downarrow \\ \{0\} & \hookrightarrow & \{0,1\} \end{array} \qquad \begin{array}{ccc} \emptyset & \longrightarrow & 1 \\ \downarrow & & \downarrow \\ 1 & \longrightarrow & 1 \end{array}$$

C.3 Common Preservation Properties

In this section, we study a number of preservation properties that are commonly assumed about set functors, and we see whether the functors from Example C.1.1 have or do not have the property. Table C.1 on page 569 summarizes the properties.

Pullbacks We first study preservation of pullbacks. It turns out that this is rare, at least as far as our examples go.

Remark C.3.1. (1) A commutative square

$$\begin{array}{ccc} P & \xrightarrow{p_2} & X_2 \\ {\scriptstyle p_1}\downarrow & & \downarrow{\scriptstyle f_2} \\ X_1 & \xrightarrow{f_1} & Y \end{array} \tag{C.2}$$

in Set is a pullback iff for every pair $x_i \in X_i$, $i = 1, 2$, with $f_1(x_1) = f_2(x_2)$, there is unique $z \in P$ such that $p_i(z) = x_i$ for $i = 1, 2$.

(2) Using the above characterization, we see that an arbitrary product or coproduct of pullbacks in Set is again a pullback.

(3) We deduce that products and coproducts of set functors preserving pullbacks also preserve them.

Example C.3.2. (1) Polynomial set functors preserve pullbacks by Remark C.3.1(3).

(2) The following subfunctor F of $(-)^\omega$ preserves pullbacks: FX is the set of all sequences (x_n) that are eventually constant: there exists k with $x_n = x_k$ for all $n \geq k$. This functor preserves pullbacks since $(-)^\omega$ does, and since a sequence in $X \times Y$ is eventually constant iff both of its projections are.

(3) The power-set functor $\mathscr{P}$ does not preserve pullbacks. For example, let X be a two-element set, and consider the following pullback of product projections

$$\begin{array}{ccc} X \times X & \xrightarrow{\pi_2} & X \\ {\scriptstyle \pi_1}\downarrow & & \downarrow{\scriptstyle !} \\ X & \xrightarrow{\ !\ } & 1 \end{array} \qquad \text{(C.3)}$$

Then $\mathscr{P}\pi_1$ and $\mathscr{P}\pi_2$ are not collectively monic.

(4) Analogously for $\mathscr{P}_\mathsf{f}$: it does not preserve pullbacks.

(5) The distribution functors $\mathscr{D}$ and $\mathscr{D}_\mathsf{f}$ do not preserve pullbacks. Again we consider (C.3), and this time we have two elements $\mu, \mu' \in \mathscr{D}_\mathsf{f}(X)$ such that $\mathscr{D}_\mathsf{f}\pi_i(\mu) = \mathscr{D}_\mathsf{f}\pi_i(\mu')$ for $i = 1, 2$. We can take μ to be constant with value $\frac{1}{4}$, and μ' takes the diagonal to $\frac{1}{2}$ and the other elements to 0.

(6) The bag functor $\mathscr{B}$ does not preserve pullbacks. Once again, we consider (C.3), and we check that $\mathscr{B}\pi_1$ and $\mathscr{B}\pi_2$ are not collectively monic. This is demonstrated by the bagso $b, b' \in \mathscr{B}(X \times X)$, where b is constant with value 1, and b' takes elements of the diagonal to 2 and the remaining ones to 0. We have $\mathscr{B}\pi_i(b) = \mathscr{B}\pi_i(b')$ for $i = 1, 2$.

Weak pullbacks We have seen that preservation of pullbacks is rare (at least in our examples), but the weaker property of preserving weak pullbacks which we turn to next is much more common. It also suffices for many applications in coalgebra, as we indicate near the end of our discussion.

Remark C.3.3. (1) A *weak pullback* of morphisms $f_i\colon X_i \to Y$ in a category is a commutative square as in (C.2) such that for every P', p'_1, and p'_2 such that

$f_1 \cdot p'_1 = f_2 \cdot p'_2$, there is a *not necessarily unique* $k \colon P' \to P$ so that $p'_i \cdot k = p_i$ for $i = 1, 2$.

(2) In Set a commutative square (C.2) is a weak pullback iff for every pair $x_i \in X_i$, $i = 1, 2$, with $f_1(x_1) = f_2(x_2)$, there is a (not necessarily unique) $z \in P$ such that $p_i(z) = x_i$ for $i = 1, 2$.

(3) We deduce that products and coproduts of set functors preserving weak pullbacks also preserve them.

Lemma C.3.4. *In a category with pullbacks, a commutative square is a weak pullback iff its factorizing morphism through the corresponding pullback is a split epimorphism.*

Proof Let us consider a commutative square as shown below:

$$\begin{array}{ccc} W & \xrightarrow{w_2} & X_2 \\ {\scriptstyle w_1}\downarrow & & \downarrow{\scriptstyle f_2} \\ X_1 & \xrightarrow{f_1} & Y \end{array}$$

Let (C.2) be the corresponding pullback, and denote by $h \colon W \to P$ the unique morphism such that $p_1 \cdot h = w_i$ for $i = 1, 2$.

(1) If the above square is a weak pullback, then we also have a morphism $k \colon P \to W$ with $w_i \cdot h = p_i$ for $i = 1, 2$. Then $p_i \cdot (h \cdot k) = p_i$ for $i = 1, 2$. Since P is a pullback, $h \cdot k = \mathrm{id}_P$. Thus, h is a split epimorphism.

(2) Conversely, suppose that h is a split epimorphism and let $k \colon P \to W$ satisfy $h \cdot k = \mathrm{id}_P$. Then the above square is a weak pullback: given $w'_i \colon W' \to X_i$ with $f_1 \cdot w'_1 = f_2 \cdot w'_2$, let $h' \colon W' \to P$ be the unique morphism such that $p_i \cdot h = w'_i$. Then $k \cdot h'$ is as desired since $w_i \cdot k \cdot h' = p_i \cdot h \cdot k \cdot h' = p_i \cdot h' = w'_i$. □

Corollary C.3.5. (1) *Every functor preserving pullbacks also preserves weak pullbacks.*

(2) *A functor preserves weak pullbacks iff it weakly preserves pullbacks (i.e. it maps pullbacks to weak pullbacks).*

In both items, we use that every functor preserves split epimorphisms.

Example C.3.6. (1) The power-set functor $\mathscr{P}$ preserves weak pullbacks. To see this use Remark C.3.3(2). Consider a weak pullback square (C.2) suppose that $\mathscr{P}f_1(M_1) = \mathscr{P}f_2(M_2)$ holds for $M_i \in \mathscr{P}X_i$. We are to find $M \in \mathscr{P}P$ with $M_i = \mathscr{P}p_i(M)$. Put $M = \{x \in P : p_i(x) \in M_i \text{ for } i = 1, 2\}$. We verify $\mathscr{P}p_1(M) = M_1$. The inclusion $\mathscr{P}p_1 \subseteq M_1$ is clear. For the reverse inclusion, given $y \in M_1$, we use that $\mathscr{P}f_1(M_1) = \mathscr{P}(M_2)$ implies that $f_1(y) \in \mathscr{P}f_2(M_2)$. Hence, we can find $z \in M_2$ such that $f_1(y) = f_2(z)$. Then there exists some

$x \in P$ such that $p_1(x) = y$ and $p_2(x) = z$. Thus, $x \in M$ and $y \in \mathscr{P}p_1(M)$. The equality $\mathscr{P}p_2(M) = M_2$ is proved analogously.

(2) Analogously for $\mathscr{P}_\mathsf{f}$: it preserves weak pullbacks.

(3) The functor $\mathscr{P}^+$ does not preserve weak pullbacks: consider the empty pullback of the two coproduct injections of $1 + 1$.

(4) The Aczel–Mendler functor $(-)^3_2$ does not preserve weak pullbacks [4, p. 363f]. Consider (C.3) where $X = \{x, y\}$. The elements (x, x, y) and (x, y, x) of $(X)^3_2$ are merged by $(!)^3_2$. However, there is no $z \in (X \times X)^3_2$ such that $(\pi_1)^3_2(z) = (x, x, y)$ and $(\pi_2)^3_2(z) = (x, y, x)$.

(5) The functor R does not preserve weak pullbacks; see Example C.3.8(6) further below.

(6) The filter functor $\mathscr{F}$ preserves weak pullbacks [149, Thm. 3.5].

(7) The bag functor $\mathscr{B}$ and the finite probability distribution functor preserve weak pullbacks. For $\mathscr{D}_\mathsf{f}$ this was first shown by Rutten and de Vink [269, Appendix A], using the graph-theoretic max flow-min cut theorem, and independently by Moss [239], using a fact about the non-negative real numbers, which is mentioned as part of a general result in Lemma C.4.3. We shall obtain weak pullback preservation for both functors as a consequence of a more general result for monoid-valued functors (Example 2.4.3) due to Gumm and Schröder [152]: a monoid-valued functor $M^{(-)}$ preserves weak pullbacks iff the monoid M is positive and refinable; we will prove this in Corollary C.4.7.

(8) The probability distribution functor $\mathscr{D}$ preserves weak pullbacks. A proof was presented by Sokolova [280, Lemma 3.5.6], based on a generalization of the Row/Column Theorem [239, Thm. 36] (cf. Lemma C.4.3 below) to the case of countably infinite sets.

If F is a set functor which preserves weak pullbacks then the category of coalgebras for F has many pleasant properties:

(1) There is a largest bisimulation relation on every coalgebra.

(2) The family of bisimulations on every coalgebra is closed under composition.

(3) A coalgebra C has proper quotients iff it has a non-trivial bisimulation (Definition 2.6.6), i.e. one which is not a subrelation of the diagonal.

For more on this topic, see e.g. [4, 153, 154, 168, 264].

Inverse images Our next preservation property concerns inverse images of subsets under functions.

Remark C.3.7. (1) Given a morphism $f: A \to B$ in a category, the *inverse* *image* of a subobject $m\colon B' \rightarrowtail B$ *under* f is the subobject $m'\colon A' \rightarrowtail A$ in the following pullback:

$$\begin{array}{ccc} A' & \xrightarrow{f'} & B' \\ {\scriptstyle m'}\big\downarrow & \lrcorner & \big\downarrow{\scriptstyle m} \\ A & \xrightarrow{f} & B \end{array} \tag{C.4}$$

Accordingly, a functor F *preserves inverse images* if it preserves pullbacks of pairs one of whose morphisms is monic.

(2) Since m is monic, so is m'. Thus, f' is a domain-codomain restriction of m. Moreover, m' is the largest subobject of A in the poset $\mathsf{Sub}(A)$ (Notation 7.2.6) for which such a domain-codomain restriction exists. This follows from the universal property of the pullback. Thus, the above definition provides a category-theoretic description of the set-theoretic notion of an inverse image.

(3) In Set, we identify B' with its image in B, and so we can take A' to be $f^{-1}[B']$ and $m'\colon A' \hookrightarrow A$ to be the inclusion map.

Example C.3.8. (1) The collection of set functors preserving inverse images is closed under composition, products, and coproducts. Indeed, a product or coproduct of inverse images in Set is again an inverse image.

(2) A polynomial set functor H_Σ preserves inverse images.

(3) The functors $\mathscr{P}$, $\mathscr{B}$, and $\mathscr{D}$ preserve inverse images since they preserve weak pullbacks (we shall prove this implication in Lemma C.3.17 further below).

(4) The functor $(-)^3_2$ preserves inverse images [154, Ex. 2.8]. To see this, suppose we are given an inverse image (C.4) and elements $(a_1, a_2, a_3) \in (A)^3_2$ and $(b_1, b_2, b_3) \in (A')^3_2$ with $f(a_i) = f(b_i)$ for $i = 1, 2, 3$. There exist $c_1, c_2, c_2 \in C$ such that $a_i = m'(c_i)$ and $b_i = f'(c_i)$ for all i. Since m' is monic and the elements a_i are not pairwise distinct, the same holds for the elements c_i. Thus $(c_1, c_2, c_3) \in (C)^3_2$ is as desired.

(5) A monoid-valued functor $M^{(-)}$ (Example 2.4.3) preserves inverse images iff the monoid M is positive (we will prove this in Proposition C.4.6 further below).

(6) The functor R does not preserve inverse images [40, Ex. 7.3]. Consider the pullback below:

$$\begin{array}{ccc} \emptyset & \longrightarrow & \{0\} \\ \big\downarrow & \lrcorner & \big\downarrow \\ \{0,1\} & \xrightarrow{\mathsf{const}_1} & \{0,1\} \end{array}$$

The elements $(0, 1) \in R\{0, 1\}$ and $d \in R\{0\}$ are mapped to d, but there is no element in $R\emptyset$ mapped to them.

(7) The functor $C_{2,1}$ does not preserve inverse images since it does not preserve monomorphisms (Example C.3.16 and Lemma C.3.17).

Intersections In Chapters 8 and 9, we worked with set functors which *preserve intersections*. This means that wide pullbacks of arbitrary sets of injective functions $m_i \colon A_i \rightarrowtail B$ are preserved. If each m_i is an inclusion, then the pullback in Set may be taken to be the intersection $\bigcap_i A_i$.

Remark C.3.9 [150, Lemmas 4.1 and 4.7]. (1) Let F be any set functor, and suppose that $m_A \colon A \hookrightarrow X$ and $m_B \colon B \hookrightarrow X$ are inclusion maps. If $m_A \leq m_B$, then $Fm_A[FA] \subseteq Fm_B[FB]$. To see this, let $j \colon A \hookrightarrow B$ be the inclusion map. Then $m_B \cdot j = m_A$, and so $Fm_A[FA] = Fm_B[Fj[FA]] \subseteq Fm_B[FB]$.

(2) For a set functor F preserving monomorphisms, and a monomorphism $m \colon X \rightarrowtail Y$, the map Fm represents the same subobject of FY as the inclusion $i \colon Fm[FX] \hookrightarrow FY$. Indeed, for each $x \in Fm[FX]$ there exists a unique $x' \in FX$ with $x = Fm(x')$. This defines an isomorphism $(-)' \colon Fm[FX] \to FX$ whose composite with Fm is i.

Proposition C.3.10 [150, Cor. 4.8]. *A set functor F preserves intersections iff F preserves monomorphisms and for every set X and every element $x \in FX$ there is a smallest subset $m \colon M \hookrightarrow X$ such that $x \in Fm[FM]$.*

Proof Let F preserve intersections. As we shall see in Lemma C.3.17, F preserves monomorphisms. Fix X and $x \in X$. We take the intersection of all subsets $m_A \colon A \hookrightarrow X$ such that $x \in Fm_A[FA]$. We may represent this intersection as a subset $m \colon M \hookrightarrow X$. By Remark C.3.9(1), $x \in Fm[FM]$. Clearly M is the least subset with this property.

For the converse implication, take a family of monomorphisms $m_i \colon A_i \rightarrowtail X$. Each m_i represents the same subobject as its image, and so we may assume that each is an inclusion $m_i \colon A_i \hookrightarrow X$. Let $A^* = \bigcap A_i$, so that the inclusion $m^* \colon A^* \to X$ is the pullback of the m_i. We prove that Fm^* is the pullback of the monomorphisms Fm_i. For this, it is enough to prove that $Fm^*[FA^*] = \bigcap Fm_i[FA_i]$. By Remark C.3.9(1), $Fm^*[FA^*] \subseteq \bigcap Fm_i[FA_i]$. For the reverse inclusion, given $x \in \bigcap Fm_i[FA_i]$, let $m' \colon M' \hookrightarrow X$ be the smallest subset of X such that $x \in Fm'[FM']$. Hence, we have $M' \subseteq \bigcap_i A_i = A^*$. This implies that $x \in Fm'[FM'] \subseteq Fm^*[FA^*]$. □

Example C.3.11. Most of our examples in Example C.1.1 straightforwardly do have the two properties mentioned in Proposition C.3.10, and so they preserve intersections.

(1) The functor R preserves intersections. It is easy to see directly that it preserves monomorphisms. If $x_1 \neq x_2$ are elements of X, then $\{x_1, x_2\} \in RX$. In this case, the smallest $m\colon M \hookrightarrow X$ has $M = \{x_1, x_2\}$. For $d \in RX$, we take $M = \emptyset$.

(2) Every monoid-valued functor $M^{(-)}$ (Example 2.4.3) preserves intersections: it preserves monomorphisms, and for every $m\colon X \to M$ in $M^{(X)}$, the smallest subset in Proposition C.3.10 is given by the support of m.

(3) The functors $C_{0,1}$ and $C_{2,1}$ do not preserve intersections (they do not even preserve finite intersections, and for that matter, $C_{2,1}$ does not preserve monomorphisms, see Example C.3.16): consider the empty intersection of the coproduct injections of $1 + 1$.

(4) The filter functor $\mathscr{F}$ does not preserve intersections [292, Prop. VI.11]. To see this consider the Fréchet filter $\mathbb{F}$ on $\mathbb{N}$ consisting of all subsets with a finite complement. For this element $\mathbb{F} \in \mathscr{F}\mathbb{N}$ there exists no smallest element $m\colon M \hookrightarrow \mathbb{N}$ with $\mathscr{F}m[\mathscr{F}M]$ (i.e. with $M \in \mathbb{F}$).

Remark C.3.12. Recall from Definition 4.3.1 that a set functor is *finitary* if it preserves directed colimits. We know that every finitary set functor F preserves nonempty intersections (Theorem 4.4.3). The Trnková hull $\bar{F}$ is finitary and preserves all intersections (including empty ones), as we shall see in Corollary C.6.11.

Finite intersections A functor *preserves finite intersections* if it preserves pullbacks of pairs of monomorphisms. Recall from Theorem C.2.2 that every set functor preserves nonempty finite intersections. Thus, for a set functor, preserving *all* finite intersections is equivalent to preserving finite intersections which are empty. This is easy to see for the filter functor $\mathcal{F}$, for example.

Remark C.3.13. If a functor preserves inverse images, then it preserves finite intersections: every pullback (C.2) of monomorphisms f_1, f_2 yields also an inverse image of f_1 under f_2. So all positive items in Example C.3.8 are functors preserving finite intersections.

Example C.3.14. (1) The set functor R preserves finite intersections but not inverse images (see Example C.3.11(1)). Likewise, every monoid-valued functor preserves finite intersections but not all of them preserve inverse images (see Proposition C.4.6 further below).

(2) The functor $C_{0,1}$ does not preserve finite intersections (Example C.2.4). Similarly, $C_{2,1}$ does not preserve finite intersections since it does not even preserve monomorphisms (Example C.3.16 and Lemma C.3.17).

Monomorphisms The weakest preservation property is preservation of monomorphisms. We have already mentioned that every set functor preserves monomorphisms whose domain is nonempty. Thus, preserving *all* monomorphisms is equivalent to the property of sending the empty inclusions $\emptyset \hookrightarrow A$ to injective maps.

Remark C.3.15. If a functor preserves finite intersections, then it preserves monomorphisms. This follows from the fact that a morphism $m\colon X \to Y$ is a monomorphism iff the following square is a pullback:

$$\begin{array}{ccc} X & \xrightarrow{\mathsf{id}} & X \\ {\scriptstyle \mathsf{id}}\downarrow & \lrcorner & \downarrow{\scriptstyle m} \\ X & \xrightarrow{m} & Y \end{array}$$

Hence, for a monomorphism the above pullback is the intersection of m with itself.

Example C.3.16. The functor $C_{2,1}$ does not preserve empty monomorphisms, since for $X \neq \emptyset$ and $e_X\colon \emptyset \to X$, $C_{2,1}(e_X)$ is the map $2 \to 1$. All of the other functors of Section C.1 and all monoid-valued functors $M^{(-)}$ preserve monomorphisms.

Our next aim is to relate all the preservation properties we have seen so far in this section.

Lemma C.3.17. *Let F be an endofunctor on a category with pullbacks. In the following list each statement implies the next:*

(1) *F preserves pullbacks.*

(2) *F preserves weak pullbacks.*

(3) *F preserves inverse images.*

(4) *F preserves finite intersections.*

(5) *F preserves monomorphisms.*

Proof For (1) $\Rightarrow$ (2) see Corollary C.3.5, for (3) $\Rightarrow$ (4) see Remark C.3.13, and for (4) $\Rightarrow$ (5) see Remark C.3.15.

For (2) $\Rightarrow$ (3), we first immediately see that F preserves monomorphisms: given a monomorphism $m\colon X \rightarrowtail Y$, the square below is a weak pullback whence a pullback (since its projections are monic):

$$\begin{array}{ccc} FX & \xrightarrow{\mathsf{id}_{FX}} & FX \\ {\scriptstyle \mathsf{id}_{FX}}\downarrow & & \downarrow{\scriptstyle Fm} \\ FX & \xrightarrow{Fm} & FY \end{array}$$

This implies that F preserves the pullback in (C.4): applying F to it, we obtain the following weak pullback

$$\begin{array}{ccc} FA' & \xrightarrow{Ff'} & FB' \\ {\scriptstyle Fm'}\downarrow & & \downarrow{\scriptstyle Fm} \\ FA & \xrightarrow{Ff} & FB \end{array}$$

Since Fm' is a monomorphism, this is a pullback. □

The gap between weak pullback preservation and inverse image preservation can be filled by preservation of *weak kernel pairs*, that is, weak pullbacks of a map along itself. That is, a functor preserving (weak) pullbacks, preserves (weak) kernel pairs. This property will also play an important role in Section C.4.

Example C.3.18. (1) All set functors preserving weak pullbacks of course preserve weak kernel pairs.

(2) The functor R preserves weak kernel pairs. Suppose we are given $f\colon X \to Y$ and we form the kernel pair $p_1, p_2\colon K \rightrightarrows X$ of f. Represent K as $\{(x_1, x_2) \in X \times X : f(x_1) = f(x_2)\}$. If $X = \emptyset$, then $Y = \emptyset = K$. In this case, $RX = RY = RK = \{d\}$, and our result is easy. So assume that all of these sets are nonempty. Let $a, b \in RX$ fulfil $Rf(a) = Rf(b)$; we must find $z \in RK$ with $Rp_1(z) = a$ and $Rp_2(z) = b$. There are several cases here, and one representative case is when $a = (x_1, x_2)$ with $x_1 \neq x_2$ and $f(x_1) = f(x_2)$, whereas $b = d$. Then $z = ((x_1, x_2), (x_2, x_2)) \in RK$ fulfils $Rp_1(z) = (x_1, x_2)$ and $Rp_2(z) = d$, as desired.

(3) The functor $C_{0,1}$ preserves weak kernel pairs. For $X \neq \emptyset$, a weak pullback of $f\colon X \to Y$ along itself must also be nonempty and our result is clear. For $X = \emptyset$, the only weak kernel pair is $\mathrm{id}_\emptyset, \mathrm{id}_\emptyset$.

(4) The functor $(-)^3_2$ does not preserve weak kernel pairs: Example C.3.6(4) presents a weak kernel pair not preserved by the functor. But this functor preserves monomorphisms, since $(\emptyset)^3_2 = \emptyset$. It also preserves all intersections.

Proposition C.3.19 [154, Thm. 2.7]**.** *A set functor preserves weak pullbacks iff it preserves inverse images and weak kernel pairs.*

The proof below works in every category with pullbacks and coproducts having monomorphic injections.

Proof For 'only if' use Lemma C.3.17 and the fact that weak kernels are special weak pullbacks.

For 'if', given a pair of maps $f\colon X \to Z$ and $g\colon Y \to Z$ we obtain their pullback by forming a square of four pullbacks as follows

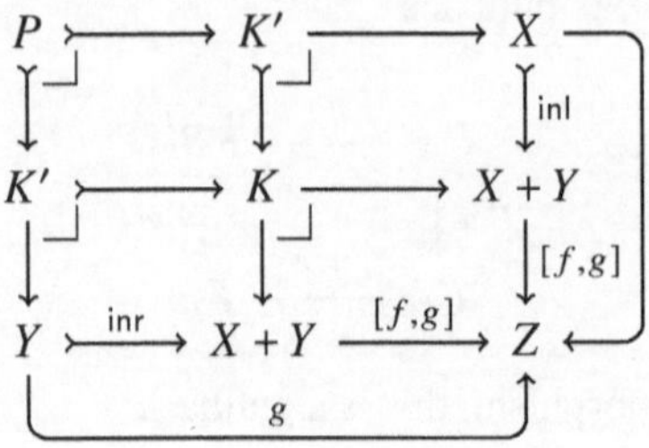

The lower right-hand square is the kernel pair of $[f, g]$, and the remaining squares are inverse images because the coproduct injections inl and inr are monic. Since F weakly preserves kernel pairs and inverse images, we see that it weakly preserves the outside pullback. □

For the record and for use in Section C.4, we mention a preservation property which is equivalent to weak kernel preservation for set functors preserving monomorphisms.

Definition C.3.20. A (weak) *epi-pullback* is a (weak) pullback of two epimorphisms.

In Set, by the stability of epimorphisms under pullback, both pullback projections are then epic, too.

The following result is due to Gumm [151, Cor. 5]. We present a simpler proof.

Proposition C.3.21. *A set functor preserves weak kernel pairs iff it preserves weak epi-pullbacks and monomorphisms.*

Proof (1) For the 'if' part, suppose that F preserves weak epi-pullbacks and monomorphisms. Then it preserves weak kernel pairs of epimorphisms, since these are special weak epi-pullbacks. Given a morphism $f\colon X \to Y$, factorize it as an epimorphism e followed by a monomorphism m. Then f and e have the same weak kernel pairs since they merge the same parallel pairs $P \rightrightarrows X$. Since $Ff = Fm \cdot Fe$ and Fm is monic by assumption, Ff and Fe also merge the same parallel pairs $P \rightrightarrows FX$ and therefore have the same weak kernel pairs. It follows that F preserves weak kernel pairs of f.

(2) For the 'only if' part, monomorphisms are clearly preserved: a morphism $f\colon X \to Y$ is monic iff $\mathsf{id}_X, \mathsf{id}_X$ is a weak kernel pair of it.

Now given epimorphisms $e_i\colon X_i \twoheadrightarrow Y$ and their weak pullback $p_i\colon P \twoheadrightarrow X_i$ $(i = 1, 2)$, we prove that the Fp_i form a weak pullback of Fe_1, Fe_2. We use the fact that F weakly preserves the kernel pair $k_1, k_2\colon K \rightrightarrows X + Y$ of $[e_1, e_2]\colon X_1 + X_2 \to Y$.

Let us choose maps $m_i\colon Y \rightarrowtail X_i$ such that $e_i \cdot m_i = \mathsf{id}_Y$. Since $[e_1, e_2]$ merges k_1 and k_2, there exists a map $k\colon K \to P$ such that the diagram below commutes:

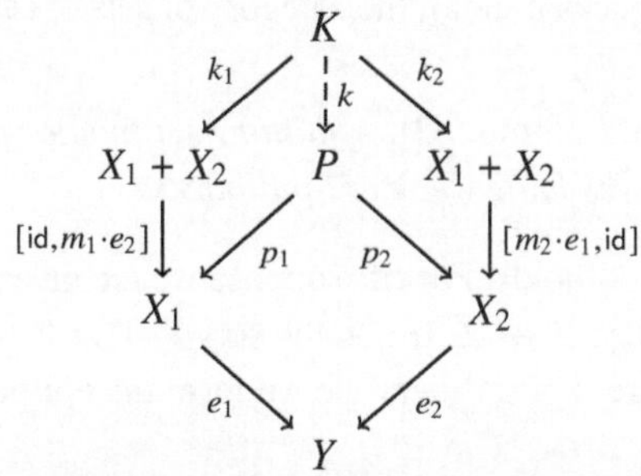

(C.5)

We now prove that every pair of maps $a_i\colon A \to FX_i$ $(i = 1, 2)$ which satisfies $Fe_1 \cdot a_1 = Fe_2 \cdot a_2$ factorizes through Fp_1, Fp_2. The above equality implies

$$F[e_1, e_2] \cdot (F\mathsf{inl} \cdot a_1) = Fe_1 \cdot a_1 = Fe_2 \cdot a_2 = F[e_1, e_2] \cdot (F\mathsf{inl} \cdot a_2). \quad \text{(C.6)}$$

Since F weakly preserves kernel pairs, we know that Fk_1, Fk_2 is a weak kernel pair of $F[e_1, e_2]$. Thus there exists a map $a\colon A \to FK$ such that

$$Fk_1 \cdot a = F\mathsf{inl} \cdot a_1 \qquad \text{and} \qquad Fk_2 \cdot a = F\mathsf{inl} \cdot a_2.$$

Finally, the composite $Fk \cdot a\colon A \to FP$ has the property $Fp_i \cdot (Fk \cdot a) = a_i$ for $i = 1, 2$:

$$\begin{aligned} Fp_1 \cdot Fk \cdot a_1 &= F[\mathsf{id}_{X_1}, m_1 \cdot e_2] \cdot Fk_1 \cdot a && \text{by (C.5)} \\ &= F[\mathsf{id}_{X_1}, m_1 \cdot e_2] \cdot F\mathsf{inl} \cdot a_1 && \text{by (C.6)} \\ &= a_1, \end{aligned}$$

and similarly for a_2. □

Remark C.3.22. (1) Preservation of monomorphisms cannot be dropped in Proposition C.3.21; indeed, the functor $C_{2,1}$ obviously preserves epi-pullbacks but not weak kernel pairs, since it does not preserve monomorphisms (Example C.3.16).

(2) The above proof shows that in every category with regular epi-mono factorizations, a functor preserving monomorphisms and (weak) regular epi-pullbacks preserves (weak) kernel pairs. For the converse implication, note that for a category with finite coproducts a functor preserving weak kernel pairs preserves weak pullbacks of split epis.

For use in Proposition C.4.5 below we record the following lemma for finitary set functors (Definition 4.3.1). By a *finite epi-pullback* we mean an epi-pullback in the category of finite sets.

The following lemma and its proof easily generalize to λ-accessible set functors and epi-pullbacks in Set_λ, the category of sets of cardinality less than λ.

Lemma C.3.23 [275, Lemma 3.2]. *A finitary set functor preserves weak epi-pullbacks iff it preserves finite weak epi-pullbacks.*

Proof 'Only if' clearly holds. For 'if' suppose we are given a pair of surjective maps $f : X \twoheadrightarrow Z$ and $g : Y \twoheadrightarrow Z$. If one the sets X, Y or Z is empty, then so are all of them, and we are done. Otherwise we form the epi-pullback

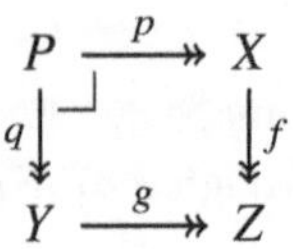

and prove that its F-image is a weak pullback. Suppose that $x \in FX$ and $y \in FY$ satisfy $Ff(x) = Fg(y)$. Since F is finitary we have finite subsets $m : X' \hookrightarrow X$ and $n : Y' \hookrightarrow Y$ as well as elements $x' \in FX$ and $y' \in FY'$ such that $x = Fm(x')$ and $y = Fn(y')$. Without loss of generality we assume that X' and Y' are nonempty and that $m[X'] = n[Y']$ (otherwise, we may suitably enlarge X' and Y'). Denote the inclusion of $m[X']$ by $o : Z' \hookrightarrow Z$. Then f and g have surjective domain-codomain restrictions $f' : X' \to Z'$ and $g' : Y' \to Z'$, respectively:

$$f \cdot m = o \cdot f' \qquad \text{and} \qquad g \cdot n = o \cdot g'.$$

We form the epi-pullback of f' and g' and denote its projections by $p' : P' \to X'$ and $q' : P' \to Y'$. We can choose this pullback so that P' is a finite subset of P, via the inclusion map $s : P' \hookrightarrow P$, say, and p', q' are restrictions of p, q. Since F weakly preserves finite epi-pullbacks, the outside of the diagram below is a weak finite epi-pullback

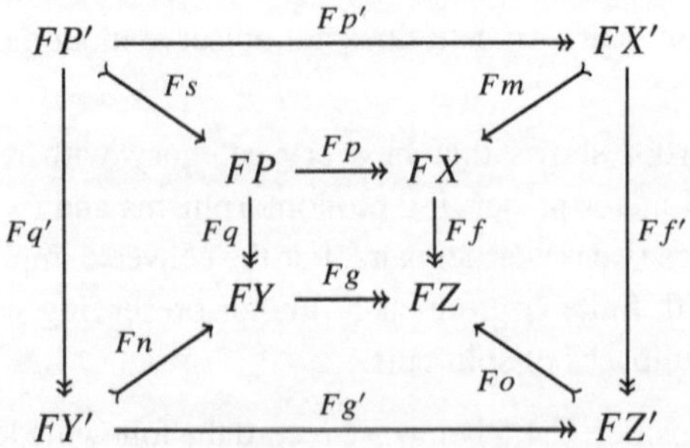

Note that Fo is monic since $o\colon Z' \hookrightarrow Z$ is a monomorphism with $Z \neq \emptyset$. Hence, we have $Ff'(x') = Fg'(y')$ since this holds when post-composed by Fo:

$$Fo \cdot Ff'(x') = Ff \cdot Fm(x') = Ff(x) = Fg(y) = Fg \cdot Fn(y') = Fo \cdot Fg(y').$$

Thus, there exists some $z' \in FP'$ such that $Fp'(z') = x'$ and $Fq'(z') = y'$. Then $z = Fs(z')$ is the desired element of FP such that $Fp(z) = x$ and $Fq(z) = y$:

$$Fp(z) = Fp \cdot Fs(z') = Fm \cdot Fp'(z') = Fm(x') = x,$$

and similarly for y and q. □

Here is a summary of the results noted in this section. See Table C.1 for the preservation properties of our examples.

Proposition C.3.24. *For set functors we have the following implications of preservation properties:*

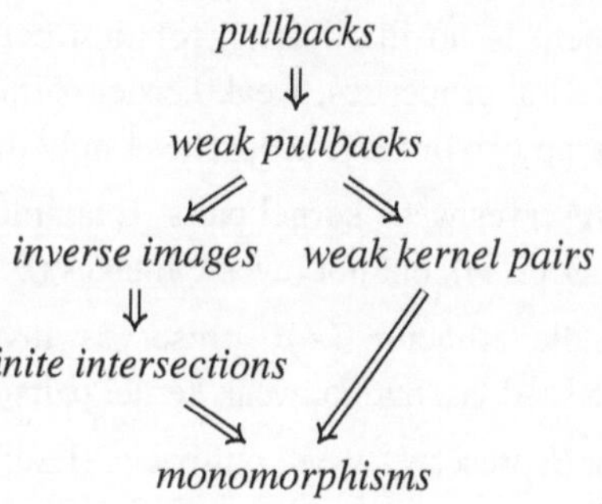

In addition the following equivalences of preservation properties hold:

$$\textit{weak pullbacks} \iff \textit{inverse images and weak kernel pairs,}$$
$$\textit{weak kernel pairs} \iff \textit{monomorphisms and weak epi-pullbacks.}$$

No implication is missing above; in particular none of the implications can be reversed. This can be seen from the examples we have previously mentioned and which are summarized in the picture below:

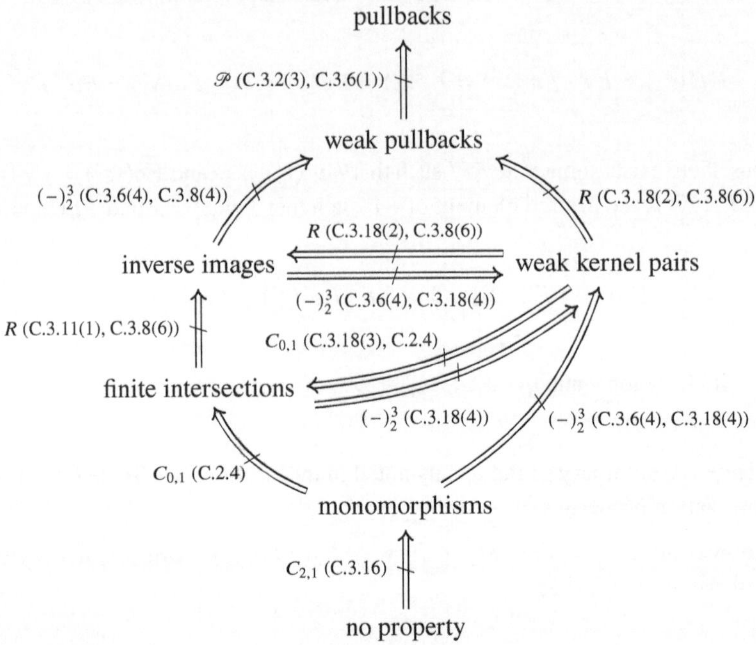

Remark C.3.25. There is no implication relation between any of the following three preservation properties: weak kernel pairs, inverse images, and intersections. Indeed, no two of these properties imply the third:

(1) The functor R preserves weak kernel pairs (Example C.3.18(2)) and intersections (Example C.3.11(1)), but not inverse images (Example C.3.8(6)).

(2) The Aczel–Mendler functor $(-)^3_2$ preserves inverse images (Example C.3.8(4)) and intersections, but not weak kernel pairs (Example C.3.18(4)).

(3) The filter functor $\mathfrak{F}$ preserves weak pullbacks (Example C.3.6(6)), hence weak kernel pairs, and inverse images. As we saw in Example C.3.11(4), it does not preserve intersections.

C.4 Monoid-Valued Functors

Here we turn to the monoid-valued functors $M^{(-)}$ introduced in Example 2.4.3 and we relate some of their preservation properties we have seen in Section C.3 to properties of the monoid M. In particular, we show that $M^{(-)}$ preserves weak pullbacks iff M is positive and refinable (Remark 4.5.24). Positivity alone

	pullbacks	weak pullbacks	inverse images	intersections	finite intersections	monomorphisms
H_Σ	+	+	+	+	+	+
$\mathscr{P}, \mathscr{P}_\lambda$	−	+	+	+	+	+
$\mathscr{D}, \mathscr{D}_\mathrm{f}$	−	+	+	+	+	+
$\mathscr{B}$	−	+	+	+	+	+
$\mathscr{F}$	−	+	+	−	+	+
$(-)^3_2$	−	−	+	+	+	+
R	−	−	−	+	+	+
$C_{0,1}$	−	−	−	−	−	+
$C_{2,1}$	−	−	−	−	−	−

Table C.1 Set functors from Example C.1.1 and various constructions which they do (+) or do not (−) preserve.

corresponds to inverse image preservation, while refinability is equivalent to preserving weak kernel pairs (equivalently epi-pullbacks). These result are due to Gumm and Schröder [152]. The proof we present is essentially due to Seifan et al. [275].

Remark C.4.1. (1) Recall from Example 2.4.3 the monoid-valued functor $M^{(-)}$ for a commutative monoid $(M, +, 0)$. It assigns to a set X the set of functions $\mu: X \to M$ with finite support. For $f: X \to Y$, the map $M^{(f)}: M^{(X)} \to M^{(Y)}$ is given by $M^{(f)}(\mu)(y) = \mu(f^{-1}(y))$; similarly as for distributions, we write $\mu(S)$ for $\sum_{s\in S}\mu(s)$ for every subset $S \subseteq X$, noting that this is always a finite sum.

(2) The bag functor $\mathscr{B}$ is the monoid-valued functor for the monoid $\mathbb{N}$ of natural numbers with addition (Example 2.4.3(3)). The finite distribution functor $\mathscr{D}_\mathrm{f}$ is clearly a subfunctor of the monoid-valued functor for the monoid $\mathbb{R}_{\geq 0}$ of all non-negative real numbers with addition (Example 2.4.3(4)).

(3) Further recall from Remark 4.5.24 the following notions: a monoid M is

(a) *positive*: $a + b = 0$ implies $a = 0 = b$, and

(b) *refinable*: $r_1 + r_2 = c_1 + c_2$ implies that there exists a 2×2 matrix with row sums r_1, r_2 and columns sums c_1, c_2, respectively.

Example C.4.2. (1) Every abelian group is refinable. Indeed, given $r_1 + r_2 =$

$c_1 + c_2$ we obtain the following 2×2 matrix

$$\begin{array}{cc|c} c_1 & r_1 - c_1 & r_1 \\ 0 & r_2 & r_2 \\ \hline c_1 & c_2 & \end{array}$$

(2) Moreover, for every ordered abelian group G, the positive cone $G^+ = \{g \in G : 0 \leq g\}$ is positive and refinable. Refinability is shown as in the previous item since by swapping rows and columns one may assume that $r_1 \geq c_1$. For the positivity, suppose that $a \geq 0$, $b \geq 0$, and $a + b = 0$, equivalently $b = -a$. Thus $-a \geq 0$. Since both a and $-a$ are ≥ 0, we have $a = 0$. It then follows that $b = a + b = 0$ as well.

(3) In particular, the monoids $\mathbb{R}_{\geq 0}$ and $\mathbb{N}$ with addition are positive and refinable.

(4) Here is an example of a monoid which is positive but not refinable [152, after Lemma 5.11]. Consider $(\mathbb{N} \setminus \{0, 2\}, \cdot, 1)$. The positivity is clear, since $n \cdot m = 1$ in this monoid implies $n = m = 1$. The failure of refinability comes from $5 \cdot 6 = 3 \cdot 10$.

As shown by Gumm and Schröder [152, Prop. 5.10], refinability may equivalently be formulated for $m \times n$ matrices for all $m, n \geq 1$. We now prove this property. This is a converse to the following obvious fact, true for every monoid: given a matrix $(a_{i,j})$, the row sums $r_i = \sum_j a_{i,j}$ and column sums $c_j = \sum_i a_{i,j}$ sum up to the same value.

Lemma C.4.3. *A monoid is refinable if and only if for every $m, n \geq 1$, given elements $r_1, \ldots, r_m$ and $c_1, \ldots, c_n$ with $\sum_{i=1}^m r_i = \sum_{j=1}^n c_j$, there exists an $m \times n$ matrix with row and column sums given by the r_i and c_j, respectively.*

Proof The 'if' direction is obvious. For 'only if', note that for $m = 1$ or $n = 1$, the statement trivially holds. For $m, n \geq 2$ one proceeds by an easy induction on $m + n$. As a hint we follow Gumm and Schröder and show the induction step for $m = 3$ and $n = 2$. Given r_1, r_2, r_3 and c_1, c_2 with $r_1 + r_2 + r_3 = c_1 + c_2$ first use the induction hypothesis to obtain a 2×2 matrix $(a_{i,j})$ with row and column sums as shown below on the left:

$$\begin{array}{cc|c} a_{1,1} & a_{1,2} & r_1 \\ a_{2,1} & a_{2,2} & r_2 + r_3 \\ \hline c_1 & c_2 & \end{array} \qquad \begin{array}{cc|c} b_{1,1} & b_{1,2} & r_2 \\ b_{2,1} & b_{2,2} & r_3 \\ \hline a_{2,1} & a_{2,2} & \end{array} \qquad \begin{array}{cc|c} a_{1,1} & a_{1,2} & r_1 \\ b_{1,1} & b_{1,2} & r_2 \\ b_{2,1} & b_{2,2} & r_3 \\ \hline c_1 & c_2 & \end{array}$$

Then use the induction hypothesis again to obtain the 2×2 matrix in the middle above. Combining these numbers yields the desired 3×2 matrix on the right.

It is straightforward to generalize this argument to an induction in $m + n$; in the induction step one uses the induction hypothesis twice: either for $m-1$ and n (with row sums $r_1, \ldots, r_{m-2}, r_{m-1}+r_m$) as well as for $m = 2$ and n (with row sums r_{m-1}, r_m), or for m and $n-1$ as well as for m and $n = 2$. □

Remark C.4.4. (1) Every monoid-valued functor $M^{(-)}$ is finitary (Definition 4.3.1). Indeed, it easy to see that $M^{(-)}$ is finitely bounded (Proposition C.5.2): given $\mu \in M^{(X)}$, take the finite set $X' = \mathsf{supp}(\mu)$, and let $\mu' \colon X' \to M$ be the restriction of μ. Then we clearly have $M^{(m)}(\mu') = \mu$ for the inclusion map $m \colon X' \hookrightarrow X$.

(2) Monoid-valued functors map finite coproducts to finite products. Indeed, we have a canonical isomorphism $M^{(X+Y)} \cong M^{(X)} \times M^{(Y)}$ mapping $\mu \colon X+Y \to M$ to the pair formed by $\mu \cdot \mathsf{inl} \colon X \to M$ and $\mu \cdot \mathsf{inr} \colon Y \to M$.

Proposition C.4.5. *For every monoid $(M, +, 0)$ the following are equivalent:*

(1) *M is refinable,*

(2) *$M^{(-)}$ preserves weak epi-pullbacks, and*

(3) *$M^{(-)}$ preserves weak kernel pairs.*

The equivalence (1) ⇔ (3) is due to Gumm and Schröder [152, Thm. 5.13(ii)]. The equivalence (1) ⇔ (2) was established by Seifan et al. [275, Lemma 3.7].

Proof The implication (2) ⇒ (3) follows from Proposition C.3.21 since every monoid-valued functor clearly preserves monomorphisms.

(3) ⇒ (1) Since $M^{(-)}$ preserves weak kernel pairs, it maps the following pullback to a weak pullback

$$\begin{array}{ccc} 2 \times 2 & \xrightarrow{\;p_1\;} & 2 \\ {\scriptstyle p_2}\downarrow & \lrcorner & \downarrow{\scriptstyle !} \\ 2 & \xrightarrow{\;\;!\;\;} & 1 \end{array}$$

Given $r_0 + r_1 = c_0 + c_1$ in M, take $\mu, \nu \in M^{(2)}$ with $\mu(i) = r_i$ and $\nu(i) = c_i$ for $i = 0, 1$. Then $M^{(!)}(\mu) = M^{(!)}(\nu)$ (and this is $\sigma \colon 1 \to M$ given by the sum). Therefore we have an element $\pi \in M^{(2 \times 2)}$ such that $M^{(p_1)}(\pi) = \mu$ and $M^{(p_2)}(\pi) = \nu$. This means that π corresponds to a 2×2 matrix with row sums r_0, r_1 and column sums c_0, c_1. Thus M is refinable.

(1) ⇒ (2) Analyzing the above argument and using Lemma C.4.3, we see that refinability of the monoid M is clearly equivalent to $M^{(-)}$ weakly preserving

pullbacks of the following form

$$\begin{array}{ccc} m \times n & \xrightarrow{p_1} & m \\ {\scriptstyle p_2}\downarrow & \lrcorner & \downarrow{\scriptstyle !} \\ n & \xrightarrow{!} & 1 \end{array} \tag{C.7}$$

where m, n denote finite ordinals. Using Lemma C.3.23 and that $M^{(-)}$ is finitary (Remark C.4.4(1)), it suffices to show that $M^{(-)}$ preserves finite epi-pullbacks. Suppose that $f\colon X \twoheadrightarrow Z$ and $g\colon Y \twoheadrightarrow Z$ are surjective maps between finite sets and denote their pullback by $p\colon P \twoheadrightarrow X$ and $q\colon P \twoheadrightarrow Y$. For every $z \in Z$ define $X_z = f^{-1}(z)$, $Y_z = g^{-1}(z)$, and $P_z = X_z \times Y_z$. We thus have the following finite epi-pullbacks

$$\begin{array}{ccc} P_z & \xrightarrow{p_z} & X_z \\ {\scriptstyle q_z}\downarrow & \lrcorner & \downarrow{\scriptstyle !} \\ Y_z & \xrightarrow{!} & \{z\} \end{array}$$

These pullbacks are of the form in (C.7) (up to isomorphism), and therefore $M^{(-)}$ weakly preserves them. Observe that the pullback of f and g is the coproduct of all these pullbacks (using that coproducts and pullbacks commute in Set). Thus, from Remark C.4.4(2) we see that $M^{(-)}$ maps the pullback of f and g to the product of the following weak pullbacks

$$\begin{array}{ccc} M^{(P_z)} & \xrightarrow{M^{(p_z)}} & M^{(X_z)} \\ {\scriptstyle M^{(q_z)}}\downarrow & \lrcorner & \downarrow{\scriptstyle M^{(!)}} \\ M^{(Y_z)} & \xrightarrow{M^{(!)}} & M^{(\{z\})} \end{array}$$

Since products of weak pullbacks are weak pullbacks, we are done. □

Proposition C.4.6. *The functor $M^{(-)}$ preserves inverse images if and only if the monoid M is positive.*

Proof (1) Suppose that $M^{(-)}$ preserves inverse images. Then it preserves the pullback

$$\begin{array}{ccc} \emptyset & \xrightarrow{\mathsf{id}} & \emptyset \\ {\scriptstyle u}\downarrow & \lrcorner & \downarrow{\scriptstyle v} \\ 2 & \xrightarrow{!} & 1 \end{array}$$

Now assume that $a_0 + a_1 = 0$ and define $\mu \in M^{(2)}$ by $\mu(i) = a_i$. Then for the unique $\nu \in M^{(\emptyset)}$ we have $M^{(!)}(\mu) = \mathsf{const}_0 = M^{(v)}(\nu)$. By the universal property of the pullback we thus have $M^{(u)}(v) = \mu$, which shows that μ must be const_0, too. Thus $a_0 = 0 = a_1$, whence M is positive.

(2) Suppose that M is positive and consider a pullback

$$\begin{array}{ccc} P & \xrightarrow{f'} & S \\ {\scriptstyle p}\downarrow & \lrcorner & \downarrow{\scriptstyle s} \\ X & \xrightarrow{f} & Y \end{array}$$

Without loss of generality we assume that $s: S \hookrightarrow Y$ and $p: P \hookrightarrow X$ are inclusion maps of subsets. Given $\mu \in M^{(X)}$ such that $M^{(f)}(\mu)$ lies in the image of $M^{(s)}$, we know that $M^{(f)}(\mu)(y) = \sum_{x \in f^{-1}(y)} \mu(x) = 0$ for all $y \notin S$. By the positivity of M, this implies that for every $y \notin S$ and $x \in f^{-1}(y)$, we have $\mu(x) = 0$. In other words, $\mu(x)$ can be non-zero only if x lies in $f^{-1}(S) = P$. This means that μ lies in the image of $M^{(p)}$, as desired. □

Corollary C.4.7. *The functor $M^{(-)}$ preserves weak pullbacks iff the monoid M is positive and refinable.*

Indeed, this follows from the above corollary and Proposition C.3.19.

Example C.4.8. The bag functor $\mathcal{B}$ preserves weak pullbacks.

Proposition C.4.9. *The finite distribution functor $\mathcal{D}_{\mathrm{f}}$ preserves weak pullbacks.*

Proof Use that $\mathcal{D}_{\mathrm{f}}$ is a subfunctor of the monoid-valued functor ${\mathbb{R}_{\geq 0}}^{(-)}$. Since $\mathbb{R}_{\geq 0}$ is positive and refinable (Example C.4.2), ${\mathbb{R}_{\geq 0}}^{(-)}$ preserves weak pullbacks due to Corollary C.4.7. Thus, given a pullback

$$\begin{array}{ccc} P & \xrightarrow{p} & X \\ {\scriptstyle q}\downarrow & \lrcorner & \downarrow{\scriptstyle f} \\ Y & \xrightarrow{g} & Z \end{array}$$

and two distributions $\mu \in \mathcal{D}_{\mathrm{f}}X$ and $v \in \mathcal{D}_{\mathrm{f}}Y$ such that $\mathcal{D}_{\mathrm{f}}f(\mu) = \mathcal{D}_{\mathrm{f}}g(v)$, we obtain $\pi \in {\mathbb{R}_{\geq 0}}^{(P)}$ such that ${\mathbb{R}_{\geq 0}}^{(p)}(\pi) = \mu$ and ${\mathbb{R}_{\geq 0}}^{(q)}(\pi) = v$. Using that the above pullback is given by $P = \{(x, y) : f(x) = g(y)\}$, the first equation yields that

$$\sum_{(x,y) \in P} \pi(x, y) = \sum_{x \in \mathsf{supp}(\mu)} \mu(x) = 1.$$

Indeed, positivity ensures that for every $x \in \mathsf{supp}(\mu)$ there is some $y \in \mathsf{supp}(v)$ such that $f(x) = g(y)$; this yields '$\geq$'. Similarly, there is no $(x, y) \in \mathsf{supp}(\pi)$ such that $\mu(x) = 0$, which yields '$\leq$'. □

C.5 Finitary Set Functors

Recall from Definition 4.3.1 that a set functor is *finitary* if it preserves directed colimits. Equivalently, one can use filtered colimits (see Definition 10.1.7) rather than the directed ones (these are more special), or one can use chains as defined in Remark 6.0.5 (even more special):

Theorem C.5.1 [23, Cor. 1.5 & 1.7]. *A category has colimits of nonempty chains iff it has filtered colimits. For such a category $\mathcal{A}$ the preservation of the following types of colimits are equivalent for every functor $F\colon \mathcal{A} \to \mathcal{B}$:*

(1) *filtered colimits,*

(2) *directed colimits, and*

(3) *colimits of nonempty chains.*

In the case of set functors, an important equivalent concept is to be finitely bounded (see Definition 4.3.3):

Proposition C.5.2 [52, Cor. 3.3]. *A set functor is finitely bounded iff it is finitary.*

Proof (1) If F is finitary, then it is finitely bounded because every set X is a directed colimit of its finite subsets $m\colon M \hookrightarrow X$. More precisely, the diagram of all finite subsets of X and inclusion maps between them is directed, and its colimit is formed by the cocone of all inclusion maps m above. Since F preserves this colimit, the cocone of all $Fm\colon FM \to FX$ is collectively epic, which means that F is finitely bounded.

(2) Let F be finitely bounded. Given a directed diagram $D\colon I \to \mathcal{A}$, we prove that F preserves its colimit. We denote the connecting morphisms by $d_{ij}\colon Di \to Dj$ for every $i \leq j$ in the directed poset I. The colimit cocone, denoted by $c_i\colon Di \to C$, is easily seen to have the following two properties (cf. Remark 10.1.16(5)):

(a) the colimit injections form a jointly epic family: $C = \bigcup_{i\in I} c_i[Di]$, and
(b) given $i \in I$ and elements $x, y \in Di$ with $c_i(x) = c_i(y)$, there exists $j \geq i$ in I such that $d_{ij}(x) = d_{ij}(y)$.

Moreover, these properties characterize directed colimits in Set. This means that a cocone $c_i\colon Di \to C, i \in I$, of D satisfying them is a colimit of D. Indeed, given any cocone $c'_i\colon Di \to C'$ there exists a unique $f\colon C \to C'$ with $c'_i \cdot f = c_i$ for every $i \in I$: due to (a) we have to define $f(x) = c'_i(y)$ for every $x \in C, i \in I$ and $y \in Di$ with $c_i(y) = x$. It follows from (a) and (b) that f is a well-defined function. Moreover, we clearly have $c'_i \cdot f = c_i$.

We are prepared to prove that the diagram FD has the colimit cocone $Fc_i\colon FDi \to FC$ $(i \in I)$. To this end, we only need to verify that it fulfils conditions (a) and (b).

Ad (a). Given $x \in FC$, since F is finitely bounded, there exists a finite set $m: M \hookrightarrow C$ and $x' \in FM$ with $Fm(x') = x$. Since I is directed and M is finite, there exists $i \in I$ such that M is a subset of $c_i[Di]$: use condition (a) for the cocone c_i. Thus, $m = c_i \cdot m'$ for some $m': M \to Di$. We conclude that $x = Fc_i(Fm'(x'))$, whence $FC = \bigcup_{i \in I} Fc_i[FDi]$.

Ad (b). Given $x, y \in FDi$ with $Fc_i(x) = Fc_i(y)$ we will exhibit $j \geq i$ such that $Fd_{ij}(x) = Fd_{ij}(y)$. Choose a finite subset $m: M \hookrightarrow Di$ such that both x and y lie in $Fm[FM]$. Condition (b) (for the cocone c_i) implies that every pair $u, v \in M$ merged by c_i is merged by some connecting morphism. Because $M \times M$ is finite, we can choose one connecting morphism for all such pairs u, v. This means that there exists $j \geq i$ such that for all $u, v \in M$ we have that $c_i(u) = c_i(v)$ implies $d_{ij}(u) = d_{ij}(v)$. In other words, the kernel equivalence of $c_i \cdot m: M \to C$ is contained in that of $d_{ij} \cdot m: M \to Dj$. Consequently, $d_{ij} \cdot m$ factorizes through $c_i \cdot m$, that is, we obtain a map $f: C \to Dj$ with $f \cdot c_i \cdot m = d_{ij} \cdot m$. Recall that we have $x = Fm(x')$ and $y = Fm(y')$ for some $x', y' \in FM$. Therefore, we obtain the desired property as follows:

$$\begin{aligned} Fd_{ij}(x) &= F(d_{ij} \cdot m)(x') \\ &= F(f \cdot c_i \cdot m)(x') \\ &= Ff(Fc_i(x)) \\ &= Ff(Fc_i(y)) \\ &= F(f \cdot c_i \cdot m)(y') \\ &= F(d_{ij} \cdot m)(y') \\ &= Fd_{ij}(y). \end{aligned}$$

□

Our next result is a characterization of finitary set functors in terms of their pre-fixed points. It depends on results concerning *almost disjoint families* and the *cofinality operation* on ordinals. We review the definitions and results which we need. For further background, see a set theory text such as [176].

Definition C.5.3. The *cofinality* of a limit ordinal λ is the smallest ordinal i such that there is an i-chain of ordinals $c_i < \lambda$ such that $\lambda = \bigvee_{j<i} c_j$. Equivalently, i is the least ordinal so that there is a cofinal map (Definition 6.0.4(1)) $f: i \to \lambda$. We write $\mathrm{cf}(\lambda)$ for the cofinality of λ.

Remark C.5.4. For every limit ordinal λ, the cofinality $\mathrm{cf}(\lambda)$ is an infinite regular cardinal. An infinite cardinal λ is regular (Definition 6.0.4(2)) iff $\mathrm{cf}(\lambda) = \lambda$.

Example C.5.5. (1) The cardinal $\aleph_\omega$ has cofinality ω. In general, for λ a limit ordinal, $\mathrm{cf}(\aleph_\lambda) = \mathrm{cf}(\lambda)$.

(2) If i and j are limit ordinals, then $\mathrm{cf}(i + j) = \mathrm{cf}(j)$, where $i + j$ is the ordinal sum (Definition 6.0.1).

(3) Given an infinite cardinal λ, for every infinite cardinal $\kappa > \lambda$,

$$\mathrm{cf}(\aleph_{\kappa+\lambda}) = \mathrm{cf}(\kappa + \lambda) = \mathrm{cf}(\lambda).$$

(Here we again use $\kappa + \lambda$ for the ordinal sum.) Since $\aleph_{\kappa+\lambda} > \aleph_\kappa \geq \kappa$, this shows a fact which we use below: for every ordinal λ there are arbitrarily large cardinals with the same cofinality as λ.

Definition C.5.6. Let λ be an infinite cardinal. Given a set X, a family $\mathcal{S}$ of subsets of X is *almost λ-disjoint* if

$$|S \cap T| < \lambda \qquad \text{for } S \neq T \text{ in } \mathcal{S}.$$

A classical result due to Sierpiński (see [176, Lemma 9.23] or [214, Prop. 5.28]) states that if an infinite set X has cardinality λ, then there exists an almost λ-disjoint family $\mathcal{S}$ with $|\mathcal{S}| > \lambda$. We need a stronger statement proved by Baumgartner [76].

Proposition C.5.7 [76, Thm. 2.8 and 2.9]**.** *Let $\lambda \leq \kappa$ be infinite cardinals with the same cofinality. Let X be a set of cardinality κ. Then there is an almost λ-disjoint family $\mathcal{S}$ of subsets of X of cardinality λ with $|\mathcal{S}| > \kappa$.*

Lemma C.5.8 [98]**.** *Let Z be an infinite pre-fixed point of a set functor F. Let X be any set such that $|X| \leq |Z|$ and these cardinals have the same cofinality. For every $x \in FX$ there is a subset $m\colon Y \hookrightarrow X$ such that $|Y| < |X|$ and $x \in Fm[FY]$.*

Proof Let $\kappa = |Z|$ and $\lambda = |X|$. Fix a family $\mathcal{S}$ of subsets of Z as in Proposition C.5.7. By adding one and the same point to each $S \in \mathcal{S}$, we may assume that $S \cap T \neq \emptyset$ for all $S, T \in \mathcal{S}$. Choose bijections $b_S\colon X \to S$ for $S \in \mathcal{S}$ using $|S| = |X|$. Let $m_S\colon X \to Z$ be $j_S \cdot b_S$, where $j_S\colon S \hookrightarrow Z$ is the inclusion.

Fix $x \in FX$. For all $S \in \mathcal{S}$, $Fm_S(x) \in FZ$. Since Z is a pre-fixed point of F and $|\mathcal{S}| > \kappa \geq |FZ|$, the elements $Fm_S(x)$ cannot be pairwise distinct. Let $S \neq T$ be such that $Fm_S(x) = Fm_T(x)$. Put $Y = S \cap T$, so $|Y| < |X|$. Write $i_S\colon Y \hookrightarrow S$, and $i_T\colon Y \hookrightarrow T$ for the inclusion maps. Consider the intersection on the left below:

$$\begin{array}{ccc} Y & \xrightarrow{i_T} & T \\ {\scriptstyle i_S}\downarrow & \lrcorner & \downarrow{\scriptstyle j_T} \\ S & \xrightarrow{j_S} & \kappa \end{array} \qquad\qquad \begin{array}{ccc} FY & \xrightarrow{Fi_T} & FT \\ {\scriptstyle Fi_S}\downarrow & \lrcorner & \downarrow{\scriptstyle Fj_T} \\ FS & \xrightarrow{Fj_S} & F\kappa \end{array}$$

That square is a nonempty intersection, and so by Theorem C.2.2 it is preserved by F, that is, the square on the right above is a pullback. Moreover, $Fj_S(Fb_S(x)) = Fj_T(Fb_T(x))$. Thus there is some $y \in FY$ such that $Fi_S(y) = Fb_S(x)$. For $m = b_S^{-1} \cdot i_S$, we obtain $Fm(y) = Fb_S^{-1}(Fi_S(y)) = x$. □

Theorem C.5.9 [45]. *A set functor F is finitary iff all sufficiently large sets are pre-fixed points of F. That is, there is a cardinal κ so that if $|X| \geq \kappa$, then $|FX| \leq |X|$.*

Proof First, assume that F is finitary. By Proposition 4.3.13, F has a finitary presentation $\varepsilon\colon H_\Sigma \twoheadrightarrow F$. Let κ be an infinite cardinal at least as large as the cardinality of the signature Σ. For all sets X with $|X| \geq \kappa$ we conclude $|\Sigma| \times |X| = |X|$. Also $|X|^n = |X|$ for all $n \in \mathbb{N}$. Thus, X is a pre-fixed point:

$$|FX| \leq |H_\Sigma X| = \sum_{\sigma \in \Sigma} |X|^{\mathsf{ar}(\sigma)} \leq |\Sigma| \times |X| = |X|.$$

Conversely, fix κ as in our statement. We prove that F is finitely bounded and therefore finitary by Proposition C.5.2. To this end we show by transfinite induction that for all infinite cardinals λ the following holds for every set X with $|X| = \lambda$: given $x \in FX$, there is a finite subset $m\colon X_0 \hookrightarrow X$ so that $x \in Fm[FX_0]$. By the induction hypothesis, we may assume our result holds for all infinite cardinals below λ. By Example C.5.5(3), we know that there are arbitrarily large cardinals with the same cofinality as λ. In particular, there is a cardinal $\mu \geq \kappa$ with the same cofinality as λ; hence μ is a pre-fixed point of F by assumption. By Lemma C.5.8, there is some subset $m\colon Y \hookrightarrow X$ with $|Y| < |X|$ so that $x \in Fm[FY]$. If Y is finite, we are done. Otherwise, take $y \in FY$ so that $Fm(y) = x$. We apply the induction assumption to $|Y| < \lambda$: there is a finite set $n\colon U \hookrightarrow Y$ so that $y \in Fn[FU]$. Then $m \cdot n\colon U \hookrightarrow X$ is a finite subset, and the image $F(m \cdot n)[FU]$ contains x. □

Corollary C.5.10. *Let F be a set functor not essentially constant on nonempty sets (see Definition 6.4.2 and Theorem 6.4.3(2)). Then the following are equivalent:*

(1) *F is finitary,*

(2) *F is finitely bounded, and*

(3) *All sufficiently large sets are fixed points of F.*

This is immediate from Theorems C.5.9 and 6.4.3.

For set functors F preserving subset inclusion maps, the concept of being *super-finitary* was introduced by Adámek and Trnková [25, Def. VII.2.4] as follows: F preserves finite sets, and there is a natural number n such that for every set X, $FX = \bigcup_{T \in \mathscr{P}_n(X)} FT$, where $\mathscr{P}_n X$ is the set of n-element subsets of X. Dropping the condition that F preserve inclusions, we obtain the following:

Definition C.5.11. A set functor F is *super-finitary* if it preserves finite sets and there exists a natural number n such that for every $X \neq \emptyset$, the cocone of $Ff: Fn \to FX$ (for all $f: n \to X$) is collectively epic.

Proposition C.5.12. *A set functor is super-finitary iff it is a quotient of a polynomial functor H_Σ for some finite, finitary signature.*

Proof (1) Let F be super-finitary. Let the signature Σ have $\Sigma_0 = F\emptyset$ and $\Sigma_n = Fn$ for all natural numbers $n \geq 1$. (That is, the constants of Σ are the elements of $F\emptyset$ and the n-ary function symbols are the elements of Fn.) Thus

$$H_\Sigma X = F\emptyset + X^n \times Fn.$$

Define $\varepsilon_X: H_\Sigma X \to FX$ by assigning to $t \in F\emptyset$ the value $F!_X(t)$ for the empty map $!_X: \emptyset \to X$ and to $(f, s) \in X^n \times Fn$ the value $Ff(s)$ (where $s \in Fn$ and $f: n \to X$).

These map are clearly surjective. We now verify they yield a natural transformation $\varepsilon: H_\Sigma \to F$. Given a map $h: X \to Y$ we show that the square below commutes:

$$\begin{array}{ccc} H_\sigma X & \xrightarrow{\varepsilon_X} & FX \\ {\scriptstyle H_\Sigma h}\downarrow & & \downarrow{\scriptstyle Fh} \\ H_\Sigma Y & \xrightarrow{\varepsilon_Y} & FY \end{array}$$

For each $t \in F\emptyset$ we have $H_\Sigma h(t) = t$, thus $\varepsilon_Y \cdot H_\Sigma h(t) = F!_Y(t)$. Moreover, $\varepsilon_X(t) = F!_X(t)$. Hence $!_Y \cdot h =!_X$ implies $Fh \cdot \varepsilon_X(t) = \varepsilon_Y \cdot H_\Sigma h(t)$. Finally, for each $(f, s) \in X^n \times Fn$ we have $H_\Sigma h(f, s) = (h \cdot f, s)$, thus

$$\varepsilon_Y \cdot H_\Sigma h(f, s) = F(h \cdot f)(s) = Fh \cdot Ff(s) = Fh\varepsilon_X(f, s).$$

(2) Let F admit an epi-transformation $\varepsilon: H_\Sigma \twoheadrightarrow F$ with Σ finite and finitary. It is easy to verify that H_Σ is super-finitary: take n to be the largest arity in Σ. For each set $X \neq \emptyset$, the cocone of all $H_\Sigma f$ is collectively epic: given $\sigma(x_1, \ldots, x_k) \in H_\Sigma(X)$, let $f: n \to X$ be given by $f(i) = x_i$ for $i = 1, \ldots, k$. Then $H_\Sigma f$ maps $\sigma(1, \ldots, k)$ to $\sigma(x_1, \ldots, x_k)$.

The functor F preserves finite sets since $|F(X)| \leq |H_\Sigma(X)|$. For every $X \neq \emptyset$, the cocone of Ff for $f: n \to X$ is collectively epic: given $t \in FX$ chose $t_0 \in H_\Sigma X$ such that $\varepsilon_X(t_0) = t$. We know that there exists $f: n \to X$ with $t_0 = H_\Sigma f(s_0)$ for some $s_0 \in H_\Sigma n$. The element $s = \varepsilon_n(s_0)$ has the desired property that $t = Ff(s)$ because

$$Ff(s) = \varepsilon_X H_\Sigma f(s_0) = \varepsilon_X(t_0) = t. \qquad \square$$

In the next result we recall the concept of a *basic equation* from Definition 4.3.16.These are equations between terms of the form $\sigma(x_1, \ldots, x_k)$ where σ is a k-ary function symbol in Σ, or a bare variable x.

Corollary C.5.13. *A set functor is super-finitary iff it is finitary and has a presentation by finitely many operations and finitely many basic equations.*

Proof Indeed, given $\varepsilon\colon H_\Sigma \twoheadrightarrow F$ with Σ finitary and finite, we know that F is presented by basic equations. Since there are are only finitely many such terms (for $x_i, x \in \{0, \ldots, k-1\}$), there are only finitely many basic equations.

Conversely, given finitely many operations and finitely many basic equations E, the corresponding set functor H_Σ/E is clearly super-finitary. □

Corollary C.5.14 [49, Cor. 3.34]. *In the category of finitary set functors, the finitely presentable objects are precisely the super-finitary functors.*

Note that there is a slightly different definition of super-finitary in op. cit.: the condition $X \neq \emptyset$ was not included. But the proof is correct.

C.6 Trnková Hull

We have stated in Proposition 8.1.13 that for every set functor F its Trnková hull $\bar{F}$ is the unique set functor which agrees with F on all nonempty sets (and functions) and which preserves finite intersections (including empty ones). We now show how to construct $\bar{F}$ and to prove that proposition.

In Chapters 2 and 3 we have used $!\colon \emptyset \to X$ to denote the empty map. In this section, we denote empty maps by $e_X\colon \emptyset \to X$.

The construction in Definition C.6.1 follows Barr [72, Proof of Thm. 3.2] and is based on a simple idea: recall from Proposition 4.4.1 that every set functor preserves finite nonempty intersections. Then consider the two subobjects $\mathsf{t}, \mathsf{f}\colon 1 \to 2 = 1+1$ whose intersection is the empty function $e_1\colon \emptyset \to 1$. Since $\bar{F}$ must preserve that intersection it is clear that $\bar{F}e_1$ is injective and forms (not only a pullback but also) an equalizer of $\bar{F}\mathsf{t} = F\mathsf{t}$ and $\bar{F}\mathsf{f} = F\mathsf{f}$.

Definition C.6.1. Let F be a set functor. We define a functor $\bar{F}$ as follows: For nonempty sets A, let $\bar{F}A = FA$, and for every map $k\colon A \to B$ with nonempty domain, let $\bar{F}k = Fk$. For the empty map $e_1\colon \emptyset \to 1$ we obtain $\bar{F}\emptyset$ and $\bar{F}e_1\colon \bar{F}\emptyset \to \bar{F}1$ as the following equalizer

$$\bar{F}\emptyset \xrightarrow{\bar{F}e_1} \bar{F}1 = F1 \underset{F\mathsf{f}}{\overset{F\mathsf{t}}{\rightrightarrows}} F2. \tag{C.8}$$

To define $\bar{F}$ on morphisms, put $\bar{F}e_\emptyset = \mathrm{id}_{\bar{F}\emptyset}$, and for $A \neq \emptyset$ let $a\colon 1 \to A$ be an element, and put

$$\bar{F}e_A = (\bar{F}\emptyset \xrightarrow{\bar{F}e_1} \bar{F}1 \xrightarrow{\bar{F}a} \bar{F}A). \tag{C.9}$$

This functor $\bar{F}$ is the *Trnková hull* of F.

Remark C.6.2. Let us check that $\bar{F}$ is a well-defined functor. First, $\bar{F}e_A$ is independent of the choice of a. For suppose that $a'\colon 1 \to A$ is given. Let $k = [a, a']\colon 2 = 1 + 1 \to A$; that is, $k \cdot \mathsf{t} = a$ and $k \cdot \mathsf{f} = a'$. Then

$$\bar{F}a \cdot \bar{F}e_1 = Fk \cdot F\mathsf{t} \cdot \bar{F}e_1 = Fk \cdot F\mathsf{f} \cdot \bar{F}e_1 = \bar{F}a' \cdot \bar{F}e_1.$$

To show that $\bar{F}$ preserves composition, it suffices to consider $h \cdot e_A$, where $h\colon A \to B$. We use the fact above to see that

$$\bar{F}(h \cdot e_A) = F(h \cdot a) \cdot \bar{F}e_1 = Fh \cdot Fa \cdot \bar{F}e_1 = \bar{F}h \cdot \bar{F}e_A.$$

Example C.6.3. (1) Every polynomial functor, the power-set functor (and in fact almost all set functors we have used so far) actually preserve finite intersections, thus, they are their own Trnková hulls.

(2) For every set M denote by C_{M1} the set functor assigning 1 to nonempty sets and M to the empty one. Its Trnková hull is the constant functor with value 1. Thus the Trnková hull adds an element to C_{01}, whereas for sets M with more than one element $C_{M1}\emptyset$ is reduced to one element.

(3) The subfunctor $\mathscr{P}^+$ of $\mathscr{P}$ assigns to a set the set of all its nonempty subsets. Its Trnková hull is $\mathscr{P}$.

(4) Let B be the same as R (Example C.1.1(7)) except that $B\emptyset = \emptyset$. Note that $R\emptyset = \{d\}$. The Trnková hull $\bar{B}$ adds a new element to $B\emptyset$ which $\bar{B}e_A$ maps to the equivalence class of the diagonal of $A \times A$ (for all nonempty sets A). Thus, $\bar{B}$ is the functor R. This shows that in general, the Trnková hull does not preserve weak pullbacks nor inverse images.

Trnková's original construction [292] made use of the functor C_{01} from Example C.1.1(8), mapping the empty set to itself and all nonempty sets to 1. She defined $\bar{F}\emptyset$ to be the set of natural transformations $\tau\colon C_{01} \to F$. We now show that this construction is equivalent to Definition C.6.1.

Proposition C.6.4. *The collection N of all natural transformations $\tau\colon C_{01} \to F$ forms a set: the function $i\colon N \to F1$ with $i(\tau) = \tau_1$ is monic, indeed it is the equalizer of $F\mathsf{t}, F\mathsf{f}\colon F1 \rightrightarrows F2$.*

Proof Note first that every natural transformation $\tau\colon C_{01} \to F$ is uniquely determined by its component $\tau_1\colon 1 \to F1$. Indeed, $\tau_\emptyset\colon \emptyset \to F\emptyset$ must be the empty map, and for $A \neq \emptyset$ the naturality of τ yields that we have (independently of the choice of $a\colon 1 \to A$):

$$\tau_A = \tau_A \cdot \mathsf{id}_1 = \tau_A \cdot C_{01} a = Fa \cdot \tau_1. \tag{C.10}$$

This implies that the natural transformations $C_{01} \to F$ form a set N. Moreover, for the map $i\colon N \to F1$ above, we have $F\mathsf{t} \cdot i = F\mathsf{f} \cdot i$, since

$$F\mathsf{t} \cdot i(\tau) = F\mathsf{t} \cdot \tau_1 = \tau_2 = F\mathsf{f} \cdot \tau_1 = F\mathsf{f} \cdot i(\tau).$$

For every $y \in F1$ such that $F\mathsf{t}(y) = F\mathsf{f}(y)$, let $z \in \bar{F}\emptyset$ be the unique element with $\bar{F}e_1(z) = y$, which we obtain since $\bar{F}e_1$ is the equalizer of $F\mathsf{t}$ and $F\mathsf{f}$. We will show that there is a natural transformation $\tau\colon C_{01} \to F$ with $e(\tau) = \tau_1 = y$. Moreover, the latter equation uniquely determines τ, as we have seen above. Put $\tau_\emptyset = e_\emptyset$, and for every nonempty set A put

$$\tau_A = (1 \xrightarrow{z} \bar{F}\emptyset \xrightarrow{\bar{F}e_A} \bar{F}A = FA).$$

We clearly have $\tau_1 = \bar{F}e_1(z) = y$. So let us check the naturality of τ. The only non-trivial naturality squares are those determined by a map $k\colon A \to B$, where $A \neq \emptyset \neq B$; we use that $k \cdot e_A = e_B$ clearly holds due to initiality of $\emptyset$:

$$Fk \cdot \tau_A = \bar{F}k(\bar{F}e_A(z)) = \bar{F}(k \cdot e_A)(z) = \bar{F}e_B(z) = \tau_B = \tau_B \cdot \mathsf{id}_1 = \tau_B \cdot C_{01}k.$$

This proves that $e\colon N \to F1$ is the equalizer of $F\mathsf{t}$ and $F\mathsf{f}$ as desired. □

Remark C.6.5. Note that with the alternative description of $\bar{F}\emptyset$ in Proposition C.6.4 we have that $\bar{F}e_A\colon \bar{F}\emptyset \to \bar{F}A = FA$ maps a natural transformation $\tau\colon C_{01} \to F$ to the element $\tau_A\colon 1 \to FA$ for every nonempty set A. Indeed this follows by the naturality of τ. Using that $e_A = a \cdot e_1$ holds for every $a\colon 1 \to A$, we get

$$\bar{F}e_A(\tau) = \bar{F}a(\bar{F}e_1(\tau)) = Fa \cdot \tau_1 \overset{\text{(C.10)}}{=} \tau_A.$$

The following result was originally proved by Trnková [292, Propositions III.5 and II.4]; for a more direct proof see Adámek and Trnková [25, Theorem III.4.5]. We present a new proof using Proposition C.6.4.

Theorem C.6.6. *The functor $\bar{F}$ of Definition C.6.1 coincides with F on nonempty sets and functions and preserves finite intersections (whence all monomorphisms). Thus, $\bar{F}$ is the Trnková hull of F.*

Proof We begin with the observation that for every $A \neq \emptyset$, $\bar{F}e_A$ is injective. This is because both $\bar{F}a$, for every $a\colon 1 \to FA$, and $\bar{F}e_1$ are injective, and $\bar{F}e_A = \bar{F}a \cdot \bar{F}e_1$.

Recall from Theorem C.2.2 that every functor preserves nonempty intersections. So we need only consider a pullback of the form

$$\begin{array}{ccc} \emptyset & \xrightarrow{e_A} & A \\ {\scriptstyle e_B}\downarrow & \lrcorner & \downarrow{\scriptstyle g} \\ B & \xrightarrow[h]{} & X \end{array} \tag{C.11}$$

with g and h monic and $A, B \neq \emptyset$. Let $a \in \bar{F}A = FA$ and $b \in \bar{F}B = FB$ be such that $\bar{F}g(a) = \bar{F}h(b)$. It is our task to find some $z \in \bar{F}\emptyset$ so that $\bar{F}e_A(z) = a$ and $\bar{F}e_B(z) = b$. Uniqueness is clear since $\bar{F}e_A$ is monic.

For the existence, we first prove that for every nonempty set Y and every pair of maps $g' \colon A \to Y$ and $h' \colon B \to Y$ we have $Fg'(a) = Fg'(b)$. Indeed, since $g \colon A \rightarrowtail X$ and $h \colon B \rightarrowtail X$ are disjoint, we can find a map $k \colon X \to Y$ such that $k \cdot g = g'$ and $k \cdot h = h'$. (Identify A and B with their images $g[A]$ and $h[B]$ in X and define $k(x) = g(x)$ for $x \in A$, $k(x) = h(x)$ for $x \in B$, and let $k(x) \in Y$ be arbitrary else.) Then we have

$$Fg'(a) = Fk \cdot Fg(a) = Fk \cdot Fh(b) = Fh'(b).$$

This proves that $Fg'(a) = Fh'(b)$ is independent of the choice of the pair g', h'. (For if we choose g', h', then we have $Fg'(a) = Fh'(b) = Fg'(a)$ and $Fh'(b) = Fg'(a) = Fh'(b)$.)

Consequently, for every set Y we can define an element of FY as a function $z_Y \colon 1 \to FY$ as follows:

$$z_Y = Fg'(a) = Fh'(b) \colon 1 \to FY \qquad \text{for some } g' \colon A \to Y.$$

The above argument proves that z_Y is independent of the choice of the maps g', h'. Moreover, from that independence it follows that for every map $u \colon Y \to Z$,

$$Fu(z_Y) = Fu(Fg'(a)) = F(u \cdot g')(a) = z_Z.$$

Thus, the elements z_Y are the components of a natural transformation $z \colon C_{01} \to F$, which means that we have the corresponding element $z \in \bar{F}\emptyset$ (Proposition C.6.4). According to Remark C.6.5 we also have

$$\bar{F}e_A(z) = z_A = F\mathsf{id}_A(a) = a \qquad \text{and} \qquad \bar{F}e_B(z) = z_B = F\mathsf{id}_B(b) = b$$

by using the above independence once again. □

Notation C.6.7. We denote by $r \colon F \to \bar{F}$ the natural transformation with components given as follows: $r_X = \mathsf{id}_{FX}$ for nonempty X, and $r_\emptyset \colon F\emptyset \to \bar{F}\emptyset$ is uniquely determined by commutativity of the triangle below, using the universal property of the equalizer (C.8) and that we clearly have $F\mathsf{t} \cdot Fe_1 = F\mathsf{t} \cdot Fe_1$:

$$
\begin{array}{ccc}
F\emptyset & & \\
{\scriptstyle r_\emptyset}\downarrow & \searrow^{Fe_1} & \\
\bar{F}\emptyset & \xrightarrow{\bar{F}e_1} & F1
\end{array}
$$

The naturality squares for r obviously commute for nonempty functions. For every empty map e_A with $A \neq \emptyset$, we have the following diagram, which commutes since $e_A = a \cdot e_1$ for every $a \colon 1 \to A$:

$$
\begin{array}{ccccc}
F\emptyset & & \xrightarrow{\quad r_\emptyset \quad} & & \bar{F}\emptyset \\
 & \searrow^{Fe_1} & & \swarrow^{\bar{F}e_1} & \\
{\scriptstyle Fe_A}\downarrow & & F1 & & \downarrow{\scriptstyle \bar{F}e_A} \\
 & & & \searrow^{Fa} & \\
FA & & \overset{r_A}{=\!=\!=\!=} & & \bar{F}A
\end{array}
$$

Theorem C.6.8 [45, Cor. VII.2]. *For every set functor its Trnková hull is a reflection in the category of set functors preserving finite intersections.*[1]
In more detail, for every set functor F, the natural transformation $r \colon F \to \bar{F}$ (Notation C.6.7) is universal: given a natural transformation $s \colon F \to G$, where $G \colon \mathsf{Set} \to \mathsf{Set}$ preserves finite intersections, there exists a unique natural transformation $\bar{s} \colon \bar{F} \to G$ such that $\bar{s} \cdot r = s$.

Proof The given functor G preserves the empty intersection e_1 of $\mathsf{t}, \mathsf{f} \colon 1 \rightrightarrows 2$, in other words, Ge is the equalizer of $G\mathsf{t}$ and $G\mathsf{f}$. Hence, we obtain a unique map $\bar{s}_\emptyset$ as displayed below:

$$
\begin{array}{ccccc}
\bar{F}\emptyset & \xrightarrow{\bar{F}e_1} & F1 & \overset{F\mathsf{t}}{\underset{F\mathsf{f}}{\rightrightarrows}} & F2 \\
{\scriptstyle \bar{s}_\emptyset}\downarrow & & \downarrow{\scriptstyle s_1} & & \downarrow{\scriptstyle s_2} \\
G\emptyset & \xrightarrow{Ge_1} & G1 & \overset{G\mathsf{t}}{\underset{G\mathsf{f}}{\rightrightarrows}} & G2
\end{array}
$$

Together with $\bar{s}_X = s_X$ for all $X \neq \emptyset$ this defines a natural transformation $\bar{s} \colon \bar{F} \to G$ with $\bar{s} \cdot r = s$. It is obvious that $\bar{s}$ is unique with this property. □

Remark C.6.9. Every element $a \in F\emptyset$ defines a natural transformation $\tau^a \colon C_{01} \to F$ whose component $\tau^a_X \colon C_{01}X = 1 \to FX$ for $X \neq \emptyset$ is $Fe_X(a) \in FX$, where $e_X \colon \emptyset \to X$ is the empty map.

Coming back to Trnková's original construction of $\bar{F}\emptyset$ as the set of natural transformations $C_{01} \to F$ (Proposition C.6.4), the components of the reflection

[1] This category does not have small hom-sets, but this is not an issue for the following proof of the universal property.

$r\colon F \to \bar{F}$ are the identity maps for $X \neq \emptyset$, and $r_\emptyset\colon F\emptyset \to \bar{F}\emptyset$ is given by $a \mapsto \tau^a$.

Definition C.6.1 also gives some further properties of the Trnková hull $\bar{F}$:

Corollary C.6.10. *Let F be a set functor.*

(1) *If $F\emptyset \neq \emptyset$, then also $\bar{F}\emptyset \neq \emptyset$.*

(2) *Fe_A factorizes through $\bar{F}e_A$ for every $A \neq \emptyset$.*

(3) *If F is finitary, then so is $\bar{F}$.*

(4) *F and $\bar{F}$ have the same algebras and the same nonempty coalgebras.*
In particular, $\mathsf{Alg}\, F \cong \mathsf{Alg}\, \bar{F}$, *and* $\mathsf{Coalg}\, F \cong \mathsf{Coalg}\, \bar{F}$.

Proof (1) This follows since we have the map $r_\emptyset\colon F\emptyset \to \bar{F}\emptyset$ in Notation C.6.7.
(2) The desired factor is $r_\emptyset\colon F\emptyset \to \bar{F}\emptyset$, since $\bar{F}e_A \cdot r_\emptyset = Fa \cdot \bar{F}e_1 \cdot r_\emptyset = Fa \cdot Fe_1 = Fe_A$.
(3) This follows since $\bar{F}$ agrees with F on nonempty sets and since finitary functors are, equivalenty, finitely bounded (see Proposition 4.3.4).
(4) The functors F and $\bar{F}$ clearly have the same algebras, due to (1). Concerning their coalgebra categories, the isomorphism takes $\alpha\colon A \to FA$ to $r_A \cdot \alpha\colon A \to F'A$. The inverse takes $(\emptyset, e_{\bar{F}\emptyset})$ to $(\emptyset, e_{F\emptyset})$; otherwise it is obvious. The only point requiring a verification is that the inverse preserves coalgebra homomorphisms $(\emptyset, e_{\bar{F}\emptyset}) \to (A, \alpha)$ when $A \neq \emptyset$, and this is an easy diagram chase. □

Corollary C.6.11. *The Trnková hull of a finitary set functor is finitary and preserves all intersections.*

Indeed, this follows from Theorem 4.4.3 and Corollary C.6.10(3).

Remark C.6.12. Recall from Remark 6.1.5(3) that the initial-algebra chain is unique up to natural isomorphism. When we claim below that two endofunctors have the same initial-algebra chains, we mean that those can be chosen so that they coincide. Analogously for the terminal-coalgebra chains.

Proposition C.6.13. *Let $\bar{F}$ be the Trnková hull of a set functor F.*

(1) *The terminal-coalgebra chains for F and $\bar{F}$ coincide.*

(2) *If $F\emptyset \neq \emptyset$, then the initial-algebra chains for F and $\bar{F}$ coincide from ω onwards.*

Proof (1) If $F1 \neq \emptyset$, then the terminal-coalgebra chain for F consists of nonempty set V_i by Proposition C.2.1(4). By an obvious induction, the terminal-coalgebra chains of F and $\bar{F}$ coincide. If $F1 = \emptyset$, then F is the constant $\emptyset$. Therefore $F = \bar{F}$ and we are done.

(2) Suppose that $F\emptyset \neq \emptyset$. We prove that the sets W_i in the initial-algebra chain for F can be chosen to be the same as the sets $\overline{W}_i$ in the initial-algebra chain for $\bar{F}$ for all ordinals $i \geq \omega$.

We have the functions $q = r_\emptyset \colon \bar{F}\emptyset \to F\emptyset$ (see Theorem C.6.8) and $p = \bar{F}e_{F\emptyset} \colon \bar{F}\emptyset \to \bar{F}F\emptyset = FF\emptyset$. We check that the diagram below commutes (note that $\bar{F}\bar{F}\emptyset = F\bar{F}\emptyset$ holds by definition of $\bar{F}$ since $\bar{F}\emptyset \neq \emptyset$):

$$
\begin{array}{ccc}
F\emptyset & \xrightarrow{\;Fe_{F\emptyset}\;} & FF\emptyset \\
{\scriptstyle q}\downarrow & \overset{p}{\nearrow} & \downarrow{\scriptstyle Fq} \\
\bar{F}\emptyset & \xrightarrow{\;\bar{F}e_{\bar{F}\emptyset}\;} & \bar{F}\bar{F}\emptyset = F\bar{F}\emptyset
\end{array}
$$

On the right, this is straight from the definitions. On the left, this follows from the naturality of $r \colon F \to \bar{F}$:

$$p \cdot q = \bar{F}e_{F\emptyset} \cdot r_\emptyset = r_{F\emptyset} \cdot Fe_{F\emptyset} = \mathrm{id}_{F\emptyset} \cdot Fe_{F\emptyset} = Fe_{F\emptyset}.$$

Consequently, the initial-algebra ω-chain for F (given by all $F^n e_{F\emptyset}$) and for $\bar{F}$ (given by all $\bar{F}^n e_{\bar{F}\emptyset}$) are connected by natural transformations with components $F^n p$ and $F^n q$ as follows:

$$
\begin{array}{ccccccccc}
\emptyset & \xrightarrow{\;e_{F\emptyset}\;} & F\emptyset & \xrightarrow{\;Fe_{F\emptyset}\;} & F^2\emptyset & \xrightarrow{\;F^2e_{F\emptyset}\;} & F^3\emptyset & \xrightarrow{\;F^3e_{F\emptyset}\;} & \cdots \\
 & & {\scriptstyle q}\downarrow & \overset{p}{\nearrow} & \downarrow{\scriptstyle Fq} & \overset{Fp}{\nearrow} & \downarrow{\scriptstyle F^2q} & \overset{F^2p}{\nearrow} & \\
\emptyset & \xrightarrow{\;e_{\bar{F}\emptyset}\;} & \bar{F}\emptyset & \xrightarrow{\;\bar{F}e_{\bar{F}\emptyset}\;} & \bar{F}^2\emptyset = F\bar{F}\emptyset & \xrightarrow{\;\bar{F}^2e_{\bar{F}\emptyset}\;} & \bar{F}^3\emptyset = F^2\bar{F}\emptyset & \xrightarrow{\;\bar{F}^3e_{\bar{F}\emptyset}\;} & \cdots
\end{array}
$$

Consequently, the colimit morphisms $\operatorname{colim} F^n p \colon \overline{W}_\omega \to W_\omega$ and $\operatorname{colim} F^n q \colon W_\omega \to \overline{W}_\omega$ are inverse to each other. Since colimits are unique up to isomorphism, this implies that we can choose $\overline{W}_\omega = \operatorname{colim} F^n\emptyset = W_\omega$ with the following colimit injections for $n \geq 1$:

$$\overline{w}_{n,\omega} = \big(\bar{F}^n\emptyset = F^n\emptyset \xrightarrow{\;F^{n-1}p\;} F^{n+1}\emptyset \xrightarrow{\;w_{n+1,\omega}\;} W_\omega\big). \qquad \text{(C.12)}$$

We prove that with this choice the next morphism

$$\overline{w}_{\omega,\omega+1} \colon W_\omega \to \bar{F}W_\omega = FW_\omega = W_{\omega+1}$$

with respect to $\bar{F}$ is equal to $w_{\omega,\omega+1}$. Recall from Definition 6.1.4 that $\overline{w}_{\omega,\omega+1}$ is defined as the unique map such that $\overline{w}_{\omega,\omega+1} \cdot \overline{w}_{n+1,\omega} = \bar{F}\overline{w}_{n,\omega}$ for all $n < \omega$. To establish the desired equality it suffices to show that $w_{\omega,\omega+1}$ has the same property for all $n \geq 1$:

$$\begin{aligned} w_{\omega,\omega+1} \cdot \overline{w}_{n+1,\omega} &= w_{\omega,\omega+1} \cdot (w_{n+1,\omega} \cdot F^n p) && \text{by (C.12)} \\ &= w_{n+1,\omega+1} \cdot F^n p && w_{i,j} \text{ compose} \\ &= F w_{n,\omega} \cdot F F^{n-1} p && \text{def. of } w_{n+1,\omega+1} \\ &= F(w_{n,\omega} \cdot F^{n-1} p) && F \text{ a functor} \\ &= F \overline{w}_{n,\omega} && \text{by (C.12)} \\ &= \overline{F} \overline{w}_{n,\omega}, \end{aligned}$$

where in the last step we use that the domain of $\overline{w}_{n,\omega}$ is nonempty for $n \geq 1$. From the equality $W_\omega = \overline{W}_\omega$ and $w_{\omega,\omega+1} = \overline{w}_{\omega,\omega+1}$ it follows (by an easy transfinite induction) that the two intial-algebra chains coincide from ω onwards as desired. □

Corollary C.6.14. *For every set functor F all the connecting morphisms of the initial-algebra chain are monic from step ω onwards.*

Proof If F preserves monomorphism, then all $w_{i,j}$ are monic; this is easy to prove by transfinite induction: $w_{0,1} : \emptyset \to F\emptyset$ is monic, if $w_{i,j}$ is monic, then so is $F_{w_{i,j}}$, and colimits of chains of monomorphisms in Set are formed by monomorphisms. If F does not preserve monomorphisms, then clearly $F\emptyset \neq \emptyset$, and we can apply Proposition C.6.13(2). □

C.7 Standard Functors

We have seen in Section C.6 that when a set functor F does not already preserve injective maps, it can easily be turned into one that does (and in fact preserves all finite intersections) by modifying its value at the empty set. Going a step further, when working set-theoretically it is often convenient that a given functor preserves inclusions in the following sense.

Definition C.7.1. A set functor F *preserves inclusions* if given sets $A \subseteq B$ with the inclusion map $i : A \hookrightarrow B$, FA is a subset of FB via the inclusion map Fi.

Example C.7.2. (1) The power-set functor preserves inclusions, and so do its subfunctors $\mathscr{P}_\mathsf{f}$ and $\mathscr{P}_\lambda$.

(2) The hom-functor $\mathsf{Set}(1, -)$ does not preserve inclusions: if $X \subsetneq Y$, then the sets $\mathsf{Set}(1, X)$ and $\mathsf{Set}(1, Y)$ are disjoint, as mentioned at the end of Section 1.2. However, $\mathsf{Set}(1, -)$ is naturally isomorphic to identity functor, which does preserve inclusions.

Definition C.7.3. A set functor is *standard* if it preserves inclusions and finite intersections.

Remark C.7.4. The original formulation of standardness in [9, Def. in Sec. III.4.5] required that F preserve inclusions and that every natural transformation from $C_{0,1}$ to F has a unique extension to C_1. This last property holds of $\bar{F}$, as we have seen in Proposition C.6.4. Going the other way, if this condition on natural transformations holds, then F is naturally isomorphic to $\bar{F}$. To see this, we check that $r_\emptyset$ is an isomorphism, where $r\colon F \to \bar{F}$ is the natural transformation in Notation C.6.7. In Remark C.6.9, we defined $r_\emptyset$ by $a \mapsto \tau^a$, where $\tau^a\colon C_{01} \to F$ is a natural transformation defined in the same remark. Proposition C.6.4 tells us that each $\tau^a \in N$ corresponds to a unique $\widehat{a} \in F\emptyset$, and every $\widehat{a} \in F\emptyset$ comes from a unique $a \in F\emptyset$.

The point is that the original formulation of standardness is equivalent to the one we adopt.

Theorem C.7.5 [9, Prop. in Sec. III.4.5]. *Every set functor preserving finite intersections is naturally isomorphic to a standard set functor.*

Proof sketch Denote by $\mathcal{C}$ the class of elements of F, that is, pairs (X, x) where X is a set and $x \in FX$. Given a pair of subsets $X, Y \subseteq Z$, two elements (X, x) and (Y, x) are equivalent if $Fi(x) = Fj(y)$ for the inclusion maps $i\colon X \to X \cup Y$ and $j\colon Y \to X \cup Y$. This defines an equivalence relation on $\mathcal{C}$, and we denote by $[X, x]$ the equivalence class of (X, x).

Define a set functor F' on a set X and a map $f\colon X \to Y$, respectively, by

$$F'X = \{[X, x] : x \in FX\} \qquad \text{and} \qquad F'f([X, x]) = [Y, Ff(x)].$$

Then F' is a well-defined standard set functor. Moreover, F' preserves intersections because F does. We have a natural isomorphism $\nu\colon F \to F'$ defined by $\nu_X(x) = [X, x]$ for every $x \in FX$. □

Example C.7.6. (1) The filter functor $\mathscr{F}$ preserves finite intersections, but it is not standard. However, as usually done in topology, for a subset $X \subset Y$, any given filter $\mathcal{U}$ on X is considered to be equivalent to the filter on Y consisting of all $S \subseteq Y$ such that $S \cap X \in \mathcal{U}$. The functor $\mathscr{F}'$ mapping a set X to the set of equivalence classes of filters on X in this sense is standard and naturally isomorphic to $\mathscr{F}$.

(2) The distribution functor $\mathscr{D}$ also preserves finite intersections but is not standard. For a subset $Y \subseteq X$ one can regard a probability distribution $b\colon Y \to [0, 1]$ as equivalent to its extension to X (mapping all elements of $X \setminus Y$ to 0). This again yields a standard functor naturally isomorphic to $\mathscr{D}$. This functor

sends X to the set of subsets $s \subseteq X \times [0,1]$ which are set-theoretic functions whose image sets sum to 1.

(3) Pertaining to the definition of standardness: not every functor which preserves inclusions preserves intersections. For this, consider $C_{0,1}$, and recall Example C.2.4.

Corollary C.7.7. *The restriction of every set functor to nonempty sets is naturally isomorphic to the restriction of a standard set functor.*

Remark C.7.8. We note a few additional properties enjoyed by standard set functors F.

(1) *F preserves images*: for every map $f\colon X \to Y$ we have $Ff[FX] = F(f[X])$. Indeed, for the inclusion map $i\colon f[X] \hookrightarrow Y$ we have the commutative diagram on the left below:

$$X \xrightarrow{\;g\;} f[X] \xrightarrow{\;i\;} Y, \quad f = i \cdot g \qquad\qquad FX \xrightarrow{\;Fg\;} F(f[X]) \xrightarrow{\;Fi\;} FY, \quad Ff = Fi \cdot Fg$$

The desired equality follows from the diagram on the right above since Fg is surjective and Fi is the inclusion map.

(2) If $f\colon X \to Y$ and $X' \subseteq X$, then $F(f|_{X'}) = (Ff)|_{FX'}$.

(3) If $f\colon X \to Y$ is injective and $y \subseteq Y$, then $F(f^{-1}[y]) = (Ff)^{-1}[Fy]$.

Proposition C.7.9. *Let F be a standard set functor. The initial-algebra chain (W_i) can be chosen so that for all ordinals $i \leq j$ we have $W_i \subseteq W_j$, where $w_{i,j}$ is the inclusion map.*

Proof We prove by transfinite induction on i that $w_{i,j}\colon W_i \hookrightarrow W_j$ can be chosen to be an inclusion map for every $j \geq i$.

(1) The base case $i = 0$ is clear since $W_0 = \emptyset$.

(2) Suppose that the desired result holds for i. We prove that it also holds for $i+1$ by transfinite induction on $j \geq i+1$. For $j = i+1$ the statement is trivial. The isolated step is immediate since F is standard and $w_{i+1,j+1} = Fw_{i,j}$. Let j be a limit ordinal. Then W_j is the colimit of all W_k for $i+1 \geq k < j$. By induction hypothesis, we know that $w_{k,\ell}\colon W_k \hookrightarrow W_\ell$ is an inclusion map for every $i+1 \leq k, \ell < j$. Then the union $W_j = \bigcup_{i \leq k < j} W_k$ is a colimit, where the colimit cocone is given by inclusion maps. This is the colimit we choose.

(3) For a limit ordinal i, we know that the connecting morphism $w_{i,j}$ is the unique factorization of the cocone given by $w_{k,j}$ for $k < i$. Each $w_{k,j}$ is an inclusion map of $W_k \subseteq W_j$. It follows that so is $w_{i,j}$. □

C.8 Accessible and Bounded Functors

Recall from Definition 11.2.2 the concept of a λ-directed colimit, for λ an arbitrary infinite regular cardinal. In the present section, we compare the following properties of a set functor F; the last one was introduced by Kawahara and Mori [186]:

(1) *λ-accessible*: F preserves λ-directed colimits (Definition 11.2.13).

(2) *λ-bounded*: every element of FX lies in $Fm[FM]$ for some $m\colon M \hookrightarrow X$ with $|M| < \lambda$ (Definition 6.5.7).

(3) *bounded by sets of size $\leq \lambda$*: given a coalgebra $\alpha\colon A \to FA$, every subset $B \subseteq A$ of size $\leq \lambda$ is included in the carrier set of a subcoalgebra of size $\leq \lambda$ (Definition 11.1.2).

When $\lambda = \aleph_1$, we speak of *countably accessible* and *countably bounded* set functors.

Assumption C.8.1. Throughout this section λ denotes an infinite regular cardinal.

Proposition C.8.2. *A set functor is λ-accessible iff it is λ-bounded.*

Proof We have seen this for $\lambda = \aleph_0$ in Proposition C.5.2. For general λ, the same arguments work. Every set is the colimit of its subsets of size at most λ. This leads to the fact that λ-accessibility implies λ-boundedness. In the opposite direction, the only time finiteness was used in our proof of Proposition C.5.2 was when we asserted that since M is a finite set, so is $M \times M$. The same fact holds for arbitrary infinite cardinalities. □

In the next lemma, we recall that for every infinite cardinal λ, the next infinite cardinal is denoted λ^+. It is always regular.

Lemma C.8.3. *A set functor is λ^+-bounded iff it is bounded by sets of size $\leq \lambda$.*

Proof Suppose that F is λ^+-bounded; we verify boundedness by sets of size at most λ. For every $x \in B$, choose a subcoalgebra $\alpha_x : C_x \to FC_x$ such that $|C_x| < \lambda^+$ and $x \in C_x$:

$$\begin{array}{ccc} C_x & \xrightarrow{\alpha_x} & FC_x \\ {\scriptstyle m_x}\big\downarrow & & \big\downarrow{\scriptstyle Fm_x} \\ A & \xrightarrow{\alpha} & FA \end{array}$$

So $|C_x| \leq \lambda$. We may assume that each m_x is the inclusion map $C_x \hookrightarrow A$. Let $C = \bigcup_x C_x$. Clearly $B \subseteq C$. Then $|C| \leq |B| \cdot \lambda \leq \lambda^2 = \lambda$. We define a coalgebra structure $\gamma\colon C \to FC$ by $\gamma(x) = Fm'_x \cdot \alpha_x(x)$, where $m'_x\colon C_x \to C$ is the inclusion map. This gives a subcoalgebra of $\alpha\colon A \to FA$ (via the inclusion map) containing B. Thus, F is bounded by sets of size $\le \lambda$.

In the opposite direction, let F be bounded by sets of size $\le \lambda$. Let X be a set, and $x \in FX$. We present a subset $m\colon M \hookrightarrow X$ of size $\le \lambda$ such that $x \in Fm[FM]$. Let $\alpha\colon A \to FA$ be the constant map with value x. We have a nonempty subcoalgebra $m\colon (M,\gamma) \to (A,\alpha)$ with $|M| \le \lambda$; in fact, A is the union of such subcoalgebras. Let $y \in M$ be arbitrary. Then $Fm \cdot \gamma(y) = (\alpha \cdot m)(y) = x$. Thus, $x \in Fm[FM]$. □

By Proposition 4.3.13 finitary set functors are precisely the quotients of the polynomial functors H_Σ for finitary signatures Σ. The statement (as well as the proof) immediately generalize to λ:

Proposition C.8.4. *A set functor is λ-accessible iff it is a quotient of a polynomial functor H_Σ, where every operation symbol in Σ has arity smaller than λ.*

Theorem C.8.5. *The collection of all λ-accessible set functors is closed under*

(1) *coproducts,*

(2) *products of less then λ functors,*

(3) *subfunctors, and*

(4) *quotients.*

Proof (4) follows from Proposition C.8.4.

(1) A set functor $F = \coprod_{i\in I} F_i$ preserves the colimit of every diagram preserved by each F_i (because colimits commute with colimits). Thus, if every F_i is λ-accessible, then so is F.

(2) A set functor $F = \prod_{i\in I} F_i$ where $|I| < \lambda$ preserves the colimit of every λ-directed diagram preserved by each F_i; this holds because colimits of λ-directed diagrams commute with products of less than λ objects in Set [23, Prop. 1.59]. Consequently, the category of set functors also has this property.

(3) Let $m\colon F \rightarrowtail G$ be a subfunctor where G is λ-accessible. Form the *amalgam* of m, the pushout of m along itself

$$\begin{array}{ccc} F & \overset{m}{\rightarrowtail} & G \\ {\scriptstyle m}\downarrow & & \downarrow{\scriptstyle f} \\ G & \underset{g}{\longrightarrow} & H \end{array}$$

Then we have a quotient $[f,g]\colon G + G \twoheadrightarrow H$. Thus, H is λ-accessible by items (1) and (4).The square above is a pullback since Set has the property that all amalgams of subobjects are pullbacks. Hence, the category of set functors has this property, too.

Pullbacks commute with filtered colimits in Set [23, Prop. 1.59]. Hence, the same holds for set functors. Given a λ-directed diagram, since G and H preserve its colimit, so does F. □

Corollary C.8.6. *The collection of set functors which are bounded by sets of size $\leq \lambda$ is closed under quotients.*

Indeed, this follows from Proposition C.8.2, Lemma C.8.3, and Theorem C.8.5.

Remark C.8.7. Recall from Example 2.4.2(6) that Moore automata with the input and output alphabets Σ and Γ, respectively, are coalgebras for the functor $G_{\Sigma,\Gamma}$ assigning to a set X the set $\Gamma \times X^{\Sigma}$.

The following theorem stems from Gumm and Schröder [153, Thm. 4.7].

Theorem C.8.8. *For every set functor F, the following are equivalent:*

(1) *F is bounded by sets of size $\leq \lambda$.*

(2) *F is λ^+-accessible (= λ^+-bounded).*

(3) *F is a quotient of the functor $G_{\Sigma,\Gamma}$ for sets Γ and Σ satisfying $|\Sigma| \leq \lambda$.*

Proof We have already seen that (1) and (2) are equivalent (Proposition C.8.2 and Lemma C.8.3).

Here is a proof that (2) ⇒(3). Let Σ be any set of size λ, put $\Gamma = F(\Sigma)$ and let $\varepsilon \colon G \to F$ be given by $\varepsilon_X \colon \Gamma \times X^{\Sigma} \to FX$ defined as $\varepsilon_X(c, f) = Ff(c)$ for all $f \colon \Sigma \to X$ and $c \in \Gamma$. Then ε is easily seen to be natural; see Proposition 4.3.13 for a similar argument. Let us check that each component ε_X is surjective. Let $x \in FX$. By λ^+-boundedness, there is a subset $m \colon Y \hookrightarrow X$ with $|Y| < \lambda = |X|$ such that $x \in Fm[FY]$. Pick $d \in FY$ such that $Fm(d) = x$. Let $s \colon \Sigma \to Y$ be surjective, so that Fs is also surjective. Choose $c \in \Sigma$ such that $Fs(c) = d$, and put $f = m \cdot s$. Then $\varepsilon_X(c, f) = Ff(c) = Fm(Fs(c)) = Fm(d) = x$.

To prove that (3) ⇒ (1), we show that $G_{\Sigma,\Gamma}$ is bounded by sets of size $\leq \lambda$, and then we recall Corollary C.8.6. Consider a coalgebra $\alpha \colon A \to \Gamma \times A^{\Sigma}$ and an element $a \in A$. Think of the coalgebra as an automaton with a the initial state, and take the subcoalgebra whose state set consists of states reachable from a. The size of this coalgebra is at most $|\Sigma^*| = |\Sigma| = \lambda$. □

Preaccessible set functors need not be accessible We end this section with a technical example. Recall from Definition 11.4.5 the notion of a preaccessible functor. We now provide an example of a set functor which is not accessible, although it preserves colimits of ω-chains (which implies that it is ω-preaccessible). We use the assumption $\neg$(M) on the set theory we work with that we explain first.

An *ultrafilter* $\mathcal{U}$ on a set X is a collection of nonempty subsets which is closed under supersets and finite intersections and for every partition $X = S \cup (X \setminus S)$ it contains S or $X \setminus S$. The ultrafilter is *σ-complete* if for every countable partition $X = \bigcup_{i=0}^{\infty} S_i$ it contains S_i for some i. More generally, given an infinite cardinal λ, $\mathcal{U}$ is *λ-complete* if for every partition $X = \bigcup_{i \in I} S_i$ with $|I| < \lambda$, $\mathcal{U}$ contains some S_i. For example, every element $x \in X$ defines the *principal* ultrafilter of all subsets of X containing x. It is λ-complete for every λ.

We recall that a cardinal κ is *measurable* if it is uncountable and on a set of cardinality κ there is a κ-complete non-principal ultrafilter. If κ is the smallest uncountable cardinal with a σ-complete ultrafilter, then κ is measurable.

Definition C.8.9. The axiom (M) states that there exists an infinite cardinal λ_0 such that for every set X of cardinality $\lambda > \lambda_0$, every λ-complete ultrafilter on X is principal.

For a discussion of the axiom (M) the reader may consult Pultr and Trnková [254, Sec. II.6]. This axiom (M) is consistent with the usual set theory. However, the negation of (M) cannot be disproved (since (M) cannot be proved).

Notation C.8.10. We denote by G the set functor which maps a set X to the set of all σ-complete ultrafilters on X. For a function $f\colon X \to Y$ the function Gf is defined by

$$\mathcal{U} \mapsto \{S \subseteq Y : f[T] \subseteq S \text{ for some } T \in \mathcal{U}\}.$$

It is easy to verify that G is a well-defined subfunctor of the filter functor $\mathcal{F}$.

Lemma C.8.11. *Let $X \subseteq Y$ and $f\colon X \hookrightarrow Y$ be the inclusion map. Then for every ultrafilter $\mathcal{U}$, $Gf(\mathcal{U})$ consists of all $S \subseteq Y$ with $S \cap X \in \mathcal{U}$.*

Proof If S lies in $Gf(\mathcal{U})$, then we have $T \in \mathcal{U}$ such that $T \subseteq S$. Then $T \subseteq X \cap S$ which shows that $X \cap S \in \mathcal{U}$.

Conversely, if $T = S \cap X$ lies in $\mathcal{U}$, then $f[T] = T \subseteq S$, whence S lies in $Gf(\mathcal{U})$. □

Proposition C.8.12. *The functor G is ω-preaccessible. Assuming $\neg$(M), it is not accessible.*

Proof (1) The functor G is ω-preaccessible because for every ω-chain of subsets X_n $(n < \omega)$ of some set, G preserves the colimit $X = \bigcup_{n<\omega} X_n$. For this, recall that colimit maps $i_n\colon X_n \hookrightarrow X$ are injective, and hence so are the maps Gi_n. It suffices to prove that every element $\mathcal{U}$ of GX lies in the image of some Gi_n. By Lemma C.8.11, we need to prove that $X_n \in \mathcal{U}$ holds for some n.

Consider the partition

$$X = \bigcup_{n<\omega} Z_n \qquad \text{where} \qquad Z_n = X_n \setminus \bigcup_{k=0}^{n-1} X_k.$$

Since $\mathscr{U}$ is σ-complete, it contains some Z_n. Since $Z_n \subseteq X_n$, we have $X_n \in \mathscr{U}$.

(2) The functor G is not λ_0-accessible for any regular cardinal λ_0, assuming $\neg$(M). This assumption means that there exists a cardinal $\lambda > \lambda_0$ and a set X of cardinality λ on which a λ-complete, non-principal ultrafilter $\mathscr{U}$ can be found. Suppose that G is λ_0-accessible. We will derive a contradiction.

Let $i_t : X_t \hookrightarrow X$ $(t \in T)$ denote the collection of all subsets of X of cardinality smaller than λ_0. Then we have a λ_0-directed colimit $X = \bigcup_{t \in T} X_t$ which G preserves. Consequently, the element $\mathscr{U}$ of GX lies in the image of Gi_t for some $t \in T$. By Lemma C.8.11, this means that $X_t \in \mathscr{U}$. We have the following partition

$$X = (X \setminus X_t) \cup \bigcup_{x \in X_t} \{x\}$$

consisting of λ_0 subsets. Since $\mathscr{U}$ is λ-complete and $\lambda_0 < \lambda$, we have that $\mathscr{U}$ contains a member of that partition. This yields the desired contradiction: since $X_t \in \mathscr{U}$, we know that $X \setminus X_t \notin \mathscr{U}$, and since $\mathscr{U}$ is non-principal, we also know that $\{x\} \notin \mathscr{U}$. □

C.9 Accessible Coreflection

Every set functor F has a largest finitary subfunctor $F_\mathfrak{f}$ which is its finitary coreflection, as we show in this section. For example, for the power-set functor $\mathscr{P}$ this is $\mathscr{P}_\mathfrak{f}$ (Example 2.2.7(4)), and for the distribution functor $\mathscr{D}$ this is the finite distribution functor $\mathscr{D}_\mathfrak{f}$ (Example 2.4.3(4)). For every infinite regular cardinal λ, we introduce the corresponding λ-accessible coreflection F_λ:

Definition C.9.1. For every set functor F its *λ-accessible coreflection* is the largest λ-accessible subfunctor $c : F_\lambda \rightarrowtail F$.

We write $F_\mathfrak{f}$ in lieu of $F_{\aleph_0}$.

We shall see in Corollary C.9.6 below that every set functor indeed has a λ-accessible coreflection. Moreover, it is constructed as explained below:

Construction C.9.2. We define a set functor F_λ such that $F_\lambda X \subseteq FX$ for every set X by putting

$$F_\lambda X = \{x : x \in Fm[FM] \text{ for some subset } m : M \hookrightarrow X \text{ with } |M| < \lambda\}$$

and where $F_\lambda f$ is the restriction of Ff for every map f. We denote by $c\colon F_\lambda \rightarrowtail F$ the natural transformation whose components c_X are inclusion maps.

Lemma C.9.3. *The assignment F_λ gives a λ-accessible subfunctor of F via c.*

Proof We first verify for every map $f\colon X \to Y$ that Ff restricts to $F_\lambda f\colon F_\lambda X \to F_\lambda Y$. That is, $x \in F_\lambda X$ implies $Ff(x) \in F_\lambda Y$. To see this, choose a subset $m\colon M \hookrightarrow X$ such that $|M| < \lambda$ and $x = Fm(x')$ for some $x' \in FM$. Factorize $f \cdot m$ into a surjective map e followed by an injective one m':

$$\begin{array}{ccc} M & \overset{m}{\rightarrowtail} & X \\ {\scriptstyle e}\downarrow & & \downarrow{\scriptstyle f} \\ M' & \overset{m'}{\rightarrowtail} & Y \end{array}$$

Then $|M| < \lambda$ implies $|M'| < \lambda$, and we have $Ff(x) = F(m' \cdot e)(x') = Fm'(Fe(x')) \in F_\lambda Y$. It is now obvious that F_λ is functorial, and the inclusion maps form a natural transformation $c\colon F_\lambda \to F$.

For every set M with $|M| < \lambda$ we clearly have $FX = F_\lambda M$. Thus, F_λ is λ-bounded, whence λ-accessible by Proposition C.8.2. □

Example C.9.4. (1) For the power-set functor $\mathscr{P}$ we obtain the subfunctor $\mathscr{P}_\lambda$ of all subsets of cardinality less than λ.

(2) For the filter functor $\mathscr{F}$ the λ-accessible coreflection $\mathscr{F}_\lambda$ is the functor of all λ-small filters (Example 6.5.14(1)). If $\lambda = \aleph_0$, then we obtain $\mathscr{F}_\mathrm{f} = \mathscr{P}_\mathrm{f}$; indeed, every finite set $M \subseteq X$ defines the filter $\mathbb{F}_M$ of all supersets of M. Conversely, every filter $\mathbb{F} \in \mathscr{F}_\mathrm{f} X$ contains a finite set; it follows that $\mathbb{F} = \mathbb{F}_M$ where M is the least member of $\mathbb{F}$.

(3) Let Σ be an infinitary signature. Then $(H_\Sigma)_\lambda$ can be identified with the set of flat terms $\sigma(x_i)_{i<n}$ where σ is an n-ary symbol and the cardinality of the set $\{x_i : i < n\}$ is less than λ.

We now show that the name λ-accessible coreflection is justified:

Proposition C.9.5. *For every natural transformation $t\colon G \to F$, where G is λ-accessible, there exists a unique natural transformation $\bar{t}$ such that the triangle below commutes:*

$$\begin{array}{ccc} F_\lambda & \xrightarrow{c} & F \\ {\scriptstyle \bar{t}}\uparrow & \nearrow{\scriptstyle t} & \\ G & & \end{array}$$

Proof We shall show that every component $t_X\colon GX \to FX$ has a codomain restriction $\bar{t}_X\colon GX \to F_\lambda X$. Given $x \in GX$, since G is λ-bounded by Proposition C.8.2, we can choose a subset $m\colon M \hookrightarrow X$ satisfying $|M| < \lambda$ and an element $x' \in GM$ such that $x = Gm(x')$. By naturality of n we obtain

$$t_X(x) = t_X(Gm(x')) = Fm(t_M(x')) \in F_\lambda X.$$

It is clear that the codomain restrictions $\bar{t}_X$ of the components t_X form a natural transformation $\bar{t}\colon G \to F_\lambda$. Moreover, $\bar{t}$ is clearly unique such that $c \cdot \bar{t} = t$. □

Corollary C.9.6. *The functor F_λ is the largest λ-accessible subfunctor of F.*

Indeed, since c is monic, we see that $\bar{n}$ is monic if n is so.

Remark C.9.7. Denote by $J\colon \mathsf{Set}_\lambda \hookrightarrow \mathsf{Set}$ the inclusion of the full subcategory given by all sets of cardinality less than λ. We clearly have $FJ = F_\lambda J$, that is, for every set X of cardinality less than λ we have $FX = F_\lambda X$.

Proposition C.9.8. *The functor F_λ is the left Kan extension of FJ along J; in symbols: $F_\lambda = \mathrm{Lan}_J(FJ)$.*

This means that for every set functor G and every natural transformation $t\colon FJ \to GJ$ there exists a unique natural transformation $\bar{t}\colon F_\lambda \to G$ with $t = \bar{t}J$.

Proof We first verify that $\bar{t}$ is unique. Given $x \in F_\lambda X$, we have a subset $m\colon M \hookrightarrow X$ and an element $x' \in FM$ such that $|M| < \lambda$ and $x = Fm(x')$. By the naturality of $\bar{t}$ we have $\bar{t}_X \cdot Fm = Gm \cdot \bar{t}_M$. Using this and $\bar{t}J = t$ (i.e. $\bar{t}_M = t_M$), we obtain that

$$\bar{t}_X(x) = \bar{t}_X(Fm(x')) = Gm(\bar{t}_M(x')) = Gm(t_M(x')).$$

Finally, it is easy to verify that the last equation does define a function $\bar{t}_X\colon F_\lambda X \to GX$ and that this yields a natural transformation $\bar{t}$ such that $\bar{t}J = t$. □

Theorem C.9.9. *If a set functor F preserves weak pullbacks, then so do all functors F_λ.*

Proof Given a weak pullback

$$\begin{array}{ccc} W & \xrightarrow{w_1} & A_1 \\ {\scriptstyle w_2}\downarrow & & \downarrow{\scriptstyle f_1} \\ A_2 & \xrightarrow{f_2} & B \end{array}$$

we know that for every pair $x_i \in FA_i$, $i = 1, 2$, such that $Ff_1(x_1) = Ff_2(x_2)$ there exists an element in FW which is mapped by Fw_i to x_i. We shall show that such an element may be found in $F_\lambda W$ provided that x_i lies in $F_\lambda A_i$ for $i = 1, 2$.

Choose subsets $m_i\colon M_i \hookrightarrow A_i$ and elements $x_i' \in FM_i$ such that $|M_i| < \lambda$ and $x_i = Fm_i(x_i')$ for $i = 1, 2$. Now form the pullback of the maps $f_i \cdot m_i$:

$$\begin{array}{ccccc}
V & & \xrightarrow{\quad v_1 \quad} & & M_1 \\
 & \searrow^{h} & & & \downarrow{\scriptstyle m_1} \\
{\scriptstyle v_2}\downarrow & & W & \xrightarrow{w_1} & A_1 \\
 & & {\scriptstyle w_2}\downarrow & & \downarrow{\scriptstyle f_1} \\
M_2 & \xrightarrow{m_2} & A_2 & \xrightarrow{f_2} & B
\end{array}$$

We have a unique map h such that the diagram commutes, and we factorize it as a surjection $e\colon V \twoheadrightarrow M$ followed by an inclusion map $m\colon M \hookrightarrow W$. Since F weakly preserves the above pullback and $F(f_1 \cdot m_1)(x_1') = F(f_2 \cdot m_2)(x_2')$, there exists $y \in FV$ such that $x_i' = Fv_i(y)$. Since e is surjective and the pair v_1, v_2 is collectively monic, we have $|M| \leq |V| \leq |M_1 \times M_2| < \lambda$. Hence, the element $z = Fh(y) = Fm(Fe(y))$ lies in $F_\lambda W$. Moreover, we have

$$Fw_i(z) = Fw_i(Fh(y)) = Fm_i(Fv_i(y)) = Fm_i(x_i') = x_i. \qquad \square$$

Remark C.9.10. (1) The same theorem holds for pullbacks, inverse images, and (finite) intersections. The proofs are analogous.

(2) The only set functors preserving pullbacks exhibited in Section C.3 are the polynomial functors. Now we obtain many more, for example, the λ-accessible coreflections $(H_\Sigma)_\lambda$ as well as coproducts of these coreflections.

Finally, we note the following property of the λ-accessible coreflection (cf. Remark 8.1.12 and (8.4)).

Lemma C.9.11. *The coreflection $c\colon F_\lambda \rightarrowtail F$ is* subcartesian*: for every injective map $s\colon X \rightarrowtail Y$ the naturality square below is a pullback:*

$$\begin{array}{ccc}
F_\lambda X & \xrightarrow{c_X} & FX \\
{\scriptstyle F_\lambda s}\downarrow & \lrcorner & \downarrow{\scriptstyle Fs} \\
F_\lambda Y & \xrightarrow{c_Y} & FY
\end{array}$$

Proof Without loss of generality we may identify X with its image in Y and assume that s is the inclusion map. If $X = \emptyset$, then we are done since $F_\lambda\emptyset = F\emptyset$ (Remark C.9.7) and c_Y is injective. So let us assume that X is nonempty, whence so is Y. Given $y \in F_\lambda Y \subseteq FY$ such that $y = Fs(x)$ for some $x \in FX$ we need to show that $x \in F_\lambda X$. Pick $m\colon M \hookrightarrow Y$ and $y' \in FM$ such that $|M| < \lambda$ and $y = Fm(y')$. We may assume that $M \cap X$ is nonempty; for otherwise replace M by $M \cup \{\bar{x}\}$, for some $\bar{x} \in X$, and y' by $Fi(y')$ where $i\colon M \hookrightarrow M \cup \{\bar{x}\}$ is the inclusion map. We have $Fs(x) = y = Fm(y')$. Since F preserves the nonempty intersection of s and m, we obtain $x' \in F(M \cap X)$ such that $Fp(x') = x$ and $Fq(x') = y'$ where $p\colon M \cap X \hookrightarrow X$ and $q\colon M \cap X \hookrightarrow M$ are the inclusion maps. Clearly $|M \cap X| < \lambda$, and thus $x \in F_\lambda X$ as desired. □

Remark C.9.12. Note that the naturality square for c is in general not a pullback for arbitrary maps m. For a counterexample consider $c\colon \mathscr{P}_\mathsf{f} \to \mathscr{P}$. Let X be infinite, and let $m\colon X \to 1$. Then any infinite subset $S \in \mathscr{P}X$ is mapped by $\mathscr{P}m$ to $\{0\} \in \mathscr{P}1 = \mathscr{P}_\mathsf{f}1$ but is not in the image of c_X.

References

[1] Abramsky, Samson. 2005. A Cook's tour of the finitary non-well-founded sets. Pages 1–18 of: *We Will Show Them: Essays in Honour of Dov Gabbay*, vol. 1. College Publications.

[2] Abramsky, Samson, and Jung, Achim. 1994. Domain theory. Pages 1–168 of: *Handbook of Logic in Computer Science*, vol. III. Clarendon Press.

[3] Aczel, Peter. 1988. *Non-Well-Founded Sets*. CLSI Lecture Notes, vol. 14. CLSI Publications.

[4] Aczel, Peter, and Mendler, Nax. 1989. A final coalgebra theorem. Pages 357–365 of: *Proc. Conference on Category Theory and Computer Science (CTCS)*. Lecture Notes Comput. Sci., vol. 389. Springer.

[5] Aczel, Peter, Adámek, Jiří, Milius, Stefan, and Velebil, Jiří. 2003. Infinite trees and completely iterative theories: a coalgebraic view. *Theoret. Comput. Sci.*, **300**(1–3), 1–45. Fundamental Study.

[6] Adámek, Jirí, Milius, Stefan, and Moss, Lawrence S. 2023. On Kripke, Vietoris and Hausdorff polynomial functors. Pages 21:1–21:20 of: Baldan, Paolo, and de Paiva, Valeria (eds), *Proc. 10th Conference on Algebra and Coalgebra in Computer Science (CALCO)*. LIPIcs, vol. 270. Schloss Dagstuhl. Full version available at `https://arxiv.org/abs/2303.11071`.

[7] Adámek, Jiří. 1974. Free algebras and automata realizations in the language of categories. *Comment. Math. Univ. Carolin.*, **15**(4), 589–602.

[8] Adámek, Jiří. 1977. Colimits of algebras revisited. *Bull. Aust. Math. Soc.*, **17**(3), 433–450.

[9] Adámek, Jiří. 1983. *Theory of Mathematical Structures*. Reidel.

[10] Adámek, Jiří. 1995. Recursive data types in algebraically ω-complete categories. *Inform. Comput.*, **118**(2), 181–190.

[11] Adámek, Jiří. 1997. A categorical generalization of Scott domains. *Math. Struct. Comput. Sci.*, **7**(5), 419–443.

[12] Adámek, Jiří. 2002. Final coalgebras are ideal completions of initial algebras. *J. Logic Comput.*, **12**(2), 217–242.

[13] Adámek, Jiří. 2003. On final coalgebras of continuous functors. *Theoret. Comput. Sci.*, **294**(1–2), 3–29.

[14] Adámek, Jiří. 2020a. Approximate coalgebra homomorphisms and approximate

solutions. Pages 11–31 of: Petrişan, Daniela, and Rot, Jurriaan (eds), *Proc. Coalgebraic Methods in Computer Science (CMCS)*. Lecture Notes Comput. Sci., vol. 12094. Springer.

[15] Adámek, Jiří. 2020b. On free iterative algebras. Pages 7:1–7:21 of: Fernández, Maribel, and Muscholl, Anca (eds), *Proc. 28th Conference on Computer Science Logic (CSL)*. LIPIcs, vol. 152. Schloss Dagstuhl.

[16] Adámek, Jiří, and Koubek, Václav. 1979. On the least fixed point of a set functor. *J. Comput. Syst. Sci.*, **19**(2), 163–178.

[17] Adámek, Jiří, and Koubek, Václav. 1995. On the greatest fixed point of a set functor. *Theoret. Comput. Sci.*, **150**(1), 57–75.

[18] Adámek, Jiří, and Merzenich, Wolfgang. 1984. Fixed points as equations and solutions. *Canad. J. Math.*, **36**(3), 495–519.

[19] Adámek, Jiří, and Milius, Stefan. 2006. Terminal coalgebras and free iterative theories. *Inform. Comput.*, **204**(7), 1139–1172.

[20] Adámek, Jiří, and Milius, Stefan. 2019. On functors preserving coproducts and algebras with iterativity. *Theoret. Comput. Sci.*, **763**, 66–87.

[21] Adámek, Jiří, and Porst, Hans-E. 2004. On tree coalgebras and coalgebra presentations. *Theoret. Comput. Sci.*, **311**(1–3), 257–283.

[22] Adámek, Jiří, and Reiterman, Jan. 1994. Banach's fixed-point theorem as a base for data-type equations. *Appl. Categ. Struct.*, **2**, 77–90.

[23] Adámek, Jiří, and Rosický, Jiří. 1994. *Locally Presentable and Accessible Categories*. Cambridge University Press.

[24] Adámek, Jiří, and Rosický, Jiří. 2022. Approximate injectivity and smallness in metric-enriched categories. *J. Pure Appl. Algebra*, **226**(6). Article 106974.

[25] Adámek, Jiří, and Trnková, Věra. 1990. *Automata and Algebras in Categories*. Mathematics and Its Applications, vol. 37. Kluwer Academic Publishers.

[26] Adámek, Jiří, and Trnková, Věra. 2011. Initial algebras and terminal coalgebras in many-sorted sets. *Math. Struct. Comput. Sci.*, **21**(2), 481–509.

[27] Adámek, Jiří, and Trnková, Věra. 2012. Relatively terminal coalgebras. *J. Pure Appl. Algebra*, **216**(8–9), 1887–1895.

[28] Adámek, Jiří, Koubek, Václav, and Pohlová, Věra. 1972. The colimits in the generalized algebraic categories. *Acta Univ. Carolinæ – Math. et Phys.*, **13**(2), 29–40.

[29] Adámek, Jiří, Nelson, Evelyn, and Reiterman, Jan. 1982. Tree constructions of free continuous algebras. *J. Comput. System Sci.*, **24**(1), 114–146.

[30] Adámek, Jiří, Milius, Stefan, and Velebil, Jiří. 2003. Free iterative theories: a coalgebraic view. *Math. Struct. Comput. Sci.*, **13**(2), 259–320.

[31] Adámek, Jiří, Milius, Stefan, and Velebil, Jiří. 2004. On coalgebras based on classes. *Theoret. Comput. Sci.*, **316**(1–3), 3–23.

[32] Adámek, Jiří, Milius, Stefan, and Velebil, Jiří. 2006a. Elgot algebras. *Log. Methods Comput. Sci.*, **2**(5), 4:1–4:31.

[33] Adámek, Jiří, Milius, Stefan, and Velebil, Jiří. 2006b. Iterative algebras at work. *Math. Struct. Comput. Sci.*, **16**(6), 1085–1131.

[34] Adámek, Jiří, Lücke, Dominik, and Milius, Stefan. 2007. Recursive coalgebras of finitary functors. *Theor. Inform. Appl.*, **41**(4), 447–462.

[35] Adámek, Jiří, Milius, Stefan, and Velebil, Jiří. 2007. What are iteration theories? Pages 240–252 of: Kučera, Luděk, and Kučera, Antonín (eds), *Proc. 32nd Symposium on Mathematical Foundations of Computer Science (MFCS)*. Lecture Notes Comput. Sci., vol. 4708. Springer.

[36] Adámek, Jiří, Börger, Reinhard, Milius, Stefan, and Velebil, Jiří. 2008a. Iterative algebras: how iterative are they? *Theory Appl. Categ.*, **19**, 61–92.

[37] Adámek, Jiří, Bloom, Stephen L., and Milius, Stefan. 2008b. On algebras with iteration. *J. Logic. Comput.*, **18**(6), 1047–1085.

[38] Adámek, Jiří, Herrlich, Horst, and Strecker, George E. 2009. *Abstract and Concrete Categories: The Joy of Cats*. 3rd edn. Dover Publications. Reprint published in *Reprints in Theory and Applications of Categories*, No. 17, 2006, `http://www.tac.mta.ca/tac/reprints/articles/17/tr17.pdf`.

[39] Adámek, Jiří, Milius, Stefan, and Velebil, Jiří. 2010. Equational properties of iterative monads. *Inform. Comput.*, **208**(12), 1306–1348. `http://dx.doi.org/10.1016/j.ic.2009.10.006`.

[40] Adámek, Jiří, Gumm, H. Peter, and Trnková, Věra. 2010. Presentation of set functors: a coalgebraic perspective. *J. Logic Comput.*, **20**(5), 991–1015.

[41] Adámek, Jiří, Rosický, Jiří, and Vitale, Enrico. 2011. *Algebraic Theories*. Cambridge University Press.

[42] Adámek, Jiří, Milius, Stefan, and Velebil, Jiří. 2011a. Elgot theories: a new perspective of the equational properties of iteration. *Math. Struct. Comput. Sci.*, **21**(2), 417–480.

[43] Adámek, Jiří, Milius, Stefan, and Velebil, Jiří. 2011b. On second-order iterative monads. *Theoret. Comput. Sci.*, **412**(38), 4969–4988.

[44] Adámek, Jiří, Milius, Stefan, and Velebil, Jiří. 2011c. Semantics of higher-order recursion schemes. *Log. Methods Comput. Sci.*, **7**(1), 15:1–15:43.

[45] Adámek, Jiří, Bowler, Nathan, Levy, Paul B., and Milius, Stefan. 2012. Coproducts of monads on Set. Pages 45–54 of: *Proc. 27th Symposium on Logic in Computer Science (LICS)*. IEEE Computer Society.

[46] Adámek, Jiří, Haddadi, Mahdieh, and Milius, Stefan. 2014a. Corecursive algebras, corecursive monads and Bloom monads. *Log. Methods Comput. Sci.*, **10**(3), 19:1–19:51.

[47] Adámek, Jiří, Milius, Stefan, Moss, Lawrence S., and Sousa, Lurdes. 2014b. Well-pointed coalgebras. *Log. Methods Comput. Sci.*, **9**(2), 2:1–2:51.

[48] Adámek, Jiří, Levy, Paul B., Milius, Stefan, Moss, Lawrence S., and Sousa, Lurdes. 2015a. On final coalgebras of power-set functors and saturated trees. *Appl. Categ. Struct.*, **23**(4), 609–641. `http://dx.doi.org/10.1007/s10485-014-9372-9`.

[49] Adámek, Jiří, Milius, Stefan, Moss, Lawrence S., and Urbat, Henning. 2015b. On finitary functors and their presentations. *J. Comput. System Sci.*, **81**(5), 813–833. `http://dx.doi.org/10.1016/j.jcss.2014.12.002`.

[50] Adámek, Jiří, Koubek, Václav, and Palm, Thorsten. 2016. Fixed points of set functors: how many iterations are needed? *Appl. Categ. Struct.*, **24**(5), 649–661.

[51] Adámek, Jiří, Milius, Stefan, Sousa, Lurdes, and Wißmann, Thorsten. 2019a. Finitely presentable algebras for finitary monads. *Theory Appl. Categ.*, **34**, 1179–1195.

[52] Adámek, Jiří, Milius, Stefan, Sousa, Lurdes, and Wißmann, Thorsten. 2019b. On finitary functors. *Theory Appl. Categ.*, **34**, 1134–1164.

[53] Adámek, Jiří, Milius, Stefan, and Moss, Lawrence S. 2020. On well-founded and recursive coalgebras. Pages 17–36 of: König, Barbara, and Goubault-Larrecq, Jean (eds), *Proc. 23rd Conference on Foundations of Software Science and Computation Structures (FoSSaCS)*. Lecture Notes Comput. Sci., vol. 12077. Springer.

[54] Adámek, Jiří, Milius, Stefan, and Moss, Lawrence S. 2021a. Initial algebras without iteration. Pages 5:1–5:20 of: Gaducci, Fabio, and Silva, Alexandra (eds), *Proc. 9th Conference on Algebra and Coalgebra in Computer Science (CALCO)*. LIPIcs, vol. 211. Schloss Dagstuhl.

[55] Adámek, Jiří, Milius, Stefan, and Urbat, Henning. 2021b. On the behaviour of coalgebras with side effects and algebras with effectful iteration. *J. Logic Comput.*, **31**(6), 1429–1481.

[56] Adámek, Jiří, Hušek, Miroslav, Rosický, Jiří, and Tholen, Walter. 2023. Smallness in topology. *Questiones Mathematicae*, **46**, 13–39.

[57] Alessi, Fabio, Baldan, Paolo, and Bellé, Gianna. 1995. A fixed point theorem in a category of compact metric spaces. *Theoret. Comput. Sci.*, **146**(1–2), 311–320.

[58] Allouche, Jean-Paul, and Shallit, Jeffrey. 2003. *Automatic Sequences: Theory, Applications, Generalizations*. Cambridge University Press.

[59] America, Pierre, and Rutten, Jan. 1989. Solving reflexive domain equations in a category of complete metric spaces. *J. Comput. System Sci.*, **39**(3), 343–375.

[60] Applegate, Harry. 1965. *Acyclic Models and Resolvent Functors*. Ph.D. thesis, Columbia University.

[61] Arnold, André, and Nivat, Maurice. 1980. The metric space of infinite trees. Algebraic and topological properties. *Fund. Inform.*, **3**(4), 445–476.

[62] Baader, Franz, and Tinelli, Cesare. 2002. Deciding the word problem in the union of algebraic theories. *Inform. Comput.*, **178**, 346–390.

[63] Balan, Adriana, and Kurz, Alexander. 2011. Finitary functors: from Set to Preord and Poset. Pages 85–99 of: Corradini, Andrea, Klin, Bartek, and Cîrstea, Corina (eds), *Proc. 4th Conference on Algebra and Coalgebra in Computer Science (CALCO)*. Lecture Notes Comput. Sci., vol. 6859. Springer.

[64] Balan, Adriana, Kurz, Alexander, and Velebil, Jiří. 2015. Extension of functors from Set to $\mathscr{V}$-cat. Pages 16–32 of: Moss, Lawrence S., and Sobociński, Paweł (eds), *Proc. 6th Conference on Algebra and Coalgebra in Computer Science (CALCO)*. LIPIcs, vol. 35. Schloss Dagstuhl.

[65] Baldan, Paolo, Bonchi, Filippo, Kerstan, Henning, and König, Barbara. 2018. Coalgebraic behavioral metrics. *Log. Methods Comput. Sci.*, **14**(3), 20:1–20:61.

[66] Banach, Stefan. 1922. Sur les opérations dans les ensembles abstraits et leurs applications aux équations intégrales. *Fund. Math.*, **3**(1), 133–181.

[67] Barlocco, Simone, Kupke, Clemens, and Rot, Jurriaan. 2019. Coalgebra learning via duality. Pages 62–79 of: Bojańczyk, Mikołaj, and Simpson, Alex (eds), *Proc. 22nd Conference on Foundations of Software Science and Computation Structures (FoSSaCS)*. Lecture Notes Comput. Sci., vol. 11425. Springer.

[68] Barnsley, Michael F. 1993. *Fractals Everywhere*. 2nd edn. Morgan Kaufmann.

[69] Barr, Michael. 1970a. Coequalizers and free triples. *Math. Z.*, **116**, 307–322.

[70] Barr, Michael. 1970b. Relational algebras. Pages 39–55 of: *Reports of the Midwest Category Seminar IV*. Lecture Notes Math., vol. 137. Springer.

[71] Barr, Michael. 1992. Algebraically compact functors. *J. Pure Appl. Algebra*, **82**(3), 211–231.

[72] Barr, Michael. 1993. Terminal coalgebras in well-founded set theory. *Theoret. Comput. Sci.*, **114**(2), 299–315.

[73] Bartels, Falk. 2004. *On Generalized Coinduction and Probabilistic Specification Formats*. Ph.D. thesis, Vrije Universiteit Amsterdam.

[74] Barwise, Jon, and Moss, Lawrence S. 1996. *Vicious Circles*. CSLI Lecture Notes, no. 60. CSLI Publications, Stanford.

[75] Bauer, Andrej, and Lumsdaine, Peter LeFanu. 2013. On the Bourbaki-Witt principle in toposes. *Math. Struct. Comput. Sci.*, **155**(1), 87–99.

[76] Baumgartner, James E. 1976. Almost-disjoint sets, the dense set problem and the partition calculus. *Ann. Math. Logic*, **9**(4), 401–439.

[77] Bergman, George. 2008. On lattices and their ideal lattices, and posets and their ideal posets. *Tbilisi Math. J.*, **1**, 89–103.

[78] Berstel, Jean, and Reutenauer, Christophe. 1988. *Rational Series and Their Languages*. Springer-Verlag.

[79] Bhattacharya, Prasit, Moss, Lawrence S., Ratnayake, Jayampathy, and Rose, Robert. 2014. Fractal sets as final coalgebras obtained by completing an initial algebra. Pages 146–167 of: *Horizons of the Mind. A Tribute to Prakash Panangaden*. Lecture Notes Comput. Sci., vol. 8464. Springer.

[80] Bielecki, Michal, Hidders, Jan, Paredaens, Jan, Tyszkiewicz, Jerzy, and den Bussche, Jan Van. 2002. Navigating with a browser. Pages 764–775 of: Widmayer, Peter, Ruiz, Francisco Triguero, Bueno, Rafael Morales, Hennessy, Matthew, Eidenbenz, Stephan, and Conejo, Ricardo (eds), *Proc. 29th Symposium on Automata, Languages and Programming (ICALP)*. Lecture Notes Comput. Sci., vol. 2380. Springer.

[81] Bird, Gregory J. 1984. *Limits in 2-Categories of Locally-Presentable Categories*. Ph.D. thesis, University of Sydney.

[82] Birkedal, Lars, Støvring, Kristian, and Thamsborg, Jacob. 2010. The category-theoretic solution of recursive metric-space equations. *Theoret. Comput. Sci.*, **411**(47), 4102–4122.

[83] Bloom, Bard, Istrail, Sorin, and Meyer, Albert R. 1995. Bisimulation can't be traced. *J. ACM*, **42**(1), 232–268.

[84] Bloom, Stephen L., and Ésik, Zoltán. 1993. *Iteration Theories: The Equational Logic of Iterative Processes*. EATCS Monographs on Theoretical Computer Science. Springer.

[85] Bloom, Stephen L., and Ésik, Zoltán. 1995. Some equational laws of initiality in 2ccc's. *Internat. J. Found. Comput. Sci.*, **6**(2), 95–118.

[86] Bojańczyk, Mikołaj. 2010. Automata for data words and data trees. Pages 1–4 of: Lynch, Christopher (ed), *Proc. 21st Conference on Rewriting Techniques and Applications, (RTA)*. LIPIcs, vol. 6. Schloss Dagstuhl.

[87] Bojanczyk, Mikolaj, Klin, Bartek, and Lasota, Slawomir. 2014. Automata theory in nominal sets. *Log. Methods Comput. Sci.*, **10**(3), 4:1–4:44.

[88] Bollig, Benedikt, Habermehl, Peter, Leucker, Martin, and Monmege, Benjamin. 2014. A robust class of data languages and an application to learning. *Log. Methods Comput. Sci.*, **10**(4), 19:1–19:23.

[89] Bonsangue, Marcello M., Milius, Stefan, and Silva, Alexandra. 2013. Sound and complete axiomatizations of coalgebraic language equivalence. *ACM Trans. Comput. Logic*, **14**(1), 7:1–7:52.

[90] Borceux, Francis. 1994a. *Handbook of Categorical Algebra: Volume 1, Basic Category Theory*. Encyclopedia of Mathematics and Its Applications. Cambridge University Press.

[91] Borceux, Francis. 1994b. *Handbook of Categorical Algebra: Volume 2, Categories and Structures*. Encyclopedia of Mathematics and Its Applications. Cambridge University Press.

[92] Borceux, Francis. 1994c. *Handbook of Categorical Algebra: Volume 3, Categories of Sheaves*. Encyclopedia of Mathematics and Its Applications. Cambridge University Press.

[93] Bourbaki, Nicolas. 1949. Sur le théoréme de Zorn. *Arch. Math.*, **2**(6), 434–437.

[94] Buchholz, Peter. 2008. Bisimulation relations for weighted automata. *Theoret. Comput. Sci.*, **393**(1–3), 109–123.

[95] Bunge, Marta, and Carboni, Aurelio. 1995. The symmetric topos. *J. Pure Appl. Algebra*, **105**(3), 233–249.

[96] Calude, Cristian, Calude, Elena, and Khoussainov, Bakhadyr. 1997. Deterministic automata simulation, universality and minimality. *Ann. Pure Appl. Logic*, **90**(1–3), 263–276.

[97] Campbell, Colin M., Robertson, Edmund F., Ruškuc, Nik, and Thomas, Richard M. 1996. On subsemigroups of finitely presented semigroups. *J. Algebra*, **180**(1), 1–21.

[98] Cancila, Daniela, Honsell, Furio, and Lenisa, Marina. 2006. Some properties and some problems on set functors. Pages 67–84 of: Adámek, Jiří (ed), *Proc. Coalgebraic Methods in Computer Science (CMCS)*. Electron. Notes Theor. Comput. Sci., vol. 164. Elsevier.

[99] Cantor, Georg. 1891. Über eine elementare Frage der Mannigfaltigkeitslehre. *Jahresbericht der Deutschen Mathematiker-Vereinigung*, **1**, 75–78.

[100] Capretta, Venanzio, Uustalu, Tarmo, and Vene, Varmo. 2006. Recursive coalgebras from comonads. *Inform. Comput.*, **204**(4), 437–468.

[101] Capretta, Venanzio, Uustalu, Tarmo, and Vene, Varmo. 2009. Corecursive algebras: a study of general structured corecursion. Pages 84–100 of: Oliveira, Marcel Vinícius Medeiros, and Woodcock, Jim (eds), *Formal Methods: Foundations and Applications*. Lecture Notes Comp. Sci., vol. 5902. Springer Berlin Heidelberg.

[102] Carboni, Aurelio, Kelly, Max, and Wood, Richard J. 1991. A 2-categorical approach to change of base and geometric morphisms I. *Cah. Topol. Géom. Différ. Catég.*, **32**(1), 47–95.

[103] Carboni, Aurelio, Lack, Steve, and Walters, Robert F. C. 1993. Introduction to extensive and distributive categories. *J. Pure Appl. Algebra*, **84**(2), 145–158.

[104] Cîrstea, Corina. 2010. Generic infinite traces and path-based coalgebraic temporal logics. Pages 83–103 of: Jacobs, Bart, Niqui, Milad, Rutten, Jan, and Silva, Alexandra (eds), *Proc. Coalgebraic Methods in Computer Science (CMCS)*. Electron. Notes Theor. Comput. Sci., vol. 264. Elsevier.

[105] Cîrstea, Corina. 2011. Maximal traces and path-based coalgebraic temporal logics. *Theoret. Comput. Sci.*, **412**(38), 5025–5042.

[106] Cîrstea, Corina. 2017. From branching to linear time, coalgebraically. *Fund. Inform.*, **150**(3–4), 379–406.

[107] Courcelle, Bruno. 1983. Fundamental properties of infinite trees. *Theoret. Comput. Sci.*, **25**(2), 95–169.

[108] Crafa, Silvia, and Ranzato, Francesco. 2011. A spectrum of behavioral relations over LTSs on probability distributions. Pages 124–139 of: Katoen, Joost-Pieter, and König, Barbara (eds), *Proc. Conference on Concurrency Theory (CONCUR)*. Lecture Notes Comput. Sci., vol. 6901. Springer.

[109] Dacar, France. 2009 (January). *The join-induction principle for closure operators on dcpos*. draft, `https://dis.ijs.si/france/notes/join-induction.pdf`.

[110] Davies, Roy O., Hayes, Allan, and Rousseau, George. 1971. Complete lattices and the generalized Cantor theorem. *Proc. Amer. Math. Soc.*, **27**(2), 253–258.

[111] de Bakker, Jaco W., and Zucker, Jeffery I. 1982. Processes and the denotational semantics of concurrency. *Inform. Control*, **54**(1–2), 70–120.

[112] Desharnais, Josée, Edalat, Abbas, and Panangaden, Prakash. 2002. Bisimulation for labelled Markov processes. *Inform. Comput.*, **179**(2), 163–193.

[113] Droste, Manfred, Kuich, Werner, and Vogler, Heiko (eds). 2009. *Handbook of Weighted Automata*. Monographs in Theoretical Computer Science. Springer.

[114] Eilenberg, S. 1974. *Automata, Languages, and Machines, Vol. A*. Pure and applied mathematics, vol. 59. Academic Press.

[115] Elgot, Calvin C. 1975. Monadic computation and iterative algebraic theories. Pages 175–230 of: Rose, Harvey E., and Sheperdson, John C. (eds), *Logic Colloquium '73*, vol. 80. North-Holland Publishers.

[116] Elgot, Calvin C., Bloom, Stephen L., and Tindell, Ralph. 1978. On the algebraic structure of rooted trees. *J. Comput. System Sci.*, **16**, 362–399.

[117] Eppendahl, Adam. 1999. Coalgebra-to-algebra morphisms. Pages 42–49 of: *Proc. Conference on Category Theory and Computer Science (CTCS)*. Electron. Notes Theor. Comput. Sci., vol. 29. Elsevier.

[118] Escardó, Martín. 2003. Joins in the frame of nuclei. *Appl. Categ. Struct.*, **11**, 117–124.

[119] Ésik, Zoltán, and Kuich, Werner. 2001. A generalization of Kozen's axiomatization of the equational theory of regular sets. Pages 99–114 of: Ito, Masami, Păun, Gheorghe, and Yu, Sheng (eds), *Words, Semigroups and Transductions: Festschrift in Honor of Gabriel Thierrin*. World Scientific.

[120] Ésik, Zoltán, and Maletti, Andreas. 2010. Simulation vs. equivalence. Pages 119–122 of: *Proc. 6th Int. Conf. Foundations of Computer Science*. CSREA Press.

[121] Ésik, Zoltán, and Maletti, Andreas. 2011. Simulations of weighted tree automata. Pages 321–330 of: *Proc. Conference on Implementation and Application of Automata (CIAA)*. Lecture Notes Comput. Sci., vol. 6482. Springer.

[122] Feys, Frank, Hansen, Helle Hvid, and Moss, Lawrence S. 2018. Long-term values in Markov decision processes, (co)algebraically. Pages 78–99 of: Cîrstea, Corina (ed), *Proc. 14th Workshop on Coalgebraic Methods in Computer Science (CMCS)*. Lecture Notes Comput. Sci., vol. 11202. Springer.

[123] Fine, K. 1975. Normal forms in modal logic. *Notre Dame J. Form. Log.*, **16**(2), 229–237.

[124] Fiore, Marcelo, Plotkin, Gordon D., and Turi, Daniele. 1999. Abstract syntax and variable binding. Pages 193–202 of: *Proc. 14th Symposium on Logic in Computer Science (LICS)*. IEEE Computer Society.

[125] Fliess, Michel. 1974. Sur divers produits de séries formelles. *Bull. Soc. Math. Fr.*, **102**, 181–191.

[126] Frank, Florian, Milius, Stefan, and Urbat, Henning. 2022. Coalgebraic semantics of nominal automata. Pages 45–66 of: Hansen, Helle Hvid, and Zanasi, Fabio (eds), *Proc. 16th Workshop on Coalgebraic Methods in Computer Science (CMCS)*. Lecture Notes Comput. Sci., vol. 13225. Springer. Full version available at `https://arxiv.org/abs/2202.06546`.

[127] Freyd, Peter. 1968. Rédei's finiteness theorem for commutative semigroups. *Proc. Amer. Math. Soc.*, **19**(4), 1003.

[128] Freyd, Peter. 1991. Algebraically complete categories. Pages 95–104 of: Carboni, Aurelio, Pedicchio, Maria Cristina, and Rosolini, Guiseppe (eds), *Category Theory: Proceedings of the International Conference held in Como*. Lecture Notes in Math., vol. 1488. Springer.

[129] Freyd, Peter. 1992. Remarks on algebraically compact categories. Pages 95–106 of: Fourman, Michael P., Johnstone, Peter T., and Pitts, Andrew M. (eds), *Applications of Category Theory in Computer Science: Proceedings of the London Mathematical Society Symposium, Durham 1991*. London Mathematical Society Lecture Note Series, vol. 177. Cambridge University Press.

[130] Freyd, Peter. 2008. Algebraic real analysis. *Theory Appl. Categ.*, **20**(10), 215–306.

[131] Freyd, Peter J. 1964. *Abelian Categories: An Introduction to the Theory of Functors*. Harper and Row.

[132] Gabbay, Murdoch, and Pitts, Andrew M. 1999. A new approach to abstract syntax involving binders. Pages 214–224 of: Longo, Giuseppe (ed), *Proc. 14th Symposium on Logic in Computer Science (LICS)*. IEEE Computer Society.

[133] Gabbay, Murdoch J., and Pitts, Andrew M. 2002. A new approach to abstract syntax with variable binding. *Form. Asp. Comput.*, **13**, 341–363.

[134] Gabriel, Peter, and Ulmer, Friedrich. 1971. *Lokal präsentierbare Kategorien*. Lecture Notes Math., vol. 221. Springer.

[135] Ghani, Neil, and Uustalu, Tarmo. 2004. Coproducts of ideal monads. *Theor. Inform. Appl.*, **38**(4), 321–342.

[136] Ghani, Neil, Lüth, Christoph, De Marchi, Federico, and Power, John. 2003. Dualizing initial algebras. *Math. Struct. Comput. Sci.*, **13**(2), 349–370.

[137] Ghani, Neil, Hancock, Peter, and Pattinson, Dirk. 2009. Continuous functions on final coalgebras. Pages 3–18 of: Abramsky, Samson, Mislove, Michael, and Palamidessi, Catuscia (eds), *Proc. 25th Conference on Mathematical Foundations of Programming Science (MFPS)*. Electron. Notes Theor. Comput. Sci., vol. 249. Elsevier.

[138] Gierz, Gerhard, Hofmann, Karl Heinrich, Keimel, Klaus, Lawson, Jimmie D., Mislove, Michael W., and Scott, Dana S. 2003. *Continuous lattices and domains*. Encyclopedia of Mathematics and Its Applications, vol. 93. Cambridge University Press.

[139] Ginali, Susanna. 1979. Regular trees and the free iterative theory. *J. Comput. System Sci.*, **18**(3), 228–242.

[140] Giry, Michelle. 1982. A categorical approach to probability theory. Pages 68–85 of: *Proc. Categorical Aspects of Topology and Analysis*. Lecture Notes Math., vol. 915. Springer.

[141] Gleason, Andrew M., and Dilworth, Robert P. 1962. A generalized Cantor theorem. *Proc. Amer. Math. Soc.*, **13**(5), 704–705.

[142] Glimming, Johan, and Ghani, Neil. 2005. Difunctorial semantics of object calculus. *Electron. Notes Theor. Comput. Sci.*, **138**(2), 79–94.

[143] Goguen, Joseph A., Thatcher, James W., and Wagner, Eric G. 1973. *A Junction Between Computer Science and Category Theory, I: Basic Concepts and Examples (part I)*. IBM Thomas J. Watson Research Center.

[144] Goncharov, Sergey, Milius, Stefan, and Silva, Alexandra. 2014. Towards a coalgebraic Chomsky hierarchy (extended abstract). Pages 265–280 of: *Proc. 8th Conference on Theoretical Computer Science (TCS)*. Lecture Notes Comput. Sci., vol. 8705. Springer.

[145] Goncharov, Sergey, Milius, Stefan, and Silva, Alexandra. 2020. Towards a uniform theory of effectful state machines. *ACM Trans. Comput. Log.*, **21**(3), 23:1–23:63.

[146] Goubault-Larrecq, Jean. 2013. *Non-Hausdorff Topology and Domain Theory: Selected Topics in Point-Set Topology*. New Mathematical Monographs, vol. 22. Cambridge University Press.

[147] Goubault-Larrecq, Jean. 2021. *Bourbaki, Witt, and Dito Pataraia*. `https://projects.lsv.ens-cachan.fr/topology/?page_id=176`.

[148] Grabmayer, Clemens, Endrullis, Jörg, Hendriks, Dimitri, Klop, Jan Willem, and Moss, Lawrence S. 2012. Automatic sequences and zip-specifications. Pages 335–344 of: *Proc. 25th Symposium on Logic in Computer Science (LICS)*. IEEE Computer Society.

[149] Gumm, H. Peter. 2001. Functors for coalgebras. *Algebra Universalis*, **45**(2), 135–147.

[150] Gumm, H. Peter. 2005. From T-coalgebras to filter structures and transition systems. Pages 194–212 of: *Proc 1st Conference on Algebra and Coalgebra in Computer Science (CALCO)*. Lecture Notes in Comput. Sci., vol. 3629. Springer.

[151] Gumm, H. Peter. 2020. Free-algebra functors from a coalgebraic perspective. Pages 55–67 of: Petrişan, Daniela, and Rot, Jurriaan (eds), *Proc. 15th Workshop on Coalgebraic Methods in Computer Science (CMCS)*. Lecture Notes Comput. Sci., vol. 12094. Springer.

[152] Gumm, H. Peter, and Schröder, Tobias. 2001. Monoid-labeled transition systems. *Electron. Notes Theor. Comput. Sci.*, **44**(1), 185–204.

[153] Gumm, H. Peter, and Schröder, Tobias. 2002. Coalgebras of bounded type. *Math. Struct. Comput. Sci.*, **12**(5), 565–578.

[154] Gumm, H. Peter, and Schröder, Tobias. 2005. Types and coalgebraic structure. *Algebra Universalis*, **53**(2), 229–252.

[155] Hardy, Godfrey H., and Wright, Edward M. 1960. *An Introduction to the Theory of Numbers*. Fourth edn. Oxford University Press.

[156] Hasuo, Ichiro. 2008. *Tracing Anonymity with Coalgebras*. Ph.D. thesis, Radboud University Nijmegen.

[157] Hasuo, Ichiro, and Jacobs, Bart. 2005. Context-free languages via coalgebraic trace semantics. Pages 213–231 of: *Proc. 1st Conference on Algebra and Coalgebra in Computer Science (CALCO)*. Lecture Notes Comput. Sci., vol. 3629. Springer.

[158] Hasuo, Ichiro, Jacobs, Bart, and Sokolova, Ana. 2007. Generic trace semantics via coinduction. *Log. Methods Comput. Sci.*, **3**(4), 11:1–11:36.

[159] Hausdorff, Felix. 1914. *Grundzüge der Mengenlehre*. Veit & Comp.

[160] Hausmann, Daniel, Milius, Stefan, and Schröder, Lutz. 2021. A linear-time nominal μ-calculus with name allocation. Pages 58:1–58:18 of: Bonchi, Filippo, and Puglisi, Simon (eds), *Proc. 46th Symposium on Mathematical Foundations of Computer Science (MFCS)*. LIPIcs, vol. 202. Schloss Dagstuhl.

[161] Hennessy, Matthew. 2002. A fully abstract denotational semantics for the pi-calculus. *Theoret. Comput. Sci.*, **278**, 53–89.

[162] Hermida, Claudio, and Jacobs, Bart. 1998. Structural induction and coinduction in a fibrational setting. *Inform. Comput.*, **145**(2), 107–152.

[163] Hinze, Ralf, Wu, Nicolas, and Gibbons, Jeremy. 2015. Conjugate hylomorphisms - or: the mother of all structured recursion schemes. Pages 527–538 of: Rajamani, Sriram K., and Walker, David (eds), *Proc. 42nd Symposium on Principles of Programming Languages (POPL)*. ACM SIGPLAN Notices, vol. 50(1). ACM.

[164] Hoffmann, Dirk, and Nora, Pedro. 2020. Hausdorff coalgebras. *Appl. Categ. Struct.*, **28**(5), 773–806.

[165] Hofmann, Dirk, Neves, Renato, and Nora, Pedro. 2019. Limits in categories of Vietoris coalgebras. *Math. Struct. Comput. Sci.*, **29**(4), 552–587.

[166] Hughes, Jesse, and Jacobs, Bart. 2004. Simulations in coalgebra. *Theoret. Comput. Sci.*, **327**(1–2), 71–108.

[167] Hyland, Martin, Plotkin, Gordon D., and Power, John. 2006. Combining effects: sums and tensor. *Theoret. Comput. Sci.*, **357**(1–3), 70–99.

[168] Ihringer, Thomas. 2003. *Allgemeine Algebra. Mit einem Anhang über Universelle Coalgebra von H. P. Gumm*. Berliner Studienreihe zur Mathematik, vol. 10. Heldermann Verlag.

[169] Iwamura, Tsurane. 1944. A lemma on directed sets. *Zenkoku Shijo Sugaku Danwakai*, **262**, 107–111. In Japanese.

[170] Jacobs, Bart. 2002. The temporal logic of coalgebras via Galois algebras. *Math. Struct. Comput. Sci.*, **12**(6), 875–903.

[171] Jacobs, Bart. 2004. Trace semantics for coalgebras. Pages 167–184 of: Adámek, Jiří, and Milius, Stefan (eds), *Proc. Coalgebraic Methods in Computer Science (CMCS)*. Electron. Notes Theor. Comput. Sci., vol. 106. Elsevier.

[172] Jacobs, Bart. 2010. Convexity, duality and effects. Pages 1–19 of: Calude, Cristian S., and Sassone, Vladimiro (eds), *Proc. 6th Conference on Theoretical Computer Science (TCS)*. IFIP Adv. Inf. Commun. Technol., vol. 323. Springer.

[173] Jacobs, Bart. 2016. *Introduction to Coalgebra. Towards Mathematics of States and Observation*. Cambridge University Press.

[174] Jacobs, Bart, Silva, Alexandra, and Sokolova, Ana. 2015. Trace semantics via determinization. *J. Comput. System Sci.*, **81**(5), 859–879.

[175] Jeannin, Jean-Baptiste, Kozen, Dexter, and Silva, Alexandra. 2017. Well-founded coalgebras, revisited. *Math. Struct. Comput. Sci.*, **27**(7), 1111–1131.

[176] Jech, Thomas. 1978. *Set Theory*. Academic Press.

[177] Johnstone, Peter T. 1975. Adjoint lifting theorems for categories of algebras. *Bull. London Math. Soc.*, **7**(2), 294–297.

[178] Johnstone, Peter T. 1986. *Stone Spaces*. Cambridge University Press.

[179] Joyal, André. 1981. Une théorie combinatoire des séries formelles. *Adv. Math.*, **42**(1), 1–82.

[180] Joyal, André. 1986. Foncteurs analytiques et espèces de structures. *Lecture Notes in Math.*, **1234**, 126–159.

[181] Kaminski, Michael, and Francez, Nissim. 1994. Finite-memory automata. *Theoret. Comput. Sci.*, **134**(2), 329–363.

[182] Kantorovich, Leonid V. 1939. Mathematical methods of organizing and planning production. *Management Science*, **6**(4), 366–422.

[183] Kantorovich, Leonid V. 1942. On the translocation of masses. *Dokl. Akad. Nauk SSSR*, **37**, 227–229.

[184] Karazeris, Panagis, Matzaris, Apostolos, and Velebil, Jiří. 2011. Final coalgebras in accessible categories. *Math. Structure Comput. Sci*, **21**(5), 1067–1108.

[185] Katsumata, Shin-ya, and Sato, Tetsuya. 2015. Codensity liftings of monads. Pages 156–170 of: Moss, Lawrence S., and Sobociński, Paweł (eds), *Proc. 6th Conference on Algebra and Coalgebra in Computer Science (CALCO)*. LIPIcs, vol. 35. Schloss Dagstuhl.

[186] Kawahara, Yasuo, and Mori, Masao. 2000. A small final coalgebra theorem. *Theoret. Comput. Sci.*, **233**(1–2), 129–145.

[187] Kelly, Max. 1980. A unified treatment of transfinite constructions for free algebras, free monoids, colimits, associated sheaves, and so on. *Bull. Austral. Math. Soc.*, **22**(2), 1–83.

[188] Kelly, Max. 1982. *Basic concept of enriched category theory*. London Mathematical Society Lecture Notes, vol. 64. Cambridge University Press. Reprint published in *Reprints in Theory and Applications of Categories*, No. 10, 2005, `http://www.tac.mta.ca/tac/reprints/articles/10/tr10.pdf`.

[189] Kerstan, Henning, and König, Barbara. 2013. Coalgebraic trace semantics for continuous probabilistic transitions systems. *Log. Methods Comput. Sci.*, **9**(4), 16:1–16:34.

[190] Khinchin, Aleksandr Ya. 1964. *Continued Fractions*. University of Chicago Press.

[191] Kőnig, Dénes. 1927. Über eine Schlussweise aus dem Endlichen ins Unendliche. *Acta Sci. Math.*, **3**(2–3), 121–130.

[192] Klin, Bartek. 2011. Bialgebras for structural operational semantics: An introduction. *Theoret. Comput. Sci.*, **412**(38), 5043–5069.

[193] Klin, Bartek, and Sassone, Vladimiro. 2013. Structural operational semantics for stochastic and weighted transition systems. *Inform. Comput.*, **227**, 58–83.

[194] Knaster, Bronislav. 1928. Un théorème sur les fonctions d'ensembles. *Ann. Soc. Pol. Math.*, **6**, 133–134.

[195] Knijnenburg, Peter W. 1993. Algebraic domains, chain completion, and the Plotkin powerdomain contruction. *Technical Report RUU-CS-93-03*.

[196] Kock, Anders. 1970. Monads on symmetric monoidal closed categories. *Arch. Math. (Basel)*, **21**, 1–10.

[197] Kock, Anders. 1972. Strong functors and monoidal monads. *Arch. Math. (Basel)*, **23**, 113–120.

[198] Koubek, Václav. 1971. Set functors. *Comment. Math. Univ. Carolin.*, **12**(1), 175–195.

[199] Koubek, Václav, and Reiterman, Jan. 1975. Automata and categories – input processes. Pages 280–286 of: Bečvář, Jiří (ed), *Proc. 4th Symposium on Mathematical Foundations of Computer Science (MFCS)*. Lecture Notes Comput. Sci., vol. 32. Springer.

[200] Koubek, Václav, and Reiterman, Jan. 1979. Categorical constructions of free algebras, colimits, and completions of partial algebras. *J. Pure Appl. Algebra*, **14**(2), 195–231.

[201] Kozen, Dexter. 2009. Coinductive proof principles for stochastic processes. *Log. Methods Comput. Sci.*, **3**(4), 8:1–8:14.

[202] Kozen, Dexter. 2011. Realization of coinductive types. Pages 237–246 of: *Proc. 27th Conference on Mathematical Foundations of Programming Semantics (MFPS)*. Electron. Notes Theor. Comput. Sci., vol. 276. Elsevier.

[203] Kozen, Dexter, Mamouras, Konstantinos, Petrişan, Daniela, and Silva, Alexandra. 2015. Nominal Kleene coalgebra. Pages 286–298 of: *42nd Colloquium on Automata, Languages, and Programming, (ICALP)*. Lecture Notes Comput. Sci., vol. 9135. Springer.

[204] Kupke, Clemens, and Rutten, Jan. 2012. On the final coalgebra of automatic sequences. Pages 149–164 of: *Logic and Program Semantics – Essays Dedicated to Dexter Kozen on the Occasion of his 60th Birthday*. Lecture Notes Comput. Sci., vol. 7230. Springer.

[205] Kupke, Clemens, Kurz, Alexander, and Venema, Yde. 2004. Stone coalgebras. *Theoret. Comput. Sci.*, **327**(1–2), 109–132.

[206] Kürtz, Klaas, Küsters, Ralf, and Wilke, Thomas. 2007. Selecting theories and nonce generation for recursive protocols. Pages 61–70 of: Ning, Peng, Atluri, Vijay, Gligor, Virgil, and Mantel, Heiko (eds), *Formal Methods in Security Engineering (FMSE)*. ACM.

[207] Kurz, Alexander. 2000. *Logics for Coalgebras and Applications to Computer Science*. Ph.D. thesis, Ludwig-Maximilians-Universität München.

[208] Kurz, Alexander, and Velebil, Jiří. 2016. Relation lifting, a survey. *J. Log. Algebr. Program.*, **85**(4), 475–499.

[209] Kurz, Alexander, Petrişan, Daniela, Severi, Paula, and de Vries, Fer-Jan. 2013. Nominal coalgebraic data types with applications to calculus. *Log. Methods Comput. Sci.*, **9**(4), 20:1–20:51.

[210] Lambek, Joachim. 1968. A fixpoint theorem for complete categories. *Math. Z.*, **103**, 151–161.

[211] Larsen, Kim G., and Skou, Arne. 1991. Bisimulation through probabilistic testing. *Inform. Comput.*, **94**(1), 1–28.

[212] Lehmann, Daniel J., and Smyth, Michael B. 1981. Algebraic specification of data types: A synthetic approach. *Mathematical Systems Theory*, **14**(1), 97–139.

[213] Leinster, Tom. 2011. A general theory of self-similarity. *Adv. Math.*, **226**.

[214] Lévy, Azriel. 1979. *Basic Set Theory*. Springer.

[215] Levy, Paul. 2011. Similarity quotients as final coalgebras. Pages 27–41 of: Hofmann, Martin (ed), *Proc. 14th Conference on Foundations of Software Science and Computation Structures (FoSSaCS)*. Lecture Notes Comput. Sci., vol. 6604. Springer.

[216] Loya, Paul. 2017. *Amazing and Aesthetic Aspects of Analysis*. Springer.

[217] Mac Lane, Saunders. 1998. *Categories for the Working Mathematician*. 2nd edn. Springer.

[218] Makkai, Michael, and Paré, Robert. 1989. *Accessible Categories: The Foundation of Categorical Model Theory*. Contemporary Math., vol. 104. American Mathematical Society.

[219] Manes, Ernest G. 1976. *Algebraic Theories*. Graduate Texts in Math., vol. 26. Springer.

[220] Manes, Ernest G., and Arbib, Michael A. 1986. *Algebraic Approaches to Program Semantics*. Springer.

[221] Manna, Zohar, and Pnüeli, Amir. 1992. *The Temporal Logic of Reactive and Concurrent Systems: Specification*. Springer-Verlag.

[222] Markowsky, George. 1976. Chain-complete posets and directed sets with applications. *Algebra Universalis*, **6**(1), 53–68.

[223] Marti, Johannes, and Venema, Yde. 2015. Lax extensions of coalgebra functors and their logic. *J. Comput. System Sci.*, **81**(5), 880–900.

[224] Martin, Keye. 2013. Nothing can be fixed. Pages 195–196 of: *Computation, Logic, Games, and Quantum Foundations*. Lecture Notes Comput. Sci., vol. 7860. Springer.

[225] Michael, Ernest. 1951. Topologies on spaces of subsets. *Trans. Amer. Math. Soc.*, **71**(1), 152–182.

[226] Milius, Stefan. 2005. Completely iterative algebras and completely iterative monads. *Inform. Comput.*, **196**(1), 1–41.

[227] Milius, Stefan. 2010. A sound and complete calculus for finite stream circuits. Pages 449–458 of: *Proc. 25th Symposium on Logic in Computer Science (LICS)*. IEEE Computer Society.

[228] Milius, Stefan. 2018. Proper functors and fixed points for finite behaviour. *Log. Methods Comput. Sci.*, **14**(3), 22:1–22:32.

[229] Milius, Stefan, and Moss, Lawrence S. 2006. The category theoretic solution of recursive program schemes. *Theoret. Comput. Sci.*, **366**(1–2), 3–59. Fundamental Study.

[230] Milius, Stefan, and Wißmann, Thorsten. 2015. Finitary corecursion for the infinitary lambda calculus. Pages 336–351 of: *Proc. 6th Conference on Algebra and Coalgebra in Computer Science (CALCO)*. LIPIcs, vol. 35. Schloss Dagstuhl.

[231] Milius, Stefan, Palm, Thorsten, and Schwencke, Daniel. 2009. Complete iterativity for algebras with effects. Pages 34–48 of: Kurz, Alexander, Lenisa, Marina, and Tarlecki, Andrzej (eds), *Proc. 3rd Conference on Algebra and Coalgebra in Computer Science (CALCO)*. Lecture Notes Comput. Sci., vol. 5728. Springer. full version avaible at `www8.cs.fau.de/ext/milius/publications/files/mps_lambda_revised_full.pdf`.

[232] Milius, Stefan, Moss, Lawrence S., and Schwencke, Daniel. 2010. CIA structures and the semantics of recursion. Pages 312–327 of: Ong, C.-H. Luke (ed), *Proc. 13th Conference on Foundations of Software Science and Computation Structures (FoSSaCS)*. Lecture Notes Comput. Sci., vol. 6014. Springer.

[233] Milius, Stefan, Bonsangue, Marcello M., Myers, Robert S. R., and Rot, Jurriaan. 2013. Rational operation models. Pages 257–282 of: Mislove, Michael (ed),

Proc. 29th Conference on Mathematical Foundations of Programming Science (MFPS). Electron. Notes Theor. Comput. Sci., vol. 298. Elsevier.

[234] Milius, Stefan, Schröder, Lutz, and Wißmann, Thorsten. 2016. Regular behaviours with names: On rational fixpoints of endofunctors on nominal sets. *Appl. Categ. Struct.*, **24**(5), 663–701.

[235] Milius, Stefan, Pattinson, Dirk, and Wißmann, Thorsten. 2020. A new foundation for finitary corecursion and iterative algebras. *Inform. Comput.*, **271**. Article 104456.

[236] Milner, Robin. 1980. *A Calculus of Communicating Systems.* Lecture Notes Comput. Sci., vol. 92. Springer.

[237] Milner, Robin. 1989. *Communication and Concurrency.* International Series in Computer Science. Prentice Hall.

[238] Moggi, Eugenio. 1991. Notions of computations and monads. *Inform. Comput.*, **93**(1), 55–92.

[239] Moss, Lawrence S. 1999. Coalgebraic logic. *Annals of Pure and Applied Logic*, **96**(1), 277–317.

[240] Moss, Lawrence S., and Noquez, Victoria. 2022. Corecursive algebras in nature. In: Hansen, Helle Hvid, and Zanasi, Fabio (eds), *Proc. 16th Workshop on Coalgebraic Methods in Computer Science (CMCS).* Lecture Notes Comput. Sci. Springer.

[241] Mulry, Philip S. 1994. Lifting theorems for Kleisli categories. Pages 304–319 of: Brookes, Stephen, Main, Michael, Melton, Austin, Mislove, Michael, and Schmidt, David (eds), *Proc. 9th Conference on Mathematical Foundations of Programming Semantics (MFPS).* Lecture Notes Comput. Sci., vol. 802. Springer.

[242] Nelson, Evelyn. 1983. Iterative algebras. *Theoret. Comput. Sci.*, **25**(1), 67–94.

[243] Neven, Frank, Schwentick, Thomas, and Vianu, Victor. 2004. Finite state machines for strings over infinite alphabets. *ACM Trans. Comput. Log.*, **5**(3), 403–435.

[244] Niven, Ivan, Zuckerman, Herbert S., and Montgomery, Hugh L. 1991. *An Introduction to the Theory of Numbers.* Fifth edn. John Wiley & Sons.

[245] Osius, Gerhard. 1974. Categorical set theory: a characterization of the category of sets. *J. Pure Appl. Algebra*, **4**(1), 79–119.

[246] Pataraia, Dito. 1997 (November). *A constructive proof of Tarski's fixed-point theorem for dcpo's.* Lecture presented at the 65th Peripatetic Seminar on Sheaves and Logic (PSSL), Aarhus.

[247] Pavlović, D., and Pratt, V. 2002. The continuum as a final coalgebra. *Theoret. Comput. Sci.*, **280**(1–2), 105–122.

[248] Pavlović, Dusko, and Escardó, Martín H. 1998. Calculus in coinductive form. Pages 408–417 of: *Proc. 13th Symposium on Logic in Computer Science (LICS).* IEEE Computer Society.

[249] Petrişan, Daniela. 2011. *Investigations into Algebra and Topology over Nominal Sets.* Ph.D. thesis, University of Leicester.

[250] Pitts, Andrew M. 2013. *Nominal Sets: Names and Symmetry in Computer Science.* Cambridge Tracts in Theoretical Computer Science, vol. 57. Cambridge University Press.

[251] Plotkin, Gordon. 1977. LCF considered as a programming language. *Theoret. Comput. Sci.*, **5**(3), 223–255.

[252] Plotkin, Gordon D., and Turi, Daniele. 1997. Towards a mathematical operational semantics. Pages 280–291 of: *Proc. 12th Symposium on Logic in Computer Science (LICS)*. IEEE Computer Society.

[253] Pompeiu, Dimitrie. 1905. Sur la continuité des fonctions de variables complexes. *Annales de la Faculté des Sciences de la Université de Toulouse pour les Sciences Mathématiques et les Sciences Physiques*, **2ième Série, 7**(3), 265–315.

[254] Pultr, Aleš, and Trnková, Věra. 1980. *Combinatorial, Algebraic and Topological Representations of Groups, Semigroups and Categories*. North-Holland.

[255] Pumplün, Dieter. 1984. Regularly ordered Banach spaces and positively convex spaces. *Res. Math.*, **7**, 85–112.

[256] Pumplün, Dieter. 2003. Positively convex modules and ordered normed linear spaces. *J. Convex Analysis*, **10**(1), 109–127.

[257] Rabin, Michael O. 1963. Probabilistic automata. *Inform. Control*, **6**(3), 230–245.

[258] Rabin, Michael O., and Scott, Dana. 1959. Finite automata and their decision problems. *IBM Journal of Research and Development*, **3**(2), 114–125.

[259] Rédei, Lásló. 1965. *The Theory of Finitely Generated Commutative Semigroups*. Pergamon.

[260] Ribes, Luis, and Zalesskii, Pavel. 2010. *Profinite Groups*. 2nd edn. Ergebnisse der Mathematik und ihrer Grenzgebiete, vol. 40. Springer.

[261] Robinson, Derek J. S. 1993. *A Course in the Theory of Groups*. Graduate Texts in Mathematics, vol. 80. Springer.

[262] Rodríguez-Lopez, Jesús, and Romaguera, Salvador. 2002. The relationship between the Vietoris topology and the Hausdorff quasi-uniformity. *Topology Appl.*, **124**(3), 451–464.

[263] Rutten, Jan. 1998. Relators and metric bisimulations. *Electron. Notes Theor. Comput. Sci.*, **11**, 252–258.

[264] Rutten, Jan. 2000. Universal coalgebra: a theory of systems. *Theoret. Comput. Sci.*, **249**(1), 3–80. Fundamental Study.

[265] Rutten, Jan. 2005. A coinductive calculus of streams. *Math. Struct. Comput. Sci.*, **15**(1), 93–147.

[266] Rutten, Jan. 2006. Algebraic specification and coalgebraic synthesis of Mealy automata. *Electron. Notes Theor. Comput. Sci.*, **160**, 305–319.

[267] Rutten, Jan. 2008. Rational streams coalgebraically. *Log. Methods Comput. Sci.*, **4**(3), 9:1–9:22.

[268] Rutten, Jan. 2019. *The Method of Coalgebra: Exercises in Coinduction*. CWI. https://ir.cwi.nl/pub/28550/rutten.pdf.

[269] Rutten, Jan, and de Vink, Erik. 1999. Bisimulation for probabilistic transition systems: a coalgebraic approach. *Theoret. Comput. Sci.*, **221**, 271–293.

[270] Schröder, Lutz, Kozen, Dexter, Milius, Stefan, and Wißmann, Thorsten. 2017. Nominal automata with name binding. Pages 124–142 of: Esparza, Javier, and Murawski, Andrzej (eds), *Proc. 20th Conference on Foundations of Software Science and Computation Structures (FoSSaCS)*. Lecture Notes Comput. Sci., vol. 10203. Springer.

[271] Schützenberger, Marcel Paul. 1961. On the definition of a family of automata. *Inform. Control*, **4**(2–3), 275–270.

[272] Schwencke, Daniel. 2010. Coequational logic for accessible functors. *Inform. Comput.*, **208**(12), 1469–1489.

[273] Scott, Dana. 1972. Continuous lattices. Pages 97–136 of: Lawvere, F. William (ed), *Toposes, Algebraic Geometry and Logic*. Lecture Notes in Math., vol. 274. Springer. Proceeding of the conference held at Dalhousie University, 1971.

[274] Segoufin, Luc. 2006. Automata and logics for words and trees over an infinite alphabet. Pages 41–57 of: Ésik, Zoltán (ed), *Proc. 15th Conference on Computer Science Logic (CSL)*. Lecture Notes Comput. Sci., vol. 4207. Springer.

[275] Seifan, Fatehmeh, Schröder, Lutz, and Pattinson, Dirk. 2017. Uniform interpolation in coalgebraic modal logic. Pages 21:1–21:30 of: Bonchi, Filippo, and König, Barbara (eds), *Proc. 7th Conference on Coalgebraic Methods in Computer Science (CALCO)*. LIPIcs, vol. 72. Schloss Dagstuhl.

[276] Silva, Alexandra, Bonsangue, Marcello M., and Rutten, Jan. 2010. Nondeterministic Kleene coalgebras. *Log. Methods Comput. Sci.*, **6**(3), 23:1–23:39.

[277] Silva, Alexandra, Bonchi, Filippo, Bonsangue, Marcello M., and Rutten, Jan. 2011. Quantitative Kleene coalgebras. *Inform. Comput.*, **209**(5), 822–849.

[278] Silva, Alexandra, Bonchi, Filippo, Bonsangue, Marcello M., and Rutten, Jan. 2013. Generalizing determinization from automata to coalgebras. *Log. Methods Comput. Sci.*, **9**(1), 9:1–9:27.

[279] Smyth, M. B., and Plotkin, G. D. 1982. The category-theoretic solution of recursive domain equations. *SIAM J. Comput.*, **11**(4), 761–783.

[280] Sokolova, Ana. 2005. *Coalgebraic Analysis of Probabilistic Systems*. Ph.D. thesis, Eindhoven University of Technology.

[281] Sokolova, Ana, and Woracek, Harald. 2015. Congruences of convex algebras. *J. Pure Appl. Algebra*, **219**(8), 3110–3148.

[282] Sokolova, Ana, and Woracek, Harald. 2018. Proper semirings and proper convex functors. Pages 331–347 of: Baier, Christel, and Dal Lago, Ugo (eds), *Proc. 21st Conference on Foundations of Software Science and Computation Structures (FoSSaCS)*. Lecture Notes Comput. Sci., vol. 10803. Springer.

[283] Sprunger, David, and Moss, Lawrence S. 2017. Precongruences and parametrized coinduction for logics for behavioral equivalence. Pages 23:1–23:15 of: Bonchi, Filippo, and König, Barbara (eds), *Proc. 7th Conference on Algebra and Coalgebra in Computer Science (CALCO)*. LIPIcs, vol. 72. Schloss Dagstuhl.

[284] Tarski, Alfred. 1955. A lattice-theoretical fixpoint theorem and its applications. *Pacific J. Math.*, **5**(2), 285–309.

[285] Taylor, Paul. 1994. *The limit-colimit coincidence for categories*. Ppreprint, `http://www.paultaylor.eu/domains/limccc.pdf`.

[286] Taylor, Paul. 1995–6. *Towards a unified treatment of induction I: the general recursion theorem*. Preprint, `www.paultaylor.eu/ordinals/#towuti`.

[287] Taylor, Paul. 1999. *Practical Foundations of Mathematics*. Cambridge University Press.

[288] Taylor, Paul. 2021 (April). *Well Founded Coalgebras and Recursion*. Preprint, `https://www.paultaylor.eu/ordinals/welfcr.pdf`.

[289] Tiuryn, Jerzy. 1980. Unique fixed points vs. least fixed points. *Theoret. Comput. Sci.*, **12**(3), 229–254.

[290] Trnková, V., Adámek, J., Koubek, V., and Reiterman, J. 1975. Free algebras, input processes and free monads. *Comment. Math. Univ. Carolin.*, **16**(2), 339–351.

[291] Trnková, Věra. 1969. Some properties of set functors. *Comment. Math. Univ. Carolin.*, **10**(2), 323–352.

[292] Trnková, Věra. 1971. On a descriptive classification of set functors I. *Comment. Math. Univ. Carolin.*, **12**(1), 143–174.

[293] Trnková, Věra. 1980. General theory of relational automata. *Fund. Inform.*, **3**(2), 189–234.

[294] Turner, David, and Winskel, Glynn. 2009. Nominal domain theory for concurrency. Pages 546–560 of: Grädel, Erich, and Kahle, Reinhard (eds), *Proc. 18th Conference on Computer Science Logic (CSL)*. Lecture Notes Comput. Sci., vol. 5771.

[295] Urabe, Natsuki, and Hasuo, Ichiro. 2015. Coalgebraic infinite traces and Kleisli simulations. Pages 320–335 of: Moss, Lawrence S., and Sobociński, Paweł (eds), *Proc. 6th Conference on Algebra and Coalgebra in Computer Science (CALCO)*. LIPIcs, vol. 35. Schloss Dagstuhl.

[296] Urabe, Natsuki, Shunsuke, Shimizu, and Hasuo, Ichiro. 2016. Coalgebraic trace semantics for Büchi and parity automata. Pages 24:1–24:15 of: Desharnais, Josée, and Jagadeesan, Radha (eds), *Proc. 27th Conference on Concurrency Theory (CONCUR)*. LIPIcs, vol. 59. Schloss Dagstuhl. Full version available at `https://arxiv.org/abs/1606.09399`.

[297] Urbat, Henning. 2017. Finite behaviours and finitary corecursion. Pages 24:1–24:15 of: *Proc. 7th Conference on Algebra and Coalgebra in Computer Science (CALCO)*. LIPIcs, vol. 72. Schloss Dagstuhl.

[298] Urbat, Henning, and Schröder, Lutz. 2020. Automata learning: an algebraic approach. Pages 900–914 of: *Proc. 35th Symposium on Logic in Computer Science (LICS)*. IEEE Computer Society.

[299] Urbat, Henning, Hausmann, Daniel, Milius, Stefan, and Schröder, Lutz. 2021. Nominal Büchi automata with name allocation. Pages 4:1–4:16 of: Haddad, Serge, and Varacca, Daniele (eds), *Proc. 32nd Conference on Concurrency Theory (CONCUR)*. LIPIcs, vol. 203. Schloss Dagstuhl.

[300] Uustalu, Tarmo, and Vene, Varmo. 2002. Similarity quotients as final coalgebra. Pages 99–110 of: Hammond, Kevin, and Curtis, Sharon (eds), *Proc. Trends in Functional Programming 3*. Intellect.

[301] Uustalu, Tarmo, and Vene, Varmo. 2011. The recursion scheme from the cofree recursive comonad. Pages 135–157 of: Capretta, Venancio, and McBride, Conor (eds), *Proc. 2nd Workshop on Mathematically Structured Functional Programming (MSFP)*. Electron. Notes Theor. Comput. Sci., vol. 229(5). Elsevier.

[302] van Breugel, Franck. 2004. De Bakker-Zucker processes revisited. *Inform. Comput.*, **188**(1), 68–76.

[303] van Breugel, Franck, and Worrell, James. 2005. A behavioural pseudometric for probabilistic transition systems. *Theoret. Comput. Sci.*, **331**(1), 115–142.

[304] van Breugel, Franck, Costantini, Camillo, and Watson, Stephen. 2004. Isometries between a metric space and its hyperspace, function space, and space of measures. *Topology Appl.*, **137**(1–3), 51–57.

[305] van Breugel, Franck, Hermida, Claudio, Makkai, Michael, and Worrell, James. 2007. Recursively defined metric spaces without contraction. *Theoret. Comput. Sci.*, **380**(1–2), 143–163.

[306] Vaseršteĭn, Leonid N. 1969. Markov processes over denumerable products of spaces, describing large systems of automata. *Problemy Peredači Informacii*, **5**(3), 64–72.

[307] Veldhoven, Joep. 2018. *Automata Extended to Nominal Sets*. Bachelor thesis, Radboud University Nijmegen.

[308] Velebil, Jiří. 1998. *Categorical Domain Theory*. Ph.D. thesis, Czech Technical University Prague. `http://math.feld.cvut.cz/velebil/research/papers.html`.

[309] Vietoris, Leopold. 1922. Bereiche zweiter Ordnung. *Monatsh. Math. Phys.*, **32**(1), 258–280.

[310] Villani, Cédric. 2009. *Optimal Transport – Old and New*. Grundlehren der mathematischen Wissenschaften. Springer.

[311] Winter, Joost, Bonsangue, Marcello, and Rutten, Jan. 2013. Coalgebraic characterizations of context-free languages. *Log. Methods Comput. Sci.*, **9**(3), 14:1–14:39.

[312] Winter, Joost, Bonsangue, Marcello, and Rutten, Jan. 2015. Context-free coalgebras. *J. Comput. System Sci.*, **81**, 911–939.

[313] Wißmann, Thorsten. 2020. *Coalgebraic Semantics and Minimization in Sets and Beyond*. Ph.D. thesis, Friedrich-Alexander-Universität Erlangen-Nürnberg (FAU).

[314] Wißmann, Thorsten, Milius, Stefan, Dubut, Jérémy, and Katsumata, Shin-ya. 2019. A coalgebraic view on reachability. *Comment. Math. Univ. Carolin.*, **60**(4), 605–638.

[315] Witt, Ernst. 1951. Beweisstudien zum Satz von M. Zorn. *Math. Nachr.*, **4**, 434–438.

[316] Worrell, James. 1999. Terminal sequences for accessible endofunctors. Pages 24–38 of: *Proc. Coalgebraic Methods in Computer Science (CMCS)*. Electron. Notes Theor. Comput. Sci., vol. 19. Elsevier.

[317] Worrell, James. 2000. Coinduction for recursive data types: partial orders, metric spaces and omega-categories. *Electron. Notes Theor. Comput. Sci.*, **33**, 337–356.

[318] Worrell, James. 2005. On the final sequence of a finitary set functor. *Theoret. Comput. Sci.*, **338**(1–3), 184–199.

[319] Zermelo, Ernst. 1904. Beweis, daß jede Menge wohlgeordnet werden kann. *Math. Ann.*, **59**(4), 514–516.

List of Categories

$\mathscr{A}_\lambda$, 367, 389
$\mathscr{A}^E$, 154
$\mathscr{A}_{\mathsf{fg}}$, 348
$\mathscr{A}_{\mathsf{fp}}$, 325
Alg F, 4, 376
BiCMet, 525
BiMet, 524
BiP, 79, 524, 537
Class, 313
CLat, 136, 137
CMet, 71, 151, 152, 179, 320, 322, 366, 394, 401, 458, 520, 537
CMet^E, 154
CMet$_*$, 152
Coalg F, 5
Coalg$_\lambda$ F, 380
Coalg$_{\mathsf{p}}$ F, 298
C/P, 340
CPO, 16, 135, 137, 176, 179, 275, 322, 367, 399, 400, 458, 537
CPO$_\perp$, 17, 135, 143
CPO$_*$, 148
DCPO, 367, 537
DCPO$_\perp$, 135, 143
Gra, 204, 320, 348
Haus, 515, 537
KHaus, 515, 517, 537
KMet, 161
K-Vec, 180, 189, 191, 281, 320, 321, 348, 366, 384, 537
Met, 71, 73, 179, 320, 322, 366, 375, 384, 394, 510, 520, 537
Nom, 18, 28, 322, 324, 348, 354, 395, 417, 537, 539
Ord, 168
Pfn, 135–137, 143, 179, 189, 191
PMet, 404, 537
Pos, 70, 320, 321, 348, 384, 512, 537
Pred, 281
Rel, 49, 80, 136, 137, 143, 146, 147, 190, 426
Set, 22, 152, 189–191, 348, 364, 366, 384, 459, 537
Set$^{\mathscr{C}^{\mathsf{op}}}$, 322
Set$^{\mathscr{F}}$, 324, 349, 354, 539
Set^S, 15, 191, 198
$\mathbb{S}$-Mod, 42, 353, 354, 433, 455, 456
Sub(A), 226
Sub$_{\mathcal{M}}(A)$, 178
Top, 384, 514, 515, 537, 539

Index

abstraction set, 19
accessible functor, 113, 368, 373, 375
Aczel–Mendler functor $(-)^3_2$, 105, 107, 553, 558, 559, 563, 568
adjoint, 392
algebra, 4, 12
  carrier, 12
  complete Elgot, 242
  completely iterative (cia), *see* completely iterative algebra
  completely metrizable, 235
  continuous, 16
  corecursive, *see* corecursive algebra
  double, 501
    for a monad, 503
  Eilenberg–Moore, *see* Eilenberg–Moore algebra
  finitely generated, 349
  for a monad, 428
  free, *see* free algebra
  homomorphism, 4, 13
  initial, *see* initial algebra
    iterative, 343
  iterative, 336
  minimal, *see* minimal algebra
  over nominal sets, 18
  over sets in context, 17
  projective, 352
  quotient, 107
  Σ-algebra, 14, 178
  sub-, 21
algebraically complete, 80, 172
almost λ-disjoint, 576
analytic functor, 78, 107, 111, 121, 122, 347, 414, 415
arity, 14
  binding, 18
  S-sorted, 15
automaton
  accepting state, 37
  Büchi, 426
  deterministic, *see* deterministic automaton
  linear weighted, 256, 272
  Mealy, 39
  Moore, 39
  nofa, 417
  non-deterministic, *see* non-deterministic automaton
  observable, 293, 309
  parity, 426
  reachable, 293
  regular nominal, 419
  terminal, 48
  weighted, 42, 43, 353
Büchi automaton, 425
bag, 40
  functor $\mathscr{B}$, 40, 78, 103, 105, 111, 233, 558, 568
Banach Fixed Point Theorem, 152, 234, 550
bar string, 355
base, 514
  case, 168
  category, 12
basic equation, 106
Beck's monadicity theorem, 376
behavioural equivalence, 60
binary
  decision diagram (BDD), 317
  decision tree, 75
  tree, 24
binding
  arity, 18

polynomial functor, 18, 19, 422, 539
signature, 18
bipointed
metric space, 524
set, 79, 524
bisimulation, 58, 97, 132, 147, 295, 309, 310, 458, 558
functional, 60
graph ~, 97
tree ~, 126
weighted, 40
bounded, 363
totally, 129
Brzozowski derivative, 48
canonical
cocone, 67
extension $\bar{F}$, 415
filtered diagram, 325
fixed point, 133, 144, 149, 150, 153, 161, 201, 210–212, 415, 423, 462
graph, 259
morphism, 189, 462
Cantor's Theorem, 22, 510
cardinal, 168, 169
measurable, 592
regular, 169
singular, 169
strong limit, *see* strong limit cardinal
cardinality, 167, 191
carrier, 12
cartesian
closed, 135
subcoalgebra, 254
category
base, 12
CMet-enriched, 151
cocomplete, 93
complete, 32
CPO-enriched, 135
Kleisli, 408, 409
λ-filtered, 365
λ-presentable, 375
left-strict, 148
locally
λ-presentable, *see* locally λ-presentable category
presentable, 367, 537
pointed, 134
presheaf, 275, 322
Scott complete, 150
strict
CMet-enriched, 157, 538
CPO-enriched, *see* strict CPO-enriched
well-powered, 21
Cauchy completion, 71, 394, 461
chain, 169, 193
i-, 169
Ord-, 169
-complete poset, 169, 250, 545
convergence ordinal, 193
initial-algebra, *see* initial-algebra chain
ω-, 66
terminal-coalgebra, *see* terminal-coalgebra chain
cia, *see* completely iterative algebra
co-well-powered, 93
coalgebra, 2, 4, 36
I-pointed, 308
canonical graph of, 259
cofree, 63, 483
corecursive
parametrically, 217
finitely presentable, 381
homomorphism, 3, 4, 36
locally
finite, 327
finitely generated (lfg), 357
finitely presentable (lfp), 327
pointed, 298
quotient, 97, 274
reachable, 298
recursive, *see* recursive coalgebra
parametrically, *see* parametrically recursive
relatively terminal, 488–489
simple, 295–297
terminal, *see* terminal coalgebra
weakly terminal, *see* weakly terminal coalgebra
well-founded, *see* well-founded coalgebra
well-founded part, 270
well-pointed, *see* well-pointed coalgebra
coalgebra-to-algebra morphism, 216, 244
cocomplete, 93
cocone, 66
cofinal, 169, 511
cofinality, 575
cofree, 63
coinduction, 3
principle, 60, 112
cointersection, 294
colimit, 67

create, 91
filtered, *see* filtered colimit
comma category, 340
commutative monad, 410, 413, 414
compactly branching tree, 129
compatible sequence, 124
complete
algebraically, *see* algebraically complete
category, 32
Elgot algebra, 242
lattice, 543
metric space, 71, 151, 464, 550
partial order (cpo), 16, 541
ideal, 44
lifting, 16
completely iterative algebra (cia), 217, 231, 232, 234
free, 483
initial, 238
completely metrizable, 235
completion
Cauchy, *see* Cauchy completion
ideal, *see* ideal completion
congruence, 97, 295
constant
functor, 101, 141, 191, 366
essentially, 191
symbol, 14
continued fraction, 529, 533
continuous
algebra, 16
function, 16
functor, 475
contracting
chain, 155
function, 152
Contraction Coinduction, 551
converge, 172, 193
in exactly λ steps, 172, 193
convergence ordinal, 172
convolution product, 353
corecursion, 3
primitive, 57
corecursive
algebra, 218, 227, 228, 533, 535
free algebra, 223
initial algebra, 221
parametrically, 217
coreflective subcategory, 274
covarietor, 63, 120, 518, 524
cpo, *see* complete partial order
CPO-enriched, 135, 181, 210
strict, 143
create
colimits, 91
limits, 93
cutting, 99, 123, 461
δ-discrete, 511
depth, 24
deterministic automaton, 37, 159, 256, 298, 309, 315
diagonal fill-in property, 224
diagram, 91, 101
filtered, 322
λ-filtered, 365
scheme, 91
dinatural transformation, 464
direct image operator, 264
directed
graph, 38
poset, 101, 546
discrete
metric, 71, 464, 521
order, 82
distributive law, 410, 414
duality, 44
edge, 38
Eilenberg–Moore algebra, 242, 361, 428
Elgot algebra, 361
complete, 242
embedding, 136, 143
-projection pair, 136
isometric, *see* isometric embedding
enriched category
CMet, 151
CPO, 135
strict CMet, 157
strict CPO, 162, 211
epi-pullback, 564
weak, 564, 571
epi-transformation, 102, 103
epimorphism
extremal, 294
split, 6, 102
strong, *see* strong epimorphism
equation
basic, 106
morphism, 217
equivalence
kernel ~, 97
equivariant map, 18
essentially

constant, 191
uniquely, 321
evaluation function, 404
eventually periodic, 332
extension, 394
left Kan, 325, 335
extensional
tree, 75, 100, 126, 127
strongly extensional, *see* strongly extensional tree
extensional quotient, 75
extremal epimorphism, 294
F-algebra, 4, 12
homomorphism, 4, 13
F-coalgebra, 4, 36
homomorphism, 4, 36
factorization
(strong epi, mono), 225
factorizes essentially uniquely, 321
factorizing morphism, 67
filter, 114
α-based, 197
functor $\mathscr{F}$, 114, 197, 553, 568, 592
λ-small, 197
filtered, 322
colimit, 195, 277, 322, 348, 380, 574
diagram, 322
final functor, 340, 342, 379
finitary
flat equation morphism, 336
presentation, 103, 104
signature, 14, 262, 325, 347, 362, 478
finite
distribution functor $\mathscr{D}_f$, 40, 558
hereditarily, 23
permutation, 18
power-set functor $\mathscr{P}_f$, 23, 75, 76, 86, 102, 103, 105, 122, 125, 233, 316, 323, 332, 347, 366, 402, 433, 469, 558
finite-to-one, 121, 122
finitely
bounded functor, 101, 104, 574
generated, 348
presentable, 320
supported, 324
fixed point, 8, 22, 187, 198, 272, 284, 335
canonical, *see* canonical fixed point
corecursive, 218
locally ffg, 359
locally finite, 357
minimal, 188, 284
post-, *see* post-fixed point
pre-, 542
rational, *see* rational fixed point
recursive, 244
unique, *see* unique fixed point
flat
equation morphism, 230
finitary, 336
term, 216
formal
language, 48, 159
power series, 49, 53, 353
free
algebra, 30, 314, 477–483
finitely generated, 352
iterative, 340
cia, 239, 400
cia monad, 400, 403
completely iterative
monad, 241
corecursive algebra, 223
corecursive-algebra chain, 223
finitely generated, 352
iterative algebra, 340
fully abstract, 330
function
continuous, 16
inflationary, 545
non-expanding, *see* non-expanding function
strict continuous, 17
functional bisimulation, 60
functor
accessible, *see* accessible functor
Aczel–Mendler, *see* Aczel–Mendler functor
analytic, *see* analytic functor
bag, *see* bag functor
binding polynomial, *see* binding polynomial functor
bounded, 363
by sets of size $\leq \lambda$, 589
constant, *see* constant functor
continuous, *see* continuous functor
contracting, 159
locally, 161
contravariant power-set $\mathsf{Set}(-, 2)$, 395
discrete probability measure $\mathscr{D}_f$, 568
distribution $\mathscr{D}$, 466
essentially constant, 191
filter $\mathscr{F}$, *see* filter functor
final, *see* final functor
finitary, 101, 113, 303, 574

set, 103, 106, 346, 577
finite power-set $\mathscr{P}_f$, *see* finite power-set functor
finitely bounded, *see* finitely bounded functor
finitely-supported power-set $\mathscr{P}_{fs}$, 324
grounded, 467, 475
Hausdorff $\mathcal{H}$, *see* Hausdorff functor
hom-contracting, 158
iteratable, 239, 403
Kantorovich $\mathscr{K}$, 368
Kripke polynomial, *see* Kripke polynomial functor
λ-accessible, 195, 589
λ-bounded, 194, 589
λ-preaccessible, 385
left strict, 538
locally
contracting, *see* locally contracting functor, 538
locally continuous, *see* locally continuous functor
monoid-valued, 40, 568
polynomial, *see* polynomial functor
power-set $\mathscr{P}$, *see* power-set functor
power-set $\mathscr{P}_\lambda$, 553, 568
preaccessible, 385, *see* preaccessible
precontinuous, 465, 475
presentation of, 103
preserves inclusions, 586
proper, 359, 453
quotient, 102, 206
R, 245, 262, 307, 310, 553, 568
stable, 210, 538
standard, 239, 587
strict CPO-enriched, 210
super-finitary, 578
Vietoris $\mathscr{V}$, *see* Vietoris functor
General Recursion Theorem, 279
Generalized Continuum Hypothesis (GCH), 199
generates, 327
generating set, 363
global element, 298
graph, 38, 266
~ bisimulation, 97
canonical, 259
directed, 38
pointed, 100
reachable, *see* reachable graph
well-founded, 267
Hartogs' Lemma, 170, 545
Hausdorff
functor $\mathcal{H}$, 153, 368, 521
metric, 153, 236
polynomial functor, 523, 539
HC, 177
height, 174
hereditarily finite, 23, 177
HF, 177
homomorphism, 14
algebra, 4, 13
coalgebra, 3, 4, 36
Σ-algebra, 14
hyperspace, 514
i-chain, 169
ideal, 44, 142, 474
completion, 142, 473, 475
idempotent, 210
semiring, 356, 435, 436
split, 210
image, 225
inclusion map, 552
Ind-completion, 328
infinitary signature, 176
infinite word, 83
inflationary, 545
initial
algebra, 4, 21, 67, 68, 108, 187, 245, 250, 282, 314, 373, 394
iterative, 343
theorem, 187
corecursive algebra, 221
iterative algebra, 343
object
simple, 227, 274
strict, 226
initial-algebra chain, 66, 171, 173, 187, 247, 274, 386, 463
intersection, 113
finite, 114, 554
nonempty, 113, 554
preservation of, 113, 560
inverse
image, 559
operator, 264, 275
preservation of, 559
isolated step, 168
isometric
cone, 522
embedding, 74, 151, 179, 511
iteration theory, 242
iterative algebra, 336

Iwamura's Lemma, 546
kernel equivalence, 97
kernel pair, 455
  weak, 563, 571
Kleene's Theorem, 65, 305, 542
Kleisli
  category, 408, 409, 414
  extension, 241
  triple, 241
Knaster–Tarski Theorem, 80, 170, 228, 276, 424, 543
Kripke
  polynomial functor, 119, 263, 374, 402, 466
  structure, 118
labelled
  transition system, 39, 302
  tree, 25
λ-accessible, 195
λ-calculus, 165
λ-presentable, 366
λ-small filter, 197
λ-term, 354
Lambek's Lemma, 22
language
  bar, 419
  context-free, 435
  finitely supported, 355
  formal, 48
  weighted, 53, 353
lattice, 543
  complete, 543
leaf, 24
left adjoint, 32, 392
left-monotone, 548
lifting, 16, 147, 210, 394, 398
  contracting, 152, 159
limit
  create, 93
  ordinal, 168
  step, 168
  strong ~ cardinal, 201
limit–colimit coincidence, 134, 140, 156
linear weighted automaton, 256, 272
locally
  continuous functor, 141, 143, 144, 162, 163, 210
  contracting functor, 152, 153, 157, 158, 166, 234, 403
  ffg fixed point, 359
  finite, 327
  finite fixed point, 357
  finitely generated, 357
  finitely presentable, 321, 327
  λ-presentable category, 367, 368, 381, 492, 495
  presentable category, 367, 537
lower Vietoris space, 520
$\mathcal{M}$-object, 178
$\mathcal{M}$-subfunctor, 204
many-sorted set, 86
map
  equivariant, 18
  strict continuous, 17
Markov chain, 40
meet, 226
metric
  Hausdorff, *see* Hausdorff metric
  maximum, 73
  space, 71
    bipointed, 524
    complete, *see* complete metric space
    discrete, 73
    totally bounded, 129
minimal
  algebra, 6, 29, 188, 284
  automaton, 293, 309, 346
monad, 31, 400, 403, 408, 410, 414, 478
  commutative, *see* commutative monad
  Kleisli category, 408, 414
  locally contracting, 403
  rational, 360
monoid
  positive, 132, 569
  refinable, 132, 569
  -valued functor, 40, 568
monomorphism
  nonempty, 226
  regular, 179
  smooth, 178
  strong, 384
  universally smooth, 276, 277
morphism
  canonical, 462
  coalgebra-to-algebra, 216, 244
  corecursively specified, 55
  factorizing, 67
  flat equation, 230
  recursively specified, 33
  tree, 124
multigraph, 41
multiset, 40
mutual recursion, 88

natural numbers object (NNO), 24
next-time operator, 265
nofa, 417
nominal set, 18
non-deterministic automaton, 39, 132, 147, 256, 433
non-expanding function, 71, 151
nullary symbol, 14

observable automaton, 293, 309
ω-chain, 66
ω-colimit, 66
operation symbols, 14
orbit, 78, 322
  -finite, 322, 348
Ord-chain, 169
ordered tree, 24
ordinal, 167
  limit, 168
  successor, 168

parametrically recursive, 245, 279, 281
parity automaton, 425
partial order
  complete (cpo), 541
Pataraia
  Induction Principle, 549
  Theorem, 548
permutation
  finite, 18
point, 298
pointed
  category, 134
  coalgebra, 298
  graph, 100
  set, 456
  subcoalgebra, 298
polynomial functor, 14, 16, 84, 88, 141, 152, 176, 195, 219, 552
  binding, *see* binding polynomial functor
  Hausdorff, 523, 539
  Kripke, *see* Kripke polynomial functor
  Vietoris, 518, 539
poset
  chain-complete, *see* chain-complete poset
  directed, *see* directed poset
positive, 132, 569
post-fixed point, 191, 192, 513, 549
power-set
  finite ~ functor, *see* finite power-set functor
  functor $\mathscr{P}$, 14, 102, 255, 298, 399, 553
pre-fixed point, 22, 66, 67, 101, 170, 186–189, 191, 192, 228, 250, 251, 272, 285, 542, 577
preaccessible, 537, 591, 592
precontinuous, 462
prefixed-closed, 25
prelimit, 465
presentation, 347
  finitary, 103
presheaf
  category, 17, 198, 275, 283, 539
  of λ-terms, 17
previous-time operator, 301
primitive
  corecursion, 57
  recursion, 34
projection, 136, 143
projective, 352
proper functor, 359, 453
pseudometric, 404

quadratic irrationals, 333
quotient, 206, 274
  algebra, 107
  coalgebra, 97
  functor, 102
  object, 96
    order on, 96
  strong, 294, 306
    coalgebra, 274

rational fixed point, 329, 346, 347, 357, 534
reachable
  automaton, 293, 309
  coalgebra, 298, 303, 304, 306
  graph, 298, 303
  part, 299
recursion, 2, 33
  mutual, 88
  primitive, 34
  Theorem, 279
recursive
  coalgebra, 244, 246, 285, 485
    parametrically, *see* parametrically recursive
    terminal, 245
recursively specified morphism, 33
refinable, 132, 569
regular
  cardinal, 169, 575
  monomorphism, 179
relation, 38, 58
  well-founded, 244
retraction-embedding pair, 151

right adjoint, 32, 392
root, 24
$\mathbb{S}$-semimodule, 42
S-sorted signature, 15
saturated tree, 130
Scott induction, 66, 250
Scott-complete, 150
semimodule, 42
semiring, 42
set
  abstraction, 19
  bipointed, 79, 524
  hereditarily countable, 177
  hereditarily finite, 23, 177
  in context, 17
  many-sorted, 86
  nominal, 18
  non-well-founded, 243
  S-sorted, 15
set functor
  accessible, 538
  finitary, 466
  grounded, 467
  precontinuous, 462
Σ-algebra, 14
Σ-term, 25
  graph, 317
Σ-tree, 25, 27, 176
  rational, 331
signature, 84, 88, 102, 103, 107, 176, 478
  binding, 18
  finitary, *see* finitary signature, 75
  infinitary, 85, 176
  many-sorted, 86, 89
  S-sorted, 15
simple, 227, 274
  coalgebra, 295–297
singular, 169
smooth, 205
  class, 178, 502
  class of monomorphisms, 502
  monomorphism, 178
  universally, 276
solution, 216, 218, 219
sorted
  set, 15
  signature, 15
split
  epimorphism, 6, 102
  idempotent, 210
stable, 210, 538
state transition graph, 256
Stone space, 520
stream, 47, 52, 55, 74, 83, 159, 243, 356, 499
  automatic, 332
  causal ~ function, 49
  eventually periodic, 332, 353
  rational, 353
  real valued, 83
strict, 17
  CMet-enriched, 157, 538
  CPO-enriched, 143, 210, 538
  initial object, 226, 277
strong
  epimorphism, 224, 294, 326
  limit cardinal, 201, 482
  monomorphism, 384
  quotient, 274, 306
  quotient coalgebra, 294
strongly extensional
  quotient, 127
  tree, 126, 127, 132, 233, 402, 471, 521
subalgebra, 6, 21
subbase, 514
subcartesian, 596
subcoalgebra, 46, 128, 364
  cartesian, 254–255
  pointed, 298
subfunctor, 204
subobject, 21
  classifier, 287
  smooth, 178
successor ordinal, 168
super-finitary, 578
support, 18
systems with termination, 37
term, 25
  flat, 216
terminal
  automaton, 48
  coalgebra, 5, 46, 103, 199, 238, 296, 312, 364, 373, 394, 461, 475
  recursive coalgebra, 245
terminal-coalgebra chain, 81, 193–195, 198, 376, 463
theorem
  Banach's Fixed Point, *see* Banach Fixed Point Theorem
  Beck's monadicity, 376
  Cantor's, *see* Cantor's Theorem
  Freyd's Iterated Square, 485
  Freyd's Special Adjoint Functor, 364

Initial Algebra, 187
Kleene's, *see* Kleene's Theorem
Knaster–Tarski, *see* Knaster–Tarski Theorem
Makkai and Paré's Limit, 376
Markowsky's, 547
Pataraia's, 548
Taylor's General Recursion, 279
Zermelo's, 170, 545
totally bounded, 129
transfinite induction, 168
transition system
labelled, 39
weighted, 40
tree
~bisimulation, 126
binary, 24
binary decision, 75
compactly branching, 129
complete n-ary, 25
cutting, *see* cutting
extensional, *see* extensional tree
strongly, 126
height, 174
labelled, 25
leaf, 24
morphism, 124
ordered, 24
saturated, 130
Σ-tree, 25, 27, 176
strongly extensional, 126, 127
quotient, 127
tupling, 25, 73
with multiplicities, 77
unordered, 24
well-founded, 174
triangular identities, 32, 392
Trnková hull, 261, 561, 579–581, 583, 584

ultrafilter, 592
σ-complete, 592
principal, 592
unique fixed point, 152, 158, 160
unit interval $\mathbb{I}$, 525, 533
universally smooth, 276
unordered tree, 24

varietor, 31, 32, 174, 191, 192, 201, 377, 389, 478, 480–482, 519, 524
variety, 313, 367, 428
finitary, 321, 349
locally finite, 321
vertex, 38
Vietoris
functor $\mathscr{V}$, 514, 517, 539
polynomial functor, 518, 539
space, 514
lower, 520

weak
kernel pair, 563
pullback, 60, 97, 132, 406, 426, 455, 556, 562, 563
weakly terminal, 95, 98, 103
weighted
automaton, 43, 433
bisimulation, 40, 434
language, 53, 434
transition system, 40
well-founded, 174
coalgebra, 255, 267, 272, 281, 282, 284, 285, 314
terminal, 282
graph, 267
graph vs. relation, 254
part, 270, 271
well-ordered set, 167
well-pointed
coalgebra, 309, 312, 314, 345
modification, 310
well-powered
category, 21
$\mathcal{M}$-, 178

zero object, 134